ósiah = Sarah
qwood I Wedgwood
30–95 1734–1815

John Bartlett Allen = Eli___
1733–1803 1738–90

sannah Josiah II = Elizabeth
765–1817 1769–1843 (Bessy)
1764–1846

Catherine (Kitty) = Sir James
1765–1830 Mackintosh
1765–1832

Caroline = Edward
1768–1835 Drewe
1756–1810

John Hensleigh
1769–1843

John = Louisa Jane
1766–1844 (Jane)
1771–1836

Lancelot Baugh
1774–1845

Harriet
1776–1847

Jessie = J.C. de
1777–1853 Sismondi
1773–1842

homas
1–1805

therine
kitty)
1774–1823

arah
zabeth (Sarah)
78–1856

Octavia
1779–1800

Emma
1780–1866

Frances
(Fanny)
1781–1875

Sarah Elizabeth
(Eliza)
1795–1857

John Allen
1796–1882

Thomas Josiah
1797–1862

Sarah
Elizabeth
(Elizabeth)
1793–1880

Caroline
1799–1825

Caroline
Sarah = Josiah III
1800–88 1795–1880

Charles
1800–20

Charles = charlotte
Langton 1797–1862
1801–86

Henry = Jessie
Allen 1804
(Harry) –72
1799–1885

Robert = Frances Crewe
1806–80 d.1845

Elizabeth
(Bessy)
1799–1823

Frances = Francis
Mosley (Frank)
d.1874 1800–88

Hensleigh = Frances (Fanny)
1803–91 1800–89

Frances (Fanny)
1806–32

Charles = Emma
Robert 1808–96
1809–82

Robert
1806–64

THE CORRESPONDENCE OF
CHARLES DARWIN

Editors
FREDERICK BURKHARDT   SYDNEY SMITH

Editorial Staff
CHARLOTTE BOWMAN   JANET BROWNE
ANNE SCHLABACH BURKHARDT   STEPHEN POCOCK
MARSHA L. RICHMOND   ANNE SECORD
NORA CARROLL STEVENSON

This edition of the Correspondence of Charles Darwin is sponsored by the American Council of Learned Societies. Its preparation is made possible by the co-operation of the Cambridge University Library and the American Philosophical Society.

Advisory Committees for the edition, appointed by the Council, have the following members:

United States Committee	*British Committee*
Whitfield J. Bell Jr	Nora Barlow
John C. Greene	W. F. Bynum
Sandra Herbert	Owen Chadwick
Ernst Mayr	Dennis Crisp
Duncan M. Porter	Peter J. Gautrey
Robert C. Stauffer	Desmond King-Hele

Support for editing has been received from the National Endowment for the Humanities, the National Science Foundation, the Andrew W. Mellon Foundation, the Alfred P. Sloan Foundation, the Royal Society of London, and the British Academy. The National Endowment's grants (Nos. RE-23166-75-513, RE-27067-77-1359, RE-00082-80-1628, RE-20166-82, and RE-20480-85) were from its Program for Editions; the National Science Foundation's funding of the work was under grants Nos. SOC-75-15840, SES-7912492, and SES-8517189. Any opinions, findings, conclusions or recommendations expressed in this publication are those of the authors and do not necessarily reflect the views of the grantors.

Asa Gray. Photograph, 1865.
(Courtesy of the Gray Herbarium of Harvard University.)

THE CORRESPONDENCE OF
CHARLES DARWIN

VOLUME 6   1856–1857

The right of the
University of Cambridge
to print and sell
all manner of books
was granted by
Henry VIII in 1534.
The University has printed
and published continuously
since 1584.

CAMBRIDGE UNIVERSITY PRESS

CAMBRIDGE

NEW YORK   PORT CHESTER

MELBOURNE   SYDNEY

Published by the Press Syndicate of the University of Cambridge
The Pitt Building, Trumpington Street, Cambridge CB2 1RP
40 West 20th Street, New York, NY 10011, USA
10 Stamford Road, Oakleigh, Melbourne 3166, Australia

First published 1990

Printed in Great Britain at the University Press, Cambridge

British Library cataloguing in publication data
Darwin, Charles *1809–1882*
The correspondence of Charles Darwin.
Vol. 6, 1856–1857
1. Organisms. Evolution. Darwin, Charles, 1809–1882
I. Title   II. Burkhardt, Frederick H. (Frederick Henry)
1912–   III. Smith, Sydney *1911–1988*
575.0092

Library of Congress cataloguing in publication data
Darwin, Charles, 1809–1882.
The correspondence of Charles Darwin.
1. Darwin, Charles, 1809–1882.   2. Naturalists—
England—Correspondence.   I. Burkhardt, Frederick,
1912– .   II. Smith, Sydney 1911–88.   III. Title.
QH31, D2A33   1985   575′.0092′4[B]   84–45347

ISBN 0 521 25586 4

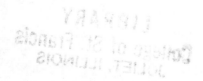

CONTENTS

List of illustrations	vii
List of letters	ix
Introduction	xiv
Acknowledgments	xx
List of provenances	xxiii
Note on editorial policy	xxv
Darwin/Wedgwood genealogy	xxx
Abbreviations and symbols	xxxii
THE CORRESPONDENCE, 1856–7	I
Appendixes	
I. Translations	518
II. Chronology	522
III. Dates of composition of Darwin's manuscript on species	526
Manuscript alterations and comments	528
Bibliography	548
Biographical register and index to correspondents	581
Index	635

ILLUSTRATIONS

Asa Gray   frontispiece

James Dwight Dana   facing p. 192

William Bernhard Tegetmeier   193

Henrietta Emma Darwin   224

Down House   225

Philip Henry Gosse   352

Alphonse de Candolle   353

Henri Milne-Edwards   416

Thomas Henry Huxley   417

LIST OF LETTERS

The following list is in the order of the entries in the *Calendar of the correspondence of Charles Darwin*. It includes all those letters that are dated in the *Calendar* within the date range covered by this volume of the *Correspondence*. Alongside the *Calendar* numbers are the current dates ascribed to every item. Some letters have been redated since the publication of the *Calendar*, so this list is necessary to enable users of the *Calendar* to locate such letters in the *Correspondence*. In the list, a date printed in italic type indicates that the item either appears only as a summary in this volume or has been omitted from it entirely.

1026. 17 November [1856–7]
1028. 20 November [1856–7]
1332. 25 May [1856]
1451. 19 August [1856]
1614. 18 December 1857
1617. [before 9 May 1856]
1619. [before 11 September 1857]
1620. [29 July 1856]
1622. [16 November 1856]
1632. 20 January [1856]
1646. 8 March [1856]
1663. 10 April [1856]
1691. *[1846–7]*
1804. [26 February 1856]
1805. [17 February 1857]
1806. [February 1856]
1807. [after 4 June 1856]
1808. *[1858]*
1809. *[1856–9]*
1810. 18 [August 1856 – January 1858]
1811. *[22 October 1855]*
1812. *[December 1855]*
1813. 1 January [1856]
1814. 2 January [1856]
1815. 3 January [1856]
1816. 3 January [1856]
1817. 8 January [1856]
1819. [before 8 January 1856]
1820. 14 January [1856]
1821. [1 February 1856]

1822. 18 January [1856]
1823. 22 January [1856]
1824. 23 January 1856
1825. 23 January 1856
1826. *See* 1824
1827. 26 January [1856]
1828. 28 January [1856]
1829. 6 [February 1856]
1830. 7 February 1856
1831. *14 [July 1855]*
1832. 23 February 1856
1833. 26 February 1856
1834. 29 February [1856]
1835. 1 March 1856
1836. 7 March 1856
1837. 8 March 1856
1838. 8 March 1856
1839. 8 March [1857]
1840. 9 March 1856
1841. 11 March [1856]
1842. 15 March [1856]
1843. 15 March [1856]
1844. 20 March [1856]
1845. [c. 22 March 1856]
1846. *2 April [1857–65]*
1847. 2 April [1856]
1848. 3 April [1856]
1849. [3 April 1856]
1850. [6 April 1856]
1851. 8 April [1856]

1852. 9 April [1856]
1853. 13 April [1856]
1854. 16 April 1856
1855. 21 April [1856]
1856. 21 April [1856]
1857. 21 April [1856]
1858. 23 April [1856]
1859. 24 April [1856]
1860. 25 April [1856]
1861. [May 1856]
1862. 1–2 May 1856
1863. 2 May [1856]
1864. 2 May 1856
1865. [before 3 May 1856]
1866. 3 May [1856]
1867. 3 May [1856]
1868. 4 May [1856]
1869. 7 May 1856
1870. 9 May [1856]
1871. 9 May [1856]
1872. 10 May [1856]
1873. 11 May [1856]
1874. 11 May [1856]
1875. 15 May [1856]
1876. 21 [May 1856]
1877. 27 May [1856]
1878. 27 May [1856]
1879. 27 May [1856]
1880. *28 [1857–65]*
1881. *28 May [1854]*
1882. 31 May [1856]
1883. 31 May 1856
1884. [14 January 1856]
1885. 1 June [1856]
1886. 3 June [1856]
1887. 4 June [1856]
1888. 4 June [1856]
1889. 4 June 1856
1890. [after 4 June 1856]
1891. 5 June 1856
1892. 5 June [1856–9]
1893. 6 June [1856]
1894. 8 June [1856]
1895. 8 [June 1856]
1896. [8 June 1856]

1897. [September–October 1856]
1898. 10 June 1856
1899. [after 10 June 1856]
1900. 12 [June 1856]
1901. 14 June [1856]
1902. 16 [June 1856]
1903. 16 June [1856]
1904. 17–18 [June 1856]
1905. 17 June 1856
1906. 18 June [1856]
1907. 20 June 1856
1908. 22 June [1856]
1909. 24 June [1856]
1910. 25 June [1856]
1911. [26 June or 3 July 1856]
1912. [27 June 1856]
1913. [July 1856]
1914. 1 July [1856]
1915. 1 July [1856]
1916. 2 July 1856
1917. 5 July [1856]
1918. 5 July [1856]
1919. 5 [July 1856]
1920. 8 July [1856]
1921. 8 [July 1856]
1922. 8 July [1856]
1923. 10 July 1856
1924. 13 July [1856]
1925. 14 July [1856]
1926. 14 July [1856]
1927. 15 July 1856
1928. [15 July 1856]
1929. 17 July 1856
1930. 18 July 1856
1931. 18 July 1856
1932. 19 July [1856]
1933. 30 July [1856]
1934. [early August 1856]
1935. [15–22 August 1856]
1936. 2 August 1856
1937. 4 August 1856
1938. 5 August [1856]
1939. 6 August [1856]
1940. 7 August [1856]
1941. 14 August [1856]

1942. 21 August [1856]
1943. 23 August [1856]
1944. 24 August [1856]
1945. 26 [July 1856]
1946. 27 [August 1856]
1947. 30 August [1856]
1948. 31 August [1856]
1949. 5 September [1856]
1950. 8 September [1856]
1951. 8 September 1856
1952. 9 September [1856]
1953. 10 September 1856
1954. 11 September [1856]
1955. [18 September 1856]
1956. 20 September 1856
1957. 21 September [1856]
1958. 22 September [1856]
1959. 23 September 1856
1960. 23 September [1856]
1961. 24 September [1856]
1962. 28 September 1856
1963. 28 September [1856]
1964. 29 September [1856]
1965. [16 October 1856]
1966. [early December 1856]
1967. 3 October [1856]
1968. 5 October [1856]
1969. 5 October [1856]
1970. 8 October – 7 November 1856
1971. 9 October [1856]
1972. 9 October [1856]
1973. 12 October [1856]
1974. [early December 1856]
1975. 15 October [1856]
1976. 19 October [1856]
1977. [19 October 1856]
1978. 20 October [1856]
1979. 27 October [1856]
1980. [1 November 1856]
1981. 3 November [1856]
1982. 4 November 1856
1983. 9 November 1856
1984. 10 November [1856]
1985. 10 November 1856
1986. 11–12 November [1856]

1987. 13 [November 1856]
1988. 14 November 1856
1989. 15 November [1856]
1990. 18 November [1856]
1991. 18 November [1856]
1992. 19 November [1856]
1993. 19 November [1856]
1994. 19 November 1856
1995. 22 November 1856
1996. 22 November 1856
1997. 23 November [1856]
1998. 23 November 1856
1999. 24 November [1856]
2000. 25 [November 1856]
2001. 26 November [1856]
2002. 26 November 1856
2003. 26 November [1856]
2004. 29 November [1856]
2005. 30 November [1856]
2006. [after 6 December 1856]
2007. *[13 August 1855]*
2008. 1 December [1856]
2009. 1 December 1856
2010. 3 December [1856]
2011. 4 December [1856]
2012. [before 6 December 1856]
2013. [11 or 18 December 1856]
2014. 7 December 1856
2015. [8 December 1856]
2016. 8 December 1856
2017. 9 December [1856]
2018. 10 December [1856]
2019. 10 [December 1856]
2020. 13 [December 1856]
2021. 23 December [1856]
2022. 24 December [1856]
2023. [28 December 1856]
2024. 29 December 1856
2025. [after 16 February 1857]
2026. *[March–April 1846]*
2027. *[24 March 1863]*
2028. *[March–April 1858]*
2029. *[1857–67]*
2030. [early November 1856]
2031. [before 3 November 1856]

2032. [before 5 February 1857]
2033. [after 20 January 1857]
2034. 1 January [1857]
2035. 3 January 1857
2036. *4 January [1858]*
2037. 4 January [1857]
2038. 10 January 1857
2039. [16 January 1857]
2040. 17 January [1857]
2041. 17 January [1857]
2042. 20 January [1857]
2043. *26 January [1858]*
2044. 26 January 1857
2045. 3 February [1857]
2046. 4 February [1857]
2047. 5 February 1857
2048. 6 February [1857]
2049. 8 February [1857]
2050. 11 February [1857]
2051. 11 February [1857]
2052. 11 February 1857
2053. 16 February 1857
2054. 18 February [1857]
2055. *21 February [1857]*
2056. 22 February 1857
2057. 22 February [1857]
2058. 23 February [1857]
2059. 26 February 1857
2060. [after 15 March 1857]
2061. [after 28 February 1857]
2062. 10 March [1857]
2063. 10 March 1857
2064. 12 March 1857
2065. 13 March 1857
2066. 15 March [1857]
2067. [21 March 1857]
2068. 28 [September 1856]
2069. 31 March 1857
2070. [before 29 September 1857]
2071. 4 April 1857
2072. 5 April [1857]
2073. 8 April [1857]
2074. [11 April 1857]
2075. 12 April [1857]
2076. [12 April 1857]
2077. 13 April [1857]

2078. 14 April 1857
2079. 19 April [1857]
2080. *[21 April 1857]*
2081. 23 April 1857
2082. 27 April [1857]
2083. 27 April 1857
2084. [29 April 1857]
2085. [30 April 1857]
2086. 1 May 1857
2087. [2 May 1857]
2088. [3 May 1857]
2089. 9 May [1857]
2090. 12 [May 1857]
2091. 13 May [1857]
2092. 16 [May 1857]
2093. 18 May [1857]
2094. 25 May [1857]
2095. [November 1857]
2096. [before 29 December 1857]
2097. 21 [July 1857]
2098. 1 June 1857
2099. 2 June [1857]
2100. 2 June [1857]
2101. 3 June [1857]
2102. 5 June [1857]
2103. 9 June [1857]
2104. [c. 24 May 1857]
2105. [before 13 June 1857]
2106. 14 June [1857]
2107. *15 June [1851]*
2108. [18 June 1857]
2109. 18 June [1857]
2110. 23 June [1857]
2111. 25 [June 1857]
2112. 25 June [1857]
2113. 26 [June 1857]
2114. [27] June 1857
2115. [19 July 1857]
2116. 1 July [1857]
2117. 5 July [1857]
2118. 5 July [1857]
2119. 7 July 1857
2120. 7 July 1857
2121. 7 July [1857]
2122. 9 July [1857]
2123. 14 [July 1857]

2124. 14 July [1857]
2125. 20 July [1857]
2126. 22 [July 1857]
2127. [before 25 July 1857]
2128. 27 July [1857]
2129. [August 1857]
2130. 1 August [1857]
2131. 2 August [1857]
2132. 10 August [1857]
2133. [November–December 1857]
2134. 22 August [1857]
2135. [before 29 October 1857]
2136. 5 September [1857]
2137. 6 September [1857]
2138. 8 September [1857]
2139. 8 September [1856]
2140. 11 September [1857]
2141. 15 September [1857]
2142. 25 September 1857
2143. 26 September [1857]
2144. [before 3 October 1857]
2145. [27 September 1857]
2146. 29 September [1857]
2147. 29 [October 1857]
2148. 30 September [1857]
2149. [22 November 1857]
2150. 3 October [1857]
2151. 4 October [1857]
2152. *13 October [1858]*
2153. 14 October [1857]
2154. 18 October [1857]
2155. 18 October [1857]
2156. 20 October [1857]
2157. [23 October 1857]
2158. 23 October 1857
2159. 26 October 1857
2160. 26 October 1857
2161. 30 October [1857]
2162. *See* 2156
2163. *11 [February 1858]*
2164. 2 November [1857]
2165. 4 November 1857
2166. [before 12 November 1857]
2167. 10 November 1857
2168. 12 November [1857]
2169. [before 12 November 1857]

2170. 14 [November 1857]
2171. 14 November 1857
2172. 17 November 1857
2173. 21 November [1857]
2174. 21 November [1857]
2175. 23 November 1857
2176. 29 November [1857]
2177. 1 December [1857]
2178. [2 December 1857]
2179. 3 December 1857
2180. 4 December [1857]
2181. [6 December 1857]
2182. 9 December [1857]
2183. 14 December [1857]
2184. 15 December [1857]
2185. 16 December [1857]
2186. [16 or 17 December 1857]
2187. 17 December [1857]
2188. [17–23 December 1857]
2189. 18 December [1857]
2190. 20 December [1857]
2191. *[after 20 December 1857]*
2192. 22 December 1857
2193. 22 December 1857
2194. 25 December [1857]
2198. [June 1857]
2206. 24 January [1857]
2241. 16 March [1857]
2329. 25 September [1857]
2378. 8 December [1857]
2392. [6 March 1857]
2396. 12 [August 1857]
2481. 11 August [1857]
4598. 23 August [1856]
13801. *[April–June 1849]*
13873. *23 January [1856–61]*
——. 10 January 1856
——. 14 January [1856]
——. 19 January [1856]
——. 6 February [1856]
——. [12 March 1856]
——. 12 April 1856
——. [1 July 1856]
——. 28 September 1856
——. [2 August 1857]

INTRODUCTION

On 14 May 1856, Charles Darwin recorded in his journal that he 'Began by Lyell's advice *writing* species sketch' ('Journal'; Appendix II). For the next two years and more, his working life was completely dominated by the preparation of this manuscript. Although advised by Lyell to publish only a brief outline— probably more for the sake of priority than anything else—Darwin was reluctant to squeeze his expansive material into such a small compass and soon abandoned Lyell's idea in favour of a full-length work that would do greater justice to his views. He had been pondering the question of species for nearly twenty years, gathering vast quantities of information, pursuing his own experiments in a variety of different areas, analysing and altering his arguments in the light of recent results across the spectrum of nineteenth-century natural history, but never relinquishing the belief that his theory of natural selection could explain the structure of the living world. The moment had come, he accepted, to marshall his facts and put his theory before the public. His book was to be called *Natural selection.*

Determined as he was to publish, Darwin nevertheless still felt cautious in expressing his views before a large scientific audience and anxious to ensure that his facts were correct, his conclusions valid. As always, his correspondents played a crucial role in this process. Still prominent in his immediate circle were Charles Lyell and Joseph Dalton Hooker, who were joined in 1856 by Hooker's friend the American botanist Asa Gray and then by the specialist in Madeiran entomology, Thomas Vernon Wollaston. Darwin also came to rely on the caustic yet considered botanical opinions of Hewett Cottrell Watson. Similarly—with only occasional protestations about his argumentative habits—he increasingly valued the views of Thomas Henry Huxley, at that time a somewhat precariously placed lecturer and palaeontologist in the School of Mines in London.

Not all of Darwin's manuscript on species has been preserved and not all of it found its way into *On the origin of species* (1859). His letters are often the only source of information about his preoccupations during 1856 and 1857. They reveal little noticed aspects of his work and particularly highlight the importance of his investigations of pigeons, poultry, and other domesticated animals. As Darwin explained to Lyell, his studies, particularly those on pigeons, were intended to provide an illustration of how selection might work in nature (letter from Charles Lyell, 1–2 May 1856, n. 10). He was surprised that no naturalist had thought of comparing wild and domesticated species in this way before. 'How very odd it is that no zoologist sh^{d} ever have thought it worth while to look to the real structure of varieties', he remarked to Hooker (letter to J. D. Hooker, 8 September [1856]).

William Bernhard Tegetmeier continued to help Darwin acquire much of the material for his practical work. This work was supported by considerable research in published and unpublished sources. Edward Blyth needed little encouragement to send prolix memorandums on domestic animals in India and elsewhere. William Darwin Fox supplied information about cats, dogs, rabbits, and geese, as well as his customary anecdotes about domestic ducks and poultry; Thomas Campbell Eyton was questioned about pigs, dogs, and cattle; Carl Johann Andersson about native Swedish ponies; Hugh Falconer about Tibetan mastiffs. The disparate facts were correlated and checked by Darwin, who adroitly used letters, published sources, and his own experiments at Down to confirm each detail. 'I am like Crœsus overwhelmed with my riches in facts,' he told Fox, '& I mean to make my Book as perfect as ever I can.' (letter to W. D. Fox, 8 February [1857]).

Darwin also attempted to test ideas emerging from his work on domestic animals by conducting experiments on plants. Expanding projects set up during 1855 and 1856 (see *Correspondence* vol. 5), he tried to 'break the constitution of plants' by altering the external conditions under which plants grew to see whether he could induce hereditable changes. He carried out comparative studies of germination under different conditions and interbred garden species with their wild congeners.

Many of Darwin's conclusions about the variation of animals and plants under domestication were written up in the first two chapters of his species book, completed by October 1856 ('Journal'; Appendix II). Unfortunately, these chapters are not extant. It seems likely that Darwin used the manuscript when compiling *The variation of animals and plants under domestication* (1868) and that it was destroyed or lost during the process. *Variation* has, until now, been the only published source for many of Darwin's views on domestic animals and plants and this, since it was composed so many years later, is not a safe guide to his pre-*Origin* thinking. The letters, on the other hand, are richly informative and cast a great deal of new light on the role that these ideas were intended to play in Darwin's formal exposition.

Since natural selection could not act without varieties to act upon, Darwin wanted to know where, how, and in what way variations appeared in animals and plants. Making the fullest possible use of his botanical friends, Darwin cross-examined them on different aspects of the question. Did naturalised plants, he asked Asa Gray, vary in the United States (letter to Asa Gray, 2 May 1856)? What about weeds? Did they vary? Plants introduced by early travellers or horticulturists were roughly comparable to domestic animals: did they ever run wild? What correlations could be established between variations and conditions of existence? One useful example that Darwin intended to include in his book was the apparent tendency of alpine plants to be more hairy than their lowland relatives. But a last-minute check with Hooker revealed that Darwin was mistaken: 'You have shaved the hair off the alpine plants pretty effectually' complained Darwin in 1857 (letter to J. D. Hooker, [2 May 1857]).

The largest project Darwin undertook along these lines was an arithmetical survey of the frequency of plant varieties. Using published catalogues of the floras of particularly well-worked countries, he attempted to show that varieties—as identified by the authors of the catalogues—tended to occur more frequently in species and genera that were geographically widespread or otherwise abundant. He wrote at length about this project, his ups and downs with the calculations and different ways of working, in letters to Hooker, Gray, and Watson. The results seemed to corroborate his general point that variation, providing abundant raw material for natural selection, led to adaptation and thence to more successful, individually numerous, and widely spread species. The significance that he placed on these results is clearly demonstrated by the shock he acknowledged when told by his neighbour and young protégé John Lubbock that his method of calculation was wrong (letter to John Lubbock, 14 July [1857]). Darwin thought his results showed that variable species occurred more frequently in large genera than in small, but all they actually showed was the self-evident fact that a large genus was more likely to contain a variable species than was a small one. Rather than relinquish the results achieved after so much effort, Darwin began the whole laborious project again, using a statistically valid method explained to him by Lubbock.

Such perseverance is perhaps the key to this period in Darwin's life. He brought the same quality of doggedness to the resolution of the single most important question revealed by his work on varieties and variation. This was the origin and function of sex in nature. Darwin had always been intrigued by the theoretical issues raised by the existence of two sexes and had long held the idea that the variations seen in individual organisms arose predominantly as the result of sexual reproduction between the parents. For that reason he believed that all organisms, even hermaphrodites, must occasionally be cross-fertilised by other individuals. Darwin sought information on this problem from many different quarters, turning to Huxley to ask him whether there were any necessarily self-fertilising hermaphrodites among marine invertebrates. His request led Huxley to make a note for future reference, 'Darwin, an absolute & eternal hermaphrodite' (letter to to T. H. Huxley, 1 July [1856]), which became a source of amusement in later letters. But Huxley could not give a categorical answer. Nor could the botanists that Darwin asked about plants whose flowers seemed consistently to fertilise in the bud or whose structure seemed to preclude crossing. The possibility of the cross-fertilisation of plants that grew under water was an equally difficult problem that he took in turn to Watson, Hooker, George Bentham, and the Belfast botanist George Dickie. Darwin's theoretical notions also encouraged him to predict that trees would tend to show a separation of the sexes, a proposal that Asa Gray and Hooker confirmed during the course of 1856.

Darwin's researches into the purpose and results of sexual reproduction were pursued, as in all his lines of investigation, by experiments at Down. His letters show that he was involved with many different experiments on plants through the summers of 1856 and 1857, particularly with garden vegetables like peas and beans. These plants posed great theoretical difficulties in that none of the available sources, published or practical, could agree on whether the structure of papilionaceous flowers would allow for cross-fertilisation. Darwin carried out his researches with relish and published a short notice about the problem in *Gardeners' Chronicle and Agricultural Gazette* in October 1857, to be followed by a second notice in 1858.

The Leguminosae question was never fully resolved to Darwin's satisfaction. Indeed, several of his experiments failed to give him the results he wanted: seeds would not germinate; beans failed to cross; pigeon crosses did not yield conclusive results; newly-hatched molluscs refused to do what he hoped. One set of experiments, in particular, exhibits the steely inner resources that Darwin brought to his work and his indomitable will to succeed in wresting facts out of nature. Having long argued against the existence of former land connections between oceanic islands and their neighbouring continents (see *Correspondence* vol. 3), he had begun in 1855 a series of researches designed to explain how animals and plants might have been transported to islands without needing to travel over dry land. The centre-piece of his study was the series of experiments begun in 1855 based on soaking a wide variety of seeds in salt water in order to show that they could be transported across oceans without losing their ability to germinate. Even when he noticed, late into the project, that seeds usually sank and therefore could not be floated out to islands by ocean currents, the set-back was only temporary. Darwin rethought the question and began alternative experiments, including feeding seeds to fish. When the fish refused to eat their designated meals, he filled up the carcases of small animals with seeds and fed them to ocean-going birds in the zoological gardens in London. As he cheerfully explained to Hooker: 'I must tell you another of my *profound* experiments. Franky said to me, "why sh^d not a bird be killed (by hawk, lightning, apoplexy, hail &c) with seeds in crop, & it would swim." No sooner said, than done: a pigeon has floated for 30 days in salt water with seeds in crop & they have grown splendidly & to my great surprise even tares (Leguminosæ, so generally killed by sea-water) which the bird had naturally eaten have grown well.' (letter to J. D. Hooker, 10 December [1856]).

His faith in his ideas and his unfailing ingenuity led him to speculate on various other possible means of dispersal. He felt that the mud on birds' feet probably had a role to play in the distribution of seeds and carried out some unusual studies that required Joseph Parslow, the butler, to shoot partridges after a heavy rainfall so that Darwin could count the number of seeds in the earth between their toes. During outings in his carriage, Darwin also made frequent stops at local ponds and ditches to collect scoops of mud to analyse for seeds. Similar experiments were carried out on the possible means of transport of live ova or newly hatched animals such as land snails. Further to these, Darwin also investigated the length of time that seeds could ordinarily retain their ability to germinate, as, for example, in the case of seeds long-buried under the roots of trees (see letters to William Erasmus Darwin, [26 February 1856] and to Charles Lyell, 3 May [1856]).

Darwin's family was amused and patient about these researches, following his successes and failures with interest. It is clear from the correspondence published here that life at Down revolved around Darwin's work. Darwin emerges as a typical Victorian patriarch, retiring to his study for the day's work on his book, emerging only to visit the greenhouse and walk in the garden, the children accompanying him on the daily round of experimental plots, the pigeon houses, and the 'Sandwalk'. Like any father of his wealth and social position he was also anxious that his boys should be adequately schooled and worried about their future professions. His concern is well evidenced in the encouragement and occasional parental reprimand given to his eldest son William, then aged 17 and coming to the end of his time at Rugby school. 'I am very glad indeed to hear that you are in the sixth; & I do not care how difficult you find the work: am I not a kind Father?' Darwin wrote in 1857, soon followed by the complaint 'You want a jobation about your handwriting—dreadfully bad & not a stop from beginning to end!' (letters to W. E. Darwin, [17 February 1857] and 21 [July 1857]). The problem of careers for his six boys (Charles Waring Darwin, the sixth and last, was born on 6 December 1856) was a constant worry, particularly since Darwin seems to have felt that the children were poorly equipped to compete in the crowded professions of Britain. He looked to the space and opportunities of the colonies with some regret and envy: 'What a much better prospect you have for your sons . . . compared to what they could have been in this old burthened country, with every soul struggling for subsistence', he wrote to Syms Covington in New South Wales (letter to Syms Covington, 9 March 1856).

Many other topics, intellectual and personal, surface in this volume. Darwin exerted a considerable influence on the awarding of the Royal Society's medals and he took time and trouble trying to get friends and relations into the Athenæum Club. Several letters touch on the publication of John Tyndall's theory concerning the dynamics of glacial flow, with which Huxley was closely involved. Darwin discussed the geological phenomenon of cleavage, still unresolved in 1856, with John Phillips and entered into a useful correspondence with Samuel Pickworth Woodward on the variability and geographical distribution of fossil shells, stimulated by an appreciative reading of Woodward's *Manual of the Mollusca* (1851–6). The letters also reveal some of the views that contemporary naturalists held about Darwin.

Most significant in terms of Darwin's future, however, was the beginning of his correspondence with Alfred Russel Wallace. The first few letters exchanged between the two have been lost, but it is clear from the surviving correspondence that Darwin initially wrote to Wallace in order to obtain specimens of Malaysian fowls. Seen in the context of other letters written at this time, Darwin's remarks to

Wallace about his work on species and the preparation of his manuscript (letter to
A. R. Wallace, 1 May 1857) seem innocuous and hardly the veiled threat that
some historians have read into his words. Darwin may indeed have been writing
in part to establish his priority in this area, for Charles Lyell thought that
Wallace's 1855 paper implied some kind of belief in transmutation (see Wilson ed.
1970, pp. 54–5), but Darwin had no reason to suspect that Wallace was in any
way a direct competitor or about to pre-empt his views on natural selection. All
the available material seems to indicate that it was Lyell rather than Darwin who
feared the transmutationist implications of Wallace's paper.

Lyell's advisory role in the public presentation of Darwin's theories is also
clarified by the publication of their correspondence. Darwin's manuscript on
species was begun only after Lyell had urged him to publish a preliminary sketch,
but it seems that this famous advice may have been given on an occasion other
than the one previously supposed. Charles and Mary Elizabeth Lyell certainly
visited the Darwins at Down House for several days in April 1856, and Darwin
took this opportunity to explain his theory of natural selection to Lyell. Yet the
suggestion of composing a preliminary sketch was apparently first made in a letter
written by Lyell from London on 1–2 May 1856. Darwin took the suggestion
seriously and went up to London to see Lyell to discuss it further (letter to Charles
Lyell, 3 May [1856]). It was after this meeting that Darwin wrote to Hooker to
say that Lyell had pressed him to write up his views (letters to J. D. Hooker, 9
May [1856]).

Darwin had also received unsolicited encouragement from rather different
quarters in the interval between these meetings with Lyell. At a second weekend
party held at Down on 26 and 27 April 1856, he had discussed the question of
species at length with his guests, Hooker, Huxley, and Wollaston. Hearing about
the party afterwards, Lyell reported in a letter to his brother-in-law that, 'When
Huxley, Hooker, and Wollaston were at Darwin's last week, they (all four of
them) ran a tilt against species farther I believe than they are deliberately
prepared to go. Wollaston, least unorthodox. I cannot easily see how they can go
so far, and not embrace the whole Lamarckian doctrine.'(letter from Charles
Lyell, 1–2 May 1856, n. 7). The excitement and intellectual stimulation of that
weekend, combined with Lyell's astute advice, undoubtedly convinced Darwin
that he should begin writing.

The labour of composition, the search for authenticated information, the zest
for practical researches, and above all the determination to succeed are all vividly
displayed in this volume of correspondence. By the end of 1857, Darwin was well
on the way towards completing his manuscript and felt confident enough in his
views to explain them in explicit detail in a long letter to Asa Gray (letter to Asa
Gray, 5 September [1857]). From this letter it is evident that Darwin had brought
together all the diverse intellectual components of his theory, including the
relatively new concept of a 'principle of divergence' which Darwin believed
necessary to account for the diverging, tree-like relationships of organisms. With
the theory complete, he was anxious to press on with its exposition.

ACKNOWLEDGMENTS

The editors are grateful to Mr George Pember Darwin for permission to publish the Darwin letters and manuscripts. They also thank the Syndics of the Cambridge University Library and other owners of the manuscript letters who have generously made them available.

The work for this edition has been supported by grants from the National Endowment for the Humanities (NEH) and the National Science Foundation (NSF). The Alfred P. Sloan Foundation and the Andrew W. Mellon Foundation provided grants to match NEH funding, and the Mellon Foundation in 1981 and 1984 awarded grants to Cambridge University that have made it possible to put the entire Darwin correspondence into machine-readable form. Research and editorial work has also been supported by timely grants from the Royal Society of London and the British Academy, and the Royal Society has helped meet the costs of publication. The Pilgrim Trust provided funds to offset the effects of fluctuations in the dollar–pound exchange rate. The late Bern Dibner and Mary S. Hopkins provided financial assistance for the purchase of books and other necessary project expenses.

The Cambridge University Library and the American Philosophical Society have generously made working space and many services available to the editorial staff.

Since the project began in 1975, we have been fortunate in benefiting from the interest, experience, and practical help of many people in many places, and the editors hope that they have adequately expressed their thanks to them individually as the work proceeded. There are some, however, who have helped over so long a period that it would be ungracious not to thank them personally in the work which has profited so much from their co-operation.

Over the years, Kathy Fuller, assistant director of the Division of Research Grants of NEH, and Ronald J. Overmann, program director for History and Philosophy of Science of NSF, have given far more time and attention to the project than their formal duties required. The editors also appreciate the interest and encouragement they received from John E. Sawyer, former president of the Andrew W. Mellon Foundation, and his colleague, James Morris.

Without the expert help of John L. Dawson of the Literary and Linguistic Computing Centre of Cambridge University, the computerisation of the correspondence would not have been possible, for the work on both the *Calendar* and the *Correspondence* required the solution of many novel technical problems.

The computer of the Cambridge University Library, as well as that of the University, was used for initial work on this edition. W. D. S. Motherwell, head of automation at the Library, arranged for the installation of terminals in the project's office in the Manuscripts Department and solved the operational problems that arose.

This book was typeset at the Oxford University Computing Service on a Monotype Lasercomp. We are grateful to Susan Hockey for enabling the typesetting of the volume to be carried out, to Stephen Miller for advice and assistance with the Lasercomp, and to the other Oxford staff involved.

The late Sir Hedley Atkins and Philip Titheradge, formerly curator of the Darwin Museum at Down House, Downe, Kent, welcomed the editors on numerous visits and responded most generously to frequent requests for information and for material from the collections at Down. Solene Morris of the British Museum (Natural History) has enabled our close relationship with Down to continue. Professor Richard Darwin Keynes kindly made available Emma Darwin's diaries and other Darwin family material in his possession.

Libraries all over the world have given help that was literally indispensable by making available photocopies of Darwin correspondence and other manuscripts in their collections. The institutions and individuals that furnished copies of letters for this volume can be found in the List of provenances. The editors are extremely grateful to them all. We are also grateful to the many people who have transmitted information regarding the whereabouts of particular letters and, in many cases, have generously provided copies of such letters for the project's use.

Among the librarians whose help has been of especial importance, the foremost is Peter J. Gautrey who, until 1989, had charge of the Darwin Archive at the Cambridge University Library. His great knowledge of Darwiniana was always readily shared with the editors, as it was with countless other Darwin scholars. His kindness and invaluable assistance will be missed in Cambridge. The editors make daily use of the incomparable facilities of the Cambridge University Library and have benefited greatly from its services and the help and expertise of its staff, particularly the staff of the Manuscripts Department. We are especially grateful to the University Librarian, Frederick W. Ratcliffe, and the under-librarian of manuscripts, Arthur Owen, for their generous support. Other members of the library's staff who have frequently responded to the editors' requests are: David Hall, Elisabeth Leedham-Green, Margaret Pamplin, Jayne Ringrose, William Noblett, Gerry Bye, Godfrey Waller, Valerie Hall, Vicki Newman, Roger Fairclough, Cynthia Webster, and Paul Emmines.

At the American Philosophical Society Library, a splendid collection of Darwiniana and works in the history of science has been continuously available to the editorial staff since the project was started. Whitfield J. Bell Jr, secretary of the society until 1983, serves on the U.S. Advisory Committee for the project and has done his utmost to further its work. The editors have also benefited from the co-operation of Edward Carter II, Roy C. Goodman, Carl F. Miller, Elizabeth Carroll Horrocks, and Bertram Dodelin, all of the APS Library.

Douglas W. Bryant, formerly of the Harvard University Library, Rodney Dennis of the Houghton Library, Constance Carter of the Science Division of the Library of Congress, Lothian Lynas of the New York Botanical Garden Library, Judith Warnement of the Gray Herbarium of Harvard, and the staff members of the New York Public Library have all been exceptionally helpful in providing material from their great collections.

In Britain, the editors often had recourse to assistance from Keith Moore, archivist of the Royal Society; Rex E. R. Banks and Dorothy Norman of the British Museum (Natural History) Library; Gina Douglas, librarian of the Linnean Society of London; John Thackray of the Geological Society Archives and the British Museum (Natural History); Sylvia FitzGerald and Leonora Thompson of the Library of the Royal Botanic Gardens, Kew; Anthony M. Carr, Miss Williams, and Mrs Ion of the Local Studies Library, Shrewsbury; Richard G. Williams and Mrs Felton of Imperial College of Science, Technology and Medicine; Christine Fyfe, archivist of Keele University; K. A. Joysey, director of the Cambridge University Museum of Zoology; and Dr and Mrs Courtney of Christ's College, Cambridge.

Among the others who have advised and assisted the editors in their work are Mario di Gregorio, Julius Held, Madeleine Ly-Tio-Fane, Garry J. Tee, Hugh Torrens, and Leonard Wilson. Fredson Bowers, and G. Thomas Tanselle were frequently called upon for advice on editorial method. We are also glad to acknowledge the invaluable support of Desmond King-Hele, Owen Chadwick, and Maldwyn Rowlands.

Continued thanks are due to William Montgomery, former associate editor with this project, for his assistance in editing the letters in the collection of the American Philosophical Society. We are especially grateful to Hedy Franks for her skill and patience in deciphering and keyboarding complex manuscript material and to Peter Saunders for thorough and valuable research assistance.

Finally, with deep regret, we record the deaths of two members of our British Advisory Committee. Lady Nora Barlow's life and work guided all of us to a better understanding of her grandfather. Dennis Crisp's enthusiasm for Darwin and incomparable knowledge of cirripedes helped to make our work easier and better.

LIST OF PROVENANCES

The following list gives the locations of the original versions of the letters printed in this volume. The editors are grateful to all the institutions and individuals listed for allowing access to the letters in their care.

Alexander Turnbull Library, Wellington, New Zealand
American Philosophical Society, Philadelphia, Pa., USA
British Library, The Manuscript Collections, Great Russell Street, London, England
British Museum, Great Russell Street, London, England
British Museum (Natural History), South Kensington, London, England
Buckinghamshire Record Office, County Hall, Aylesbury, Bucks., England
Cambridge University Library, Cambridge, England
Christ's College Library, Cambridge, England
Cleveland Health Sciences Library, Cleveland, Ohio, USA
CUL *see* Cambridge University Library
DAR *see* Cambridge University Library
Deutsche Staatsbibliothek, Berlin, German Democratic Republic
Down House, Downe, Kent, England
D. J. McL. Edmondston (private collection)
Edinburgh University Library, Edinburgh, Scotland
Gardeners' Chronicle and Agricultural Gazette (publication)
Gloucestershire Record Office, Clarence Row, Gloucester, England
Gosse 1890 (publication)
Gray Herbarium of Harvard University, Cambridge, Mass., USA
J. Hancock 1886 (publication)
Houghton Library, Harvard University, Cambridge, Mass., USA
Imperial College of Science, Technology and Medicine, South Kensington, London, England
Leeds University Library, Leeds, England
Lincolnshire Archives Office, The Castle, Lincoln, England
Linnean Society of London, Burlington House, London, England
Lady Lyell (private collection)
John Murray (publishers), London, England
New York Botanical Garden Library, Bronx, N.Y., USA
Oxford University Museum, Oxford, England

I. J. Pincus (private collection)
Praeger 1935 (publication)
Royal Society, London, England
Shrewsbury Local Studies Library, Castle Gates, Shrewsbury, Shropshire, England
Sotheby's, London, England (dealers)
Suffolk Record Office, Raingate Street, Bury St Edmunds, Suffolk, England
Sydney Mail (publication)
University of Birmingham, Birmingham, England
University of British Columbia, Vancouver, British Columbia, Canada
University College London, London, England
University of Texas at Austin, Harry Ransom Humanities Research Center, Austin, Tex., USA
Wellcome Institute for the History of Medicine Library, London, England
Yale University, New Haven, Conn., USA

A NOTE ON EDITORIAL POLICY

The first and chief objective of this edition is to provide complete and authoritative texts of Darwin's correspondence. Insofar as it is possible, the letters have been dated, arranged in chronological order, and the recipients or senders identified. Darwin seldom wrote the full date on his letters and, unless the addressee was well known to him, usually wrote only 'Dear Sir' or 'Dear Madam'. After the adoption of adhesive postage stamps in the 1840s, the separate covers that came into use with them were usually not preserved and thus the dates and the names of many recipients of Darwin's letters have had to be derived from other evidence. The notes made by Francis Darwin on letters sent to him for his editions of his father's correspondence have been helpful, as have matching letters in the correspondence, but many dates and recipients have had to be deduced from the subject-matter or references in the letters themselves. These tasks, together with the deciphering of the handwriting of Darwin and others, have presented the most troublesome problems.

Whenever possible, transcriptions have been made from manuscript. If the manuscript was inaccessible but a photocopy or other facsimile version was available, that version has been used as the source. Other copies, published letters, or drafts have been transcribed when they provide texts that are otherwise unavailable.

The method of transcription employed in this edition is adapted from that described by Fredson Bowers in 'Transcription of manuscripts: the record of varients', *Studies in Bibliography* 29 (1976): 212–64. This system is based on accepted principles of modern textual editing and has been widely adopted in literary editions.

The case for using the principles and techniques of this form of textual editing for historical and non-literary documents, both in manuscript and print, has been forcefully argued by G. Thomas Tanselle in 'The editing of historical documents', *Studies in Bibliography* 31 (1978): 1–56. The editors of the *Correspondence* followed Dr Tanselle in his conclusion that a 'scholarly edition of letters or journals should not contain a text which had editorially been corrected, made consistent, or otherwise smoothed out' (p. 48), but they have not wholly subscribed to the statement made earlier in the article in which he says, 'In the case of notebooks, diaries, letters and the like, whatever state they are in constitutes their finished form, and the question of whether the writer "intended" something else is irrelevant' (p. 47). The editors have preserved the spelling, punctuation, and grammar of the original, but they have found it impossible to set aside entirely the question of authorial intent. One obvious reason is that in reading Darwin's writing, there must necessarily be reliance upon both context and intent. Even when Darwin's general intent is clear, there are cases in which alternative readings are, or may be, possible and therefore the transcription decided upon must to some extent be conjectural. In this work, when the editors were uncertain of their transcription they have enclosed the doubtful text in italic square brackets.

A major editorial decision was to adopt the so-called 'clear-text' method of transcription, which so far as possible keeps the text free of brackets recording deletions, insertions, and other alterations in the places at which they occur. Darwin's changes are, however, recorded in the back matter of the volume as alteration notes keyed to the printed text by paragraph and line number. All lines above the first paragraph of the letter (i.e., date, address, or salutation) are referred to as paragraph 'o'. Separate paragraph numbers are used for subscriptions and postscripts. These notes enable the reader who wishes to do so to reconstruct the manuscript versions of Darwin's letters while furnishing printed versions that are uninterrupted by editorial interpolations. They record all alterations made by Darwin in his letters and any editorial amendments made in transcription. For copies and drafts of Darwin letters included in the correspondence no attempt has been made to record systematically all alterations to the text, but ambiguous passages in copies and major revisions of drafts are noted. The editors believe it would be impracticable to attempt to go further without reliable information about the texts of the original or final versions of the letters involved. Letters to Darwin have been transcribed without recording any of the writers' alterations unless they reflect significant changes in substance or impede the sense; in such cases footnotes will bring them to the reader's attention.

Misspellings have been preserved, even when it is clear that they were unintentional as, for instance, 'lawer' for 'lawyer'. Such errors often indicate excitement or haste and may exhibit, over a series of letters, a habit of carelessness in writing to a particular correspondent or about a particular subject.

Capital letters have also been transcribed as they occur except in certain cases, such as 'm' and 'c', which are frequently written somewhat larger than others as initial letters of words. In these cases the normal practice of the writers has been followed. If there is doubt about Darwin's intention, these letters have been transcribed as capitals.

In some instances that are not misspellings in a strict sense, editorial corrections have been made. In his early manuscripts and letters Darwin consistently wrote 'bl' so that it looks like 'lb' as in 'albe' for 'able', 'talbe' for 'table'. Because the form of the letters is so consistent in different words, the editors consider that this is most unlikely to be a misspelling but must be explained simply as a peculiarity of Darwin's handwriting. Consequently, the affected words have been transcribed as normally spelled and no record of any alteration is given in the apparatus. Elsewhere, though, there are misformed letters that the editors have recorded because they do, or could, affect the meaning of the word in which they appear. The main example is the occasional inadvertent crossing of 'l'. When the editors are satisfied that the intended letter was 'l' and not 't', as, for example, in 'stippers' or 'istand', then 'l' has been transcribed but the actual form of the word in the manuscript has been given in an alteration note.

Editorial interpolations in the text are in square brackets. Italic square brackets enclose conjectured readings and descriptions of illegible passages. To avoid confusion, in the few instances in which Darwin himself used square brackets, they have been altered by the editors to parentheses with the change recorded in the back matter. In letters to Darwin, square brackets have been changed to parentheses silently.

Material that is irrecoverable because the manuscript has been torn or damaged is indicated by angle brackets; any text supplied within them is obviously the responsibility of the editors. Occasionally, the editors are able to supply missing sections of text by reference to transcripts or photocopies of manuscript material that has suffered further damage after the copy was made.

Words and passages that have been underlined for emphasis are printed in italics in accordance with conventional practice. When the author of a letter has indicated greater emphasis by underlining a word or passage two or more times, then the greater emphasis is indicated by printing the text in bold type.

Paragraphs are often not clearly indicated in the letters. Darwin and others sometimes marked a change of subject by leaving a somewhat larger space than usual between sentences; sometimes Darwin employed a longer dash. In these cases, and in very long stretches of text when the subject is clearly changed, a new paragraph is started by the editors without note. The start of letters, valedictions, and postscripts are also treated as new paragraphs regardless of whether they appear as such in the manuscript. Special manuscript devices delimiting sections or paragraphs, e.g., blank space left between sections of text and lines drawn across the page, are treated as normal paragraph indicators and are not specially marked or recorded unless their omission leaves the text unclear.

Additions to a letter that run over into the margins or are continued at its head or foot, are transcribed at the point in the text at which the editors believe they were intended to be read. The placing of such an addition is recorded in a footnote if it seems to the editors to have some significance or if the position at which it should be transcribed is unclear.

Occasionally, punctuation marking the end of a clause or sentence is not present in the manuscript, but the author has made his or her intention clear by allowing, for example, extra space or a line break to function as punctuation. In such cases, the editors have inserted an extra space following the affected sentence or clause to set it off from the following text.

Some Darwin letters and an occasional letter to Darwin are known only from entries in the catalogues of book and manuscript dealers or mentions in other published sources. Whatever information these sources provide about the content of such letters has been reproduced.

For every Darwin letter, the text that is available to the editors is always given in full. Some other items, however, are not printed in their entirety. Most memorandums and other documents that are not letters but are relevant to the correspondence have been summarised, as have those letters to Darwin that the editors consider can be presented adequately in shortened form.

The format in which the transcriptions are printed in the *Correspondence* is as follows:

1. *Order of letters.* The letters are arranged in chronological sequence. A letter that can be dated only approximately is placed at the earliest date on which the editors believe it could have been written. The basis of a date supplied by the editors is given in a footnote unless it is derived from a postmark, watermark, or endorsement as recorded in the physical description of the letter (see section 4, below). Letters with the same date, or with a range of dates commencing with that date, are printed in the alphabetical order of their senders or recipients unless their contents dictate a clear alternative order. Letters dated to a year or a range of years precede letters that are dated to a particular month or range of months and these, in turn, precede those that are dated to a particular day or to a range of dates commencing with a particular day.

2. *Headline.* This gives the name of the sender or recipient of the letter and its date. The date is given in a standard form, but those elements not taken directly from the letter text are supplied in square brackets.

3. *The letter text.* The transcribed text follows as closely as possible the layout of the source, although no attempt is made to produce a type-facsimile of the manuscript, that is, word-spacing and line-division in the running text are not adhered to. Dates and addresses given by authors are transcribed as they appear, except that if both date and address are at the head of the letter they are always printed on separate lines with the address first, regardless of the exact manuscript order. If no address is given on a letter by Darwin, the editors have supplied one, when able to do so, in square brackets at the head of the letter. Addresses on printed stationery are transcribed in italics, often in a simplified form. Addresses, dates, and valedictions have been run into single lines to save space, but the positions of line-breaks in the original are marked by vertical bars.

4. *Physical description.* All letters are complete signed autograph letters unless otherwise indicated. If a letter was written by an amanuensis, or exists only as a draft or a copy, or is incomplete, or is in some other way unusual, then the editors provide the information needed to complete the description. Postmarks, endorsements, and watermarks are recorded only when they are evidence for the date or address of the letter.

5. *Source.* The final item provides the provenance of the text. Some sources are given in abbreviated form (e.g., DAR 140: 18), but are listed in full in the List of provenances unless the source is a published work. References to published works are given in author–date form, with full titles and publication details supplied in the bibliography at the end of the volume.

6. *Darwin's annotations.* Darwin frequently made notes in the margins of the letters he received, scored significant passages, and crossed through details that were of no further interest to him. These annotations are transcribed or described following the letter text together with details of their location on the manuscript and the writing medium. They are keyed to the letter text by paragraph and line numbers. Most notes are short, but occasionally they run from a paragraph to several pages sometimes written on separate sheets appended to the letter. Extended notes relating directly to a letter and physically associated with it are transcribed whenever practicable.

Quotations from Darwin manuscripts in footnotes and elsewhere, and the text of his annotations and notes on letters, are transcribed in 'descriptive' style. In this method the alterations in the text are recorded in brackets at the places in which they occur. For example:

'See Daubeny ['vol. 1' *del*] for *descriptions of volcanoes in [*interl*] S.A.' *ink*

means that Darwin originally wrote in ink 'See Daubeny vol. 1 for S.A.' and then deleted 'vol. 1' and inserted 'description of volcanoes in' after 'for'. The asterisk before 'descriptions' marks the beginning of the interlined phrase, which ends at the bracket. The asterisk is used when the alteration applies to more than one preceding word. The final text can be read simply by skipping the material in brackets. Descriptive style is also used in the alteration notes in the transcription of deleted passages from Darwin's letters.

Volumes of the *Correspondence* are published in chronological order. Each volume is self-contained, having its own index, bibliography, and biographical notes. A comprehensive index is planned for the final volume. References are supplied for all persons and subjects mentioned, even though some repetition of material in earlier volumes is involved.

If the name of a person mentioned in a letter is incomplete or incorrectly spelled, the full correct form is given in a footnote. Brief biographies of every identifiable person mentioned in a volume and dates of each correspondent's letters to and from Darwin are given in the Biographical register and index to correspondents.

The editors use the abbreviation 'CD' for Charles Darwin throughout the notes. A list of all abbreviations used by the editors in this volume is given on p. xxxii. For references to Darwin's books, articles, and collections of his letters, short titles are used (e.g., *Descent*, *Collected papers*, *LL*). Short titles are also used for some standard reference works (e.g., *Alum. Cantab.*, *Wellesley index*, *DNB*). For all other works, author–date references are used. The full titles of all the books referred to are given in the bibliography; the short titles and author–date references are listed together in alphabetical order.

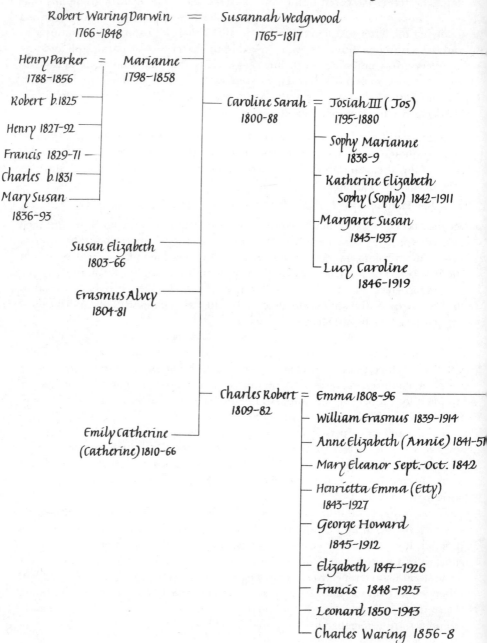

Robert Waring Darwin = Susannah Wedgwood
1766-1848   1765-1817

Henry Parker = Marianne
1788-1856   1798-1858

Robert b.1825

Henry 1827-92

Francis 1829-71

Charles b.1831

Mary Susan
1836-93

Caroline Sarah = Josiah III (Jos)
1800-88   1795-1880

Sophy Marianne
1838-9

Katherine Elizabeth
Sophy (Sophy) 1842-1911

Margaret Susan
1843-1937

Lucy Caroline
1846-1919

Susan Elizabeth
1803-66

Erasmus Alvey
1804-81

Charles Robert = Emma 1808-96
1809-82

William Erasmus 1839-1914

Anne Elizabeth (Annie) 1841-51

Mary Eleanor Sept.-Oct. 1842

Henrietta Emma (Etty)
1843-1927

George Howard
1845-1912

Elizabeth 1847-1926

Francis 1848-1925

Leonard 1850-1943

Charles Waring 1856-8

Emily Catherine
(Catherine) 1810-66

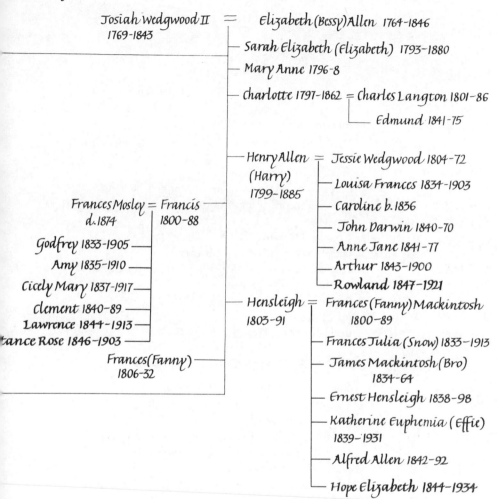

Josiah Wedgwood II 1769-1843 = Elizabeth (Bessy) Allen 1764-1846

— Sarah Elizabeth (Elizabeth) 1793-1880
— Mary Anne 1796-8
— Charlotte 1797-1862 = Charles Langton 1801-86
        └─ Edmund 1841-75

— Henry Allen (Harry) 1799-1885 = Jessie Wedgwood 1804-72
        — Louisa Frances 1834-1903
        — Caroline b. 1836
        — John Darwin 1840-70
        — Anne Jane 1841-77
        — Arthur 1843-1900
        └─ Rowland 1847-1921

Frances Mosley = Francis
d. 1874   1800-88
Godfrey 1833-1905
Amy 1835-1910
Cicely Mary 1837-1917
Clement 1840-89
Lawrence 1844-1913
...ance Rose 1846-1903

Frances (Fanny)
1806-32

— Hensleigh = Frances (Fanny) Mackintosh
1803-91   1800-89
        — Frances Julia (Snow) 1833-1913
        — James Mackintosh (Bro)
             1834-64
        — Ernest Hensleigh 1838-98
        — Katherine Euphemia (Effie)
             1839-1931
        — Alfred Allen 1842-92
        └─ Hope Elizabeth 1844-1934

ABBREVIATIONS

A	autograph, i.e., in the hand of the sender
L	letter
mem	memorandum
S	signed by the sender
AL	autograph letter
ALS	autograph letter signed
LS	letter in hand of amanuensis, signed by sender
LS(A)	letter in hand of amanuensis with additions by author
CD	Charles Darwin
CUL	Cambridge University Library
DAR	Darwin Archive, Cambridge University Library
del	deleted
illeg	illegible
interl	interlined
underl	underlined

TRANSCRIPTION CONVENTIONS

[some text]	'some text' is an editorial insertion
⌈some text⌉	'some text' is the conjectured reading of an ambiguous word or passage
⌈some text⌉	'some text' is a description of a word or passage that cannot be transcribed, e.g, *3 words illeg*
⟨ ⟩	word(s) destroyed
⟨some text⟩	'some text' is a suggested reading for a destroyed word or passage
⟨*some text*⟩	'some text' is a description of a destroyed word or passage, e.g., *3 lines excised*

THE CORRESPONDENCE OF CHARLES DARWIN
1856–1857

To William Bernhard Tegetmeier 1 January [1856][1]

Down Bromley Kent
Jan 1st

My dear Sir

I write one line to thank you very sincerely for the card for the Philo-peristeron;[2] which I will certainly attend if I *possibly* can: if I do not come, I will send the cranium per post;[3] but I hope that I shall be able.— I am sorry to hear that you have been harassed in what seems a most just cause.

Your's sincerely | C. Darwin

Very many & sincere thanks for keeping me in mind about Pigeons.

New York Botanical Garden Library (Charles Finney Cox collection)

[1] Dated by CD's attendance at the annual show of the Philoperisteron Society (see n. 2, below).
[2] Tegetmeier was secretary of the Philoperisteron Society, a club of gentlemen pigeon-fanciers that met in the Freemason's Tavern, London. See Secord 1981, p. 172, and E. W. Richardson 1916, pp. 100–2. CD attended the meeting of 8 January (*Cottage Gardener*, 15 (1855–6): 301), and became a member of the society on 14 October 1856 (Secord 1981, p. 176).
[3] Probably the head of the jungle fowl from India that CD had offered to lend Tegetmeier (see *Correspondence* vol. 5, letter to W. B. Tegetmeier, 6 December [1855]).

To John Maurice Herbert 2 January [1856][1]

Down Bromley Kent
Jan. 2d—

My dear Herbert

I thank you sincerely for your most friendly letter.—[2] I shall have to send at end of week to Athenæum[3] & will there get the volume of Poems, which Mrs Herbert & you have so kindly sent us, & which I have no doubt we shall like much.[4] My wife desires to be most kindly remembered to Mrs Herbert, & she has often regretted that we have been so completely separated. Some few months ago I was looking over a few old & valued letters which I keep, & amongst others there were some of yours & which were so pleasant that they quite warmed my old heart.[5] I began a letter to you on the strength of old remembrances, but as I was interrupted I afterwards destroyed it, having nothing particular to say, except what I am sure you will believe, that I shall keep to my dying day as unfading remembrance of the many pleasant hours (especially at Barmouth) which we have spent together.—[6]

My health will, I fear, never be fully reestablished, & though anyone would think me a strong man, this is very far from the case; & hence I never go from home, so that I much fear there is little chance of our being able to visit you at Rocklands,[7] though I am very Sure it would give us much pleasure.— If ever you have time to spare when in London it would give us real pleasure, if you could pay us a visit here.—

With our united very kind remembrances to you both, My dear Herbert | Your old & affect.^e friend | Charles Darwin

American Philosophical Society (121)

[1] The year in which the letter was written is conjectured from an 1855 watermark (Carroll 1976).

[2] The letter has not been found. Herbert had been one of CD's close friends during their undergraduate days at Cambridge University. See *Correspondence* vol. 1.

[3] CD used his London club as an address for the delivery of books and parcels.

[4] The volume of poetry has not been identified. In 1877, Herbert edited a book of his wife's poems entitled *Poems by the late Mary Anne Herbert* (London).

[5] For the extant correspondence between CD and Herbert, see *Correspondence* vols. 1 and 2.

[6] In the summer of 1828, CD spent two months with Herbert and Thomas Butler in Barmouth on an 'Entomo-Mathematical expedition' (*Correspondence* vol. 1, letter to W. D. Fox, [30 June 1828], and Appendix I).

[7] Herbert's home in Herefordshire. Since 1847, Herbert had been a county court judge on the Herefordshire, Radnorshire, and Monmouthshire circuit (*Alum. Cantab.*).

To William Darwin Fox 3 January [1856]

Down Bromley Kent
Jan. 3^d

My dear Fox

Thanks for your letter: I had your name on my list to write to soon to tell you how I got on in the Cock & Hen line of business. I have got nothing but promises as yet, & there are few or none like you, who do what they promise, though I fully believe they intend at the moment to do so.

I sh^d be very glad of old & good Cochin, Dorking & Malay Call Drake;[1] though I think it possible that I may get Cochin elsewhere; if I succeed I will instantly write to you.— Indeed almost any cock w^d be very valuable to me, if of good breed; but I do not want to be unreasonable, after the immense trouble you took for me in regard to Chickens.[2] I have as yet not tried energetically for Poultry, but I have for dead Pigeons, & my success has been not great, so that I do not want throw away any chance of good Bird from any quarter.

I offer payment for dead Pigeons, & Ducks & Rabbits to Mr Baker.—[3] Young Baily, from whom I have bought live Pigeons, I found not at all obliging about dead birds.[4] If I fail by other channels, I will apply to the Father Baily & use Pulleine's name,[5] or get him to write a note to him.— I have since I wrote last greatly extended my scheme; & I have now written above 20 letters to every great quarter of the world to professional skinners, & others to get me collection of Poultry & Pigeons skins;[6] if I succeed I think it will be a very curious collection.—

With respect to the Athenæum I have f^d it so dull that I have for some time left off taking it in;—I quite forgot to mention this to you.—[7]

With respect to seeds, my few remarks were made, on account of Hookers, extreme obstinacy, (as it appears to me) in not believing that seeds can live even a few years in the ground.[8] Your Isle of Wight case w^d have been interesting if written at the time with certainty about the species; but without every particular given Hooker & Bentham, sneer at every account.—[9]

I am particularly obliged by the particulars on the eggs & colour of the Muscovy Ducks & sheep: all such facts are *valuable* to me.[10] I have heard something analogous about crossing Cochin Chinas.— Did you ever see the Poultry Chronicle; one of the best contributor B. P. Brent, lives not far off, & is very kind in giving me information: unfortunately he has given up keeping many Birds.—[11]

Farewell, we are not in a very flourishing condition; as many of us are rather poorly, & my wife for a couple of months has been suffering much from headachs.— I congratulate & condole with you on your 12th child:[12] in my own case, I sh^d have wished only for condolence.

Yours very affectionately | C. Darwin

P.S. | Would you blow for me an *average* egg of the White & Slate-coloured or other var. of Musk Duck.— I sh^d very much like to have these.

Postmark: JA 3 1856
Christ's College Library, Cambridge (Fox 86)

[1] The Cochin, Dorking, and Malay are breeds of fowl; the 'Call Drake', which CD wrote above 'Malay' without changing the punctuation, is a breed of domestic duck.

[2] In 1855, Fox had provided CD with specimens of young chickens and ducks for his study of variation and the particular question of whether the young of different domestic breeds differ from each other as much as their parents do. See *Correspondence* vol. 5, especially letters to W. D. Fox, 23 May [1855] and 22 August [1855].

[3] Either Samuel C. or Charles N. Baker, dealers in poultry, Chelsea.

[4] John Baily, poultry dealer, was a regular contributor on poultry-keeping to the *Cottage Gardener*. From 1855 onwards, there are frequent entries for 'Baily Pigeons' in CD's Account book (Down House MS). Baily did eventually supply CD with skins and other specimens, as noted in *Variation* 1: 132 n. 2.

[5] Robert Pulleine, rector of Kirkby-Wiske, Yorkshire, had been a friend of CD's at Cambridge. He was well known as a judge at poultry shows (see *Cottage Gardener* 15 (1855–6): 208 and 227, and letter to W. B. Tegetmeier, 14 January [1856]). The elder John Baily, of Mount Street, Grosvenor Square, London, was also an established judge of poultry.

[6] See *Correspondence* vol. 5, CD memorandum, [December 1855], and subsequent letters in December 1855.

[7] CD used to read the *Athenæum* and then send it on to Fox (see *Correspondence* vol. 4, letter to W. D. Fox, 4 September [1850]).

[8] CD refers to his recent letters to the *Gardeners' Chronicle* (see *Correspondence* vol. 5, letter to *Gardeners' Chronicle*, 13 November [1855], and the first letter to *Gardeners' Chronicle*, [before 29 December

1855]), concerning the vitality of ancient and long-buried seeds. Joseph Dalton Hooker had written to the same journal (*Gardeners' Chronicle and Agricultural Gazette*, 8 December 1855, pp. 805–6) about CD's report on charlock seeds that had germinated after being buried for at least eight years, stating: 'my objection to placing complete confidence in this case, is the want of evidence that Charlock seed will withstand the destroying effects of moisture for any number of years' (p. 806).

[9] Fox visited the Isle of Wight for his health (see *Correspondence* vol. 1, letter from W. D. Fox, 1 November 1834). His account of seeds from the island has not been traced. George Bentham had expressed a view similar to Hooker's (see n. 8, above) in an article on the vitality of charlock seed in *Gardeners' Chronicle and Agricultural Gazette*, 10 November 1855, pp. 741–2.

[10] See *Variation* 1: 248–9.

[11] Bernard P. Brent then lived at Bessel's Green, Riverhead, Kent. In *Variation* 1: 132 n., CD wrote: 'Mr. B. P. Brent, well known for his various contributions to poultry literature, has aided me in every way during several years'. Brent was a member of the Columbarian Society. The *Poultry Chronicle* published three volumes between 1854 and 1855 before being subsumed by the *Cottage Gardener, and Country Gentleman's Companion*. Brent contributed a regular column to the former entitled 'Colombiary' and continued to write articles for the poultry section of the *Cottage Gardener*.

[12] Frederick William Fox was born on 8 December 1855 (*Darwin pedigree*).

To John Stevens Henslow 3 January [1856][1]

Down Bromley Kent
Jan.ʸ 3ᵈ

My dear Henslow

I have received your letter, the Report & pamplet,[2] for all of which very many thanks.— Your letter has been of *real* use to me, in deciding what to do, which will be, in consequence, very little.—

What trouble Government does give about Clubs.— I have been two whole days in drawing up the annual & Quinquennial Returns!

Farewell with many thanks.— Pray remember to let me have the case of the Canada Geese with the seed in crop, if ever you shᵈ meet with it; as the *means* of distribution is, at present, a great hobby with me.—[3]

Farewell | C. Darwin

Your servants relation. failed, I presume, in getting the wild Carnation seed from Rochester.—

DAR 93: 106–7

[1] Dated by the relationship to an earlier letter to Henslow about local insurance clubs and societies (*Correspondence* vol. 5, letter to J. S. Henslow, 26 December [1855]).

[2] The report and pamphlet were sent in reply to CD's request for advice on 'various worrisome questions' about the Down Friendly Club, of which he was treasurer. See *Correspondence* vol. 5, letter to J. S. Henslow, 26 December [1855].

[3] See *Correspondence* vol. 5, letter to J. S. Henslow, 26 March [1855], n. 5.

From Charles Wade Crump to Edward Blyth [before 8 January 1856][1]

Facts regarding Lions in Central India[2] and Wild Cattle in Southern India | Chas Crump. | Artillery[3]

Wild Cattle in the South.

There is as I have always heard a breed of wild cattle of a dull brick red colour in the Tinneyvelly District, who it is the tradition of the country sprang originally from the tame village cows who were driven off to the jungles by the people of the country when Tippo Sahib invaded the Raja of Travancore about the year 1788—[4] So much I know only by hearsay— the following is of my own knowledge— On my way down from the Hills (Neilgheeries) in 47—I stayed with our 49th at Vellore, there, one of the Officers (Oliver Butler I think)[5] told me of a strange herd of wild cattle to be met with on the great plains about Rajas Choultry on the road to Madras and shewed me the skin and horns of one a bull which he had himself killed there a month or so before— this skin was dull brick red colour the horns same shape and as far as I remember rather thicker than common bullocks horns—

I halted a day at Rajas Choultry and went out after this herd but could not find them, however they were perfectly well known to all the people thereabouts and the account I received was as follows—

About 4 years before that time a wild bull of a red colour had appeared on the great plains which stretch from Rajas Choultry to Arcot, (and the people all said he came from the jungles *below* Trichinopoly)— this bull fed by himself but at times consorted with the tame herds fought and drove away the tame bulls and then walked into the cows; from this intercourse in process of time sprung calves— two at first, red like their father—who when they had done sucking forsook the tame beasts and lived apart with their sire, and so the herd went on *he* bulling the Village cows, and the calves forsaking the tame herds for the wild one till the wild cattle were 7 in number. viz—the original old bull the young bull Butler shot, and 5 cows.— this young bull was driven out of the herd by his Father for taking improper liberties with the cows took up his position close to a village where he molested cows, men, and everybody the Potail[6] of the place came to the travellers Bungalow informed B of the fact and requested him to bring his gun and rid the village of the nuisance, young bull was under a tree, wouldnt budge but when B drew near, charged and was so shot.—

I heard afterwards that the whole race was exterminated by Kennedy of our 1st L. C. when he was stationed at Arcot—[7]

I have heard from Jerden of another breed of wild cattle living in the salt water lagoons and islands above Nellore, he has been at the killing of some, and describes them as being of a dirty grey white with dark points.[8]

DAR 98: 114–16

CD ANNOTATIONS
3.3 but at times . . . the cows; 3.5] *scored brown crayon*
3.5 sprung calves . . . tame beasts 3.7] *scored brown crayon*

[1] Dated by the relationship to the following letter. Blyth forwarded this letter to CD with his own letter of 8 January 1856 (see following letter).

[2] Blyth here added: 'These I have retained, EB.' before he sent Crump's letter on to CD. Crump had presumably supplied Blyth with information on Asiatic lions for his article in the *Calcutta Sporting Review* (see letter from Edward Blyth, 8 January 1856).

[3] Blyth added 'Madras' before 'Artillery'. Crump wrote this note on the cover of his letter.

[4] Tippoo Sahib, sultan of Mysore, ravaged the territories of the raja of Travancore in 1789 (*EB*).

[5] John Olive Buttler was a lieutenant in the Madras forty-ninth regiment native infantry stationed at Vellore from March 1847 (*East-India register and army list, for 1848*).

[6] An alternative spelling of patel, a village head-man in south and central India (*OED*).

[7] Lord David Kennedy was a lieutenant in the Madras first regiment light cavalry stationed at Arcot from May 1847 (*East-India register and army list, for 1848*).

[8] Later described by Thomas Claverhill Jerdon of the Madras service in Jerdon 1867, p. 301: 'Near Nellore, in the Carnatic, on the sea-coast, there is a herd of cattle that have been wild for many years. . . . Their horns were very long and upright, and they were of large size. I shot one there in 1843, but had great difficulty in stalking it, and had to follow it across one or two creeks.'

From Edward Blyth 8 January [1856][1]

Calcutta,
Jan.Y 8/55—

My dear Sir,

I have two letters of yours to answer, respectively dated Nov.r 5th & 22nd—[2] but I have already penned three sheets of "Notes for M.r D." & I have also an interesting notice for you of wild humped cattle in S. India,[3]—and finally a few queries for M.r Moore, (D.r Horsfield's assistant in the India house).[4] Will you kindly send to him the scrap of paper containing these last, and ask him to let me have the replies as soon as ever he conveniently can. Tell him that I am greatly obliged to him for his comm.n about the Finches, &c; and that I will write to him by an early opportunity. My paper on Orangs is out, but I cannot send it to you by this mail;[5] and also 2 papers of mine in the 'Calcutta Sporting Review', respectively on *Asiatic Lions* & *wild Asses*. These I recommend to your attention, but can only send you the first; so apply for the number to Lepage & C°, the London publishers,—no 44, for Dec.r 1855.[6] I am preparing several elaborate articles for this work, & have just sent one in on the Tiger; & I am writing also the gallinaceous tribes of India, which will probably furnish 3 or 4 articles. So I think you will not be disappointed if you take this Review henceforth. I get tolerably well paid for these articles, & therefore send them to this 'Sporting Review'; but I endeavour to make them suggestive, & they may be the means of eliciting no small amount of information from the more observant of Indian sportsmen.

Now for your letters. Many thanks for your kind endeavours to procure specimens of the British Crustacea. Horns. Those of Red Deer are what I most want, & if on the skull or frontlet, of course so much the better; but unless fine specimens with branching 'crown', they are scarcely worth sending. Good Fallow-Deer frontlets would also be acceptable; and I should like a good pair of horns of *C. virginianus*, should such be procurable.

You have evidently quite misunderstood me about *Treron* and *Coracias*. Different species of *Treron* seem to interbreed, & ditto *Coracias*; but not *Treron* with *Coracias*![7] That would, indeed, be "an astounding fact". I will treat further on this subject when I have got through the rest of your letters. My friend Bashford has recently returned to Bengal, & to his silk factory.[8] You are doubtless right about the attitudes assumed by insects, spiders, &c, when *shamming*; not so, however, with beasts & birds when pulled about, & *"possuming"*![9] I cannot recollect where I met with that anecdote about the cattle near a railway; but think it was in the London Athenæum, in the review of some continental tourist's work. No! I seem to fancy that it was headed 'Communication of ideas among animals,' or some such title; & it appeared probably in the Illustrated News, 3 or 4 years ago, & was certainly quoted from the work of some tourist in Germany. I am extremely sorry now that I did not note down the authority.[10] You mention a breed of cattle "of American origin almost certainly, but I cannot make out the name—'*Niata*', it looks like.—[11]

There is a Chinese breed of silky fowls, with white feathers & black skin (& periostæum of course); and also a Malayan race of silky fowls with white skin & feathers. The latter has the ordinary comb & wattles; but the other is a most singular bird about the head, having (as near as I can remember) an *even comb* as if cut, & no lateral wattles (as usual), but a tumid throat, with much blue about the naked skin of the face. A friend of mine here received a pair, which were presented to him as *Eagles*! The hen soon died, but the cock became the sire of some chicken by a Chittagong hen, which showed *scarcely any trace* of the peculiarities of their dad! I can send you the skin of a youngish cock of this mixed race. Our other black-skinned *hens* have nought else remarkable about them than the melanism of the skin, *comb*, &c Those you speak of, in England, "with hair-like feathers", are probably some intermixture, perhaps of the Chinese silky fowl with black-skinned ordinàry fowls. Vide Griffiths edition of the 'animal Kingdom' VIII, 222, for notice of black-skinned fowls in S. America, by Azara![12] Also *feral*, in p. 177!—[13] I doubt much any breed of Canaries having intermixture of Siskin blood. Look to shape of beak, & length of tail. All domestic Canaries, so far as I know, remain quite true to the particular African type, exemplified by certain species which Ruppell refers to *Serinus*. His *S. melanocephalus* for instance.[14] I much wish that you could get hold of & study the true wild Canary, & note its *song*; coloured figures of a pair of them are worthy of publication in your forthcoming work. The song of a Siskin-hybrid would, for certain, be much modified: & how curious it is that the Goldfinch and other mules should have even the *song* intermediate! There are some ten or a dozen species of true Siskin in both Americas; one only of which (*Stanleyi*) resembles the Himalayan *spinoides* by its thicker bill; & the Pr. of Canino makes a particular division of these two.—[15]

Camels. Those which I referred to as having bred in the Z. Gardens, were (both parents) of the large dark-coloured & 1-humped breed, known as the Armenian or Caramanian Camel, and which is stated by every author to be a mixed breed between the 1-humped & the 2-humped. Vide my note, in *J. As. B.* XV, 162, (to which I referred you before,)[16] also Hutton's remarks there,[17] & Chesney, 'Jl of Euphrates Expedition', I, 582 & 82.[18] In Hutton's Scripture Geology book, he figures the 2-humped Camel *pale*;[19] it is usually dark-coloured, & the 1-humped is always pale *in India*. The reference you cannot make out—"Bur—?" must be Burckhardt, 'Travels in Nubia', p. 222; but see my note as above.—[20]

The *Jumni Pari* Goat = the Syrian, a very remarkable race with long legs and excessively elongated pendent ears.— Have not *all* domestic Cats a very slight *Lyncine* tuft on the ears? As for the marks on Donkey's legs, I have sufficiently gone into this subject in my paper on Wild Asses, to which I therefore refer you.[21] I do not think that the stripes imply Zebra intermixture, but are merely what we see *in so very many instances* in various classes, of markings (often strongly pronounced) in the young, which disappear more or less in adults. Vide Lion & Puma cubs, reptiles innumerable, very many fish also (as the young of *Salmonidæ*), &c &c &c. Are not some caterpillars even more intensely marked and coloured when small? Some of your other queries I have noticed by anticipation in my 'notes for Mr Darwin'.[22] The rest when I can manage it.

Yours truly ever | E. Blyth.

P.S. Coracias, 3 Indian species,— *garrula* in the N.W. only,— *indica* throughout India, replaced E. of the Bay by *affinis*. The 2 last interbreed, & shew every possible intermediate gradation, *i.e.* where they come into contact,—each race remaining pure in its proper region. Whether *C. garrula* & *indica* also interbreed, I am unaware.

Treron. 3 yellow-footed races,— *chlorigaster* in the Indian peninsula & Ceylon,— *phænicoptera* in Bengal & all Upper India,—& *viridifrons* in Burma. The first and second seem to intergrade where they come in contact.[23]

Turtur suratensis of India & *T. tigrina* of the Malay countries, do. N.B. The latter is distinct from *chinensis*, with which the Pr. of Canino seems to confound it.[24] But the *Columbidæ* present *very numerous* cases of species of different regions *barely separable*,—or what many would call distinct local races of the same species, but there is no knowing where to stop when this principle is once admitted. The *C. livia* group affords one instance out of *very many*: still I cannot help thinking that *Carpophaga ænea* of the Nicobars, as compared with *ænea* of all the surrounding countries, is just such a local race as *Lagopus scoticus* & *Lepus hibernicus*, which I have treated of in my 'notes for Mr D.' now forwarded.

The Scandinavian Bottletit has no markings on the head; & compare, if you can, other Scandinavian Tits with British specimens.

For hybrid Kallij Pheasants, & the intermediate races, vide *J. As. B.* XVII, 694.[25]

For representative species or races, differing only in certain details of colouring,—& which are never found in the intermediate country,—cite *Mustela Gwatkinsii* of the Nilgiris,—*M. flavigula* of the entire Himalaya & even Arakan Mountains,—& *M. ——?* (hitherto *flavigula*, var.,) of the Malayan peninsula.

The *Mydaus meliceps* thus occurs in Java, only on certain elevated table-lands, & *never* in the intermediate country; & the species seems absolutely the same in the Malayan peninsula! See Horsfield's *Zool. Res. in Java.*[26]

Have you studied Corse's paper on the Indian Elephant & its varieties in the As.ᶜ Researches?[27]

The groups exemplified by *Sciurus maximus* & *Pteromys petaurista* are especial puzzles, as to what are to be considered species & what varieties!

Now for another matter. Is not the *Kual-Kole* (however should it be spelt?) or 'turnip-rooted Cabbage' quite a modern *variety*? Originating I think in S. Africa, famous for fat-rumped sheep & Hottentots, & to which the broad-sterned Dutch race have taken kindly! No particular analogy, but the Cape climate is *likely to have originated* the *Kual Kole*[28]

[Enclosure]

Notes for M.ʳ Darwin.

Distribution of domesticated *Cairina moschata*. In Crawfurd's 'Mission to Siam & Cochinchina' (*A.D.* 1821), p. 434, we read— "The Muscovy Duck (*Anas moschata*), now very generally found throughout the east, although a native of America, is bred in small numbers about Bangkok. Its name *Pet Manilla*, or the 'Duck of Manilla', indicates the direction from whence it reached the country".—[29] *N.B.* In enumerating the domestic animals derived from America, do not omit the Musk Duck.

Gosse, I see, in his 'Birds of Jamaica', has a note about the Spaniards having derived the domestic Turkey *from the Antilles*, & argues that the species was probably found there *wild* of yore.—[30] See his remarks on the *Guinea-fowl*, which abounds wild in Jamaica, & has done so for more than 150 years![31] Yet Edwards, in his 'Birds' (not a century ago), Vol VII, p. 269, Volume dated 1764, states that when he was a boy, the 'Guiney-hens' were shewn *as rarities* in *England*; but that now they were commonly bred all over the country, & already exhibited the full amount of variation attained up to this present time (1856). Those first brought, says he, *were all of the wild colour*! Now their variation corresponds to that of the *Pheasant*, white, partially white, or pale—the 'Bohemian Pheasant' of the dealers.[32]

I have been investigating the *history* of domestic Guinea-fowl, & come to sundry conclusions.— Firstly, our race *is not descended from the Roman*, inasmuch as it is clear to me that the *Meleagris* of the Romans (& doubtless of Aristotle) is the E. *African* bird, *N. ptilorhyncha*, Lichtenstein; an inhabitant of Abyssinia, Sennar, & Kordofan, which might have been received through *Nubia*;—& query, is (or was) there not confusion between the names *Nubia* & *Numidia*, when the latin authors termed it the *Numida* or *Numidian fowl* instead of *Nubian*? There are none in Barbary! Now the often quoted passage from Columella, which it has been thought indicates that 2 species were known to the Romans, appears to me to refer to *the two sexes* of *Ph. ptilorhyncha*! And the words *"paleam et cristam"* will apply to *ptilorhyncha*, & not to the modernly termed *N. meleagris*;[33] the former having a *hairy frontal crest* additional to the bony coronal peak,—& the latter not a trace of such crest! Then the very name *Guinea*-fowl indicates the immediate source of the modernly domesticated species, which is widely diffused over *W. Africa* Such name would never have been attached to it, if we had derived our bird from the old Roman stock, prior to the opening of the modern Guinea trade!; & a friend tells me that he had always understood that our bird was first introduced into Europe by the famous Prince John of Portugal,—[34] nothing more likely; but there were probably many importations, in Britain & elsewhere, & those which have gone wild in the Antilles may have been taken *direct from Africa* in the slower ships, & at all events must originally have been the near descendants of wild-caught African birds, as uncontaminated by farm-yard influences as the English Pheasants and Partridges which are now fast multiplying in N. Zealand. Indeed, considering that they are enumerated among the wild game of Jamaica 150 years ago, it may even be that they were taken there in the slavers, and introduced *from the W. Indies* into England! Nevertheless, Latham describes the "white-breasted Guinea-fowl from Jamaica,", which shews that particoloured birds occur (or did occur) there.[35] Ask Gosse about this.[36] I find that Buffon has anticipated me a good deal,[37] but I think that I now *first prove* the Roman *meleagris* to have been the *ptilorhyncha*, & therefore *impossible* to have been the progenitors of our "Guinea"-fowl, which must date from the opening of the modern trade with the W. Coast of Africa. Referring to the *Encyclopædia Brittanica* on the subject, I see it remarked that Pennant seems to prove that there were domestic Guineafowl in England so early as the 13th century! "At least prior to 1277"!!! Unfortunately I have not Pennant to refer to, *so you must see to this*![38] A third species of Guinea-fowl with the bony coronal knob is that of S. Africa & also Madagascar, *N. mitrata*, which instead of the lateral wattles has a sort of medial throat-wattle not unlike that of a Turkey; & there are two species without the knob—*cristata* & *vulturina*—& probably more remain to be discovered.— There is one (perhaps *ptilorhyncha*) in Arabia, vide Niebuhr, Descr. de l'Arabie, I, 234.[39]

I think it is a common notion that Alexander introduced the Peafowl from India into Greece; but how about Juno & *her emblem the Peacock*! Aristophanes also Long enough before, to say nought of King Solomon's merchants bringing 'Apes and Peacocks'. By the way, the Cape Dutchmen call the large Bustards by the name 'Pauw', which has given rise to the stories about wild Peafowl in Africa; and the Levantine Europeans call the Bustard the "wild Turkey", which has similarly originated the notion of 'wild Turkeys' inhabiting *Syria*. The Dutch name perhaps hints the old Roman pronunciation of the word *Pavo*, imitative of the cry of the bird! How came we by the name *Turkey*?; surely not because the *strut* recals to mind the haughtiness of the "grand Turk"! A queer conceit, rather! In the narrative of a sporting excursion in China, I saw an undescribed species of Bustard (clearly), frequently denominated "wild Turkey".— As *Otis tarda* abounds in

Syria (in the dominions of the Grand Turk), could it even have been known, to the Crusaders for instance, as the Turkey-fowl, or Turkey-cock? If so, the name might well have passed to the American *Dindons*!

Another word about Guinea-fowl. What Gosse remarks of their power of flight does not accord with my observation;[40] for I have seen a *covey* rise in England from the midst of a stubble-field, as strongly as any Partridges; nor are they worse organized for flight than the generality of other Partridges (for a group of Partridges they undoubtedly are); & looking to the skeleton, I find that the sternal crest is even considerably deeper, & the inner emargination of the sternum much more filled up, than in other typical poultry birds, as Pheasants, Partridges, &c &c.

On the subject of *permanent varieties*, I very much incline to the opinion that the *Lepus hibernicus* is a permanently *coloured* variety of *L. variabilis*; and *Lagopus scoticus* ditto of *L. albus* (et *lapponicus*, Gm., *subalpinus*, Nilsson, and *saliceti*, Tem.) Otherwise, is it not remarkable that the Variable Hare should not occur on the Irish mountains; nor *L. albus* in Scotland? Both having *so very extensive a distribution elsewhere*![41] But then *L. timidus* has only recently been *introduced* into Ireland; and if I remember rightly, the Squirrel, Dormouse, & Mole are unknown there! Still it is most particularly worthy of notice that *hibernicus* should **represent** *variabilis*, & *scoticus* **represent** *albus*! Is it true that there is a difference *in the colour of the flesh* of the Irish & Alpine Hares? That of the former being *dark*, as in *L. timidus*; and of the other *white*, as in a Rabbit? See also to the so called *Perdix montana* (*Tetrao montanus*, Gm.), which I believe is a mere variety of the *cinerea*; but is it true to its particular distinctions? Hodgson has lately picked up a true congener to the English Grey Partridge, in a much finer bird received from Tibet, to which (*more suo*), of course, he gives a new generic name.— [42] I have been working much at the gallinaceous order lately, and consider the *Ammoperdix* type (of Afghanistan, Persia, & Arabia,) to come nearest to true *Perdix*, & to be intermediate to this and the *Caccabis* or red-legged group: & of this last, you should endeavour to find whether the Himalayan *Chukar* does not *grade imperceptibly* into the *C. græca* of the Alps, Pyrenees, Greece, Syria, &c. I much suspect *that it does so*; & after all, the difference is very slight indeed, between the eastern & western birds. The *chukar* has a wide range over central Asia; & Ruppell notes it from *Mt. Sinai*, calling it "*græca*, of the variety figured by Hardwicke"![43]

See also to intermixture of *Phasianus colchicus* & *Ph. torquatus* in the British islands. If you see rows of Pheasants hanging up at the poulterers, you will always find *green* predominate upon the rumps of the cock-birds which have any white feathers about the collar; and *coppery-red* on the rumps of those which have no trace of the white collar— This I have often remarked in the London-markets. There is considerable difference in the *markings* of the typical English bird (without the white collar), and true *torquatus* of China; especially on the breast-feathers (vide samples);[44] the flanks of the Chinese bird are much paler and *contrasting*; and the tail is shorter, with the bars on it *very much broader*. Now I have some considerable suspicion that the European Pheasant is, after all, *indigenous*; & some time ago met with a notice of it in the Anglo-Saxon times in England(!), which I must endeavour to hunt up. This, however, is consistent with *Roman agency*! Moreover I suspect that the Colchian (i.e. Mingralian) bird, which (I suppose is the one that) abounds along the wooded Elbury chain (which skirts the S. extremity of the Caspian),—*i.e.* Hyrcania of old, renowned for its Tigers (which *still exist there*),— wants looking to. I have seen a hen bird *in the masculine plumage* from the vicinity of Herát! And from peculiarities about this specimen suspect that it is not of the European species, however nearly affined! Has not Gould obtained such a Pheasant from Káfferistán or thereabouts?[45] Enquire about this. For some notice of the region inhabited by this Pheasant, refer to A. Conolly's Journey, Vol I, 289.[46] My impression is, that the true or *unmixed* British Pheasant will prove to be more nearly affined to the Chinese *torquatus*, than is that of middle Asia! But even if so, which is the Hyrcanian & the true Colchian Pheasant? And *is* the British species *indigenous?*

Now for an etymological argument, however! The words *Phasianus* & Pheasant, or 'Faisan', are redolent of the name *Phasis*, on the banks of which river the bird is said to have been originally found!

For information on the breeds of Indian cattle, vide 'Transactions of the Agricultural & Hort! Soc.Y of India', VII, 112.—[47] The *feral humped cattle* of Oudh are noticed by Cautley in the *Journ. As. Soc. B.* 1840, 623;[48] & again by D.r Butter, in his 'Outlines of the Topography & Statistics of the Southern districts of Oudh', &c, p. 29.—[49] These works you will see at the India-house Library. I have learned that there are some feral cattle also in S. India, which *tradition* dates from the time of Tippoo Sultán! But correctly so? These, I hear, are uniformly coloured, brown; & I am promised some details respecting them which I have now the pleasure to send you.—[50]

A letter just rec.d from Sir J. Brooke informs me that— "The wild cattle are here (in Borneo) called *Tabadan*; & it is the Banteng (so far as I know) of Java and the peninsula. The Burmese species I remember to have seen many many years ago, & that is very similar. But there are, it is affirmed, two species in the peninsula of Malacca, one much larger than the other."—[51] True, & the larger is *Bos gaurus*; but is the smaller *B. sondaicus?* I am not quite satisfied that it is so; & the *Bantengs* are termed 'Báli cattle' at Singapore.

I believe that I have now hit upon the true origin of the name 'Turkey', as applied to the bird; my former suggestion being considerably more *plausible than probable*. The Guinea-fowl appears to have been denominated the 'Turkey hen' in former days; and as the Portugueese discoveries along the W. coast of Africa preceded those of the Spaniards in America, there is reason to believe that our British ancestors became acquainted with the Guinea-fowl long prior to their knowledge of the Turkey; and the English trade being then chiefly with the Levantine countries, our ancestors must have fancied that it came from thence. Referring to a curious old Dictionary in my possession (published in 1678) for the word *Meleagris*, I find it translated "a Guinny or Turkey hen".— "*Gallinæ Africanæ sive Numidicæ*, Var. sive quæ vulgò *indicæ*".[52] Again, *Numidica guttata* (Martial) is rendered "a Ginny or Turky hen". Looking also into an English and Spanish Dictionary of 1740, I find *Gallipavo* rendered "a Turkey or Guinea Cock or Hen". Well, our British forefathers must have derived the Turkey probably from Spain, or France (certainly not direct from America); and meanwhile have learned the true habitat of the Guinea-fowl; and therefore have supposed the former to be the true *Turkey* fowl, as distinguished from the *Guinea*-fowl,—the latter bearing both names previously. In France it was *le coq d'Inde*, now corrupted into *Dindon*, from a similar mistake In 1764, Edwards figures the Australian *Talegalla* as "the Turkey-Pheasant, or *Phaisan-Dindon*"!;[53] and the latin-sounding name *Gallipavo* seems to be of Spanish origin, and obtains among the Spaniards to this day.— I don't know whether I ever mentioned to you another curious derivation which I traced some years ago. Poor Strickland asked me if I knew the origin of the name *Amadavat*, applied to the little eastern Finch which you so often see alive in England, & which has regularly gone wild in Malta. In Sheridan's 'School for Scandal' (Act V, Sc. 1), brought out in 1777, "amadavats" are more than once mentioned! They actually take the name from the city of Ahmadabád in Guzerat! Vide the following passage from "a New Account of East India and Persia", by John Fryer, M.D., *Cantabrig*. (1698). Among other natural curiosities brought to Surat, were—"Milk-white Turtles from Bussorah, Cockatooas, and Newries (Lewries or Lories, commonly pronounced Loories), as also a Cassowar that digests iron. From **Amidavad** small birds, who, besides that they are spotted with white and red no bigger than measles, the principal chorister beginning, the rest in concert, fifty in a cage, make an admirable chorus". Thus the name of the place from which they were brought to Surat, has become transferred to themselves; as in the more familiar instances of *Bantam* & *Canary*, and *Turkey* by mistake! While the specific name *amandava*, L., and the generic name *Amadina*, Swainson, are further derivations from the same source!!! I have a sort of impression that I have told you about this before;[54] but *n'importe*, only it swells the postage.

Looking over some recent Nos. of the 'Illustrated News', in the Supplement to No. 769, I observed a representation of some paintings on the interior of a Greek tomb, & inter alia a *cock's head*.[55] Now I have a strong impression that all the antique representations of fowls which I have seen (Etruscan, Greek, & Roman) exemplify *the same type*, which comes near that of the wild bird; and so far as I have been able to trace, the Greeks and Romans had no *marked races of fowls* What breeds of fowls are there now in the 'Isles of Greece', especially in Cyprus, & in out-of-the-way parts of Turkey? The old Greek race is likely enough to have come down to us!! The history of which in Europe should therefore be sought during the middle ages. The Crested Polish seems to be of European origin.— Does the name 'Polish' refer to their crested *polls*? Oh! It is also a most noteworthy fact, that no *satisfactory and certain* notice of the domestic fowl occurs throughout the Old Testament! (Vide Dixon)[56] Nor in Homer!! And it seems that Aristophanes termed it the *Persian bird*, thus indicating the direction from which it came to Greece. According to Chesney, (I, 82), there are Jungle-fowl in the eastern parts of Persia;[57] but I doubt it. *Domestic* fowls have several Sungskrit names, & were formerly reared by Hindus!!! It is certain that Cæsar found them in Britain! And Cook and others in the S. Seas &c! You should con well old Aldrovand,[58] to whom I have not access; and I will study Buffon, Bachart,[59] &c & try to find out the earliest notices of the fowl in oriental works. I hear that Solomon's "Peacocks" are *probably* a mistranslation, & that instead of 'Apes & Peacocks', *two sorts of Apes* are probably meant. See to this; & bear in mind *Juno* & her Peacocks! *Aristophanes* of course has the Peafowl!— I have been looking over Col. Chesney's 'Jl. of the Euphrates Expedition', & have extracted therefrom its reliable Zoology. Curious that he should not have known the Flamingo! (Vide I, 732).[60] I recommend you to see what he says about *Camels*, in I, 582 *et seq.*, & also 82;[61] & you may also consult Niebuhr's '*Descr. de l'Arabie*', I, 582.[62] Still only what I have told you before! Of the mammalia common to Asia & Africa, it is interesting to trace them through the intermediate country, as especially the Leopard, Hunting Leopard, & Caracal; but we hear nought of the Ratel! *Humped cattle* of small size in Arabia! (I, 586); & Buffalos wherever there is water (in that parched & riverless land!) *ibid.* Buffalos found *in a wild state* towards the shores of the Black Sea! (I, 362). Can these be of the distinct race now in Italy? And *aboriginally wild*? Wild Horses *and* Wild Asses in Arabia (p. 581); the former probably *Ghorkhurs* Termed *Zebra* by Kinnier. Fact! (vide p. 108, where the *hemione* is called "more properly the wild Horse"! If so, is the other the true Ass, *with humeral cross*? Perhaps!) Green Parrots common *in Syria* (I, 443, 537). Is then *Palæornis torquatus* identical, after all, on the two continents? "*Phasianus colchicus* & another" (Appendix, No 4). The Stag of the Taurus is doubtless the Persian *Maral*, & not the *elaphus*! And "further south the Fallow Deer, *C. dama*"![63]

I have just seen Gould's 2^d Supplementary No. to his 'Birds of Australia',[64] & much wish that he would give us the rest of the birds of the N. Zealand group, & for that matter those of the S. S. islands generally, & also of N. Guinea, N. Ireland, &c, which lead on to the Moluccas. Those of the Philippes might do for a distinct work. How splendidly he would get up the *Paradiseidæ*! We have not even his 'Birds of Asia', which I often want to consult, & I have not seen the last 2 or 3 Nos.[65] Now the Court of Directors subscribe for 40 copies, for distribution of course, & I wish you would mention this to Col. Sykes when you see him.[66] Also try to induce Gould to go on with the 'Birds of N. Zealand'.

Mem. The Arabs at Jiddha (or Jeudda, &c) not only eat Turkeys, but prize them very much. This too close upon Mecca! So, also, do the Egyptian Musalmans, in opposition to those of India.

DAR 98: 110–11, 121, 119–20, 112–13, 117–18

CD ANNOTATIONS

0.1 Calcutta . . . tourist's work. 3.9] *crossed pencil*
1.10 *wild Asses*] *underl brown crayon*
1.11 so apply . . . 1855. 1.12] *scored brown crayon*
1.12 no 44] *underl brown crayon*
3.1 You have . . . tourist's work. 3.9] *crossed brown crayon*
3.13 You mention . . . looks like.— 3.15] *crossed brown crayon*
4.1 There is] 'Fowls' *added brown crayon*
4.1 There is . . . of tail. 4.16] *scored brown crayon*; 'Fowls' *added brown crayon*
4.19 & note . . . modified 4.21] '*Canaries.*' *added brown crayon*
4.24 the Pr. of Canino . . . manage it. 6.11] *scored brown crayon*
8.1 the N.W. . . . *separable,*— 10.4] *scored brown crayon*; 'Interbreeding' *added brown crayon*
12.1 For hybrid . . . 694. 12.2] *scored brown crayon*; '10'⁶⁷ *added brown crayon*
13.1 For representative . . . Mountains, 13.4] *double scored brown crayon*
14.1 The *Mydaus* . . . *Java.* 14.3] *double scored brown crayon*
Top of first page: '(Fr. Moore Esqᶜ)' *pencil*; '10' *brown crayon*

[Enclosure]
2.6 the 'Guiney-hens' . . . dealers. 2.11] *scored brown crayon*; 'Consult Sloane'⁶⁸ *added pencil*
2.10 'Bohemian Pheasant'] *underl brown crayon*
3.1 I have . . . crest! 3.13] 'Guinea Fowl' *added brown crayon*
3.22 the English . . . N. Zealand. 3.23] *double scored brown crayon*
4.1 I think . . . *Dindons!* 4.15] 'Peacocks Names' *added brown crayon*
6.1 On . . . *varieties,*] '10' *added brown crayon*
6.1 On . . . *montanus*, Gm.), 6.12] *scored brown crayon*
6.1 On . . . Ireland; 6.6] *double scored brown crayon*
6.13 Hodgson . . . name.— 6.15] *crossed brown crayon*
7.1 See also . . . islands. 7.2] *double scored brown crayon*
9.1 For information . . . 112.— 9.2] *double scored brown crayon*
9.5 I have . . . send you. 9.9] 'Cattle' *added brown crayon*
11.1 I believe . . . postage. 11.39] *crossed brown crayon*
11.1 I believe . . . previously. 11.18] 'Name of Turkeys' *added brown crayon*
11.25 which has . . . Malta.] *underl brown crayon*
11.25 In Sheridan's . . . mistake! 11.36] 'Names' *added brown crayon*
12.3 Now I have . . . us!! 12.9] *double scored brown crayon*
12.16 *Domestic* . . . Britain! 12.17] *scored brown crayon*
12.24 Curious . . . 732). 12.25] *double scored brown crayon*
13.1 I have . . . Zealand'. 13.9] *crossed brown crayon*
Top of first page: 'Musk Duck' *pencil*

---

[1] Although Blyth dated his letter 1855, this was clearly a mistake for 1856. The content of the letter indicates that it follows Blyth's letters written in 1855 (see *Correspondence* vol. 5). CD's numbering of Blyth's letters (see CD's annotations and n. 67, below) also indicate the letter was written in 1856.

[2] CD began his correspondence with Blyth in 1855 (see *Correspondence* vol. 5), but none of the letters from CD to Blyth during the period that Blyth was curator of the museum of the Asiatic Society of Bengal has been found.

[3] For Blyth's notes, see the enclosure following this letter. For the notice of the wild cattle of India, see the letter from C. W. Crump to Edward Blyth, [before 8 January 1856].

[4] Thomas Horsfield, keeper of the East India Company's Museum, Leadenhall Street, London, was assisted by Frederic Moore. Moore published several papers in the *Proceedings of the Zoological Society*

of London describing specimens in the museum. The part of Blyth's letter directed to Moore was evidently sent on by CD. It has not been located.

[5] Blyth 1855b (see *Correspondence* vol. 5, letter from Edward Blyth, [30 September or 7 October 1855] and nn. 27 and 28). A lightly annotated copy is in the Darwin Pamphlet Collection–CUL.

[6] It has not been possible to locate the *Calcutta Sporting Review*. None of the papers referred to by Blyth as having been published in this periodical are in the Darwin Pamphlet Collection–CUL.

[7] See *Correspondence* vol. 5, letter from Edward Blyth, 7 September [1855].

[8] F. Bashford had earlier sent information on the interbreeding of different races of silkworms to Blyth for CD (*Correspondence* vol. 5, letter from F. Bashford and Edward Blyth, [after 3 July 1855]). Bashford had been in England since 3 July 1855.

[9] Blyth had written a paper on the counterfeiting of death by animals wishing to escape danger (Blyth 1837). See *Correspondence* vol. 5, letter from Edward Blyth, [22 September 1855] and n. 19.

[10] Blyth mentioned this anecdote in a previous letter (*Correspondence* vol. 5, letter from Edward Blyth, [22 September 1855]). The article has not been located in the *London Illustrated News*.

[11] 'Niata' oxen, a South American breed of cattle with a curious skull formation. CD later described this 'monstrous breed' in *Variation* 1: 89–91.

[12] E. Griffith *et al.* 1827–35, 8: 222:

> M. d'Azara, in his essays on the natural history of the quadrupeds of Paraguay [Azara 1801], says . . . at Buenos Ayres, and in the range of the Andes, there are also hens, whose feathers, feet, crest, barbs, and skin, are black . . . It is singular that no mention is made of these birds in M. d'Azara's book on the ornithology of those countries.

In *Variation* 2: 209, CD referred to Azara 1801, 2: 324, on the black-skinned fowl of Paraguay. This book is in the Darwin Library–CUL and was annotated by CD.

[13] E. Griffith *et al.* 1827–35, 8: 177.

[14] The reference to *Serinus melanocephalus* has not been found in any work by Wilhelm Peter Eduard Simon Rüppell.

[15] Bonaparte 1850–7, 1: 514–15, describes *Chrysomitris spinoides* of Asia and *C. stanleyi* of America as closely related species.

[16] See *Correspondence* vol. 5, letter from Edward Blyth, [22 September 1855], n. 37. The reference is to Hutton 1846 (p. 162 n. 60) for which Blyth provided the footnotes.

[17] Hutton 1846, pp. 162–8.

[18] In Chesney 1850, 1: 582–4, Francis Rawdon Chesney described various kinds of camels and dromedaries, including 'a mule breed between [the Bactrian camel] and the Arabian animal, with a hump which can neither be called single nor double . . . This is a large, useful, and highly prized animal . . . but the creature is short-lived, and the Arabs do not breed from him; giving as a reason, that the progeny are intractable, and bad-tempered.' (p. 584). On p. 82, Chesney mentioned a 'mule breed [of camel], between the Arabian and Bactrian, with a single hump, but much larger than that on the back of the former'.

[19] The two-humped camel, *Camelus bactrianus*, is figured as the frontispiece of Hutton 1850. This work is in the Darwin Library–CUL and was annotated by CD.

[20] In his reference to John Lewis Burckhardt's 'Travels in Nubia' in Hutton 1846, p. 162 n. 60, Blyth gives page 232. The passage concerning camels, however, has not been located on page 232 or page 222 in either Burckhardt 1819 or Burckhardt 1822. However, Burckhardt 1830, 1: 195, states: 'The Anatolian breed is produced between an Arab she-camel, and the double-humped male dromedary imported from the Crimea.'

[21] In the *Calcutta Sporting Review*. See n. 6, above.

[22] See the enclosure following this letter.

[23] In his abstract of this letter (DAR 203), CD noted: 'Last Page, important on varieties crossing when ranges meet.— Coracias, Treron.—' CD used this information when writing his species book: 'in India reputed species of Coracias, as I am informed by Mr. Blyth, intermix & blend on the confines of their range.' (*Natural selection*, p. 259).

[24] Bonaparte 1855a, p. 17.

[25] Hutton 1848.

[26] Horsfield 1824:

> The Mydaus meliceps . . . is confined exclusively to those mountains which have an elevation of more than 7000 feet above the level of the ocean; on these it occurs with the same regularity as many plants. The long-extended surface of Java, abounding with conical points which exceed this elevation, affords many places favourable for its resort.

[27] The only paper published by John Corse in the *Asiatic Researches* is devoted to describing the methods by which wild elephants are caught (Corse 1799a). It would appear that Blyth intended to refer to Corse 1799b in the *Philosophical Transactions*, to which he had earlier directed CD (see *Correspondence* vol. 5, letter from Edward Blyth, [22 September 1855] and n. 53).

[28] Blyth is presumably referring to the Kohlrabi or choux-raves. In *Variation* 1: 323, CD referred to the 'recently formed new race of choux-raves . . . in which the enlarged part lies beneath the ground like a turnip.'

[29] Crawfurd 1828, p. 434. Blyth's date, 1821, is an error for 1828. CD had read Crawfurd 1828 in March 1844 (*Correspondence* vol. 4, Appendix IV, 119: 15b).

[30] P. H. Gosse 1847, p. 329. Philip Henry Gosse stated that 'the turkey is, as far as European knowledge is concerned, indigenous to the greater Antilles, having been found by the Spanish discoverers, already domesticated by the Indians'. He went on to maintain that 'the European domestic breed is descended from West Indian, and not from North American parentage. This would perhaps tend to confirm, what has been suspected, that the domestic Turkey is specifically distinct from the wild Turkey of North America.'

[31] P. H. Gosse 1847, pp. 325–7. Gosse began his discussion of the guinea-fowl by stating: 'In a country whose genial climate so closely resembled its own . . . the well-known wandering propensities of the Guinea-fowl would no doubt cause it to become wild very soon after its introduction. It was abundant in Jamaica as a wild bird, 150 years ago' (p. 325). In his reading notebook, CD noted: 'Gosse Birds of Jamaica— account of wild Guinea Fowls— C^d he get specimen. read', and he also recorded having read the work on 11 May 1856 (*Correspondence* vol. 4, Appendix IV, *128: 161; 128: 18). He later used the information on wild guinea-fowls in *Variation* 1: 190 and 294, having obtained further information about Jamaican guinea-fowls from Richard Hill, Gosse's collaborator (*Variation* 1: 294 nn. 43 and 44; see also letter from Richard Hill, 10 January [1857]).

[32] Edwards 1758–64, 3: 269.

[33] Columella, *De re rustica* 8. 2. 2–3. Columella distinguished an African fowl, called 'Numidian' with a red helmet and crest, from 'Meleagris' with a blue helmet and crest. Blyth has mistakenly used *palea* (wattles) instead of *galea* (helmet) in quoting Columella.

[34] During the reign of John I of Portugal (1357–1433), his son, Prince Henry 'the navigator', with other Portuguese navigators began exploring the area designated Guinea (*EB*).

[35] Latham 1821–8, 8: 147. CD recorded having read volume eight 'on Pigeons & Fowls' in March 1856 (*Correspondence* vol. 4, Appendix IV, 128: 16).

[36] Gosse had lived in Jamaica and studied its fauna. See n. 31, above.

[37] Buffon 1793, 2: 144–68.

[38] Blyth has confused his sources. In *The London encyclopaedia* (London, 1829), 16: 16, it is stated that: 'Mr. Pennant contends, and seems to prove, that the pintadoes had been early introduced into Britain, at least prior to 1277'. No such statement by Thomas Pennant has, however, been found.

[39] Niebuhr 1779, 1: 234.

[40] P. H. Gosse 1847, pp. 326–7: 'Flight cannot be protracted by them, nor is it trusted to as a means of escape, save to the extent of gaining the elevation of a tree: the body is too heavy, the wings too short and hollow, and the sternal apparatus too weak, for flight to be any other than a painful and laborious performance.'

[41] In his abstract of this letter (DAR 203), CD here noted: 'Quote Blyth on relation of range to question of Red Grouse being species.' CD later stated in *Origin*, p. 49:

Several most experienced ornithologists consider our British red grouse as only a strongly-marked race of a Norwegian species, whereas the greater number rank it as an undoubted species peculiar to Great Britain. A wide distance between the homes of two doubtful forms leads many naturalists to rank both as distinct species . . .

[42] Hodgson 1856.

[43] Rüppell 1845, p. 106, in which Rüppell recorded that the *Chacura graeca* 'varietas, Gray Indian Zoology Vol. I Taf. 54' was found 'paarweise am Sinai'. The reference is to J. E. Gray [1830–5], which was illustrated from the collection of Thomas Hardwicke.

[44] Attached to the letter are two feathers, labelled 'China' and 'English', differing in their black, marginal markings. Next to the feather labelled 'China' Blyth has added, 'The figure of this bird in Griffiths' 'Animal Kingdom' is **atrocious**!'. See E. Griffith *et al.* 1827–35, 8, facing p. 232.

[45] J. Gould 1850–83, 7: pl. 28, which figures '*Pucrasia castanea*, Kafiristan Pucras Pheasant'. This plate was first published in July 1854 in part 6 of *The birds of Asia*.

[46] Conolly 1834, 1: 289, describes the country between Meshed and Heraut.

[47] Little 1840, pp. 111–14.

[48] Cautley 1840, p. 623: 'the natives of Hindostan . . . have in their affection for the cow and ox, given rise to a race of wild cattle perfectly distinct from those of the forest. . . . in the province of Oude, large herds of black oxen are . . . found in the wild and uncultivated tracts'.

[49] Butter 1839.

[50] See letter from C. W. Crump to Edward Blyth, [before 8 January 1856].

[51] James Brooke was raja of Saráwak, Borneo.

[52] Littleton 1678.

[53] Edwards 1758–64, 3: pl. 337.

[54] Blyth had indeed previously given this information to CD. See *Correspondence* vol. 5, letter from Edward Blyth, 4 August 1855. His references are to Sheridan 1781, act 5 scene 1, and Fryer 1698, p. 116.

[55] *Illustrated London News*, 10 November 1855, p. 564, figures 'Paintings on a Greek tomb lately found near Pæstum.'

[56] Dixon 1848, pp. 173–4. 'It is true that there is no mention of Fowls by name in the Old Testament, except a doubtful allusion in the Vulgate translation of the Book of Proverbs (xxx. 31), which is lost in the authorised version' (p. 173).

[57] Chesney 1850, 1: 82.

[58] Aldrovandi 1599–1603. CD recorded having read this work in March 1856 (*Correspondence* vol. 4, Appendix IV, 128: 16).

[59] Bochart 1675. Volume two of this work included Samuel Bochart's *Hierzoicon*, which treated of the animals of Scripture and was first published in 1663. See also letter from Edward Blyth, [*c*. 22 March 1856] and n. 2.

[60] Chesney 1850, 1: 731–2: 'Description of the bird called . . . "The Magnanimous Bird." '

[61] See n. 18, above.

[62] Blyth has inadvertently repeated the page number of his previous reference. The correct reference is Niebuhr 1779, 1: 229–30.

[63] All these references are to the first volume of Chesney 1850. The last reference to *Cervus elaphus* is in Chesney 1850, 1, appendix 3, p. 728.

[64] The seven volumes of John Gould's *The birds of Australia* had been issued in 1848 (J. Gould 1848). From this date, parts of a supplement to the work were published. The supplement was completed in 1869.

[65] J. Gould 1850–83. By January 1856, seven parts of this work had been published.

[66] William Henry Sykes was the chairman of the court of directors of the East India Company. The museum of the Asiatic Society of Bengal was under the jurisdiction of the company.

[67] The brown crayon numbers that CD wrote on Blyth's letters indicate their chronological sequence and relate to CD's abstracts of the letters (DAR 203), which he also numbered.

[68] Sloane 1707–25.

From John Davy 10 January 1856

[Reports the results of experiments on the vitality of impregnated fish ova carried out in November and December 1855. The ova of Charr were exposed to temperatures ranging from 70 °F to 84 °F for various periods of time. Ten experiments to ascertain the maximum temperature at which the ova will stay alive were carried out. A second series of experiments was also performed in which a trial of the vitality of the ova was made by packing them in wet wool and sending them a distance by post. Considers it as proved that the power of resisting an undue temperature is possessed in a higher degree by the ova in an advanced, than in an early stage of development; and that the power of retaining life in moist air in like degree increases with age. It may further be inferred that the ova of other species of the Salmonidæ, were they similarly exposed, would afford similar results.][1]

Proceedings of the Royal Society of London 8 (1856–7): 27–33.

[1] The letter was intended to be communicated by CD to the Royal Society of London in the same way that CD had sent an earlier letter from Davy to be published in the *Philosophical Transactions of the Royal Society of London* (see *Correspondence* vol. 5, letter from John Davy, 21 March 1855).

To John Edward Gray 14 January [1856][1]

Down Bromley Kent
Jany 14th

My dear Gray

You have often helped me,[2] will you be so kind as to help me this time in regard to the enclosed memorandum, with Mr Birch.[3] It is my only imaginable channel by which I can ever learn anything about the varieties of our domesticated animals & plants in China.—[4] Do pray use your interest for me with Mr Birch; I could not ask myself.—

My dear Gray | Yours very truly | C. Darwin

[Enclosure]

Is there any translation of any Chinese work, ancient or modern, descriptive, or even simply enumerative, of the varieties of *domestic* Pigeons & Fowls or Ducks kept by the Chinese; & likewise of the Dogs, sheep, cattle &c; but I care more about the former even than the latter.— And the same in regard to the varieties of cultivated plants, but more especially of tobacco & maize; for these latter plants, the work, of course, must not be ancient.—

If any such Chinese agricultural work or Encyclopædia exists in the British Museum but has not been translated, would it be possible for Mr Birch, & would he be so very kind as to take the trouble as to look at it (& as probably saving him a little trouble) & let me be present to note down names of any varieties mentioned, if such are specified.. This would be of extreme interest to me; but I hardly know how great a favour I am asking, for Chinese seems to be so wonderfully difficult to read.—

  C. Darwin

British Museum (Department of Western Asiatic Antiquities, correspondence 1826–67: 1490, 1488)

[1] Dated by the relationship to subsequent letters to Samuel Birch and William Darwin Fox (see n. 4, below).

[2] In particular, Gray had assisted CD in obtaining permission to borrow the British Museum's collection of Cirripedia while CD was working on his monograph of the group (see *Correspondence* vol. 4, letters to J. E. Gray, [18 December 1847], 18 December 1847, and [5 or 6 February 1848]).

[3] Samuel Birch was assistant keeper of the department of antiquities, British Museum. Gray was keeper of the zoological collections.

[4] See letters to Samuel Birch, 6 February [1856] and [12 March 1856], and to W. D. Fox, 15 March [1856]. In both *Origin* and *Variation*, CD cited Birch as having translated for him passages from ancient Chinese, Japanese, and Egyptian texts that described breeds of pigeons and fowls. See *Origin*, pp. 27–8, and *Variation* 1: 205, 230, 238, 246 n. 33, and 247.

To John Lubbock [14 January 1856][1]

                                                              Down
                                                             Monday
Dear Lubbock

  Very many thanks for the Books; I had meant to have sent you a line on Sunday, but quite forgot it myself.— Indeed we are all sick & miserable, & I hardly care even for Pigeons, so may guess what a condition I am in! Nevertheless, I have life left in me to ask whether you ever saw the Chinese M.ͬ Smith:[2] *pray* do not trouble yourself to write, if you have to send a negative; but if affirmative I would write to him, if you think there is any chance of his helping me in the domestic Bird line.—

  Yours most truly | C. Darwin

  Did you ever give orders to preserve corpses of Sebright Bantams?—[3]

  Forgive so much trouble.— | Adios

Endorsement: 'Jan. 1856'
Down House

[1] The endorsement is confirmed by the reference to CD's ill health: according to the following letter, he had been ill for the past week. The first Monday in January was the 7th, but it is unlikely that this letter was written then because CD was well enough to attend the Philoperisteron Society Show on 8 January (see letter to W. B. Tegetmeier, 1 January [1856], n. 2).

[2] Probably George Smith, bishop of Victoria, Hong Kong, who was the son-in-law of the rector of Beckenham, Kent.

[3] John Saunders Sebright had crossed a common bantam with a Polish fowl, then recrossed the offspring with a hen-tailed bantam to obtain the famous Sebright bantam, a small fowl in which the cock lacks male plumage (see *Variation* 2: 54).

To W. B. Tegetmeier 14 January [1856]

Down Bromley Kent
Jan. 14th.—

My dear Sir

I have been unwell for a week, otherwise I sh^d not have left so many days elapse without thanking you very sincerely for your most kind offer of buying for me old Cocks at Stephens.—[1] I have only *one* skeleton as yet, of a good Spanish Cock, so that I sh^d be glad of anything or everything, which you consider a distinct breed. I sh^d be willing to go to 5^s per bird.— My old friend the Rev. R. Pulleine (whose name, I daresay you have heard as a good Poultry judge) sent me a message the other day that he was sure that Mr Baily would at his request send me anything;[2] but I believe your scheme is more sure & I will not as yet try Baily. I am in no hurry. If I succeed in my attempts to get the *skins* of Poultry from all quarters of the world,[3] I shall want skins of the breeds of England for comparison; so if you stumble on a bird *in good plumage*, I wish you would have its neck broken, instead of cut, & then I shall understand that you think it worth skinning, instead of skeletonising. Should I *ultimately* succeed in making good collection of skins & skeletons of our domestic birds, I shall give whole to British Museum.[4]

I have sent a few addresses, as possibly saving you a very little trouble.—

I do not think I shall come up to London for a few weeks, but when I do I shall ask permission to visit either Mr Wickings or Bults collection, & I will inform you, so as to know what hour will suit you, if you are inclined to come & can put off your visit till I do come, as it was evident I had better come as your companion, if you think it worth while to inspect these collections.—[5] But I will write again nearer the time.

With very sincere thanks | Your's truly | C. Darwin

Mr Bult, I sh^d think knew most, & it must be near to you, though far for me.—

Endorsement: 'Jany 14/56'
New York Botanical Garden Library (Charles Finney Cox collection)

[1] CD refers to John Crace Stevens, auctioneer at 38 King Street, Covent Garden. Advertisements appeared occasionally in the *Gardeners' Chronicle and Agricultural Gazette* of auctions of poultry at Stevens's address. CD wanted old cocks of different breeds of fowl for his study of their osteological differences.

[2] Robert Pulleine and John Baily (see letter to W. D. Fox, 3 January [1856] and n. 5).

[3] See *Correspondence* vol. 5, CD memorandum, [December 1855].

[4] CD donated his collection of sixty domestic pigeons and six ducks to the British Museum in 1867 (British Museum (Natural History) 1904–6, 2: 256 and 336).

[5] Matthew Wicking and Benjamin Edmund Bult were members of the Philoperisteron Society, and CD probably met them at the society's meeting of 8 January (see letter to W. B. Tegetmeier, 1 January [1856], n. 2). In *Natural selection*, p. 258, CD referred to Wicking as having 'kept a larger stock of various breeds [of pigeons] together than any man probably in Britain' and in *Variation* 1: 201, as having had 'more experience than any other person in England in breeding pigeons of various colours'. In *Variation* 1: 208, CD described Bult as 'the most successful breeder of Pouters in the world'. See letter to W. B. Tegetmeier, [1 February 1856], in which a meeting with Bult was arranged and letter to W. E. Darwin, [26 February 1856], in which CD mentions a visit to Wicking.

To John Phillips 18 January [1856]¹

Down Bromley Kent
Jan. 18ᵗʰ—

Dear Phillips

I have been looking over my Chapter on Cleavage & Foliation in my Geolog. Observ. on S. America 1846, & with that candour so characteristic of authors I really think it worth your looking over.— Some remarks in first part of Chapt. p. 140, are, perhaps, worth skimming over; but the concluding remarks p. 162 give my results.— I see I give one case p. 147 of cleavage not strictly coincident with mountain range. I may remark that I think you may trust my observations, as I made a vast number *always* with compass (corrected for var.) & note-book, as I was at time deeply interested in subject, & astounded at cleavage being quite distinct from stratification. The case p. 144 of the confused cleavage, where two *great* series crossed each other; the separate hillocks running in different directions, each with its own folia parallel to its longer axis seems interesting.

The foliation of the Chonos Group p. 157 seems to me a grand case, & here it is *not* parallel to line of coast: the apparent crossing (p 159) of a subsequently formed chain is worth notice I think.— In my general conclusions, I allude to well-known fact of cleaved clay-slate when metamorphosed by neighbouring granite, becoming foliated in the planes of cleavage; & I *now* suppose that this is the explanation of most of the cases of widely extended foliated rocks having the same strike, described by me, for instance that of the Chonos group; at the same time, I think, it should never be forgotten that rocks which have been liquefied by heat, sometimes have their crystallized materials so arranged, as almost to deserve to be called foliated, of which I saw grand instance in Eastern Cordillera of Chile.—²

In a brief description of the Falkland Islᵈ. in Journal of Geolog. Soc. (read in March 1846) Vol. 2 (?) p. 270,³ I give a case of a range of stratified quartz, changing its course, & with it the cleavage of the clay-slate at its base: I remember making numerous careful observations on this head.— At p. 271, I give from Capt. Sulivan⁴ (a careful observer) a case, which I have never myself seen; of cleavage in a set of folded beds, in some vertical, in others at right angles to each bed.—

I ought to apologise for troubling you with so long a note more especially as I do not know how much you are concerned with the vaguely denominated foliated rocks.—

Yours very sincerely | Ch. Darwin

The remark p. 163 of Geolog. Observat. in S. America of difference in Mineralogical composition in the planes of cleavage; I do not remember to have seen noticed: I am sure of its accuracy, & presume it is the first step towards metamorphosism or the segregation of separate minerals in the planes of cleavage.—

One other remark,—for years & years the existence of grauwacke with Clay-Slates in various parts of world has perplexed me; as clay-slates seem to have been formed in deep & tranquil seas— Do you think that the same pressure which causes the cleavage, & some movement along the planes of cleavage, can in *part* have actually broken up the rock & mingled different varieties together,, like fragments of ice in a glacier, & like these subsequently recemented together by pressure.— Do just think of this.—

My Geolog. Observ. in S. America must be in Bodleian Library[6]

If you care to have copy of my 3 vols. together, I w^d with pleasure give you order for one. on Smith & Elder.[7]

American Philosophical Society (122)

[1] Dated on the basis that Phillips was preparing his report on cleavage for the meeting of the British Association for the Advancement of Science to be held in August 1856 (Phillips 1856).

[2] Phillips's intention in Phillips 1856 was to distinguish the phenomenon of cleavage in rocks from that of stratification, a distinction that CD had also taken pains to make in *South America* (see also *Correspondence* vol. 4, letter to Charles Lyell, [on or before 20 January 1847], n. 2). In his discussion of the relationship between the planes of cleavage and the inclination of strata, Phillips cited CD's observations (Phillips 1856, pp. 375–6), and in commenting on the symmetry between cleavage planes and axes of elevation he quoted from *South America*, p. 162: 'The cleavage laminæ range over wide areas with remarkable uniformity, being parallel in strike to the *main axes* of elevation, and generally to the outlines of the coast' (Phillips 1856, p. 375).

[3] 'On the geology of the Falkland Islands,' *Quarterly Journal of the Geological Society of London. Proceedings of the Geological Society*, pt 1, 2 (1846): 267–74 (*Collected papers* 1: 203–12). The page reference is to a passage on quartz.

[4] Bartholomew James Sulivan, lieutenant in the *Beagle*, made two surveying voyages to the Falkland Islands, 1837–9 and 1842 6.

[5] The Bodleian is the University Library of Oxford. Phillips was keeper of the Ashmolean Museum at Oxford and reader in geology. He became professor of geology in 1857.

[6] In 1851, Smith, Elder & Co. had published a combined edition of *Coral reefs*, *Volcanic islands*, and *South America* entitled *Geological observations on coral reefs, volcanic islands, and on South America* etc. (*Geology of the 'Beagle'*).

To J. E. Gray 19 January [1856]

Down Bromley Kent
Jan. 19th

My dear Sir

I write one line to say how very sincerely obliged I feel to you, for so very kindly acceding to my troublesome request.—[1]

Believe me | Yours truly obliged | Charles Darwin

British Museum (Department of Western Asiatic Antiquities, correspondence 1826–67: 1491)

[1] See letter to J. E. Gray, 14 January [1856].

From George Gulliver 20 January [1856][1]

Mount Alton, Templeogue. [Dublin]
Jan. 20, 1855.

My dear Sir,

It is not singular that the Blood-Discs of different genera of birds should be similar, for widely distinct Families have those corpuscles not distinguishable in size, shape, or structure. When a marked difference of shape occurs it seems to be merely aberrant, as in the Snowy Owl, Passenger Pigeon, Snow-Bunting, Great Butcher Bird, Java Sparrow, &c. And so of certain saurian Reptiles. In my Appendix to Gerber's Anatomy, 8°. Lond. 1842,[2] where the details are fuller than in the Notes to Hewson.[3] Indeed, the difference of the corpuscles in the entire class of Birds is not more than may be found in a single Family of Mammals.

All I recollect of the cows and goats that I examined is, that they were those most easily got at in London, & therefore probably the most common ones.[4]

The measurements of the Dog's corpuscles $\frac{1}{3542}$th, of the Dingo $\frac{1}{3395}$th, & of the Wolf $\frac{1}{3600}$th of an inch,[5] seem much more different in figures than in fact, and no greater than might be obtained by reducing the averages of different sets of measurements of the corpuscles of the very same individual, provided the measurements were not confined to a single dried & invariable specimen of blood.

The measurement in the Dog in Hewson was from a little mongrel. I have examined those of a fox-hound and of other good breeds without noting any marked difference. But the corpuscles of the Fox, after many comparative trials, have always proved *very slightly smaller* than the corpuscles of the Dog.

I am, | Yours very truly, | George Gulliver.

If you want to see a marked difference of size in the corpuscles of a single family of mammals, compare the comparatively minute corpuscles of the smallest Ruminants or Rodents with the larger corpuscles of the largest species of the same order.

CUL (Darwin Library: bound in Gulliver ed. 1846)

CD ANNOTATIONS
1.3 When a . . . Pigeon, 1.4] *scored pencil*
4.1 I have examined . . . difference. 4.3] *double scored pencil*

[1] Although the letter is dated 20 January 1855, it seems that this was a mistake. The letter answers queries arising from CD's reading of Gulliver ed. 1846, a book that he received from Gulliver only after 18 December 1855 (see *Correspondence* vol. 5, letter to George Gulliver, 18 December [1855]). CD's copy of Gulliver ed. 1846 is in the Darwin Library–CUL and contains annotations by CD.
[2] Gulliver's appendix to Friedrich Gerber's anatomy (Gerber 1842) included tables of measurements of the blood corpuscles of mammals and birds.
[3] Gulliver ed. 1846 comprises a collection of the writings of the physiologist William Hewson, to which Gulliver added extensive notes relating to his own microscopical observations.
[4] CD had marked the measurements of the blood corpuscles of several species and varieties of goats, sheep, and cattle in his copy of Gulliver ed. 1846, p. 238, and wrote in the margin: "What var."
[5] CD had also marked these measurements (Gulliver ed. 1846, p. 238).

To J. S. Henslow 22 January [1856][1]

Down Bromley Kent
Jan. 22[d]

My dear Henslow

I write merely to thank you for your note, though my former one did not require an answer.

I have entirely forgotten (& it is stupid of me) that you had told me about the wild carnation seed.—[2]

M[r] Tollet, (W. Clive's father in law) *is* dead.—[3] Have you seen A. de candolle's Geographie Botanique; it strikes me as a quite wonderful & admirable work.—[4]

I saw in the Times the death of your mother, but at so venerable an age that life can hardly be to any worth much further prolongation.[5] In one sense I never knew what this greatest of losses is, for I lost my mother in very early childhood.—[6]

My dear Henslow | Yours most truly | Charles Darwin

P.S. | I have been sowing some of the seeds from Hitcham this morning.[7]

DAR 93: 108–9

[1] Dated by the reference to the death of Henslow's mother (see n. 5, below).
[2] See letter to J. S. Henslow, 3 January [1856].
[3] George Tollet of Betley Hall, Staffordshire, a close friend of Emma Darwin's father Josiah Wedgwood II, had died in 1855. William Clive, vicar of Welshpool, Montgomeryshire, had been an exact contemporary of Henslow's at St John's College, Cambridge (1813–18). He had married Marianne Tollet in 1829.
[4] A. de Candolle 1855.
[5] Frances Henslow, née Stevens, had died on 15 January 1856 at the age of 80.
[6] Susannah Darwin had died on 15 July 1817, when CD was 8½ years old. In an autobiographical fragment written in 1838 (*Correspondence* vol. 2, Appendix III), CD stated that he scarcely remembered his mother's death and had few recollections of her.
[7] CD intended to begin experiments designed to 'break the constitution' of plants (see *Correspondence* vol. 5, letter to J. S. Henslow, 29 October [1855]). Several small girls from Henslow's parish of Hitcham, Suffolk, had collected seeds from local plants for CD in 1855. In his Experimental book (DAR 157a) on 22 January 1856, CD recorded sowing Henslow's seeds. The experimental results were recorded in May and June (see letter to J. S. Henslow, 16 June [1856]).

From Edward Blyth 23 January 1856

Calcutta,
Jan[y] 23/56.

My dear Sir,

Last night I was roused about midnight by the arrival of my home letters,—a sufficiently formidable array of correspondence, which it is utterly impossible that I can do justice to by the present out-going mail. I am pleased to find that Owen does not oppose my views regarding the "great Orang-utan question".[1] I have a long letter from you undated, but mentioning in a *P.S.* that you had just rec[d] mine of Oct[r] 8th & 22[nd][2] Since my last, I have had no time for penning notes; but must

138,711

call your attention to an article in the 'Echo du monde savant,' No 98, Jan.ʸ 24/36, which I only know of by the reference to it in *Rev. Zool. de la Soc. Cuv.* 1841, p. 33.[3] The said article describes "une belle et rare variété du *Cyprinus carpio*, L., dont la couleur était d'un beau *rouge aurore*". This at once led me to suspect that *C. auratus* is just such a variety of an affined species, which John Chinaman had carefully bred from in the first instance, though now become so common![4] If so, there should be a grey wild Chinese Carp, of which the 'Gold & Silver Fish' our cultivated varieties,—the curious triple-tailed *C. macropthalmos*, however, being probably an abnormal variety of another species indigenous to China. As these are the only very marked *cultivated varieties* in the class of *Pisces*, the subject merits investigation. Since I last wrote, I have been following up my enquiries respecting gallinaceous birds in general, and have embodied a great deal of curious matter in an article for the 'Calcutta Sporting Review', which you will see in due course;[5] & you will learn from it that you at present much underrate our actual knowledge of the wild gallinaceous birds all the world over; and at once comprehend the grounds for my *very decided opinion*, that we may seek in vain for wild types of *G. giganteus*, &c. I don't wish to seem dictatorial; but feel that my knowledge is now about as complete as it well can be, regarding the wild types of *Gallus*. Should however one still remain unknown to me, which I think most unlikely, un-doubtedly Cochin China, Cambogia, & Siam, are the countries of S Asia least known to zoologists; but then for ages past they have been more or less connected with China, & the wealthy Chinese are fond of keeping pheasants, &c &c, & pay such high prices for rarities that our first knowledge of various Malayan species was derived from the inmates of Chinese aviaries. I have some skins of Jungle-fowl for you, illustrative of the variation observable among them; & the tarsus I find varies remarkably in length, as you will see.[6]

By the way, do you know the positively wild *Numida meleagris* from *Guinea*, as distinguished from Ogilby's *N. Rendallii*,[7] said to be the ordinary species of the Gambia? Another species which I should like to know about, is the *N. coronata* (in addition to *N. mitrata*, if not also *N. cristata*) in S. Africa. Have you seen Albin's figure of the lost breed of crested Turkeys?[8] There seems no doubt about them; but Dixon's supposed "wild Crested Turkeys" of Central America are clearly *Cracidæ*, from the notice cited of the nest & (two) eggs of one of them.[9] A true Turkey would lay more eggs than could be accommodated in a *tree-nest*; & all the *Pavonidæ* without exception nestle on the ground. By *Pavonidæ* I mean the united *Phasianidæ* & *Tetraonidæ*, Auct., which are *empyrical* & artificial divisions, the *very types* of which (*Ph. colchicus* & *T. tetrix*) are so nearly affined as not unfrequently to interbreed *in the wild state*!

Upon sound anatomical distinctions, I divide the gallinaceous birds into 5 essentially distinct families, which do not intergrade,—viz. *Cracidæ*, *Megapodiidæ*, *Syrrhaptidæ*, *Pavonidæ*, and *Tinamidæ* (including *Turnix*). Thus arranged, we can generalize a good deal to some purpose. It is not unlikely that any treatise on the Gallineaceæ in the 'Calcutta Sp. Rev.' may grow in time as Prichard's original

Essay on the Human races expanded into a big work,[10] but in the meantime, I trust that its present form will elicit the information necessary for completion.[11] You mention the Turkey's tuft as a curious 'abnormity' (as it were); especially, I may add, as the Ocellated Turkey does not possess it. Bear in mind, however, that this tuft consists not of bristles, but *bristle-like plumes*, which are annually moulted & developed like other true feathers; as are also the *eyelashes* of various birds, &c &c.— See Leadbeater's fine specimen of this rare bird,[12] & notice (what has never been described) the curious structure of the appendage over the bill, & all the little warts & caruncles. I was not aware of what you mention concerning the *muscular foundation* of the tuft in the Polish (*Polled?*) fowls. I have been trying to hunt up a notice I remember reading some time ago, in some French work, respecting the wild range of the Golden Pheasant extending I think to Orenbourg; a very remarkable fact, which may account for the ancients having some knowledge of it, however vague, which Cuvier connects with the old descriptions of the *Phœnix*!![13]

Yesterday I had the great pleasure of seeing a particularly fine living specimen of the exceedingly interesting (to me) animal, *Canis primævus* of Hodgson;[14] which I suppose you will make acquaintance with by & bye, in the London Zool! Gns. It was much more Fox-like than I expected, but hunts in packs as you know; a most particularly agile, graceful & *game* animal, & the original *Canis* **aureus** no doubt. This animal has an immense range, from the Altai, Tibet, &c, over all altitudes of the Himalaya, to central India, the Nilgiris, &c, also in Burma, Malacca, Sumatra & Java; for I strongly lean to the opinion that the *C. javanensis, sumatrensis, dukhanensis, primævus,* &c of authors refer all to the same animal, the best name for which is *rutilus* of Temminck, since *aureus* has been transferred to the Jackal. For many years past I wished much to see this species alive, & now that I have done so, I begin to feel that I understand it.— As you receive specimens from Madeira, kindly procure me some skins of the wild Canary; also ascertain which is the 'Red-legged Partridge' of Madeira & also of the Canary isles,—*P. petrosa* I suppose. According to Widdrington, this is not found in Spain,[15] but is the only species in Sardinia, as in Barbary. I have no specimen of it. I should like also to see the peculiar Chaffinch of Madeira.

About the Seychelles, I know of no conspectus of the terrene fauna of those islands, but am aware that peculiar species exist, especcially of land-shells, & also a peculiar Chamæleon. The *Helix monodon* is a fine and remarkable species, found always on the *Cocos de Mer* which is indigenous only to two or three of the islets, & does not thrive on the others! Consult the well known conchologist, M. Liénard of the Mauritius, respecting the zoology of the Seychelles;[16] & if you can get at the publications of the Nat. Hist. Society of the Mauritius, which was an active body during the secretaryship of Julien Desjardins, you are likely to find what you want.[17] Did I tell you, that since writing my article on wild Asses,[18] I have come to the conclusion that if the real *asinus* still exists anywhere in the wild state, it will be in the southern districts of Arabia![19] I find that the Hindu prejudice against domestic fowls is not very ancient, & that fowls were reared in great numbers by the ancient Hindus.[20] Consult Horace Hayman Wilson at the India-house.—[21] See a notice of a curious breed of Indian cattle at Dacca, in Capt. R. Tytler's paper on the zoology of Dacca, published a year or two back in the *Ann. Mag. N.H.*—[22] I now refer to your letter, *seriatim*, having thus far cited it from memory. 'Rock Pigeons'. Remember that the *Pterocles* genus is here so called by sportsmen & others; whence the term may be misunderstood by correspondents in this country. You must have misunderstood me about numerous *races* of fowls in Negroland, at least I think I must have said that fowls were reared numerously by the Negros of Africa. They certainly now are so, vide Niger Expedition, &c &c;[23] but old Barbot tells us that neither the common poultry nor ducks are natural to Guinea, any more than the Turkey; and that very few Turkeys are to be met with there, & those only in the hands of the chiefs of the European forts; the Negros declining to breed any on account of their tenderness. I quote from the Encyclopædia Britt[a]., & cannot get at the date of Barbot's work; but it is curious, from the notice of the Turkey so early in Africa.[24] Let me know the date, if you can do so without overmuch trouble. "Domestic fowls with double spur".[25] According to my observation, when birds that typically bear a single spur have also a second, the latter grows from the base of the other underneath. In all the genera of typically double-spurred *Pavonidæ*, the spurs are curiously irregular; but whenever a third occurs, it is situate, in like manner, at the base underneath of one of the normal spurs

We have a common Jungle-hen, well spurred.—

I dont remember what I said about the origin of *Bantams*, but probably referred merely to that of the name.[26] The appellation "Himalayan Rabbit" must necessarily be a misnomer.[27] What could I have said about varieties of Fallow Deer, beyond new colouring?[28] For my articles on the Elk & Reindeer, you must hunt up Vols. 8 & 10 of the 'Calcutta Sp. Rev.'.[29] I must congratulate you in getting a good Indian 'Pigeon fancier' correspondent in Cap[t] Vine; he will be able to assist you more than I can with Indian domestic Pigeons.[30] Valuable birds rarely die in fine condition; & no native could be made to understand the scientific value attached to a dead bird, which is sure to be pitched away; and what few Indo-portuguese bird-stuffers we have are no better. I will do what I can, which after all is not promising much.

I am obliged to finish somewhat abruptly, even thus, | & remain | Ever truly Yrs, | E Blyth

C. Darwin Esq

DAR 98: 122–5

CD ANNOTATIONS

0.1 Calcutta . . . notes; 1.6] *crossed brown crayon*
1.7 'Echo . . . 36,] *scored brown crayon*
1.26 Cochin China . . . zoologists; 1.27] *double scored brown crayon*
1.27 connected . . . will see. 1.32] *scored brown crayon*
2.9 By *Pavonidæ* . . . state! 2.12] *scored brown crayon*
3.8 I may add . . . possess it. 3.9] *scored brown crayon*
4.4 much more . . . *Canis* **aureus** 4.5] *scored brown crayon*
4.8 the opinion . . . same animal, 4.9] *scored brown crayon*
5.11 the Hindu . . . numbers 5.12] *scored brown crayon*
5.14 in Capt. R. Tytler . . . *N.H.*—5.16] *scored brown crayon*
5.17 'Rock Pigeons' . . . & others; 5.18] *scored brown crayon*
5.21 They certainly . . . Turkey; 5.23] *double scored brown crayon*
5.25 I quote . . . work; 5.26] *scored brown crayon*
6.4 For my articles . . . Rev.'. 6.5] *double scored brown crayon*
Top of first page: '11'[31] *brown crayon*
End of last page: '11' *brown crayon, circled brown crayon*

[1] See letter from Edward Blyth, 8 January [1856]. Blyth had previously discussed this paper (Blyth 1855b) with CD (*Correspondence* vol. 5, letter from Edward Blyth, [30 September or 7 October 1855] and n. 28). Blyth refers to a section in Richard Owen's paper on the anthropoid apes (Owen 1855a, p. 31) in which Owen stated that there seemed to be two species of orang-utan in Borneo. This point had also been discussed in Blyth 1855b. There are offprints of both papers in the Darwin Pamphlet Collection–CUL.

[2] See *Correspondence* vol. 5, letters from Edward Blyth, 8 October 1855 and [22 October 1855]. CD's letter has not been located.

[3] Hérétieu 1841, p. 33 n. 1.

[4] In his abstract of this letter (DAR 203), CD noted: 'suspects Carp a Golden var.' Later, in *Variation* 1: 296, CD stated: 'Mr. Blyth suspects from the analogous variation of other fishes that golden-coloured fish do not occur in a state of nature.'

[5] See letter from Edward Blyth, 8 January 1856, n. 6.

[6] In his abstract of this letter (DAR 203), CD noted: 'variation of wild Gallus Bankiva'. In his discussion of *Gallus bankiva* in *Variation* 1: 235, CD noted that in the Indian *G. bankiva*, 'Mr. Blyth finds the tarsus remarkably variable in length.'

[7] Ogilby 1835, p. 103–4.

[8] Albin 1731–8, 2: pl. 33. In February 1856, CD recorded having read 'E. Albin's Nat. Hist. of Birds 1734' (*Correspondence* vol. 4, Appendix IV, 128: 16). The second volume of Eleazar Albin's work is dated 1734.

[9] Dixon 1851, pp. 277–8. This work is in the Darwin Library–CUL and was annotated by CD.

[10] Prichard 1843.

[11] Blyth never published a larger work on gallinaceous birds.

[12] Probably a reference to John Leadbeater, who was a London bird dealer and ornithologist to Queen Victoria. Blyth's 'Ocellated Turkey' was probably the Honduras turkey. A pair of these birds, 'long desired in European collections', was presented by the Queen to the Zoological Society's gardens in 1856 (Scherren 1905, p. 117).

[13] The work to which Blyth refers is Ajasson de Grandsagne 1829–33, with notes by Georges Cuvier in volume seven. In this volume (p. 368), Cuvier stated that the description of the phoenix by Pliny was that of a real bird, the golden pheasant.

[14] Hodgson 1833.

[15] Widdrington 1844, 1: 397: 'Another great mistake of Temminck, is the statement, that the Perdix petrosa is extremely abundant in the mountains of Spain; whereas it most certainly does not exist there, nor any other but the P. rufa.'

[16] Elizée Liénard was a Mauritian notary and naturalist.

[17] CD had twice noted the proceedings of the Natural History Society of Mauritius in the 'books to be read' section of his reading notebooks (*Correspondence* vol. 4, Appendix IV, *119: 6v., 16v.). Julien François Desjardins founded and served as secretary of the Natural History Society of Mauritius from 1829 to 1840. During this period the *Rapports annuels sur les travaux de la Société d'Histoire Naturelle de l'Ile Maurice* (1835–8) were published. For CD's contact with the Natural History Society of Mauritius, see letters from Victor de Robillard, 20 September 1856 and 26 February 1857.

[18] The article has not been located. See letter from Edward Blyth, 8 January [1856], n. 6.

[19] For Blyth's evidence for this view, see letter from Edward Blyth, 23 February 1856 and n. 16.

[20] See letter from Edward Blyth, [*c.* 22 March 1856] and n. 16.

[21] Horace Hayman Wilson was director of the Royal Asiatic Society in London.

[22] Tytler 1854, p. 177.

[23] Allen and Thomson 1848, 1: 387. See *Correspondence* vol. 5, letter from Edward Blyth, [22 October 1855], for Blyth's earlier references to fowl in Africa.

[24] In discussing the coasts of South Guinea in Barbot 1732, p. 217, John Barbot stated: 'The several sorts of tame-fowl, consist properly in hens, ducks, turkeys and pigeons; the two former whereof are not common to the *Blacks*, but only to be found in or about the *European* forts and factories.' He further added about turkeys: 'There are only a few in the hands of the chiefs of the *European* forts . . . The *Blacks* breed none at all, perhaps because they are very tender, and require much care to bring them up.' (p. 217).

[25] See *Correspondence* vol. 5, letter from Edward Blyth, [30 September or 7 October 1855], in which Blyth stated: 'It is remarkable that there is no *Double-spurred* race of domestic fowls', and described the positions of spurs in wild fowl.

[26] *Correspondence* vol. 5, letter from Edward Blyth, 4 August 1855: 'the name of that town has become transferred to themselves, as in the more familiar instances of *Canary* & *Bantam*!' According to the *OED*, bantams were named from Bantam in the north-west of Java, from whence they were supposed to have been introduced into Europe. CD, following John Crawfurd, believed they came originally from Japan (*Variation* 1: 230).

[27] In his letter of 4 August 1855 (*Correspondence* vol. 5), Blyth told CD that in India hares were indigenous whereas rabbits were introduced. In *Variation* 1: 108–11, CD gave an account of the origin of the 'so called Himalayan rabbits'.

[28] In *Correspondence* vol. 5, letter from Edward Blyth, [22 October 1855], Blyth discussed the fallow deer but made no reference to varieties.

[29] See *Correspondence* vol. 5, letter from Edward Blyth, [30 September or 7 October 1855], in which Blyth quoted from 'an article on the Rein Deer which I wrote some years ago.' For the difficulty in locating the *Calcutta Sporting Review*, see letter from Edward Blyth, 8 January 1856, n. 6.

[30] See *Correspondence* vol. 5, CD memorandum, [December 1855], for a list of correspondents, including William Vine, to whom CD had 'written to for Pigeon & Poultry Skins'. Vine was an officer in the Madras cavalry.

[31] CD's numbering of Blyth's letters.

To Walter Elliot 23 January 1856

> Down Bromley Kent
> Jan. 23. /56

My dear Sir

I hope that you will not have quite forgotten a call, most pleasant to me, which you were so good as to make on me in Glasgow.[1] You there mentioned two or three things, which on reflection have interested me so much, that I hope you will excuse me troubling you in regard to them. You said that you had measurements of tigers, showing differences in their proportions & that you could let me have a copy, this I sh^d value much.—[2]

You, also, referred to some work in an Eastern language, with remarks on domestic Pigeons (& Poultry?);[3] if the extracts are short, I sh^d be extremely grateful for any, or even for any enumeration of the breeds or races. I should mention that I have heard that such exist in the Ayin Akbaree in Persian (I know not whether I have spelt this right) but as this work is translated I can consult it in the India House.[4]

Lastly I want to beg a very great favour of you, if in your power to grant it, & I think your wonderful zeal for Nat. History will lead you to help me if you can.— I am trying energetically to get the skins of all the *domestic* varieties of Pigeons & Poultry from all parts of the world, in order to study the amount of variation.— Now if there is any one, who can for payment skin birds, will you aid me by making enquiries of any *natives* for the names of *any* varieties which are supposed to have been long bred in the country, & then direct the bird skinner to purchase such birds & skin them for me; the skinner might buy very *old* birds, which would be best for me & cheapest.— I could easily, permit me to say, repay you.—[5]

The only caution required would be not to get birds *recently* introduced from Europe.— This would be an enormous assistance to me.— I sh^d like to have the native name, & a notice whether any of the varieties are Tumblers or Carrier.— There are Tumblers in India with most curious habits.[6] The birds sh^d be adult or old; a characteristic specimen sh^d be selected; & in skinning the whole of the bones of wing & leg sh^d be left in, & as much as possible of the skull.— In Poultry Cock & Hen sh^d be selected; but in Madras itself, probably all the Poultry will be imported breeds.— Now can you forgive me asking you to take so much trouble? I fear you will think me very unreasonable & I have nothing to trust to, but your goodnature.

Pray believe me, with many apologies. My dear Sir | Yours very faithfully | Ch. Darwin

I fear my address is very incorrect; but living in country, I have no one to ask

American Philosophical Society (123)

[1] CD had attended the British Association for the Advancement of Science meeting in Glasgow in September 1855 (see *Correspondence* vol. 5, Appendix I). CD had first heard of Elliot's natural history collections from Joseph Dalton Hooker (see *Correspondence* vol. 4, letter from J. D. Hooker, 20 February – 16 [March] 1848). Elliot was a prominent Indian civil servant and a member of the council of the governor of Madras.

[2] A table of measurements of tigers is preserved in DAR 205.10 (Letters). On the verso CD wrote: 'Madras | Mr Walter Elliot | June 10th 1856'. The table is marked with a '3', the number of CD's portfolio on variation and varieties, and CD has noted in pencil: 'I have not thought worth using'. Elliot is cited several times in *Variation*, but the information on tigers was not used in that work nor in *Natural selection* or *Origin*.

[3] A Persian treatise by Sayzid Mohammed Musari. Elliot is thanked in *Variation* 1: 141, for providing CD with a translation of this work.

[4] Gladwin trans. 1783–6. Edward Blyth had first drawn CD's attention to this source (see *Correspondence* vol. 5, letters from Edward Blyth, 4 August 1855 and 8 December 1855). India House was the former office of the East India Company in London. It subsequently became the headquarters of the British government's India Office.

[5] In *Variation* 1: 132 n. 1, CD wrote: 'I am deeply indebted to Sir Walter Elliot for an immense collection of skins from Madras, with much information regarding them.' CD recorded payments to Walter Elliot in November 1856 and September 1857 (CD's Account book (Down House MS)).

[6] Elliot sent CD specimens of the *Lotan*, or Indian ground tumbler, which 'present one of the most remarkable inherited habits or instincts which have ever been recorded.' (*Variation* 1: 150). When gently shaken and then placed on the ground, the pigeons begin tumbling head over heels.

To Katharine Murray Lyell 26 January [1856]

Down Bromley Kent
Jan[y] 26th

My dear M[rs] Lyell

I shall be very glad to be of any sort of use to you in regard to the Beetles. But first let me thank you for your kind note, & offer of specimens to my children: my Boys are all butterfly-hunters, & all young & ardent lepidopterists despise from the bottom of their souls coleopterists.—[1]

The simplest plan for your end & for the good of entomology, I should think, would be to offer the collection to D[r] J. E. Gray for British Museum, on condition that a perfect set was made out for you. If the collection was at all valuable I should think he would be very glad to have this done.—[2] Whether any third set would be worth making out, would depend on value of collection: I do not suppose that you expect the insects to be named for that would be a most serious labour.— If you do not approve of this scheme, I sh[d] think it very likely that M[r] Waterhouse would think it worth his while to set a series for you, retaining duplicates for himself, but I say this only on a venture.[3] You might trust M[r] Waterhouse implicitly, which I fear, as rumour goes, is more than can be said for all entomologists.—

I presume, if you thought of either scheme, Sir Charles Lyell could easily see the gentlemen & arrange it; but if not, I could do so when next I come to town, which however will not be for 3 or 4 weeks.—

With respect to giving your children a taste for Natural History, I will venture one remark, viz that giving them specimens, in my opinion, would tend to destroy such taste. Youngsters must be themselves collectors to acquire a taste; & if I had a collection of English Lepidoptera, I would be systematically most miserly & not give my Boys half-a-dozen butterflies in the year. Your eldest Boy[4] has the brow of an observer, if there be the least truth in phrenology.—

We are all better, but we have been of late, a poor household.—

Pray give my kind remembrances to Colonel Lyell[4] & believe me, my dear M[rs] Lyell | Yours truly obliged | Charles Darwin

Postmark: JA 27 1856
American Philosophical Society (124)

[1] For William and George Darwin's interest in Lepidoptera, see *Correspondence* vol. 5, letter to G. R. Waterhouse, 8 July [1855]. In 1859, Francis, Leonard, and Horace Darwin, aged 7, 5, and 4, respectively, reported their collecting activities in the *Entomologist's Weekly Intelligencer* (see *LL* 2: 140).

[2] The accessions list of the British Museum (Natural History) does not record any donation by Katharine Murray Lyell (British Museum (Natural History) 1904–6, vol. 2).

[3] George Robert Waterhouse was preparing a monograph on the Coleoptera (Waterhouse 1858).

[4] Leonard Lyell, aged 5.

[5] Katharine Murray Lyell, Mary Lyell's sister, was the wife of Colonel Henry Lyell, the brother of Charles Lyell.

To John Phillips 28 January [1856][1]

Down Bromley Kent
Jan. 28th

My dear Phillips

I have received this morning your very kind present;[2] but I am almost ashamed to accept so beautiful a book in return for my old & dry works on Rocks.—[3] With very sincere thanks. Believe me. Yours very truly | Ch. Darwin

Oxford University Museum (Geological collections)

[1] Dated by the relationship to the letter to John Phillips, 18 January [1856].

[2] Phillips 1855. CD's copy is in the Darwin Library–Down. The work includes a series of views of Yorkshire taken from sketches by Phillips.

[3] *Geology of 'Beagle'.* See letter to John Phillips, 18 January [1856], n. 6.

From Thomas Vernon Wollaston [February 1856][1]

left Funchal, & too late for me to procure another pair (of really wild ones) alive,—sic transit gloria mundi!—

I was very much vexed about this; however it could not be helped, & I have put them (being full grown) into spirits (along with the other pair, which were shot for me by a friend), & I trust that they will not be altogether useless to you.[2] I was sorry that I was not able to get *even the others* prepared, after all: so that everything I have for you is in pickle. The brace of P? S? rabbits (+ an *eel*,—M! Lowe's offering to you, though in reality found by me in a stream in P? S?, I having given it to him)[3] are in a small cask; &, in addition to these, I have got you a bottle of Frogs (from Madeira proper, & which have been introduced into the island *within* 50 years, though whether from the Canaries or England appears to be doubtful,— a point however which the *species* will probably prove), which have increased so prodigiously of late years as to have become a literal nuisance. I do not know whether they will be of any use to you; but as I remember your remark about the general absence of the Frog family (as however aborigines, I am aware) in islands far removed in the oceans, I thought perhaps you might like to have some.—[4] If you will kindly tell me *how* I am to forward the cask & bottles to you, I will

AL incomplete
DAR 205.3 (Letters)

CD ANNOTATIONS
1.1 left Funchal . . . small cask; 2.7] *crossed pencil*
2.11 I do not know . . . I will 2.16] *crossed pencil*
Top of last page: '19'[5] *brown crayon, circled brown crayon*

[1] The date is conjectured from the recent return of Wollaston from a seven-month visit to Madeira and from a letter written by Charles Lyell, 19 February 1856, mentioning that he had recently met Wollaston (K. M. Lyell ed. 1881, 2: 208–9).

[2] In his chapter on domestic rabbits in *Variation*, CD devoted several pages to the discussion of the feral rabbits of Porto Santo, Madeira. He stated that 'Mr. Wollaston, at my request, brought home two of these feral rabbits in spirits of wine' (*Variation* 1: 113). The rabbits of Porto Santo were known to be descended from those put ashore in 1418 or 1419 from a Portuguese ship. Since their introduction they had varied so greatly from the common European rabbit that they would have been ranked as a distinct species had their history not been known (see *Variation* 1: 112–15). CD had sent Wollaston a long list of queries concerning the natural history of Madeira early in 1855 (see *Correspondence* vol. 5, letter from T. V. Wollaston, 2 March [1855]).

[3] Richard Thomas Lowe, rector of Lea, Yorkshire, was a friend of Wollaston's and the two made frequent trips to Madeira. Lowe had been chaplain on Madeira, 1832–52, during which time he had become an expert on its natural history. See letter from R. T. Lowe, 12 April 1856.

[4] In the notes which Charles Lyell made of a conversation with CD on 13 April 1856 during a visit to Down, he wrote: 'Frogs are not found in volcanic islands. Even New Zealand only provides one species lately discovered in island & Darwin finds frogs' spawn to be very easily killed by salt water.' (Wilson ed. 1970, p. 53). See also *Correspondence* vol. 5, letter to J. D. Hooker, 10 June [1855].

[5] The number of CD's portfolio of notes on the geographical distribution of animals.

To W. B. Tegetmeier [1 February 1856][1]

57 Queen Anne St | Cavendish Sq^e
Friday morning

My dear Sir

I received a note from M^r. Bult late last night, in which he very courteously offers to show me his Pigeons, at 11 oclock tomorrow morning, & tells me how best to get to his house per omnibus. He says "if Mr Tegetmeier comes with you it will add to the pleasure". I hope you may, but I see it is quite a long walk.[2]

Pray believe me, Your's very sincerely | Ch. Darwin

New York Botanical Garden Library (Charles Finney Cox collection)

[1] Dated by the relationship to the letter to W. B. Tegetmeier, 14 January [1856]. CD was in London, staying at the house of his brother Erasmus Alvey Darwin, to attend a council meeting of the Royal Society of London on 31 January 1856 (Royal Society council minutes).

[2] Benjamin Edmund Bult's business premises were at 25 New Quebec Street, only about half a mile from Queen Anne Street, but presumably he resided elsewhere. Tegetmeier lived about eight miles distant in Wood Green, near Tottenham (*Post Office London directory* 1855; CD's Address book (Down House MS)).

To Samuel Birch 6 February [1856][1]

Down Bromley Kent
Feb. 6th

My dear Sir

I am most sincerely obliged for your note, & for the very great trouble you are so kindly willing to take for me.—[2] I shall be in London in a fortnight's time, when I will call, previously letting you know when I will come.[3] I hope that this little delay will not be inconvenient to you.—

With my best thanks | Your's truly obliged | Ch. Darwin

British Museum (Department of Western Asiatic Antiquities, correspondence 1826–67: 1492)

[1] Dated by the relationship to the letter to J. E. Gray, 14 January [1856].
[2] CD had asked Birch, through John Edward Gray, to check ancient Chinese sources for references to varieties of domesticated animals and cultivated plants (see letter to J. E. Gray, 14 January [1856]).
[3] See letter to Samuel Birch, [12 March 1856].

To William Yarrell 6 [February 1856]

Down Bromley Kent
6th.

My dear Mr Yarrell

The Pigeons are all quite well, vigorous, & in good spirits.— They are really quite beautiful.— I have now 15 kinds of Pigeons!—[1]

Pray give my very sincere & cordial thanks to Mrs Cotton.—[2] I send with this the Book & your Cage.—

Pray believe me, | Yours most truly obliged | Charles Darwin

Endorsement: 'Feby 7th 1856'
Cleveland Health Sciences Library (Robert M. Stecher collection)

[1] Early in 1855, when CD began his study of pigeons, Yarrell had advised him about breeds and how to procure them. See *Correspondence* vol. 5, letters to W. D. Fox, 19 March [1855] and 27 March [1855].
[2] Mrs Cotton has not been identified but according to his Catalogue of Down specimens (Down House MS), CD acquired red tumbler pigeons and trumpeters from her. CD's Account book (Down House MS) has regular entries, beginning in June 1855, recording payments made for the purchases of pigeons. By the end of June 1856, he had acquired eighty-nine birds (see letter to W. B. Tegetmeier, 24 June [1856]).

From Charles James Fox Bunbury 7 February 1856

Mil⟨denh⟩all
February 7. 1856

My dear Darwin,

As I know you are much interested in the questions relating to *varieties* & *species* among plants as well as ⟨an⟩imals, I will mention to you a ⟨c⟩ase which has excited my curiosity a good deal. In the Journal of the Horticultural Society, a few years ago, (5th vol. p. 31 & 32,) Lindley described & figured what he believed to be a very extraordinary variety of *Colletia spinosa* (a South American shrub),[1] said to have been raised from seed of that species in Lady Rolle's garden in Devonshire.[2] The deviation from the original type, as shortly as I can state it, is mainly this: that ⟨   ⟩ spines, or spiny-pointed branches ⟨in⟩stead of being nearly cylindrical, are so excessively dilated vertically, as to form nearly right-angled triangles, with ⟨   ⟩ base (along the branch) often equal to the perpendicular. And this is a real extension of the wood, not merely of the cellular substance.[3] Now ⟨the⟩ curious thing is, that this same variety (if such it be) grows wild 'in South America. Lindley's description & figure agree perfectly (tho' he seems to have quite overlooked this,) with those which Sir W. Hooker gives, in the first volume of the Botanical Miscellany, of his Colletia cruciata,[4] found by Dr Gillies on sand hills near Maldonado.[5] I find in Mr Fox's collection, several fine dried specimens of the same, gathered also at Maldonado.[6]

I wonder whether you met with it there? *If* the origin of this plant in cultivation from seeds of Colletia spinosa be really certain, it is a very curious case of a seminal variety having ⟨   ⟩ the appearance of a distinct species, & occurring in the wild state as well as in cultivation but I confess the evidence as to its origin does not appear to me quite satisfactory.[7] You probably have the means of referring to the Journal of the Hort. Soc.,—I wish you would tell me what you think.

I was told last summer that you were becoming a believer in the unlimited mutability of species,—almost to the extent of the "Vestiges of Creation."[8] I suspect this is not strictly correct. I should be very glad to hear any thing you may think fit to tell me about your researches into the laws of species, a subject on which I hope you will one day enlighten us very much; also I should like to know whether you have obtained any further results as to the germination of seeds ex⟨pos⟩ed to salt water.[9] I am afraid I have not yet any remarkable facts to send you in reference to Cape plants.[10] I forget whether I mentioned *Frankenia* to you as one of those genera which have their headquarters in Europe, but which have some truly indigenous species at the Cape of Good Hope. One species indeed, Frankenia lævis, appears to be common to Europe & the Cape; whether it is found in any intermediate country I do not know; but there are one or two peculiar to the Cape, with a very strong likeness certainly to those of the northern hemisphere, but distinct, as specific distinctness is generally understood. Dianthus I think I mentioned to you before: it is an interesting case, because I believe the genus exists nowhere in the southern hemisphere except at the Cape, & the genus is a peculiarly natural & well-marked one.[11] We have a vast deal yet to learn with respect to the limits of species; the excessive differences in the views of different ⟨na⟩turalists on this point intro⟨duces⟩ c⟨onf⟩usion & uncertainty into ⟨ ⟩ reasonings on the geogra⟨phy of⟩ plants & animals. When ⟨ ⟩ that what according to Duval's views are nearly 50 different species of Solanum,[12] are considered by Bentham as all referable to the one Solanum nigrum; & that another botanist has made 12 species out of our common White Water-lily; it is rather bewildering. I am myself quite ready to believe that the range of variation of species may be greater than even the most cautious botanists at present allow for; but I should be slow to believe that it is unlimited.

Ever yours very sincerely | C J F Bunbury

DAR 160: 374, DAR 205.4 (Letters)

CD ANNOTATIONS
2.4 does not . . . Cape plants. 3.8] *crossed pencil*
3.8 I mentioned . . . well-marked one. 3.17] '20' *added brown crayon*; 'Bunbury' *added pencil*[13]
3.24 range of variation . . . unlimited. 3.26] *crossed pencil*
Top of first page: '1' *brown crayon, del brown crayon*; '3' *brown crayon*

[1] Bunbury refers to Lindley 1850.
[2] Louisa Barbara Rolle of Bicton House, Devonshire. Her gardener, James Barnes, provided John Lindley with an account of the *Colletia* (Lindley 1850, p. 29).
[3] Lindley considered this a case of transmutation without parallel, which threw 'greatest doubt upon the original distinction of numerous plants now admitted into books as species' (Lindley 1850, p. 32).
[4] [W. J. Hooker] 1830a, pp. 152–3.
[5] John Gillies, a naval surgeon, had collected plants in South America.
[6] Henry Stephen Fox was Bunbury's uncle. He had collected plants while British minister in Buenos Aires and Rio de Janeiro, 1831–3. On his death in 1846, Bunbury inherited his herbarium (R. Desmond 1977).
[7] Lindley's informant, James Barnes (see n. 2, above), had not at first been certain that the plant had been raised from seed (Lindley 1850, p. 29).
[8] CD had told Bunbury as early as November 1845 that he was 'to some extent' a transmutationist (see *Correspondence* vol. 3, letter from Charles Lyell, [after 2 August 1845], n. 5). Bunbury refers to the anonymous evolutionary work *Vestiges of the natural history of creation* ([Chambers] 1844).
[9] CD was nearing the end of a series of experiments begun in 1855 to test the vitality of seeds after submersion in salt water. See letter to M. J. Berkeley, 29 February [1856], and n. 1.
[10] CD had previously written to Bunbury requesting information on Cape Colony plants (*Correspondence* vol. 5, letter from C. J. F. Bunbury, 10 April 1855).
[11] CD included this information in *Natural selection*, p. 552 n. 3: 'Mr. Bunbury thinks that the genera Dianthus, Franklinia Statice are the most striking cases of northern genera having representative species at the Cape.' 'Franklinia' was a mistake for 'Frankenia'.
[12] A reference to the French gardener known only as jardinier Duval. CD had read three of his works in July 1854 (see *Correspondence* vol. 4, Appendix IV, 128: 8). Bunbury's particular reference is to Duval 1852, a treatise on the potato family. There is an extract from this work in the Darwin Library–CUL.
[13] CD cut the letter into two pieces. These annotations were to label the portion placed in portfolio 20, which contained CD's notes on the geographical distribution of plants. The '3' on the remainder was used for notes on marked varieties.

From Edward Blyth 23 February 1856

Calcutta,
Feb. 23/56.

My dear Sir,

My last to you was an exceedingly brief and hurried production, but got safe to the post-office.[1] I now write a little more at leisure, but have not overmuch time to go into details, *i.e.* the minute discussion of various subjects.— Imprimis, I send you 4 copies of my printed letter, which is now in circulation among the Council of our Society: a roughish proof of it I sent you by ⟨l⟩ast mail.[2] One of the 4 now sent, kindly give to Col. Sykes; another to Prof. Owen; a third to the Comm. in Chief, Visc. Hardinge, & make sure that he gets it;[3] & the 4th keep yourself, or do as you think best with it. If I meet with any more unworthy opposition from the old quarter (that *medical clique* who have uniformly opposed me always),[4] I certainly shall not mince matters at all; but republish and circulate widely, to the discredit of the Asiatic Society, a correspondence on the same subject which passed about ten years ago; respecting which our present Secretary,[5] who has just read it, writes me word that he thinks the conduct of the Council then to me was "most illiberal and narrow-minded". Not only that, but I will review my reviewers a bit, and criticize their claims to sit in judgement upon me, by analysing their own very humble contributions to science—an exposé they little dream of! "Oh that mine enemy would write a book" said somebody;[6] & it is little enough that my opponents have done in the writing line, but that little will abundantly suffice; & heaven knows that I have been patient enough hitherto, & will have been fairly provoked to retaliation, if it does come to that! The fact is, my mind has always been too much occupied with other matters, to allow it to dwell on one subject, even though it concern my comfort so nearly. But I am lapsing into the style of "the man with a grievance", of which *jam satis*. The present however is an auspicious time to agitate the matter, as there is a grand movement just now in India for museums & even Zool. Gns, of which latter one is now being established at Madras![7] Calcutta is likely to follow, under the auspices of Ld & Lady Canning, who will have personally seen not only what is doing at Madras, but at Bombay also; & her ladyship I hear takes much interest in the matter.—[8]

Three days ago I indulged in my first holiday this cold weather, & had a glorious ramble over the botanic gn, myself & wife only; for it was the 2nd anniversary of our wedding day, & we kept it thus; both having pretty much the same tastes.[9] Would not you too have enjoyed it? *Bougainvillia* in all its glory (you may remember what Humboldt says of it in his *Cosmos*);[10] & a week or two hence the *Amherstia* (quite a grove of it) will be in its glory, & superb beyond expression: lots of orchids, too, &c &c; & the ever beautiful palms, *Cycadeæ*, &c, which I never tire of feasting my eyes on. I imagine that not many feel so intense a pleasure as I do in contemplating the grand and beautiful forms of vegetation; & the more so, as I so seldom give myself the chance. I have just had the pleasure of seeing the *Gallus varius* (vel *furcatus*) *alive* for the first time. This, as you know, has an unserrated comb, Single medial throat-wattle, & broad scale-like nuchal feathers, in lieu of hackles: but the colouring of the naked parts surprised me.

a, pale bluish-lake
b Red
c. Bright yellow
d —Blue—

The *G. œn⟨eus⟩* of Temminck is a hybrid between this & a common hen, often enough raised in captivity in Java.—[11] Well, I have begun to work in earnest at domestic Pigeons, and am surprised to find how cheap many beautiful varieties of them are, which I will tell you more about another time; and I shall accordingly be able to send you a good supply *alive* £10, which you proposed to send, would more than suffice; but good (*rat-proof*) cages are expensive here, and I might otherwise lay out a part of the money to advantage. But I would *very much rather* have the value in hardy living creatures; as especially Maccaws,—and Marmozets if you could but procure some. I should be glad of any number of Maccaws, and would willingly share with you *the costs and the profits* of a few speculations of the kind; the latter being somewhat inordinate; but this need not be published to the world! It is a fact, that I can always get £50 a pair for Maccaws, the cost in London being £3 or 4; & they are particularly hardy on the voyage: the best plan being to send them in charge of a ship's butcher, promising him a reward from me for all that arrive in health. For one or more pairs of Marmozets, I think that I could now get £100 per pair, without difficulty; & they go into *Zenanas*[12] where nobody sees them or is even likely to hear about them. Natives of enormous wealth arc the purchasers, who care not what they give for what they particularly fancy; and the money is too often much worse *mispent* by them. Well, what say you to the spec, in a quiet way? Bartlett formerly supplied me, but since I advanced to him a considerable sum for purchases (borrowed too at high interest), I can get no answer to my letters from him, nor any account of animals which ⅃ know to have been safely delivered to him from me, the last being a fine Tibetan Bear.[13] I may yet be compelled to resort to harsh measures in his case, which I am loth in the extreme to do; but not getting the Maccaws, &c, which I paid for in advance, & fully calculated on receiving and then immediately disposing of, has caused me a vast deal of trouble and annoyance, which is entirely due to Bartlett's misconduct. Perhaps you may think such traffic *infra dig.*; but you might manage it by an agent; & the Commercial road is a great place to pick up Maccaws, &c. *cheap*; or they might be purchased of Herring & other regular dealers.—[14]

But now for your letter of Dec.ʳ 16/– & its queries. About the fine Asses of the province of Omán in Arabia, I cannot recal to mind my authority for the *genealogies*; but I am certain that I have read it somewhere; perhaps in one of the late Lt. Wellsted's papers, or in Burckhardt?[15] But the authorities on Arabia are not numerous. You will, I suppose, have read my paper on 'Wild Asses', entitled *Nat. Hist. Queries*, by *Scrap-collator*, in the last No. of the *Indian Sporting Review*, to which I called your attention.[16] Here you will find some notice of those superb donkeys; and I much incline to think now that if the *genuine* wild Ass still exists anywhere, it will be in S. Arabia, where Chesney speaks of a wild Ass, in addition to his "wild Horse", the latter seeming to be the Hemione.—[17] Ditto about the infertility of Irish & Devon Red Deer; my *impression* is, that I saw this remarked in an article in one of the English sporting periodicals; or it may have been in Chambers's Jl, or some such work: but there is no doubt at all of my having met with the remark, & that it applied more especially to the Devon Deer. Why not try the 'Notes & Queries'? Or make enquiry, as a correspondent of one of the sporting periodicals,—about as likely a way as any to elicit the truth. If you can get the 'India Sporting Review', Vol. VI, 252, you will find an interesting notice of experiments with canine hybrids:[18] but the writer's suggestion about the Hyæna is of course nonsense, inasmuch as mos eorum copulandi mos canum non est! By the way, the so called 'Wild Dog' I told you of is dead,[19] & transferred (temporarily at least, if not permanently,), to our museum—: Hodgson saw it at Darjiling, & considers it to be totally distinct from his *primævus*; but I will let you know more about these rufous wild dogs (so called) after further enquiry. I am having both skin & skeleton prepared, &, as usual, the second lower tuberculous molar is wanting. What chiefly surprises me in this animal is the length of the tail, which is fully as *long* (as well as bushy) as any Fox's; and there is much long hair about the *jowl*. I am not at all satisfied regarding the alleged facts respecting the infertility of canine hybrids in the third generation. Unless you have two or more sets of them, from different pairs of parents, the question of *breeding-in-& in* comes to disturb your conclusions: my own views on which question I have before explained to you— However, supposing that you had a litter of Jackal-hybrids which bred *inter se*, it would be easy to parallel all experiments by subjecting ordinary puppies of the same litter to precisely the same conditions; & more than one set of them, the more the better of course, to ascertain if the Jackal-hybrids really presented exceptional phenomena.[20]

A mere luxurious life is inimical to propagation, alike in the lower animals & in so many wealthy human families You remember the Irish medico's remark, that sterility was hereditary in some families!! ! And also in many choice cultivated plants! The prolificacy of hybrids being of course liable to be similarly affected, & the more so as being rare and taken much care of, they are apt to be a good deal pampered. Infertility is thus a negative result, & the non-fertility of some hybrids is not necessarily due to their mixed (or mule) origin; whether the admixture be of pure species or of undoubted varieties. Nevertheless there can be no doubt of the broad fact that the *tendency* of hybrids is to be sterile, and especially that the semen of the males is often unprolific, & has been found (in the equine hybrid) deficient of spermatozoa. And there can be no mistake regarding my experiments with the ½ bred *Gallus Sonneratii* of both sexes, which were infertile *inter se*, and either of them with other fowls; although the hen produced many eggs, & the cock was particularly salacious; for the eggs in which either was concerned never would hatch, although other eggs placed with them hatched as usual: & surely this instance of infertility may be fairly ascribed to the hybridity.[21] Perhaps, however, other hybrids of the very same kind might be more or less prolific as some individuals of pure species are sterile.? I am going over old ground I fear; & shall finish it by telling you that I have procured some hybrid skins of the *Coracias*, & also of the two parent species, *C. indica* & *C. affinis* pure. For Skua Gull in both hemispheres, consult Gould's 'Birds of Australia'.[22] It is the most striking instance of the kind I know of. The *Thalassidromi Leachii* was considered another; but I think this has lately been observed within the tropics. Is Gould's Australian crested Grebe really different from that of the N. hemisphere (which I have obtained here)? I fancy not; & that several of Gould's species want knocking on the head, *ex. gr.* his *Sylochelidon strenuus*, which = *S. caspius*, of which European, Indian, & Australian specimens are in our museum here, absolutely undistinguishable. *Totanus glottoides*, Vigors & Gould, of India & Australia is just the winter dress of *T. glottis*, most abundant here, & the specimens totally undistinguishable from British. The Turnstone is the most universally distributed of all birds, being found literally on all (habitable) sea-coasts; & *Cynthia cardui* seems to be an universally diffused Butterfly. See to Horsfield's description of *Mydaus* in the *Zool. Res. in Java*; this species also inhabiting a certain elevation in the Malayan pen^a—.[23]

Ever truly Yrs, E Blyth—

DAR 98: 128–32

CD ANNOTATIONS
0.3 My dear . . . the vegetation; 2.9] *crossed pencil*
2.10 I have just had . . . Java.— 2.15] '12' *added brown crayon, circled brown crayon*
2.15 Well, . . . suffice; 2.19] '12' *added brown crayon*
2.15 Well, . . . regular dealers.— 2.43] *crossed pencil*
3.17 Vol. VI,] *underl pencil*
4.19 I have procured . . . pure. 4.20] '12' *added brown crayon*
4.20 For Skua . . . know of. 4.22] *double scored pencil*
4.32 See to Horsfield's . . . pen^a—. 4.34] *double scored pencil*
Top of first page: '12'^24 *brown crayon*

[1] Letter from Edward Blyth, 23 January 1856.
[2] This probably refers to a memorial Blyth wrote to the court of directors of the East India Company petitioning for an increase in salary and a pension. It was submitted to a general meeting of the

Asiatic Society of Bengal on 7 May 1856, which moved to forward it to the government of Bengal 'with the expression of the high sense entertained by the Society of the value of Mr. Blyth's labours in the department of Natural History' (*Journal of the Asiatic Society of Bengal* 25 (1856): 237–9). Blyth received no official reply to his memorial. By 1862, when Blyth had to leave India due to failing health, he had still not received a pension in spite of a second memorial prepared on his behalf by the council of the Asiatic Society of Bengal (*Journal of the Asiatic Society of Bengal* 31 (1862): 60). Blyth was finally awarded a pension of £150 per annum after his return to England (*Journal of the Asiatic Society of Bengal* 33 (1864): 73). In comparison, the usual pension allowed to members of the Bengal Medical Establishment was £300 per annum (see *Correspondence* vol. 4, letter from J. D. Hooker, 13 October 1848, n. 3).

[3] William Henry Sykes, chairman of the court of directors of the East India Company; Richard Owen, Hunterian professor at the Royal College of Surgeons; Henry Hardinge, Viscount Hardinge of Lahore, general commander-in-chief of the forces during the Crimean War, 1854–6. The latter was receiving a pension of £5000 a year from the East India Company and £3000 a year from the British government, in gratitude for his services as governor-general of India, 1844–7 (*DNB*).

[4] Presumably the Bengal Medical Establishment of the East India Company.

[5] William Henry Atkinson was secretary of the Asiatic Society of Bengal in 1856.

[6] 'Oh . . . that mine adversary had written a book!' (Job 31: 35).

[7] The Asiatic Society of Bengal was exploring the possibility of founding an Imperial Museum in Calcutta in which the whole of its collections would be housed. The plan was rejected in 1858. See Mitra 1885, p. 44. A National Museum was eventually established in 1865 (see letter from Edward Blyth, 26 February 1856, n. 2).

[8] Charles John Canning, governor-general of India, assumed the government of India on the last day of February 1856, having visited Bombay and Madras en route to Calcutta with his wife Charlotte Canning (*DNB*).

[9] According to Grote 1875, p. x, Blyth's married life was extremely happy, and it was a severe blow to him when his wife died in December 1857.

[10] Humboldt 1846–58. Blyth's reference has not been located.

[11] In his abstract of this letter (DAR 203), CD noted: 'Description of Gallus *varius vel [*interl*] furcatus G. œnæus is a hybrid.'

[12] Zenana: 'In India and Persia, that part of a dwelling-house in which the women of a family are secluded; an East Indian harem.' (*OED*).

[13] Abraham Dee Bartlett was superintendent of the natural history department of the Crystal Palace (*Modern English biography*).

[14] William Herring was a bird and animal dealer at 34 FitzRoy Terrace, New Road (*Post Office London directory* 1855). Both New Road and Commercial Road are in the East End of London.

[15] John Lewis Burckhardt had not travelled as far as Oman in his Arabian expedition, but in his book (Burckhardt 1829) he included an appendix detailing the route taken by pilgrims to Mecca. A 'fine breed of mules and asses' is mentioned in volume 2, p. 379.

[16] See letter from Edward Blyth, 8 January [1856], in which he tells CD that his article on wild asses is in number 44, December 1855, of the *Calcutta Sporting Review*. It has not been possible to locate either the *Calcutta Sporting Review* or the *Indian Sporting Review* for 1855.

[17] Chesney 1850, 1: 581, 586.

[18] See n. 16, above.

[19] See letter from Edward Blyth, 23 January 1856.

[20] See *Correspondence* vol. 5, letter from Edward Blyth, 7 September [1855], on the prolificacy of the hybrids from two or three generations of intermixing dogs and jackals.

[21] In his abstract of this letter (DAR 203), CD noted: 'Repeat on infertility of G. Sonneratii'. See *Correspondence* vol. 5, letter from Edward Blyth, 21 April 1855, for Blyth's earlier mention of this experiment. For CD's later discussion of the fertility of such hybrids, see *Variation* 1: 234–5.

[22] In an earlier letter to CD, 8 October 1855 (*Correspondence* vol. 5), Blyth had stated: 'Take the Skua

Gull of Australia as compared with that of the north; & this bird *has never been seen within the tropics.*'
In his abstract of this letter (DAR 203), CD noted: 'Refers about Skua Gull not in Tropics, but in
Australia, to Gould; & I *have* consulted him.' John Gould provided CD with information on
northern birds found in Australia but not in the tropics (*Natural selection*, p. 554); the skua gull was
not listed as one such bird. However, the skua gull is described in J. Gould 1848, 7: pl. 21, with the
comment: 'So little difference is observable between the examples of the Southern Ocean and those
found in our own seas, that I have been compelled to consider them to belong to the same species.'
[23] See letter from Edward Blyth, 8 January 1856, n. 26.
[24] CD's numbering of Blyth's letters.

From Edward Blyth 26 February 1856

Calcutta,
Feb.? 26/56.

My dear Sir,

A mail viâ Bombay is advertized for today, and you will probably get this letter
by the same steamer which conveys my last, of a few days back.[1] Now it is quite as
well to anticipate contingencies, & to take time by the forelock; and a long
experience of India and its doings makes one uncommonly vigilant and what is
commonly termed *wide-awake.* Now that a movement is in contemplation for the
establishment of a Presidency or Gov.^t museum here, to which our Asiatic Society's
Collections are likely to be transferred, it is probable that the whole thing will be
done on a liberal and efficient scale, & that a decent salary will be attached to the
Curatorship.[2] In that case, the office will undoubtedly be sought, and (if we don't
look sharp) very possibly obtained by some member of the medical body here,
who will assuredly look upon it as virtually appertaining to their service, and I am
quite sure will try for it: & I, who have so long borne the burden & heat of the day,
may find myself placed in a subordinate position! That I have not performed
impossibilities with insufficient means will then doubtless be attributed to
incapacity, & so forth. Now, it is really a fact that I have been able to do very little
of late, for want of the most necessary aid in the museum; our only taxidermist and
general assistant in my department having been laid up for more than a fortnight
past, & prior to that having been much away from his post all this year; so that I
can get nothing done that I want done. Judging from past experience, I know full
well that no allowance will be made for all this by & bye, when a good salary is to
be striven for; people will be *interested* in misrepresenting the matter, whom I too
well know are not at all scrupulous; & so because I can't get on as I could wish
without proper assistance, it will be made out that I am incompetent to take
official charge of a proper establishment; & this you must kindly manage to
represent properly to Col. Sykes. I enclose a portion of a letter just rec.^d from our
Secretary, which will shew you that my fears are not unfounded.—[3]

Looking up materials for an article on the *Leopard* (for 'Feline Animals of India',
No 3),[4] I have just hit upon some important conclusions, which I feel satisfied are
sound. 1, That the ancients never descriminated the *Panther* and *Leopard* of modern
zoologists, respecting the distinctions of which no two original observers seem to

be agreed even now.— 2, that they were acquainted, however, with two large spotted cats, viz. the *Nimmer* and the *Fáádh* of the Arabs, both of which are common enough in Asia minor to this day. These two being the *Guepard* or 'Harting Leopard' and the true *Pard*.— 3, that the *Panther* of the Greeks was the *Guepard*; and their *Pardus* and *Pardalis* the two sexes of the other.— 4, that the *Pardus* of the Romans signified the male, and their *Panthera* [var] *Varia* the female, of the *Pard*; and wanting a name for the other, when they came to know it in after-times, they christened it the *Leo-pardus*, *i.e.* 'maned' or 'Leonine Pard', in allusion to the lengthened fur upon the nape; whence also the name *F. jubata*. Hence, it appears, that the Guepard was the original πανθηρ of the Greeks, & also the original *Leo-pardus* of the Romans; both of which names have since been transferred to the other! While the name *Cheeta* now applied to it in zoological works belongs properly to the other, the *Chita-bagh* (*i.e.* 'Spotted Tiger') of the natives here!! Now I should hardly have hit upon this éclaircissement, had I not recently learned from Chesney's work how common the *jubata* is in Syria;[5] and if you look to the authorities cited in Cuvier's *Ossemens fossiles*,[6] I think you will agree with me in these interpretations.—

Are you aware that *in China* Ducks are artificially hatched on a grand scale? In an elaborate article 'on the diet of the Chinese', published in the 'Chinese Repositary', Vol. III, 463, it is stated— "Ducks are reared in great numbers. The eggs are hatched by artificial heat much in the same manner as in Egypt; and the young are kept in boats, which are provided with coops and railings".[7] We have read often enough of the 'De'il take the hindmost' style in which the Ducks in China rush back into their boat at night, the last being touched up with a whip! Well, my idea some time ago was, or at least my suggestion, that the Egyptian artificial hatching in the olden time may have referred to *Geese*. That *fowls* are *now* so hatched there, I am well aware; and for highly interesting and elaborate notice thereof vide our *Journ. As. Soc.* Vol. VIII, pp.38 *et seq.*—[8]

Yours ever most truly, E Blyth.

DAR 98: 126–7

CD ANNOTATIONS
1.24 & this . . . Col. Sykes. 1.25] *cross added brown crayon*
3.11 Vol. VIII,] *underl brown crayon*
Top of letter: 'Nothing for me except Hatching eggs in Ægypt.' *pencil*; 'Nothing for me' *ink*; 'Nothing' *brown crayon*
Last page: 'Nothing' *brown crayon*

[1] Letter from Edward Blyth, 23 February 1856.
[2] See letter from Edward Blyth, 23 February 1856, n. 7. Blyth's salary as curator of the Asiatic Society's museum was 250 rupees a month and was never increased during his time of service. When the National Museum was eventually established in Calcutta in 1865, the new curator was paid as a government official. According to Blyth, he received 'just double what I had after more than twenty years' work, with an additional £50 yearly, and house accommodation!' (Grote 1875, p. xiii).

[3] William Henry Atkinson was secretary of the Asiatic Society of Bengal. The portion of the letter referred to has not been preserved.

[4] Part of the article published in the *Calcutta Sporting Review*. See letter from Edward Blyth, 8 January 1856.

[5] Chesney 1850, 1: 442.

[6] Cuvier 1834–6, 7: 390–412.

[7] This quotation is taken from the article entitled 'Diet of the Chinese: little known of their domestic life; grains, garden vegetables, fruits, and other plants cultivated for food; fish extensively used for the same purpose; also domesticated and wild animals; beverages of the Chinese; modes of cooking, and eating; cost of living' (*Chinese Repository* 3 (1835): 457–71).

[8] Demas 1840.

To William Erasmus Darwin [26 February 1856][1]

[Down]
Tuesday Evening

My dear old Willy

I was very glad to get your letter this morning, but I wish I could hear that your leg was quite healed: be sure tell us particularly how it goes on.— I am glad to hear of your sixth-form power;[2] it is good to get habit of command & discretion in commanding; & you unfortunate wretch, how you will enjoy reading the prayers, & keeping the accounts; as for carving you will cut a good figure.— You know Mamma is at Hartfield with the 3 little chaps;[3] I enclose a note from Lenny. He sent such a funny one lately to Leith Hill: it began "Baby has a shag coat, but it is brown.— I have bought some sealing wax & I have bought some note paper: it is quite true.— Is not this a jolly letter?." & so on for 4 pages.—[4] Snow, the dog has come back, very fat & is just as much at home as before.—

We have today cut down & grubbed the big Beech tree by the roundabout: I find by the rings it is 77 years old:[5] I am going to try whether there are any seeds in the earth from right under it, for they must have been buried for 77 years.—[6] I am getting on splendidly with my pigeons; & the other day had a present of Trumpeters, Nuns & Turbits; & when last in London, I visited a jolly old Brewer, who keeps 300 or 400 most beautiful pigeons & he gave me a pair of pale brown, quite small German Pouters:[7] I am building a new house for my tumblers, so as to fly them in summer.—[8]

I am sorry to say that I have had to strike out your name for Athenæum Club, as you cannot be entered till 18 years old.[9] Several members mistook you for me & Lord Overstone[10] called here to say that he should propose me to be elected by the Committee, who have power of electing 8 members every year, so that I have had a deal of bother on the subject.— I sh^d like to hear what you do in Chemistry.— Good night, my dear old man,

Your affect. father | C. D.—

I will send on your letter to Hartfield. We are such a little party at home, as I never remember.—

DAR 210.6

[1] An entry in CD's Experimental book, p. 3 (DAR 157a) gives 26 February 1856 as the date on which the beech tree with seventy-seven rings referred to in the letter was cut down.

[2] William was a pupil at Rugby School. He was evidently given senior 'powers' before he was actually in the sixth form, which he did not enter until February 1857 (see letter to Syms Covington, 22 February 1857).

[3] Emma Darwin was in Hartfield, Sussex, where her sisters Sarah Elizabeth (Elizabeth) Wedgwood and Charlotte Langton lived at The Ridge and Hartfield Grove, respectively, from 21 February to 1 March 1856 (Emma Darwin's diary). Since William and George were at school, only Francis, Leonard, and Horace, in addition to Henrietta and Elizabeth, accompanied her.

[4] Leonard Darwin was 6 years old. His childhood sayings were a frequent source of amusement to the family (see *Correspondence* vol. 4, Appendix III). Leith Hill Place was the home of Caroline and Josiah Wedgwood III.

[5] See n. 1, above. The results of CD's experiment were recorded in his Experimental book, p. 3 v.

[6] There are several entries for 1856 in CD's Experimental book (DAR 157a) describing similar kinds of experiments. For CD's earlier interest in the vitality of seeds long buried, see *Correspondence* vol. 2, Appendix VI, and *Correspondence* vol. 5, letter to *Gardeners' Chronicle*, 13 November [1855].

[7] Matthew Wicking of the brewers Jenner, Wicking and Jenner, 153 Southwark Bridge Road (*Post Office London directory* 1856).

[8] Throughout 1855 and 1856, CD paid John Lewis, the Down carpenter, each half-year for work on pigeon houses (CD's Account book (Down House MS)).

[9] See also letter to John Lubbock, [6 March 1857], in which the same rule is discussed in relation to Lubbock's election.

[10] Samuel Jones Loyd, Baron Overstone, was a member of the committee of the Athenæum Club; he was one of the trustees from 1857 to 1883 ([Cowell] 1975). William Darwin was not elected to the Athenæum Club until 1884 (Waugh [1888]).

To Miles Joseph Berkeley   29 February [1856][1]

> Down Bromley Kent
> Feb. 29th

My dear Sir

At Hookers suggestion I am going to send to Linnean Soc, an account of the few experiment which I made on salting seeds,[2] & I want to know whether you will permit me to tabulate your results with mine.[3] I thought of arranging the genera in their nat. Families—

I had intended trying many more experiments but my ardor was damped, by finding that plants would float for so short a time in salt-water:[4] I am, however, not quite sure that I tried this part of the affair quite fairly, so shall try the floating again— I kept the plants in the water in my room & therefor too warm & in nearly the dark, & this might have hastened their decay. Will you be so kind as to tell me the proper name of the Aubergine & Corn Salad & of the Kidney Bean; I tried the *dwarf*.— was yours the tall or dwarf?

I hope you will excuse me troubling you & believe me Dear Sir | Yours sincerely & obliged | Charles Darwin

Do you think it of any consequence how I arrange the Families: shd I follow Lindley or put the Families by mere chance, keeping, perhaps, Endogens &

Exogens apart?—[5] It will take me some little time to find out order of the Natural Families—

I planted the curious Peas you were so kind as to send me near some other kinds, but unfortunately they were gathered before they were quite ripe, so that I c^d not tell positively whether they had been affected.[6] But another lot of "Pois sans parchemin", which were ticketed as your purest seed, were sown separately & have come nearly true, yet some few of the peas were not mottled with brown, & I do not think they could have been impregnated with the pollen of any other variety.— A neighbour has a curious pea with black pods, & these when planted by themselves sometimes come false. This leads me to ask whether you do not think that the change in the seed may be due to mere variation, & not to the anomalous & direct action of the pollen.—[7]

I have forgotten, indeed, Gærtner's experiments, which certainly seem to prove direct action of the pollen.—[8] Yet on other hand how strange it is that nurserymen take no pains to prevent crossing in the Peas which they raise from seed close together—[9] And in Sweet Peas they *certainly* come true if planted *quite close* together. Have you tried any other experiments? Can you throw any light on this point which interests me extremely?[10]

I have a small collection of Pois sans Parchemin from Vilmorin,[11]—all his varieties of this sub-class: Have you any wish for a sample of each kind?— One word more, there seems to me capital evidence *for* & *against* the *natural* crossing of Pisum & Lathyrus, & I am completely puzzled.—[12]

Shrewsbury Local Studies Library

[1] Dated by CD's reference to collating the results of his seed-soaking experiments for a paper (see n. 3, below). The year 1856 was a leap year.

[2] Joseph Dalton Hooker, initially doubtful of the ability of most seeds to withstand salt water, was interested in the results of CD's seed-soaking experiments (see *Correspondence* vol. 5, letter to J. D. Hooker, 13 April [1855]).

[3] Berkeley had reported his results of experiments on immersing seeds in sea-water in *Gardeners' Chronicle and Agricultural Gazette*, 1 September 1855, p. 278. CD's paper summarising his results and those of Berkeley was read at the Linnean Society of London on 6 May 1856. The paper was entitled 'On the action of sea-water on the germination of seeds' (*Journal of the Proceedings of the Linnean Society (Botany)* 1 (1857): 130–40; *Collected papers* 1: 264–73).

[4] See *Correspondence* vol. 5, letter to *Gardeners' Chronicle*, 21 November [1855], in which CD described why his experiments were 'of little or no use (excepting perhaps as negative evidence) in regard to the distribution of plants by the drifting of their seeds across the sea'. He later found that dry fruits and capsules floated for much longer (see *Origin*, p. 359).

[5] See letter from M. J. Berkeley, 7 March 1856.

[6] See *Correspondence* vol. 5, letters to M. J. Berkeley, 7 April [1855] and 11 April [1855].

[7] In 1854, Berkeley contributed a regular column on 'Vegetable pathology' to the *Gardeners' Chronicle and Agricultural Gazette*. In the number for 24 June 1854, he stated his opinion that the colour of the skin of peas is sometimes different from the colour exhibited by the peas of either of the parent plants. He attributed this effect to the direct action of the pollen on the external coat of the ovule (*Gardeners' Chronicle and Agricultural Gazette*, 24 June 1854, p. 404).

8 Gärtner 1849. In *Variation* 1: 397, CD summarised Karl Friedrich von Gärtner's experiments on this
   point as follows: 'Gärtner . . . selected the most constant varieties, and the result conclusively
   showed that the colour of the skin of the pea is modified when pollen of a differently coloured variety
   is used.'

9 See *Correspondence* vol. 5, letter to William and Julius Fairbeard, [October 1855 – May 1856], in
   which CD asked these nurserymen specifically about this point.

10 CD eventually decided that different varieties of peas growing together rarely crossed because they
   were generally self-fertilised before being pollinated by visiting insects. See also n. 12, below.

11 Pierre Philippe André Lévêque de Vilmorin, French botanist and horticulturist, was a partner in
   the plant-breeding company Vilmorin-Andrieux.

12 The problem was discussed in *Natural selection*, pp. 69–71 and *Variation* 1: 329–30.

From John Morris 1 March 1856

Kensington
Mar 1. 56

My dear Sir

The only evidence on the existence of living land Mollusc in England is the *Helix labyrinthica* Say, of the eocene Hordwell deposits & considered by Mr. Wood[1] to be identical with the American species of Say.[2]

The identity of the Melaniopsis buccinoidea of the Woolwich (lower eocene) series is probably doubtful with the *Nile* form. although I believe considered so by Ferrussac & Deshayes.[3] So that is but little evidence of many existing forms in the older tertiary period.

Nor is there much evidence respecting the greater duration of life of the fresh water-testaceæ to the marine gasteropoda none of the eocene species ranging into the upper strata

I can only call to mind at present the genus, *Strophostoma* or *Ferrusina* of the miocene which is an extinct genus—

The relation of the gasteropoda found in the Mammalian beds at Grays[4] &c &c to the marine deposits, is not definitely settled—*ie* whether they are equivalent to or *superior* to some of the deposits of the glacial or pleistocene epoch

There is a notice (I think in the journal (geological)) of some deposits in which I believe it is mentioned that a stratum of extinct freshwater shells, cover a deposit of marine mollusca of still existing species. I believe it is alluded to to by Sir R. Murchison in his Alps paper, but I have not the reference.[5] I will look for it & others and write you again—

The best evidence is (as you are aware) that the *freshwater genera* as *genera* have survived many mutations of the surface, and outlived many of the marine genera, of course I allude to the still living genera of *Physa* Planorbis, Melanopsis, Cyrena, Paludina, &c in the Purbeck & Wealden—so in the lower and middle tertiary, &c

Yours sincerely | John Morris

C Darwin Esq

DAR 205.2 (Letters)

CD ANNOTATIONS
Top of first page: 'On age of Land & F.W. & marine Mollusca.—' *pencil*; '18'[6] *brown crayon, circled brown crayon*

[1] Searles Valentine Wood was an expert on the fossil Mollusca of the East Anglian Crag.
[2] Thomas Say.
[3] Férussac and Deshayes 1820–51.
[4] Grays near Thurrock, Essex, the site of fluviatile deposits from the Crag period.
[5] Morris refers to Murchison 1847, in which the order and implications of the fossiliferous strata of Œningen are discussed. Roderick Impey Murchison reconsidered his conclusions in Murchison 1849, pp. 233–7, having decided that the Alps had experienced geological upthrust so violent that some strata had been completely overturned.
[6] The number of CD's portfolio of notes on the means of geographical dispersal of plants and animals.

From M. J. Berkeley 7 March 1856

My dear Sir/

All that I did about the seeds was done at the suggestion of Dr Hooker entirely to forwar⟨d⟩ your views, and the little that I was able to ascertain is entirely at your service.[1] I think I should arrange the families as they appear in Lindleys Vegetable Kingdom.[2] It is best to have some definite order. The Corn salad consists of different species of Fedia. I cannot tell you what particular species or supposed species. The Aubergine is a variety of Solanum Melongena. The Kidney bean I tried was the dwarf Belgian Phaseolus vulgaris. It is very possible that the change of seed in the Sugar Peas may be due to mere variation and not to impregnation.[3] The subject is not capable of solution from one or two Experiments. The mottled seeds last year produced grey, red & other colored peas. A few only came quite true. But I cannot say that the change is from impregnation of their neighbours. The white sugar peas come more true than the mottled ones, but they vary in tint. I have not got Gærtners book to refer to.[4]

I am so busy this year with my Introduction to Crypt. Bot that I have no time for gardening.[5]

Do you know the Rawsons at Bromley?[6] Mrs Rawsons brother lives here and has married my neice.

I am very truly yours M J Berkeley.—

I should be glad of two or three of the Black Peas

King's Cliff | March 7. 1856.

Two or three red peas sowed by themselves produced mottled peas.

DAR 160: 174

CD ANNOTATIONS
0.1 My dear . . . vulgaris. 1.7] *crossed pencil*
1.9 The subject . . . Experiments. 1.10] *cross added brown crayon*

2.1 I am . . . 1856. 6.1] *crossed pencil*
Top of first page: '14'[7] *brown crayon*; '[illeg]' *brown crayon*
Verso of last page: 'Vibrio in Agrostis'[8] *pencil*

[1] Berkeley refers to his experiments on the immersion of seeds in sea-water. See letter to M. J. Berkeley, 29 February [1856].

[2] Lindley 1846. In his paper, CD stated: 'I have arranged the families in accordance with Lindley's "Vegetable Kingdom." ' (*Collected papers* 1: 268).

[3] See letter to M. J. Berkeley, 29 February [1856] and n. 7.

[4] Gärtner 1849.

[5] Berkeley 1857.

[6] Arthur and Charlotte Elizabeth Rawson are not listed in CD's Address book (Down House MS).

[7] The number of one of CD's portfolios of notes.

[8] See *Correspondence* vol. 5, letter to J. S. Henslow, 7 July [1855].

From W. D. Fox 8 March [1856][1]

Delamere | Northwich
March 8

My dear Darwin

I have anxiously inspected My Dorking & Cochin friends Yards for an old Cock of Each, & written to Captain Hornby[2]—but I fear you have not received any yet. There must be some die before long I think.

Have you a Sebright Bantam yet?[3] If not I have an old Gentleman I will send you shortly. You should have an old *White* Dorking also, as they are quite distinct from the other in form.

I forget whether I ever told you that I had long considered the Scotch Deer Hound a mongrel, par Excellence. Dont tell any Scotch so, or I shall be murdered. It has long been a pet idea of mine, & I have often said I could breed them without any Deer Hound blood in them. I have also always thought the Irish Deer or Wolf Dog, was merely a cross with the Scotch & a Mastiff.

Some months ago in a conversation on this head with a M[r] Lister near here,[4] he told me to my great delight that he had a Bitch ½ Deer Hound & ½ Mastiff. On looking at her it is wonderful how little the ½ Mastiff is recognisable in her. On minutely examining however, you find her mastiff Blood in neck & shoulder. I much wished this Bitch crossed back with Deer Hound. This has been done, & the result, as shown in a splendid Bitch puppy, is to *completely* restore the Scotch Deer Hound. I dined there last week, & met a stranger who was enthusiastic about Scotch Dogs—of which by the way he gave a pretty story as having happened to himself. Walking one day in Regent S[t] he felt something cold in his hand, & on looking, found a Scotch Dogs nose there, who had been with him Deer stalking &c 2 years before in the Highlands, & was then walking in London with his Master.

Lister, rather spitefully introduced me to this Captain Warren[5]—as being one who believed in the Scotch Dogs being mongrels. Of course I maintained my ground, when, to my intense amusement he (after warning me not to go to Scotland & especially Badenoch,[6] with such views) quoted *the* puppy as an Example of pure blood, as might be seen by any one, & which he said was well worth 40£. He was so enthusiastic that I was obliged to break the fact by degrees, "that her Grandfather was a Mastiff."

I am trying now to get the ½ breed Scotch & Mastiff Bitch put to a pure Mastiff—& I expect either the produce of that—or the next cross at all events, to be the Irish Wolf Dog.

I would defy any Scot to detect the false blood in this puppy Bitch— She is quite a perfect Scotch Deer Hound.

I see Tegetmeyer—or some such name, who doctors all the Fowls in England— says he is engaged with you in examining the anatomy of Fowls.[7] He seems to know a great deal about them from his letters in Cottage Gardener—but I often think his prescriptions rather foolish. You are not meddling with Geese I think yet, are you.

Tell me how M[rs] Darwin & your little ones all are—also Susan Catherine & M[rs] Wedgwood[8] & Believe me always Yours affec[ly] W D. Fox.

DAR 164: 174

CD ANNOTATIONS
0.3 My dear . . . shortly. 2.2] *crossed brown crayon*
7.1 I would . . . Hound. 7.3] *scored brown crayon*
8.1 I see . . . Fox. 9.2] *crossed pencil*
Top of first page: '16'[9] *brown crayon*

[1] The letter is dated by its relationship to the letter to W. D. Fox, 15 March [1856].
[2] Windham W. Hornby of Knowsley, Lancashire, was a prominent breeder of Dorking fowl.
[3] CD had asked John Lubbock for specimens of Sebright bantams (letter to John Lubbock, [14 January 1856]).
[4] The *Post Office directory of Cheshire* (1857) lists 'E. Lister, esq., Marston, Northwich'.
[5] Captain Warren has not been identified.
[6] A district near Inverness, Scotland.
[7] In the report on the 'Annual grand show of the Philo-Peristeron Society', *Cottage Gardener* 15 (1855–6): 301, it was stated that 'Mr. Yarrell, whose name is a "household word" with all zoologists, and Mr. Darwin, whose "Naturalist's Voyage round the World" is known all over the world, were present, and with our old correspondent, Mr. Tegetmeier, were examining bird after bird, with a view to ascertain some of those differences on which the distinction between species or varieties depend.' It is likely that William Bernhard Tegetmeier, who wrote a regular column for this journal on the diseases of poultry, was the author.
[8] CD's sisters, Susan Elizabeth Darwin, Emily Catherine Darwin, and Caroline Sarah Wedgwood.
[9] The number of CD's portfolio of notes on hybridism.

From Thomas Hutton 8 March 1856

Mussooree
8th March 1856

My dear Sir,

When in my former letter[1] I stated that the children descended from mixed European and Native blood would often in the third and fourth generations show symptoms of a return to the dark complexion of the Grandmother or Great Grandmother, even when their own immediate parents showed no traces of it, you will perhaps feel inclined to ask why this should be, and why, when the *White* had appeared to have superseded the *Black* complexion, there should be a retrograde movement to the latter.— I am tempted to reply that the reason is to be found in the fact that Nature abhors what is called civilization as much as she does a vacuum.— That is to say in other words, that Nature remains true to herself, and endeavours to avoid all that is *artificial.*— The European is decidedly a highly cultivated and therefore an artificial breed;—the native of India on the other hand, is almost a child of nature— Hence when these two races intermix Nature makes violent efforts to retain the simple and true, and to reject the artificial and false blood. It is only after the repeated addition of European blood that these efforts cease,—because the race has then become wholly artificial by the predominance of the father's blood.—

This is seen still more clearly in the case of the breed obtained between a highly bred European dog, and the common village cur of India.— If we cross the latter with an English Greyhound, Foxhound or Bulldog, all of which are highly bred and artificial, the pups will partake of the characteristics of both parents, but most so of the village dog; precisely as between the first cross of European man and Native woman, the children more strongly resemble the mother's complexion— If these pups are again crossed with European blood, the characteristics of the village dog will gradually disappear as before with the man and woman,—with an occasional effort to return to the least artificial stock of the two.— But if the pups are left to breed *inter se* their produce will reject the artificial European blood, and become wholly native,—and so would the Half Caste children produced from the European and Native.—

We call this a tendency to degenerate, unless the breed is kept up to an artificial standard, by the addition of civilized blood,—but it is in reality the very reverse of degeneracy,—being a return to a natural type only.— The civilization or domestication of a species soon renders it artificial, and gradually obliterates the original appearance.— Nature abhors such artifice and invariably endeavours to return to herself, and would always succeed were we not to counteract her influence by keeping up the artificial form by the repeated admixture of similar blood;—thus, as like breeds like,—the desired artificial form is preserved only by frequently recrossing it with similar artificial stock.—

On my return from Afghanisthan in 1841, I brought with me two half bred male Goats,—the Mother having been a domestic Goat and the father the wild Capra Elgagrus.— These more strongly resembled the wild, than the tame breed.— They were again crossed upon domestic goats, and the produce altho' now only $\frac{1}{4}$ wild, was still much more closely allied to the *wild* than to the tame breed; and the reason is that Nature avoided the artificial and clung to the wild blood.—[2]

I think I may say that you will perceive the same thing in regard to the common Mule.— A breed is produced between the High bred and artificial horse and the poor uncultivated donkey;—the foal is always, where both parents are equally healthy,—more Assinine in appearance than Equine, because the blood of the one is more natural than that of the other!— In our farm yards the Duck altho' running into all sorts of colours as the effects of ⟨do⟩mestication, yet constantly returns, more especially among the Drakes to the original Mallard.—

With regard to our breed of Geese, I find that I can obtain no trustworthy information upon the points you require to know,—but this much I can venture to say from my own casual observation, viz that the breed dispersed generally over the country is a hybrid race between the Chinese or knobbed Goose (A. Cygnoides.) and the common Grey Lag.— The progeny is prolific *inter se* & often becomes pure white.—[3] I will however make further inquiries on the subject though truth to tell where nobody thinks of a goose until they see it *roasted on the table*, it is not easy to procure any information at all.—

Pigeons are a source of great amusement to the Mahomedan population of India, and are of several varieties, all of which are probably known to pigeon fanciers in Europe— The only wild breed that is occasionally domesticated among these is the common Blue Pigeon of India generally, and known as *Columba intermedia.*— This is common throughout the North Western parts of the country where it breeds in communities in large dry wells (*unbricked*) scooping out holes for the purpose in the earthy sides.— It is likewise found in Afghanisthan where it breeds in similar situations, and among old buildings and rocks— It is easily tamed, and breeds very readily with all[4]

AL incomplete
DAR 166: 283

CD ANNOTATIONS

0.3 My dear . . . blood.— 1.16] *crossed pencil*; 'Geese Hybrids' *added brown crayon, del ink*; 'Q' *added ink, circled ink*; 'Nothing except Pigeons very last Page' *added ink, circled ink*
2.4 the pups . . . woman, 2.6] *scored brown crayon*
2.9 But if . . . Native.— 2.12] *scored brown crayon*
4.1 On my return . . . blood.— 4.7] *scored brown crayon*
6.3 viz . . . white.— 6.6] *scored brown crayon*; 'Q' *added pencil, circled pencil*
7.7 It is . . . all 7.9] 'Copied' *added pencil*
7.8 It is . . . all 7.9] *triple scored brown crayon*
Top of first page: 'Marked with' *pencil* 'Red' *brown crayon, underl brown crayon*; '(5)' *brown crayon*; '17'[5] *brown crayon*; 'Breeds taking [on] wild forms' *pencil*; 'Geese' *brown crayon, underl brown crayon*; 'Pigeon' *brown crayon, underl brown crayon*

[1] This letter has not been found, nor has it been possible to identify Hutton. CD had originally corresponded with Hutton at the suggestion of Edward Blyth. See *Correspondence* vol. 5, letter from Edward Blyth, 22–3 August 1855, for extracts from a letter from Hutton to Blyth responding to some of CD's queries.
[2] CD cited Hutton to this effect in *Natural selection*, p. 486.
[3] See *Natural selection*, p. 439, where CD gives information from both Hutton and Edward Blyth on these and other points about Indian geese.

[4] In *Variation* 1: 185, CD stated: 'In India, as Captain Hutton informs me, the wild rock-pigeon is easily tamed, and breeds readily with the domestic kind'.
[5] CD's numbering refers to his portfolios on varieties and hybrids, respectively.

To George Henry Kendrick Thwaites 8 March 1856

Down Bromley Kent
March 8th /56

My dear Sir

Though I have nothing very particular to say I must thank you cordially for the extremely kind manner with which you have received my letter.[1] I remember at Oxford that you had attended to many of the points on which I was then & am now so much interested.[2] I hope that you will publish some of the facts on variation to which you allude: I shd be particularly glad to see in print or M.S. some particulars in regard to the species from different elevations, which show different degrees of capacity for cultivation at a new level: Hooker has published a similar case in regard to the Himmalaya Rhododendrums.[3]

As you have received my letter in so very friendly a spirit, I will mention one or two other points on which I am much interested; viz in regard to the distribution of alpine plants; have you collected at the greater heights in Ceylon, & is there anything new in relation to the vegetation at these heights in comparison with the Himmalaya, Neilgherries or other mountains? Again have you observed whether the introduced & *perfectly naturalised* plants tend to vary much in Ceylon? The courrse of my work makes me more & more sceptical on the eternal immutability of species; yet the difficulties on the other theory of common descent seem to me frightfully great. In my work, which I shall not publish for 2 or 3 or perhaps more years; it is my intention to give, as far as I can & that will be very imperfectly, all the arguments & facts on *both* sides of the case, stating which side seems to me to preponderate.—

You cannot possibly render me more material assistance than by getting me any skins of Indian or Ceylon (or any breed except English) breeds of Pigeons;[4] for I have concluded it would be better to work carefully at the varieties of a few animals, than compile brief notices on all our domestic animals. I have now all English breeds of Pigeons alive, & am carefully observing them, making skeletons & crossing them. There are some remarkable Tumblers in India.—[5] Have you any domestic Ducks? or Rabbits These I mean to work at, as well as at Poultry. Any skeleton of Ducks wd be very valuable.— Dr Kellaart has offered to help me in regard to Poultry, for I met him accidentally at Brit. Museum, after writing to you;[6] but I do not know how far he will be so kind as to take trouble for me.—

When I began, I meant merely to thank you; but when a beggar once begins to beg he never knows when to stop![7]

Pray accept my very cordial thanks & good wishes & believe me, My dear Sir | Your's very sincerely | Ch. Darwin.

American Philosophical Society (125)

¹ See *Correspondence* vol. 5, letter to G. H. K. Thwaites, 10 December 1855. Thwaites's reply has not been found.

² CD had attended the meeting of the British Association for the Advancement of Science in Oxford in 1847 at which Thwaites read a paper announcing his discovery of conjugation in the Diatomaceae (Thwaites 1847).

³ Probably J. D. Hooker 1852, pp. 69–71. CD quoted Thwaites's observations on the cultivation of varieties from different elevations in *Natural selection*, p. 286: 'Mr. Thwaites, the curator of the Botanic Garden at Ceylon, . . . writes to me, that he finds "that individuals of the same species are acclimatised to different elevations,—being more & more impatient of cultivation at any station, according as they have been transported to it, from stations of greater & greater altitude." ' See also *Correspondence* vol. 7, letter to G. H. K. Thwaites, 7 February [1858].

⁴ Thwaites's name is not among those in *Natural selection* or *Variation* who provided CD with specimens of birds or rabbits from Ceylon.

⁵ See letter to Walter Elliot, 23 January 1856.

⁶ Edward Frederick Kelaart is described in *Variation* 1: 259, where his name is spelled 'Kellaert', as one who had 'closely studied the birds of Ceylon' and is cited on the native fowl of that island (*Variation* 1: 234).

⁷ That CD wrote more than he intended is indicated by the fact that the last page was written on the verso of the first. To clarify the order, he headed it 'Last Page'.

To Syms Covington[1] 9 March 1856

> Down Bromley, Kent.
> March 9, '56.

Dear Covington,—

I was very glad to get a month or six weeks ago your letter of the 4th of September, with its interesting account of the state of the Colony and your own affairs, which I am most truly glad are so prosperous. You did a wise thing when you became a colonist.[2] What a much better prospect you have for your sons, bringing them up as farmers—the happiest and most independent career a man can almost have—compared to what they could have been in this old burthened country, with every soul struggling for subsistence. I have lately been talking a good deal on the subject with Captain Sulivan, who has four boys, and who often seems half-inclined to start for some Colony and make his boys farmers.[3] Captain Sulivan, owing to all his practice in the old Beagle (I have heard that our old ship is now a collier),[4] was the right hand of the fleet in the Baltic, and had all the difficult work to do in placing the ships in the bombardment of Sweabourg. I heard of a letter from a seaman in the fleet, but not in Captain Sulivan's ship, who said he was the best sailor in the whole lot, and that if the men could elect their Admiral they would elect him.[5] Captain Stokes is married again, to a widow, and will never, I believe, go afloat again.[6]

I have finished my book on the barnacles (in which you so kindly helped me with the valuable Australian specimens).[7] I found out much new and curious about them, and the Royal Soc. gave me their great gold medal[8] (quite a nugget, for it weighs 40 sovereigns), chiefly for my discoveries in regard to these shells, which are not perfect shells, but more allied to crabs.[9]

My health is better, but I have a few bad days almost every fortnight, and cannot walk far or do any hard work. I am now employed on a work on the variation of species, and for this purpose am studying all about our domestic animals and am keeping alive all kinds of domestic pigeons, poultry, ducks. Have you ever noticed any odd breeds of poultry, or pigeons, or ducks, imported from China, or Indian, or Pacific islands? If so, you could not make me a more valuable present than a skin of such. But this, I know, is not at all likely.

My children, thank God, are all well, and one gets, as one grows older, to care more for them than for anything in this world. With every good wish for the health and happiness of yourself and family, believe me, dear Covington, yours sincerely, CHARLES DARWIN.

Sydney Mail, 9 August 1884

[1] The manuscript of CD's letter has not been found. This transcript has been taken from the earliest known printed source; another version published in de Beer 1959, pp. 24–5, differs in some respects from the one given here. See also *Correspondence* vol. 5, letter to Syms Covington, 14 March 1852, n. 1.

[2] Covington, CD's servant and assistant for eight years during and after the *Beagle* voyage, had emigrated to Australia in 1839 and become postmaster at Twofold Bay, New South Wales. Covington's letter has not been found.

[3] Bartholomew James Sulivan had been a lieutenant on board the *Beagle*. Following surveying work in the Falkland Islands, he had farmed on the islands, 1848–51.

[4] From 1847 to 1870, when it was sold for scrap, the *Beagle* served as a coast guard watch vessel on the river Roach (Freeman 1978). For the history of the *Beagle* after CD's voyage, see Thomson 1975.

[5] Sulivan had recently seen active service in the Baltic as captain of the surveying vessel HMS *Lightning*. The naval attack on Sveaborg, in the Gulf of Finland, was an unsuccessful minor diversionary gambit during the siege of Sevastopol in 1855 (Sulivan ed. 1896).

[6] John Lort Stokes, mate and assistant surveyor in the *Beagle* with CD, was subsequently engaged on coastal surveys of Australia and New Zealand. In 1851 he returned to England and remained on half-pay until 1860.

[7] *Living Cirripedia* (1854). Covington had sent CD cirripede specimens from Twofold Bay (see *Correspondence* vol. 4, letter to Syms Covington, 23 November 1850).

[8] The Royal Medal of the Royal Society of London was awarded to CD in November 1853 for his work on geology and for his monograph on the Cirripedia. See *Correspondence* vol. 4, Appendix II.

[9] Prior to the 1830s, cirripedes had been classified as molluscs. One of the accomplishments of CD's study was to show how, according to their developmental history and anatomy, the barnacles were related to the Crustacea.

To Herbert Spencer 11 March [1856][1]

> Down Bromley Kent
> March 11th. [1856]

M^r. Darwin presents his compliments to M^r. Herbert Spencer, & begs leave to thank him very sincerely for his extremely kind present of the Principles of Psychology.—[2] M^r. Darwin may remark, in order to show how acceptable M^r. Spencer's present has been, that only about a fortnight since he wrote down in his list of books to be read, the name of M^r. Spencer's work.[3]

Copy
DAR 147

[1] Dated by reference to the publication date of Spencer 1855 (see n. 2, below).
[2] Spencer 1855, which had been published in September 1855 (*Publishers' Circular*, 1 September 1855, p. 341). The presentation volume is in the Darwin Library–CUL. It is lightly annotated, with notes by CD on the end-papers.
[3] Spencer 1855 is not listed in the books to be read section in CD's reading notebooks (*Correspondence* vol. 4, Appendix IV).

To Samuel Birch [12 March 1856][1]

57 Queen Anne St | Cavendish Sq^{re}.
Wednesday

My dear Sir
I will take advantage of your most kind offer[2] & call on you on Friday about 11 oclock; if I do not hear to contrary I will venture to assume that this will suit you.—
Pray believe me, | Yours very truly obliged | Ch. Darwin

British Museum (Department of Western Asiatic Antiquities, correspondence 1826–67: 1489)

[1] CD was in London between 10 and 14 March 1856 (see the following letter) staying with his brother, whose address he gives. He attended a council meeting of the Royal Society of London on Thursday, 13 March (Royal Society council minutes).
[2] See letters to J. E. Gray, 14 January [1856], and to Samuel Birch, 6 February [1856]. Birch translated for CD passages referring to pigeons from an ancient Chinese encyclopaedia and on fowls from ancient Egyptian and Japanese sources. See *Origin*, 27–8, and *Variation* 1: 205, 230, 247. For a list of CD's Chinese sources, see Pan 1984.

To W. D. Fox 15 March [1856]

Down Bromley Kent
March 15th

My dear Fox.
I was very glad to get your note & congratulate you on your triumph about the Scotch Deer Hound, which however I do not know by sight.—[1] How I wish I further knew (can you find out for me) whether the real Scotch Deer Hounds breeds true; but I suppose this must be the case, whether or no your mongrel would do so.[2] This seems to **very valuable** case for me, for it would be a most bold hypothesis to imagine that the real Scotch Deer Hound was a pure & distinct aboriginal race, but that your mongrel,, though identical in appearance, was *essentially* different: I do not even know what a common Deer Hound is.—
Many thanks for your continued remembrance of me & my poultry skeletons: I am making some progress & have been working a little at their ancient History & was yesterday in the British Museum getting old Chinese Encyclopedias

translated.[3] This morning I have been carefully examining a splendid Cochin Cock sent me (but I sh^d be glad of another specimen) & I find several important differences in number of feathers in alula,[4] primaries & tail, making me suspect quite a distinct species.—[5] I am getting on best with Pigeons, & have now almost every breed known in England alive: I shall find, I think great differences in skeleton for I find extra rib & *dorsal* vertebra in Pouter.—[6]

I have just ordered the Cottage Gardener:[7] Mr Tegetmeier is a very kind & clever little man; but he was not authorised to use my name in any way, & we cannot be said to be working at all together; for our objects are very different, & he began on skulls before I had thought on subject: I have not yet looked at our pickled chickens & hardly know when I shall, for I have my hands very full of work; but they will come in some day most useful, as will a large series of young Pigeons, which I have myself killed & pickled.—[8]

I sh^d be very glad of old Sebright Bantam.—

I have been in London nearly all this week, working at Books[9] & we had at Erasmus's a very pleasant dinner & sat between M^r. & M^rs. Bristowe & was charmed with both.[10] Bristowe often so reminds me of you some 25 years ago in certain expression of face & manner. They told me you had been far from well: why did you not mention yourself? Do not I always prose at good length about myself & pursuits. So I will say that my stomach has been better for some months than average, & I am able decidedly to work harder.— My sisters are pretty well: you heard of D^r. Parkers release about 2 months ago.—[11]

How I do wish I had you nearer to talk over & benefit by your opinions on the many odds & ends on which I am at work. Sometimes I fear I shall break down for my subject gets bigger & bigger with each months work.—

My dear old friend | Most truly yours | Ch. Darwin

Endorsement: 'C Darwin March 18/56'
Christ's College Library, Cambridge (Fox 97)

[1] See letter from W. D. Fox, 8 March [1856], in which Fox related his view that the Scottish deerhound was actually a 'mongrel' breed.
[2] The Scottish deerhound had been established as a distinct breed by the sixteenth century (*EB*).
[3] CD visited the British Museum to consult Samuel Birch, whose help he sought concerning ancient references to breeds of pigeons and poultry. See preceding letter.
[4] The alula is the 'bastard-wing' of birds, consisting of three or four large quill-like feathers carried on the 'thumb' (*EB*).
[5] In wild birds, the number of feathers is generally constant and is used as a character in classification. The Cochin China cock had been sent to CD by Bernard P. Brent (see letter to W. B. Tegetmeier, 20 March [1856]).
[6] The differences in the number of ribs is discussed in *Variation* 1: 165–6. CD recorded that he was not sure that he had designated the vertebrae correctly; he stated the dorsal vertebrae were always eight in number (*Variation* 1: 165 and n.).
[7] *Cottage Gardener, and Country Gentleman's Companion*. See letter to W. D. Fox, 8 March [1856] and n. 7.
[8] CD's series of young pigeons was discussed in the chapter on embryology in *Origin*, pp. 445–6. He also used this material in his discussion of the differences between breeds of fowls in *Variation* 1: 248–50.

[9] There are entries in CD's reading notebooks for March 1856 relating to early ornithological works. See *Correspondence* vol. 4, Appendix IV, 128: 16.

[10] Probably Henry Fox Bristowe, a London barrister, and his wife Selina. Bristowe was Fox's nephew. CD was staying with Erasmus Alvey Darwin.

[11] Henry Parker, physician to the Shropshire Infirmary and CD's brother-in-law, had died in January 1856 (*Eddowes Salopian Journal*, 16 January 1856, p. 5).

To W. B. Tegetmeier 15 March [1856][1]

Down Bromley Kent
March 15th

My dear Sir

I am going to beg a rather unreasonable favour of you.— I called yesterday at M.ʳ H. Weirs[2] & he told me that some Leghorn Runts are to be sold on Tuesday at Stevens.[3] Would you be so kind as to purchase a pair for me (or single bird) that is if the Birds have short tail, long legs & long neck, in short appear of *a different shape* from your Smyrna Runts, which *in shape* are like what I already possess.—

I would go as far 20ˢ, or 25, or even 30, if they appear very distinct in shape; but M.ʳ Weir thought that perhaps they would go for under 20ˢ—

I presume that you could purchase a basket & send them with the enclosed address by some trustworthy Porter to Golden Cross, Charing Cross by 3 oclock.[4] & I could repay you immediately for these incidental expences & whatever you paid for them.— But I am well aware that I am asking you to take a very scandalous amount of trouble.—

If by any odd chance a pair of black Carriers, very long in the Beak & narrow in the Head, were to be sold for 20ˢ (I do not care much for wattle) I sh.ᵈ be glad of them. But the Leghorn Runts are important, as I sh.ᵈ never be able to pick up skeleton.—[5]

Forgive me if you can & believe me | Yours sincerely | Ch. Darwin

New York Botanical Garden Library (Charles Finney Cox collection)

[1] Dated by the relationship to the following letter.

[2] Harrison William Weir was a noted pigeon fancier, animal painter, and member of the Philoperisteron Society who lived in Peckham, Kent. He illustrated Tegetmeier's *Poultry book* (Tegetmeier ed. 1856–7) and was to have provided the text for the pigeon and rabbit sections of the work if publication had not ceased prematurely. Weir and Tegetmeier collaborated on a work on pigeons in 1868 (Tegetmeier and Weir 1868). CD acknowledged his indebtedness to Weir in *Variation* 1: 132 n. 2.

[3] John Crace Stevens. See letter to W. B. Tegetmeier, 14 January [1856], n. 1.

[4] The departure point for the coach that called at Farnborough, two miles from Down (*Post Office London directory* 1858).

[5] CD described the Leghorn runt as a breed of pigeons that was probably extinct in Europe in *Variation* 1: 144. He knew of it only through the description and illustration in Aldrovandi 1599–1603 and in [J. Moore] 1765 (see *Correspondence* vol. 4, Appendix IV).

To W. B. Tegetmeier 20 March [1856][1]

Down Bromley Kent
March 20th

My dear Sir

Many thanks for your note & kind offers of assistance.—[2] I sh[d] be very glad indeed if you could procure me the Carriers as you propose. I sh[d] considerably prefer black Carriers, as I have dun Dragons & the colour will be useful in distinguishing crosses.[3] And I thank you for not buying the Runts: I have had a sick Scanderoon & the skeleton is now making.[4] I was aware that the Leghorn Runt is very like the Scanderoon, but yet according to the German Pigeon Book,[5] there seems some difference, the neck being not so long or so curved, & the body larger.—

If ever you sh[d] stumble on any *odd* breed of Pigeon at Stevens, not very dear I sh[d] be glad to purchase.

By the way I must mention that M[r] Brent has sent me a splendid Cochin Cock, so that I shall not want such.[6]

What an excellent table you have sent me, as a specimen; I cannot doubt you will make a first-rate work on your subject.—[7]

I saw quite lately (& have just now been relooking over my papers, but cannot refind the note) that Pallas in his Spicilegia Zoologica (I think vol. 2.) has described the protuberance on the skull of the Polish Fowl.—[8] This, perhaps, would be worth your looking to.—

I have had a most unfortunate, & curious in medical point of view, accident this morning; viz in 3 of my best, old Pigeons dying & 2 or 3 others ill, from overeating Bay Salt: & what makes it odder, they have been accustomed to it; but have not had any for 2 or 3 weeks: I noticed that they ate very much, but I never dreamed of its making them ill; but in 3 or 4 hours I had half-a-dozen very ill & 3 are now dead. I now remember before that it seemed to make them sick on a former occasion; this seems to me *very surprising*; but I cannot doubt in the least, that it was the salt & nothing but the salt.—[9] The deaths happened in two houses.—

Pray believe me, Yours very sincerely | Ch. Darwin

New York Botanical Garden Library (Charles Finney Cox collection)

[1] The year is established by the incident with the pigeons overeating salt, which is reported in the letter and was subsequently published by Tegetmeier (see n. 9, below).

[2] See preceding letter.

[3] Notes recording the results of CD's crosses between various races of pigeons are preserved in DAR 205.7: 166–89 and in CD's Catalogue of Down specimens (Down House MS).

[4] Scanderoons are described in *Variation* 1: 142–4. There is an entry in CD's Account book (Down House MS) for 10 March 1856 that reads: 'Townsend: Skeletons'. There are several subsequent entries to the same effect. CD had previously prepared his own skeletons (see *Correspondence* vol. 5, letter to T. C. Eyton, 26 November [1855]).

[5] Neumeister 1837. CD's annotated copy is in the Darwin Library–CUL.

[6] Bernard P. Brent was one of CD's chief sources of information on poultry.

[7] CD possibly refers to some preliminary work for Tegetmeier's *Poultry book* (Tegetmeier ed. 1856–7), the first number of which was issued in May, or to Tegetmeier 1856 (see n. 8, below).

8 Pallas 1767–80, pt 4, pp. 18–23 and Plate III. One of Tegetmeier's specialities was the breeding of Polish fowl, which are characterised by a skeletal protuberance on the crown of the head exaggerated by colourful crest feathers. Tegetmeier exhibited live specimens and skeletons of this breed at a meeting of the Zoological Society of London on 25 November 1856. In the written report (Tegetmeier 1856) he mentioned Pyotr Simon Pallas's earlier description but corrected Pallas's opinion that the breed was a product of a cross between a domestic fowl and the guinea-fowl.

9 Tegetmeier published an account of the death of CD's pigeons. He stated it was 'communicated to me by one of the most distinguished naturalists of the day, whom I am most proud to number on my list of friends' (*Cottage Gardener* 16 (1856): 73–4). Tegetmeier concluded that 'it is necessary to limit the quantity of salt given to birds, after they have been deprived of it for any time.'

From Edward Blyth [*c.* 22 March 1856][1]

Notes for M![r] Darwin.

Have you studied yet Bocharts great work? More especially Vol. 2 of his '*Omnia opera*', where fowls and Pigeons are treated of.[2] If you have not, do so; albeit there seems to me a vast deal of uncertainty respecting the correct application of many of the names. As regards fowls, I see that Capt. Allen's notice of feral fowls in the isle of Annobono has been made the text of a paper read before the *Académie des Sciences*, on Oct![r] 29th last; which see.[3] I have not Allen's work now by me; but there is something very suspicious in the Frenchman's words—"Les Poules d'Annobono resemblent aux *Pintades*".—[4] By reference to my paper on a zool![l] collection, from the Somáli country, you will see that a kind of Francolin is considered "the wild hen" east of the Red Sea;[5] & in the S. of Arabia, Chesney talks of "Jungle-fowl" & "Pheasants".[6] Bosh of course! Guinea-fowl there are (*N. ptilorhyncha?*), as long ago stated by Niebuhr,[7] and latterly by D![r] Nicholson in *P. ZS.* 1851, p. 128;[8] and here I may remark that D![r] N's supposed new Francolin (*Fr. yimensis*) is evidently the *Perdix melanocephala* of Rüppell, a species of *Caccabis* (or *red-legged P.*), not a Francolin! Among the birds of almost general distribution, bear in mind *Squatarosa helvetica*.[9] I hear that there is a good collection of skins of Arabian birds in the Bombay museum, and have accordingly written thither to make enquiries about these *Gallinaceæ* of S. Arabia. I have just rec![d] a large batch of the *Proc. Z. Soc.*; and find a *Gallus Temminckii* described by Gray (& it would seem also *figured*).[10] I have no faith in it; suspecting it *very strongly* to be a hybrid of some kind,[11] probably a cross between male *varius* (v. *furcatus*) and hen of the large Malayan breed of domestic fowls; while *G. æneus*, Tem., as we are assured by Schlegel, is mixed *varius* & (small?) common hen.—[12] Owen, I see, has been identifying a skull found in the S. of England with the 'Musk Ox' of the Barrengrounds of Arctic America![13] Good! He before had *Rein Deer* from Devon; & these serve as additional evidence for Agassiz's glacial period: but whatever does Owen mean by calling it the "Musk *Buffalo*"? *Bubalus moschatus*!!! Beyond a certain resemblance in the flexure of the horns to *B. caffer*, it seems to me about the furthest remove from the true Buffalos of all the Bovines; & it has the *ovine* nose, &

I believe 2 teats only; but the former the Yak also has, in reference to climate,—also the Rein Deer! Buffalos proper are creatures which pass much of their lives in water, with just the eyes & nostrils above the surface, have thick & scantily clad *pachydermatous* hides, & feed on coarse aquatic plants in the hottest climates!!! The style of animal too, is altogether dissimilar; judging from what looks to be a capital portrait of the 'Musk Ox', published by one of the Arctic navigators, somewhat thus (from memory on my part),

or with the head still more hanging, if possible: the *characteristic* attitude of all the true Buffalos is with the head held out *straight*: observe too the short tail of the *Ovibos*; but *jam satis*—

At last I have fairly taken up the subject of domestic Pigeons; & shall send you many living specimens, if not some that will carry high prizes as novelties in England; What particularly interests me is, that I have already obtained nearly all the distinct races procurable here, of the typical or wild colour, *i.e.* slaty, with 2 black bars on wing, &c. Would not a *fantail* of that hue be a novelty in England? Well, the common wild blue Pigeon of India (*intermedia* of Strickland)[14] chiefly differs from the European *livia* in having no white on the rump; & I find that the typically coloured *carriers* here are white-rumped, & therefore probably of European extraction; whereas all the other races which I have seen typically coloured as yet, have ashy rumps, & are therefore probably of Asiatic (not Indian) origin![15] Whether 'fancy Pigeons' are bred by the Chinese, I have not yet been able to ascertain, *but will do so*! If so, probably only of late years, & chiefly about the foreign ports: for I cling to the suspicion that to the old Semitic civilizations of W. Asia, we are indebted for these curious products of domestication and *careful breeding*. I now more than ever suspect that 'fancy pigeons' were *unknown in India prior to the Muhammedan conquest*! For there appear to be no Sungskrit names denoting the different races; and in Kamandáki's 'Elements of Polity', a work of the 3rd. Century *B.C.*, there is no mention made of curious Pigeons, although it notices almost all animals brought to the use of man at that time, so far of course as known to the author. Moreover Kamandáki notices tame fowls, but says nothing of different breeds of them;[16] and the domestic fowl is prohibited to be eaten, although the wild is permitted, so long back as by Mánu, or 12 Centuries *B.C.* (according to Sir W. Jones);[17] and therefore tame fowls were familiarly known *so far back* in this country![18] Mind that you keep a good look out among the Assyrian monuments— Also consult H. H. Wilson, Rawlinson, & others.[19] Looking over Dixon's book on Pigeons, I observe an interesting extract concerning the *breeding & rearing* of young Cormorants in China.[20] And now about our Indian breeds of

Pigeons. In general, each is to be had both *bare-legged* and *feather-legged*; & the latter divide into those which have long feathers growing from the outer side of the toes, & those which have the shank feathered—but few or no plumes on the toes. All three such varieties of *fantails* are to be had, and each with an occipital *toupet*; & there are also fantails *with smooth occiput*, as I think are all those of England. Those *fantails*, however, with long feathers on toes are rare and somewhat difficult to obtain, and all such that I have seen yet are *parti-coloured*. The others are commonly pure white, or parti-coloured; sometimes black, brown, or ashy; and there is a strong tendency to have the tail black or ashy, while all the rest (save a patch on each flank) is pure white! I have obtained an oddity of this kind, which is a most strange looking bird, as it struts about, with its black tail spread over the back so as to meet in front, and the neck thrown back, & head perking out from behind, or where the tail ought to be! I have also obtained one typically coloured fantail, slaty with bars on wing, but with the blemish of some white feathers on one wing. I am trying to mate both, and ere long shall succeed no doubt. Would not pure white fantail with black tail be a novelty in England? And still more those with long feathers on toes, & occipital *toupet* also? Likewise the following, termed *Parpaun*. A handsome medium-sized Pigeon, with very long feathers on toes, and very proud (remarkably fine) occipital crest of reverted feathers, with a whorl on either side of the crest— Colour pure white, or dark cinnamon, or black (finely iridiscent on neck), or *normal*, or (of course) parti-coloured. Have you such a Pigeon in England?— Another is somewhat larger (on the average), with same variety of colouring, & long feathers on toes, but smooth occiput, & some carunculated skin round eye,—also a very handsome Pigeon. The *Lotun* or *ground-tumbler* has little to distinguish it, save an occipital toupet, & shank feathered $\frac{1}{2}$ way down. Often white, or variously pied. I have obtained one white with black crown & tail! The *Powters* seem to resemble those of England; but I have not hitherto met with fine specimens, with *long* feathers on toes, such as I remember in England. The *Suragie* (or *Shirazi*, from Shiraz in Persia,) is a *Runt*, always bred to have the throat (forehead & cheeks *sometimes*), front of neck, underparts, rump & tail, pure white, the rest of any uniform colour, as black, ruddy, slaty, or delicate pearl-grey (like *Larus argentatus* or even *L. glaucus*), bare-legged, or with feathers more or less, but I have never seen long feathers on the toes. These ought to be *very large*; but the common run are not particularly so, though specimens of astounding size are bred at Lucknow, Delhi, &c., & these sell at high prices. *Carriers* (termed *Oran*) are distinguished by their general figure, long bill, strong feet (with short feathers occasionally), & more or less caruncle around eyes & at nostrils, sometimes enormously developed.— Colours black, &c, & often *normal* (but with white rump). A sort of *Bald-pate* is common, & is a pretty Pigeon, remarkably true to its colouring, which is black with fine reddish-purple glosses, with contrasting white cap & 2 or 3 primaries on each side, & feet generally bare, though sometimes with short plumes, in which case those on toes are often white: an occipital toupet, & a sort of hood. Some have the dark parts ruddy, more or less pale. These Baldpates (termed *Mukkhi*) are also a good deal distinguished by their manners; they are particularly active, & saucy towards other Pigeons; & they shake the head, somewhat like the 'broad-tailed shaker' or *fantail*. This is the only race which I have not seen with the wild colouring. *Sky-tumblers* (*Gira-baz*) I have not seen much of as yet; but they seem to = the European; but *unimproved*. *Turbits* and *Jacobins* I do not remember to have seen.— I think the irides *vary* in colour in every breed, "pearl", "gravelly", &c; but I have not hitherto paid sufficient attention to this matter.[21] I purpose attempting several curious *crosses*, to mark the results. Have you been trying experiments of this kind?[22]

—March 22/— Since last writing, it has struck me that our wild Pigeons (*intermedia*) have dark feet, whereas the tame have generally pink feet; but though I have 4 times crossed the Calcutta esplanades, going & coming to the museum, I have not, strange to say, seen a single flock; though they are generally there numerous, & keep rising before the horse, & settling again *just in the way*, ½ doz. times in succession. This "important observation" must now, therefore, be deferred for a future opportunity.—

I suppose, ere this reaches you, that you will have read my article on wild Asses.[23] Strange to say, the very day of its publication, the subject was engaging the attention of the Académie des Sciences in Paris, as I see by the *Comptes Rendus*; on the occasion of the presentation of two animals from the Syrian desert (as you will see by a note), which Is. Geoffroy distinguishes as a new species, *E. hemippus*, & I think he is quite right.[24] But my suspicions are confirmed regarding the indigenous abode of the Ass; and I now feel satisfied that it is an *African* rather an *Asiatic* quadruped! Nor do I see what reason the Prince Canino can find for *pronouncing* the troops which range the deserts & also mountains of N. E. Africa, to be descendants of tame Donkeys (like the S. American);[25] seeing that they are of pre-historic antiquity; & moreover that the finest breeds of tame Donkeys inhabit now (as from the remotest traceable period) the adjacent countries, and, as a general rule, degenerate as we recede from them. I also now suspect that Chesneys "wild Horses" in N. Arabia refer to the *hemippus*, & his "wild Asses" in S. Arabia to *asinus ferus* (vel *Onager*),[26] also that Wellsteads Sacotran "wild Asses" are *asinus* aboriginally wild & not 'feral'![27] Already I have prepared another paper on the subject, in which I have gone sufficiently into details—[28] Is. St. Hilaire, I perceive, confirms what I say regarding the voice of the Indian *Ghor-Khur*; & is of the same opinion as myself respecting the identity of the Indian animal with the *hemionus* of N. Asia, which is undoubtedly the Tibetan *Kyang*.[29] But now comes the question regarding the relative distribution of these animals in S. Asia. The Mesopotamian is, in all probability, *hemippus*, or may not different species occur in the same region, to a greater or less extent? The younger Gmelin's Mongolian *Wild Ass* I take to be a *Hemione* with incipient shoulder-stripe; but that of Pallas must be a distinct & peculiar species, *nameless*![30] Certainly not true *asinus*; & Kerr Porter's 2 individuals *sans* even dorsal streak may possibly yet prove to be another.—[31]

I shall not have time to write to you today, so believe me here to be, as ever, Yrs very truly, | E. Blyth.

Extract from an able communication on the resources of Oudh, published in the Calcutta 'Englishman' newspaper for March 13/56— Under the heading of "hides, horns, & tips,"—we read "There are hundreds of thousands of horned cattle, almost wild, to be found in the *tarai* of the hill forests, from the borders of the Pillibhit and Shajahánpur districts, down to the Gorukhpur one; and there thousands of hides, & hundreds of thousands of horns are said to be rotting for want of a demand, which is no more than the consumption of the locality".— *Old story, of want of roads, &c.*

DAR 98: 133–9

CD ANNOTATIONS

1.4 As regards . . . which see. 1.6] *scored brown crayon*
1.19 and find . . . Gray] *double scored brown crayon*
1.21 probably . . . fowls; 1.22] *double scored brown crayon*
1.22 while G. . . . common hen.— 1.23] *double scored brown crayon*
2.2 carry high . . . wing, &c. 2.5] *double scored brown crayon*
2.7 & I find . . . white-rumped, 2.8] *scored brown crayon*
2.10 ashy rumps] *underl brown crayon*
2.15 *unknown . . . B.C.* 2.17] *scored brown crayon*
2.22 so long back . . . monuments— 2.25] *double scored brown crayon*
2.27 And now . . . the toes. 2.30] 'Pigeons' *added brown crayon*
2.31 All three . . . there numerous 3.5] *scored brown crayon*
4.2 Strange . . . *Rendus;* 4.3] *scored brown crayon*
4.9 of N.E. Africa, . . . from them. 4.13] 'Asses' *added brown crayon*
Top of letter: 'Received May 6/56/' *ink*; '13'[32] *brown crayon*
Verso of last page: '13' *brown crayon*

---

[1] The date '22 March' is given half-way through the letter, at which point Blyth indicates that the preceding part was written earlier. His valedictory sentence implies that the letter was completed after 22 March.

[2] Bochart 1675, published in two volumes, includes Samuel Bochart's work on the animals of scripture. It went through several subsequent editions.

[3] Dureau de la Malle 1855, pp. 690–2, discusses William Allen's description of the wild fowl of Annobon Island.

[4] This exact statement does not occur in Dureau de la Malle 1855, but see p. 689 n. 3 of this article.

[5] Blyth 1855a, p. 304: 'PTERNESTES RUBRICOLLIS . . . or "wild hen" . . . represents the domestic fowl in E. Africa; and its flight and run resemble those of the Guinea-fowl.'

[6] Chesney 1850, 1: 588: 'It is understood that in Nedjd and the southern parts of the territory, the pheasant, the jungle-fowl . . . are met with'.

[7] Niebuhr 1779, 1: 234.

[8] Nicholson 1851.

[9] This sentence was added by Blyth in the margin without indicating where it was to be read.

[10] G. R. Gray 1849.

[11] In *Variation* 1: 235, CD stated: 'Mr. Blyth and others believe that the *G. Temminckii* (of which the history is not known) is a . . . hybrid.'

[12] See letter from Edward Blyth, 23 February 1856 and n. 11. Hermann Schlegel was director of the National Museum of the Netherlands in Leiden.

[13] Owen 1856. The fossil ox was discovered by John Lubbock (see *Correspondence* vol. 5, letter to John Lubbock, 19 [July 1855]).

[14] Blyth believed that Hugh Edwin Strickland's *Columba intermedia* was 'simply a local race of *livia*' (*Correspondence* vol. 5, letter from Edward Blyth, 21 April 1855).

[15] In his abstract of this letter (DAR 203), CD noted: 'Has got all the breeds of Pigeons f^d in India of the Slate colour with bars. & on colour of Rumps.'

[16] An English edition of Kamandaki's *The elements of polity* was published in Calcutta in 1849, translated and edited by Rajendralal Mittra.

[17] Jones trans. 1796, pp. 124–5, lists among items forbidden in the Hindu diet 'the town cock'.

[18] In his abstract of this letter (DAR 203), CD noted: 'on antiquity of Fowls in India'.

[19] Horace Hayman Wilson was professor of Sanskrit at Oxford University and director of the Royal Asiatic Society of London; Henry Creswicke Rawlinson, who deciphered a Persian cuneiform inscription and excavated many sculptures in Babylonia, had returned to England in 1855.

[20] Dixon 1851, pp. 415–19.

[21] In his abstract of this letter (DAR 203), CD recorded that Blyth devoted two sheets to 'Description of Indian Pigeons domestic varieties.—'

[22] CD began keeping pigeons for experimental purposes in April 1855 (see *Correspondence* vol. 5, letter to W. E. Darwin, [25 April 1855]). In August 1856, he began crossing all his kinds 'to see whether crosses are fertile & for the fun of seeing what sort of creatures appear.—' (letter to W. B. Tegetmeier, 30 August [1856]). The records of his crosses are in DAR 205.7: 166–89.

[23] See letter from Edward Blyth, 8 January [1856].

[24] Geoffroy Saint-Hilaire 1855a.

[25] Bonaparte 1855b.

[26] Chesney 1850, 1: 581.

[27] Wellsted 1840, 2: 294.

[28] Blyth 1859. There is an annotated copy of this paper in the Darwin Pamphlet Collection–CUL.

[29] Geoffroy Saint-Hilaire 1855b.

[30] Johann Georg Gmelin's wild ass is described in Schreber 1778–1846, 6: 154–66 and Pl. CCCXII; that of Pyotr Simon Pallas is in Pallas 1774, Pl. VII, and Pallas 1798, Pl. V, fig. 2. See *Correspondence* vol. 5, letter from Edward Blyth, 21 April 1855, for Blyth's first mention to CD of these wild asses.

[31] Robert Ker Porter described a wild ass in Porter 1821–2, 1: 459–61; on p. 460, he noted: 'No line whatever ran along his back, or crossed his shoulders, as are seen on the tame species with us.'

[32] CD's numbering of Blyth's letters.

To Thomas Henry Huxley 2 April [1856][1]

Down Bromley Kent
April 2$^{\underline{d}}$—

My dear Huxley

The Hookers are coming here on the 22$^{\underline{d}}$ & I believe Mr Wollaston & H. C. Watson, & I want to persuade you to join us;[2] I know from your lectures you could not come on that day,[3] but could you not spare a few days, & come here the *next* day, or if that be not possible on Saturday the 26th, when the Hookers at least will be here.— I do not know how Mrs Huxley now is; but do you not think a little change might be beneficial to her. My wife desires me to say (& would have written herself, had I not said I was sure that you would excuse the form) that she could be as quiet as she liked, & sh^d have a comfortable arm-chair in her bedroom, so as to live upstairs as much as she liked. Do think of the possibility of all this. A coach leaves Silver cross Charing Cross (& another from the Golden

Cross) at 3¼ & would drop you at Farnborough about 13 miles, & there I would have vehicle to bring you the two miles on here.—

A little change is good for man woman & beast, so do reflect favourably, & come if you can, & believe me | Your's most sincerely | Charles Darwin

Many thanks for the Abstract your lecture, which I was *very glad* to see

Imperial College of Science, Technology and Medicine Archives (Huxley 5: 46)

[1] The year is established by CD's invitation to Huxley to stay at Down (see n. 2, below).

[2] According to Emma Darwin's diary, Joseph Dalton Hooker and his wife Frances arrived at Down on 22 April 1856. Thomas Vernon Wollaston came on 25 April and the Huxleys on 26 April. No mention is made of Hewett Cottrell Watson.

[3] Huxley, Fullerian professor at the Royal Institution, delivered a course of lectures on physiology and comparative anatomy in 1856 (L. Huxley ed. 1900, 1: 138). One, entitled 'On Physiology and Comparative Anatomy', was advertised for 22 April 1856 (*Medical Times & Gazette* 12 (1856): 404). CD was subsequently invited to attend the lecture (see letter to T. H. Huxley, 9 April [1856]).

From Edward Blyth [3 April 1856][1]

Notes for M.ʳ Darwin.

The following remarks on Indian Pigeons are transcribed from a letter, rec.ᵈ from D.ʳ David Scott, of the Hurriána Light Infantry, stationed at Hansi (about 60 miles, distant from the desert):[2] He lost his right hand not long ago, & is therefore obliged to correspond with his left; & his present writing is not a little difficult to decypher. A gun accident; Scott being still a most indefatigable sportsman, and an excellent Nat. Hist. observer, the *J. J.* of the 'Indian Sporting Review', under which signature you will find some capital articles of their kind. Poor fellow, his calligraphy is indeed such, that I must transcribe it in order to understand it myself.—

"For the fancy Pigeons, I put in one column the English names, with what I believe to be the native names opposite; and I take them according to size".— Vide overleaf. *Mem.* That our wild Pigeons here *have* pink feet; & that I have now obtained *Bald-pates* of the typical colouring, but with *white rump*.

"English	Hindustáni
Runts.—	1. *Kábūli Kabuta* (literally Kábul Pigeon).
"	2. *Si maab*
"	3 *Shirazi*
Powter	*Gul phula* (literally *full gullet, gula!*)—
Fantails	*Lukkha*
Tumblers	*Girra-baz*
" Almond	*Choa-chundun*
" Ground	*Lotun*
Carrier	(native name unknown).
Nuns	*Mūki*
	Gáhu

"Two other native names I know nothing of, viz.—(illegible).

"The three first I believe are all what are called *Runts* at home.— As its name betokens, the first comes from Kábul; and I had a lot of them at Pesh/wur, where I remember paying four rupees for a pair. They are very large, generally blue-speckled, with more or less white. Often like what is called 'feather-about'. They are heavy clumsy birds, and very bad breeders; for I never remember above one young in a nest brought up". (This is evidently what is here known as the Bághdád breed!)

"The *Simaab* or 'silver-water' is, I think, merely so called from the colour resembling quicksilver. The only pair I ever saw came from Calcutta." (An ash-coloured *Shirázi*, apparently; here common enough, with upper-parts silvery-grey, more or less dark, the same as in most Gulls and Terns.)

"The *Shirázi* is very common, of various colours,—black & white, blue & white, or red & white. They are large heavy birds, also bad breeders, & useless brutes" (*sic*, after much trouble to make it out! Well, our Indian *runts* bear names which indicate a western origin, so far as this country is concerned; *i.e.* Shiráz, Bághdád, & Kábul).

"The Powter I have never seen in India, though it does exist, introduced I think from home".—

, "The Fantails are precisely like those at home, but generally indifferently bred. I have seen the blue bar-winged ones at home, but not here". (*N.B.* I have got another white fantail with black tail, making *a pair*. Are there English fantails with occipital top-knot, & with long feathers on toes?) Scott adds, "I have seen Fantails generally bare-legged".

"*Palmon* (?) is a name applied to all kinds having feathered legs. Some here have handsome tufts on head, like $\frac{1}{2}$ bred 'Ruffs'. The *Yábu* is like a $\frac{1}{2}$ bred 'Ruff', but has a few feathers like a tuft just over the beak. I have only seen one.

"The *Mūkis* are, I believe, simply 'Nuns'. They are small, and black, blue, or slate-colour, with white heads and white flight-feathers. They are handsome, but bad breeders.

"The first kind of *Tumbler* is that kept for flying. They are generally blue with light-coloured eyes (Here dark-legged, black bill, pearl eye, & small & delicately made). They are of every colour here.

"The *Choa Chundun* are, I think, but am not sure, 'Almond Tumblers'. They are of various colours, with short heads, and very small bills; & the bird itself is small. The *Lotuns* are a kind that roll about on the ground.

"What is called the *Carrier* here is not the real kind, but has got the name from having merely a fleshy knob on beak, and the eyes also surrounded by a fleshy margin. Those I have seen were thick-necked brutes, a shade better than common Pigeons.

"Delhi & Lucknow are the places for Pigeons, and I have never been stationed at either." (*N.B.* Both *Musálman* capitals!)

Thus much *verbatim et libratim.* You will observe that Dr Scott's observations in Upper Hindustán coincide very nearly with mine here. However, our 'Carriers' are not so bad; & show, conspicuously, a much longer bill than other Pigeons, with more or less caruncle at base & surrounding eyes.— Turn to Dixon's book on Pigeons, p. 101,—[3] the black 'Barb' there figured, I could here match exactly; the 'Jacobine' figured has the colouring of our *Mūki*, except that the head only should be white; which latter (*ours*) is clearly the *Nonnain Maurin* of Temminck, vide p. 100;[4] and the 'Nun' there figured has the same crest as our *Mūki*, but its colouring is like that of some *Lotuns* or 'Ground-tumblers': these last have the same occipital crest, but no reverted plumes lower down. The 'Jacobein' looks to be a further modification of our *Mūki*; & the latter is not the 'Baldpate' figured at p. 86.[5] I have now obtained *Mūki* with smooth occiput, *i.e.* true 'Bald-pates'. I have not yet seen 'Runts' (*Shirázi*) of the typical colouring, slaty, with 2 black bars on wing, &c; but the *fantail* so coloured, & also our feather-footed with occipital reverted feathers, ditto, have no white band on rump; whereas the 'Carriers' & also the 'Nuns' (*Mūki*), *have it*; wherefore the latter should derive from the European *livia*, the former from the Asiatic *intermedia*! I must look to this matter as regards 'Powters'.— It strikes me that Egypt (& Cairo—*Kahirá*—in particular) would be a good place where to study the races of tame Pigeons; also Turkey & Syria. Have you no friends in those countries, who might assist your enquiries?— With reference to my remark that the *Numida* or *Meleagris* or *Gallina numidica* (mistake for *nubica*) of the old Romans was *N. ptilorhyncha* of E. Africa, tamed no doubt, but scarcely *domesticated*,—I call your attention to a passage in G. W. Browne's 'Travels in Africa', &c (1792 to 1798), p. 264. We here learn, that so late as the close of the last century (if not still), "Guinea-fowl"—meaning of course the *ptilorhynca*—"are found in great numbers in Dar-fûr; whence they are carried, *as a profitable commodity*, to Kahirá, where, however, in a domestic state it is said that *they seldom or never breed*".[6] Mitchell should endeavour to procure a few pairs from Cairo.[7] They are of a conspicuously distinct species from the modernly denominated *N. meleagris*; with hair-like frontal crest, and no vinous or plum-colour about the neck & breast. But who can now doubt the source of the Roman supply? A traffic doubtless continued uninterruptedly even to our times!— The more I see of domestic Pigeons, the more obvious becomes *the conclusion*, according to my judgement, of their having *all* descended from the *livia* (or barely separable *intermedia*, &c); and I think the voice attests this, quite as decidedly as with domestic fowls. Some difference of voice there undoubtedly is, among the races; but what is that amount of difference, compared with the distinct *coo* of any really different species? It is indeed surprising what varieties of *cooing* there are among the various wild or true species of *Columbidæ*; in all cases an unmistakeable Pigeon's voice, yet how different one species from another, & comparatively how very similar the coos of the different domestic races! Then their pairing so indiscriminately, however different the races!

DAR 98: 140–3

CD ANNOTATIONS

1.1 The following . . . size".— 2.2] *crossed pencil*
1.2 Hurriána] *underl brown crayon*
1.2 Hansi] *underl brown crayon*
1.2 about 60 miles 1.3] *underl brown crayon*
2.7 *Si maab*] 'mere colour' *added pencil*
2.14 Carrier] 'seems different from English' *added pencil*
2.15 Nuns] 'White heads & white flight— Bald Pates' *added pencil*
6.3 Well, . . . Kábul). 6.5] *double scored brown crayon*
9.2 The *Yábu* . . . seen one. 9.3] *double scored brown crayon*; '—Distinct' *added pencil*
9.3 beak.] 'beak' *added pencil*
15.2 However, . . . eyes.— 15.4] *scored brown crayon*
15.5 the black . . . exactly;] *scored brown crayon*
15.18 It strikes . . . times!— 15.32] *crossed pencil*
15.32 The more . . . races! 15.42] *scored brown crayon*
Top of letter: 'April 3. 1856' *pencil*; '14'[8] *brown crayon, circled brown crayon*
Top of last page: '14' *brown crayon, circled brown crayon*

[1] The date is based on CD's annotation.
[2] David Scott of the Bengal Medical Establishment was assistant surgeon to the Hurrianah light
infantry battalion stationed at Hansi (*East-India register and army list, for 1856*).
[3] Dixon 1851, p. 101, illustrates a nun, a barb, and a jacobin.
[4] Dixon 1851, p. 100, quotes Coenraad Jacob Temminck's description of the most beautiful specimens
of nuns as 'those which are black, but have the quill feathers and the head white: they are called
Nonnains-Maurins.'
[5] Dixon 1851, p. 86.
[6] W. G. Browne 1799, p. 264: 'This beautiful bird [guinea-fowl] is found in great numbers in Fûr . . .
They are carried as a profitable commodity to Kahira, where however, in a domestic state, it is said
they seldom or never breed.' CD recorded having read this work in 1838 (*Correspondence* vol. 4,
Appendix IV, 119: 4a).
[7] David William Mitchell had been secretary of the Zoological Society of London since 1847.
[8] In his abstract of this letter, CD wrote only: 'Letter 14— Notes on domestic Pigeons' (DAR 203).

To Walter Baldock Durrant Mantell   3 April [1856][1]

Down Bromley Kent
April 3[d.]

My dear Sir

When I saw you in London, you were so kind as to say that I might remind you
about the note & sketch, which you said you would give me, on the position,,
Latitude, approximate size, height above sea, & degree of separation by valleys
from any higher land, &c, of the quartz boulders, which you saw in New
Zealand;[2] & likewise on the icebergs with fragments of rock, which you saw in the
southern Ocean.— I am really *most anxious* to hear about these, whenever it may
be most convenient to you to spare the time; though I fear that you will think me a
very troublesome person.—

I am tempted to put 2 or 3 other questions, if you will be so kind as to answer
them, *if in your power*, on a separate piece of paper & you can write answers at foot.

These questions you will think very ridiculous ones; but they all bear in some degree on the origin & history of the cultivation of plants, & domestication of animals.—[3]

With hopes, that you will forgive me I remain | My dear Sir | Your's sincerely | Charles Darwin

Alexander Turnbull Library (Mantell 83.268)

[1] Information referred to in this letter and the letter to W. B. D. Mantell, 10 April [1856], was included in the chapter on geographical distribution in *Natural selection*, the first draft of which was completed in July 1856 (see *Natural selection*, pp. 532–4). Mantell was in England between 1856 and 1859.

[2] See *Correspondence* vol. 5, letter to W. B. D. Mantell, 17 November 1854, in which CD asked whether there were erratic boulders in New Zealand. In *Natural selection*, p. 546, CD noted: ' Mr. W. Mantell has shown me sketches of great fragments of quartz, lying on tertiary strata, which probably are erratic boulders'. CD was anxious to establish that New Zealand, as other countries, had experienced glaciation so that he could postulate a worldwide cool period to explain the distribution of recent animals and plants. See *Natural selection*, pp. 544–57, and letter to C. J. F. Bunbury, 21 April [1856].

[3] CD's questions and Mantell's answers have not been found, but see letter to W. B. D. Mantell, 10 April [1856]. In *Variation* 2: 161 n. 65, CD stated: 'the New Zealanders, as Mr. Mantell informs me, kept various kinds of birds.'

From Charles John Andersson [6 April 1856][1]

with regard to the *beau ideal* satisfactorily; for amongst us Europeans it is well known that one man selects his partner for a handsome face, whilst another make his choice of a good figure without much regard to beauty. All I can say is that savages generally select their partners more for any attraction the body may posses than for beauty of face.[2]

By the by have you not described somewhere the flesh of the Puma as palatable[3] Did you ever taste the flesh of lions? I find opinions vary greatly with regard to the flavour of its flesh—

With kindest Regards | Yours very truly | Ch.ˢ J. Andersson

Ch.ˢ Darwin, Esq.ʳᵉ

Incomplete
DAR 85: 102

CD ANNOTATIONS
2.1 By the . . . Esqʳᵉ 4.1] *crossed pencil*
Bottom of second page: 'Apr 6—1856' *pencil*

[1] Dated by an endorsement in CD's hand, presumably repeating the date given on the missing page of the letter (see CD's annotations). This fragment was preserved in a portfolio of notes on sexual selection used in preparing *Descent*.

[2] Andersson was a Swedish-born explorer of Africa. CD also asked Walter Baldock Durrant Mantell the same question about sexual selection among indigenous populations (see letter to W. B. D. Mantell, 10 April [1856]). Neither Mantell nor Andersson was cited in *Descent*.

[3] In *Journal of researches*, p. 135, CD reported having eaten puma meat on a journey from Patagones to Buenos Aires and pronounced it 'remarkably like veal in taste'.

To Joseph Dalton Hooker 8 April [1856][1]

Down Bromley Kent
April 8th

My dear Hooker

I have been *particularly* glad to get your splendid eloge of Lindley.[2] His name had been lately passing through my head & I had hoped that Miers would have proposed him for the Royal medal.[3] But I most entirely agree that the Copley is more appropriate; & I daresay he would not have valued the Royal.—[4] From skimming through many Botanical works & from often consulting the Vegetable Kingdom,[5] (I had ignorant as I am), formed the highest opinion of his claims as a Botanist.

If Sharpey will stick up strong for him, we sh[d] have some chance;[6] but the Natural Sciences are but feebly represented in the Council.[7] Sir P. Egerton, I daresay, would be strong for him.—[8] You know Bell is out.—[9] Now my only doubt is, & I hope that you will consider this, is, that the Natural Sciences being weak on Council, & (I fancy) the *most powerful man* on council, Col. S.[10] being strong against Lindley, whether we sh[d] have any chance of succeeding, it would be so easy to name some eminent man, whose name would be well known to *all* the Physicists. Would Lindley *hear of*, & dislike being proposed for Copley & not succeeding?[11] Would it not be better on this view to propose him for Royal? Do think of this.— Moreover if Lindley is *not* proposed for Royal, I fear both Royal medals would go Physcicists; for I, for one, sh[d] not like to propose another zoologist, though Hancock w[d] be a very good man.[12] & I fancy there would be feeling against medals to *two* Botanists.—[13] But for whatever Lindley is proposed, I will do my best.— We will talk this over here.—[14]

Your's ever | C. Darwin

P.S. | Has Falconer appeared in world yet;[15] if so & you know his address, I wish you would let me have it.— If I do not hear I shall understand you do not know.—

I have written the following in answer to M[rs] Hooker's note to my wife.—

Our carriage & a Fly shall be ready at 12°. 15′ at Croydon on the 22[d]: I am vexed to see that you must go to Vauxhall & wait so long there; I had fancied as a matter of course that you could have stopped at Wimbledon, Croydon is same distance as Sydenham St. from us. viz 9½ or 10 miles.—

It is very good of your coming for really it is an awful task.[16] A Railway is actually making to Beckenham, which will save 2 miles.[17]

Do bring some work with you so as not to cut your visit very short.—

M[r] & M[rs] Huxley come on Saturday 26th & return on 28th—

DAR 114.3: 160, 159B

[1] Dated by the reference to Hooker's forthcoming visit to Down House (see n. 16, below).

[2] Hooker had apparently written to CD in support of John Lindley as a candidate for the Copley Medal of the Royal Society of London. CD was vice-president and a member of the council of the society.

[3] John Miers, engineer and botanist, was also a member of the Royal Society council in 1856. In 1855, he had seconded Thomas Bell's nomination of Lindley for a Royal Medal. The medal, however, had been awarded to John Obadiah Westwood.

[4] The Copley Medal was considered to be 'the highest scientific distinction' that the Royal Society had to bestow. The two Royal Medals awarded each year were, on the other hand, given for what were deemed to be 'the two most important contributions to . . . Natural Knowledge' published within the preceding ten years (Royal Society of London 1940, pp. 12, 116–17).

[5] Lindley 1846.

[6] William Sharpey was one of the secretaries of the Royal Society.

[7] The naturalists on the council in 1856 were, apart from CD, John Miers (see n. 3, above), William Benjamin Carpenter, and Philip de Malpas Grey-Egerton. There were also two medical members of council: Neil Arnott and Benjamin Collins Brodie. The full list is given in *Athenæum*, 1 December 1855, p. 1403.

[8] See n. 7, above.

[9] Thomas Bell, secretary of the Royal Society from 1848 to 1853, had been a member of council in 1854 and 1855.

[10] Edward Sabine, as treasurer of the Royal Society, was 'Senior Vice-President *de facto* although not *de jure*' (Hall 1984, p. 132).

[11] Lindley had twice been an unsuccessful candidate for a Royal Medal. As well as being nominated by Miers in 1855 (see n. 3, above), he had been proposed by Hooker in 1853 (*Correspondence* vol. 5, letter from J. D. Hooker, [4 November 1853]), and letter to J. D. Hooker, 5 November [1853]).

[12] CD had first discussed the possibility of a Royal Medal for Albany Hancock in *Correspondence* vol. 5, letter to T. H. Huxley, 31 March [1855]. Hancock was awarded the Royal Medal in 1858.

[13] Since 1853, one of the two Royal Medals awarded annually went to an author of a work in the physical sciences and the other to an author in the biological sciences. There was, moreover, an attempt within the latter category, to alternate between zoology and botany. CD apparently felt that the council would be reluctant to award both the Copley Medal and one of the Royal Medals to botanists in the same year. In the end, both the Copley and a Royal Medal were awarded to zoologists (see n. 14, below).

[14] Lindley was not proposed for either medal. Rather, Carpenter proposed, and CD seconded, the nomination of Henri Milne-Edwards for the Copley Medal, which he received in November 1856. CD successfully nominated, seconded by Sabine, John Richardson, a zoologist, for the Royal Medal (see letter to Edward Sabine, 23 April [1856]). Lindley received a Royal Medal in 1857 (see letters to J. D. Hooker, 2 June 1857 and 5 June [1857], and Royal Society council minutes).

[15] Hugh Falconer had retired from the Indian civil service in the spring of 1855. Upon arriving in England, he immediately began work on his study of the vertebrate fossils from the Siwalik Hills, for which he visited 'almost every museum in Western Europe' (*DNB*).

[16] Hooker and his wife visited Down from 22 to 28 April 1856 (Emma Darwin's diary).

[17] The Mid-Kent Railway was completed in 1858. The Beckenham junction was completed on 1 January 1857 (see letter to W. D. Fox, 8 February [1857]).

To T. H. Huxley 9 April [1856][1]

Down Bromley Kent
Ap. 9th

My dear Huxley

We are *very glad* that M[rs] Huxley & yourself will come to us on the 26th.—
I am heartily sorry at so poor an account of your health.

I will tell you a pleasanter way for M[rs] Huxley to come, viz by Croydon & Epsom line from London Bridge to Sydenham Station, where my phaeton shall be & bring you here, 9–10 miles,— I will send to meet Train which leaves London at 3°. 15′, supposing that you would wish to have best part of day in London; but if you like to come earlier, so much the better, & fix earlier Train.—

We dine at 7. & shall perhaps have few visitors besides those in House.—[2]

Very many thanks for your card, which I fear I shall not take advantage of, as the Lecture of 22[d] was what I most wanted to hear: but I sh[d] be very glad to hear any lecture, having heard of your lecturing powers.—[3]

Your's Ever C. Darwin

Imperial College of Science, Technology and Medicine Archives (Huxley 5: 33)

[1] Dated by the relationship to the letter to T. H. Huxley, 2 April [1856] and the preceding letter.
[2] See letter to T. H. Huxley, 2 April [1856], n. 2, for the members of the party staying at Down House. According to Emma Darwin's diary, John Lubbock joined them for dinner on 26 April.
[3] See letter to T. H. Huxley, 2 April [1856], n. 3.

To W. B. D. Mantell 10 April [1856][1]

Down Bromley Kent
Ap. 10th

My dear Sir

I am very much obliged to for finding time, amidst all your avocations, to answer my questions so fully.—

With regard to the erratic blocks, the most suspicious circumstance in regard to their truly erratic character, is, as it strikes me, their all consisting of the quartz rock.— Generally, but certainly not always, one finds several kinds of rock; nor do I quite understand that you are sure that they are separate fragments, & not rock in situ peeping out. Did M[r] Harris refer to them as loose or separate blocks?[2] if so I sh[d] think the evidence was in favour of their belonging to the so-called erratic class.—[3]

Perhaps when you send me the iceberg sketch, you will answer this about M[r] Harris.—

If I have not utterly exhausted your patience, I sh[d] be *particularly* obliged if you would inform me whether you think the evidence is really good that there formerly existed some animal (*with hair?*) like an otter or Beaver: I am much surprised at this. Could it not have been any water bird or reptile?[4]

Lastly, I fear you cannot answer my question whether the beau ideal of beauty amongst the less civilised natives (ie those least influenced by being accustomed to European faces) would agree with ours; viz whether we & they would pick out the same kind of beauty.—[5] Forgive me if you can, & believe me,

Your's truly obliged | Charles Darwin

Alexander Turnbull Library (Mantell 83.268)

[1] See letter to W. B. D. Mantell, 3 April [1856], n. 1, for the basis of the date.
[2] Possibly John Williams Harris, who in 1837 had discovered in New Zealand bones of *Dinornis*, the large extinct moa later described by Richard Owen (*DNZB*).
[3] See letter to W. B. D. Mantell, 3 April [1865], n. 2.
[4] Before the arrival of Polynesian settlers, the only native land mammals in New Zealand were two species of bats (*Encyclopedia of New Zealand* 2: 380).
[5] CD wanted to compare the criteria affecting selection of mates among different species (see also letter from C. J. Andersson, [6 April 1856]). In *Descent* 2: 369, he wrote: 'Until recently, as I hear from Mr. Mantell, almost every girl in New Zealand, who was pretty, or promised to be pretty, was *tapu* to some chief.' 'Tapu' is a Maori variant of taboo, meaning set apart for a special use or purpose or restricted to the use of a god or chief (*OED*).

From Richard Thomas Lowe[1]   12 April 1856[2]

12 April '56

The Flora of Porto S^{to} may be stated at from 270 to 280 species (indig^s and perfectly naturalized).[3]

1. Of these, 7 or 8 are certainly, and one or 2 others doubtfully endemic—the doubt arising from difficulty of ascertaining identity of species— 3 or 4 of the 7 or 8 are sufficiently striking and abundantly growing plants. 2 or 3 represent Mad^n endemic sp: 2 or 3 are more of the nature of "weeds".
2. Of plants common to P^o S^o & Mad: but not hitherto found elsewhere, there are 20 or 21, perfectly certain, and 3 others doubtful.
3. Again, of plants common to P^o S^o and other countries (mainland) of Europe, but not *found at all* in Mad^a there are 25 certain & 4 doubtful.
4. Lastly, of plants common to P^o S^o and other countries (mainland) of Europe, but *very rare* in Mad^a there are 6, to which may perhaps be added 2, which are indeed only at this day occasional garden plants in Mad^a, whence they were introduced in 1834, into P^o S^o which they have now completely overspread! One of these is a Tamarix, (*T. orientalis* L.?); and the other is the Hottentot Fig, (*Mesembryanthemum edule* L.)

Of classes 3 & 4, almost all are common European "weeds", proving of course little any way. And let me add, that the very peculiar nature of the soil & climate of P^o S^o as compared with Mad^a accounts of itself in great measure for the very different character & aspect generally of the vegetation in the 2 Islands. Many plants common in one Island, *can not* be made by any efforts to grow in the other.

This may be useful as a caution against attributing too much in this particular case to other possible modifying general causes or influences.

Hooker has I think considerably underrated the number of *good* endemic species in Madeira,[4] which exclusive entirely of P? S^{tan} & Dez^{tan} plants will be found I think rather to exceed than fall short of 100. But it would take a good deal more time & research than I can just now afford to speak positively & with accuracy on this head.

There are *very many common* endemic species in Madeira not occurring in P? S?

In the Dezertas there are 3 very remarkable *endemic common* plants, one forming a new *genus* of Umbelliferae;[5] another a new genus of Gramineae;[6] the 3^d a new shrubby Chrysanthemum, representing *C. pinnatifidum* L. fil. of Mad^a.[7]

There is no freshwater fish in P^o S^o (as in Mad^a) but the Eel (*Ang. latirostris* Yarr.) I believe. Wollaston & I caught one ourselves in a puddle last spring.

Contemporary copy
Lady Lyell

[1] The original manuscript of this letter has not been found, but there is a copy by Charles Lyell in his 'scientific journals' (Wilson ed. 1970, pp. 53–4). Lyell visited CD at Down House from 13 to 16 April 1856, and the 'Migration of Plants & Shells' was one of the topics discussed, with particular reference to Madeira, which greatly interested Lyell (Wilson ed. 1970, pp. xlviii–xlvi, 52–5). CD's notes on their conversation about Madeiran shells, dated 16 April 1856, are in DAR 205.3 (Letters).

[2] The date as given by Lyell.

[3] CD's query probably concerned a point he had noted in his copy of A. de Candolle 1855, 2: 801:

Could I get list of Naturalised Plants from Lowe for Madeira . . . This w^d be important as showing means of distribution, & as showing inhabitants of islands not well adapted.

Lowe was preparing a short catalogue of new species of plants found on Madeira for *Hooker's Journal of Botany and Kew Garden Miscellany* (Lowe 1856). The list includes species from Porto Santo, part of the Madeiran group of islands.

[4] W. J. Hooker and J. D. Hooker 1847.

[5] The genus was called *Monizia* by Lowe and described in Lowe 1856, pp. 295–6. It comprised only one species, *Monizia edulis*, found only on the island of Dezertas.

[6] The genus was given the name *Arthrochortus* and described in Lowe 1856, pp. 301–2. Lowe considered the single species *A. loliaceus* to be intermediate between the genera *Lolium* and *Lepturus*.

[7] *Chrysanthemum haematomma*, described in Lowe 1856, p. 296. It was distinguished from its nearest ally, *C. pinnatifidum* of Madeira, by the 'dark blood-coloured florets'.

To Henry Tibbats Stainton 13 April [1856][1]

Down Bromley Kent
April 13th

Private

Dear Sir

I am much obliged to you for so courteously sending me a copy of "Entomologist Weekly Intelligencer".—[2] I do not suppose that I ought to mention anything which passed on the Council. But I may say that I individually have nothing to object to in your remarks.—[3] I can see, however, that apparently you are not aware of a most important change made 3 years ago, with consent of the Queen, in the distribution of the Royal Medals; before that time it was compulsory in the Council to give it to men for publications in the Transactions, & this will explain, if you take the trouble to look at the names, the cause of many of the awards.— It was, I think, an *extremely* bad rule. Hence I think you will perceive why, **except to Mʳ Newport,** no medal was given to an Entomologist.[4]

Since the rule was changed, the 3 recipients have been myself, (when I was not on council) Dʳ Hooker, & Mʳ Westwood.[5] As I have been a recipient, of course I can say nothing whether or no the awards have been well, or atrociously ill made; but this I can say that the Council take great trouble in deciding, & a most difficult & disagreeable task it is to perform.—

Pray believe me | Dear Sir | Yours very faithfully | Ch. Darwin

British Museum (Natural History) (General Library MSS/DAR: 16)

[1] Dated by the reference to the *Entomologist's Weekly Intelligencer*, which began publication on 5 April 1856.

[2] Stainton was the founder and editor of the *Entomologist's Weekly Intelligencer*. He had previously corresponded with CD about another entomological publication of his, the *Entomologists' Annual* (see *Correspondence* vol. 5, letter to H. T. Stainton, 20 October [1855]).

[3] The second number of the *Entomologist's Weekly Intelligencer* (12 April 1856) opened with an unsigned editorial, clearly written by Stainton, entitled 'Why did Mr. Westwood get the Royal Medal?'. John Obadiah Westwood had been awarded the medal in 1855 (see n. 5, below). Stainton thought the award was long overdue and complained that Westwood's most famous contribution to science had been published many years previously (Westwood 1839–40). He concluded that the recent recognition of Westwood's merits was because entomology was now 'more prominent than formerly' ([Stainton] 1856, p. 9–10).

[4] George Newport was awarded the Royal Medal in 1851, not for his entomological work but for his physiological investigations in Newport 1851, which had been published in the *Philosophical Transactions of the Royal Society*. None of Westwood's papers had appeared in the *Philosophical Transactions*.

[5] It had been CD who had nominated Westwood for the Royal Medal in 1855 (see *Correspondence* vol. 5, letter to T. H. Huxley, 31 March [1855]). In his nomination text, CD had cited Westwood's 'Introduction' (Westwood 1839–40) and his 'various Monographs and Papers on Entomology' (Royal Society council minutes). However, since Westwood 1839–40 had been published more than ten years previously, under the new rules it had to be excluded from the citation.

From C. J. F. Bunbury 16 April 1856

Mildenhall
April 16, 1856

My dear Darwin,

I hardly know how to account for my long silence, after the very interesting letter which I received from you upwards of two months ago, & which I assure you I value highly.[1] But in truth I felt that I had nothing to say which could be at all an adequate return for your letter, & I waited (rather vainly) in hopes of "something turning up", that is, in hopes of striking out some remark which might be worth sending.

About the *Colletia*, however, I *have* something to say.[2] I wrote to Joseph Hooker about it, & he wrote in reply:—"You are quite right about the Colletia; it was a hasty affair of Lindley's, & would have gone beyond even my notions of variability." I presume he means that Lindley was misled as to the garden origin of the supposed variety, or too hasty in believing it.

I am exceedingly interested by all you tell me about your researches & speculations on species & variation & distribution, & am delighted that you are going on working at the subject. I trust that you will not on any account give up the idea of publishing your views upon it; tho' neither you nor any one else may be able to unravel the whole mystery, or to command the universal assent of naturalists, still the researches of one who has studied the whole question so long, & with such extensive knowledge & in so philosophical a spirit, cannot fail to be of very great advantage to science. The whole subject,—I mean every thing connected with the geography of plants & animals, including all the questions of distribution & variation, is to me particularly interesting & delightful; but how much we have yet to learn upon it! The difficulties which appear to attend upon each & every one of the theories,—of specific centres, of *multiple* creation, & of transmutation,—are so many, that what is most clear to me is the necessity of caution & candour, of avoiding dogmatism, & of giving a fair consideration to every fact & argument on any side.[3] I say this, because the theory to which *you* lean is the most remote from that to which *I* incline, & yet I am quite ready to admit that your notion *may* be the right one. Certainly, on the "specific centre" theory, it is very difficult to account for the fact that (to take one instance only,) certain plants, not in the least likely to have been introduced by man or by accident, are common to Europe & the Cape, & *not* found (as far as we know) in tropical Africa; such are Typha latifolia, Scirpus lacustris, & Scirpus fluitans. To be sure, one might cut the knot boldly, by saying that they *may* exist in the intermediate countries, tho' not yet found there; but this would not be very satisfactory. How can one account, by the way, for the well ascertained fact that, of those phanerogamous plants which are common to very distant countries, (*excluding* those introduced thro' cultivation or other human agencies,) the majority belong to the division of glumaceous monocotyledons (Grasses & Sedges), & most of the remainder are aquatic or marsh plants? Are there any freshwater shells or insects common to Europe & the Cape?

The only flowering plant which is at once North American & British, & at the same time *not* found on the continent of Europe, is a water plant, Eriocaulon septangulare; & it is one which does not seem to have any peculiar facilities for migration. How did it cross over from Canada to Connemara & the Isle of Skye? Many & puzzling are the questions of this sort which occur to one.

Again, as to the Cape:—poor E. Forbes, I remember, thought that there was an axis of high land running up through central Africa, connecting the plateau of the Cape with that of Abyssinia;[4] & it is true that there are some remarkable botanical analogies (but I believe no case of specific identity) between those two regions.[5]

Still, this would help us but very little way towards an overland route for plants between Europe & the Cape; for between Abyssinia, & the Mediterranean there is a huge gap, the arid deserts of Nubia & the low burning valley of the Nile. There are indeed but few species common to Europe & the Cape, but those few are very difficult to account for.

I want to know whether you agree in the opinion which I have expressed in my Cape book,[6] that the *zoology* of that country is more analogous to that of the rest of Africa, than its *botany*? I do not at all know how far it may hold with respect to the invertebrate animals, but (as far as I can depend upon my very slight knowledge of zoology,) it strikes me as holding good to a certain extent with the quadrupeds, birds, & even the serpents. I wish I could visit Natal, which is now far more accessible than when I was at the Cape. It seems clear that as far as the Flora of the Cape is at all connected with that of tropical Africa, it is so in that direction, by the east coast & the Quathlamba mountains; whereas towards the N.W. & N. it is entirely cut off by the deserts. It appears from Krauss's account (which you will find in the Ray Soc. volume for 1849, p. 387,)[7] that at Natal there are many tropical forms on the coast, with Aloes & tall Euphorbias on the hills, & several species of peculiar Cape genera in the mountain meadows. I am sure I have read, tho' I cannot at this moment recollect *where*, that the *Doum* Palm of Upper Egypt & Nubia extends down to that south-eastern coast of Africa,—I think to Delagoa Bay.— That it is the heat & aridity of the deserts which stops the extension of the peculiar Cape Flora towards the north, seems certain. Prof. Piazzi Smyth[8] told me that, in travelling towards the interior, north-westward from Cape Town, after crossing a vast extent of *Karroo* country, without any vegetation but a few scattered & stunted succulents, he ascended to the high table-land of the Bokkeveld, & found again Heaths, Proteas, & other characteristic forms of the Cape Flora, in striking contrast to the productions of the Karroo below. But why some families of plants, which abound in the *western* parts of the colony, are scantily represented in the same parallel in the *eastern* part, is by no means evident.

I have run on with this monstrous long prose about Cape botany, because I remember that, when I saw

AL incomplete
DAR 205.2 (Letters)

CD ANNOTATIONS
1.1 I Hardly . . . sending. 1.6] *crossed pencil*
2.1 About the . . . reply:— 2.2] *scored brown crayon*
3.1 I am . . . to science. 3.8] *crossed pencil*
3.13 the necessity . . . side. 3.15] *scored brown crayon*
3.21 such are Typha . . . boldly, 3.22] *scored brown crayon*
3.26 the majority . . . aquatic 3.28] *scored brown crayon*
4.2 *not* found . . . septangulare; 4.3] *scored brown crayon*
5.5 an overland route . . . account for. 5.9] *scored brown crayon*
5.9 account for.] 'Must have been an isl^{d}' *added pencil*

6.1 expressed . . . *botany?* 6.3] *scored brown crayon*
6.6 serpents.] *underl brown crayon*
6.9 east coast] *underl brown crayon*
Top of first page: '18'⁹ *brown crayon*

[1] The letter, which has not been found, was CD's reply to the letter from C. J. F. Bunbury, 7 February 1856.
[2] See letter from C. J. F. Bunbury, 7 February 1856.
[3] For Bunbury's further views on CD's theories, see letter to C. J. F. Bunbury, 21 April [1856], n. 9, and letter from Charles Lyell, 1–2 May 1856, n. 7.
[4] E. Forbes 1854.
[5] Discussed by Bunbury in C. J. F. Bunbury 1848, p. 219. See also *Natural selection*, p. 552.
[6] C. J. F. Bunbury 1848, p. 218.
[7] Krauss 1844–6 was discussed in August Heinrich Rudolf Grisebach's report on the progress of geographic botany for 1844, published by the Ray Society (Grisebach 1849). Ferdinand Krauss published a more complete account of the flora of the Cape and of Natal in 1846 (Krauss 1846).
[8] Charles Piazzi Smyth had worked at the Royal Observatory at the Cape of Good Hope from 1835 to 1845 (*DNB*).
[9] The number of CD's portfolio of notes on the means of dispersal of animals and plants.

To C. J. F. Bunbury 21 April [1856][1]

Down Bromley Kent
April 21

My dear Bunbury

You are quite right, I do take a very great interest about the Cape Flora & Fauna, & I thank you much for your letter,[2] which, as all yours do, has pleased & instructed me much.— I have lately been especially attending to Geograph. Distrib, & most splendid sport it is,—a grand game of chess with the world for a Board. The fact you allude to about the zoology (at least mammifers) of the Cape not being nearly so peculiar as the Botany has often struck me much: I think the most probable **hypothetical** explanation is that it was long a group of islands, since united with the continent allowing the vertebrata to enter.—

Thank you about the Colletia, I called on Lindley, but c^d extract nothing & wrote to the Gardener who raised the seed, (but have not, & shall not receive any answer) to ask whether he ever had seed from S. America of any kind; undoubtedly the common form was in the Garden.[3]

I am very glad to hear you are still thinking of Madeira; there seems to me much to be done there yet; but I hear from Mr Lowe, he is going to publish a Flora, & he has sent me a curious account of vegetation of P. Santo.[4] A careful comparison of the Floras of Madeira, Azores, & Canary Is^d would, I cannot doubt, lead to some very curious results.

You speak in **far** too flattering a way about my work, in which I will persevere; & I will endeavour (eheu how difficult) "to be cautious & candid & avoid dogmatism". My determination to put difficulties, as far as I can see them, on both sides is a great aid towards candour; because I console myself, when finding some great difficulty, in endeavouring to put is as forcibly as I can.— I am trying many *little* experiments, but they are hardly worth telling, though some I am sure will bear on distribution & I think on *aquatic* plants.[5]

As you say you like scientific chat, & your kind letter makes me sure that you will not think me an egotistical bore, I will tell you of a theory I am maturing (by the way please do not mention it to anyone, for 2 directly opposite reasons, viz whether valueless or valuable). As glacial action extended over whole of Europe, & in Himmalaya, on *both* sides of N. America & *both* sides of Southern S. America & I believe in N. Zealand, within very late times (existence of recent species); I cannot but think the *whole* world must have been rather colder during the Glacial Epoch: (I know I ought to be able to show that the glacial action was *actually* & *absolutely* coincident in North & South, & this I cannot do, nor can I here enter in details to show *how far* I can show them coincident)[6]

At this period I look at the intertropical plants as somewhat distressed, but not (or only a few) exterminated.— Under these conditions I consider it probable that some of the warmer temperate plants would spread into the Tropics, whilst the arctic plants reached the foot of the Alps & Pyrenees. (according to poor Forbes' view; by the way I had this part of the theory *written out.*, 4 years before Forbes published!)[7]

Some, I consider it possible might cross the Tropics & survive at C. of Good Hope, T. del Fuego & S. Australia; but within the Tropics, when warmth returned, all would be exterminated, except such as crawled up mountains, as in Ceylon, Neilgherries, Java, Organ mountains in Brazil. This theory, I conceive, explains certain aquatic productions in S. hemisphere &c &c. (& European Fish at C. of Good Hope)— But on the view that species change, it throws, I think, far more light on the analogous, but not identical species, on the summits of the above named mountains. Of course I cannot enter in details (& you would not care to hear them) on this subject, which I am sure in some degree would render the view more probable than it will seem to you at first.—

You will probably object, why have so many more Northern species & forms gone to the south, than southern forms come to the north; I can explain this only on a pure hypothesis of cold having come on first from the north; but there has been *some* migration from south to north, as of Australian forms on *Mountains* of Borneo. And I am sure I have notes of a few S. African forms, as wanderers across the Tropics, into N. Africa & Europe: is not this so with Gladiolus, Stapelia(?). Can you help me in this, either identical species, or allied forms, of well marked S. African forms? By the way I look at Abyssinia, *during the cold period*, as the channel of communication; for some, (as I know from Richard) very northern temperate species of plants are found there; & some S. African forms likewise.—[8]

There, I am sure, you will agree that I have prosed enough on my own doctrines; which I may have to give up, but I strongly suspect that the theory is a sound vessel & will hold water. I look at the vegetation of the Tropics, during the cold period, as having been somewhat like the vegetation described by Hooker at foot of Himmalaya, as essentially Tropical, but with an odd mixture of Temperate forms & even identical species, before they became mostly modified.—

What will you say to such a dose of speculation! You will exclaim, "he is a pretty fellow to talk of caution"!—

Pray believe me | Your's very sincerely | Charles Darwin

If at any time you are inclined to write pray attack my doctrine.—[9]

With respect to diffusion of water plants in *very distant* regions, it seems, as far as my doctrine is concerned, sufficient answer that the same species of water plants in the same continent are very widely diffused, & whatever the means of diffusion may be, the same means w^d tend to carry them to the most distant parts during the cold period.— The same argument is applicable to the Glumaceæ to some extent; but Decandolle thinks that certain lowly organised phanerogams, which are very widely diffused (I forget whether he includes Glumaceæ, which I think some authors consider the highest of the monocots?) are diffused owing to such species having been *very anciently* created & therefore having had *more time* to become diffused.[10] I doubt whether he has any grounds for his belief, without it be a very feeble analogy of the greater duration of mammifers compared with Molluscs.—

Suffolk Record Office, Bury St Edmunds

[1] Dated by the relationship to the letter from C. J. F. Bunbury, 16 April 1856.

[2] Letter from C. J. F. Bunbury, 16 April 1856.

[3] See letters from C. J. F. Bunbury, 7 February 1856 and 16 April 1856.

[4] Lowe 1856. See letter from R. T. Lowe, 12 April 1856. Richard Thomas Lowe was also at work on *A manual flora of Madeira* (Lowe 1857[-72]).

[5] In 1856, CD began experiments on the transport of plants, seeds, and ova by birds. Experiments on the germination of seeds from pond mud were commenced on 6 April 1856 (DAR 157a).

[6] CD gave a fuller discussion of this theory in *Natural selection*, pp. 534–54, and *Origin*, pp. 365–82.

[7] Edward Forbes published this view in E. Forbes 1846. CD gave a similar explanation in the 1842 sketch of his species theory, expanded in the essay of 1844 (*Foundations*, pp. 31 and 165–9). For a discussion of the views of Forbes and CD on Arctic–alpine floras, see J. Browne 1983, pp. 117–27.

[8] Richard [1847].

[9] No further letters from Bunbury about CD's theories in 1856 have been found, but Bunbury did record the substance of a conversation on species on 20 June 1856, when CD visited Bunbury in London (F. J. Bunbury ed. 1891–3, *Middle Life* 2: 413–14). With reference to the topics discussed in this letter, Bunbury wrote (p. 414):

> He spoke of his theory (which he had before mentioned to me by letter) that an interchange of plants, to a certain degree, may have taken place between Europe and the Cape of Good Hope, during the glacial period, when a much colder climate than the present existed beyond certain parallels of latitude, in both hemispheres. In accordance with his doctrine of the mutability of species, he supposes that the representative species in either hemisphere (*e.g.* the species of Dianthus at the Cape, and those of Gladiolus in Europe), may be modified forms originating from the opposite hemisphere in which their respective genera have their head quarters. His chief difficulty is, that there are so many more representatives of northern forms at the Cape than representatives of southern forms in Europe; as if the migration had taken place chiefly from north to south.

[10] A. de Candolle 1855, 1: 499–500 and 604.

To Charles Lyell 21 April [1856][1]

Down Bromley Kent
April 21st

My dear Lyell

Your case seems to me most perplexing;[2] I have never seen or heard of anything like it, & have been puzzling my brains quite in vain. One thing I quite see & agree with you, that it is a *strong* argument against the angle of (say) 12° having been produced by upheavement, for it is manifestly impossible that the columns in a flat field of lava could have been all formed at angles inclined from the centre of the island, so as to become vertical when the beds were upturned 12°. The cause must (I sh[d] think) be connected with the lava not having absolutely ceased its downward tendency when the shrinking which produced the columns & fissures was first superinduced.

Would it not be worth while to ask Hopkins or some such man,[3] whether in a body of very viscid matter just before all motion ceased, whether there could be any vertical sliding movement in closely approximate planes **tending** to make (not really making) steps like these;

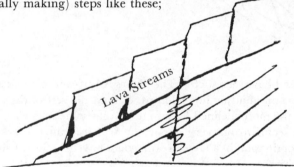

for if so I can well fancy such tendency might have *guided* the planes of division of the columns.— I know I have seen some columnar lava with columns directed in manner for which I could conceive no cause, but I cannot remember where.—

I have seen **very** narrow, **very** convex streams of lava, (which must have been very viscid) with the proper radiating columnar structure beautifuly displayed.—

Have I not read of Glaciers, coming down a very steep slope, breaking up into vertical steps?

Do make the *local* areas crowded with parallel dikes a prominent feature, though I daresay you are very wise & safe in allowing some central elevation of the whole. What a deal of interesting work your Madeira expedition has given you![4]

Yours most truly | Ch. Darwin

I supposed you observed your facts in Madeira & not from drawings alone.— I have been drawing the angles on paper & in *thin* beds of lava, I sh^d have thought the difference w^d not have been conspicuous; in *thick* beds it undoubtedly would

Has not Forbes,[5] or Hopkins or some one given figure of some very viscid substance flowing down slope, with plane of division, something like this? —

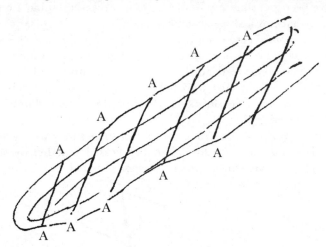

I have some *vague* idea I have seen such— The planes AA, if ever so obscure, might well guide the formation of adjoining columns, & then the rest of the columns.

The more I think the more inclined I am to think that some such view must be cause of your, as it seems to me very important & extraordinary fact.— The longitudinal planes of division in a lava-stream, could have no effect, I sh^d think, on the dip of the columns.—

American Philosophical Society (126)

[1] Dated by the subject of the letter, a topic that was discussed during Lyell's visit to Down House, 13–16 April 1856 (see n. 2, below).

[2] Lyell had recently visited CD at Down House (Wilson ed. 1970, p. xliii). An unpublished letter from Lyell to Georg Hartung, dated 11 and 15 April 1856, indicates that Lyell and CD had discussed the geology of Madeira in great detail, with particular reference to the angle at which molten lava could solidify (the editors thank Lady Lyell and Leonard G. Wilson for this information). The point was crucial in the debate, at that time still continuing, between Lyell and Christian Leopold von Buch and Jean Baptiste Armand Louis Léonce Élie de Beaumont. Lyell held the view that volcanic cones were built up gradually, layer by layer, through the accumulated outpourings of lava, a theory proposed in detail in C. Lyell 1850; Leopold von Buch considered that lava could not solidify on slopes and that the beds must at first have been horizontal and only subsequently elevated and inclined. CD had discussed Lyell's argument at length in 1849 (see *Correspondence* vol. 4, letter to Charles Lyell, [1 November 1849], and subsequent letters).

[3] William Hopkins was an expert in mathematical geology (C. Smith 1989).

[4] Charles and Mary Elizabeth Lyell, accompanied by Frances Joanna and Charles James Fox Bunbury, visited Madeira and the Canary Islands in the winter of 1853–4 to study the volcanic

phenomena of the islands. Some of the results of Lyell's observations were incorporated into the fifth edition of his *Manual* (C. Lyell 1855, pp. 515–22). The paper on the volcanic geology of Madeira and the Canaries, which Lyell and Georg Hartung were preparing, was ultimately abandoned. Lyell discussed the question, in part, in C. Lyell 1858. Hartung later published, with Lyell's permission, the results of their work (Hartung 1864).

5 James David Forbes had analysed the physical properties of glaciers in 1845 and considered that they moved primarily though acting as a viscous mass. CD had corresponded with him about the similarity of movement in glaciers and lava (see *Correspondence* vol. 3, letters to J. D. Forbes, 11 October [1844] and [November? 1844]). Forbes was at that time engaged in a bitter controversy with William Hopkins, who had opposed the viscous theory with his own theory of motion based on sliding. CD's reference in the letter indicates that he meant Forbes's work rather than Hopkins's (C. Smith 1989). There are several diagrams similar to the one in this letter in J. D. Forbes 1845, pp. 368–436.

To Godfrey Wedgwood 21 April [1856][1]

Down Bromley Kent
Ap. 21

My dear Godfrey

I am extremely much obliged to you for all the trouble which you have taken about the Rabbits & for your clear Report. I fear the case has broken down, except so far as knowing that such a breed has run wild for some years at Sandon.[2] The only chance of hearing further particulars would be from the agent of the Estate or any old farmer, & such might know that the breed has been there for at least some given number of years. But perhaps it is not worth taking more trouble about; unless you sh^d come across any one residing at or near Sandon.—

I am very much obliged to you for all your trouble—

Believe me | Yours affectionately | Charles Darwin

P.S. | I have omitted most important part, viz that it is not worth sending the bodies, as the history of the Breed is so obscure.—

I. J. Pincus

[1] Dated from CD's interest in domesticated animals and their reversion to the wild state.

[2] CD was interested in comparing feral breeds with their domesticated ancestors to see the extent to which adaptation to the wild had caused changes. It was generally thought that the wild conditions would cause reversion to the primitive state, a view that CD thought was unproved (see *Origin*, pp. 14–15, and *Variation* 1: 111–15). CD mentioned the case of the half-wild rabbits of Sandon Park (about ten miles south of the Wedgwood potteries at Etruria, Staffordshire) in *Variation*. He later received further information from a gamekeeper at Sandon Park, who told him that the rabbits had been wild for a considerable time and had originated 'from variously-coloured domestic rabbits which had been turned out.' (*Variation* 1: 122 and n. 26).

To Edward Sabine 23 April [1856][1]

Down Bromley Kent
Ap. 23^d

My dear Colonel Sabine

As I probably shall not get to the speech of you today, I write to ask you what you think of Sir John Richardson, which occurred to Hooker (who is staying here) & myself whilst considering the Royal medals.— It strikes us a most appropriate award considering the high merit of the Fauna Boreali-Americana & of his late work on Fishes.[2] Several of his later works, as the Zoology of the Voyage of the Herald come within the 10 years.[3] His whole career will, also, tell forcibly.—

Forgive me for suggesting, that if you would propose him, I sh^d be proud to second him.—[4]

As I shall be soon off the Council, & as you take much interest in Royal Soc. permit me just to mention a few other names, which after much consideration, seem to me eminently well qualified for the Royal medal: viz (according to Hooker, & even what little I know) Bentham,[5] Prestwich for his admirable Tertiary Geology[6]—& Albany Hancock for Zoology;[7] with respect to latter, Huxley was inclined to think that he had higher claims than Westwood.—[8] By the way from remarks made by foreigners, I am quite easy that the award to Westwood was a sound one:[9] I got frightened about it at one time & vowed to myself that I never would interfere again, & now I want to do so only indirectly.

Your's very sincerely | Ch. Darwin

Royal Society (Sa: 387)

[1] Dated by the reference to Joseph Dalton Hooker being at Down House. Hooker and his wife were guests together with Thomas Henry Huxley and his wife and Thomas Vernon Wollaston (Emma Darwin's diary and letter to J. D. Hooker, 8 April [1856]).

[2] J. Richardson 1829–37 and 1848.

[3] J. Richardson 1852. The Royal Medal was awarded to authors for works published within the preceding ten years.

[4] In the event, CD nominated John Richardson and Sabine seconded the nomination (Royal Society council minutes). Richardson was awarded a Royal Medal at the annual meeting of the Royal Society, 1 December 1856, for 'his contributions to Natural History and Physical Geography'. CD was the author of the announcement of the award, published in the *Proceedings of the Royal Society of London* 8 (1856–7): 257–8.

[5] George Bentham was awarded the Royal Medal in 1859. He was elected FRS in 1862.

[6] Joseph Prestwich was recognised as an expert on the Tertiary geology of Britain through his papers on the Coalbrookdale coalfield in Shropshire (Prestwich 1840) and his work on the stratigraphy of the beds underlying London (Prestwich 1854). He received the Royal Medal in 1865.

[7] Albany Hancock was awarded the Royal Medal in 1857. For CD's earlier suggestion of Hancock's name, see letter to J. D. Hooker, 8 April [1856].

[8] See *Correspondence* vol. 5, letters to T. H. Huxley, 31 March [1855] and 18 April [1855]. John Obadiah Westwood had been awarded the Royal Medal in 1855, having been nominated by CD and seconded by Huxley.

[9] See letter to H. T. Stainton, 13 April [1856]. In his 12 April 1856 editorial in the *Entomologist's Weekly Intelligencer*, Henry Tibbats Stainton had stated that an eminent German entomologist remarked 'that by giving the Royal Medal to Mr. Westwood the honour had been conferred on the [Royal] Society', implying that London scientific circles had been slower to recognise the value of Westwood's work than had some Europeans ([Stainton] 1856, p. 9).

To John Lubbock 24 April [1856][1]

Down
April 24th

Dear Lubbock

Before I come to the purpose of my note, let me give you my very sincere congratulations on your marriage;[2] & I do not believe that any of your many friends can send you more cordial & true good wishes.

Can you get "leave of absence" & come & dine here on Saturday at 7 oclock to meet D.^r & Mrs Hooker, Mr Wollaston, M.^r & Mrs Huxley.—[3]

Mrs Darwin had intended (but is not at present quite well)[4] to have ventured on a very early call on Mrs Lubbock in hopes of persuading her to come with you, but I fear there would have been but very little chance of the party at High Elms[5] letting her leave home so soon.—

Dear Lubbock | Yours very sincerely | Ch Darwin

Down House

[1] Dated by the reference to Lubbock's marriage (see n. 2, below).
[2] Lubbock had married Ellen Frances Hordern, only daughter of the late Peter Hordern of Chorlton-cum-Hardy, at Rostherne, Cheshire, on 10 April 1856 (*Gentleman's Magazine* (1856) pt 1: 640).
[3] Lubbock dined at Down House with CD and his visitors on 26 April (Emma Darwin's diary).
[4] On 17 April, Emma Darwin recorded in her diary: 'I began to be bad.' Charles Waring Darwin was born on 6 December 1856.
[5] The Lubbock family residence, a short distance from Down House.

To W. B. Tegetmeier 25 April [1856][1]

Down Bromley Kent
Ap. 25th.—

My dear Sir

You could not possibly have done me a greater kindness than in bespeaking for me the a pair of Indian laughing Pigeons.—[2] And as you mention that they were brought by a Mr Wooler,[3] I presume that you authentically know that they did really come from India.— I shall be, also, very much obliged if you will bear me in mind in regard to Carriers (or any Cocks.) at Stevens.—[4]

Upon my life you let me trespass **too much** on your kindness, all Fanciers, I find sell their *choice* eggs, & why not sell me a sitting of your Polish;[5] but if you will not, I should be very glad of 2 or 3 eggs, that I might have opportunity of seeing them alive; but I foresee that you will work out the Poultry so well, that I shall be gladly saved the trouble. Nevertheless, I intend to keep a few of each sort of Fowls, that I may better appreciate their differences.—

The very next time I write to my Bookseller I shall order your Poultry Book,[6] & much I am sure I shall find in it, very interesting to me. I am not very well today so no more.—

Your's very sincerely | Ch. Darwin

P.S | About sending the eggs, the best way would be to send them by Deliverance Co? (or if not safe, by hand, & I can repay you) addressed exactly as follows.

C. Darwin Esq.
care of G. Snow
Nag's Head
Borough.

Mr Snow is our Carrier & leaves the Nag's Head every Thursday *morning*. This wd. be safer than per Coach to Bromley.—

New York Botanical Garden Library (Charles Finney Cox collection)

[1] Dated by the reference to Tegetmeier ed. 1856–7 (see n. 6, below).
[2] Indian laughing pigeons are not mentioned in *Variation*. CD stated that he kept two varieties of laughers, both of which came from Arabia (*Variation* 1: 155).
[3] Possibly W. A. Wooler, who provided information for CD on Himalayan rabbits (see *Variation* 1: 109).
[4] See letter to W. B. Tegetmeier, 28 [September 1856], in which CD stated that carriers were the only variety of pigeons he had not yet examined.
[5] See letter to W. B. Tegetmeier, 20 March [1856], n. 8.
[6] The first number of Tegetmeier's *Poultry book* was published in May 1856. This and the following ten parts of Tegetmeier ed. 1856–7 (after which publication was terminated) are in the Darwin Library–CUL. CD did not actually acquire parts of the work until 1857 (see letter to W. B. Tegetmeier, 18 May [1857]).

To Miss Holland[1] [May 1856][2]

[Down]

My dear Miss Holland

Fortunately for my entomological credit, a first-rate Entomologist has been staying with me, to whom I showed the pupa which you sent me, & he says it would turn into one of the Lackey moths, probably the Eriogaster lanestris.—[3] Last summer moths & butterflies abounded in an unprecedented degree, & entomologists attribute this abundance, I believe rightly, to the great destruction of birds during the winter previous to that just passed;[4] & the scarcity of birds saved many caterpillars which otherwise would have been devoured, & hence the numerous cocoons on your Hawthorn Hedges.—

Emma demands the rest of this note,[5] so pray believe me, dear Miss Holland, yours very sincerely | Charles Darwin

American Philosophical Society (Getz 1861)

[1] Miss Holland has not been identified. The Darwins and Wedgwoods were related to the Holland family.

[2] Dated by the reference to Thomas Vernon Wollaston, who visited the Darwins from 25 to 28 April 1856 (Emma Darwin's diary).

[3] See n. 2, above. In the manuscript, 'Mr Wollaston' was interlined in a different hand, possibly Emma Darwin's, after 'first-rate Entomologist' and, in the same hand, 'Eriogaster lanestris' was written above CD's text.

[4] The weather in January and February 1855 had been particularly severe (see *Correspondence* vol. 5, letter to W. D. Fox, 19 March [1855] and n. 7). CD refers to the large-scale destruction of birds in his garden during that winter in *Origin*, p. 68.

[5] Emma Darwin's note has been excised.

From Charles Lyell 1–2 May 1856

May 1. 1856.

My dear Darwin

As I sent you a list of the land-shells picked up at Down when we were last there[1] I wish you, in case you have kept it to put ?? to Helix carthusiana which I still think is among them— But the shell which I meant by that name is I find called by Forbes & Hanley Helix Cantiana Mont. as being the prior name—[2] In my own list & perhaps in the one I sent you I find I had added this name of Montagues[3] as a synonym—, for I had made no mistake about the species which I meant—

I also find amongst them another shell which I suppose to be Helix rufescens? and the Bulimus which I left unnamed must I think be B. obscurus— The Clausilia which I called C. plicatula ought to have been C. nigricans, according to Forbes & Hanley.

I have met with such a remarkably conical variety of Helix aspersa from Charing Kent, in Woodward collection[4] that I cannot regard that shell as quite so constant in England as I told you—& as it is reputed to be. I sh^d like to show it to you as I have got this variety

I have just heard from Woodward that his friend Mr C. Prentice of Cheltenham caught that large & most powerful of our water beetles Hydrobius piceus with an ancylus fluviatilis adhering to him! & treating him as he would a stone after the beetle was out of the water.—[5] Here is a new light as to the way by which these sedentary mollusks may get transported from one river basin to another— That species of Ancylus seems to have got into Madeira before Man—[6] How far can an Hydrobius fly with a favourable gale?

I hear that when you & Hooker & Huxley & Wollaston got together you made light of all species & grew more & more unorthodox—[7]

Heer of Zurich has given us a capital essay on the old Atlantis in a new paper just out on the fossil leaves of Madeira—[8] When I go abroad I will lend it to you for now I am always using it.[9] He makes out a good case in favour of the old union provided one believes in specific centres. According to any other hypothesis I cannot as yet very well see how to bring the geograph! facts to bear one way or the other— I wish you would publish some small fragment of your data *pigeons* if you please & so out with the theory & let it take date—& be cited—& understood.[10]

With my love to Mrs Darwin & the children ever truly Y | Cha Lyell

To encourage you to get up J. E. Gray's subdivision of the Genus Helix I may mention that Macandrew brought back from the Salvages a Helix allied to (some think only a var of) Helix pisana a British species[11] it was named by some one H. Macandrei & by Lowe who had received it first H. ustulata—[12]

Gray enters into his catalogue

> Nanina ustulata Salvages
> N—— Macandrei d°—

making two species, & on being asked why he called them Nanina said, because it is the form which belongs to islands in the *Pacific*! He had imagined the Salvages to be in the Pacific otherwise he should have certainly had some other subgenus assigned to it—! some Atlantic genus of Helicidæ![13]

May 2. This letter was not posted yesterday by accident. I went to see poor Curtis the entomologist who being nearly blind by amaurosis was run over 10 days ago by three Cabs & broke a shoulder bone but is doing surprisingly well.[14]

He told me that a french entomologist brought him a Dytiscus marginalis the large water beetle (which flies about freely) with a small bag of eggs of a water spider under his wings, evidently put there by the parent. It was not a parasitic insect which had done this— He could not look up the case because of his illness & blindness but some day I will get it perhaps—

What unexpected means of migration will in time be found out—[15]

The multiple creation of Agassiz will one day rank with spontaneous generation but Madeira seems to me to favour the single specific birth-place theory & I long to see your application of any modification of the Lamarckian species-making modification system— Representative forms in closely adjoining islands are great puzzles where the conditions are one would think so nearly the same—[16]

ever truly yrs | Cha Lyell

C Darwin Esq

DAR 205.3 (Letters)

CD ANNOTATIONS
0.1 May . . . got this variety 3.4] *crossed pencil*
4.6 How far . . . gale? 4.7] 'I have taken Colymbeste 45 miles fr Land' *added pencil*
5.1 I hear . . . unorthodox— 5.2] *crossed pencil*
6.1 Heer . . . Madeira— 6.2] *scored pencil*
6.2 When I . . . understood. 6.7] *crossed pencil*
10.1 May 2. . . . well. 10.3] *crossed pencil*
13.1 The multiple . . . Lyell 14.1] *crossed pencil*
On first page: '18[17] & Salvages Helix' *brown crayon*

[1] The Lyells had visited the Darwins from 13 to 16 April 1856 (Emma Darwin's diary). Lyell's list of shells has not been found.
[2] E. Forbes and Hanley [1848–] 1853.
[3] George Montagu was the author of *Testacea Britannica* (Montagu [1803–8]).

[4] The personal collection of Samuel Pickworth Woodward.

[5] See also Wilson ed. 1970, p. 83. There is a note on this in DAR 205.2: 139.

[6] Thomas Vernon Wollaston had given Lyell a manuscript 'Catalogue of Madeira & Porto Santo land Mollusca, living and fossil' in February 1856. *Ancylus fluviatilis* was included among the living species. See Wilson ed. 1970, pp. 19–27.

[7] Joseph Dalton Hooker, Thomas Henry Huxley, and Wollaston had visited CD during the last week of April (see letter to T. H. Huxley, 2 April [1856], n. 2). Lyell had evidently heard about the weekend discussions from Hooker and Huxley, for he recorded other remarks from them about species in his 'Scientific journal' (Wilson ed. 1970, pp. 56–7). In a letter of 30 April 1856 to Charles James Fox Bunbury, Lyell stated: 'When Huxley, Hooker, and Wollaston were at Darwin's last week, they (all four of them) ran a tilt against species farther I believe than they are deliberately prepared to go. Wollaston least unorthodox. I cannot easily see how they can go so far, and not embrace the whole Lamarckian doctrine.' (K. M. Lyell ed. 1881, 2: 212). CD made notes, dated 28 April 1856, on his conversation with Wollaston. These are preserved in DAR 197.2, together with further notes dated 8 May 1856. In a letter to Lyell dated 12 May 1856 (F. J. Bunbury ed. 1891–3, *Middle life* 2: 391–2), Bunbury expressed surprise that Hooker was unorthodox and commented:

> Darwin goes much further in his belief of the variability of species, than I am disposed to do, but even he, I imagine, would not assert an *unlimited* range of variation: he would hardly, I conceive, maintain that a Moss may be modified into a Magnolia, or an oyster into an alderman; though he seems to hold that all the different forms of each natural group may have sprung from an original stock, even (for instance) that the Ericas of Europe and of the Cape may have had a common origin: which I am not disposed to believe.

[8] Oswald Heer, Swiss palaeontologist and botanist, had visited Madeira for his health in 1850. He had sent Lyell his paper (Heer 1855) on the fossil plants of Madeira (see K. M. Lyell ed. 1881, 2: 210–11). To account for the similarity in flora between Madeira and Europe, Heer had suggested that Madeira was at one time connected to North America and at a later date joined to Europe. Lyell also entertained the possibility that a land-bridge had previously connected Madeira and Europe (see K. M. Lyell ed. 1881, 2: 212, and Wilson ed. 1970, pp. 92–3).

[9] The Lyells left for a tour of the Continent late in July. CD recorded having read Heer 1855 on 7 August 1856 (*Correspondence* vol. 4, Appendix IV, 128: 20).

[10] During Lyell's visit to Down in April, CD had told Lyell about his theory of natural selection as a mechanism for species change. Although Lyell had long known about CD's investigation of the species question, there is no evidence that CD had communicated the substance of natural selection before this date. Lyell's notes on this conversation, dated 16 April 1856 and headed 'With Darwin: On the Formation of Species by Natural Selection', cited, among other things, the example of pigeons. He wrote: 'The young pigeons are more of the normal type than the old of each variety. Embryology, therefore, leads to the opinion that you get nearer the type in going nearer to the foetal archetype & in like manner in Time we may get back nearer to the archetype of each genus & family & class.' (Wilson ed. 1970, p. 54). In a letter to Bunbury, 30 April 1856 (see n. 7, above), Lyell again discussed pigeons in connection with CD's transmutation theory: 'Darwin finds, among his fifteen varieties of the common pigeon, three good genera and about fifteen good species according to the received mode of species and genus-making of the best ornithologists, and the bony skeleton varying with the rest!' (K. M. Lyell ed. 1881, 2: 213).

[11] Robert McAndrew, Liverpool merchant and yachtsman, had joined the dredging committee of the British Association for the Advancement of Science in 1844. He had spent a month in the spring of 1852 collecting molluscs in the Canary Islands and the Madeira group, including the Salvages, later described in McAndrew 1854. CD's annotated copy of a reprint of this paper is in the Darwin Pamphlet Collection–CUL.

[12] Richard Thomas Lowe.

[13] In his published catalogue of Pulmonata, John Edward Gray, keeper of the zoological department of the British Museum, listed *Nanina MacAndrewiana* as a species from the Pacific islands (J. E. Gray

1855, p. 125). He also mistakenly gave its habitat as the 'Great Savage Island'. His species descriptions apparently were not well trusted by contemporary naturalists (see Gunther 1975, pp. 178–9).

14 John Curtis.

15 CD and Lyell had discussed the topic of migration and the various means of dispersal during Lyell's visit to Down in April (see Wilson ed. 1970, pp. 52–3). On his return to London, Lyell began his own experiments on the power of seeds to endure salt water, evidently intending to verify CD's work in relation to the diffusion of plants (see F. J. Bunbury ed. 1891–3, *Middle life* 2: 389–91).

16 A reference to Lyell and CD's discussion of the distinctness of the fauna on individual islets in archipelagos such as the Madeira group and the Galápagos. In his notes relating to this discussion, Lyell recorded: 'In the Galapagos isles the land shells collected were not numerous, but so far as the evidence went, it corroborates the Mad.ª & P.º S.º case in regard to the distinctness of the fauna in each island.' (Wilson ed. 1970, p. 52)

17 The number of CD's portfolio of notes on the means of geographical dispersal of animals and plants.

To Asa Gray 2 May [1856][1]

Down Bromley Kent
May 2ᵈ—

My dear Dᵣ Gray

I have received your very kind note of Ap. 8ᵗʰ [2] In truth it is preposterous in me to give you hints;[3] but it will give me real pleasure to write to you just as I talk to Hooker, who says my questions are sometimes suggestive owing to my comparing the ranges &c in different Kingdoms of Nature. I will make no further apologies about my presumption; but will just tell you (though I am *certain* there will be **very** little new in what I suggest & ask). the points on which I am very anxious to hear about.— I forget whether you include Arctic America, but if so for comparison with *other parts of world*, I would exclude the Arctic & Alpine-arctic, as belonging to a quite distinct category.[4] When excluding the naturalised, I think Decandolle must be right in advising the exclusion, giving list, of plants exclusively found in cultivated land, even when it is not known that they have been introduced by man.—[5] I would give list, of temperate plants (if any) found in Eastern Asia, China & Japan, & not elsewhere.— Nothing would give me a better idea of Flora of U.S. than the proportion of the genera to all the genera, which are confined to America; & the proportion of genera, confined to America & Eastern Asia with Japan; the remaining genera wᵈ be common to America & Europe & rest of world; I presume it wᵈ be impossible to show any especial affinity in genera (if ever so few) between America & *Western* Europe.[6] America might be related to Eastern Asia (always excluding Arctic forms) by a genus having the *same* species confined to these two regions; or it might be related by the genus having different species, the genus itself not being found elsewhere. The relation of the genera (excluding identical species) seems to me a most important element in geographical distribution often ignored, & I presume of more difficult application in plants than in animals, owing to the wider ranges of plants; but I find in N. Zealand (from Hooker) that the consideration of genera with representative species tells the story of relationship even plainer than the identity of the species with different parts of the world.—[7]

I should like to see the genera of the U. States, say 500 (excluding Arctic & Alpine) divided into 3 classes, with the proportions given, thus,

$\frac{100}{500}$ American genera

$\frac{200}{500}$ Old World genera, but not having any identical species in common.—

$\frac{200}{500}$ old world genera, but having some identical species in common: supposing that these 200 genera included 600 U.S. plants; then the 600 would be the denominator to the fraction of the species in common to the Old World.—[8]

But I am running on at a foolish length.—

There is an interesting discussion in Decandolle (about p. 503 to 514) on the relation of the size of Families to the average range of the individual species; I cannot but think from some facts which I collected long before De Candolle appeared,[9] that he is on wrong scent in having taken Families (owing to their including too great a diversity in the constitution of the species), but that if he had taken Genera, he would have found that the individual species in large genera range over a greater area than do the species in small genera:[10] I think if you have materials that this would be *well* worth working out; for it is a very singular relation.—[11]

With respect to naturalised plants; are any *social* with you, which are not so in their parent country?[12] I am surprised that the importance of this, has not more struck Decandolle. Of these naturalised plants are any or many more variable in your opinion, than the average of your U.S. plants: I am aware how very vague this must be: but De Candolle has stated that the naturalised plants do not present varieties;[13] but being very variable & presenting distinct varieties seems to me rather a different case: if you would kindly take the trouble to answer this question, I sh^d be very much obliged, whether or no, you will enter on such points in your Essay.—

With respect to such plants, which have their southern limits within your area, are the individuals ever or often stunted in their growth or unhealthy: I have in vain endeavoured to find any Botanist who has observed this point; but I have seen some remarks by Barton on the *trees* in U.S.[14] Trees seem in this respect to behave rather differently from other plants.—

It would be a very curious point, but I fear you would think it out of your Essay, to compare the list of European plants in Tierra del Fuego (in Hooker)[15] with those in N. America; for without multiple creation, I think we must admit that all now in T. del Fuego, must have travelled through N. America, & so far they do concern you.—

The discussion on Social plants (vague as the term & facts are) in De candolle strikes me as the best, which I have ever seen:[16] two points strike me as eminently remarkable in them; that they should ever be social close to their extreme limits; & secondly that species having an *extremely* confined range, yet sh^d be social where they do occur: I sh^d be infinitely obliged for any cases *either by letter* or publickly on these heads, more especially in regard to a species remaining *or ceasing* to be social, on the confines of its range.

There is one other point, on which I individually sh^d be extremely much obliged, if you could spare the time to think a little bit & inform me; viz whether there are any cases of the same species being more variable in U.S. than in other countries in which it is found: or in different parts of U.S. Wahlenberg says generally that the same species in going S. becomes more variable than in extreme north.[17] Even still more am I anxious to know whether any of the genera, which have most of their species horribly variable (as Rubus or Hieracium &c) in Europe, or other part of world, are less variable in U. States: or the reverse case, whether you have any odious genera with you, which are less odious in other countries. Any information on this head, would be a real kindness to me.—

I suppose your Flora is too great; but a simple list in *close columns* in *small type* of all the species, genera, & Fam^s, each consecutively numbered, has always struck me as most useful; & Hooker regrets that he did not give such list in introduction to N. Zealand & other Flora.—

I am sure I have given you a larger dose of questions than you bargained for, & I have kept my word & treated you just as I do Hooker. Nevertheless if anything occurs to me during the next two months, I will write freely, believing that you will forgive me & not think me very presumptuous.

My dear Dr Gray | Your's very sincerely | Charles Darwin

I have reread this letter & it really is not worth sending, *except for my own sake*.: I see I forgot in beginning to state that it appeared to me that the 6 heads of your Essay included almost every point which could be desired, & therefore that I had little to say.—

How well De Candolle shows the necessity of comparing nearly equal areas for proportion of Families![18]

(Excuse this poor paper I am run short)

Gray Herbarium of Harvard University

[1] Dated by the suggestions CD makes in this letter, which were incorporated into A. Gray 1856–7 (see n. 3, below).

[2] Gray's letter has not been found. CD's previous letter to Gray had been written in August 1855 (see *Correspondence* vol. 5, letter to Asa Gray, 24 August [1855]).

[3] Gray was preparing the first part of his paper 'Statistics of the flora of the northern United States' (A. Gray 1856–7). In the published version, Gray acknowledged, without mentioning his name, CD's contribution to the work: 'While engaged in the preparation of a second edition of the *Manual of the Botany of the Northern United States*, I was requested by an esteemed correspondent, upon whose judgment I place great reliance, to exhibit, in a compendious and convenient form, the elements of flora I was occupied with.' (A. Gray 1856–7, p. 204).

[4] Gray generally followed this recommendation and devoted a separate section to a discussion of the alpine and sub-alpine flora of the United States (A. Gray 1856–7, pp. 62–76).

[5] A. de Candolle 1855, 2: 642–5, 703–4. Gray agreed with this point, stating that the admission of introduced species into the comparison 'seriously vitiates our conclusions' (A. Gray 1856–7, p. 212).

[6] This information is tabulated in A. Gray 1856–7, pp. 215, 217–24, and 226–9.

[7] Gray attempted a 'Comparison of the flora of the northern United States with that of Europe in respect to the similar or related species', in spite of the difficulties such an enterprise entailed 'owing to the impossibility of estimating the degrees of resemblance among species, or at least of expressing them in any precise or definite way, or bringing shades of difference to any common standard.' (A. Gray 1856–7, p. 78).

[8] Gray stated that he found this comparison problematic. He therefore settled on taking species rather than genera for his comparison and did not attempt 'anything beyond an enumeration' of species in the three classes (A. Gray 1856–7, pp. 80–3).

[9] A. de Candolle 1855.

[10] Bound in CD's copy of A. de Candolle 1855, 1: pp. 528–9, is a note, dated 1 May 1856, made by CD when studying Alphonse de Candolle's discussion of the 'area of species according to families' (A. de Candolle 1855, 1: 501–32). After working through his own calculations made on data for the Cape flora of South Africa, CD concluded by stating: 'So that nothing can be inferred safely from these results. Families ['Genera' *del*] being too large.—' See J. Browne 1980.

[11] Gray followed this suggestion explicitly. After mentioning that Candolle's calculations were founded upon families, he stated: 'To be of any value, at least upon our limited scale, the comparison should be made with genera, as Mr. Darwin suggests; and from some investigations of his own, this sagacious naturalist inclines to think that species in large genera range over a wider area than the species of small genera do.' (A. Gray 1856–7, p. 77). For the most part, Gray's calculations tended to confirm CD's generalisation, particularly when the woody plants were examined, although Gray noted that the converse of the proposition 'does not tell in the same way' (A. Gray 1856–7, p. 381).

[12] Gray addressed this question only indirectly. As he stated, 'properly to discuss this and kindred topics would require a great amount of detailed investigation, and would expand these articles into a treatise.' (A. Gray 1856–7, p. 389). CD had explained his interest in this point in his letter to J. D. Hooker, 14 November [1855] (*Correspondence* vol. 5).

[13] A. de Candolle 1855, 2: 607–8. The passage is marked in CD's copy in the Darwin Library–CUL.

[14] Probably Barton 1809.

[15] European plants in Tierra del Fuego are not separately listed in *Flora Antarctica* (Hooker 1844–7), but presumably CD hoped Gray would extract them from the Fuegian flora details.

[16] A. de Candolle 1855, 1: 457–73.

[17] Wahlenberg 1824–6 on which CD commented in his reading notebook: 'most curious passage about species not varying in North.' (*Correspondence* vol. 4, Appendix IV, *128: 169).

[18] A. de Candolle 1855, 2: 1154.

From Samuel Pickworth Woodward 2 May 1856

40. U. Park St! Barnsbury
May 2 1856.

Dear Sir

Your question respecting the relative value of species in times past & present, can only, I fear, be answered hypothetically—[1]

1. *The proportion of species to genera* was only 7 to 1 in the Silurian Period; it rose to 14 in the Jurassic, & 16 in the Eocene Tertiary. At the present day there are 40 sp. to each genus of *shells*, on the average, according to my estimate (p. 417)[2] Now as shells are very simple contrivances presenting few points of comparison, (for in this inquiry we must set aside *colour*) the difficulty of determining species must increase very rapidly with the number in each genus. There are 3 or 400 sp. of Conus & 4

or 500 Pleurotomas—so that any new species possessing the *essential* characters of either genus will probably run very close to some sorts already known— The number of *possible* variations of these types must be well-nigh exhausted.

It does not follow from this that the species are *bad*—ie varieties of a common stock.

On the other hand, fossil shells present difficulties, on account of their *condition*, which usually more than make up for their comparative fewness.

When suites of fossils can be obtained in good condition—such as the Bayeux fossils & the U. Greensand sp. of Mans, which are like recent shells—they present just the same degrees of specific variation as their living representatives

In Deshayes Fossils of the Paris Basin about *half* the species are good, according to my notion of species.[3] In his catalogues of recent shells (*Veneridæ*, *Cycladidæ*, *Tellinidæ*) perhaps one fourth are good for something.[4]

The accurate Dr Louis Pfeiffer seems to have grown weary, or disgusted with the task of discriminating *Helicidæ* after describing **3000** species, & tossed in another odd 900 species sans discretion—perhaps to please Mr Cuming.[5]

Mr Gray says only $\frac{1}{10}$ of the reported species of shells are really distinct;[6] & you know the rest of the joke. Henceforth the definition of "**Species**" will be—"a little group of Genera".— "**Genus**; the smallest term in Zoology; something less than a variety".

2. *Your question may be answered by* appealing to Geographical Distribution. Dr Hooker says "the species in isolated islands are generally well-defined; this is in part the consequence of a law that genera in islands bear a large proportion to the species" (Antarctic Botany.)[7] I believe this to be true, & that the Island faunas are older (generally) than those of Continents. Still it cannot be denied that the *Helicidæ* of Madeira exhibit a high degree of specific variation— Even in the more ancient S! Helena have we not Bulimus auris-vulpinus & B. Darwini? And if you appeal to the Galapagos Ids I must almost admit the converse of the above proposition!

3. Lastly we may compare the relative value of species in Genera of different degrees of antiquity. The worst genus of all—*Clausilia* is at least of Eocene date— its 280 species are restricted to the Lusitanian Province, save a few which range through the Germanic to China.[8]

The *Terebratulæ* of the Oolites are not more variable than the *Terebratellæ* of the present seas— Mr Buckman's attempt to demolish some of them was founded on a comparison of young & undeveloped specimens—[9] Some of these have *internal* characters with which he was unacquainted. In every case the *period of maximum* development of a genus may be taken as the time at which the diagnosis of its species is most difficult—& the *entire fauna* of each period presents the same aspect of a combination of groups, some commencing, others culminating, the rest declining. It is no use pursuing this further, as I cannot give you a matter-of-fact answer. The truth is that the species of every group, country & time, cannot be distinguished *empirically*, but require judgment & experience—& this will be no new truth to you—

Pray when you can, let me know what points in my book I ought to re-consider!
Yrs sincerely | S. P. Woodward

DAR 181

CD ANNOTATIONS

1.1 Your . . . genus. 2.6] *crossed pencil*
2.6 There are . . . known—2.8] *scored pencil*
2.8 The number . . . proposition! 9.9] *crossed pencil*
5.1 When suites . . . representatives 5.3] *scored pencil*
10.1 3. Lastly] '5'[10] *added brown crayon*
11.1 The *Terebratulæ* . . . seas— 11.2] *double scored pencil*
11.5 may be . . . difficult— 11.6] *double scored pencil*

[1] CD's reply (letter to S. P. Woodward, 27 May 1856) makes it clear that CD had asked Woodward whether the variability of species within a particular genus was different at different periods.
[2] Woodward 1851–6, 3: 417, a table illustrating the 'Development of Families, Genera, and Species, in Time'. It is possible that Woodward enclosed an advance copy of the third part of his book on shells with this letter: the volume was received by CD before 15 May (see letter to S. P. Woodward, 15 May [1856]). CD also recorded having read the entire work on 5 June 1856 (*Correspondence* vol. 4, Appendix IV, 128: 18). In any case, Woodward could be confident that CD would have access to this reference in the very near future: the third volume was published in May (see letter to S. P. Woodward, 15 May [1856]).
[3] Deshayes 1824–37.
[4] Deshayes 1853–4.
[5] Ludwig Georg Karl Pfeiffer compiled catalogues of various Mollusca; the *Helices* are described in Pfeiffer 1848. Hugh Cuming had made one of the most extensive collections of shells of the time, which he gave to the British Museum. Pfeiffer and Gérard Paul Deshayes had both prepared catalogues for the British Museum of various families from Cuming's collection.
[6] John Edward Gray himself had a reputation for needlessly coining new specific and generic names (*DNB*). See also letter from Charles Lyell, 1–2 May 1856.
[7] J. D. Hooker 1844–7, 2: 217 n.
[8] Woodward divided the globe into twenty-seven land provinces and eighteen marine provinces. As he stated in Woodward 1851–6, 3: 382: 'The Land Provinces represented on the map are the principal Botanical Regions of Prof. Schouw, as given in the Physical Atlas of Berghaus'. He refers to Joachim Frederik Schouw's map of botanical distribution in Heinrich Karl Wilhelm Berghaus's *Physikalischer Atlas* (Berghaus 1845–8). For the marine provinces, Woodward followed Edward Forbes's work (E. Forbes 1846 and 1856).
[9] James Buckman studied the palaeontology and stratigraphy of the Jurassic series of the Cotswolds.
[10] The number of CD's portfolio of notes on variation.

From Laurence Edmondston [before 3 May 1856][1]

still they may not be decisive for as you may know, the vaunted conjugal fidelity of the ark bird has like most general rules had its exceptions[2]

I have no idea that the wild pigeons feed far from their breeding places, but believe that they remain— each tribe in the vicinity of their *coves*, & as to crossing to the Mainland of Scotland that is a feat of migration I should never give them credit for—[3]

They have always been very numerous on these islands, & always present the usual uniform colour of the Rock Dove—

I am delighted to hear you are occupied in the effort to throw light on the interesting & difficult subject of *Varieties* of Species, & I beg you will freely command me when you suppose I may be of use in your researches—

Your query as to drifted Trees I can answer in the affirmative—but instances of them are less frequent now than formerly—& for this obvious enough reasons present themselves—

M^r McGillivray[4] was one of my earliest & most respected scientific friends, &

AL incomplete
DAR 205.2 (Letters)

CD ANNOTATIONS
2.1 I have . . . for— 2.4] *crossed pencil*
4.1 I am delighted . . . researches— 4.3] *crossed pencil*
5.1 Your query . . . themselves— 5.3] 'Drifted Trees' *added brown crayon*; '18'[5] *added brown crayon*; 'Shetland Is^d' *added pencil*
6.1 M^r McGillivray . . . friends, &] *crossed pencil*
Bottom of last page: 'L. Edmondston' *pencil*

[1] The correspondent is identified by CD's annotation. The letter is dated by its relationship to the following letter.
[2] The dove, particularly the turtle-dove, has long been considered the ideal of conjugal love and fidelity.
[3] Edmondston resided on the Shetland Islands, to the north-west of Scotland.
[4] William Macgillivray, Scottish naturalist and author of *A history of British birds* (1837–52), had died in 1852. A copy of his work is in the Darwin Library–CUL.
[5] The number of CD's portfolio of notes on the means of geographical dispersal of plants and animals.

To Laurence Edmondston   3 May [1856][1]

Down Bromley Kent
May 3^d.

Dear Sir

I beg to thank very truly for your kind & very interesting answer to my queries.— The fact, which you communicate to me of a cock Rock Pigeon, having come to your Dove Cot & paired with a domestic Bird is of value to me.—[2] I had fancied from the several statements in poor M^r. Macgillvray's work vol I p. 378 etc,[3] that the taming of the Rock Dove was a much commoner event than it appears to be. Mr James Barclay certainly seems to believe that they have paired & bred in domestication.[4] If you should ever hear of any cases, I sh^d. be extremely much obliged if you would kindly take the trouble to inform me.

I am making a collection of skeletons of all the domestic kinds of Pigeons, but I have failed in getting a *real wild adult* Rock Dove. Should you think me very unreasonable to beg you to do me the great favour to send one, in strong paper, or light Box per post: but I ask this on one condition that you will permit me to return you the 3^s or 4^s worth of Stamps. This would be a *real assistance* for I would skin it & keep skin with wing & leg on one side, & make skeleton of the rest.— To make the favour quite perfect, I sh^d. like just the head of a second specimen, as I cannot make skeleton & skin of this part from one specimen.—[5] But I much fear that you will think me exorbitant in my request & that I trespass on your *very kind* offer of assistance.

I thank you for the information in regard to the drifted trees, it adds one more archipelago to my list.— If any fact should ever occur to you in regard *even to any very slight variation or difference or Habits* in regard to any of the domesticated animals of the Shetland Isl^{ds}. I sh^d. be most grateful for the information.— I sh^d. very much like to hear whether the bones of any large quadrupeds have ever been found **deep** in the peat of the Shetlands; for I suppose Peat is dug.—

You ask about myself; I have devoted my whole life to do what little I could for our favourite pursuit of Natural History, & I volunteered my Services on board H.M.S. Beagle in her circumnavigation, & did my best during our long voyage of five years, & published an account of it.—

With my cordial thanks, pray believe me, my dear Sir, Yours' truly obliged | Ch. Darwin

D. J. McL. Edmondston

[1] Dated on the basis of CD's interest in acquiring skeletons of pigeons and other birds.
[2] Cited in *Variation* 1: 185 n. 15.
[3] Macgillivray 1837–52.
[4] In *Variation* 1: 185 n. 15, CD stated: 'Mr. James Barclay and Mr. Smith of Uyea Sound, both say that the wild rock-pigeon can be easily tamed; and the former gentleman asserts that the tamed birds breed four times a year.'
[5] Edmondston sent CD two wild rock-doves from the Shetlands (*Variation* 1: 134 n. 5). A drawing of one of them is in *Variation* 1: 135. See also letter to L. D. Edmondston, 11 September [1856].

To Charles Lyell 3 May [1856][1]

Down Bromley Kent
May 3^d.

My dear Lyell

It was very very good of you to write me so long & very interesting a letter; but I wish you had mentioned whether you see any further into your very odd case of the vertical divisions of the lava-streams.—[2]

I have kept your list of the land-shells & made the corrections: I had no idea how wonderfully learned you were on the subject.— I shall be very glad to borrow Heer, when you go abroad.—[3] Your cases of possible transportal beat all that I have ever heard of; & if any body had put such cases hypothetically I sh^d. have laughed at them.[4] I have known Colymbetes fly on board Beagle 45 miles from land,[5] which, by the way, surprised Wollaston much. We had much to me most interesting conversation, when he & the others were here: Wollaston strikes me as quite a first-rate man & very nice & pleasant into the bargain. It is really striking (but almost laughable to me) to notice the change in Hookers & Huxley's opinions on species during the last few years.——[6]

With respect to your suggestion of a sketch of my view; I hardly know what to think, but will reflect on it; but it goes against my prejudices. To give a fair sketch would be absolutely impossible, for every proposition requires such an array of facts. If I were to do anything it could only refer to the main agency of change, selection,——& perhaps point out a very few of the leading features which countenance such a view, & some few of the main difficulties. But I do not know what to think: I rather hate the idea of writing for priority, yet I certainly sh^d be vexed if any one were to publish my doctrines before me.——[7] Anyhow I thank you heartily for your sympathy. I shall be in London next week,[8] & I will call on you on Thursday morning for one hour precisely so as not to lose much of your time & my own: but will you let me this one time come as early as 9 oclock, for I have much which I must do, & the morning is my strongest time.

Farewell | My dear old Patron | Yours | C. Darwin

By the way *three* plants have now come up out of the earth **perfectly** enclosed in the roots of the trees.——[9] And 29 plants in the table-spoon-full of mud out of little pond:[10] Hooker was surprised at this, & struck with it, when I showed him how much mud I had scraped off one Duck's feet.——

If I did publish a short sketch, where on earth should I publish it?

If I do *not* hear I shall understand that I may come from 9–10 on Thursday.——

American Philosophical Society (127)

[1] Dated by the relationship to the letter from Charles Lyell, 1–2 May 1856.

[2] See letter to Charles Lyell, 21 April [1856].

[3] Heer 1855.

[4] See letter from Charles Lyell, 1–2 May 1856.

[5] CD mentioned this case in *Origin*, p. 386. Lyell also recorded it in his 'scientific journal' (Wilson ed. 1970, p. 83).

[6] Thomas Vernon Wollaston, Joseph Dalton Hooker, and Thomas Henry Huxley had recently visited CD at Down House, where they 'made light of all Species & grew more & more unorthodox' (letter from Charles Lyell, 1–2 May 1856 and n. 7).

[7] CD further debated the pros and cons of publishing a preliminary outline of his theory with J. D. Hooker (see letters to J. D. Hooker, 9 May [1856] and 11 May [1856]) before he 'Began by Lyell's advice *writing* species sketch' on 14 May ('Journal'; Appendix II).

[8] CD went to London on 5 May 1856. On Tuesday, 6 May, he read a paper at the Linnean Society summarising the results of his and Miles Joseph Berkeley's experiments on 'The action of sea-water on the germination of seeds' (*Collected papers* 1: 264–73). On Thursday, 8 May, he attended a council meeting at the Royal Society (Royal Society council minutes).

[9] Since February, CD had been interested in the possibility of seeds being found in earth enclosed in tree roots (see letter to W. E. Darwin, [26 February 1856]). The case referred to here was of seeds found under an old oak tree, recorded in CD's Experimental book, p. 6 (DAR 157a). It was cited in

Origin, p. 361, as an example of possible means of transport in the geographical distribution of plants:

> I find on examination, that when irregularly shaped stones are embedded in the roots of trees, small parcels of earth are very frequently enclosed in their interstices and behind them,—so perfectly that not a particle could be washed away in the longest transport: out of one small portion of earth thus *completely* enclosed by wood in an oak about 50 years old, three dicotyledonous plants germinated: I am certain of the accuracy of this observation.

[10] See *Origin*, pp. 386–7, where CD described another of his 'several little experiments' attempting to germinate seeds contained in pond mud.

To T. H. Huxley 4 May [1856][1]

Down Bromley Kent
May 4th

My dear Huxley

I have been thinking over the Ray Soc. business,[2] & it seems to me so improper that my advances of £117, during the last 2–3 years to pay Mr Sowerby should never have been mentioned to the Council, that I hope that you will bring forward the case, & you can say very truly that I asked you when I was going to be paid & was surprised to find that you knew nothing about it.—[3]

Moreover it strikes me that it rather deeply concerns you personally, for if you were to get Royal Soc. money & the Ray Soc became bankrupt it w^d be an awkward affair for you.—[4] There may be other debts besides mine.—

As I feel under great obligations to the Ray Soc, which I beg you to express on my part, I would not for the world do anything troublesome or unfair towards the Society.—

I enjoyed your visit here *extremely*, & I much hope that Mrs Huxley did not suffer.

Yours most truly | Ch. Darwin

P.S. I have this minute received your note in regard to the Chinese Pots; & very much obliged I am to you for so kindly remembering the subject: I am very sorry to hear about Mrs Huxley.[5]

C. ·D.

Imperial College of Science, Technology and Medicine Archives (Huxley 5: 35)

[1] Dated by the reference to Huxley having recently visited CD at Down (see letter to T. H. Huxley, 2 April [1856]).
[2] Huxley, a member of the council of the Ray Society, had apparently mentioned to CD during his visit to Down in April that owing to the failure of a number of members to pay their subscriptions and to over-publication, the society was facing bankruptcy. In 1857, however, the situation began to improve through withdrawing the names of non-paying members and increasing the number of new members ('Report of the council of the Ray Society', 2 September 1857, bound into the copy of T. H. Huxley 1859b in the Cambridge University Library).
[3] George Brettingham Sowerby Jr had drawn the figures for *Living Cirripedia* (1851) and (1854), published by the Ray Society. CD had paid him on behalf of the Ray Society (see *Correspondence* vol. 5, letters to J. S. Bowerbank, Ray Society, 28 September [1851], and to John Lubbock, 10

[September 1853], n. 3). The Ray Society repaid CD £67 on 21 November 1857. An entry in his Account book (Down House MS) reads: 'Ray Society. Part repayment 50£ due.'

4 Huxley had made repeated attempts, after returning from the voyage of the *Rattlesnake*, to obtain funds from the Admiralty for the publication of his work on marine invertebrates. Only after resigning his commission in 1854 was he awarded £300 from the government grant administered by the Royal Society. The Ray Society undertook to publish the work, but this was delayed until 1859 owing to Huxley's 'heavy official duties' and the great number of other publications undertaken by the Ray Society in the intervening years. See the preface to T. H. Huxley 1859b and L. Huxley ed. 1900, 1: 71–6.

5 Henrietta Anne Huxley was pregnant.

From J. D. Hooker   7 May 1856

Kew
May 7ᵗʰ/56

My dear Darwin

Nymphæa seeds—we can get you plenty in Autumn if you will remind me.[1]

Ascension plants. I found only 3 or 4 native phænogams, & as many Ferns— I was wrong about the Rubiaceous genus & should have said *Hedyotis* (a very common trop. genus)— the plants are

> Euphorbia endemic
> Hedyotis— do
> Aristida do. (grass.
> & perhaps one or two Cyperaceæ

The Ferns are generally *not* S.ͭ Helena Species, but W. Indian— this is unaccountable.

Had the winds transported seeds at all there ought to be many common trop. weeds on the Mt. top—but I believe that every non endemic flowering plant of Ascension has been imported by man— *all* may be traced to the garden & houses, as far as I could detect.[2]

Scabiosa & Cnestis—the observations no doubt refer to species and not to individuals. I should rebut that the amount of albumen (perisperm) would be constant in the individuals of the species,—at least in some other exalbuminous orders in *some few* species of which I have found a little albumen (Compositæ & Halorageæ) I do not find the quantity variable[3]

Bentham did not cultivate the Dandelions (to which his observations refer) he watched the effects of light, shade, seasons &c on the forms (species of Dl.) growing about him at Montpellier.[4]

Grasses

Seed of Glyceria fluitans is abundantly collected (from the wild plants) & eaten in Holland & Russia & exported also, being the *true* "Manna Croup".—[5] Zizania was & is eaten abundantly by N. Am Indians, but is not cult.[6]

Dréges Verseidinity is not a mere catalogue, but elaborate table of distributions hideously closely printed, & diabolically complicated—[7] I doubt if you can get it in the Linnean, but I will lend you the Kew copy for a fortnight or month if you would care for it

in Russia

Primrose—Caucasus only
Oxlip—Caucasus to lat of Moscow.—
Cowslip—Caucasus to lat. of St Petersbugh,
(Compare English distribution)

Forbes map sent to Royal Society.[8]
Lyallia Kerguelensis— I have lost clue to the American affinity—probably a myth.[9]
In Herefordshire where both Oxlips & Cowslips grow, Bentham finds plenty of intermediates
Bentham would not have altered his Targioni review[10] for ADC^s book[11]—but I think that ADC. would for Bentham's review—
Will you let me know when you have spoken to Lord Overton[12] about Huxley & Athenæum[13] & I will back it up through Lord O^s bosom friend Strzelecki.[14]

AL
DAR 100: 94–5

CD ANNOTATIONS
1.1 Nymphæa . . . me.] *crossed pencil*
2.6 Aristida do.] 'do.' *del pencil*; 'PS.' *added pencil*
5.1 Scabiosa . . . for it 9.4] *crossed pencil*
9.3 I will lend . . . for it 9.4] *cross added pencil*
10.2 Primrose . . . Petersbugh, 10.4] 'Q' *added pencil, circled pencil*;
'[Petersb] 59, 56
$\frac{55\ \ 54}{}$' *added pencil*
11.1 Forbes . . . Strzelecki. 15.2] *crossed pencil*
Top of first page: '20'[15] *brown crayon*

---

[1] CD had apparently sent Hooker a list of questions relating to their discussions during Hooker's visit to Down House from 22 to 28 April (Emma Darwin's diary). The *Nymphaea* seeds were for CD's seed-soaking experiments (Experimental book, p. 14 (DAR 157a)).

[2] Hooker had visited Ascension Island and St Helena in 1843 during the voyage of H.M.S. *Erebus* (L. Huxley ed. 1918, 1: 53). He had first mentioned the dissimilarity in their floras, despite their proximity, in *Correspondence* vol. 3, letter from J. D. Hooker, 29 November 1844.

[3] The amount of albumen in seeds was thought by Hooker to be related to their ability to survive immersion in sea-water. He had earlier suggested that CD test the germination of albuminous seeds after immersion in sea-water (*Correspondence* vol. 5, letter to J. D. Hooker, 19 April [1855]).

[4] The observations have not been found. George Bentham had lived in Montpellier from 1816 to 1826 (*DNB*).

[5] 'Manna-croup' is the name given both to the coarse residue from grinding flour and to a similar meal made from the seeds of the manna-grass, *Glyceria fluitans* [hence the *true* manna-croup] (*OED*).

[6] *Zizania aquatica*, so-called wild rice, was once a staple food of North American Indians.

[7] Jean François Drège had made an extensive collection of South African plants during an eight-year visit to the Cape Colony. Hooker may be referring to Drège's 'Geographische Vertheilung', pp. 38–43, in Drège 1843, which is a table listing the numbers and areas of distribution of families of South African plants. See also the following letter.

[8] Probably Edward Forbes's 'Map of the distribution of marine life, illustrated chiefly by fishes, mollusca, and radiata; showing also the extent & limits of the homoiozoic belts', published in Alexander Keith Johnston's *Physical atlas* (E. Forbes 1856).

[9] In a manuscript list headed 'Kerguelen Lands List of Plants of' bound into CD's copy of the introductory essay of J. D. Hooker 1853–5 (Darwin Library–CUL), CD wrote next to this species: 'Hooker says (May 13/56) that he has lost trace of relation of this plant to some Cordillera plant'.

[10] Bentham 1855, a review of Targioni Tozzetti 1850.

[11] Hooker refers to Alphonse de Candolle's long discussion of plants naturalised in Europe, which makes reference to Antonio Targioni Tozzetti's work on the same subject (A. de Candolle 1855, 2: 800).

[12] Samuel Jones Loyd, Baron Overstone.

[13] Hooker planned to propose Thomas Henry Huxley for membership of the Athenæum Club (see letter to J. D. Hooker, 9 May [1856]). Huxley was not elected, however, until 1858 (L. Huxley ed. 1900 1: 150).

[14] Paul Edmund de Strzelecki was a member of the election committee of the Athenæum. Lord Overstone had assisted him in becoming a British subject (*DNB*).

[15] The number of CD's portfolio of notes on the geographical distribution of plants.

To C. J. F. Bunbury [before 9 May 1856][1]

List of Plants, common to Europe, observed by Meyer in Drége's Cape collection: Flora B. 2. 1843. "Zwei pflanzengeograph. Documente." s. 9.[2]

N. Certainly or probably naturalised by man's intervention.

A. Aquatic or Marsh Plants.

S sea-side Plants

(?) Plants of which Mr Bunbury knows nothing.

Those unmarked will be the most striking cases of specific identity with Europe.[3]

A memorandum
DAR 73: 159

[1] The date is established by CD's reference in the following letter to a 'list'. See also letter to C. J. F. Bunbury, 21 April [1856], in which CD asked for Bunbury's assistance in comparing the floras of South Africa and Europe.

[2] Drège 1843 included an introduction by Heinrich Friedrich Meyer.

[3] There follows a list of ninety-six plant species copied by an amanuensis from Drège 1843. The list has been annotated by Bunbury using the symbols listed by CD. Ten species also have had '(S. Am.)' written next to them by Bunbury. At the bottom of the list, Bunbury added the following comments:

> I have added the mark (S. Am.) to those which I know to be found also in South America.—
> CB.

Addenda
to the above list. CB.

Typha latifolia—A.
Polypogon Monspeliensis.
Lappago racemosa
    (= Cenchrus racemosus, L. Rubus fruticosus (according to Ecklon & Zeyher, but I will not
    undertake to answer for its absolute identity with the British plant.)

The reference is to Ecklon and Zeyher 1835–7. The list was also annotated by CD after it was
returned by Bunbury. CD referred to this list and to Bunbury's analysis in *Natural selection*, p. 552,
where he stated:

> Some are littoral plants which may possibly have travelled by the coast; about 14 are aquatic or
> marsh plants which seem to have, as we have seen, some special means of diffusion; but 30
> plants apparently do not come under either of these categories & I should infer (if really not
> naturalized by man's agency) had migrated through the tropics during the cold period.

To C. J. F. Bunbury 9 May [1856][1]

Down Bromley Kent
May 9th

My dear Bunbury

I am extremely much obliged to you for the list,[2] which is capital & gives me as
good an idea of the subject as one who is not a Botanist can have, & that I very
deeply feel at the best is but a poor idea.— I shall be *particularly* glad, whenever
you may have time & inclination, to hear anything which you may have to say on
representative species (but I have not yet read Heer) & on my supposed cold
mundane period.—[3]

When I most loosely spoke of all Europe, I was thinking of it in an E. & W.
sense, which concerns me especially as showing the wide extension of the cold. I
cannot prove this cold period geologically, I can only show that it is in some
degree probable, & then if it explains a *good many* facts in distribution (45 algæ in
New Zealand, common to the north & not found in Tropics), then I think it may
be admitted as probable hypothesis. The several northern species in T. del Fuego,
which during a cold period may have travelled down the Cordillera is one of the
strongest cases. So many plants, some European, common to Himmalaya,
Neilgherries, Ceylon, Java (I believe) & S. Australia. But My notions absolutely
require some greater means of dispersal than A. Decandolle & Hooker are
inclined to admit, but I cannot believe that we know $\frac{1}{10}$ of means of dispersal. I
stated on last Tuesday at Linnean Soc. (& I saw it made considerable impression
on the cautious Bentham & on Lyell) that I had removed earth **perfectly**
enclosed within roots of trees & in this earth (with every precaution taken) 3 seeds
germinated; & I enumerated the oceanic islands on which I *know* trees (some with
stones in roots) are cast up.[4]

But I shall weary you with my speculations & facts, so adios with many thanks |
C. Darwin

Suffolk Record Office, Bury St Edmunds

[1] The year is established by CD's reference to not yet having read Heer 1855. This work was recorded in his reading notebook on 7 August 1856 (*Correspondence* vol. 4, Appendix IV, 128: 20). See also letter from Charles Lyell, 1–2 May 1856.

[2] See preceding letter.

[3] See letter to C. J. F. Bunbury, 21 April [1856]. CD refers to Heer 1855 (see letter from Charles Lyell, 1–2 May 1856).

[4] CD's remarks must have been made after the paper he delivered at the meeting of the Linnean Society on 6 May 1856. They are not included in the published version (*Collected papers* 1: 264–73). See also letter to Charles Lyell, 3 May [1856].

To J. D. Hooker   9 May [1856][1]

Down Bromley Kent
May 9th

My dear Hooker

Read & return the enclosed from Consul Crowe (a friend of Col. Sabine) & send me whatever answer you think fit & I will write civilly to M.r Crowe.—[2]

With respect to Huxley,[3] I was on point of speaking to Crawfurd & Strzlecki (who will be on committee of Athenæum) when I bethought me of how Owen would look & what he would say.[4] Cannot you fancy him, with a red face, dreadful smile & slow & gentle voice, asking, "Will M.r Crawfurd tell me what M.r Huxley has done, deserving this honour; I only know that he differs from, & disputes the authority of Cuvier, Ehrenberg & Agassiz as of no weight at all".—[5] And when I began to consider what to tell M.r Crawfurd to say, I was puzzled, & could refer him only to some *excellent* papers in R. Trans. for which the medal had been awarded.[6] But I doubt *with an opposing faction*, whether this would be considered enough, for I believe real scientific merit is not thought enough, without the person is generally well known; now I want to hear what you *deliberately* think on this head: it would be bad to get him proposed & then rejected; & Owen is very powerful.—

Lastly, & of course especially, about myself; I very much want advice & *truthful* consolation if you can give it. I had good talk with Lyell about my species work, & he urges me strongly to publish something.[7] I am fixed against any periodical or Journal, as I positively will *not* expose myself to an Editor or Council allowing a publication for which they might be abused.

If I publish anything it must be a *very thin* & little volume, giving a sketch of my views & difficulties; but it is really dreadfully unphilosophical to give a resumé, without exact references, of an unpublished work. But Lyell seemed to think I might do this, at the suggestion of friends, & on the ground which I might state that I had been at work for 18 years, & yet could not publish for several years, & espccially as I could point out difficulties which seemed to me to require especial investigation. Now what think you?. I sh.d be really grateful for advice. I thought of giving up a couple of months & writing such a sketch, & trying to keep my judgment open whether or no to publish it when completed.[8] It will be simply impossible for me to give exact references; anything important I sh.d state on authority of the author generally; & instead of giving all the facts on which I ground any opinion, I could give by memory only one or two. In Preface I would state that the work could not be considered strictly scientific, but a mere sketch or outline of future work in which full references &c sh^d be given.— Eheu, eheu, I believe I sh^d sneer at anyone else doing this, & my only comfort is, that I *truly* never dreamed of it, till Lyell suggested it, & seems deliberately to think it adviseable.

I am in a peck of troubles & do pray forgive me for troubling you.—

Yours affecti^y | C. Darwin

Emma desires her best thanks to Mrs Hooker for her kind note received this morning.

DAR 114.3: 161

[1] Dated by the relationship to the letter from J. D. Hooker, 7 May 1856.

[2] John Rice Crowe was British consul-general in Norway, 1843–75. He and CD had previously corresponded about seeds washed up on the coast of Norway (*Correspondence* vol. 5, letter from J. R. Crowe, 27 September 1855, and letter to J. R. Crowe, 9 November 1855). Edward Sabine was treasurer of the Royal Society of London.

[3] Hooker was attempting to get Thomas Henry Huxley elected to the Athenæum (see letter from J. D. Hooker, 7 May 1856).

[4] John Crawfurd, Paul Edmund de Strzelecki, and Richard Owen. Huxley's relationship with Owen had sharply deteriorated in recent months (A. Desmond 1982).

[5] Huxley had roundly criticised Louis Agassiz's theory of 'progressive development' of living forms through geological time and Georges Cuvier's application of his principle of 'the physiological correlation or coadaptation of organs' in two Friday evening lectures held at the Royal Institution (T. H. Huxley 1855 and 1856a). He had previously described Christian Gottfried Ehrenberg's researches as 'wonderful monuments of intense and unremitting labour, but at least as wonderful illustrations of what zoological and physiological reasoning should *not* be' (T. H. Huxley 1851, p. 436).

[6] Huxley had received the Royal Medal of the Royal Society in 1852 for his paper on the anatomy and physiology of Medusae (T. H. Huxley 1849).

[7] CD had made an appointment to see Lyell on 8 May (see letter from Charles Lyell, 1–2 May 1856, and letter to Charles Lyell, 3 May [1856]).

[8] A note inserted in the front of CD's copy of A. de Candolle 1855, volume 1 (Darwin Library–CUL), indicates CD's intentions: 'When this read skim over (make index) Review Hooker N. Zealand &c & Fl. Antarctica *& Galapagos [*added pencil*] Skim my own portfolio Then read my own old sketch, & write Essay'.

To Henry Ambrose Oldfield 10 May [1856]

Down Bromley Kent

May 10th

Dear Sir

I hope that you will forgive the liberty which I take in addressing you;[1] but Mr Vaux has told me that I might use his name as an introduction.—[2] When looking at the Assyrian drawings, Mr Vaux told me that you had remarked that the Dogs there represented, were like the Thibetan Dogs, with which you were familiar in Nepaul.—[3] I am *greatly* interested in regard to all domesticated animals, & their ancient history, & sh.d feel very much obliged if you would inform me whether this resemblance is close.—

In drawings which I have seen, (but perhaps they were poor ones) of Thibetan Mastiffs, the ears were longer & the chops depended much lower than in the Assyrian, drawings, giving to the head more of the appearance of the Blood Hound.— As I am writing, would you be so good as to inform me whether there are other breeds of Dogs in Nepaul. If by any chance your attention sh.d have been turned to any of the domesticated animals or birds, I sh.d be most grateful for any information, more especially in regard to Poultry & the breeds of Fancy Pigeons, or Rabbits if such be kept, as I am paying especial attention to the Pigeons, & am endeavouring to make a collection from all parts of the world.—

Hoping that you will excuse this intrusion, I beg to remain | Dear Sir | Your faithful servant | Charles Darwin

Postmark: MY 11 1856
American Philosophical Society (128)

[1] Oldfield was surgeon to the British Residency in Nepal, 1850–63. He was apparently on leave in 1856, for the cover preserved with the letter is addressed to him at 8 Gloucester Gardens, Gloucester Terrace, Hyde Park, London.

[2] William Sandys Wright Vaux, of the department at antiquities, British Museum, was the author of *Handbook to the antiquities in the British Museum: being a description of the remains of Greek, Assyrian, Egyptian, and Etruscan art preserved there* (Vaux 1851).

[3] In *Variation* 1: 17, CD mentioned having seen these drawings, which he stated were from the Assyrian tomb of the son of Esar Haddon dated 'about 640 B.C.', at the British Museum. He also added: 'This dog has been called a Thibetan mastiff, but Mr. H. A. Oldfield, who is familiar with the so-called Thibet mastiff, and has examined the drawings in the British Museum, informs me that he considers them different.' (*ibid.* 1: 17 n. 4).

To J. D. Hooker 11 May [1856][1]

Down Bromley Kent
May 11.th

My dear Hooker

I am really *grateful* to you for your two very long letters, which must have cost you much time.[2] I will just allude to the first; the several facts you give me are very valuable, especially the marvellous one on Primrose & cowslip.— At some *future* time, I sh.d be particularly obliged for Drege,, ie if it be not in Linnean Soc.y—[3] By the way what a noble interest you take in Linnean: I was extremely glad to hear at Council of R. Soc. that the Linnean will get rooms in B.H.[4] (By the way I feel pretty certain of medal for Sir J. Richardson, though he has not yet been proposed.)[5]

(Many thanks for Forbes map, which I shall pick up when next in London.).—[6]
I have written to M.r Crowe & repeated what you say.

In regard to Huxley; I sh^d think his chance was so good if Owen is not on Committee (& his **real** claims I have not for a moment ever doubted about), that I will call when next in London on Ld Overstone & speak about it: if you can interest Murchison I sh^d think he w^d be invaluable.[7] I would, also, with pleasure write to Mr Crawfurd, if you think fit & I do not meet him soon.— I will, also, speak to Sir P. Egerton, when I see him next at Council of R. Soc^y.—[8]

Now for a *more important*! subject, viz my own self: I am extremely glad you think well of a separate "Preliminary Essay" i.e. if anything whatever is published; for Lyell seemed rather to doubt on this head; but I cannot bear the idea of *begging* some Editor & Council to publish & then perhaps to have to *apologise* humbly for having led them into a scrape. In this one respect I am in the state, which according to a very wise saying of my Father's, is the only fit state for asking advice, viz with my mind firmly made up, & then, as my Father used to say, *good* advice was very comfortable & it was easy to reject *bad* advice.— But Heaven knows I am not in this state with respect to publishing at all any preliminary essay. It yet strikes me as quite unphilosophical to publish results without the full details which have led to such results.

It is a melancholy, & I hope not quite true view of your's that facts will prove anything, & are therefore superfluous! But I have rather exaggerated,, I see, your doctrine. I do not fear being tied down to error, i.e. I feel pretty sure I should give up anything false published in the preliminary essay, in my larger work; but I may thus, it is very true, do mischief by spreading error, which as I have often heard you say is much easier spread than corrected. I confess I lean more & more to at least making the attempt & drawing up a sketch & trying to keep my judgment whether to publish open. But I always return to my fixed idea that it is dreadfully unphilosophical to publish without full details. I certainly think my future work in full would profit by hearing what my friends or critics (if reviewed) thought of the outline.—

To anyone but you I sh^d apologise for such long discussion on so personal an affair; but I believe, & indeed you have proved it by the trouble you have taken, that this would be superfluous.

Your's truly obliged | Ch. Darwin

Pray thank Mrs Hooker for her little note about Mr Gosse's book.—[9]

P.S | What you say (for I have just reread your letter) that the Essay might supersede & take away all novelty & value from my future larger Book, is very true; & that would grieve me beyond everything. On the other hand, (again from Lyell's urgent advice) I published a preliminary sketch of Coral Theory & this did neither good nor harm.—[10] I begin *most heartily* to wish that Lyell had never put this idea of an Essay into my head.[11]

DAR 114.3: 162

[1] Dated by the relationship to the letter from J. D. Hooker, 7 May 1856, and the letter to J. D. Hooker, 9 May [1856].

[2] Letter from J. D. Hooker, 7 May 1856. The other one, now missing, was a response to the letter to J. D. Hooker, 9 May [1856].

[3] See letter from J. D. Hooker, 7 May 1856.

[4] Burlington House, Piccadilly. The Linnean Society had lobbied the government since November 1854 with the proposal that it should be accommodated in Burlington House along with the Royal Society and the Geological, Astronomical, and Antiquarian Societies. The Linnean's request was granted by the Treasury in a letter, dated 22 May 1856, addressed to the president of the Royal Society and forwarded to Thomas Bell in time for the anniversary meeting on 24 May (Gage and Stearn 1988, pp. 50–1). CD had attended the meeting of the council of the Royal Society on 8 May (Royal Society council minutes).

[5] CD nominated John Richardson for a Royal Medal in June 1856 for his work in natural history and physical geography. Richardson received the medal in November (Royal Society council minutes).

[6] E. Forbes 1856.

[7] See letters to J. D. Hooker, 7 May 1856, n. 13, and 9 May [1856].

[8] Philip de Malpas Grey-Egerton was a member of the Royal Society council with CD.

[9] CD probably refers to P. H. Gosse 1856, a guide to the seaside town of Tenby, which had been published in April 1856 (*Publishers' Circular*, 15 April 1856, p. 164). CD was planning a visit to Tenby (see following letter).

[10] CD had presented this paper, 'On certain areas of elevation and subsidence in the Pacific and Indian Oceans, as deduced from the study of coral formations', at a meeting of the Geological Society of London on 31 May 1837 (*Collected papers* 1: 46–9). He subsequently published *Coral reefs* in 1842.

[11] See letter from Charles Lyell, 1–2 May 1856. On 14 May, CD followed Lyell's advice and began writing his 'species sketch' ('Journal'; Appendix II).

To W. B. Tegetmeier   11 May [1856][1]

Down Bromley Kent
May 11th

My dear Sir

To take your note seriatim, I am very much obliged for the authentic information in regard to the Laughing Pigeons:[2] I shall be **most particularly** glad to have the pair you refer to at 7^s. or more if you think right.

I am almost sorry I did not get the Silk fowls for 18^s though it would have been rather extravagant as one dead Bird would have done as well for me; nevertheless I am rather sorry that I was not extravagant.—

Very many thanks for the renewed offer of the Silver-spangled Poland eggs.—

With respect to the Anerley Show,[3] I have very great fears I shall not be at home, which I shall very much regret; but I am going to take my whole Family to Tenby for a month or 5 weeks,[4] & I doubt whether I shall have returned.

If you really think the committee would like my name, certainly I can have no sort of objection to the compliment thus paid me.[5]

I suppose in such cases any sort of title is adviseable, so that you can, if you think fit, append to my name "Vice Pres: Royal Society" & F. Zoolog. Soc.^{y} if you like.—

With many thanks for all your kindness, pray believe me | Your's very sincerely | Ch. Darwin

PS. With respect to Steven's sale-list if you see anything which might interest me, I sh^d. be very much obliged for a copy, but not otherwise.—6

New York Botanical Garden Library (Charles Finney Cox collection)

1 Dated by the reference to the Anerley show and to CD's proposed Tenby trip (see nn. 3 and 4, below).
2 See letter to W. B. Tegetmeier, 25 April [1856].
3 The poultry and pigeon show held from 29 July to 1 August 1856 at Anerley Gardens, near the Crystal Palace in Sydenham, Kent. CD had attended the 1855 exhibition (see *Correspondence* vol. 5, letter to W. D. Fox, 31 July [1855]). The 1856 show was the last one held because it failed to meet expenses (see Secord 1981, p. 171).
4 The Darwin family did not go to Tenby because of Emma Darwin's poor health (see letter to W. D. Fox, 8 June [1856]), so CD was able to attend the Anerley exhibition (see letter to W. B. Tegetmeier, 14 August [1856]).
5 Possibly a reference to the fact that Tegetmeier was the general manager of the poultry section for the Anerley show (*Cottage Gardener* 16 (1856): 339). No specific use of CD's name has been found in reports of the Anerley show.
6 The sale catalogue of John Crace Stevens, poultry and pigeon auctioneer in Covent Garden, London.

To S. P. Woodward   15 May [1856]¹

Down
15 May

[Returning thanks for the Supplement to his correspondent's *Manual of the Mollusca*,² eloquently praising the book and Woodward's labours in the cause of science]
'What an amount of labour is condensed in your little volume! & how marvellously cheap.— I fully believe & hope that you will reap the only reward worth having, the consciousness that you have done good service to the cause of Science.'

Sotheby (21 March 1966)

1 The date as given in Sotheby's catalogue. It is corroborated by the publication date of part three of Woodward 1851–6 (see n. 2, below).
2 The third part of Woodward's *Manual of the Mollusca* (Woodward 1851–6), entitled 'Supplement to the rudimentary treatise on recent fossil shells', was published in May 1856 (*Publishers' Circular*, 2 June 1856, p. 227). A copy of the entire work is in the Darwin Library–CUL. The third part, covering the geographical distribution of shells, is heavily annotated.

To J. D. Hooker   21 [May 1856]¹

Down Bromley Kent
21st

My dear Hooker
I have got the Lectures & have read them. The Lectures strike me as very clever. Though I believe, as far as my knowledge goes that Huxley is right, yet I

think his tone very much too vehement, & I have ventured to say so in a note to Huxley.—[2] I had not thought of these Lectures in relation to the Athenæum, but I am inclined quite to agree with you & that we had better pause before anything is said.[3] It might be urged as a real objection the way our friend falls foul of every one (N.B I found Falconer very indignant at the manner in which Huxley treated Cuvier in his R. In^n Lecture;[4] & I have gently told Huxley so.) I think we had better do nothing, to try in earnest to get a great Naturalist into Athenæum & fail, is far worse than doing nothing—

How strange, funny & disgraceful that nearly all—(Faraday, Sir J. Herschel at least exceptions) our great men are in quarrels in couplets; it never struck me before.—

I hope to meet you at Club.—[5] When there, tell me whether Leptospermum & Stylidum are confined to southern Australia, or are they, also, Tropical? Can you lend me paper on crossing of Fucus?[6]

Ever yours | C. Darwin

Also can you tell me whether the fossil Casuarina & Banksia of Flinders Is^d can be recognised as distinct species.—[7]

DAR 114.3: 163

[1] Dated by the reference to Hugh Falconer's reaction to T. H. Huxley 1856a (see n. 4, below).

[2] CD's note has not been found. He refers to the first two lectures given by Thomas Henry Huxley in his series 'Lectures on general natural history', published in the *Medical Times & Gazette* (T. H. Huxley 1856–7). The first two lectures, printed in issues dated 3 May and 17 May 1856, deliberately attacked Richard Owen's definition of life (*Medical Times & Gazette* 12 (1856): 430), Owen's apparent lack of reference to embryology (p. 432), his work on parthenogenesis (p. 482), and his classification scheme (p. 484).

[3] Huxley was not proposed for election to the Athenæum Club in 1856 but was elected in 1858 under Rule 2, which provided for the election of persons distinguished in the arts, science, or letters independently of the normal balloting by the membership (L. Huxley ed. 1900, 1: 150).

[4] Huxley had attacked Georges Cuvier's principle of the correlation or coadaptation of organs in T. H. Huxley 1856a. Hugh Falconer later criticised Huxley for his attack in the June issue of the *Annals and Magazine of Natural History* (Falconer 1856). See letter to J. D. Hooker, 17–18 [June 1856].

[5] The Philosophical Club of the Royal Society. CD was in London to attend the meeting of the council of the Royal Society on Thursday, 30 May 1856 (Royal Society council minutes). The Philosophical Club usually met once a month on Thursdays (Bonney 1919).

[6] Thuret 1854–5. Gustave Adolphe Thuret was well known for his researches on the reproduction of algae and had, in this paper, identified both spores and antherozoids in *Fucus*, thereby showing that sexual reproduction took place in seaweeds. See Farley 1982, pp. 64–5.

[7] On the back of a note headed '*Hooker*. Notes on N. Zealand Flora *In conversation May 1856 [*added pencil*]', CD wrote: 'The Casuarinæ are quite [*above del* 'very'] absent in N. Zealand & makes another strong case like Eucalyptus. There are 2 (but distinct species) of Proteaceæ in N. Zealand; no Banksia; but yet Tert. deposit, From Flinders isl^{ds} there is a Casuarina & Banksia, showing that probably these forms anciently in Australia.—' *ink* (MS inserted in CD's copy of the introductory essay of J. D. Hooker 1853–5 in the Darwin Library–CUL).

To Albany Hancock 25 May [1856][1]

Down Bromley, Kent
May 25

My dear Sir,

I am really very much obliged to you and Mr. Storey, and am quite ashamed at having caused so much trouble, but I was very curious to obtain this information.[2] My present work leads me to wish to get as accurate information as I can on what some call the economy of nature, and the point in question seemed to me deserving of attention, as aiding in shewing how far the struggle with other species checked the extreme possible northern range of any species. It seems odd that dwarfing should be so frequent on mountains, and so rare, or at least not equally conspicuous, at the extreme northern lowland limits of a species.[3]

I hope that you will be so kind whenever you see Mr. Storey to present to him my sincere thanks for all the trouble he has so kindly taken for me, and pray believe me,

My dear Sir, | Yours very sincerely, | CHARLES DARWIN.

J. Hancock 1886, pp. 277–8

[1] The letter is dated by the reference to topics discussed in chapter 5 of *Natural selection*, a draft of which was completed in February 1857 (see 'Journal'; Appendix II).

[2] John Storey worked on the flora of Northumberland and Durham and, like Hancock, resided in Newcastle upon Tyne and was a member of the Tyneside Naturalists' Field Club. CD had previously corresponded with Hancock concerning his monograph on the Cirripedia (see *Correspondence* vols. 4 and 5).

[3] CD discussed this point in his chapter on 'The struggle for existence' in *Natural selection*, p. 195, where he stated: 'as several British plants do not range beyond Northumberland & Durham, I asked Mr. Story to attend to this point for me, & he has sent me a list of 32 plants in this predicament observed by himself & friends; & it appears that only three or four of these are at all dwarfish.'

To T. H. Huxley 27 May [1856][1]

Down Bromley Kent
May 27th

My dear Huxley

I have just written & will post tomorrow two very strong notes to L[d] Overstone & Sir J. Lubbock, & I shall be heartily glad if they are of the least service to you.—[2]

I have this morning received your Lecture,[3] but have not had time yet to look at it.

Your's most truly | C. Darwin

Imperial College of Science, Technology and Medicine Archives (Huxley 5: 174)

[1] Dated by the relationship to the following letter. Both letters relate to Huxley's application for the position of examiner in physiology and comparative anatomy at London University, which fell vacant in 1856.

[2] Samuel Jones Loyd, Baron Overstone, was a member of the senate of London University (Harte 1986, p. 125). John William Lubbock, who had been the university's first vice-chancellor, 1836–42, was also a member of the senate (Harte 1986, p. 83).

[3] Probably the third of Huxley's lectures on general natural history (T. H. Huxley 1856–7), published on 24 May 1856. CD had already received the first two lectures (see letter to J. D. Hooker, 21 [May 1856]).

To J. W. Lubbock 27 May [1856][1]

Down
May 27th

Dear Sir John Lubbock

I hope that you will forgive me troubling you, & this note requires no answer.— I believe that you are interested in the University of London & have influence there.— My friend, M[r] Huxley F.R.S. is a candidate for the Examinership in Physiology & Comp: Anatomy vacant by Dr Carpenter's resignation.—[2] He has asked me to write in his favour, & I can do this most conscientiously, for his merits in these particular branches of Natural Science, are of the highest order.[3] I think if you will ask your Son he will agree with me in this.—[4]

He is Palæontologist in the Museum of Practical Geology, & they have never selected any but the best men.— He has lately been lecturing on Physiology at the Royal Institution. He has published several papers in the Phil: Transact, on comparative anatomy which papers were honoured by the Royal Medal, & have been translated into German.— His acquaintance with foreign literature in Nat. History is remarkably accurate & extensive. And lastly he gained when a medical Student in London the Gold Medal for Physiology.[5]

I hope that you will forgive my bringing M[r] Huxley's claims to your attention, & pray believe me, | Your's very sincerely | Ch. Darwin

I was extremely sorry that I was not able to come to dinner yesterday.—

Royal Society (LUB: D23)

[1] See the preceding letter for the basis of this date.

[2] William Benjamin Carpenter resigned his lectureships and other offices when he was appointed registrar of London University (*DNB*). The university was primarily an administrative and examining body (*EB*).

[3] Thomas Henry Huxley was appointed to the examinership on 9 July 1856, a position he held until 1870 (L. Huxley ed. 1900, 1: 148, and Pingree 1968, p. 59).

[4] Huxley and John Lubbock shared common interests in the morphology of invertebrates. They had recently been guests at CD's house (see letter to John Lubbock, 24 April [1856]).

[5] Huxley won a gold medal for anatomy and physiology upon his graduation in 1845 from London University (*DNB*).

To S. P. Woodward   27 May 1856[1]

Down. Bromley. Kent.
May 27th— /(56.)

Dear Sir.

I am very much obliged to you for having taken the trouble to answer my query so fully—[2] I can now be at rest, for from what you say & from what little I remember Forbes said, my point is unanswerable. The case of Terebratula[3] is to the point as far as it goes and is negative. I have already attempted to get a solution through geographical distribution by Dr Hooker's means, & he finds that same genera which have very variable species in Europe, have other very variable species elsewhere— This seems the general rule, but with some few exceptions— I see from the several reasons which you assign, that there is no hope of comparing the same genus at two different periods, & seeing whether the tendency to vary is greater at one period in such genus than at another period— The variability of certain genera or[4] groups of species strikes me as a very odd fact.

I shall have no points as far as I can remember to suggest for your reconsideration, but only some on which I shall have to beg for a little further information—[5] However I feel inclined very much to dispute your doctrine of Islands being generally ancient in comparison I presume with continents— I imagine you think that islands are generally remnants of old continents a doctrine which I feel strongly disposed to doubt— I believe them *generally* rising points, you, it seems think them sinking points—[6]

With many thanks. | Yours very sincerely. | C. Darwin.

Copy
DAR 148

[1] The date on the copy is corroborated by the reference to previous correspondence with Woodward (see n. 2, below).

[2] This letter is a response to the letter from S. P. Woodward, 2 May 1856, even though CD had written an additional letter in the interim (see letter to S. P. Woodward, 15 May [1856]).

[3] In the copy, 'Terebratula' has been substituted for the copyist's 'Tenebratula'.

[4] In the copy, 'or' has been substituted for the copyist's 'in'.

[5] See letter to S. P. Woodward, [after 4 June 1856].

[6] Woodward expressed his belief that island faunas were generally older than those of continents in the letter from S. P. Woodward, 2 May 1856. He gave a fuller discussion in Woodward 1851–6, 3: 406 n. See also Woodward 1851–6, 3: 381, where Woodward referred to insular flora and fauna as 'the survivors, seemingly, of tribes which the sea has swallowed up.' CD considered most oceanic islands to be volcanic in origin and unconnected with any continents.

From Edward William Vernon Harcourt   31 May 1856

Dear Sir.

I found your note here, on my arrival from London the other day. It would give me great pleasure to be able to answer any of your questions.[1] I passed four winters in Madeira, i.e. from Oct: 1847 – April 1848 Nov: 1848 – May 1849 Nov: 1849 – May 1850. Nov: 1850 – April 1851. So that I cannot speak from personal experience as to the habits of birds in Madeira during the months of June, July, August, and Sept: I had however the advantage of an acquaintance with the Rev.ᵈ R. T. Lowe, whose residence of 24 years in Madeira and whose habits of accurate scientific observation gave such authority to the information he most kindly and freely afforded me, as I was very fortunate to obtain.[2] I am quite aware how little dependance may be placed generally on the accounts which travellers often, *on the authority of others*, are in the habit of giving. I sh.ᵈ therefore always be especially careful in receiving the testimony of those who had not been well drilled in *habits* of observation.

I enclose for your acceptance my notes on the Madeiran birds, and shall be extremely gratified if they interest you;—[3] I have placed a pen mark against those birds which I have myself seen in the Island, it follows therefore that those have been found there between the months of October & May; I have placed *o* against those birds which I did not see my-self, but of whose presence as occasional visitants, from the reliability of my informants, I can have no doubt.[4]

Cathartes percnopterus, may perhaps be found in Madeira once in every three or four years, it appears in Berthelot's list of the birds of the Canary islands,[5] in Malherbe's list of African birds,[6] in Ignatius Asso's list of Spanish birds (Aragon)[7] as well as in Capt. Widrington's list of Spanish birds (Seville).[8] from any of these contiguous countries it might well find its way to Madeira— I may observe that during the prevalence of the Easterly winds which blow from the African coast to Madeira, (being called *Harmattan* on the W. coast of Africa and *l'Este* in Madeira) most of the visitants arrive—and as these winds do not prevail more at one time of year than another, the migration of birds to Madeira is probably involuntary and more affected by winds than seasons. This does not apply to the seabirds which occasionally visit the Island.

Falco nisus, which occurs in the lists of the birds of the Canaries, of Africa, & of Spain, is found in Madeira perhaps once in two years

Falco subbuteo, may be described in like manner.

Corvus corax, is occasionally brought over in a domestic state by the traders from Lisbon, there is therefore the chance that the only individual I heard of as having been killed in Madeira had escaped from captivity. Then, again, on the other hand, we find this species in the African and Canary lists.

Corvus corone, occurs perhaps once in two years this bird appears in the lists of African specimens as well as in those of Spain, but *not* in those of the Canary islands.

Corvus frugilegus, of which I have lately received three fine specimens from Madeira shot in June of last year by a friend, does not appear in Berthelot's list of the birds of the Canaries, altho *C. monedula* does occur in Teneriffe, which bird has never, to my knowledge, been found in Madeira.

Oriolus galbula, is of not unfrequent occurrence.

Sternus vulgaris, appears in flights of 5 or 6 together after or during strong easterly wind (I omitted to observe that *Oriolus galbula* is not mentioned in Berthelot's list!)

Turdus iliacus & musicus are both rare visitants They may be found perhaps once in two years. This is the more curious as the *T. merula* is *very* plentifully found in the island.

Sylvia hortensis, I never heard of but one specimen of this bird being shot in Madeira, I did not see the specimen, and as the bird does not appear in either the African or Canary lists, its advent to Madeira must be considered remarkable. (If my authority is as correct in his information as he is positive)

Troglodytes Europæus, has occurred in Madeira, on the Authority of the Revd. R. T. Lowe; it is found in the African lists, but not in those of the Canary islands. I never heard of but one specimen being found in Madeira, & I did not myself see that specimen.

Motacilla alba, occurs, according to my information once in every two or three years, it is included in the lists of the Canary islands.

Alauda arvensis, has been found in Madeira, I have no data for determining how often.

Fringilla chloris, may be found some two or three times a year in Madeira; this bird is not mentioned in the lists of the Canary I.!

F. domestica which has been once (as far as I know) obtained in Madeira does not appear in the lists of the Canaries or of W. Africa. It is however not uncommon in parts of Spain.

Cuculus canorus, one or two specimens of this bird are generally shot in the course of the year in Madeira.

Musophaga Africana, has been found about once in 3 or 4 years; it does not appear in the lists of the Canary Islands.

Upupa epops, appears in flocks every year, at various times.

Merops apiaster, of this bird perhaps on an average one specimen is obtained in the Course of the year.

Alcedo ispida, two or three specimens are generally taken during the year.

Hirundo urbica, one specimen may perhaps be obtained of this bird in Madeira during the course of the year. It does not appear in Berthelot's list.

H. rustica, two or three specimens a year is the utmost that the most attentive observer has remarked, tho' it is more frequently seen in Madeira than the *H. urbica*.

H. riparia. I once saw one of these birds in Madeira, and tradition goes that another had once been seen, at any rate its occurrence is very rare, and it is not known in the Canaries.

Caprimulgus Europeus, is obtained at the average of one a year, perhaps hardly so frequently.

Columba Œnas, is of rare occurrence, I only saw one specimen when I was in Madeira. It does not appear in the lists of the Canary islands.

C. turtur, occurs once or twice a year in Madeira I have seen it blown on board ship at a lat: of about 2°. north of Madeira in the month of May— And once we had contemporaneously a flight of *Hirundo rustica*, which fell like rain in a state of complete exhaustion on the deck and about the rigging of our ship; in the same lat:

Œdicnemus crepitans, occurs about once in two years.

Calidris arenaria, I never saw in Madeira, but I understand it is found every two or three years there.

Vanellus cristatus, is found in parties of 2 & 3 about every other year, at most,

Charadrius hiaticula, so common in Africa and S. Spain, has only (to my knowledge) occurred once in Madeira, and has never been noted in the Canary isles.

C. pluvialis, occurs after the same fashion as *C. cristatus*.

Strepsilas interpres, has occurred in Madeira and that is all I know about it.

Ardea Nigra, has also occurred, but I sh^d imagine it is *very* rare.

Ardea Cinera, perhaps 6 or 7 specimens on average appear in Madeira in the course of the year—

A. ralloides, makes it appearance about once in two or three years, at most.

A. russata, is of still rarer occurrence, and does not appear in the lists of the Canaries.

A. purpurea, is shot about once in two years; it is curious that this bird sh^d not find its way to the Canary islands!

A. minuta, must be very rare.

A. Stellaris. one or two a year.

A. nycticorax. Not an uncommon visitant Several in the course of the year.

Platalea leucorodia, about once in 4 years

Limosa melanura, about once a year.

Numenius arquata, two or three a year; this bird does not occur in the lists of the birds from the Canaries.

N. phæopus, found most years, but not so frequently as the *N. arquata*.

Tringa pugnax, I saw two females, I never heard of any others in Madeira, nor did I hear of the male bird of that species being found there.

T. subarquata, I am informed is found in Madeira once in every two or three years.

T. variabilis, I found twice in Madeira, I could obtain no further information concerning it.

T. cinerea, I am told by Mr Lowe has been seen in Madeira. *T. variabilis* is the only *T.* mentioned as occurring in the Canaries.

Totanus hypoleucos, is rare, occurring once in 3 or 4 years.

T. Glottis, is more common, appearing on an average once in two years; it does not appear in the lists of the Canary islands.

Scolopax gallinago, is shot three or four times in the course of year;

S. major appears in about the same numbers as the common Snipe.

Crex Baillonii, occurs once in three or four years. neither this, nor the former bird are recorded by Webb & Berthelot.

C. pratensis, is said to occur once in two or three years.

Porhpyrio allenia, the specimen I obtained of this bird was very young; I never heard of any other being found in Madeira

Gallinula chloropus, is found every year in more or less abundance.

Fulica atra, the same may be said of this bird.

Anser segetum, is shot most years in Madeira, it appears in companies.

Marecea Penelope, I never saw in Madeira, & have only the record of one having been shot there.

Anas crecca, appears in about the same numbers as *Anser segetum*.

A. boschas, I have only certain information of one having been killed in Madeira.

Sterna nigra, M.r Lowe once saw one of these birds.

S. Dougalli, appears in my list on the authority of Sir W. Jardine.[9]

Larus tridactylus, is not uncommon in Madeira It may possible breed there, but of that I have no certain evidence.

L. cataractes, is found about once in two years— As you have, doubtless, Webb & Berthelots book,[10] there is no use my remarking further upon which birds do or do not occur in the Canary islands; but, considering the propinquity of the two groups to each other, the comparison is curious.

Colymbus glacialis, the specimen I saw was a young bird, it was killed in an exhausted state; I never heard of another being taken in Madeira, wh: is far out of his accustomed beat.

Sula Alba, was once obtained by M.r Lowe.

Procellaria mollis, I never heard of but three specimens of this bird in Madeira.

P. Pacifica, and *one* specimen of this bird.

Prion brevirostris, I saw the one specimen named by M.r Gould, & exhibited by M.r Yarrell at a meeting of the Zool: Society—[11] this constitutes my entire acquaintance with the bird.

Thalassidroma pelagica, finishes the list, and I never heard of but one specimen in Madeira.

I must now apologise for the length to which I have extended my explanations, but could not well make them shorter in answer to your questions.

I have no data to go upon wh: would enable me to answer the question "whether wanderers of the *same species* of birds which permanently inhabit the island are ever blown from the Continent to Madeira". Of course, where the distance is only 250 miles, it is perfectly possible, and indeed probable.

In answer to your other question "Whether any regular migratory Birds inhabit Madeira?" *Cypselus unicolor, C. murarius*, & *Scolopax rusticola*, Breed in the island and always are found there.[12]

The variance of the recorded visits of different species of birds of the same genus to the Madeiras and the Canary islands, will doubtless strike you as very remarkable: as also, why many birds of the same genus should be distinguished from each other as *frequent* and *infrequent* yearly visitants at the same island.

I have now to the best of my power, answered the questions you put to me; I only hope the length of my answers have not tired you— For my part, it has given me great pleasure to have had an opportunity of thus making acquaintance with one whose name is so well known to all loves of Natural History.

Believe me, dear Sir, | Y^{rs} faithfully, | Edward Vernon Harcourt.

Hastings | May 31. | 1856.

DAR 166: 100

CD ANNOTATIONS

3.5 I may observe . . . seasons. 3.10] *double scored pencil*
4.2 found in Madeira] *cross added pencil*
5.1 *Falco subbuteo,*] *cross added pencil*
7.1 *Corvus corone,*] *cross added pencil*
8.1 three fine specimens] *underl pencil*
9.1 *Oriolus galbula,*] *cross added pencil*
10.1 *Sternus vulgaris, . . .* wind 10.2] *cross added pencil; double scored pencil and brown crayon*
11.1 *Turdus iliacus and musicus*] *two crosses added pencil*
14.1 *Motacilla alba,*] *cross added pencil*
16.1 *Fringilla chloris,*] *cross added pencil*
18.1 *Cuculus canorus,*] *cross added pencil*
19.1 *Musophaga Africana,*] *cross added pencil*
20.1 *Upupa epops, . . .* times.] *cross added pencil; scored brown crayon*
21.1 *Merops apiaster,*] *cross added pencil*
22.1 *Alcido ispida,*] *cross added pencil*
23.1 *Hirundo urbica,*] *cross added pencil*
24.1 *H. rustica,*] *cross added pencil*
26.1 *Caprimulgus Europeus,*] *cross added pencil*
28.1 *C. turtur,*] *cross added pencil*
28.2 And once . . . same lat: 28.4] *triple scored ink*
31.1 *Vanellus cristatus, . . .* most,] *scored brown crayon*
60.1 *Anser segetum, . . .* companies.] *scored brown crayon*
67.3 but, considering . . . curious. 67.4] *scored brown crayon*
75.1 I have no . . . tired you— 78.2] *scored brown crayon*

[1] From the letter it seems that CD had asked Harcourt about birds that were rare or occasional visitors to Madeira and whether there were any migratory species based there. Harcourt was the author of *A sketch of Madeira; containing information for the traveller, or invalid visitor* (Harcourt 1851). In the 'To be read' section of his reading notebook, CD wrote: 'Vernon Harcourt has published account of Madeira with list of Birds (*some migratory*). Yarrell has'. On 8 June 1855, CD noted that he had read the work (*Correspondence* vol. 4, Appendix IV, *128: 173; 128: 12). CD's notes on Harcourt 1851 are in DAR 71: 87–8.

[2] Richard Thomas Lowe had been chaplain on Madeira, 1832–52.

[3] Harcourt's pamphlet on the ornithology of Madeira (Harcourt 1855) is now in DAR 196.4. It is annotated by CD. On the final page (p. 8) Harcourt added the information: 'I have since received from Madeira the *Fringilla velata*, vel *Hyphantorius textor*, shot by a friend in June/55, and the *Prion brevirostris* of Gould (new to science) & *Corvus frugilegus*.'

[4] CD's copy of Harcourt 1855 has been marked, as described in the letter, by Harcourt. To one or two species he also added further information as to the age or sex of the specimen seen. The list of birds given by Harcourt in the letter follows the order given in Harcourt 1855, p. 8.

[5] Webb and Berthelot 1836–50, 2 (pt 2) Zoologie: 1–48. Horace Bénédict Alfred Moquin-Tandon was actually the author of the 'Ornithologie Canarienne'.

[6] Malherbe 1846.

[7] [Asso y del Rio] 1784.

[8] [Widdrington] 1834.

[9] William Jardine.

[10] CD did not own a copy of Webb and Berthelot. He recorded that he had read the work on 30 January 1846 with the comment: '(My notes with Hookers Copy)' (*Correspondence* vol. 4, Appendix IV, 119: 16a). See also letter to Charles Lyell, 16 [June 1856], in which CD asked to borrow Lyell's copy. Some undated notes on this work are in DAR 196.4.

[11] John Gould exhibited a new species of *Prion* 'through the kindness of Mr. [William] Yarrell' at the Zoological Society of London in 1855 (J. Gould 1855).

[12] When CD came to write up his species book, he made the following comment on migratory birds (*Natural selection*, p. 494):

> I have been much struck in the case of oceanic islands, lying at no excessive distance from the main-land, but which for reasons to be given in a future chapter, I do not believe have ever been joined to the mainland, with the fact that they seem most rarely to have any migratory Birds. Mr. E. V. Harcourt who has written on the birds of Madeira informs me that there are none at Madeira: . . .

He also discussed the information given in this letter in *Natural selection*, pp. 256–7.

To W. B. Tegetmeier 31 May [1856][1]

Down Bromley Kent
May 31st

My dear Sir

I have to thank you for manifold kindnesses.— I have received the eggs safely & put them under a Hen.—[2]

I am particularly obliged to you for mentioning the Angora Doe, for I have just lately been comparing 2 or 3 skeletons, & I find differences enough to make me wish to go on; therefore I sh[d] be extremely much obliged if you would purchase her for me, if she goes under say 15[s], but I hope she may fetch less. Could you get the Porter to stick her, for I do not want her alive, & she would get knocked about & half-starved in our cross country Roads.

I find that it ruins the skull to kill a rabbit in the ordinary way by a blow, & I sh[d] think it would be difficult to break the neck below the atlas.— I really do not wish or expect you to do so disagreeable task as to stick the poor beast, but I daresay the same Porter whom you employ to carry her (addressed to C. Darwin, care of M[r] Acton, P. Office, Bromley Kent) would do it— You must please remember that I must pay you for all little contingent expences. With respect to Malays &c, I sh[d] be glad at any time of a Cock for skeleton if one goes cheap..—

Should you ever stumble across any good natured Rabbit fancier, will you remember my wish for dead Bodies of fine specimens.—

With respect to the Laughers, I shall be in London on June 21st & shall leave 57 Queen Anne St Cavendish Sq^{re} on that day, if that would not be too long for you to keep the Birds.—[3]

Upon my life it is almost laughable the trouble I cause you, & how you can be so goodnatured, I hardly understand. I am only just in time to catch Post

Yours etc | C. Darwin

New York Botanical Garden Library (Charles Finney Cox collection)

[1] Dated by the reference to a forthcoming trip to London (see n. 3, below).
[2] Eggs from Tegetmeier's silver-spangled Polish hens (see letters to W. B. Tegetmeier, 25 April [1856] and 11 May [1856]).
[3] CD went to London on 18 June 1856 and returned on 21 June (Emma Darwin's diary). The Queen Anne Street address is that of his brother Erasmus Alvey, with whom he stayed.

To J. D. Hooker 1 June [1856][1]

Down Bromley Kent
June 1st—

My dear Hooker

I read in your note as far as "unutterable mortification" & was in despair, for I came instantly to the conclusion that probably Government had determined to give up Kew Gardens!! & you may imagine how I laughed when I came to the real Cause of your mortification.[2] It is the funniest thing in the world that you do not rejoice; for you have (& I never have) put in print that you do not believe in multiple creation, & therefore you surely sh^d rejoice at every conceivable means of dispersal. Well, I & my wife have enjoyed a jolly laugh, & all the more from fully believing for a second that some great calamity had befallen you.— If you *publish* a note on the seed, please be sure state that the seeds "were procured by the kind intervention of H. M. Consul Mr T. Carew Hunt". Unfortunately I have not been able to read the name of the man who *actually* sent them; I fancy he is acting consul & the name looks like Ives

The leaf from Old Red Sandstone is a grand fact.—[3]

Adios my dear Hooker | Yours hatefully triumphant | C. D.

The Handkerchief is Mrs Hooker.

I am very sorry to hear about Sharpe.—[4]

I really cannot get up steam for London so soon again: I was horribly knocked up by the Fireworks.—[5]

I shall be particularly glad to see your Review of Decandolle whenever I can.[6]

DAR 114.3: 164

[1] Dated by the references to the death of Daniel Sharpe and to CD's recent London trip (see nn. 4 and 5, below).

2 A reference to seeds sent to CD by Thomas Carew Hunt and evidently passed on to Hooker to be planted in Kew Gardens. The seeds had been washed up on the Azores Islands (see *Correspondence* vol. 5, letter from T. C. Hunt, 2 July 1855). Hooker's 'mortification' stemmed from his belief that seawater would mostly kill plant seeds.

3 Probably a reference to the discovery of fossil plants in Kiltorcan, Co. Kilkenny, in deposits considered the same age as the Old Red Sandstone. The fossils, the earliest plant remains known at that time, were sent to Adolphe Théodore Brongniart in Paris and subsequently announced in R. Griffith 1857. They were identified as belonging to a new genus, *Cyclostigma*, by Samuel Haughton, who gave a full description of the specimens in Haughton 1859 (originally intended to be read in conjunction with R. Griffith 1857).

4 Daniel Sharpe died on 31 May 1856 following a riding accident, only a few months after being elected president of the Geological Society.

5 CD had gone to London on 29 May 1856 (Emma Darwin's diary) and attended a council meeting of the Royal Society on 30 May (Royal Society council minutes). On the evening of 29 May, as part of the celebration of the peace treaty concluding the Crimean War, there was a great pyrotechnic display organised by Ralph Fenwick at four localities in London (*Annual Register* (1856), Chronicle, p. 116) The final display consisted of five illuminated fixed pieces, the last bearing the words 'God save the Queen', and the simultaneous firing of 10,000 rockets in red, blue, green, and yellow from each of the four stations.

6 Hooker's review of Alphonse de Candolle's *Géographie botanique raisonnée* (A. de Candolle 1855) appeared in *Hooker's Journal of Botany and Kew Garden Miscellany* in seven parts in 1856 ([J. D. Hooker] 1856). An annotated copy is in the Darwin Pamphlet Collection–CUL.

To S. P. Woodward 3 June [1856][1]

Down Bromley Kent
June 3d.

My dear Sir

I have just finished studying with all the attention of which I am capable, your Book.[2] And I, for one, am deeply indebted to you for its publication, as I have not derived for years so much solid instruction & interest, from any other book.—

I have some questions, which I am *most* anxious to ask you: I have written them down;[3] but I think I shall cause you less trouble, by reading them over with you. (if you are kindly willing to assist me) & then leaving them if necessary.— I shall be in London in about a fortnight & will then call & see whether you will give me a little help—[4]

Yours very sincerely | Ch. Darwin.

American Philosophical Society (129)

1 Dated by the reference to Woodward 1851–6 (see n. 2, below).

2 Woodward 1851–6. CD recorded having read this work on 5 June 1856 (*Correspondence* vol. 4, Appendix IV, 128: 18). See letter to S. P. Woodward, 15 May [1856].

3 Probably letter to S. P. Woodward, [after 4 June 1856].

4 Emma Darwin's diary indicates that CD was in London on 12 June 1856 and again from 18 to 21 June.

To W. D. Fox   4 June [1856][1]

Down Bromley Kent
June 4th

My dear Fox

I write one line only (for I have been writing till I am quite wearied) to say that a Dorking Cock, more like an ostrich than a simple fowl, arrived this morning; & I presume from you.—

It is a splendid acquisition, & I thank you much for it.— It will make a grand skeleton.—

I hope all is well with you.— | My dear Fox | Yours most truly | Ch. Darwin

American Philosophical Society (130)

[1] Dated by the relationship to letter to W. D. Fox, 8 [June 1856].

To W. B. Tegetmeier   4 June [1856][1]

Down Bromley Kent
June 4th

My dear Sir

The Rabbit arrived safely & I thank you heartily for all the trouble which you have so kindly taken for me.[2]

I send a P.O. (payable to *William* B. Tegetmeier) for 1£, as I thought it would be best to send even sum, & you can keep, if you kindly will, the few shillings in advance.[3] Then you will have to pay a Porter to bring the Birds to me on Saturday *morning* the 21st of June.—[4]

With my repeated thanks. | In Haste | Yours very sincerely | Ch. Darwin
How cheap the Rabbit was!

New York Botanical Garden Library (Charles Finney Cox collection)

[1] Dated by an entry in CD's Account book (Down House MS) (see n. 3, below).
[2] The Angora doe mentioned in letter to W. B. Tegetmeier, 31 May [1856].
[3] There is an entry in CD's Account book (Down House MS) for a payment to Tegetmeier on 4 June 1856. It reads: 'Tegetmeier Rabbit & Pigeon: Peas [pigeon food]'.
[4] Tegetmeier sent a pair of laughing pigeons to Erasmus Alvey Darwin's house on 21 June for CD to take home with him on his return from London (see letters to W. B. Tegetmeier, 31 May [1856] and 24 June [1856]).

From S. P. Woodward   4 June 1856

British Museum.
June 4th— 1856.

Dear Sir

I am sorry you are not coming to the Meeting to-night—as I have a paper on the highly popular subject of *Orthocerata!*—[1] Poor Mr Sharpe—the loss is great to the Societies—but greater still to those individual members who like myself enjoyed his friendship & relied upon his aid. He had recovered so far as to sit up & chip fossils! And then relapsed.[2]

You honour me greatly—but the credit belongs to those whose opinions I have collected—Waterhouse, Forbes, Hooker &c[3] I am glad you do not object to having been yourself brought in to testify to heterodox theories.[4]

If you will send me the written memoranda you refer to (ie. if it does not give you the trouble of writing them out afresh for the purpose)—I will give them my best attention—[5] I fear I am *very* slow— I seldom see the force of a joke in less than 48 hours— But I might come to some conclusion by the period of your visit, a fortnight hence.

After the rec.[t] of your last letter[6] I consulted with Mr Waterhouse again about the *Island Faunas*—& we agreed that the doctrine might be maintained against all comers. Sir Chas. Lyell is *coming* to the same opinion—tho' he would account for it metaphysically. We can start with a good case in *Tasmania*—an island which must have been separated from Australia *since* the creation of *Thylacinus* & *Dasyurus* (which occur *fossil* in the mainland) & *before* the arrival of those species peculiar to the continent. In the north of Australia additional groups occur, & faint *Asiatic symptoms* appear—and we cannot doubt that if a land way from India to Australia had remained *since* the epoch of the *Felidæ*, some of that wandering tribe would have found their way & made sad havoc with the poor Opossums— The "stream of migration" has been from Asia towards Tasmania—but has been arrested periodically, so that Tasmania represents the oldest condition—Australia the next—New Guinea & Timor next—& then there is a vast interval between them & Borneo.

Respecting the Falklands—Fuegia & Chiloe—my impression is that they have been *drowned* at no very remote period, & so lost all symptoms of the **ancient** Fauna of S. America— Their present isolation cannot date so further back than the Newer Tertiary period, in which the modern vulgar Foxes took the lead in repeopling them.

Some years ago I had a gossip with Mr Whewell in Combination-room,[7] when he cordially agreed to this view, as more probable than the Lyellian doctrine.

I should think Dr Pickering must hold something like the same notion—from his chapter on the Probable scene of the Creation of Man.[8] He hesitates between the *Area* of the Orangs, & that of the *Chimpanzees* & seems inclined to make the first man black!

I am strongly impressed with a conviction of the **oneness** of the scheme of Creation— But collecting *data* is a serious matter! Mr Waterhouse advises me not to abandon the project of a general exposition of my theory, & promises to supply the Mammalian & Insect *facts*. I want chiefly a monster-map (on the conical projection)—so as to map out the data & put them to the test of discussion— If I live another six years I may do it— Meanwhile I look eagerly for the publication of your *specific* researches!

Yours sincerely | S. P. Woodward

Chas Darwin Esqr

DAR 205.3 (Letters)

CD ANNOTATIONS
0.1 British Museum . . . metaphysically. 4.4] *crossed pencil*
4.8 and we cannot . . . Opossums— 4.10] *double scored brown crayon*
4.8 and we cannot . . . Borneo. 4.14] 'Mastodon awful difficulty' *added pencil*
6.1 Some years . . . Woodward 9.1] *crossed pencil*
Top of first page: '19'^9 *brown crayon*

[1] Woodward read his paper (Woodward 1856a) at a meeting of the Geological Society on 4 June 1856.
[2] Daniel Sharpe, president of the Geological Society and a fellow of the Royal, Linnean, and Zoological Societies, had died on 31 May 1856 as a result of a fall from his horse on 20 May. He was working on a memoir on the Mollusca of the Chalk for the Palaeontographical Society at the time of his death (*DNB*).
[3] George Robert Waterhouse and Edward Forbes were cited at length in Woodward 1851–6, 3: 349–54 and 381–3, as Woodward's sources for his general discussion of the geographical distribution of land and marine Mollusca. Joseph Dalton Hooker was cited in Woodward 1851–6, 3: 406.
[4] A reference to Woodward's belief that island faunas were usually older than those of continents (see letter from S. P. Woodward, 2 May 1856). CD had been 'inclined very much to dispute' this doctrine (see letter to S. P. Woodward, 27 May 1856). Woodward had used information from CD's *Journal of researches* on the shells of St Helena and Ascension to support his view of the antiquity of such faunas (see Woodward 1851–6, 3: 389).
[5] See following letter.
[6] Letter to S. P. Woodward, 27 May 1856.
[7] William Whewell was the master of Trinity College, Cambridge. The combination room in Cambridge colleges is the fellows' common room.
[8] Charles Pickering, author of *The races of man* (Pickering [1848]), discussed briefly the possible origins of the several races of man in chapter 20 of his book.
[9] The number of CD's portfolio of notes on the geographical distribution of animals.

To S. P. Woodward [after 4 June 1856][1]

[Down]

p. 353. Genera found in Arctic countries, & in S. Hemisphere.[2]
Are they never found in Tropics??[3]

Chrysodomus (p 109) no southern Habitat given)[4] (p. 109) Cuming[5] says in St. of Macassar

Trophon (I thought there were some aberrant Tropical forms) North. Falklands. New Zealand.[6]

Trichotropis (p. 109) (no southern habitat given.)[7]

Margarita (p 144) Greenland. Brit. Falkland Isl^d.[8]

Rhynchonella (p. 227) Melville Is^d &c. New Zealand.[9]

Crenella (p. 266) Nova Zembla. &c New Zealand.[10]

Yoldia (p 270) "Arctic & Antarctic seas— Greenland.[11] Massachussetts **Brazil**!! (no S. Habitat given)[12]

Astarte (p. 299) Behring St. Norway **Canaries** (no S. Habitat given)[13]

(I see Forbes (in new map.) gives Patina in N. & S. & *not* in Tropics.—)[14]

p. 371. Is it really certain that Monoceros is in New Zealand?[15] Was Sowerby right that one of my Patagonian fossils is a Struthiolaria?[16]

Is Bankivia confined to Cape & N. Zealand, (page wrong in Index).—[17] Do you mean that Venus Stutchburyi & Modiolarca trapezina are common to New Zealand & Kerguelen Land.—[18]

p. 382. Do you know whether any of the *same* species of F. W. shells have very wide ranges.?—[19] p. 397. Does D.[r] Gould give distribution of individual species of Land Molluscs in Pacific islands.—[20] Pfeiffer gives several in common to rather distant isl[ds].; but can he be trusted?[21]

p. 410 How do you calculate that each Geolog. period = 3 times average duration of species.[22]

p. 411. (first paragraph) do you think you have really good evidence in regard to numbers of individuals.[23]

p. 411. (3.[d] par.) & p. 419—(bottom) Longevity of Land & F.W. molluscs great: is it so compared with sea-molluscs, or land vertebrata? Is there means of comparing longevity of *land* & *F.W.* Molluscs?[24]

p. 414 Table, what meaning of names in *Italics*?[25]

p 421. Is Cardium edule *var*. rusticum, found either fossil or recent elswhere, besides in White Sea.[26]

p 421. (4.th Par) I cannot find what is distrib. of recent species of Fulgur:[27] Is it certain that there is a Gnathodon (p 309) in Moreton Bay? What is range of living Mercenaria (p 305).—[28]

A memorandum
DAR 72: 59–61

[1] Dated on the assumption that this is the list of questions referred to in the letters to S. P. Woodward, 3 June [1856], and from S. P. Woodward, 4 June 1856. It seems probable that CD posted the questions to Woodward, as requested in the preceding letter, rather than delivering them in person, as was first suggested in the letter to S. P. Woodward, 3 June [1856]. Woodward wrote his answers on the list of questions and returned it to CD, probably during CD's next visit to London on 18–21 June. Most of the questions posed in the letter were first noted in the margins of CD's copy of Woodward 1851–6, now in the Darwin Library–CUL.

[2] The page numbers refer to Woodward 1851–6. Woodward wrote notes on the list and then returned it to CD. CD subsequently wrote remarks next to Woodward's notes.

[3] CD underlined this sentence again and overwrote the second '?' in brown crayon.

[4] Woodward added '= Buccinum — antarcticum'; under 'antarcticum' he subsequently wrote '[= Pisania]'. He also wrote 'Buc. Donovani' in pencil, then crossed this out in ink. CD drew a line through 'Chrysodomus (p 109)' in pencil. At the bottom of the page, keyed to this passage, Woodward wrote:

Fusus Fontainei D'Orb. ⎫   p. 377—
— Zealandica Q & G.   ⎬   = Chrysodomus (Neptunea)   all in Brit. M.
— dilatata Q & G.   ⎭   teste Adams.

"Buccinum" antarcticum, Falklands, = Pisania–Bivon. operc. claw-like.
—— ligatum &c Cape = Cominella, Gr. operc. like Turbinella

Woodward refers to Orbigny 1835–47 and Quoy and Gaimard 1830[−4]. Next to this, CD wrote 'Used' in pencil.

[5] Hugh Cuming. CD wrote 'Ask' in pencil before Cuming's name.

[6] Woodward underlined CD's 'aberrant' and added 'aberrant' after 'New Zealand.' and 'aberrant U.S. *Fusus* cinereus (p. 380) is called *Trophon* at p. 359.'

[7] Woodward added 'a *New Zealand* shell in the Brit. Mus. is referred to this genus'. CD added '(?)' after this in pencil.

[8] Woodward underlined 'Falkland Isl^d.' and added '(D'Orb.)'. He refers to Orbigny 1835–47.

[9] Woodward underlined 'New Zealand' and wrote: 'A new sp. lately sent by McGillivray from the south (*Feejees*?)'. CD added '?' in pencil over Woodward's '?' and wrote 'ancient form' in pencil at the beginning of this section.

[10] Woodward wrote 'typ.' after 'Crenella' and then added 'A minute sp. in *Cuba* (p. 473) *Crenella* (Forbes, including Lanistes, Sw.)'. Following 'New Zealand.', Woodward added 'Brit. U. Greensand'.

[11] After this, Woodward wrote: 'Y. Eightrii (Gould) Sandwich I. ?? Y. sp. p. 270.'

[12] Woodward answered: 'The Brazilian sp. p. 379, (L. cultrata, *Cumana* & L. tellinoides) are scarcely the same genus— Adams will give them a new name—'. The reference is to Adams and Adams [1853–]1858.

[13] Woodward crossed out '(no S. Habitat given)' and wrote: 'A. longirostris, D'Orb p. 378. Falklands. *a single valve, & that lost. A. bilunulata, Florida* p. 380. He also underlined '**Canaries**' and added 'A. fusca also Medit.'

[14] CD refers to E. Forbes 1856. Woodward added after 'Patina': '(Leach) = Nacella, Schum. 1816. Celtic & Lusitanian—not *Arctic* N. cymbularia Fuegia—'. CD added, in pencil: '(A species very like at Cape.)'.

[15] Woodward responded:

No! Monoceros calcar?? ⎫
·   —— terullatus   ⎬   Dieffenbach (Gray)

CD underlined Woodward's 'No!' three times, and in his copy of Woodward 1851–6, 3: 371, CD wrote next to this genus: '**No**'.

[16] Woodward wrote: 'certainly wrong. see p. 130.' Woodward 1851–6, 1: 130, reads: 'Australia and New Zealand; where alone it occurs sub-fossil.' George Brettingham Sowerby described CD's fossil shells for *South America*.

[17] In answer to CD's question, Woodward wrote: 'Swan R—Tasmania—Sandy Cape.' Woodward underlined CD's 'page wrong' and added: 'p. 144 *is right.*'

[18] Woodward crossed out all this sentence except 'Modiolarca trapezina' and added: 'is common to Fuegia & Kerguelen'. CD later wrote in pencil: '(adheres to floating seaweed.', presumably following the description in Woodward 1851–6, 2: 266. He also added a large exclamation mark in pencil after his original query. CD later crossed the letter up to this point in pencil.

[19] Woodward wrote:

British freshwater shells in N. India. in U.S. 6 sp. p. 399. p 384, 18 British freshwater shells range to *Siberia* (Limnæidæ in N. America.)
    Brit. F. W. shells in Tibet— Limnæa stagnalis, peregra, auricularis truncatula? Valvator piscinalis — Corbicula consobrina —
    Cyclas rhomboidea, Say = Paddington Canal!
    —— partrenseia Say. Ohio = Cornea?

After 'Cornea', CD wrote 'British' in pencil.

20 Woodward responded: '**Yes**'. CD refers to Augustus Addison Gould's monograph on the Mollusca collected during the United States Exploring Expedition (A. A. Gould 1852–6).

21 CD added in pencil 'in Pacific' after 'isl[ds]'. Woodward added another '?' and commented: ' "*Helix similaris* is found wherever the coffee plant grows; *H. vitrinoides* in like manner accompanies the *Arum esculentum*" Gould.' CD refers to Pfeiffer 1848.

22 Woodward drew the following diagram to illustrate his point.

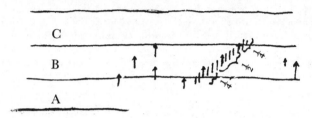

23 CD underlined 'individuals' in pencil. Woodward wrote: 'Yes! This is not my experience only—it is the conviction of all *collectors*.'

24 Woodward wrote: 'There are no recent *sea*-shells in Eocene strata—but there are land & freshwater sp. e.g. *Helix labyrinthica & Melanopsis buccinoides*. The *Genera* of recent *fr. water* shells date back furthest.' CD added in pencil: 'than even land-shells.'

25 Woodward wrote: '*Cephalopoda*—marking organic eras'.

26 Woodward answered: 'Black Sea p. 365. Britain p. 357..'

27 Woodward added: 'W. Indies p. 379. Mass.— S. Carolina p. 380.' and 'F. perversus B. of Campeachy.'

28 Woodward responded: 'The 2 *Mazatlan* sp. are correct. (see Cuming's Col.) For *Moreton B.* I have no authority but Petit de la Saussaye in Journ. Conch.' Woodward refers to the *Journal de la Conchyliologie*, the first four volumes (1850–4) of which were edited by Jean François de Paule Louis Petit de La Saussaye. After this, CD wrote in pencil: 'Fulgar thinks American. Mercenaria unimportant genus'. He also added in pencil: 'In eocene beds greater *resemblance* in F.W. & L shells than in sea-shells.' and 'Has some evidence that Aralo-Caspian was miocene'.

From S. P. Woodward [after 4 June 1856][1]

p. 399.[2] *Vitrina limpida*, Gould— Lake Superior ill-distinguished from the British sp. See also. *V. Angelicæ* of Greenland.

Marine shells of New England ill-distinguishable from British p. 358.

Psammobia fusca = Tellina solidula var?

————— sordida = calcaria?

Cryptodon flexuosus Brit. = *C. Gouldii*, Phil. of Greenland & US.

Lima sulcata, Leach of Arctic seas & US probably same as *L: subauriculata* Mont.

Terebratulina caput-serpentis Brit = T. septentrionalis, Couth. Mass.

Cancellaria buccinoides, Couth. Mass. = C. viridula, Müll. Norway.

Trichotropis conica, Möll Greenland = var. of T. Borealis Brit.

A memorandum incomplete
DAR 205.9 (Letters)

CD ANNOTATIONS

Left margin: '22'³ *brown crayon*

Bottom of page: 'Note given me by Woodward cases of representative shells doubtful whether vars or species— not many if any such cases in Mediterranean' *pencil*

¹ Dated by the relationship to the preceding letter. This note supplements the answers given by Woodward on the manuscript of the preceding letter (see CD's annotations).

² The page references are to Woodward 1851–6, where most of the species noted are referred to. The general argument in Woodward 1851–6, 3: 398–9, also alluded to on p. 358, was that the Canadian lakes contain a few freshwater molluscs identical, or nearly identical, to European species. Woodward proposed that these species 'strengthen the evidence . . . of a land-way across the north Atlantic having remained till after the epoch of the existing animals and plants.' (p. 398).

³ The number of CD's portfolio of notes dealing mainly with palaeontology and extinction.

To W. B. D. Mantell 5 June [1856–9]¹

Down Bromley Kent
June 5th

My dear Sir

I am much obliged for your kind remembrance of my wish to hear particulars in regard to the iceberg seen by you & I thank you for having sent them to me.—²

I think that you were so kind as to answer fully all my previous questions.—

I am not now at work on Barnacles, but I shall like someday to see at Brit Museum, the specimens to which you refer.—

With many thanks, Pray believe me, My dear Sir | Yours very sincerely | Ch. Darwin

P.S. | When you return to N. Zealand, I hope that you will kindly remember that the aboriginal rat³ (& the Frog)⁴ are great desiderata in Natural History.

Alexander Turnbull Library (Mantell 83.268)

¹ Mantell, normally resident in New Zealand, was in England during this period.

² See letters to W. B. D. Mantell, 3 April [1856] and 10 April [1856].

³ *Rattus exulans* Peale, a small brown rat, was evidently introduced into New Zealand at an early period since its bones have been found with those of the moa. When CD visited New Zealand, the species was disappearing rapidly as a result of the arrival of the Norway rat (see *Journal of researches*, p. 511).

⁴ In *Origin*, p. 393, after stating that frogs are never found on oceanic islands, CD wrote: 'I have, however, been assured that a frog exists on the mountains of the great island of New Zealand; but I suspect that this exception (if the information be correct) may be explained through glacial agency.'

From Hewett Cottrell Watson 5 June 1856

Thames Ditton
June 5th 1856

My dear Sir

Allow me to suppose your letter divisible into four questions, instead of two.

1ˢᵗ Sir Charles Lyell's wish or advice seems sound, that you should not postpone publication of your views to a distant date;—rather, look forward to perfect or improve at the future time.[1]

2$\underline{d}$ My own volumes on distribution of British plants, prior to Cybele Britannica, were either three forms or three editions—expanding onwards.[2] The third went very much into detail, but only one part (3 orders) was printed as an experiment, & not published. Had that third form been proceeded with, the information you want as to the species common to Europe & America would have been found. But as only 39 species were treated, the mere fragment would be practically useless. A Table or List in the second form ("Remarks") attempts to give the information, but incompletely & too condensedly.[3] At its date, the published data for America were far less ample & reliable than is now the case; & the distribution of the species beyond Britain was restricted to a single line for each.[4]

3rd America certainly possesses very few European species, which are absent from the Scandinavian Peninsula & Russia northward of 55°. Perhaps it cannot be said positively that no exceptions are found. But among the few which are reported to be so, some are highly questionable in respect to the identity of the species, or to the certain nativity on both continents. I will look out a few examples, to show what is here meant, & write them on a separate paper.[5]

4th In England some 40 or 50 species of plants may be held as restricted, or almost restricted, to tillage ground (ploughed or dug); & about as many more are prevalent chiefly on such ground, altho' more or less extending to road sides, sands, commons, old gravel pits, etc. I think these 50 or 100 species do not run into varieties more than other species of the same genera or same orders, which grow in other situations. If any difference in this respect, it may be that they are less widely variable, on an average, than are the non-agrarian species. But there are some striking instances of variability among the agrarians; and especially among those which grow also in other situations. For example, the varieties of Polygonum aviculare & Viola tricolor are numerous & wide; the agrarian forms differ much from those which grow in other situations; and the agrarian forms vary among themselves also. It would be easy to pick out three, or six, or perhaps more numerous varieties of either of these two species, which would have been received as so many good & distinct species, if they had been brought from different countries, & unaccompanied by those intermediate forms, which seemingly connect them into single species.[6]

Assuming that agrarian weeds differ less widely than others, on an average,—how far may this assumed fact be connected with another fact, that they are almost all of them Annuals? Ask of British botanists, which genera are their greatest opprobria in technical description,—in which is it most difficult to define & describe the Species, & to decide as to which forms should be called Species, & which should be called varieties?— They will answer, Salix, Rubus, Rosa, Mentha, Batrachium (sub-genus of Ranunculus), Potamogeton, Poa, Festuca, Viola, Galium,—Genera exclusively or chiefly of perennial Species.

You can make such use as you deem fit of answers to questions, which you pay me the compliment of asking;—but I fear the replies are unavoidably of that vague & unsatisfactory kind which must render them of no really available use—

Very faithfully | Hewett C. Watson

To C. Darwin Esq

Examples of alleged American & European plants, not extending far north in Europe—[7]

Chrysosplenium oppositifolium. Said to occur in middle Russia & extending eastward to Altai. Absent from Sweden, but found in Denmark & the more southern portion of Norway. In Britain plentiful, extending to Orkney, & high on the Grampian Mountains. Absent from Shetland, Faroe, & more northerly isles.

Was formerly held to be an American plant, found in various parts of Canada, &c. But in the Flo. Bor. Am. of Hooker the Eastern American plant is described as a distinct species, under name of Americanum.[8] Still, Hooker keeps a Western form as a Variety of C. oppositifolium, viz. "β Scouleri", from Columbia river; & he remarks on the resemblance of this Western-American form to the C. Nepalense of N. India, in its leaves, tho' more like C. oppositifolium by its inflorescence. Thus, we have here a species which would have been deemed common to both Continents, tho' not extended far northward in either, if the question had been asked some years ago. Now, they are either different "Species", or different "varieties". Looking beyond these two ("oppositifolium" & "β Scouleri") we find Chamisso joining to them also the C. Nepalense & C. Kamtschaticum, as other varieties of the type species C. oppositifolium;[9] thus tracing this last, but only in its alleged varieties, round the earth in latitudes which attain a culminating point northwards in Orkney.

Isnardia palustris. Less of a boreal plant than the preceding. In Europe, attaining to the South of England, North of Germany, & S.E. of Russia. In America, to the Saskatchawan, & far southward.— Perhaps a true example,—but a semi-aquatic, widely distributed.

Astragalus Hypoglottis. Absent from the Scandinavian Peninsula & N. Russia. Occurs in Scotland, Denmark, Island of Oesel, Middle Russia (say to lat 58°), Siberia. In British America, northward to the Saschatchawan river.

Thlaspi alpestre.— England, Forfarshire, Gottland, Germany, &c; Absent from Russia, & Scandinavian peninsula. A single authority for its occurrence in Canada (Hooker Flo. Bor. Am.) but whether the author had seen specimens, or recorded on statement only, does not appear. Torrie & Gray simply copy, & add "Introduced?"[10] But it is an unlikely plant to be introduced. The European "T. alpestre" has been latterly subdivided into various species by Jordan &c.[11]

Arenaria verna. Not admitted as a plant of Scandinavia, in the Summa Vegetabilium of Fries.[12] It occurs on some few hills in Scotland, northward to Aberdeenshire. Hooker (Flo. Bor. Am.) says that the specimens from Canada & Western America correspond with the A. Verna of Britain. It is to be noted, however, that Ledebour (Flora Rossica)[13] combines with this a number of so-called species, & thus makes the grouped forms, or the name, represent a species extending into arctic latitudes.

Lythrum hyssopifolium— In Europe, northward to England & Lithuania; not in Denmark, Baltic Isles, or Scandinavian peninsula. Locally in the American States, & perhaps introduced.

These half dozen species are perhaps as good examples as can be cited, of species alleged to be common to Europe & America, but having an early limit northward, comparatively with most species certainly common to both continents.

DAR 181

CD ANNOTATIONS

0.3 My dear . . . time. 2.3] *crossed pencil*
3.2 The third . . . each. 3.11] *crossed pencil*
5.9 For example, . . . situations; 5.11] *double scored pencil*
Top of first page: '1'[14] *brown crayon*; 'Agrarian plants not variable some vars. confined to Fields.— Very few plants which do not range far n. are common to America' *pencil*

---

[1] CD 'Began by Lyell's advice *writing* species sketch' on 14 May ('Journal'; Appendix II). See also letter from Charles Lyell, 1–2 May 1856, n. 10.

[2] Watson refers to the three editions of a work that he published under different titles: *Outlines of the geographical distribution of British plants* (Watson 1832); *Remarks on the geographical distribution of British plants* (Watson 1835); and *The geographical distribution of British plants* (Watson 1843). The third edition was, like the first, printed privately, and only one part was published, a copy of which is in the Darwin Library–CUL. The first volume of *Cybele Britannica* appeared in 1847 (Watson 1847–59). This work is also in the Darwin Library–CUL.

[3] There are two tables in Watson 1835 pertinent to CD's point. The 'Table showing the number of British species found in other countries' (Watson 1835, p. 113) indicates that some 480 species were common to North America and Britain. A much longer table (Watson 1835, pp. 187–258) attempted to show the 'geographical extension of British plants beyond 30 °N. latitude' and was arranged species by species, as Watson mentions in the letter. CD recorded that he had read Watson 1835 on 15 June 1856 (*Correspondence* vol. 4, Appendix IV, 128: 18).

[4] CD was writing a chapter on geographical distribution for his species book, the first draft of which was finished in July (*Natural selection*, p. 531). In this he cited the table to which Watson refers (see n. 3, above) but stated that 'Mr. Watson informs me that since his publication in 1835 our knowledge has been much increased, & that the above numbers can be considered only as approximate.' (*Natural selection*, p. 539). CD later deleted this discussion, adding in the margin: 'No give Asa Gray's facts, far more accurate' (*ibid.*, p. 539 n. 2). See letter to Asa Gray, 12 October [1856].

[5] See the second part of the letter, following the valediction.

[6] Watson's comments on *Polygonum aviculare* are cited in *Natural selection*, p. 104.

[7] CD used this information in *Natural selection*, p. 539.

[8] W. J. Hooker 1840.

[9] Chamisso 1831, pp. 557–8.

[10] Torrey and Gray 1838[–43].

[11] Jordan 1846[–9], pt 3: 1–34.

[12] Fries 1846.

[13] Ledebour 1842–53.

[14] CD numbered Watson's letters.

To T. V. Wollaston 6 June [1856][1]

Down Bromley Kent
June 6th

My dear Wollaston

I have to thank you for your note & Lyell's letter.—[2] M.ʳ Janson shall be properly attended to—[3] I have been very glad to see Lyell's letter: it is a capital one, & how well he seems to have read your Book.[4] I agree with *almost* everything which he says there as far as I have gone, which is not half through yet.[5] With respect to your Book, I may say that all which I have read has been *most* interesting (notwithstanding that I remembered well the passages quoted from I.M.[6]) & several of your facts & views have already given me quite devilish puzzles, which you ought to take as a compliment. What Lyell says about links being destroyed, I think, is very true.[7] I did snigger at your "legitimate variation" & I see I dashed the word with a (!).[8]

I have heard Unitarianism called a feather-bed to catch a falling Christian;[9] & I think you are now on just such a feather bed, but I believe you will fall much lower & lower.[10] Do you not feel that "your little exceptions" are getting pretty numerous?[11] It is a funny argument of yours that I (& other horrid wretches like me) may be right, because we are in a very poor minority! anyhow it is a comfort to believe that *some others* will soon be with me.

Adios | Your's very sincerely | C. Darwin[12]

Edinburgh University Library (Gen. 1999/1/29)

[1] Dated by CD's notes about a letter from Charles Lyell to Wollaston (see n. 2, below).

[2] The letter from Lyell to Wollaston concerned Lyell's view of Wollaston 1856. CD's notes on this letter survive in DAR 205.1: 61, 62. The notes are headed: 'Lyell letter to Wollaston. June 1856'.

[3] Perhaps a reference to CD's intention to read Edward Westley Janson's paper on the occurrence of rare Coleoptera in Britain (Janson 1848).

[4] CD's annotated copy of Wollaston 1856, presented to him by the author, is in the Darwin Library–CUL. The work was dedicated to CD, 'Whose researches, in various parts of the world, have added so much to our knowledge of Zoological geography'. CD entered it into his reading notebook on 5 June 1856 (see *Correspondence* vol. 4, Appendix IV, 128: 18).

[5] Among his comments on points raised by Lyell in Lyell's letter to Wollaston, CD made the following note (DAR 205.1: 61v.):

> Lyell says that in [*interl*] the transmutationists the weakest point is the appearance of fresh organs.— *true, but what an example [*added pencil*] (Oh how weak for he instances tail, that man might get tail by wishing, because he has rudiment—but how first get the vertebræ—oh oh—look at [*increase*] in vertebræ—if he mean how ever get vertebræ, I have nothing to do with it

[6] A reference to Wollaston's *Insecta Maderensia* (Wollaston 1854).

[7] Probably a reference to something said by Lyell in his letter to Wollaston (see n. 2, above). Wollaston had cited Lyell's views on the effects of exterminating species in Wollaston 1856, p. 179.

[8] The passage occurs in Wollaston 1856, p. 35, where Wollaston discussed his belief in the ultimate fixity of species:

We should remember, also, that the boundaries of insect instability are restricted; and, although we would advocate freedom of development within limits which are more or less comprehensive according to the species, to pass beyond them would be confusion, and such as could result from a *lapsus Naturæ* only, rather than from a power of legitimate variation.

CD underlined the last two words and added an exclamation mark in the margin.

[9] Erasmus Darwin had defined Unitarianism in this way (*LL* 2: 158).

[10] CD's expectation that Wollaston would be converted to transmutation was disappointed. Wollaston's review of the *Origin* (Wollaston 1860) was one of the more hostile ones.

[11] Wollaston had stated that 'where there is a law there must be an exception to it' (Wollaston 1856, pp. 72–3).

[12] Attached to the original of this letter in the Edinburgh University Library is one from Wollaston to Lyell, indicating that Wollaston was forwarding CD's letter to Lyell.

To W. D. Fox 8 [June 1856]

Down Bromley Kent
8th

My dear Fox

I wrote just to say the splendid Cock had arrived, & now I have got your letter.[1] I am most sincerely sorry, my dear old friend, at all the great suffering you have undergone. I have always understood that Lumbago & sciatica cause extreme pain.— How much illness you have had in your life, & at different times how much misery! It is lucky, when young, that one does not foresee the future. I do hope that the waters are so nasty that they will do you good.—[2] We are all well here, except my poor dear wife who is *wretched* (perhaps in Family way) & has been wretched for weeks; & I fear that it will stop Tenby.[3]

I have been working of late very hard & have now an enormous correspondence.— I think I shall make an interesting collection of domestic Vars. for promises are coming in from all quarters.[4] This morning I heard from Rajah Brooke with promises of energetic assistance.[5] and Hon[bl] Ch. Murray says Pigeons & Fowls are on their road for me from Persia,[6]—as are others from E. Africa.—[7]

At this present moment I am most interested about domestic Rabbits; having just sent an Angora to be skeletonised, & having compared a P. Santo, common, & Hare Rabbit skeletons & found, I believe, most remarkable differences.—[8] Ducks too are my delight: pray do not forget a Call *Drake* or Duck, if you sh[d] have the *misfortune* (oh what hypocrisy!) to lose one. Can you tell me anything about differences in habits in Rabbits? Do you know any breed of Ducks besides tufted, hook-billed, (both of which I have had from Holland); Penguin; Call; & Black Indian or B. Ayres Duck.—[9] Even sub-vars w[d] be of value to me. On account of doubt on origin I have come to care more about these & Pigeons than about Poultry; not that I shall give up poultry.—

Sir C. Lyell was staying here lately, & I told him somewhat of my views on species, & he was sufficiently struck to suggest, (& has since written so strongly to urge me) to me to publish a sort of Preliminary Essay.[10] This I have begun to do, but my work will be horridly imperfect & with many mistakes, so that I groan &

tremble when I think of it. By the way I want just to put, as an illustration of singular corelations in organisation. That "I have been assured by the Rev. W. D. Fox that he has never seen or heard of a blueish-grey Cat which was not deaf." May I quote you? or must I put in "friend"

Do you still believe in the generality of fact? Might I say that you have seen or heard of as many as 6.?[11] But please mind, do **not** answer this note, if it pains or troubles you.— I do most truly regret that you were not able to pay us a visit; I sh^d have so much enjoyed it.—

My dear old friend | Yours most truly | Ch. Darwin

I have sent you a very long note all about myself.— And now to write to Borneo & the Cape of Good Hope![12]

Postmark: JU 10 1856
University of British Columbia, Woodward Library

[1] CD's letter to W. D. Fox, 4 June [1856], was forwarded to Fox at Harrogate, where he had gone to take the waters, by his wife Ellen Sophia. The cock had been sent from Fox's rectory at Delamere, but Fox's letter was presumably sent from Harrogate. CD addressed this letter to the 'Old Parsonage | High Harrogate'.

[2] In a letter written on the back of CD's letter of 4 June [1856] (see n. 1, above), Ellen Sophia Fox mentioned Fox 'drinking the odious waters' at Harrogate.

[3] Emma Darwin was expecting their tenth child, born December 1856. CD and Emma did not go to Tenby in July as planned, judging from an entry of 29 July in Emma's diary: 'Willy came home from Tenby'.

[4] CD had sent out a letter to correspondents in different parts of the world requesting information on variation in various kinds of domestic animals (see *Correspondence* vol. 5, CD memorandum, [December 1855]).

[5] James Brooke, raja of Saráwak, Borneo, sent CD specimens of pigeons and ducks (*Variation* 1: 132 n. 1, 280).

[6] Charles Augustus Murray was envoy and minister plenipotentiary to the court of Persia. See letter to W. B. Tegetmeier, 29 November [1856].

[7] Perhaps from S. Erhardt, who provided CD with information on domestic animals in East Africa (see *Variation* 1: 25, 246 n.).

[8] CD included an analysis of the osteological characters of rabbits in *Variation* 1: 115–24.

[9] These breeds are discussed in *Variation* 1: 276–7.

[10] Charles Lyell and his wife had visited Down from 13 to 16 April 1856. See letter from Charles Lyell, 1–2 May 1856, and letter to Charles Lyell, 3 May [1856].

[11] The manuscript of the first two chapters of CD's species book, composed in 1856 and dealing with 'Variation under domestication', is not extant. CD later used much of the material, and perhaps the manuscript itself, in compiling *Variation*. In *Variation* 2: 329, after making the statement that 'white cats, if they have blue eyes, are almost always deaf', he wrote: 'The Rev. W. Darwin Fox informs me that he has seen more than a dozen instances of this correlation in English, Persian, and Danish cats; but he adds "that, if one eye, as I have several times observed, be not blue, the cat hears. On the other hand, I have never seen a white cat with eyes of the common colour that was deaf."' The information also appears in *Origin*, p. 12, but without attribution.

[12] A reference to James Brooke and Edgar Leopold Layard. The letter to Brooke has not been found, but see the following letter.

To Edgar Leopold Layard 8 June [1856][1]

Down Bromley Kent
June 8th

My dear Sir

Let me thank you cordially for your really to me *very* valuable letter.[2] The kind spirit with which you have answered my not little troublesome letter has gratified me extremely.—

I admire & honour your zeal in thinking of your Madagascar expedition, & I have no doubt, if you go, you will do much for Natural History.[3] It is such a good piece of fortune for me, as I could never have anticipated; for it is almost certain that there will be some odd breeds of domestic animals there.—

My chief object in writing now, besides thanking you, is to trouble you with two or three questions, if you can find leisure (for you seem indeed to be greatly overworked) to answer them.— Firstly I do hope that you will try & discover whether the hybrid Cats are fertile; I fear that this w[d] be very difficult inter se; but with either parent you cannot fail to discover.[4] Secondly Licktenstein asserts that the natives have a breed of domestic dogs like the C. mesomelas, & I think he asserts that they sometimes get a cross with the wild species to improve the Breed;[5] I wish you w[d] enquire in regard to this.— Thirdly I have always been curious to know, where many breeds of dogs are promiscuously crossed, whether any sort of uniformity is acquired in the mongrel race: how is this at the Cape?

Fourthly would you kindly take the trouble to ask M[r] Fry two questions;[6] whether, if he can remember, the feral Pigeons at Ascension, had black bars on wing & white rumps, or whether they were chequered like common Dove cot.—[7] Also did he ever see in N. Africa, a Grey-Hound, with a *very* short & much curled tail: such a Dog is figured on the ancient monuments, & has been said now to exist, but I cannot avoid doubting this.[8]

Lastly (& in truth I am ashamed to ask so much) I shall be most grateful for any information, from the Pigeon Fancier, mentioned to you by M[r] Fry, in regard to the Pigeons of the Cape. If any breed, it is supposed, has long been at Cape, *even if very slightly different*, I should wish *beyond all measure* for a specimen; & so with *Ducks & Poultry*.[9] Country Farm Houses w[d] offer only chance.— I am much interested about Ducks, & find some great peculiarities in their skeletons.— I venture to repeat that very slight differences interest me *greatly*.

With respect to your question about Books by which to make out sea-fowl, I have not knowledge to answer, but I will enquire when next in London: but from similar enquiries I made formerly, I much doubt, whether anything like a general synopsis is published.

With cordial thanks for your very kind & interesting letter, pray believe me | My dear Sir | Your's very sincerely | Charles Darwin

American Philosophical Society (143)

¹ Dated by CD's reference in the preceding letter to his intention to write to 'the Cape of Good Hope'. Layard was employed by the judicial branch of the civil service in Cape Town. He was also curator of the natural history museum there.

² This letter has not been found. CD had written to Layard in 1855 requesting information on domestic animals (see *Correspondence* vol. 5, letter to E. L. Layard, 9 December 1855).

³ Layard undertook a voyage to Mauritius, Mombasa, Zanzibar, Madagascar, and the ports along the southern coast of South Africa between October 1856 and March 1857 (*SADNB*).

⁴ In his discussion of domestic cats in *Variation* 1: 44, CD stated: 'In South Africa, as Mr. E. Layard informs me, the domestic cat intermingles freely with the wild *F. caffra*'.

⁵ Lichtenstein 1812–15, 2: 272.

⁶ CD cited Layard's acquaintance 'Mr. Fry' as a source of information on domestic cats and fowls in *Variation* 1: 44 and 238 n. 27. Fry has not been identified.

⁷ See *Variation* 1: 190 n. 18, where Layard is cited as CD's authority that the common dovecot pigeon had become feral on Ascension Island. No description was given because Fry was not familiar with pigeon breeds (see letter from E. L. Layard, [September–October 1856]). Fry is cited as confirming that the feral fowls of Ascension did revert to the primitive coloration of the breed (*Variation* 1: 238 n. 27).

⁸ In *Variation* 1: 17, CD wrote: 'The most ancient dog represented on the Egyptian monuments is one of the most singular; it resembles a greyhound, but has long pointed ears and a short curled tail: a closely allied variety still exists in Northern Africa'.

⁹ No information on ducks or fowls from South Africa was included in *Variation*.

To John Lubbock [8 June 1856]¹

Down.
Sunday

Dear Lubbock

Will you be so kind as to lend me, if you are not using it, the Fly-Pincers, like a pair of scissors.

Georgy has taken a passion for Bees,² & I want to see whether they will be useful to him, & if so I can get a pair made for him.—

I have been very sorry to hear how unwell you have been; but I hope you are now strong again.—

Ever yours | C. Darwin

Mʳˢ Darwin continues very poorly.—

Wollaston's book on Insect variation is published & I am reading it.—³

Down House

¹ June 1856d from the reference to CD's reading of Wollaston 1856 (see n. 3, below).

[2] George Howard Darwin was nearly 11 years old.

[3] Wollaston 1856. CD recorded it in his reading notebook on 5 June 1856 (*Correspondence* vol. 4, Appendix IV, 128: 18).

From H. C. Watson 10 June 1856

Thames Ditton
June 10th/56

My dear Sir

I know not whether a few words in reference to Mr. Forbes's views about former expanses of land, alluded to in your letter of 7th., will be worth your trouble in reading them.[1]

The plants specially common to Ireland, S.W. England, W. France, Spain or Portugal, Azores,—or to two or more of these countries, & not general to Europe,—are precisely those, unadapted to bear continental climates.[2] They thrive in mild, humid, equable climates; are killed by low temperatures, or wither under dry heat; some bearing neither. Whatever the present dissevered & isolated distribution of these plants may suggest about former connexions between their present localities, unless assumed to be by *intermediate hilly islands*, the constitution or nature of the species physiologically, seems to negative. I know not a tittle of evidence to show that the species in question are of older creation or existence on the earth, than any others in Britain or Europe.—

The idea of a connexion between Europe & America, (whether quite continuous, or partially interrupted,) in high latitudes, seems rather supported than negatived by botanical facts.[3] True, a good many plants of N.W. Europe are unknown in America; & vice versa. But (as a mere surmise, not investigated) if we could reduce the Ural chain to a few islands (like Iceland, Faroe &c.), & wash away much of the land on both sides, we should have a parallel case:

1	2	3
North	Sea	North
West	&	East
Europe	islands	Asia

Many species would be in common to 1 & 3. Many peculiar to 1 or 3. Many of the common, & some of the peculiar, would be found in the islands of 2.

Forbes was not under the trammels of a very strict Conscience. He would be more likely to find "facts" to suit a conjecture, than those calculated to test its soundness. I should deem any one unwise who adapted his theories to Forbes's statements of facts, unless those facts were verified afresh, or corroborated by others of like nature.[4]

Very faithfully | Hewett C. Watson

To | C. Darwin | Esq

DAR 181

CD ANNOTATIONS

1.1 I know . . . them. 1.3] *crossed pencil*
2.8 I know not . . . Europe.— 2.10] *scored brown crayon; crossed pencil*
Top of first page: '2'⁵ *brown crayon*

¹ CD's letter has not been found, but see the letter from H. C. Watson, 5 June 1856. Watson refers to Edward Forbes's views set out in E. Forbes 1846.
² These plants were called 'The Atlantic type' in Watson 1835, p. 86.
³ Watson listed the British species found in Europe and America in Watson 1835, pp. 91–114. CD discussed this subject in *Natural selection*, pp. 538–44.
⁴ After the publication of E. Forbes 1845 and 1846, Watson had launched a bitter attack on Forbes, claiming that he had appropriated Watson's scheme for separating and describing the various elements of the British flora (see *Correspondence* vol. 3, letter from J. D. Hooker, [before 3 September 1846], n. 2).
⁵ CD numbered this series of Watson's letters sequentially.

To H. C. Watson [after 10 June 1856]¹

And now for my question, which bears on the capital manner in which you compare the relation of Europe & N. America, with what Europe probably would have to E. Asia if intervening tracts were submerged.— It is, whether all, or nearly all (as far as you know) the plants, which are common to Europe & N. America live North of the Arctic Circle.² By rights **N.** Scandinavia ought to be excepted on account of warmth from Gulf Stream from my question, but this would much complicate my question.— I do not know whether you will care to hear what I am speculating on (one never cares for other people's speculations) but for the chance, I will write & you can slightly skim over it, if you so like.— My notions, however, are based on the belief that all the plants common to Europe & N. America do not live within the Arctic circle, (especially if northern Scandinavia be excepted) & in this, I imagine lies the difficulty in accounting for those which are in common to old & new worlds.— From reasons not worth explaining I have been led to speculate that about the middle of the pliocene period, when most of the species (at least of shells) were the same as now, & when the climate was rather warmer than now, the productions now inhabiting middle Europe & N. United States, might have lived fairly within the Arctic Circle, say between 65° & 75°. As between these Latitudes there is almost continuous land, from Scandinavia, by Asia & N. America & Greenland, would not the Flora in all probability have been pretty nearly uniform? As the climate became cooler, I imagine the productions to have been driven a little further south; & during glacial epoch much further south, & since glacial epoch to have remigrated northward, but not so far north, as during the pliocene period.— I fancy some such idea might do instead of Forbe's immense subsidences.³ But even if there be any truth in this notion, it would require more working out than I fear I am at all capable of.—⁴ And really I ought to apologise for troubling you with such vague notions. Finally to put my question in another way,—does not the difficulty in understanding the relations of the Floras of Europe & N. America, depend on the plants in common not inhabiting the Arctic circle lands, where the land is nearly continuous?

Can you forgive me!

Your's most truly obliged | Charles Darwin

Incomplete
DAR 185

[1] Dated on the assumption that this is the reply to the preceding letter.
[2] Watson had attempted a tabulation of this sort in Watson 1835, pp. 187–260.
[3] E. Forbes 1846.
[4] CD's views were fully explained in *Natural selection*, pp. 534–44, and in *Origin*, pp. 365–82. See also J. Browne 1983, pp. 114–31.

To John Lubbock   12 [June 1856][1]

Down.—
12th

Dear Lubbock

Many thanks for your note just received.— **Pray** do not get an insect-catcher made, for I do not know whether it would answer, & I c[d] get it made, if I wanted it.—[2] I had one, & my foolish Boys lost it.— But I write now chiefly to say, to prevent you having a journey for nothing, that I shall be out all Saturday, & in next week from Wednesday to Saturday inclusive.[3] After (or before) that time I hope to be at home every day.—

I am very glad that you are well again.— Have you heard that we have Small Pox in the village. We have had a cruel death in the little Boy, after only 18 hours illness, of our servant Parslow.—[4]

Your's most truly | C. Darwin

Down House

[1] Dated by the reference to CD's intended trips to London (see n. 3, below).
[2] See letter to John Lubbock, [8 June 1856].
[3] CD went to London and back on 12 June 1856 and again from 18 to 21 June (Emma Darwin's diary).
[4] Joseph Parslow was the Darwins' butler.

To W. D. Fox   14 June [1856]

Down Bromley Kent
June 14th

My dear Fox

Very many thanks for the capital information on Cats;[1] I see I had blundered greatly, but I know I have somewhere your original notes; but my notes are so numerous during 19 years collection that it w[d] take me at least a year to go over &

classify them.—² I do not intend to attend systematically to cats, there are such great doubts on origin & they have been crossed in so many countries with native Cats.—

I have bespoken the *so-called* Himmalaya Rabbit in Zoolog. Gardens.³

I am particularly obliged about Call Drake.⁴ Have you ever crossed them with common Ducks: it would be a very valuable experiment for me to know whether the half-bred are fertile inter se, or with some third breed; I am trying this extensively with Pigeons; but I am overwhelmed with subjects & work, & do so wish I was stronger.—

What you say about my Essay, I daresay is very true; & it gave me another fit of the wibber-gibbers; I hope that I shall succeed in making it modest. One great motive is to get information on the many points on which I want it. But I tremble about it, which I sh⁴ not do, if I allowed some 3 or 4 more years to elapse before publishing anything.

My dear old friend | Yours affect^y | C. Darwin

I am off in 10 minutes to a great Pigeon Fancier at Black Heath.—⁵

Postmark: JU 14 1856
Christ's College Library, Cambridge (Fox 98)

¹ See letter to W. D. Fox, 8 [June 1856]. CD cited Fox on cats in *Variation* 2: 329.
² 'I keep from thirty to forty large portfolios, in cabinets with labelled shelves, into which I can at once put a detached reference or memorandum.' (*Autobiography*, p. 137).
³ The gardens of the Zoological Society of London. CD had been a member of the society since 1839. CD was collecting skeletons of wild and domestic rabbits to enable him to make osteological comparisons (see *Variation* 1: 115–30). He cited information given to him by Fox about Himalayan rabbits in *Variation* 1: 109. See also letter to John Thompson?, 26 November [1856].
⁴ See letter to W. D. Fox, 8 [June 1856].
⁵ CD's Address book (Down House MS) gives Matthew Wicking's address as 'Clifton Villa, Blackheath Park'.

To J. S. Henslow   16 June [1856]¹

Down Bromley Kent
June 16th

My dear Henslow

You may remember sending me seed of "Myosotis repens or cæspitosa, Stowmarket".² The next time you go that way, would you be so kind as to gather me a tuft in flower & send it in letter that I may see what the aboriginal is like.— I send one of my cultivated spec⁵, (1st generation) that you may see it, not that I suppose it is anyways remarkable.—

Secondly, can you give me the address of shop in London, where, years ago, I got on your recommendation nice square strong paste-board Boxes, about 15 inches square: I cannot myself remember in the least where it was.—

Thirdly, when will you publish some little Book to show how to teach Botany on Nat. System to children:³ How I wish you would: my children are always asking me, & I have no idea how to begin. If you can't or won't publish, pray tell me what Book I had better get: Lindley's School Bot. is out of Print,[4] which Hooker recommended to me;—not that, I suppose, that would have done to teach children by.— Forgive my 3 questions, & answer them when at leisure, or rather when least busy.

My dear Henslow | Your's most truly | C. Darwin

Your Lychnis-plants are flourishing & I am dosing them & others, with Guano water, salt-petre & common salt, & intend thus to make the most wonderful transformations,—that is, if the plants have any gratitude, for they evidently much like the doctoring.—[5]

DAR 93: 110–11

[1] Dated on the basis of entries in CD's Experimental book (DAR 157a) pertaining to the plants discussed in the letter (see nn. 2 and 6, below).

[2] See letter to J. S. Henslow, 22 January [1856]. On 22 January 1856, CD recorded in his Experimental book, p. 1 (DAR 157a) having planted Henslow's seeds. Later in the year, he recorded (p. 7):

> May 21st The flowers of Myosotis repens or cæspitosa in sun in K. Garden, seem from memory twice as big as former pink flowers.— Those in shady Place behind wall are some blue & large & some pink & small. The tube of ['corolla' *del*] calyx seems certainly longer in the blue than in pink flowers, in proportion to division of calyx.— June 16th The specimens in shade have larger flowers with emarginate petals in slight degree.

[3] Henslow had been teaching botany to children in the Hitcham parish school since 1852. His 'Practical lessons in botany for beginners of all classes' were published in a series of fourteen articles in the *Gardeners' Chronicle and Agricultural Gazette* beginning in July 1856. See also Russell-Gebbett 1977, chapter 4, for an account of Henslow's school teaching. No separate primer of the natural system in botany was published by Henslow.

[4] Lindley 1854, the third edition of John Lindley's *School botany: an explanation of the characters of the principal natural classes and orders of plants* (London, 1839).

[5] See *Correspondence* vol. 5, letters to J. S. Henslow, 11 July [1855], 21 July [1855], and 28 July [1855]. On 13 June 1856, CD recorded in his Experimental book, p. 9 (DAR 157a):

> The *Lychnis (red) dioica* which has had so much salt & Guano, in the female stems are certainly less red & less hairy than the other plants [*interl*] which only occasionally have had Guano alone.

CD was attempting, by varying external conditions, to 'break' the constitution of plants.

To Charles Lyell 16 [June 1856][1]

Down
16th.—

My dear Lyell

I am going to do the most impudent thing in the world. But my blood gets hot with passion & runs cold alternately at the geological strides which many of your disciples are taking.

Here, poor Forbes made a continent to N. America & another (or the same) to the Gulf weed.—[2] Hooker makes one from New Zealand to S. America & round the world to Kerguelen Land.[3] Here is Wollaston speaking of Madeira & P. Santo "as the sure & certain witnesses" of a former continent.[4] Here is Woodward writes to me if you grant a continent over 200 or 300 miles of ocean-depths (as if that was nothing) why not extend a continent to every island in the Pacific & Atlantic oceans![5]

And all this within the existence of recent species! If you do not stop this, if there be a lower region for the punishment of geologists, I believe, my great master, you will go there. Why your disciples in a slow & creeping manner beat all the old catastrophists who ever lived.— You will live to be the great chief of the catastrophists!

There, I have done myself a great deal of good & have exploded my passion. So my master forgive me & believe me | Ever yours | C. Darwin

P.S. | When you go abroad you are to lend me Webb. & Heer, & can you add Maury ocean chart;[6] Woodward had it sometime ago.

Dont answer this, I did it to ease myself.—

American Philosophical Society (131)

[1] Dated by Lyell's reply (see following letter).
[2] E. Forbes 1846, pp. 348–50, 382–3, and 401–2.
[3] J. D. Hooker 1853–5, 1: xxiii.
[4] In Wollaston 1854, p. x, Thomas Vernon Wollaston referred to 'that ancient continent of which these Atlantic clusters are the sure witnesses.'
[5] Samuel Pickworth Woodward's letter has not been found, but CD had questioned Woodward on his belief that oceanic islands were of great antiquity in letter to S. P. Woodward, 27 May 1856. See also letter from S. P. Woodward, 4 June 1856.
[6] Webb and Berthelot 1836–50; Heer 1855 (see letter from Charles Lyell, 1–2 May 1856); and Maury 1855a.

From Charles Lyell 17 June 1856

53 Harley St., Lond.,
June 17, 1856.

My dear Darwin

I wonder you did not also mention D. Sharpe's paper just published[1] by which the Alps were submerged as far as 9000 ft. of their present elevation above the sea in the Glacial Period and then since uplifted again.[2] Without admitting this you would probably convey the Alpine boulders to the Jura by marine currents and if so make the Alps and Jura islands in the Glacial Sea. And would not the Glacial theory as now very generally understood immerse as much of Europe as I did in my original Map of Europe when I simply expressed all the area which at some time or other had been under water since the commencement of the *Eocene* Period.[3] I almost suspect the glacial submergence would exceed it.

But would not this be a measure of the movement in every other area northern (arctic) antarctic or tropical during an equal period—oceanic or continental for the conversion of sea into land would always equal the turning of much land into sea.

But all this would be done in a fraction of the Pliocene Period— The Glacial shells are barely 1 per cent extinct species?

Multiply this by the older Pliocene and Miocene epochs.

You also forget an author who by means of Attols contrived to submerge archipelagos (or continents?) the mountains of which must originally have differed from each other in height 8000 (or 10000?) feet, so that they all just rose to the surface at one level, or their sites are marked by buoys of coral.[4] I could never feel sure whether he meant this tremendous catastrophe, all brought about by what Sedgwick[5] called "Lyell's niggling operations" to have been effected during the era of existing *species* of corals. Perhaps you can tell me for I am really curious to know. At all events he did not I suspect go back as far as the faluns of Tourain for the commencement of the era of the Coral Building spec[s] of Atolls now growing[6]

Now altho' there is nothing in my works to warrant the building up of continents in the Atlantic and Pacific even since the Eocene period yet as some of the rocks in the central Alps are in part Eocene I begin to think that all Continents and Oceans may be chiefly if not all post-Eocene and Dana's "Atlantic Ocean" of the *Lower Silurian* is childish (see the Anniversary Address, 1856).[7] But how far you are at liberty to call up continents from "the vasty deep" as often as you want to convey a helix from the U.S. to Europe in *Miocene* or *Pliocene* periods is a question; for the Ocean is getting deeper of late, and Haughton says the *mean* depth is 11 miles! by his late paper on tides.[8] I shall be surprised if this turns out true by soundings.

I thought your mind was expanding so much in regard to time that you would have been going ahead in regard to the possibility of mountain chains being created in a fraction of the period required to convert a swan into a goose or vice versâ. Nine feet did the Rimutaka chain of N. Zealand gain in height in Jan. 1855 and a great earthquake has occurred in N. Zealand every 7 years for half a century nearly. The Washingtonia (Californian conifer) lately exhibited was 4000 years old,[9] so that one individual might see a chain of hills rise, and rise with it, much less a species—and those islands which J. Hooker describes as covered with N. Zealand plants 300? miles to the N.E.? of N.Z. may have been separated from the mainland 2 or 3 or 4 generations of Washingtonias ago.?[10]

If the identity of the land shells of all the hundreds of British Isles be owing to their having been united since the Glacial Period and the discordance almost total of the shells of Porto Santo and Madeira be owing to their having been separated all the newer and possibly older pliocene periods, then it gives us a conception of Time which will aid you much in your conversion of species, if immensity of Time will do all you require for the glacial period is thus shown as we might have anticipated to be contemptible in duration or in distance from us, as compared to the older Pliocene, let alone the Miocene when our contemporary species were tho' in a minority, already beginning to flourish.

The *littoral* shells according to Macandrew imply that Madeira and the Canaries were once joined to the main land of Europe or Africa but that those isles were disjoined so long ago that most of the species came in since.[11] In short the marine shells tell the same story as the land shells. Why do the plants of P° S° and Madeira agree so nearly? And why do the shells which are the same as European or African species remain quite unaltered like the Crag species which returned unchanged to the British seas after being expelled from them by Glacial cold, when 2 millions? of years had elapsed, and after such migration to milder seas. Be so good as to explain all this in your next letter.[12]

Sincerely yours | C. Lyell

Copy
DAR 146

[1] Sharpe 1856, which had been read on 5 December 1855. Daniel Sharpe had died on 31 May 1856.

[2] Sharpe 1856, pp. 118–23.

[3] The map given in the second volume of C. Lyell 1830–3, facing p. 304, entitled 'Map shewing the extent of surface in Europe which has been covered by water since the commencement of the deposition of the older Tertiary strata (strata of the Paris and London basins &c &c.)'.

[4] An allusion to CD's theory that coral reefs originated by the gradual growth of corals as the floor of the ocean subsided (*Coral reefs* (1842)).

[5] Adam Sedgwick.

[6] The faluns of Touraine was identified by Lyell as a Miocene deposit, similar to but older than the coralline Crag of Suffolk (C. Lyell 1855, pp. 176–9). Nearly all the corals identified in the faluns were different from living forms, so that if the subsidence in the Pacific had taken place during the era of existing coral species, it must have occurred after the deposition of the faluns beds.

[7] An address given by James Dwight Dana as president of the American Association for the Advancement of Science in August 1855 (Dana 1855). Dana had suggested that during the Silurian period the United States was entirely covered by shallow seas.

[8] Haughton 1856, pp. 137–9, read on 24 April 1854. Samuel Haughton also presented his results in a short preliminary notice published earlier (Haughton 1853–4).

[9] Lyell refers to the first specimen of the giant sequoia tree, sometimes known as *Washingtonia*, to have been seen in England. It was brought to London in 1853 by William Lobb, who collected plants in California and Oregon from 1849 to 1857 for the Veitch nursery (R. Desmond 1977, p. 391). Now known as *Sequoia gigantea*, this specimen was first given the name *Wellingtonia* by John Lindley (*EB*).

[10] It is not clear whether Lyell is referring to Chatham Island, the flora of which Joseph Dalton Hooker described as the same as that of New Zealand in J. D. Hooker 1853–5, 1: vii, or to Raoul Island, in the Kermadec group, which also possesses a New Zealand flora but is some 600 miles from New Zealand. Hooker wrote a paper on the botany of Raoul Island in 1857 (J. D. Hooker 1857). CD's reply indicates that he thought Lyell referred to Raoul Island (see letter to Charles Lyell, 25 June [1856]).

[11] McAndrew 1854. CD had corresponded with Robert McAndrew about the meaning of his results in 1855 (see *Correspondence* vol. 5, letter to Robert McAndrew, 6 October 1855).

[12] See letter to Charles Lyell, 25 June [1856]. Lyell evidently considered his questions to be important for he made a rough abstract of the main points in his scientific journal (see Wilson ed. 1970, pp. 104–5).

To J. D. Hooker 17–18 [June 1856][1]

Down Bromley Kent
17th

My dear Hooker

I was actually wishing *much* to hear what you thought on the two subjects, to which your note, received this morning is chiefly devoted.— I did not like to give up the time to form a very certain judgment to my own satisfaction, in Falconer v. Huxley. But the article struck me as very clever.—[2] I rather lean to the Huxley side, & without Falconer can say that he could have told, without the knowledge of habits of any bears, that the Polar bear was carnivorous & the brown bear frugivorous from structure alone, I think Huxleys argument best.—[3] But to deny all reasoning from adaptation & so called final causes, seems to me preposterous.[4] But I am most heartily sorry at the whole dispute: it will prevent two very good men from being friends.— It was a pity that Falconer ludged in Owen into Huxley's "id genus omne" & *not fair*; but I deprecate the contemptuous tone of Huxley on compilers &c. What is a Lecturer but a compiler?—[5]

I have been very deeply interested by Wollaston's book,[6] though I differ *greatly* from many of his doctrines. Did you ever read anything so rich, considering how very far he goes, as his denunciations against those who go further, "most mischievous" "absurd", "unsound". Theology is at the bottom of some of this. I told him he was like Calvin burning a heretick.—[7] It is a very valuable & clever book in my opinion.— He has evidently read very little out of his own line: I urged him to read the New Zealand Essay.[8] His Geology also is rather *eocene*, as I told him. In fact I wrote most frankly; I fear too frankly; he says he is sure that *ultra*!-honesty is my Characteristic: I do not know whether he meant it as a sneer; I hope not.—

Talking of eocene geology, I got so wrath about the Atlantic continent, more especially from a note from Woodward (who has published a **capital** book on shells) who does not seem to doubt that *every island* in Pacific & Atlantic are the remains of continents, submerged within period of existing species; that I fairly exploded & wrote to Lyell to protest & summed up all the continents created of late years by Forbes, (the head *sinner!*) *yourself*, Wollaston, & Woodward & a pretty nice little extension of land they make altogether![9] I am fairly *rabid* on the question & therefore, if not wrong already, am pretty sure to become so.—

I have just reread your note & it seems to me that there is great justness in your remarks on Huxley & the general question, being discussed as it has been discussed.—[10]

I have enjoyed your note much

Adios. | C. Darwin

P.S | 18th Lyell has written me a **capital** letter on your side,[11] which ought to upset me entirely, but I cannot say it does quite.— Though I must try & cease being rabid & try to feel humble, & allow you all to make continents, as easily as a Cook does pancakes.—

P.S. | Just think of this question & tell me at Club,[12] I believe *N.W.* portion of America is rather more temperate than middle parts of America & of Asia; are there, any plants there common to Europe & *not* found in middle parts of America & middle parts of Asia?

DAR 114.3: 170

[1] Dated by the reference to Falconer 1856 (see n. 2, below).

[2] A reference to Hugh Falconer's strongly worded defence (Falconer 1856) of Georges Cuvier's palaeontological doctrines, which had been attacked by Thomas Henry Huxley in T. H. Huxley 1856a. For CD's first response to the argument, see letter to J. D. Hooker, 21 [May 1856]. CD's copy of Falconer 1856 is in the Darwin Pamphlet Collection–CUL.

[3] Huxley had asked: 'If bears were only known to exist in the fossil state, would any anatomist venture to conclude from the skull and teeth alone, that the white bear is naturally carnivorous, while the brown bear is naturally frugivorous?' (T. H. Huxley 1856a, p. 191).

[4] Huxley had insisted that the principle of physiological correlation or coadaptation of organs was a 'utilitarian principle, [which though] valuable enough in physiology, helps us no further, and is utterly insufficient as an instrument of morphological research' (T. H. Huxley 1856a, p. 192).

[5] Huxley had referred to 'compilers, and copyers, and popularizers, and *id genus omne*' who handed down from 'book to book, that all Cuvier's restorations of extinct animals were effected by means of the principle of the physiological correlations of organs' (T. H. Huxley 1856a, p. 190). In his defence of the validity of Cuvier's methodology, Falconer cited Richard Owen's work on the Stonesfield fossil mammals as showing beyond reasonable doubt that there were both placental and marsupial forms (Falconer 1856, pp. 91–2). Richard Owen was the most prominent advocate of the principle of the physiological correlation of parts. The word 'ludged' was a mistake for 'lugged', and in the manuscript it has been altered to 'lugged' in pencil by an unidentified hand.

[6] Wollaston 1856.

[7] This remark is not in an extant letter; but see letter to T. V. Wollaston, 6 June [1856].

[8] CD refers to the introductory essay of J. D. Hooker 1853–5, in which J. D. Hooker had examined contemporary views about the variability and fixity of plant species.

[9] See letter to Charles Lyell, 16 [June 1856].

[10] In a letter to T. H. Huxley dated 'June? 1856' in L. Huxley ed. 1918, 1: 427, Hooker expressed his favourable opinion of Huxley's critique of Cuvier:

> I have been dissipating the disconsolation of my solitude (rather fine that) by reading old Quarterlies [*Quarterly Review*] . . . and find in xli. 313 a passage that will amuse you and rile Falconer.— 'Under the influence of this *delusion* "the necessary conditions of existence" the deservedly celebrated Cuvier is found asserting that any one who observes only the prints of a cloven hoof, etc., etc.,— it is worth your reading.

[11] Letter from Charles Lyell, 17 June 1856.

[12] The Philosophical Club of the Royal Society met on 19 June 1856 (Bonney 1919). CD was in London from 18 to 21 June and attended the meeting ('Journal'; Appendix II; Royal Society Philosophical Club minutes).

To Robert Everest?[1] 18 June [1856][2]

Down Bromley Kent
June 18th.

Dear Sir

Our mutual friend Dr Falconer has told me that I might use his name as an apology for troubling you.—[3] I am very much interested on the subject of the variation of our domestic animals; & I have seen in several works, statements on the degeneracy of dogs in India & that Bull-Dogs, for instance, not only lost courage, but actually changed in form.[4] I never could quite believe this, but D.[r] Falconer says he feels pretty sure that it is true.[5] He thinks that you could give me reliable information on this head; & it would be a very great kindness, if you could.—[6]

The danger of a cross seems the obvious source of error. If you would be so kind, when at leisure, to answer this note, I should feel extremely much obliged.— I have sometimes gone so far as to doubt whether climate has any direct influence even on colour.—

Pray believe me | Dear Sir | Your's faithfully & obliged | Charles Darwin

I. J. Pincus

[1] The correspondent is conjectured on the basis that CD cited Everest in *Variation* on the deterioration of European breeds in India. Everest, a keen geologist and author of papers on the geology and climate of India, was listed in the *Clergy list* 1855 as residing in Calcutta (see n. 3, below).

[2] Dated from CD's particular interest in eliciting information about domestic animals in other countries at this time (see letter to E. L. Layard, 8 June [1856]).

[3] Hugh Falconer, at the time resident in England, had been superintendent of the Calcutta botanic garden and professor of botany at the Calcutta Medical College from 1848 to 1855.

[4] In *Variation* 1: 36 n. 65, CD cited William Youatt 1845, p. 15, and 'The veterinary' (*The Veterinarian, or monthly journal of veterinarian science*) 11 (1838): 235 (a mistake for p. 535) on the degeneration of dogs in India.

[5] CD cited Falconer on the subject of the bulldog in *Variation* 1: 38.

[6] In *Variation* 1: 36, CD cited Everest on Newfoundland dogs (Everest 1834, p. 19) and two pages later stated that:

> The Rev. R. Everest informs me that he obtained a pair of setters, born in India, which perfectly resembled their Scotch parents: he raised several litters from them in Delhi, taking the most stringent precautions to prevent a cross, but he never succeeded, though this was only the second generation in India, in obtaining a single young dog like its parents in size or make; . . .

From H. C. Watson 20 June 1856

Thames Ditton
June 20. 1856

My dear Sir

The title page of my 'Remarks on Geogr. Distr. Brit. Plts.' bears the date of 1835. The published data towards compiling such a List as that termed 'Geog.[l] Extension' have been greatly increased & corrected in the past score years. So that many gaps in the List would now be filled, & sundry corrections should be made.[1] Still, the general fact remains as there imperfectly appearing, that plants common to Europe & America range far northward, in one or both, with few exceptions.[2]

Since giving my own reply I communicated your two queries to Mr. J. T. Syme, a good European botanist, & asked his opinion.[3] He writes:—

1.st "I quite agree with you that plants of cultivated ground do *not* vary more than others of the same genus, growing in other situations".

2$^{\underline{d}}$ "I agree with you that very few European species are wild in U.S. except such as pass 55°.."

Mr S. illustrates the first of these two opinions, by naming several agrestal species[4] which do not vary so much as allied species which grow in other situations than tilled ground.

In illustrating the second, he gives a list of 22 species, noted in running thro' Gray's Botany of the N. States,[5] which he thinks may not pass much north of 55 in Europe. i.e. "which do not occur in Scandinavia, or beyond 55 in Britain."

The paucity of this list is the essential point; & even in this short list of 22, some 5 or 6 may likely have passed by human agency from Europe to America, or vice versa. Two are common to America & the British Isles, without reaching European continent. Two are common to Europe & E. America, but not British. Three connect with America better by way of E. Asia & W. America, than by way of W. Europe & E. America. Two are littoral plants. Four are natant aquatics; & such plants are often difficult to discriminate specifically, & are widely diffused either in seeming or really.

I am not prepared to say whether the number of species common to E. Asia & W. America is less or more than of those common to W. Europe & E. America. I doubt whether the botany of E. Asia & W. America is sufficiently known to allow of a true comparison.— Under existing knowledge, the list for W. Europe & E. America would proby. be the more numerous.

Truly I cannot at all say whether the Americo-European species, as a prevailing rule, could have spread from one Continent to the other by way of Kamchatka & Aleoutia, with some moderate changes of the land. Perhaps there are too many of these plants not yet known on the two coasts of the upper part of the N. Pacific, & intervening islands, to allow of the hypothesis, that they did so spread.

As to the Hispano-Hibernians. These are few, & you are aware chiefly Heaths & Saxifrages. Unite Ireland, England, & France, & project a few promontories into the Oceanic Bay thus formed, or throw up some few Volcanic cones between Spain & Ireland,—breakwaters to be gradually washed away again,—& you would make the physical conditions by which those Heaths & Saxifrages might pass from Ireland to Spain, or from Spain to Ireland. The Hispano-Hibernian Heaths (Arbutus Unedo, Menziesia polifolia, Erica Mediterranea, ciliaris, Mackaiana) are less patient of cold than the Saxifrages. I suppose all are "semi-alpine" in the latitude of Spain; the Saxifrages are so in that of Ireland. And if the latter naturalize themselves in Britain, it is in the Dales of Yorkshire, about the Lakes of Engl$^{\underline{d}}$, or the borders of the Highlands. Whether they could exist on a coast cold enough to admit of Glaciers, seems doubtful & scarcely probable.—

Sorry I can give no better answers or comments on your queries.— I suspect they must be passed down to a future generation for solution—

Very truly | Hewett C. Watson

DAR 181

CD ANNOTATIONS

1.5 plants . . exceptions. 1.6] *scored brown crayon*
3.1 1ˢᵗ . . . situations". 3.2] *scored brown crayon*
4.1 European . . . 55°." 4.2] *scored brown crayon*
6.1 In . . . Britain." 6.3] *scored brown crayon*
7.1 The paucity . . . really. 7.8] *scored brown crayon*
9.3 Perhaps . . . spread. 9.6] *scored brown crayon*
10.5 make the . . . Spain; 10.9] *scored brown crayon*
Top of first page: '3'⁶ *brown crayon*

[1] Watson had begun to publish a revised edition of his *Remarks on the geographical distribution of British plants* (Watson 1835) in 1843. Only the first part was published (Watson 1843).
[2] See letter from H. C. Watson, 10 June 1856, and letter to H. C. Watson, [after 10 June 1856]. CD appended a note in the manuscript of *Natural selection* concerning this point: 'Asa Gray thinks there are not a few plants common to U.S. & Europe, which do *not* range to Arctic regions.' To this Joseph Dalton Hooker added, when commenting on the chapter, 'Certainly' (*Natural selection*, p. 539 n.).
[3] John Thomas Irvine Boswell-Syme was the co-editor, with Watson, of *The London catalogue of British plants* (Watson and Syme eds. 1853), a work used by CD in his study of variation (see *Correspondence* vol. 5).
[4] Species that grow wild in cultivated land (*OED*).
[5] A. Gray 1848.
[6] CD numbered this series of Watson's letters sequentially.

To J. D. Hooker 22 June [1856][1]

Down
June 22ᵈ

My dear Hooker

A full abstract on the Laburnum case is given in Flora Neue Reihe. VI Jahrgang. 1 B. 1848. p. 26.[2] & I have seen another abstract in Gærtner:[3] the case is worthy of your consideration: I cannot see how there well can be a fallacy; but then I am as credulous as you are sceptical,—oh, how credulous I must be!—[4]

I said I would not trouble you about the plants of N. W. northern N. America,[5] but on reflection, I shᵈ be very glad to know a little more so as to be (*probably*) able to say "Dʳ Hooker informs me that **about** a dozen (or half-dozen, or score) of plants grow there, which have not been found in Siberia or in central (or Eastern?) America.[6]

You see I do not want you to waste time in a rigorous search.— As far as I can make out the distribution of sea-shells in Arctic regions, it seems to agree very closely with that (as far as I can see) of Plants—[7]

I have just received the ticket from Mʳˢ Hooker, please give her my thanks. I daresay I can send it again.— Mʳˢ Darwin is hardly at all better; indeed she has rather gone back.—[8]

Yours very truly | C. Darwin

DAR 114.3: 165

[1] Dated by the relationship to the letter to J. D. Hooker, 17–18 [June 1856].

[2] CD refers to Christian Friedrich Hornschuch's paper on the sporting of plants (Hornschuch 1848) in which an account is given of Siegfried Reissek's case of red flowers appearing on an ordinary yellow-flowered laburnum (Hornschuch 1848, pp. 25–7). CD had recorded Hornschuch 1848 in his list of books to be read (*Correspondence* vol. 4, Appendix IV, *128: 177).

[3] Gärtner 1849, pp. 624–9. Reissek's case is described on p. 628. In his copy of Gärtner 1849, in the Darwin Library–CUL, CD wrote next to this passage:

> Case of sport in common Laburnum with flowers like C. Adami Is not this like the orchis case? Were they sterile? The sports and parent in Austrian Bramble are sterile.— (Herbert has shown are sterile in Hort. Journal)

CD refers to Herbert 1847. CD's annotated copy of the *Journal of the Horticultural Society of London* in which the paper appeared is in the Darwin Library–CUL.

[4] The appearance of different and sometimes also parti-coloured flowers on the laburnum was known to occur only on plants of mixed ancestry. CD had expressed much interest in the phenomenon in 1847 (see *Correspondence* vol. 4, letters to J. D. Hooker, [2 June 1847] and [10 June 1847]) with reference to garden plants at that time called *Cytisus adami* (*Laburnocytisus adamii*). Nevertheless, Reissek had insisted that the parent of his coloured flowers was a pure specimen of the aboriginal species of laburnum, *Cytisus laburnum*. CD's reference indicates that Hooker probably thought Reissek was mistaken about the identity of the plant. See also letter to J. D. Hooker, 8 September [1856].

[5] See letter to J. D. Hooker, 17–18 [June 1856].

[6] See letter from J. D. Hooker, [26 June or 3 July 1856].

[7] This correlation between subarctic plants and the seashells common to the shores of Europe and North America was discussed in *Natural selection*, p. 539.

[8] Emma Darwin, aged 48 on 2 May, was in the fourth month of her tenth pregnancy.

To W. B. Tegetmeier   24 June [1856][1]

Down Bromley Kent
June 24th

My dear Sir

I brought home the Laughers quite safely:[2] I sh^{d.} have written sooner, but I feared to give rather a bad Report of one, which seemed ill the next morning, but is now better, but yet not quite well.— I shall be very curious to hear their coo.—

The Scanderoons have a very large massive frame, & if, as I suspect, they are youngish Birds, will be ultimately very large.—[3] I shall be really pleased to let you have a pair, whenever they breed; but sh^{d.} they lay only one egg the first time, I shall most likely sacrifice that, as I am very anxious to compare soon all very young Pigeons: but afterwards you may rely on my keeping the very first pair for you.—

If it sh^{d.} turn out that you could spare a pair of Mr Gullivers Runts,[4] they would be very valuable to me. I have a pair of poorish Runts from Mr Baily,[5] but the Hen seems to be quite sterile; & I shall kill her before long for skeleton, if she does not improve her ways.— I am really glad to hear so good an account of your Polands: mine are Hatched, but I have not seen them yet as I was forced to send them to Farm House; we not then having a broody Hen.

Yours very sincerely | Ch. Darwin

I counted my Pigeons the other day & I have 89!

P.S. | Do you ever see Mr Gulliver, if so, I wish you would kindly remind him, what a valuable treasure to me an old dead Bird would be.— I did once venture to ask him—

New York Botanical Garden Library (Charles Finney Cox collection)

[1] Dated by the relationship to the letter to W. B. Tegetmeier, 25 April [1856], in which laughing pigeons and the possibility of acquiring the eggs of Polish fowl were mentioned.
[2] CD had been in London from 18 to 21 June 1856 ('Journal'; Appendix II). The pigeons were probably the ones mentioned in letter to W. B. Tegetmeier, 25 April [1856]. For CD's description and history of this breed, see *Variation* 1: 155, 207.
[3] See *Variation* 1: 142–3.
[4] See letter to W. B. Tegetmeier, 21 September [1856].
[5] See letter to W. D. Fox, 3 January [1856], n. 4.

To Charles Lyell 25 June [1856][1]

Down Bromley Kent
June 25th.—

My dear Lyell

I will have the following tremendous letter copied to make the reading easier, & as I want to keep a copy.—[2]

As you say you would like to hear my reasons for being most unwilling to believe in the continental extensions of late authors, I gladly write them; as, without I am convinced of my error, I shall have to give them condensed in my Essay, when I discuss single and multiple creation.[3] I shall therefore be particularly glad to have your general opinion on them. I may *quite likely* have persuaded myself in my wrath that there is more in them than there is.— If there was much more reason to admit a continental extension in any one or two instances (as in Madeira) then in other cases, I should feel no difficulty whatever. But if on account of European plants and littoral sea-shells, it is thought necessary to join Madeira to mainland, Hooker is quite right to join New Holland to New Zealand, and Aukland Isl^d (and Raoul Is^d to N.E) and these to S. America and the Falklands, and these to Tristan d'Acunha, and these to Kerguelen Land;—thus making, either strictly at the same time, or at different periods, but all within the life of recent beings an almost circumpolar belt of land.[4] So again Galapagos and Juan Fernandez must be joined to America; and if we trust to littoral sea-shells, the Galapagos must have been joined to Pacific isl^d (2400 miles distant, as well as to America; and as Woodward seems to think all the islands in the Pacific into a magnificent continent:[5] also the islands in the Southern Indian Ocean into another continent with Madagascar and Africa and perhaps India. In the N. Atlantic, Europe will stretch half way across the ocean to the Azores, and further North, right across. In short we must suppose probably, half the present ocean was land within the period of living organisms.— The globe within this period must have had a quite different aspect. Now the only way to test this, that I can see, is to consider whether the continents have undergone within this same period such wonderful permutations. In all N. and S. and central America, we have both recent and miocene (or eocene) shells, quite distinct on the opposite sides, and hence I cannot doubt that *fundamentally* America has held its place since at least the miocene period. In Africa almost all the living shells are distinct on the opposite sides of the intertropical regions, short as the distance is compared to the range of marine mollusca in uninterrupted seas; hence I infer that Africa has existed since our present species were created. Even the isthmus of Suez, and the Aralo-Caspian basin have had a great antiquity. So I imagine from the Tertiary deposits, has India. In Australia the great Fauna of extinct Marsupials, shows that before the present mammals appeared, Australia was a separate continent. I do not for one second doubt that very large portions of all these continents have undergone *great* changes of level within this period, but yet I conclude that fundamentally they stood as barriers in the sea, where they now stand; and therefore I should require the weightiest evidence to make me believe in such immense changes within the period of living organisms in our oceans; where, moreover from the great depths, the changes must have been vaster in a vertical sense.

Secondly. Submerge our present continents, leaving a few mountain peaks as islands, and what will the character of the islands be.— Consider that the Pyrenees, Sierra Nevada, Apennines, Alps, Carpathians are non-volcanic Etna and Caucausus, volcanic.— In Asia, Altai and Himmalaya, I believe non-volcanic.— In N. Africa the non-volcanic as I imagine alps of Abyssinia and of the Atlas: in S. Africa the Snow Mountains.— In Australia the non-volcanic Alps. In N. America, the White Mountains, Alleghanies, and Rocky Mountains, some of the latter alone I believe volcanic. In S. America to the East, the non-volcanic Silla of Caraccas and Itacolumi of Brazil, further South the S. Ventana; and in the Cordilleras, many volcanic but not all.— Now compare these peaks with the oceanic islands; as far as known all are volcanic, except S.t Pauls (a strange bedevilled rock) and the Seychelles, if this latter can be called oceanic in the line of Madagascar; the Falklands only 500 miles off are only a shallow bank; New Caledonia, hardly oceanic, is another exception.— This argument has to me great weight. Compare on a Geographical[6] map, islands which, we have *several* reasons to suppose were connected with mainland, as Sardinia and how different it appears. Believing, as I am inclined, that continents as continents, and oceans as oceans, are of immense antiquity, I should say that if the existing oceanic islands have any relation of any kind to continents, they are forming continents; and that by the time they could form a continent the volcanos would be denuded to their cores, leaving peaks of syenite diorite or porphyry. But have we nowhere any last wreck of a continent in the midst of the ocean? S.t Pauls rock and such are old battered volcanic islands, as S.t Helena may be; but I think we can see some reason why we should have less evidence of sinking than of rising continents (if my view in my Coral volume has any truth in it, viz: that volcanic outbursts accompany rising areas) for during subsidence there will be no compensating agent at work,— in rising areas there will be the *additional* element of out-poured volcanic matter.

Thirdly. Considering the depth of ocean, I was, before I got your letter, inclined vehemently to dispute the vast amount of subsidence, but I must strike my colours,— with respect to coral-reefs I carefully guarded against its being supposed that a continent was indicated by the groups of atolls.[7] It is difficult to guess, as it seems to me, the amount of subsidence indicated by coral-reefs; but in such large areas, as the Low Arch: the Marshall Arch. and Laccadive group, it would, judging from the heights of existing oceanic archipelagoes, be odd if some peaks of from 8000 to 10,000 had not been buried.— Even after your letter a suspicion crossed me whether it would be fair to argue from subsidences in middle of greatest oceans to continents; but refreshing my memory by talking with Ramsay in regard to probable thickness in one vertical line of the Silurian and carboniferous formations,[8] it seems there must have been *at least* 10,000 feet of subsidence during these formations in Europe and N. America, and therefore during the continuance of nearly the same set of organic beings. But even 12,000 feet would not be enough for the Azores or for Hooker's continent. I believe Hooker does not infer a continuous continent, but approximate groups of islands, with, if we may judge from existing continents, not *profoundly* deep sea between them: but the argument from the volcanic nature of nearly every existing oceanic island tell against such supposed groups of islands,—for I presume he does not suppose a mere chain of volcanic islands belting the Southern Hemisphere.

Fourthly.— The supposed continental extensions do not seem to me *perfectly* to account for all the phenomena of distribution on islands:—as the absence of mammals and Batrachians; the absence of certain great groups of insects on Madeira, and of Acaciæ and Banksias &c in New Zealand:—the paucity of plants in some cases &c.— Not that those who believe in various accidental means of dispersal can explain most of these cases; but they may at least say that these facts seem hardly compatable with former continuous land.

Finally, for these several reasons, and especially considering it certain (in which you will agree) that we are extremely ignorant of means of dispersal, I cannot avoid thinking that Forbe's Atlantis was an ill-service to Science, as checking a close study of means of dissemination. I shall be really grateful to hear as briefly as you like, whether these arguments have *any* weight with you, putting yourself in the position of an honest judge. I told Hooker I was going to write to you on this subject; and I should like him to read this; but whether he or you will think it worth time and postage remains to be proved.—

Yours most truly. | Charles Darwin

LS(A)
American Philosophical Society (132)

[1] Dated by the relationship to the letter from Charles Lyell, 17 June 1856.

2 The remainder of the letter, except for CD's signature, is in the hand of the copyist. CD corrected the copy and filled in many gaps left by the copyist, as well as making some additions. Only these alterations have been noted in the Manuscript alterations and comments section. CD's draft is preserved in DAR 50 (ser. 4): 6–11.

3 Although CD evidently intended to discuss this topic and mentioned it in the introductory paragraphs of the chapter on geographical distribution in *Natural selection* (p. 534), he did not do so until he included it in *Origin*, pp. 352–6.

4 J. D. Hooker 1853–5, 1: xxi–xxvii.

5 Samuel Pickworth Woodward expressed this view in Woodward 1851–6, 3: 406.

6 In his draft (DAR 50 (ser. 4): 8), CD had written 'geological', but he omitted to correct the copyist's 'Geographical' in the copy sent to Lyell.

7 *Coral reefs*, pp. 142–8. CD had previously discussed this issue with Lyell (see *Correspondence* vol. 2, letter to Charles Lyell, [September – December 1842]). He also commented on the subsidence required by his coral reef theory in his letter to Charles Maclaren, [15 November – December 1842] (*Correspondence* vol. 2).

8 Andrew Crombie Ramsay's researches in Wales had led both Lyell and CD to discuss with him the probable depth of deposits before erosion had taken place (see *Correspondence* vol. 4, letter to Charles Lyell, [2 September 1849], and *Correspondence* vol. 3, letters to Charles Lyell, [3 October 1846], and to A. C. Ramsay, 10 October [1846]). CD may have met Ramsay on 18 June 1856 at a special general meeting held at the Geological Society to elect a new president following the death of Daniel Sharpe.

From J. D. Hooker   [26 June or 3 July 1856][1]

Kew
Thursday N^t

Dear Darwin

I can make no story at all out of the N.W. American plants:[2] the cases I had in view have all turned up in the Altai, & especially in the Baikal Siberia & Dahuria[3], which is a very fertile nook of N. Asia: there are however a considerable number of plants absolutely peculiar to N.W. America, west of the Rocky mountains.[4] Various Asiatic & Europæan plants that advance Eastwards to Sitka & the Aleutians but no further East in Am. advance much further North on the two Pacific coasts than they do in the interior of Asia, & were long supposed (by me at any rate) to be foreign to Siberian Asia. I am however writing to Asa Gray & will ask him if he can give any information.[5]

By the greatest good luck in the world we have two seedlings of the Asiatic *Entada*, growing side by side with the Azorean, & they *are very different species*; so that you have no sooner found proof that the West Indian one will travel with unimpaired vitality to the old world, when it turns out that it has not taken advantage of its powers after all.! a beastly disgusting fact, which I hope will give a little more countenance than you will allow to my dogma, that it it is much more difficult so to wash seeds up that they shall grow, than to transport them.

When opening some *Aristolochia* flowers I find the pollen all escaped & *on the stigma before expansion*—ditto in some *Visca*, the buds being firmly closed.[6]

A Dr Radlkofer[7] tells me that Siebold has *proved* that some ♀ Bees & Butterflies are sometimes fertile without impregnation;[8] is this true?

That paper in Flora to which you allude seems to be very good,[9] I am thinking of getting it translated: would it not be much better for the Ray Club to confine its efforts to translating such things & leave it to Societies to publish original monographs, which *with proper* support the Soc.[s] could do better than the Ray Club.[10]

Thomson[11] has refound near Calcutta one of the most remarkable anomalies in Geogr: dist: of plants in *Aldrovandra*, a *most* singular & curious rare S. Europæan water plant, allied to *Drosera*, of which a drawing exists at the Calcutta Garden, but for which Griffith,[12] Wallich,[13] Falconer, Hook fil & Thomson & scores of others have hunted the length & breadth of India for in vain: but which Thomson has found abundantly in a few small ponds about 5 miles from Calcutta, It proves to be absolutely identical with S. European local plants[14]

Incomplete
DAR 104: 197

CD ANNOTATIONS
0.1 Kew . . . long 1.7] *crossed pencil*
2.1 By . . . them. 2.7] *crossed pencil and ink*
3.1 When . . . closed. 3.2] *square brackets added ink*
3.1 *Aristolochia*] 'labiosa' *added ink*
3.2 *Visca*, . . . closed.] 'No a mistake; is diœcious' *added pencil and ink*
4.1 A . . . Club. 4.5] *crossed ink*
6.1 Thomson . . . Calcutta] 'F. W.' *added brown crayon*
Top of first page: ' I think I may say this these spot *[in circle]*, at N.W. America & near ['Lake' *del*] Baikal, plants are f.[d] whereas not f.[d] in intermediate districts; suits me as well.' *ink*; '20'[15] *brown crayon*; 'Aldrovanda' *pencil*

[1] The conjectured dates are the two Thursdays between the letters to J. D. Hooker, 22 June [1856] and 5 [July 1856].
[2] See letter to J. D. Hooker, 22 June [1856]. CD was anxious to ascertain whether any of the plants that were found in both Europe and North America could also exist under Arctic or sub-Arctic conditions. He had asked Hooker and Hewett Cottrell Watson (letter from H. C. Watson, 5 June 1856) whether these species were known to range northwards. CD's point was explained in *Natural selection*, p. 539, where he attempted to show that the species that are now common to Europe and North America had formerly comprised part of a circumpolar flora that migrated southwards during the glacial period. The case of local species in north-west America, west of the Rocky Mountains, was cited by CD as an example of 'nests of species' having been left behind after the flora retreated northwards (*Natural selection*, p. 540).
[3] The area formerly inhabited by the Dahur cossacks in northern Manchuria.
[4] CD repeated this information in *Natural selection*, p. 540.
[5] Shortly after receiving this letter, CD wrote on his own account to Asa Gray asking the same question (see letter to Asa Gray, 14 July [1856]).
[6] Hooker was mistaken in his observation of *Viscum*, which has separate sexes (see CD's annotations, above, and letter from J. D. Hooker, 10 July 1856).
[7] Ludwig Radlkofer was an authority on both the sexual and asexual reproduction of plants, about which he wrote several books, including *Die Befruchtung der Phanerogamen* (Leipzig, 1856). In 1857 and 1858 he published studies on parthenogenesis.

[8] A reference to Karl Theodor Ernst von Siebold's researches on parthenogenesis (Siebold 1856). Siebold overturned Richard Owen's definition of parthenogenesis (Owen 1849) by showing that the cells from which new organisms developed were true ova and not simply pre-existing 'germinal' cells contained within the parent's body. Siebold demonstrated that these ova were capable of development without fertilisation. See Farley 1982, pp. 100–5.

[9] See letter to J. D. Hooker, 22 June [1856].

[10] See letter to T. H. Huxley, 4 May [1856], in which the financial difficulties of the Ray Society were discussed. The society had undertaken to translate and publish some foreign monographs, but most of its publications were original monographs, including both volumes of CD's *Living Cirripedia* (1851 and 1854).

[11] Thomas Thomson was superintendent of the Calcutta botanic garden and professor of botany at the Calcutta Medical College. He was co-author with Hooker of the *Flora Indica* (J. D. Hooker and Thomson 1855).

[12] William Griffith had travelled widely in Bhutan and Assam between 1835 and 1838 before becoming superintendent of the Calcutta botanic garden. He died in 1845.

[13] Nathaniel Wallich had been superintendent of the Calcutta botanic garden before Griffith and had catalogued the plants in the East India Company's museum in London.

[14] The words 'absolutely . . . plants' were added to the bottom of the page by CD and were presumably copied from the missing part of Hooker's letter.

[15] The number of CD's portfolio of notes on the geographical distribution of plants.

From T. V. Wollaston[1] [c. 27 June 1856][2]

sure that **practically** the greater number of them would be at once allowed. Thus, for instance, if a particular island (e.g. the Dezerta Grande) is found to be for the most part more productive of *large states* than islands alongside it (as may be proved by the fact of intermediate links *in* **each** of the various islands,—the *tendency of the mass* being merely, in that especial locality, to assume a gigantic bulk), it would seem unreasonable to regard a form *on that rock* as specifically distinct from its European analogue *simply because it is a* **trifle larger** *than the latter*.—[3] And so, in other cases.—

As regards the smallness of size as in some measure accounted for by the unnatural in-breeding which a minute area must of necessity entail,—I can conceive it possible that *the power of production* may continue unchecked, (so as to cause the existence of large n^{os} of *individuals*), & **yet the race deteriorate**. Will not this stand the test of analysis?

Loss of flight v. *Increase of Bulk*. The conclusion seems to me rather the other way. If either of the above are unequal (on the compensation theory), I should imagine that the loss of so essential an organ as the wings was *greater* than the gain in stature. This indeed I rather assumed throughout, & therefore expressed my belief that it was only a "partial compensation" [w^h] the

AL incomplete
DAR 205.3 (Letters)

CD ANNOTATIONS

3.1 *Loss . . . the* 3.5] *crossed pencil*
3.2 I should . . . wings 3.3] *double scored brown crayon*; 'N.B. muscles may & must abort' *added pencil*

[1] Wollaston is identified from the handwriting.
[2] Dated on the basis of the relationship to the following letter. In both letters, Wollaston answers queries that CD posed after reading Wollaston 1856. Although the letters are incomplete, differences in the paper indicate that they were not a single letter.
[3] Wollaston's point was described in detail in Wollaston 1856, pp. 80–92. He claimed that the 'annihilation of the powers of flight' in Coleoptera led to a compensatory increase in bulk (p. 88). Nevertheless, he went on to say, size could not be taken as a primary and fixed characteristic of the species; rather, it was what he called one of several 'qualifying results, from isolation' (p. 88). On page 81 of his copy of Wollaston 1856 (Darwin Library–CUL), CD noted that Wollaston 'Thinks decrease of wings increases size in some instances & so makes up for isolation which tends to reduce size'.

From T. V. Wollaston[1] [27 June 1856][2]

it s^{d} have in any way affected (for even a single night) the functions of your alimentary canal; &, could I have foreseen this, I should most certainly have given an anti-dyspeptic chapter, of a sedative tendency.—[3]

As regards your question about the *other* Orders of Madeiran insects, I believe I can say thus much: that they do **not**, apparently, include forms generally anomalous, or at all comparable with the Coleoptera. Indeed that *Atlantic region* (that **continuous** Atlantic province which it puts you in such a rage to think about) was I believe, strictly, a *Beetle-land* (what a glorious place it must have been!),—a Scarabæoideous area of radiation; numbering amongst its endemic forms few remarkable modifications (I mean **primary** modifications, of course) which were not Coleopterous. It might be defined as a Scarabæo- and Helico-metropolis;[4] & magnificent sport it must have been, to our Atlantic ancestors, to give free scope to their hunting propensities in such a land: it makes one's modern entomological blood curdle

AL incomplete
DAR 181

CD ANNOTATIONS

1.1 it s^{d} . . . tendency.— 1.3] *crossed pencil*
2.3 Indeed . . . curdle 2.11] *crossed pencil*
Top of first page: 'June 27/56'

[1] Wollaston is identified from the handwriting.
[2] Dated by CD's annotation.
[3] A reference to CD's criticism of theories of land-bridges as proposed by Edward Forbes, Joseph Dalton Hooker, Wollaston, and Charles Lyell, which he expressed in the letter to Charles Lyell, 16 [June 1856].
[4] CD discussed this view in *Natural selection*, p. 253.

To W. B. Tegetmeier [July 1856][1]

Down Farnborough | Kent

My dear Sir

I am glad to say that the Laughers are quite well; & I have heard one make a very odd note, which I suppose was laughing.[2]

You may *rely on it*, I will rear the first pair & let you have them, of the Scanderoons: they are now sitting, but rather badly & whether they will hatch I am rather doubtful.—

With respect to vertebræ, I sh^d. not like to be quoted; but as far as one casual look went, they *certainly* seemed to have an extra neck vertebra; but whether they have not one less in back, I will not say positively, for though I have now many skeletons, I really have not had time to look to them well—it is so much less trouble to do such work, when one's materials are nearly complete.—[3]

My present so called Scanderoons[4] though much larger than the diseased Bird, which I had from Mr Baker,[5] are not, I believe, nearly so well characterised.—

I shall be very glad to meet you at Anerly.[6]

With many thanks for all your assistance to me, | Yours very sincerely | Ch. Darwin

I am very glad you are going to give a Paper to Zoolog. Soc.—[7]

Your Polands (one died) are going on admirably.—

P.S. I have just looked at skeleton & I believe I have made simple blunder; but the neck has not been well cleaned & I cannot make out certainly.—

New York Botanical Garden Library (Charles Finney Cox collection)

[1] Dated by the relationship to the letter to W. B. Tegetmeier, 24 June [1856], and by CD's reference to the poultry and pigeon show at Anerley (see n. 6, below).

[2] See letter to W. B. Tegetmeier, 24 June [1856].

[3] In *Variation* 1: 165 n. 36, CD remarked that he was not sure that he had correctly designated the different kinds of vertebrae: all breeds examined had twelve cervical (neck) vertebrae. For CD's summary of the osteological characters of the various breeds of pigeons, see *Variation* 1: 162–71.

[4] One of five sub-races of runts that CD described in *Variation* 1: 142–3.

[5] Samuel C. or Charles N. Baker.

[6] The 1856 poultry show at Anerley Gardens, near Sydenham, Kent, was held from 29 July to 1 August. A report of the show is in *Cottage Gardener*, 16 (1856): 338–40.

[7] Tegetmeier 1856 was read at the meeting of 25 November 1856.

To J. E. Gray 1 July [1856][1]

Down Bromley Kent
July 1st—

My dear Gray

You once told me that you would help me in my Essay on variation. I want *much* some information on a point of Geographical Distrib., & to be allowed, to give information on your authority. It is, whether there are genera of Echinoderms, starfish &c, which have species (especially if closely related) in the Northern &

Southern seas, but have not any one species in the Tropical seas? Or whether there are any closely related & representative genera in the north & south, without any closely related genus within the Tropics?[2]

I am quite ignorant about the range of Echinoderms & perhaps all the genera have very confined ranges. Could I find information on this head in any publication?

Pray forgive, if you can, the trouble, & believe me, Yours very sincerely | Ch. Darwin

British Museum (Natural History) (General Library MSS/DAR: 69)

[1] The year is suggested by CD's interest in the distribution of marine animals (see following letter).
[2] Gray is not listed among the authorities cited in the section of *Natural selection* devoted to the distribution of marine animals in the polar and tropical seas (pp. 555-7).

To T. H. Huxley 1 July [1856][1]

Down Bromley Kent
July 1st

My dear Huxley

The Society for Prevention of Cruelty to animals ought to be at me, for troubling you, overworked & unwell as you are; but I do cruelly want one question answered; & you can lay aside *for present* the remarkable case, of "Darwin, an absolute & eternal hermaphrodite"[2] Have you published a Catalogue of the Ascidians in B. Mus;[3] if so & you could lend it me for a few days, I daresay my question would be answered.— My question is, are there any Ascidian genera, with closely allied species in the northern & southern cold or temperate seas, but such genera *not* found anywhere in the Tropical seas.—[4] I have some vague idea that there are some genera of compound Ascidians in this predicament.— But it is very likely that the subject has been so neglected that even if you knew of a genus in north & south, yet you could not form any opinion whether or no it occurred in Tropics. The best chance would be in very northern genera.— I shd like to quote you as authority

Hoping for forgiveness | Yours most truly | C. Darwin

Thanks for the last lecture,[5] which as all the others have done, has interested me much.—

You pay me a grand compliment, far more than I deserve; but this did not lessen my satisfaction.[6]

What success with Examinership?[7]

P.S. | Two closely allied genera, one in north & the other in south, with *no* closely allied in Tropics, is almost equally a case in point. I think it quite possible that Ascidians in spirits may be hardly recognizable, & if so my queries are unanswerable.—

P.S. 2d— | I see I have not answered your question about the antennæ.[8] It is mere chance whether easy or excessively difficult to detect antennæ, depending on nature of surface & amount of cement poured out. Generally young are best. It is easy in some cases for reasons I cannot explain. The best specimens are young attached to calcareous substances which can be dissolved. But you must remember these organs very small. For months, at first, I only obscurely made them out, & could never conceive what they were! How I have puzzled over them!

As you will be a good deal of sea-side for next few years,[9] I wish you would remember to observe, sh^d you chance ever to see a tree washed on shore, will you carefully observe whether any earth, ever so little, is embedded between roots—on account of transport of plants.—

P.S. 3^d | You have some slides with cement-glands of sessile cirripedes in London, please do not destroy them; as I sh^d. like sometime to have them back.—[10] I have antennæ preserved, sh^d. you ever wish to see them.

Imperial College of Science, Technology and Medicine Archives (Huxley 5: 175, 37–9)

[1] The year is indicated by CD's use of information in this letter in chapter 11 of his species book (see n. 4, below), which was completed late in the summer of 1856 (*Natural selection*, p. 531).

[2] CD had asked Huxley to give him examples of organisms that were truly hermaphrodite, that is, in which cross-fertilisation seemed anatomically impossible. (The joke apparently arose through Huxley entering this request in his diary in the form CD quoted.) Such examples would disprove CD's belief that all hermaphrodites could, and occasionally did, cross. The issue is discussed at length in *Natural selection*, pp. 44–6. CD refers to Huxley's observations on ascidians which, though hermaphrodite, were thought by Huxley to release their spermatozoa and ova at different times, thereby being functionally unisexual (*Natural selection*, p. 45). He also mentions having asked Huxley whether he knew of 'any animals whose structure was such that an occasional cross was physically impossible' (*Natural selection*, p. 46). See also letters to T. H. Huxley, 8 July [1856], and to J. D. Hooker, 13 July [1856].

[3] Huxley endeavoured to compile a catalogue of the ascidians in the collection of the British Museum (see T. H. Huxley 1852). The catalogue, however, was never completed by Huxley.

[4] Huxley's reply has not been found, but CD cited it in *Natural selection*, p. 556: 'In the Ascideae, the genus Boltenia has allied species in the arctic & antarctic seas, & Prof. Huxley thinks that the genus is not Tropical; but here again from our ignorance much caution is requisite.' See also letter to T. H. Huxley, 8 July [1856].

[5] CD refers to Huxley's lectures on general natural history, delivered at the School of Mines, that were published serially in the *Medical Times & Gazette* beginning with the issue of 3 June 1856 (T.·H. Huxley 1856–7).

[6] In Huxley's fifth lecture, published on 21 June 1856 and covering the Actinia, Huxley founded his discussion of the coral-forming actinians on CD's *Coral reefs*, warmly praising the work 'as a striking example of the manner in which Geology and Natural History may be made to elucidate one another.' (T. H. Huxley 1856–7, 12: 623).

[7] See letters to T. H. Huxley, 27 May [1856], and to J. W. Lubbock, 27 May [1856]. Huxley was appointed examiner in comparative anatomy and physiology at the University of London. The appointment was announced in the *Medical Times & Gazette* 13 (1856): 103.

[8] CD refers to the antennae of sessile cirripedes. Huxley discussed this group of Crustacea in his twelfth lecture on general natural history. For earlier correspondence between CD and Huxley concerning cirripede anatomy, see *Correspondence* vol. 5.

[9] Huxley had visited Tenby in the summers of 1854 and 1855 to carry out researches on marine invertebrates. He returned there again in September and October 1856 (L. Huxley ed. 1900, 1: 143).

[10] A set of slide specimens of cirripede parts prepared by CD is in the Cambridge Zoology Museum.

From Charles Lyell [1 July 1856]

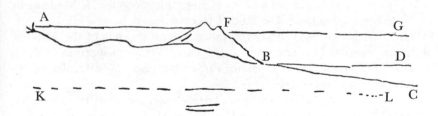

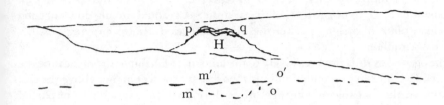

"Whether Volcanos are in areas of elevation"

Extract of a Letter from C Lyell to C Darwin July 1. 1856—[1]

Of course it is true, as you well show in your Coral volume, that the active volcanos have recent deposits with marine shells uplifted in them.[2] This is the case to some small extent in all the principal Atlantic islands except Palma, which I visited, & Palma has not been thoroughly examined & may somewhere exhibit signs of elevation Comparatively therefore, and by contrast with the Atoll areas, you may represent the volcanic as rising.

Still I have always felt a little uncomfortable at being called upon to assume that in recent & pliocene ages volcanic action has been and is connected with the growth of land. Were this the case should we not find that the continents would be the great areas of extinct Pliocene and of active volcanos, and that the latter did not affect sea-side and insular and even mid-ocean sites.

If we find active volcanos in Oceanic areas, & few or none of them in the middle of continental areas, it furnishes a primâ facie case in favour of the doctrine that the grand uplifting power acts very independently of the accidental sites of existing superficial outbreaks. An argument might even be raised in support of the theory that active volcanos are more connected with sinking on a great scale however true it may be that locally they tend to upheave as well as to form land by outpouring of lava & of ejectamenta.

Maurys last chart of the Atlantic[3] makes the Atlantis hypothesis more bold than it appeared when E. Forbes proposed it for the Canaries are separated from Africa & Europe by deep sea depressions of more than 6000 feet & Madeira by depths exceeding 12,000 feet![4] The data, I fear are scanty however.

I find in Madeira & the Canaries upraised littoral deposits of the Miocene period, in my sense of Miocene when there was a certain proportion of living species already in being. This, I think, rather increases the difficulty of the continental extension hypothesis.

But I want to ask you whether it may not be true that the bed of the Atlantic has been gradually sinking all the while the Canaries & Madeiras have been forming & that very slight local upheaval only has occurred even on the sites of these volcanic islands. I sometimes think I can dispense with all excess even of local upheaval, over & above that of the adjoining deep sea spaces.

Thus for example suppose A. B. C. to represent the original Europeo-African continent & B. D. to be the level of the Atlantic.[5] A gradual sinking down of 6000 feet takes place in a short part of the Miocene Period, (not occupying possibly above ½ a million of years.

The ocean has thus risen relatively to the land up to G. But in the meantime the volcano F. has been gradually built up 7500 f^t & is 1500 feet higher above the sea. A pause in the volcanic action takes place during which a subsidence of partial extent under I^d occurs causing F. to lose 1500 feet of its height by slow depression during which every part of the subaerial mass of F. gets submerged & covered or faced with a marine littoral deposit, full of rolled boulders and pebbles with

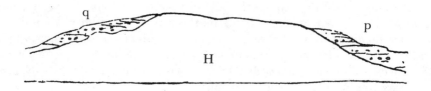

patellæ, & other littoral shells. The subjacent rocks H. all volcanic but as entirely free from marine remains as if exclusively subaerial.

We now have the original subterranean layer K. L. bending down at m, n, o. 1500 feet below the general depression of 6000 feet, & if it be then restored to the level m′ o′ we have the volcano H. pushed up again 1500 feet with the marine beds p. q., abutting against the foundation of older subaerial rocks.[6] This is what I observed in one part of the Grand Canary.

The 4000 or 5000 of additional subaerial volcanic beds may be built up & you have Madeira. In the Grand Canary I suspect most of its height was attained before the submergence of 1500 or (1100? feet.

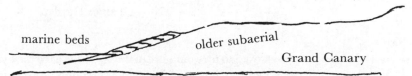

marine beds
older subaerial
Grand Canary

But my reasoning you see is the same as that which I adopted about the Atolls before you invented your theory, namely, that oscillations occurring in a sea filling up with coral or with volcanic matter may cause uplifted marine formations provided subsidence & upheaval be just equal the one to the other.[7]

Take away all the volcanic matter from Etna, Ischia &c & the marine shells could sink down below the sea level. All the marine beds in the Canaries & Madeiras are volcanic except the corals & shells themselves. If the active volcanos were connected with a continent-making power, we should see secondary and non-volcanic rocks uplifted by them.

I do not however want to contend that active & Pliocene volcanos belong to subsiding areas rather than to areas of elevation altho' half inclined to that alternative in preference to the opposite theory. But surely they are so distributed as that they seem to belong quite as much to Pliocene & recent subsidence as to upheaval during the same period.[8]

Contemporary copy
Lady Lyell

[1] The letter has not been found. The heading, date, and text given here are taken from Lyell's scientific journal 2, pp. 82–90 (Kinnordy House MS). It is also printed in Wilson ed. 1970, pp. 110–14.

[2] *Coral reefs* (1842), pp. 140–2, 'On the absence of active Volcanos in the areas of subsidence, and on their frequent presence in the areas of elevation'. Lyell had queried CD's suggestion that volcanoes were mostly associated with rising land in his scientific journal 2, p. 74, in an entry dated 30 June 1856 (Wilson ed. 1970, p. 108):

> There has always seemed to me a difficulty in reconciling two facts in Darwin's theory of volcanic & Coral areas—namely that Volcanoes are the upheaving power and yet, that nearly all the islands in the middle of great oceans are volcanic, whereas there are not many active, nor an extraordinary number of Tertiary volcanoes in continental areas.

[3] Maury 1855a, which Lyell had been studying with reference to the possibility of former land-bridges between Madeira and Africa (see Wilson ed. 1970, pp. 109–10).

[4] Lyell had previously noted the correct depth of 1200 feet in his scientific journal 2, p. 80 (Wilson ed. 1970, p. 110).

[5] Lyell refers to the first diagram given at the beginning of his extract, dated 'June 1st', in his scientific journal (Wilson ed. 1970, p. 111).

[6] Lyell refers to the second diagram given at the beginning of his extract.

[7] C. Lyell 1830–3, 2: 283–301.

[8] Following the letter, Lyell added:

P.S. not sent to Darwin

The deepness of the sea round Madeira & P°. S°. & other Atlantic Islands is against Volcanos being connected with upheaval for the upraising power wᵈ tend at least to render the sea shallow shᵈ it fail to push up dry land in the neighbourhood of oceanic volcanos. July 2ᵈ—/56

From John Henry Gurney 2 July 1856

No 24 Palace Gardens | Kensington
2 July 1856

Sir

M^r C Buxton[1] has forwarded to me your note of the 23^d June as to the hybrids of P. Versicolor which I bought at Knowsley[2] & which certainly appear to breed freely between themselves as well as with the Common Pheasant—[3] I have not however made any accurate experiments as to the degree of relationship between the hybrids so intermixing & by this time these degrees of relationship have been lost sight of

I am assured that the hybrids between the mallard & pintail are sometimes fertile inter se—which I mention as it bears on the same subject & may be interesting to you[4]

I am Sir | Yours faithfully | J H Gurney

Charles Darwin Esq

DAR 165: 259

CD ANNOTATIONS
1.2 bought . . . themselves 1.3] 'N. Q' *added pencil, circled pencil*
2.3 interesting to you] 'See Fiennes in Zoolog Proc' *added pencil*
Top of first page: '17'[5] *brown crayon*

[1] Charles Buxton was married to the eldest daughter of Henry Holland, who was a distant relative of CD's. The Buxtons and the Gurneys were noted Quaker business families.
[2] Edward Smith Stanley, Earl of Derby, had established a private menagerie at Knowsley Hall, Lancashire.
[3] The fertility of offspring of *Phasianus versicolor* crossed with *P. colchicus* is discussed in *Natural selection*, pp. 438, 440.
[4] In *Natural selection*, p. 439, CD cited two cases of a hybrid between the mallard and pintail: the first was a specimen exhibited by Twiselton Fiennes (see CD's annotations, above) at a meeting of the Zoological Society on 13 December 1831 (*Proceedings of the Zoological Society of London* pt 1 (1830–1): 158); the second was mentioned in a letter to the *Magazine of Natural History* by James H. Fennell (*Magazine of Natural History* 9 (1836): 615–16).
[5] The number of CD's portfolio of notes on hybridism.

To J. D. Hooker 5 [July 1856][1]

Down.
5th

My dear Hooker

I have to thank you for a most interesting note,[2] but I want to catch post.— I am going mad & am in despair over your confounded Antarctic isl^d flora.[3]

Will you read over the Tristan list,[4] & see if my remarks are at all accurate:[5] I cannot make out why you consider the vegetation so Fuegian:[6] I suppose you think that many of the species which bear distinct names are really Fuegian.— I have marked with red crosses the genera which seem to me most telling; & these strike me as indicating a *general* relation to southern circumpolar Flora, as much as to Fuegia. But if many of the species are identical this alters the question.— Except Chevreulia, which seems an American genus & Phylica (& partially Pelargonium) which is a Cape genus, the *generic* affinities seem mundane & S. circumpolar. To save all trouble which I can, I send envelope addressed & stamped.— A very few words would help me greatly.—

Ever yours | C. Darwin

I shall write soon

DAR 114.3: 167

[1] Dated on the basis that this is the note referred to in the second letter to J. D. Hooker, 5 July [1856], as having been written the same morning. It was followed by the longer letter in the evening.

[2] Letter from J. D. Hooker, [26 June or 3 July 1856].

[3] CD was completing his chapter on geographical distribution, which included an extensive discussion of the migration of Arctic plants through the tropics to Antarctic regions during a former cold period (see *Natural selection*, pp. 534–54). At the end of his discussion, he referred to the 'most extraordinary' cases of distribution that arose from Hooker's work on the flora of the Antarctic islands (J. D. Hooker 1844–7). CD attempted to explain the close similarities in their floras without invoking any former land-bridges between the islands (*Natural selection*, pp. 560–4).

[4] The list has not been located, but the plants of Tristan d'Acunha are discussed in *Natural selection*, pp. 560–1, 563.

[5] In *Natural selection*, pp. 560–1, CD gave a summary of the flora of Tristan d'Acunha. He considered it to be related to the flora of Tierra del Fuego rather than of South Africa, the island's closest neighbour.

[6] The character of the flora of Tristan d'Acunha is discussed in J. D. Hooker 1844–7 but CD forgot to look there (see letter to J. D. Hooker, 8 [July 1856]). In J. D. Hooker 1844–7, Hooker stated: 'Though only 1000 miles distant from the Cape of Good Hope, and 3000 from the Strait of Magalhaens, the Botany of this island is far more intimately allied to that of Fuegia than Africa.' (2: 216). Hooker went on to state that of the twenty-eight species on Tristan d'Acunha, two were allied to South African forms and seven were either natives of Tierra del Fuego or typical South American plants. This passage is marked in CD's copy of the work in the Darwin Library–CUL.

To Charles Lyell[1] 5 July [1856]

Down, Bromley, Kent
July 5th

My dear Lyell

I am very much obliged for your long letter,[2] which has interested me much. But before coming to the volcanic *Cosmogony*! I must say that I cannot gather your verdict as judge and jury (and not as advocate) on the continental extensions of late authors, which I *must* grapple with, and which as yet strikes me as quite unphilosophical, in as much as such extensions must be applied to every oceanic island, if to any one, as to Madeira; and this I cannot admit, seeing that the skeletons at least of our continents are ancient, and seeing the geological nature of the oceanic islands themselves. Do aid me with your judgment; if I could honestly admit these great extensions, they would do me good service.

With respect to *active* volcanic areas being rising areas, which looks so pretty on the coral-map,[3] I have formerly felt "uncomfortable" on exactly same grounds with you, viz; maritime position of volcanos. And *still more* from the immense thicknesses of Silurian &c volcanic strata; which thickness at first impresses the mind with idea of subsidence; if this could be proved, the theory would be smashed;[4] but in deep oceans, though the bottom were rising great thicknesses of submarine lava might accumulate. But I found after writing coral-book cases in my notes of submarine *vesicular* lava-streams in the upper masses of the Cordillera, formed as I believe during subsidence, which staggered me greatly. With respect to the maritime position of volcanos, I have long been coming to conclusion that there must be some law, causing areas of elevation, (consequently of land) and of subsidence to be paralell (as if balancing each other) and *closely approximate*: I think this from the form of continents with deep ocean on one side,—from coral-map,—and especially from conversations with you on immense subsidences of the carboniferous &c periods, and yet with continued great supply of sediment: if this be so, such areas with opposite movements would probably be separated by sets of paralell cracks, and would be the seat of volcanos and tilts, and consequently volcanos and mountains would be apt to be maritime: but why volcanos should cling to the rising edge of the cracks I cannot conjecture.— That areas with *extinct* volcanic archipelagos may subside to any extent, I do not doubt.

Your view of bottom of Atlantic long sinking with continued volcanic outburst and local elevations at Madeira, Canaries &c grates (but of course I do not know how complex the phenomena are, which are thus explained) against my judgment: my general ideas strongly lead me to believe in elevatory movements being widely extended. The notion of local subsidence under great *volcanic* piles (from shrinking &c) I think receives support from very frequent coincidence of volcanic tertiary streams and lakes or Fresh Water beds,—an idea which, I think, you might work into something, if it be not already, in your Encyclopedic Principles.— One ought, I think, never to forget that when a volcano is in *action*, we have distinct proof of an action from within outwards.— Nor should we forget, as I believe follows from Hopkins,[5] and as I have insisted in my Earthquake Paper,[6] that volcanos and mountain-chains are mere accidents resulting from the elevation of an area, and as mountain-chains are generally long so should I view areas of elevation as generally large.

Your old original view that great oceans must be sinking areas, from there being causes making land and yet there being little land, has always struck me, till lately, as very good. But in some degree this starts from assumption that within periods of which we know anything, there was either a continent in such areas, or at least a sea-bottom of not extreme depth. But my vague ideas on this head are worth absolutely nothing. By the way this letter from brevity of expression may appear as if my notions were dogmatic, which Heaven knows is far from the case.

I am *delighted* that I may say (with absolute truth) that my essay is published at your suggestion;[7] but I hope it will not need so much apology as I at first thought; for I have resolved to make it nearly as complete as my present materials allow. I cannot put in all which you suggest, for it would appear too conceited. I shall not attempt history of subject, but in one page devoted to two or three leading and opposed authorities, I had *already*, after a few remarks on the Principles, ventured on the words—"and with a degree of almost *prophetic* caution which must excite the admiration &c &c." But I shall hereafter beg you to look at what I say on Principles in this one respect.[8]

With hearty thanks | Your's most truly | Ch. Darwin

I wrote this so badly that I have had it copied to facilitate your reading, & I am sure it deserves facilitation.—

Endorsement: 'July 5. 1856'
LS(A)
American Philosophical Society (133)

[1] With the exception of the salutation, date, valediction, and postscript, this letter is in the hand of a copyist. CD corrected the copy before it was sent to Lyell. Alterations and additions made by CD have been noted in the Manuscript alterations and comments section. CD's draft of the letter is preserved in DAR 50 (ser. 4): 1–5.

[2] Letter from Charles Lyell, 1 July 1856.

[3] CD refers to his map (pl. 3) in *Coral reefs* on which fringing reefs, which he believed to be areas of elevation, are coloured red. These coincided with areas of active volcanoes, marked by vermilion spots.

[4] CD, however, found no reason in later years to change his view. The second edition of *Coral reefs*, published in 1874, contains numerous revisions in the light of new knowledge, but the conclusion that active volcanoes generally occurred in areas of elevation remains unchanged. In the preface (p. vi), he continued to maintain that 'volcanos in a state of action are not found within the areas of subsidence, whilst they are often present within those of elevation'.

[5] W. Hopkins 1835, 1836, and 1847.

[6] 'On the connexion of certain volcanic phænomena . . .' (*Collected papers* 1: 53–86). William Hopkins's early papers are cited on pp. 76–8.

[7] Lyell had urged CD to publish his views on natural selection (see letter to Charles Lyell, 3 May [1856]). From an entry in Lyell's notebooks (Kinnordy House MSS, Notebook 213), it seems that CD may have recently written to Lyell to ask whether he might dedicate the proposed book on species to him. Following the comment 'Letter Darwin dedication.', Lyell wrote:

> Y[r] anecdote of my saying that I ought in consistency to have gone for transmut[n]—that I have uniformly taken the other side in all edit[s] but have shewn more inclin[n] to appreciate the simulation of permanent varieties of the character of Species—that I have urged you to publish & set forth all that can be s[d] ag[st] me—that in no book has the gradual dying out & coming in of spec[s] been more insisted upon, nor the necessity of allowing for our ignorance & not assuming breaks in the chain because of no sequence & of admitting lost links owing to small area observed or observable—that finally you hope your book will convert me wholly or in part— To this I c[d] reply in a new Ed of Manual or P. of G— wh. w[d] act in setting the case well before the public—also that in communication with CL. he has f[d] an [approxim[n]] in some points. & knowing that I shall be a fair judge. no unnecessary intervention of unknown or hypothetical agency.

In L. G. Wilson ed. 1970, pp. xlviii–xlix, the date is given as 'June 29, 1856'. It has not been possible to verify this date, nor is it clear whether Lyell's summary served as the basis of a reply to CD. From CD's comments, it would appear that the subject was raised in the letter from Charles Lyell, 1 July 1856, and that Lyell had given permission for CD to say that his work was published at Lyell's suggestion. The section of the letter in which this was discussed is now missing.

[8] The reference to Lyell's *Principles* (C. Lyell 1830–3 and subsequent editions) may have been made in the preliminary pages, which are now missing, of CD's species book (see *Natural selection*, p. 22).

To J. D. Hooker 5 July [1856][1]

Down.
July 5th

My dear Hooker

I wrote this morning in tribulation about Tristan d'Acunha. The more I reflect on your antarctic Flora, the more I am astounded. You give all the facts so clearly & fully, that it is impossible to help speculating on the subject; but it drives me to despair, for I cannot gulp down your continent;[2] & not being able to do so gives in my eyes the multiple creationists an awful triumph. It is a wondrous case,[3] & how strange that A. Decandolle should have ignored it,[4] which he certainly has, as it seems to me.

I wrote Lyell a long geological letter about continents,[5] & I have had a very long & interesting answer;[6] but I cannot in the least gather his opinion about all you continental extensionists; & I have written again beseeching a verdict.[7]

I asked him to send to you my letter,[8] for as it was well copied it would not be troublesome to read; but whether worth reading I really do not know: I have given in it the reasons which make me strongly opposed to continental extensions.

I was very glad to get your note some days ago:[9] I wish you would think it worth while, as you intend to have the Laburnum case translated, to write to "Wien" (that unknown place) & find out how the Laburnum has been behaving since that year: it really ought to be known.—[10]

The Entada is a beast;[11] I have never differed from you about the growth of a plant in a new island being **far** harder trial than transportal, though certainly that seems hard enough. Indeed I suspect I go even further than you in this respect; but it is too long a story.—

Thanks you for the Aristolochia & Viscum cases: what species were they? I ask, because oddly these two very genera I have seen advanced as instances (I forget at present by whom, but by good man) in which the agency of insects was *absolutely* necessary for impregnation. In our British diœcious viscum, I suppose it must be necessary. Was there anything to show that the stigma was ready for pollen in these two cases? for it seems that there are many cases in which pollen is shed long before stigma is ready.

As in one Viscum insects carry sufficiently regularly for impregnation pollen from flower to flower; I sh[d] think that there must be occasional crosses even in an hermaphrodite Viscum. I have never heard of Bees & Butterflies, only Moths producing fertile eggs without copulation.— [12]

What a case of Aldrovanda!

I remember Decandolle, (the father) quotes it as wonderful case *even* in two rivers of Europe.[13]

With respect to Ray Soc. I profited so enormously by its publishing my Cirripedia, that I cannot quite agree with you on confining it to translations; I know not how else I could possibly have published.—

I have just sent in my name for £20 to Linn. Soc;[14] but I must confess I have done it with heavy groans, whereas I daresay you gave your £20 like a light-hearted gentleman.

My dear Hooker | Ever yours | C. Darwin

Wollaston speaks strongly about the intermediate grade between two varieties in insects & mollusca, being often rarer than the two varieties themselves.[15] This is obviously very important for me, & not easy to explain. I believe I have had cases from you. But, if you believe in this, I wish you would give me a sentence to quote from you on this head. There must, I think, be a good deal of truth in it; otherwise there could hardly be nearly distinct varieties under any species, for we should have instead a blending series as in Brambles. & Willows.—

DAR 114.3: 166

[1] This letter was the second written to J. D. Hooker on 5 July 1856. CD states that his first letter was written in the morning and that he had also written to Charles Lyell (see preceding letter). The original order of the letters has been preserved for clarity.

[2] Hooker had proposed a large Antarctic land mass that, at different times, linked South America, Australia, some Pacific islands, and the Antarctic islands with New Zealand (J. D. Hooker 1853–5, 1: xxi–xxiv). He had alluded to this suggestion in his earlier volume on Antarctic botany (J. D. Hooker 1844–7, 2: 210–11).

[3] See the first letter to J. D. Hooker, 5 [July 1856].

[4] A. de Candolle 1855.

[5] Letter to Charles Lyell, 25 June [1856].

[6] Letter from Charles Lyell, [1 July 1856].

[7] See preceding letter.

[8] CD reiterated this intention in his letter to J. D. Hooker, 13 July [1856]. It seems that Lyell did not forward the letter to Hooker, for CD eventually had a copy made of his own draft (the latter was retained at Down) and forwarded it to Hooker, enclosed with his letter to J. D. Hooker, 30 July [1856].

[9] Letter from J. D. Hooker, [26 June or 3 July 1856].

[10] See letter to J. D. Hooker, 22 June [1856], n. 2.

[11] See letter from J. D. Hooker, [26 June or 3 July 1856].

[12] See letter from J. D. Hooker, [26 June or 3 July 1856], n. 8.

[13] A. P. de Candolle 1820, p. 406, in which Augustin Pyramus de Candolle discussed the distribution of the freshwater plant *Aldrovanda* in the river basins of the Po and the Rhône. The passage is marked in CD's copy of the work (Darwin Library–CUL); in the margin CD wrote in pencil: '& at Calcutta Hooker'.

[14] A subscription was opened in June 1856 to pay the expenses for moving the Linnean Society's quarters from Soho Square to Burlington House (Gage and Stearn 1988, p. 51). CD's Account book (Down House MS) has an entry for this payment on 22 February 1857.

[15] Wollaston 1856, pp. 105–6.

To J. D. Hooker 8 [July 1856][1]

Down Bromley Kent

8th

My dear Hooker

I do hope that this note may arrive in time to save you trouble in one respect. I am perfectly ashamed of myself, for I find in introduction to Flora of Fuegia a short discussion on Tristan plants,[2] which though scored I had quite forgotten at time & had thought only of looking into introduction to New Zealand Flora. It was very stupid of me.—

In my sketch, I am forced to pick out the most striking cases of species which favour the multiple creation doctrine, without indeed great continental extensions are admitted. Of the many wonderful cases in your books, the one which strikes me most is that list of species, which you made for me, common to N. Zealand & America, & confined to Southern hemisphere; & in this list those common to Chile & N. Zealand seem to me the most wondrous. I have copied these out & enclosed them.[3] Now I will promise to ask no more questions, if you will tell me a little about these. What I want to know is whether any or many of these are *mountain* plants of Chile, so as to bring them *in some degree* (like the Chonos plants) under the same Category with the Fuegian plants. I see that all the genera (Edwardsia even having Sandwich Isd & Indian species) are wide ranging genera, except Myosurus, which seems extra wonderful. Do any of these genera cling to sea-side?— Are the *other* species of these genera wide rangers?

Do be a good Christian & not hate me.

Ever yours | C. Darwin

I began last night to reread your Galapagos paper,[4] & to my taste it is quite admirable: I see in it, some of the points which I thought best in A. Decandolle! such is my memory.—

Lyell will not express any opinion on continental extensions.

DAR 114.3: 168

[1] Dated by the relationship to the letters to J. D. Hooker, 5 [July 1856] and 5 July [1856].

[2] See the letter to J. D. Hooker, 5 [July 1856] and n. 6. CD refers to J. D. Hooker 1844–7, 2: 216–17. These pages are scored in pencil in CD's copy of the work in the Darwin Library–CUL.

[3] A list headed 'Plants common to New Zealand and South America but not Europœan' in J. D. Hooker's hand is bound into the back of CD's copy of the introductory essay of J. D. Hooker 1853–5 (Darwin Library–CUL). It has been extensively annotated in pencil by CD. A second list, headed 'Plants common to *Chile* *not Fuegia [*interl*] & New Zealand found only in Southern Hemisphere' in CD's hand, is also bound in the back of the same work. The latter list has been annotated by J. D. Hooker.

[4] J. D. Hooker 1851. CD's copy is in the Darwin Pamphlet Collection–CUL.

To T. H. Huxley 8 July [1856][1]

Down Bromley Kent
July 8th

My dear Huxley

I will use the Boltenia case, if I use it, cautiously.[2] I am very sorry to trouble you but I cannot read the word scored in your first page, (which please return) & therefore cannot understand the sentence. The case is really important to me & it strikes me as in itself a singular physiological fact I presume you do not think that much water is taken in by the mouth: my impression had been that much was so taken in for respiratory & digestive processes.— But I suspect that I am forgetting & that ova & spermatozoa are in a closed receptacle.[3]

The fact which you give about the Polyzoa & M[r] Hincks[4] is very curious:[5] I fancy Nordmanns case of bisexual Flustra with channel from male to female cell tells also against extraneous fertilisation.[6] Do the bisexual compound Ascidians throw any light on the point? I presume there is no such a thing as a *uni*sexual ciliograde acalephe? Does the position of ova & spermatozoa in the unisexual pulmonogrades throw any light on the possibility of crossing?[7]

With very many thanks | Yours very truly | C. Darwin

Imperial College of Science, Technology and Medicine Archives (Huxley 5: 40)

[1] Dated by the relationship to the letter to T. H. Huxley, 1 July [1856].
[2] See letter to T. H. Huxley, 1 July [1856], n. 4.
[3] Probably a reference to the reproductive mechanisms of some of the jellyfish (*Beroidae*) which, CD recorded, 'seem to offer the greatest difficulty' to occasional cross-fertilisation. In *Natural selection*, p. 46, CD gave Huxley's opinion that 'it is not positively known whether or not the eggs are discharged fertilised; & that as these animals derive their food from indrawn currents of water, which bathe the ovaria, it is certainly quite possible that the spermatozoa of other individuals might come into action.'
[4] CD refers to Thomas Hincks's paper on Bryozoa (Hincks 1852), which included a discussion of *Flustra*.
[5] CD discussed this in *Natural selection*, p. 46, in a passage immediately following the remarks quoted in n. 3, above. CD wrote: 'Again Prof. Huxley informs me that he should have thought that the hermaphrodite Bryozoa or Polyzoa (certain corallines) would have offered insuperable difficulties to an occasional cross, had it not been for Mr. Hinck's observations'.
[6] In the manuscript of *Natural selection*, CD pencilled a memorandum after the passage on jellyfish spermatozoa (see n. 3, above) referring to the work of Alexander von Nordmann: 'Nordmann & Owen on sexes separate in Flustra. L'*Institut* 1839, p. 95—on sexes in coralline allied to Flustra & on zoospermatic animalcules!' (*Natural selection*, p. 46 n. 1). The reference is to Nordmann 1839, which was cited in Richard Owen's discussion of Bryozoa in Owen 1855b, pp. 151–3. These pages in CD's copy of Owen 1855b (Darwin Library–CUL) were marked by CD.
[7] Huxley's reply has not been found. CD discussed the pairing of hermaphrodite organisms in *Natural selection*, pp. 43–6, but he made no mention of acalephs or pulmonigrades, both of which are types of jellyfish.

To Charles Lyell 8 July [1856]

Down Bromley Kent
July 8th.

My dear Lyell

Very many thanks for your two notes[1] & especially for Maury's map:[2] also for Books which you are going to lend me.

I am sorry you cannot give any verdict on continental extensions; & I infer that you think my arguments of not much weight against such extensions: I know I wish I could believe.—

I have been having a good look at Maury (which I once before looked at)[3] & in respect to Madeira & co, I must say that the chart seems to me against land-extension explaining introduction of organic beings. Madeira, the Canaries & Azores are so tied together that I shd have thought that they ought to have been connected by some bank if changes of level had been connected with their organic relation. The azores ought too to have shown, more connection with America. I had sometimes speculated whether icebergs could account for the greater number of European plants & their more northern character on the Azores compared with Madeira; but it seems dangerous until boulders are found there.[4]

One of the most curious points in Maury, as it strikes me, is the little change, which about 9000 feet of sudden elevation would make in the continent visible, & what a prodigious change 9000 feet subsidence would make! Is this difference due to denudation during elevation? Certainly 12,000 feet elevation would make a prodigious change.—[5]

I have just been quoting you in my essay on ice carrying seeds in S. hemisphere;[6] but this will not do in all the cases.— I have had a week of such labour in getting up the relations of all the antarctic floras from Hooker's admirable works. Oddly enough I have just finished in great detail giving evidence of coolness in Tropical regions during the glacial epoch, & the consequent migration of organisms through the Tropics. There are a good many difficulties, but upon the whole it explains much. This has been a favourite notion with me almost since I wrote on erratic boulders of the south.—[7] It harmonises with the modification of species, & without admitting this awful postulate the glacial epoch in the south & Tropics does not work in well. About Atlantis, I doubt whether the Canary islands are as much more related to the continent as they ought to be if formerly connected by continuous land.—

Yours most truly | C. Darwin

Hooker with whom I have formerly discussed the notion of the world or great belts of it having been cooler, though he at first saw great difficulties (& difficulties there are great enough) I think is much inclined to adopt the idea. With modification of specific forms it explains some wondrous odd facts in distribution.

But I shall never stop if I get on this subject, on which I have been at work, sometimes in triumph & sometimes in despair, for the last month.

Endorsement: 'July 8. 1856'
American Philosophical Society (134)

[1] The notes have not been found, but see n. 4, below.
[2] See letter to Charles Lyell, 16 [June 1856]. CD refers to Maury 1855a, in which there is a detailed map indicating the various depths of the 'Basin of the North Atlantic Ocean'.
[3] CD's reading notebook records a different work by Matthew FontaineMaury, also published in 1855 but concerned with sailing directions (Maury 1855b), which he also noted 'Lyell has' (*Correspondence* vol. 4, Appendix IV, *128: 163).
[4] Lyell's opinion that icebergs may have contributed to the dispersal of animals and plants was expressed in a note made in his scientific journal after receiving CD's letter of 5 July (letter to Charles Lyell, 5 July [1856]). In his journal (Wilson ed. 1970, p. 116) Lyell wrote:

Letter Darwin July 5, 1856

Icebergs & floating ice between latitudes 35 & 80 in each hemisphere may have been great agents of transporting species & have done much which is attributed to continental extension.

Take the Glacial period as a unit & multiply this by all post-miocene Time & the result may afford an amount of geograph! change capable with the additional aid of means of transport & migration of species of carrying them every where even across the line by aid of cold periods & mountainous islands & floating ice & floating timber—

[5] See letter from Charles Lyell, [1 July 1856].
[6] CD cited the ninth edition of Lyell's *Principles of geology* (C. Lyell 1853) in *Natural selection*, pp. 561 and 562.
[7] 'On the distribution of the erratic boulders and on the contemporaneous unstratified deposits of South America' (*Collected papers* 1: 145–63).

From J. D. Hooker   10 July 1856

Kew
July 10/56

Dear Darwin

We have flowered Dr Bell Salters hybrid *Epilobium* from seeds he sent us, & it is so clearly E. roseum and nothing else, having no trace of either parent that I have no faith whatever in his experiments—[1] he himself remarks in the Phytologist that it is to all appearance similar to E roseum, but differs in the slightly 4 Cleft stigma—[2] now though *E. roseum* is put in the section with entire stigmata, it is described both by Babington[3] & Hook Arnott[4] as having a *slightly 4-cleft* stigma— This is just the way, whenever I do make an experiment it is sure to end either in smoke, or disappointment, or in a disgusting opposition to some preconceived theory of my own. (As the *Entada* seeds.)[5]

Per contra Henslow is going ahead with his Ægilops & procured a sport from it very like Revet wheat after the 3^d year![6]

I have been just reading Ed. Forbes first 112 pages of his little work on the European seas, all that is which Van Voorst had printed before Forbes death:[7] would you like to see it?

I have read Huxleys response to Falconer with eminent gusto—how admirably neatly & clearly he puts the whole question.[8] I never understood the distinctions between Morphology & Physiology in their relation to systematic Zoology half so well before. I certainly only half understood the question before & I do not think that Huxleys original Lecture was particularly good at all.—[9] he has put forth his strength here & will I think startle old Falconer. I had a note from the latter a week ago from Paris, in which he alluded to the subject as if he had eaten Huxley without salt & left no bones at all; by Jove he will find this pungent

Huxley & Mrs H staid with us from Saturday till Monday last, but I had not then read his response.

I began this letter intending to tell you why I had not earlier attended to your letters which is because my father is in Scotland: so you may set your mind at ease on the score of the Tristan d'Acunha list.[10]

The said Tristan d'Acunha list is I think capable of some modification in my favor, a most remarkable and wholly distinct grass inhabiting it having been lately found in the Island of S! Pauls (N. of Kerguelens land)— I see you have a note of this I shall hunt up the Tristan d'Acunha plants & let you know.

With regard to the Chili + N.Z. list I have gone over it— *none* are mountain plants in any sense that I know of— As a rule none of the genera have wide ranging species except Epilobium though none are remarkably restricted as to the ranges of their individual species— they are nothing particular in short.[11]

Upon the whole the most wonderful cases, almost demanding double centres, are the presence of the European plants in the Australian Alps & in Tasmania as Cardamine pratensis, ˣLysimachia vulgaris, Aphanes arvensis, ˣTurritis glabra, ˣVeronica serpyllifolia—besides various Carices, & grasses plants that are not common in Australia & are found no where else in the Southern Hemisphere & are not *very* widely distrib^d in the Northern except the Card.[12]

ˣThese on the faith of D! Ferd. Mueller Govt Bot. of Victoria an able Botanist.[13]

It appears however true that multiple centres are worse for your theory than any thing else.

Lyell has not sent me your letter,[14] I wish he would.

The *Aristolochia* was *A. labiosa* or an allied species, the brute of a *Viscum* is diœcious after all, it looked Hermaphrodite, & certainly the pollen is all out before the bud opens, so *that* observation is worthless.—[15] The *Aristolochia* stigma seemed all ready for pollen but I will look again. Its anthers are on the stigmata hence the necessity of Insect or other action to shake out the grains

I shall try to get you cases of well marked varieties being common & the intermediates rare[16]—but these cases are always explained away by the assumption of hybridism— e.g. *Geum urbanum, rivale* & *intermedium*— Wheat & Ægilops— Hewett Watson would probably give you more good cases than I could.

Ever yours | Jos D Hooker

Has it ever struck you that the fact of Parnassia, Saxifraga & anthers approaching & being applied to the Stigma in pairs or in turns is against cross hybrid impregnation, in as much as that a plants own pollen shoots its tubes on its own stigma before that of another species will applied at same time & in general overcomes the latter.— You know what I mean.

DAR 100: 96–9

CD ANNOTATIONS
1.1 We have . . . roseum 1.2] *scored brown crayon*
3.1 I have . . . Australian Alps 8.2] *crossed brown crayon*
8.4 plants . . . the Card. 8.6] *scored brown crayon*
9.1 an able Botanist.] *underl brown crayon*
10.1 It appears . . . after all, 12.2] *crossed brown crayon*
13.2 but these . . . hybridism— 13.3] *double scored brown crayon*
13.4 Hewett Watson . . . could.] *double scored brown crayon*
15.1 Has it . . . impregnation 15.3] *scored brown crayon*
Top of first page: 'other facts besides Antarctic Distribution)' *pencil*

[1] For CD's interest in Thomas Bell Salter's claim to have discovered natural and completely fertile hybrids in the genera *Epilobium* and *Geum*, see *Correspondence* vol. 5, letter from T. B. Salter, 25 June 1855. Salter had promised to send seeds to CD; these had evidently been passed on to Hooker to grow at Kew.
[2] T. B. Salter 1852.
[3] Babington 1851.
[4] W. J. Hooker and Arnott 1855.
[5] See letter from J. D. Hooker, [26 June or 3 July 1856].
[6] John Stevens Henslow described his experiments on *Aegilops squarrosa*, a wild grass found in southern Europe, at the August meeting of the British Association for the Advancement of Science held in Cheltenham (Henslow 1856). He had intended to test the claim made by the French botanist Esprit Fabre (Fabre 1854) that *Aegilops ovata* was the original source of wheat (*Triticum sativum*). Henslow announced that although he had succeeded in changing the character of the experimental plants, he had not yet succeeded in obtaining all the specific features of wheat.
[7] Hooker refers to the pages printed and corrected by Edward Forbes before his death in 1854 for his book on the natural history of the European seas. This text was edited and continued by Forbes's friend Robert Alfred Cloyne Godwin-Austen and was published in 1859 (E. Forbes and Godwin-Austen 1859). The section that Hooker had read included descriptions of Forbes's proposed Arctic, Boreal, Celtic, and Lusitanian provinces.
[8] T. H. Huxley 1856b, a response to Falconer 1856. See letters to J. D. Hooker, 21 [May 1856] and 17–18 [June 1856].
[9] T. H. Huxley 1855.
[10] See letter to J. D. Hooker, 5 [July 1856].
[11] See letter to J. D. Hooker, 8 [July 1856], n. 3. The list was annotated by Hooker, giving brief descriptions of the localities inhabited by the twelve species. At the bottom, CD wrote: 'These genera, Hooker says are not particularly wide rangers; but species with restricted ranges.— Nothing particular in short.—'
[12] CD used Hooker's information in *Natural selection*, pp. 553–4.
[13] Ferdinand Jakob Heinrich von Mueller was the government botanist in Melbourne, Australia. From 1853, he issued annual reports on the vegetation of the colony.
[14] See letter to J. D. Hooker, 5 July [1856].
[15] See letter from J. D. Hooker, [26 June or 3 July 1856], and letter to J. D. Hooker, 5 July [1856].
[16] CD had asked Hooker whether he could confirm Thomas Vernon Wollaston's observations of the same point in the insect kingdom (see letter to J. D. Hooker, 5 July [1856]).

To J. D. Hooker 13 July [1856][1]

Down
July 13th

My dear Hooker

Your letter, as usual, has been most valuable to me. I am delighted at what you say about Huxley's answer & I agree most entirely: it is excellent & most clear; I thought from the first that he was right, but was not able to put it clearly to myself.— By the way do you remember Huxley's entry of "Darwin, an absolute & eternal hermaphrodite":[2] he can find no certain case, nor have I ever been able. Apropos to my asking him whether the ciliograde acalephes could not take in spermatozoa by the mouth,[3] which takes in so much water, he gives me a sentence like our case of pollen, in which nature seems to us so clumsy & wasteful. He says "The indecency of the process is to a certain extent in favour of its probability, nature becoming very *low* in all senses amongst these creatures". What a book a Devil's chaplain might write on the clumsy, wasteful, blundering low & horridly cruel works of nature! With respect to crossing, from one sentence in your letter I think you misunderstand me:[4] I am very far from believing in hybrids; only in crossing of same species or of *close* varieties. These two or 3 last days, I have been observing wheat & have convinced myself that L. Deslongchamps is in error about impregnation taking place in closed flower;[5] ie of course I can judge only from external appearances. By the way R. Brown[6] once told me that the use of brush on stigma of grasses was unknown: do you know its use?

You once asked me whether I had your Lemann's list of Madeira plants,[7] I see in Forbes Memoir in note,[8] that you lent it him, as he says; probably he never returned it—

I enclose old note of yours about Lyallia; it may refresh your memory: as for this plant & the Pringlea, I sh[d] think the Vestiges' theory that they were converted algæ, was as good as any![9] Confound & exterminate them.—

Very many thanks for the answers about Chile & New Zealand plants.

You say most truly about multiple creations & my notions; if any one case could be proved, I sh[d] be smashed: but as I am writing my Book, I try to take as much pains as possible to give the strongest cases opposed to me, & offer such conjectures as occur to me: I have been working your Books as richest (& vilest) mine against me: & what hard work I have had to get up your New Zealand Flora! As I have to quote you so often, I sh[d] like to refer to Mullers case of Australian Alps:—where is it published? Is it a Book? a correct reference would be enough for me, though it is wrong ever to quote without looking oneself.— I sh[d] like to see very much Forbes sheets, which you refer to; but I must confess (I hardly know why) I have got rather to mistrust poor dear Forbes.—

There is wonderful ill logic in his famous & admirable memoir on distribution,[10] as it appears to me, now that I have got it up so as to give the Heads in a page.— Depend on it, my saying is a true one, viz that a compiler is a *great* man, & an original man a common-place man. Any fool can generalise & speculate; but oh my Heavens to get up *at second hand* a New Zealand Flora, that is work.—

I am so glad to hear about Henslow & wheat: I do hope there was no wheat-field near: he ought to state distance & whether flowering coincides with that of wheat.—

And now I am going to beg almost as great a favour, as a man can beg of another: and I ask some 5 or 6 weeks before I want favour done, that it may appear less horrid: it is to read, *but well copied out*, my pages (about 40!!) on alpine floras & faunas arctic & antarctic floras & faunas & the supposed cold mundane period.— [11] It w^d be really an enormous advantage to me; as I am sure otherwise to make Botanical blunders. I would specify the few points on which I most want your advice. But it is quite likely that you may object on ground that you might be publishing before me (I hope to publish in a year at furthest) so that it would hamper & bother you; & secondly you may object to loss of time; for I daresay it would take hour & half to read.— It certainly would be immense advantage to me; but of course you must not think of doing it, if it would interfere with your own work.—

My dear Hooker | Ever yours | C. Darwin

I do not consider this request *in futuro*, as breaking my promise to give no more trouble for some time.

From Lyell's letters he is coming round at a Railway pace on the mutability of species, & authorises me to put some sentences on this head in my preface.[12]

I shall meet Lyell on Wednesday at Ld Stanhopes & will ask him to forward my letter to you;[13] though as my arguments have not struck him; they cannot have force, & my head must be crotchety on subject; but the crotchets keep firmly there.— I have given your opinion on continuous land, I see, too strongly.

DAR 114.3: 169

[1] Dated by the relationship to the preceding letter.

[2] See letter to T. H. Huxley, 1 July [1856], n. 2.

[3] See letter to T. H. Huxley, 8 July [1856].

[4] See the final paragraph of the preceding letter.

[5] Loiseleur Deslongchamps 1842–3. CD recorded having read this work on 5 April 1856 (*Correspondence* vol. 4, Appendix IV, 128: 18). There is a copy of the first part in the Darwin Library–CUL.

[6] Robert Brown.

[7] A reference to the manuscript flora of Madeira drawn up by Charles Morgan Lemann before his death in 1852. The list had evidently accompanied his herbarium, which was deposited at Kew (R. Desmond 1977).

[8] E. Forbes 1846, p. 401 n.

[9] A joke between CD and Hooker relating to an article about the Kerguelen Land cabbage (*Pringlea antiscorbutica*) printed in *Chambers's Edinburgh Journal* n.s. 5 (1846): 76–7. See *Correspondence* vol. 3, letter from J. D. Hooker, 1 February 1846, and *Correspondence* vol. 5, letter to J. D. Hooker, 5 June [1855].

[10] E. Forbes 1846.

[11] A reference to the manuscript pages of what CD called 'the before part of Geograph Distr.' that eventually formed the bulk of chapter 11, on geographical distribution, of his species book (see *Natural selection*, pp. 531, 534–66).

[12] See letter to Charles Lyell, 5 July [1856] and n. 7.

[13] See letter to J. D. Hooker, 5 July [1856]. CD had previously met Philip Henry Stanhope at Stanhope's family seat in Chevening, Kent, and on several occasions he went to Stanhope's London house to join 'one of his parties of historians and other literary men' (*Autobiography*, p. 111).

To James Dwight Dana   14 July [1856][1]

Down Bromley Kent
July 14th

My dear Sir

I want to beg one more favour to the many which formerly you have conferred on me.[2] I am extremely much interested in regard to the blind cave animals, described some time since in your Journal by Prof. Silliman Junr.,[3] as the subject is connected with a work of somewhat general nature, which I am endeavouring to draw up on variation & the origin of species, classification &c.—

Are the specimens at Newhaven? and if so could you get any good entomologist to look at the insects— What I want to know is, whether any of the Crustacea, spiders, insects (flies beetles, crickets &c) & Fish belong to the American type[4] (Has not Agassiz noticed the Fish?)[5] ie to genera or sections of genera, found only on the American continent.— I sh^d be most grateful for any, the least, information on this head.— All the American mice have a peculiar character in their teeth by which they can be recognised.—

Secondly I have been rereading with renewed interest your memoir on geograph. Distrib. of Crustacea[6] & I want to ask a question on this head: Botanists have remarked on several cases in which northern temperate & arctic genera have sent the same or representative species into corresponding zones of S. hemisphere.— You give several similar & striking cases; but I do not feel sure from my ignorance that these genera can be called from their general affinities & range strictly northern genera. How is this? Might they not be called southern genera, which have sent species to the North: I ask this because in plants, it is very remarkable as observed by Dr Hooker & A. Decandolle, that southern genera have *not* their represetatives or identical species in the north, though there are so many cases of northern genera which have their congeners & same individual species in the south.— Will you be so very kind as to reflect on this, & take the trouble to inform me.[7]

Lastly can you remember whether any author (I think Mr Eights whose writings I have never seen) has described fossil trees in the S. Shetland islands.—[8]

Now I am sure I have put your kindness to a severe proof, & can only beg to be forgiven.— If you have a few minutes to spare, I sh^d *very* much like to hear a little news of yourself, & whether all things go well with you. Are you at work at any particular great subject? I should expect so, though no one whatever in the world has a better right to rest on his oars than you have. I never cease being fairly astounded at the amount of labour which you have performed.— But the other day I was sitting at dinner by Prof. Miller of (our) Cambridge,[9] & he was speaking warmly on your mineralogical work.—

As for myself I live a very quiet & retired life, with a large set of very happy & good children round me, & do daily 3 or 4 hours work at Natural History; for more than which I have not, & shall never have, strength.— Our neighbour J. Lubbock, has married a young & pretty wife, & a very young couple they are reckoned in this country, & I think & hope he will be as happy as he deserves: he works away during the very little leisure which he has, at his Entomostraca, & if he could give himself up to Nat. History, he would make a capital Naturalist.[10]

Pray believe me, my dear Sir, with every good wish & sincere respect. | Yours very truly | Ch. Darwin

I have directed this to care of Prof. Silliman; as I heard some time since that you were Professor of Geology at some new place.[11]

Yale University Library (Manuscripts and archives, Dana papers)

[1] Dated from Dana's reply (see letter from J. D. Dana, 8 September 1856).

[2] CD and Dana had corresponded since 1849 about Cirripedia and the geological observations they had made on coral islands and Australia.

[3] Silliman 1851. Benjamin Silliman Jr, Dana's brother-in-law, was co-editor with Dana of the *American Journal of Science and Arts*, usually called 'Silliman's Journal' after its founder, the elder Benjamin Silliman.

[4] CD had previously asked Dana much the same question about the cave fauna (see *Correspondence* vol. 5, letter to J. D. Dana, 8 May [1852]). For Dana's opinion, see letter from J. D. Dana, 8 September 1856. See also letter from J. O. Westwood, 23 November 1856, in which John Obadiah Westwood discussed the insect genera found in the cave.

[5] Agassiz 1851. Louis Agassiz had written to Benjamin Silliman that he considered the fish 'an aberrant type of my family of Cyprinodonts' (Agassiz 1851, p. 127).

[6] Dana 1853. CD's copy is in the Darwin Library–CUL.

[7] Dana's answer to this query has not been found, but see letter to J. D. Dana, 29 September [1856].

[8] James Eights had collected in the Antarctic and published several papers on the marine Crustacea of the South Shetland Islands. He was primarily a palaeontologist. Dana evidently replied in some detail since there is a note made by CD, along with a reference to Eights 1856, in the manuscript of his species book (see *Natural selection*, p. 579 n. 3) reminding him to look at 'Dana's letter on Mr. Eights'. The letter was probably part of the letter from J. D. Dana, 8 September 1856, which is now incomplete.

[9] William Hallowes Miller was professor of mineralogy at Cambridge University. CD may be referring to the Philosophical Club dinner of 19 June that both he and Miller attended (Royal Society Philosophical Club minutes).

[10] CD had previously told Dana about John Lubbock's work on Entomostraca (see *Correspondence* vol. 5, letter to J. D. Dana, 27 September [1853]). Lubbock had married on 10 April 1856 (see letter to John Lubbock, 24 April [1856]).

[11] Dana had not become professor of geology at any 'new place' but had, in 1855, finally taken up the duties of the Yale professorship in natural history to which he had been appointed in 1849 when the elder Benjamin Silliman retired (*DAB*).

To Asa Gray   14 July [1856][1]

Down Bromley Kent
July 14th

My dear D[r] Gray

You have been so very kind in giving me information of the *greatest* use to me;[2] that I venture to trouble you with a question, which cannot cause you much trouble.— I have been reading a paper by you on plants on mountains of Carolina, (in London Journal of Botany)[3] in which you state that most are the same with the plants of the N. States & Canada.— Now what I want to know is, whether the Alleghenies are sufficiently continuous so that the plants could travel from the north in the course of ages thus far south?[4] I remember Bartram makes the same remark with respect to several trees on the Occone M[ts],—not that I know where these Mountains are.—[5]

How does your memoir on Geograph. Distrib. get on?[6] I do heartily wish it was now published; for I have been trying to make out how many plants are common to Europe, which do not range up to the Arctic shores, & they seem to be very few.—

I have just thought of one other question, connected with my subject, which I cannot resist asking.— I have seen it remarked by entomologists, that it *often* happens that the intermediate varieties connecting together two varieties (& thus showing that such are varieties) are less common or numerous in individuals, than the two varieties themselves.[7] If you can enlighten me on this head I sh[d] be **very** much obliged. I am inclined to think there must be some truth in it; otherwise varieties would not be so well marked as they often are.— I wrote some time ago, a troublesome letter, in which I begged for information on the amount of *variability* of your naturalised & your agragrian plants, as compared with other species of the same genera.—[8]

But I know I have been scandalously troublesome to you.— Can you forgive me? & believe me, Yours truly obliged | Ch. Darwin

New York Botanical Garden Library (Charles Finney Cox collection)

[1] Dated by the relationship to the letter to Asa Gray, 2 May [1856].

[2] See letter to Asa Gray, 2 May [1856], and the letters exchanged between Gray and CD in 1855 (*Correspondence* vol. 5).

[3] A. Gray 1842. CD cited the paper in his species book (*Natural selection* p. 537) but later cancelled the citation when he found more extensive data in Gray's 'Statistics of the flora of the northern United States' (A. Gray 1856–7).

[4] See *Natural selection*, p. 537.

[5] William Bartram discussed the flora of Mount Oconee (now called Stratton Mountain) in South Carolina in Bartram 1791, p. 335, which was cited by CD in *Natural selection*, p. 537 n. 2.

[6] The revised edition of Gray's *Manual of the botany of the northern United States* was published later in the year (A. Gray 1856a).

[7] This point is made in Wollaston 1856, pp. 105–6. See CD's comments in letter to J. D. Hooker, 5 July [1856].

[8] Letter to Asa Gray, 2 May [1856].

From S. P. Woodward 15 July 1856

Brit. M.
July 15th 1856.

Dear Sir

Your question was answ^d. almost by express anticipation in my paper at the Zool. Soc^y. last week,[1]—on the L. & Fr. water shells of Tibet & Kashmir,[2] but as that will not be printed just yet I will endeavour to make out the list when I return home to-night.[3]

I proposed to consider 20 of M. Deshayes[4] *Cyrenæ* (Corbiculæ) as geographical varieties of **one** species. (C. fluminalis, Müll) Since after examining Mr Cuming's collection & *ours* I can find no characters by which a miscellaneous mixture could be sorted.[5] Philip Carpenter also glanced at them—& said it was like the case he was investigating (*Calyptræa*) in which varieties of the same species were rated as members of distinct "sub-genera" by H & A. Adams.[6]

The "species" which may be most safely referred to *Cyrena fluminalis*, are

> C. ambigua, Desh. Euphrates
>   Euphratica, Bronn.
>   Cor, Lam. Nile
>   consobrina, Caillaud. Alexandrian Canal
>   Sea of Tiberius
>   triangularis, Desh. (no locality but exactly like
>   specimens from Alex. Canal)
>   Panormitana, Bivon Sicily.
>   Gemmellarii Phil. Fossil, Sicily.
>   trigonula, Searles Wood—Brit.
>   Cashmiriensis, Desh. Kashmir—
>   Scinde—Candahar.
>   radiata, Phi. Nile—India

The *C. occidens* (Benson) Sikkim, & C *Bengalensis* Desh. C. striatella, Desh. Pondicherry are only *specimens* a little more "transverse".

The same shell when found in China (which was not so far from Noah's Ark as ultima Thule) has another set of names—

> *C. fluviatilis*, Müll
>   Largillierti, Phi
>   Woodiana, Lea &c

In the Peninsula it becomes *C. Malaccensis*, Desh. in Java *C. compressa* (Mousson)[7]

Sir Chas. Lyell was at the Meeting & said that Dr Hooker would reduce the reduce the reputed Indian Flora at the rate of **19** to **1**—; and made some sharp remarks on species-making & the "present state of Conchology"—[8]

Gray,[9] who was in the chair, went still further & (Cuming being present!) denounced the greater part of the reputed shells as "dealers' species".

In the last number of the "Annals" my friend Mr Benson has drawn a very fine distinction between *Clausilia Rolphii* of Charlton & a specimen of the so-called *Cl. Mortilleti* from Charing—[10] I told him before-hand I didn't believe or understand it—& now I have sent him down a lot of the *Charlton* Shell which are *all Cl. Mortilleti*!! so that "species" is annihilated, at least for a while. However Mr Benson is a gentleman & a philosopher—& will acknowledge his error!

Lastly, *have you* the most admirable book in which the great question of "creative action" is treated—

Dr **Harvey's** "Seaside Book" (Van Voorst)[11]

Because if not, you ought to get it immediately—& having read the first & last chapters—give it to that young lady of your family who has most sympathy with "the sunshine & the calm of mute, insensate things".

I presume you are acquainted with Dr Pickering's "Races of Man"[12]—& with that chapter in which, when discussing the probable scene of the Creation of man, he speaks more respectfully of the *Orang* & *Gorilla* than Agassiz does of "our black brethren".[13] It is fortunate for those of us who respect our ancestors & repudiate even the contamination of Negro blood—that Agassiz remains, to do battle with the transmutationists

Yours sincerely | S. P. Woodward

Chas Darwin Esq.

DAR 205.3 (Letters)

CD ANNOTATIONS
0.3 Dear . . . to-night. 1.4] *crossed pencil*
2.1 I proposed . . . varieties 2.2] *double scored pencil*
3.1 The species . . . are] *scored pencil*
3.2 C. ambigua . . . India 3.14] 'What a range' *added pencil*
3.2 Euphrates] *double underl pencil*
3.4 Nile] *double underl pencil*
3.8 Alex. Canal] *double underl pencil*
3.9 Sicily.] *double underl pencil*
3.10 Sicily.] *underl pencil*
3.11 Brit.] *double underl pencil*
3.12 Kashmir.] *double underl pencil*
5.1 China] *double underl pencil*
6.1 Java] *double underl pencil*
7.1 Sir Chas . . . Woodward 12.1] *crossed pencil*
Top of first page: '19'[14] *brown crayon*; 'F. W. Shells' *brown crayon*

---

[1] CD's question was evidently similar to that posed in the letters to H. C. Watson, [after 10 June 1856], and to J. D. Hooker, 22 June [1856], in which he asked whether species common to both Europe and North America were also found in the Arctic or sub-Arctic zone. The range of shells was discussed by CD in *Natural selection*, p. 539, but the information in this letter was not cited.

[2] Woodward 1856b was read at a meeting of the Zoological Society of London on 8 July 1856. Woodward had attempted to ascertain which of the Kashmir and Tibetan species were also found in Europe.

³ See following letter.

⁴ Gérard Paul Deshayes.

⁵ Hugh Cuming owned one of the largest scientific collections of shells in Britain. Although part of his collection was already in the possession of the British Museum, the bulk of it was not purchased until after his death (British Museum (Natural History) 1904–6, 2: 727). Woodward had been on the staff of the department of geology and mineralogy in the British Museum since 1848.

⁶ Philip Pearsall Carpenter had bought a valuable and extensive collection of shells known as the Mazatlan collection in 1855; it was purchased from him by the British Museum in 1857. Henry and Arthur Adams were the co-authors of *Genera of recent Mollusca* (Adams and Adams [1853–] 1858).

⁷ The list given by Woodward corresponds to information given in Woodward 1856b, p. 186 n.

⁸ The shells described by Woodward in Woodward 1856b had been collected in India by Thomas Thomson in 1847 and 1848 and given to the British Museum by Charles Lyell and Joseph Dalton Hooker.

⁹ John Edward Gray.

¹⁰ William Henry Benson was cited in Woodward 1856b. The notice referred to was a letter from Benson on the 'Occurrence of *Clausilia Mortilleti*, Dumont, in Kent' printed in *Annals and Magazine of Natural History* 2d ser. 18 (1856): 74–5.

¹¹ Harvey 1854, published by John Van Voorst, in which William Henry Harvey attempted to demonstrate the majesty of God's design in nature. In the conclusion, he stated (Harvey 1854, p. 313):

> For though we may admit that physical laws suffice to explain the mutations of the mineral world,—the regular succession of seasons, and the irregular action of the earthquake and the storm, we cannot attribute to physical agency the existence of organic life—itself the clearest witness to a supernatural power.

CD owned a copy of the first edition of this work (Harvey 1849).

¹² Pickering [1848], subsequently published in several editions. CD owned a copy of Pickering 1850, which is in the Darwin Library–CUL. The passage referred to occurs in Pickering 1850, p. 314, where Charles Pickering stated that orangs, 'of all animals, in physical conformation and even in moral temperament, make the nearest approach to humanity'. According to a page of notes made by CD and inserted in the back of his copy, CD reread the work in October 1856.

¹³ Louis Agassiz believed in the multiple origin of the human race. See Lurie 1954.

¹⁴ The number of CD's portfolio of notes on the geographical distribution of animals.

From S. P. Woodward [15 July 1856]¹

Lusitanian Shells of wide range *beyond the Province*.²

Helix *pulchella*, Europe generally; Madeira; Caucasus; Tibet; Cape (introduced); Mass. Missouri

— *Costata*, Brit. Sweden, Russia, Caucasus; Iskardo, Tibet 7200 feet.

— *fulva*. Brit. Sweden. US. Georgia (Say) Mass. (Gould) Russ. Caucasus

— *ruderata*, Brit. (fossil) Sweden–Finland, Russia, Stanowoj Mtns, E. Siberia. = ? *striatella*, Anth. US. Vermont, Ohio.

Helicella *pura*, Alder. Brit. Germany, Switz. Russia. E. Siberia; US. Mass. (*Gould*)

— *cellaria*, Europe; (Cape; N. Zealand;) U. States. Sweden

— *nitida*, Müll. Brit. Iskardo, Tibet. Finmark, Sweden. Mass. US. (*arborea*, Say)

Zua *lubrica*, Brit. Sweden. Madeira; Iskardo. N. West Territory. Mass. US. Oregon.

Succinea *amphibia* (*varieties* of) Brit. Sweden Kashmir. U. States.

Vitrina *pellucida*. Brit. Sweden. Greenland (V. "Angelicæ") U.S. N. England. (V. "Americana")

Limnæa *stagnalis*, Brit. Sweden— Bercsov, N. Lat. 63° Irkutsk. Kashmir; speciosus, Rossm. + fragilis, Mont. = L. *jugularis*. Port Vancover, Oregon— Canada.

— *palustris*, Brit. Bernaul, Irkutsk—52° = ? *elodes*, Say. Mass.

— *truncatula*, Brit. Sweden Bercsov. 63° Tomsk. Iskardo; Candahar; Madeira— = ? *desidiosa*, Miss!— Atlantic, 35°–45° Lat.

— *peregra*   ⎫
                    ⎬ Brit. + Tibet. Siberia 63° Sweden
— *auricularia* ⎭

Physa hypnorum, Brit. Siberia 70° U. States (Ohio)

Cyclas *calyculata* (= C. partumeia, Say) Sweden. Brit.— Kamtschatka Europe— US.

— *rhomboidea*, Say. U.S.— Lake Champlain. Paddington Canal! An. N. H. June 1856—

Valvata *piscinalis*, Brit. Sweden + Kashmir.

Unio *margaritiferus*, Brit. Sweden, U. States. Labrador. Oregon

It must be observed that *Latitude* is not of so much consequence as the antiquity of the land, or its *elevation*. •

The mountain ranges of Scandinavia, & Central Asia, & the northern part of the Alleghanies, equally belong to the "Arctic Province"—& formed centres from which the species which escaped the Glacial subsidence again diffused themselves—or indeed were *driven* by the descent of the snow-line. The plain of Kashmir is 5,300 feet above the sea— Iskardo Tibet, 7,200 feet—& the Brit. *Limnæa peregra* is said to have been found at 18000 feet in Tibet.— I have not yet asked Dr Hooker about it.

The high antiquity of these wide-spread shells is proved in the case of *Helix ruderata* & *pulchella*, *Zua lubrica* & *Helicella nitida* by their occurrence in the older Pliocene, along with *Elephas merid*. Rhinoceros *leptorhinus* & *Mastodon*—animals which became extinct & were succeeded by *another set*, in their turn to disappear whilst these land-*shells* continued to exist![3]

A memorandum
DAR 205.3 (Letters)

CD ANNOTATIONS
1.2 Helix *pulchella*,] 'Ancient' *added pencil*
1.2 Madeira] *double underl pencil*
1.10 — *nitida*,] 'Ancient' *added pencil*
1.11 Zua *lubrica*,] 'Ancient' *added pencil*
1.11 Madeira] *underl pencil*
4.1 The high . . . *Elephas merid*. 4.3] *double scored pencil*
Top of first page: '19'[4] *brown crayon*

[1] According to the previous letter, Woodward intended to compile this list on the evening of 15 July. Both the list and the letter were received by CD by 18 July (see letter to S. P. Woodward, 18 July 1856).

[2] The list expands on the information given in Woodward 1856b, p. 186.

[3] The same information is given in Woodward 1856b, p. 187. CD cited Woodward to this effect in *Natural selection*, p. 539.

[4] The number of CD's portfolio of notes on the geographical distribution of animals.

From John Richardson 17 July 1856

Lancrigg— Grasmere | Westmoreland
17 July 1856

Dear Sir

The Common Pike and Salmon are the only fishes that I recollect at present which are common to the fresh-waters of Europe and North America.[1] The identity of the American Pike with the European one has been called in question but some American Ichthyologists, but I am convinced wrongly, because they had no specimen or no good specimen of the Common Pike of Rupert's Land[2] to compare with— I most minutely examined the American Salmon & the Pike with European ones and feel assured that precisely the same species of each exists in both hemispheres— There are other pikes in Rupert's land & Agassiz believes that there is a peculiar species in each river system— The Salmon also extends along the Labrador coast to the arctic sea north of America but I am not sure that it goes on to Behrings Straits— The numerous Salmon of the seas of Kamtschatka & Beerings Sea down to the Oregon seem to have much resemblance to the Asiatic ones described by Pallas but they have not been compared. Many of the *Sclerognathæ* are common to the English channel, the Baltic, the Greenland seas and american arctic seas but in Beerings sea where that family also abounds the forms are different from the European ones. The same genera of fish are common to the northern parts of the old & new worlds, with exceptions, but the species differ— The American Siluridæ are a totally distinct group from the Asiatic ones & in the temperate regions of Am[a] there are many genera not found elsewhere, of all divisions of Fresh water fishes. The Salmon speaking roundly does not go south of the 40th parallel and only a stray one enters the Mediterranean. One once appeared at the market in Malta—

Trout are natives of northern regions— D[r] Hooker found them to the north of the Himalaya's but not to the south— In Australasia & Patagonia, the Trouts are replaced by *Galaxias*, fish assuming the external appearance & beautiful spotting of the Trout but wanting the Adipose fin. There are many genera in the high southern latitudes which do not exist to the north of the equator. Certain species *Muræna helena*, Elecata and I believe some Scopelidæ or (Sauridæ as the family has been named) are also cosmopolite in the temperate & warmer seas Several sharks also—

Of sea fish there are certainly some forms common to the coasts of Australia & those of Japan but in the latter the curious northern Sclerogenidæ begin to appear & I should suppose that in the north of Japan and along the shores of Sagalien the general features of the ichthyology have changed. From the southern part of Japan down to the Indian Archipelago, westward to near the Cape of Good Hope southward to Australia and eastward to the China seas again the great mass of species is the same though there are some genera and species which are very local.

Many of the red sea genera and the fresh water fishes of the Nile are represented in the Senegal and on the Goldcoast but the species as far as I have learnt are never the same— Curious forms with lung-like air-bladders, exist on both sides of Africa but are not confined to that continent—

The Falkland Island fish resemble those of south New Zealand & islands further south but differ from the Japanese & northern forms— I do not think that any *fresh-water* species is common to the Falklands and New Zealand— Marine species are[3] The Ichthylogy of the higher southern latitudes has a peculiar character from the predominance of certain forms. I believe that ranges of fresh water species to two continents will be found chiefly among anadromous fish. The Pike descends into salt water but I have not heard of its having been taken at sea. It is said to be found in Lake Aral. Sticklebacks which are very nearly alike on both continents, (the differences being as little as those which the species assumes in one locality,) are also inhabitants of the sea & one was taken by Sir Edward Belcher beyond Wellington inlet, which would have passed for an English one, if its origin had not been known.

I have answered your questions I think but have written hurriedly, being unable at present to devote much thought to the subject—[4]

Yours faithfully | John Richardson

DAR 205.3 (Letters)

CD ANNOTATIONS

1.1 Common . . . Salmon] *underl brown crayon*
1.18 The American . . . ones] *scored brown crayon*
1.20 not go . . . Mediterranean 1.21] *double scored brown crayon*
2.1 Trout . . . south— 2.2] *scored brown crayon*
4.1 fresh water] *underl brown crayon*
4.1 fresh water . . . Goldcoast 4.2] *scored brown crayon*
5.5 of fresh . . . at sea. 5.7] *scored brown crayon*
Top of first page: '19'[5] *brown crayon*

[1] Richardson was an acknowledged expert on Arctic and North American zoology, having travelled widely in this region. His works were frequently cited by CD in *Natural selection*.
[2] The former name of the Canadian territory comprising the drainage basin of the Hudson Bay (*EB*).
[3] 'I do not think . . . species are' was added at the top of the first page for insertion here.
[4] The information given in this letter was not directly used by CD in his species book; it confirmed points that CD had already drawn from Richardson's two works on ichthyology (J. Richardson 1836 and 1845). See *Natural selection*, pp. 539 and 555.
[5] The number of CD's portfolio of notes on the geographical distribution of animals.

To the Secretary, Royal Society 18 July 1856

Down.
July 18th/56

Sir

I recommend Dr Carpenter's paper, Part II. on the Foraminifera, to be published in the Transactions.[1]

Sir | Your obed. servt | Ch Darwin

To the Sec^y | Royal Soc^y—

Royal Society (RR3: 40)

[1] The second part of William Benjamin Carpenter's paper on the Foraminifera, on *Aveolina* and other genera (Carpenter 1856, pt 2), had been read to the Royal Society on 19 June 1856. CD had previously refereed the first part of Carpenter's paper (on *Orbitolites*) when it was submitted in 1855 (see *Correspondence* vol. 5, letter to the Council of the Royal Society, 18 August 1855).

To S. P. Woodward 18 July 1856

Down. | Farnborough Kent.[1]
July 18th (1856.)

My dear Sir

Very many thanks for your kindness in writing to me at such length,[2] and I am glad to say for your sake that I do not see that I shall have to beg any further favours.

What a range & what a variability in the Cyrena! Your list of the ranges of the Land & F. W. Shells, certainly is most striking & curious; & especially as the Antiquity of four of them is so clearly shewn.[3]

I have got Harveys sea-side book, & liked it; but I was not particularly struck with it but I will reread the 1st & last chapter.

I am growing as bad as the worst[4] about species & hardly have a vestige of belief in the permanence of species left in me, & this[5] confession will make you think very lightly of me; but I cannot help it, such has become my honest conviction though the difficulties & arguments against such heresy are certainly most weighty.

Yours very sincerely | Ch Darwin.

Copy
DAR 148

[1] After October 1855, CD's postal address was usually given as Bromley, but Farnborough was occasionally used.
[2] Letters from S. P. Woodward, 15 July 1856 and [15 July 1856].
[3] See letter from S. P. Woodward, [15 July 1856].
[4] In the copy of this letter, Francis Darwin substituted 'worst' for the copyist's 'rest'.
[5] Francis Darwin here substituted 'this' for the copyist's 'the'.

To J. D. Hooker   19 July [1856][1]

Down Bromley Kent
July 19th

My dear Hooker

I thank you warmly for the *very kind* manner with which you have taken my request.[2] It will in truth be a most important service to me; for it is absolutely necessary that I sh^d. discuss single & double creations, as a very crucial point on the general origin of species, & I must confess, with the aid of all sorts of visionary hypotheses, a very hostile one.—

I am delighted that you will take up possibility of crossing;[3] no Botanist has done so which I have long regretted, & I was glad to see that it was one of A.[4] Decandolles desiderata. By the way he is curiously contradictory on subject.— I am far from expecting that no cases of *apparent* impossibility will be found; but certainly I expect that ultimately they will disappear; for instance Campanulaceæ seemed a strong case, but now it is pretty clear that they must be liable to crossing. Sweet Peas—Bee-orchis, & perhaps Hollyocks are, at present, my greatest difficulties; & I find I cannot experimentise by castrating Sweet Peas, without doing fatal injury. Formerly I felt most interest on this point as one chief means of eliminating varieties; but I feel interest now in other ways.—[5]

One *general* fact makes me believe in my doctrine, is that **no** terrestrial animal in which semen is liquid is hermaphrodite except with mutual copulation: in terrestrial plants in which the semen is dry there are many hermaphrodites.— Indeed I do wish I lived at Kew or at least so that I could see you oftener.

If you were to take up crossing, I w^d. look over my notes, which perhaps w^d. guide you to some of the most difficult cases.[6]

To return again to subject of crossing; I have been inclined to speculate so far, as to think (my!?) notion (I say **my** notion, but I think others have put forward nearly or quite similar ideas) perhaps explains the frequent separation of the sexes in *Trees*, which I think I have heard remarked, (& in looking over the mono- & diœcious Linnean classes in Persoon seems true[7]) are very apt to have sexes separated; for a tree having a vast number of flowers on same individual or at least same stock, each flower if only hermaphrodite on common plan would generally get its own pollen or only pollen from another flower on same stock,—whereas if sexes were separate there would be better *chance* of *occasional* pollen from another distinct stock. I have thought of testing this in your New Zealand Flora, but I have no standard of comparison & I found myself bothered by bushes— I sh^d. propound that some unknown causes had favoured development of trees & bushes in New Zealand, & consequent on this, there had been a development of separation of sexes to prevent too much intermarriage— I do not of course suppose the prevention of too much intermarriage the only good of separation of sexes.— But such wild notions are not worth troubling you with the reading of.—

With respect to Ægilops I subscribe to your very just & new to me remarks; I did not know Æ. triticoides was rare in wild state: what I remarked on was in relation to the French objectors.[8]

With respect to the Pringlea I am ready to admit any theory whatever; & generally I w^d observe that I would admit a continental extension in any *few* cases, when the facts required it more than in the generality of cases, but it seems to me that you will have to admit continental extensions to every island whatever, & that I cannot swallow. Indeed even one *continental* extension is an awful gulp to me. I never made a continent for my Coral Reefs.— But how I am running on: it is the greatest temptation to me to write ad infinitum to you.—

I have been much tempted by your invitation for Friday & sh^d like it extremely, but I have been having a bad 3 weeks & do not think I could stand the fatigue; but I sh^d have enjoyed such a party extremely.—

I am very sorry that so much of your time sh^d be taken up with thy young Doctors.[9]

My dear Hooker | Yours affecly | C. Darwin

My poor wife keeps as wretched as ever, but now with general oppression & not nausea.—

DAR 114.3: 171

[1] Dated by the relationship to the letter to J. D. Hooker, 13 July [1856].

[2] CD had asked Hooker to read the manuscript he had written on geographical distribution for his book on species (letter to J. D. Hooker, 13 July [1856]). Hooker's reply has not been found.

[3] This topic was discussed in the letter from J. D. Hooker, 10 July 1856, and the letter to J. D. Hooker, 13 July [1856].

[4] Alphonse de Candolle discussed the need to establish the species in which 'le croisement par des circonstances naturelles est impossible' (A. de Candolle 1855, 2: 1346). CD scored this passage in his copy of the work (Darwin Library–CUL).

[5] CD's views were written up in the chapter 'On the possibility of all organic beings occasionally crossing' (*Natural selection*, pp. 35–91).

[6] CD's notes for his chapter on crossing (see n. 5, above) and on dichogamy are in DAR 49 and DAR 75. Other notes are preserved in DAR 205.8.

[7] Either Christiaan Henrik Persoon's edition of Linnaeus's *Systema vegetabilium* (Persoon ed. 1797) or Persoon 1805–7, both of which CD owned. The latter is now in the Darwin Library–CUL.

[8] See letter to J. D. Hooker, 13 July [1856].

[9] Hooker was an examiner in botany for the East India Company medical services (see L. Huxley ed. 1918, 1: 385–7).

To J. D. Hooker 26 [July 1856][1]

Down Bromley Kent
26th—

My dear Hooker

Lyell has lent me Westminster Review telling me to forward it to you, which I will do on Monday.[2]

I sent off few days ago Forbe's sheets.—[3]

I said in former letter that the passage in Flora Antarctica about Tristan d'Acunha nearly sufficed for me;[4] but on further reflexion I find that it would be very important for me to know, whether there is as much identity of species, or more representative species as compared with Kerguelen Land, in relation both to Fuegia.— I am aware that it is *very* unlikely that you know the Flora in such detail, for if you had, you would probably have inserted it some of your Books; but I ask for chance.—

Have you alive at Kew, Montia fontana, Limosella aquatica & Callitriche verna; if so, do you not think it would be well worth trying whether the *seed-pods or heads* would float in salt-water & whether they would survive. I would gladly try.—[5] What a wonderful case of distribution these 3 plants are in your Antarctic isl^ds^!— It is a great bore that we have *no* water plants on the Chalk.— You speak in your Himalaya Journal of the number of plants which you found within arms reach.[6] I suppose you did not make list & if you did it would probably be buried in mass of papers; otherwise I sh^d^ have liked to have looked at it, to see whether many species of same genera & genera of same Family.—

I have let 3 × 4 sq^e^ feet of **old** Lawn grow up, & 18 plants in 17 genera have flowered during this summer. Exactly same numbers as in whole Keeling islands, though so many miles in length!—

Ever your's | C. Darwin

DAR 114.3: 175

[1] Dated by the relationship to the letter to J. D. Hooker, 13 July [1856], in which CD stated that he would like to see pages from Edward Forbes's unfinished manuscript (see n. 3, below).
[2] The July issue of the *Westminster Review* included an anonymous article on 'Hereditary influence, animal and human' (*Westminster Review* 66: 135–62). The author was George Henry Lewes (*Wellesley index* 3: 625).
[3] See letter from J. D. Hooker, 10 July [1856] and n. 7. The reference is to pages eventually published in E. Forbes and Godwin-Austen 1859.
[4] See letter to J. D. Hooker, 8 [July 1856].
[5] CD had to ask Hooker again for seed-pods of these plants in his letter to J. D. Hooker, 28 September [1856].
[6] Hooker stated in J. D. Hooker 1854, 2: 67, that he gathered forty-three plants 'without rising from the ground'. In a footnote to this passage, marked in CD's copy of the work (Darwin Library–CUL), he added that in England only thirty species could be collected in an equivalent space.

To John Lubbock [29 July 1856][1]

Down
Tuesday Evening

Dear Lubbock

Parslow utterly forgot to give me till within 5 minutes your note.

I am very sorry to say that I really cannot refer you to what you want: this very same point was always a trouble to me.

I am delighted to hear that you are going on well.— I have just returned from Poultry show & am very tired so no more[2]

Ever your's | C. Darwin

James Dwight Dana. Photograph, *c.* 1860.
(Courtesy of Yale University Library.)

William Bernhard Tegetmeier (left) with a fellow pigeon-fancier.
(E. W. Richardson 1916. By permission of the Syndics of the Cambridge University Library.)

Can you let me have back my $\frac{1}{10}'$ glass as I sometimes want it. The $\frac{1}{8}'$ compound I am not in a hurry for;[3] it had better be packed carefully whenever returned as a jar would hurt it.—

Has not Martin Barry in last 10 years in Phil. Transact. given drawings of germinal vesicle?[4] I think he *must* have done so.—

Down House

[1] Dated by CD's reference to the poultry show at Anerley (see n. 2, below) and to Lubbock's microscopical investigations (see n. 4, below).

[2] CD refers to the poultry show at Anerley, near Sydenham, Kent, held from Tuesday, 29 July to Friday, 1 August 1856 (see letters to W. B. Tegetmeier, [July 1856] and 14 August [1856]).

[3] The reference is to lenses, probably for the Smith and Beck simple dissecting microscopes that both CD and Lubbock owned, and for their compound microscopes (see *Correspondence* vol. 4, letter to Richard Owen, [26 March 1848], and Appendix II, p. 404).

[4] Barry 1838. Lubbock was in the process of studying the two means of reproduction in the water-flea *Daphnia*, a freshwater crustacean, to see whether the common eggs destined to develop without fertilisation (the agamic ova) were morphologically different from the so-called 'ephippial' eggs (Lubbock 1857). See letter to John Lubbock, 27 October [1856].

To J. D. Hooker 30 July [1856][1]

Down Bromley Kent
July 30

My dear Hooker

Your letter is of **much** value to me. I was not able to get definite answer from Lyell, as you will see in enclosed letters,[2] though I inferred that he thought nothing of my arguments. Had it not been for this correspondence, I sh^{d} have written sadly too strongly. You may rely on it I shall put my doubts modestly.[3] There never was such a predicament as mine; here you continental extensionists would remove enormous difficulties opposed to me, & yet I cannot honestly admit the doctrine, & must therefore says so.— I cannot get over the fact that not a fragment of secondary or palæozoic rock has been found on any isl^{d} above 500 or 600 miles from a mainland.— You rather misunderstand me when you think I doubt the *possibility* of subsidence of 20, or 30,000 feet; it is only *probability*; considering such evidence as we have independently of distribution.— I have not yet worked out in full detail the distribution of mammalia both *identical* & allied with respect to the *one element of depth of the sea*;[4] but as far as I have gone, the results are to me surprisingly accordant with my most troublesome belief in not such great geographical changes as you believe; & in Mammalia we certainly know more of *means* of distribution that in any other class.— Nothing is so vexatious to me, as so constantly finding myself drawing different conclusions from better judges than myself, from the same facts.

I fancy I have lately removed many (not geographical) great difficulties opposed to my notions, but God knows it may be all hallucination.—

Please return Lyells letters.— What a capital letter of Lyell's that to you is,[5] & what a wonderful man he is.— I differ from him greatly in thinking that those who believe that species are *not* fixed will multiply specific names:[6] I know in my own case my most frequent source of doubt was whether others would not think this or that was a God-created Barnacle & surely deserved a name. Otherwise I sh^d. only have thought whether the amount of difference & permanence was sufficient to justify a name:[7] I am, also, surprised at his thinking it immaterial whether species are absolute or not:[8] whenever it is proved that they all species are produced by generation, by laws of change what good evidence we shall have of the gaps in formations. And what a science Natural History will be, when we are in our graves, when all the laws of change are thought one of the most important parts of Natural History. I cannot conceive why Lyell thinks such notions as mine or of Vestiges, will invalidate specific centres.[9] But I must not run on & take up your time. My M.S. will not I fear be copied before you go abroad.

With hearty thanks | Ever yours | C. Darwin

Do pray keep the crossing doctrine occasionally before your mind.[10]

What a capital party you will be abroad.—[11]

After giving much condensed my argument versus continental extensions, I shall append some such sentence, as that two better judges than myself have considered these arguments & attach no weight to them.—

DAR 114.3: 172

[1] Dated by the reference to the exchange of letters between CD and Charles Lyell on continental extensions (see n. 2, below).

[2] It seems probable that CD had sent to Hooker the original manuscripts of Lyell's letters to CD (letters from Charles Lyell, 17 June 1856 and [1 July 1856]) and copies of his replies to Lyell, 25 June [1856] and 5 July [1856]. The copies are now bound in DAR 114.3 following letters 165 and 167, respectively.

[3] In the chapters on geographical distribution in *Natural selection* and *Origin*, CD made it clear that he disagreed with the theory of continental extensions because he could not admit the vast geographical changes it required, but in the main he contented himself with showing that the geographical distribution of species, particularly those on oceanic islands, could be better explained by various means of dispersal. Although he admitted that many difficult problems of distribution remained, CD maintained that none was insuperable and that a theory of migration or transportal from single centres of creation better accounted for the facts and raised fewer problems than theories invoking continuous land masses or multiple creations.

[4] The relationship is discussed in *Origin*, pp. 395–6.

[5] A reference to a letter from Charles Lyell to J. D. Hooker, dated 25 July 1856 (K. M. Lyell ed. 1881, 2: 214–17). Lyell had discussed the definition of species entailed in Hooker's introductory essay (J. D. Hooker 1853–5, 1: vii–xii) and how CD's views on geographical distribution and transmutation were difficult for Lyell to accept.

[6] Lyell had complained to Hooker that if naturalists believed that the 'boundaries of species are . . . artificial, or mere human inventions,' as CD's theory seemed to require, there would be no check to the proliferation of species in catalogues. Lyell felt that so long as species-multipliers 'feared that a species might turn out to be a separate and independent creation, they might feel checked; but once abandon this article of faith, and every man becomes his own infallible Pope.' (K. M. Lyell ed. 1881, 2: 214–15).

[7] CD refers to the difficulty he experienced in defining species in his monograph on the Cirripedia. See *Correspondence* vol. 5, letter to J. D. Hooker, 25 September [1853].

[8] Lyell had stated that: 'In truth it is quite immaterial to you or me which creed proves true, for it is like the astronomical question still controverted, whether our sun and our whole system is on its way to the constellation Hercules.' (K. M. Lyell ed. 1881, 2: 215).

[9] Lyell had written: 'I fear much that if Darwin argues that species are phantoms, he will also have to admit that single centres of dispersion are phantoms also' (K. M. Lyell ed. 1881, 2: 216). Lyell based much of his geological work on the idea that groups of species spread out from a single geographical source while remaining constant in form: deposits were considered to be the same age if their fossils were of identical species.

[10] See letter to J. D. Hooker, 19 July [1856].

[11] The Hookers were about to leave for a holiday in Switzerland (see following letter).

From Asa Gray [early August 1856][1]

besides 20 or 30 of Introd. matter[2]—(all simple & for students' use)—among which is 4 pages of *statistics*.—of which I enclose you a proof of the last page, that you may see what our flora amounts to.[3] The genera of the Crypt. Ferns—down to *Hepaticæ*—are illustrated in 14 crowded plates—so that the volume has become rather formidable as a Class-book—which is what it is intended for—

I have revised the last proofs to-day— The publishers will bring it out some time in August.— Meanwhile, I am going to have a little holiday, which I have earned, little as I can spare the time for it—and my wife and I start on Friday to visit my Mother and friends in W. New York, and on our way back I will look in upon the scientific meeting at Albany on the 20th inst. or later—just to meet some old friends there.[4]

Why could not you come over, on the urgent invitation given to European savans—& free passage provided back & forth in the steamers! Yet I believe nobody is coming. Will you not come next year, if a special invitation is sent you on the same terms?

Boott[5] lately sent me your *photograph*,[6] which (tho. not a very perfect one) I am well pleased to have.—

But there is another question in your last letter—one about which a person can only give an impression— And my impression is that—speaking of plants of a well-known flora—that what we call intermediate varieties *are generally less numerous* in individuals than the two states which they connect. That this would be the case in a Flora where things are put as they naturally should be, I do not much doubt, and the wider are your views about species (say for instance with D.[r] Hooker's very *latitudinarian* notions) the more plainly would this appear.—

But practically two things stand hugely in the way of any application of the fact or principle, if such it be, 1, Our choice of what to take as the typical forms very often is not free. We take, e.g. for one of them the particular form of which Linnæus, say, happened to have a specimen sent him, and on which established the species; and I know more than one case in which that is a rare form of a common species: the other var. will perhaps be the opposite extreme—whether the most common or not, or will be what L. or [Willd],[7] say, described as a 2^d species.— Here various intermediate forms may be the most abundant; 2^d It is just the same thing now, in respect to specimens coming in from our new western country. The form which first comes & is described & named, determines the spec. char.— And this long sticks as the type, tho, in fact it may be far from the most common form.

Yet of plants very well known in all their aspects, I can think of several of which we recognize 2 leading forms, and rarely see anything really intermediate,—such as our Mentha borealis, its hairy and its smooth varieties.

Your former query, about the variability of Naturalized plants as compared with others of same genera, I had not forgotten, but have taken no steps to answer.— I was going hereafter to take up our list of naturalized plants & consider them, it did not fall into my plan to do it yet. Off hand I can only say that it does not strike me that our introduced plants generally are more variable, nor as variable, perhaps, as the indigenous. But this is a mere guess,—

When you get my sheets of 1st part of article in Sill. Journal,[8] remember that I shall be most glad of free critical comments; and the earlier I get them, the greater use they will be to me.

Dr Hooker writes me that he is to be off on the 12th inst to Switzerland, &c.— So I cannot write to him till his return.

One more favor: Do not I pray you speak of your letters *troubling* me. I should be sorry indeed to have you stop, or write more rarely, even tho' mortified to find that I can so seldom give you the information you might reasonably expect from
Yours most | sincerely | Asa Gray

Incomplete
DAR 165: 93

CD ANNOTATIONS
1.1 besides . . . have.— 4.2] *crossed pencil*
5.1 But there . . . appear.— 5.7] *scored brown crayon*; 'Q' *added pencil*
6.1 But practically . . . free 6.3] 'Q' *added pencil*
6.3 We take . . . species.— 6.8] 'Q' *added pencil*
6.10 The form . . . intermediate,— 7.2] 'Q' *added pencil*
7.3 smooth varieties.] '(that may be jump)' *added pencil*
8.4 Off hand . . . indigenous. 8.6] *double scored brown crayon*
8.6 But . . . Gray 10.1] *crossed pencil*
Top of first page: 'Intermediate forms. & *[illeg]*' *pencil*; '5' *brown crayon, circled brown crayon*; 'B' *brown crayon, circled brown crayon*; '20'[9] *brown crayon, circled brown crayon, del brown crayon*
Bottom of last page: 'Received Aug 20th /1856/'

[1] Dated on the assumption that the mail from Boston to England would have taken about two weeks (see CD's annotations).

[2] Gray refers to the proof-sheets of the forthcoming second edition of his *Manual* (A. Gray 1856a).

[3] A. Gray 1856a, pp. xxv–xxviii, an 'Arranged list of the natural orders of the flora of the northern United States, with the number of their genera and species, the number of introduced species, and of those common to Europe'.

4 The American Association for the Advancement of Science met in Albany, New York in 1856.

5 Francis Boott, a friend of CD, Gray, and Joseph Dalton Hooker, had settled in London and, after his retirement from medical practice, devoted his time to botany.

6 This may be the photograph taken by Maull and Fox (see *Correspondence* vol. 5, facing p. 448).

7 Karl Ludwig Willdenow.

8 The first part of Gray's 'Statistics of the flora of the northern United States' (A. Gray 1856–7).

9 The '5' may refer to chapter 5 of *Origin*; '20' is the number of CD's portfolio of notes on the geographical distribution of plants.

From J. S. Henslow 2 August 1856

Hitcham, Bildeston, *Suffolk*.

2 Aug *1856*

My dear Darwin,

I go to town on Tuesday & on to Cheltenham on Thursday—[1] Are you to be there? L. Jenyns writes me word he wants to get up a discussion about the limitation of species—[2] He means to attack the subject thro' Ornithology. I have a case or two this year among plants—[3]

I have clearly reduced Centaurea nigrescens or decipiens (*of Bab.*[4]) to C. nigra, after about 4 years culture— I think I told you I have had a small bed of Ægilops squarrosa self sown for the last 3 or 4 years— This year one plant in the very midst of the patch assumed the *triticoidal* character, which has been noticed in 2 other species—[5] I have carefully netted it—& watched it but from its appearance I begin to fear it will not ripen any seed. I had fancied it to be a clear example of Ægilops passing to wheat—& certainly it is a most remarkable plant— It may however be a hybrid between the Ægilops & some revet wheat (?)— It is upright, & the ear has long straight awns & is quite downy. Would you like some seeds of the Ægilops now ripe?[6] I have manured a piece of the border next the patch, & mean to extend my experiment next year over it— You have not told me yet what seeds you would like us to get for you— We have been in a greater

AL incomplete
DAR 166: 178

CD ANNOTATIONS

0.3 My . . . plants— 1.4] *crossed pencil*

2.1 I have . . . culture— 2.2] *following* '[' *added pencil; crossed pencil*

2.10 I have . . . greater 2.12] *crossed pencil*

Top of first page: '(3)'[7] *brown crayon*

1 The British Association for the Advancement of Science meeting was held at Cheltenham, 6–13 August 1856. CD did not attend.

2 Leonard Jenyns read a paper at the British Association meeting entitled 'On the variation of species' (Jenyns 1856).

3 Described in Henslow 1856, a paper delivered at the British Association meeting in Cheltenham. See also letter from J. D. Hooker, 10 July 1856.

4 Charles Cardale Babington.

5 See letter from J. D. Hooker, 10 July 1856 and n. 6.

[6] CD had sown seeds from the second generation of Henslow's experimental *Aegilops* in the spring of 1856 (see letter to J. S. Henslow, 6 August [1856] and n. 3).

[7] Scraps marked '3' frequently concern cross- and self-fertilisation and the number may have referred to chapter 3 of *Natural selection*, 'On the possibility of all organic beings crossing'.

From J. D. Hooker 4 August 1856

Aug 4th. 1856

Dear Darwin

Thanks for Lyells letters which are very tough reading—[1] I certainly enjoy all the freedom of motion in vacuo as far as the Geological difficulties are concerned, not because I would underrate their magnitude & importance, but because we really seem to know so little about them. What the deuce are these Eocenes, Miocenes, & Pleiocenes but shadows of Geological phenomena, whose substance (in so far as that substance is the reconstruction of the Earths surface at the period) we really appear to know nothing whatever about.

If Snowdon has gone up 3000 feet since the Glacial epoch & the sea has left no trace of its presence over the thousands of square miles but the wretched little shell beds that you might almost pocket in a visit, it would really seem that we have but fixed the hazy edge of the shadow of Geological change during the period of existing species As an outside barbarian I have always thought that Geologists make too much out of their facts: though I do not think they can magnify the importance of the facts I think they can make too much of them. Have you read Austen on the possible extension of the coal fields?[2] that does appear to me outgeologizing Geology, with a vengeance

My object in now writing is not to discuss Lyells letters wh. I have not digested at all, but to tell you that two of my New Zealand *Epacrideæ* that I had thought new, are the same with Tasmanians: both are mountain Tasmanian; & one truly Alpine, both in Tasm. & N. Zealand. This prevalence of scarce Alpines common to N.Z. & V.D L. coupled with the total absence of Eucalypti & any Myrtaceæ & Acaciæ or any Legum: in common, is to my mind a damning evidence against migration.[3] The mere fact that little scarce alpine things with no apparent means of transport should be common to the two tracts of land, whilst, hundreds of shrubs & trees & Composites that clothe Tasmania & Australia like a garment (I especially allude to Acacias, Epacrid & Eucalypti & other Myrtaceæ) which have, some, millions of minute seeds, wafted aloft, others good hard seeds for transport of other kinds, others pappus & viscid seeds & when you consider that these are not only in abundance of individuals but in point of number of the genera & species they belong to **dominant** families of Tasmania & Australia—I do think that to call for migration for those rarer & more local plants that have got across is against all facts & all philosophy & is to me the most inconceivably gratuitous assumption, being further unsupported by a single fact of special adaptation to migration in the plants themselves.[4]

I am quite ready to admit the gigantic difficulties in the way of Continental Extension, & I also admit that it does not explain all the facts & is no more than an idea perhaps; but it does not fly in the face of known facts in the history of distribution & the geological arguments against it are of no *proved* value. The Continental Extension may be a retrograde step, but it is no harm done, whereas the migration strikes at the root of logical induction from known facts in distribution.

This is all very fine talking. I shall certainly have another shy at the subject when Fl. Tasmania is done.[5]

To my mind the matter stands thus;—there are facts against migration which you & every one acknowledge: & there are arguments (not facts) against the theory of Continental Extension, whose full value you do not yourself know & upon which no two Geologists agree. Under these circumstances I reject migration, & hesitatingly accept continental Extension, waiting for more light upon the subject

This is the aspect the question bears to my eyes at present. I grant that it may be susceptible of huge modifications— The difference between us lies in this—that you are more of a Geologist than Botanist & feel the real weight of the Geolog. objections. I am no Geologist at all & enjoy the aforesaid freedom of motion.

With regard to *Lyells* letters I doubt if the throws any real light upon the grand question of continental extension by niggling (as Sedgwick seems to have it)[6] at local subsidences. If there is, as you seem to show, a relation between lofty volcanic Mts & ocean,[7] & between lofty non volcanic Mts & continents, we must seek some explanation of so grand a generalization in the operation of cosmical causes & the first that occurs is hydrostatic pressure. now if one had only the gift of the gab, & cheek enough, one might keep the Geolog. Soc. in extasy with speculations of that description. Central heat would be a trifle to it.

Your argument about great Mammalia is good, but cuts both ways, why did they not migrate if the Java & Borneo &c &c &c beasts are the same—& again are not the means of destruction of large beasts on Islands tenfold greater than of small.? want of food in case of Carnivora—

I think I can quite understand what Lyell means by saying that, your or vestigial notions will invalidate specific centres;[8] & that it resides in this, that you will have to hedge about your theory with (what will appear to most people) to be gratuitous assumptions to prevent their or your quoting change of form for every centre. Thus, if Ægilops & wheat are transmutations & the duration of *Ægilops* as a species be 3 geological epochs (A. B. C.) & its spread in area to be the Northern hemisphere, you may have physical causes causing its transmutation during A & C. only—& at 3 independent spots X, Y. Z. in the area— You thus annihilate both centres & time— My strongest argument against transmutation are 1) that the vegetable world does not appear in the confusion I should expect it to be in, were transmutation the law. 2) that I do not find that repetition of species, or of forms, under closely similar physical conditions, that I should have expected transmutation to have effected. If I accept transmutation I am prepared to give up much specific centralization as an inevitable consequence.— This shakes Geological systems to the foundation & geographical distribution too.

But I must break off. I return Lyell's letters, they are very suggestive but I do not see that they much touch the question— What do you think of Tyndall's explanation of Cleavage?[9]

Ever Y[rs] | J D Hooker

DAR 100: 100–4

CD ANNOTATIONS

0.3 Dear . . . vengeance. 2.9] *crossed pencil*
7.1 This . . . time— 10.9] *crossed pencil*
10.9 1) . . . law. 10.11] 'no' *added pencil*
10.11 2) . . . effected. 10.13] 'no' *added pencil*
10.13 If . . . Hooker 12.1] *crossed pencil*

[1] See letter to J. D. Hooker, 30 July [1856].
[2] Godwin-Austen 1856 was read at a meeting of the Geological Society of London on 30 May 1855. Robert Alfred Cloyne Godwin-Austen concluded that geological evidence was sufficient 'to admit of the restoration of the original surfaces which supported the coal-vegetation' in Europe (pp. 72–3) and that there were strong *a priori* reasons to suppose there was a band of coal measures along the line of the valley of the Thames (p. 73).
[3] In *Natural selection*, p. 564, n. 2, CD suggested that seeds carried on boulders transported by icebergs might explain alpine plants common to New Zealand and Van Diemen's Land (Tasmania).
[4] In J. D. Hooker 1853–5, 1: xix, Hooker noted: 'The idea of transportation by aerial or oceanic currents cannot be entertained, as the seeds of neither [*Edwardsia grandiflora* nor *Oxalis magellanica*] could stand exposure to the salt water, and they are too heavy to be borne in the air.' He also commented that the plants showed no apparent adaptations for transportal (p. xxi).
[5] Hooker had been publishing his *Flora Tasmaniæ* in parts since 1855 (see Wiltshear 1913). The work was completed in 1860 (J. D. Hooker 1855[-60]).
[6] See letter from Charles Lyell, 17 June 1856.
[7] See letter to Charles Lyell, 5 July [1856].
[8] An allusion to the transmutationist views expressed in *Vestiges of the natural history of creation* ([Chambers] 1844). See letter to J. D. Hooker, 30 July [1856].
[9] In a lecture delivered at the Royal Institution on 6 June 1856 (Tyndall 1856), John Tyndall discussed cleavage in rocks and other solids with the intention of distinguishing between cleavage due to crystalline action and cleavage caused by mechanical agencies. The latter, argued Tyndall, was primarily the result of compression. CD had long been interested in the phenomena of cleavage and foliation (see *Correspondence* vols. 3, 4, and 5).

To J. D. Hooker   5 August [1856][1]

Down.
Aug[t] 5th

My dear Hooker

Many thanks for your Review, which I shall read with *greatest* interest.[2] I quite agree about Lyell's letters to me, which though to me interesting have afforded me no new light.— Your letters, under the *geological* point of view have been more valuable to me. You cannot imagine how earnestly I wish I c[d] swallow continental extension, but I cannot; the more I think (& I cannot get the subject out of my head), the more difficult I find it. If there were only some half-dozen cases, I sh^d. not feel the least difficulty, but the generality of the fact of all islands (except 1 or 2) having a considerable part of their production, in common with one or more mainlands, utterly staggers me.— What a wonderful case of the Epacridæ! It is most vexatious, almost humiliating to me that I cannot follow & subscribe to the way in which you strikingly put your view of the case. I look at your facts (about Eucalypti &c) as *damning* against continental extension, & if you like, also damning against migration, or at least of *enormous* difficulty.— I see the grounds of our difference (in a *letter* I must put myself on an equality in arguing) lies in my opinion that scarcely anything is known of means of distribution. I quite agree with A. Decandolles (& I daresay your) opinion that it is poor work putting together the merely *possible* means of distribution;[3] but I see no other way in which the subject can be attacked; for I think that A. Decandolles argument that no plants have been introduced into England except by man's agency, of no weight.—[4] I cannot but think that the theory of continental extension does do some little harm as stopping investigation of means of dispersal, which whether *negative* or positive seem to me of value: when negatived, then everyone, who believes in single centres, will have to admit continental extensions.—

I agree about much greater chance of extermination on isl^ds—but the (I think) (I have got lists) *invariability of rule* that on all islands there are naturalised mammifers, in some cases of some hundred years duration, in some degree weakens your argument.— I do not understand your argument why Java & Borneo beasts did not migrate; I wish I did.— Nor yet entirely about about specific centres; but I have entered both points in my note Book to ask you, when we meet; for you have of late taken enormous trouble for me.— I see from your remarks that you do not understand my notions, (whether or no worth anything) about modifications,—I attribute very little to the direct action of climate &c.—

I suppose in regard to specific centres, we are at cross purposes; I sh^d. call that kitchen garden, in which the Red Cabbage was produced, or the Farm in which Bakewell made the Short-Horn Cattle,[5] the specific centre of these *species*! and surely this is centralisation enough!—

I thank you most sincerely for all your assistance; & whether or no my Book may be wretched you have done your best to make it less wretched. Sometimes I am in very good spirits & sometimes very low about it. My own mind is decided on the question of origin of species but good Heavens how little that is worth.

I have not read Austen with attention. Tyndall's pamphlet is *capital* & has made me finally give up a cherished opinion, but which was much shaken before:[6] I cannot but think, that Tyndall underrate the effect of the actual movement of the particle, in planes parallel to the pressing surfaces which he shows does take place, as pressure without extension produces no effect.

I must now look at all the cases of foliated schists which I have described, as metamorphosed *cleaved* rocks with segregation & crystallization along the planes of cleavage, (as along the planes of division caused by movement in obsidian lava-streams); I formerly thought that,[7]—but I wont bore you—

so my dear Hooker, farewell & may you & Mrs Hooker have a delightful tour. Adios | C. Darwin

My M.S. will not be ready till your return.—

DAR 114.3: 173

[1] Dated by the relationship to the preceding letter.

[2] Hooker's review ([J. D. Hooker] 1856) of Alphonse de Candolle's *Géographie botanique raisonnée* (A. de Candolle 1855). CD's annotated copy of an offprint of the review is in the Darwin Pamphlet Collection–CUL.

[3] 'Tels sont, au premier aperçu, les moyens de transport. Il ne suffit pas d'en constater l'existence, il faut encore prouver par des faits que ces moyens ont agi. On se contente trop souvent d'indiquer les possibilités de transports, sans examiner si elles se réalisent' (A. de Candolle 1855, 2: 623). The passage is marked in CD's copy (Darwin Library–CUL).

[4] A. de Candolle 1855, 2: 704–9. Hooker refers to Candolle's conclusions about naturalised plants in [J. D. Hooker] 1856, pp. 82–8.

[5] The short-horn breed of cattle was bred by Robert and Charles Colling in County Durham after the brothers had visited Robert Bakewell at Dishley (*EB*).

[6] See preceding letter and n. 9. CD agreed with John Tyndall that cleavage was the result of pressure, but he also believed that foliation was a continuation of the same process in metamorphosed rock. Tyndall's view of crystallisation, however, was that foliation took place along lines of stratification and that when such rocks were compressed and contorted, they produced cleavage foliation.

[7] See *Correspondence* vol. 3, letter to J. D. Forbes, 11 October [1844], in which CD drew James David Forbes's attention to the laminae in obsidian that CD thought resembled those of glacier ice.

To J. S. Henslow 6 August [1856][1]

Down Farnborough Kent
Aug[t] 6th

My dear Henslow

I received your letter dated 2[d] only yesterday: I shall not come to Cheltenham, though your presence & L. Jenyns paper would be a great temptation.

I am particularly pleased to hear about the Centaurea, the seed, which you gave me did not germinate.[2] Your Ægilops has come up & has ripened seed: I forced it so as not to flower at same time with wheat: it has not varied: you formerly called it Æ. ovata now Æ. squarrosa.—[3]

The Myosotis was sown in open ground, both in sunny & shady places;[4] in former place, whence the specimens sent to you came, it was watered weekly with Guano water. Nearly all the flowers are brightish blue, & only a very few on dwarf branches are pink.— The specimens in the more shady place have the lobes of corolla slightly emarginate. The tube of corolla, in comparison with the Calyx seems to be longer in the blue than in the smaller pinkish flowers.

With respect to seeds: I sh[d] be extremely glad of any water plants; especially of Callitriche verna, Limosella aquatica & Montia fontana, (if such you have).—[5] I want, also, to try whether the ripe pods on heads of seed would float in sea-water;

if you could help me by sending a few specimens in Box by Post. I have just been correcting my paper on salting seeds for Linnean Journal.—6

My dear Henslow | Yours most truly | Ch. Darwin

The seed of Rosa tomentosa did not come up.—

DAR 93: 55–6

1 This letter was endorsed '1858', but the contents and the relationship to the letter from J. S. Henslow, 2 August 1856, indicate that 1856 is the correct year.

2 In his Experimental book, p. 3 (DAR 157a), CD recorded that he planted seeds of *Centaurea nigra* var. *decipiens* from Henslow on 1 March 1856. In May 1856 he further recorded: 'Centaurea nigra, var decipiens. Seedling 2ᵈ removed from original plant wild at Bath. From Henslow Hitcham. 1855.'

3 See letter from J. S. Henslow, 2 August 1856. CD's Experimental book, p. 1 (DAR 157a), includes the entry: 'Ægilops ovata, seed from Henslow "(2ᵈ generation in Hitcham Garden 1855".—'. On the facing page adjoining the entry, CD wrote: '(Henslow, now Augᵗ 6ᵗʰ 56 calls it Æ squarrosa)'.

4 See letter to J. S. Henslow, 16 June [1856].

5 CD had also asked Joseph Dalton Hooker, in the letter to J. D. Hooker, 26 [July 1856], for specimens of these water plants.

6 CD's paper 'On the action of sea-water on the germination of seeds', read 6 May 1856, was published in *Journal of the Proceedings of the Linnean Society (Botany)* 1 (1857): 130–40 (*Collected papers* 1: 264–73).

To J. D. Hooker 7 August [1856][1]

Down
Augᵗ 7ᵗʰ

My dear Hooker

Hearty thanks for the Tristan list:[2] it has cost you much trouble, but I trust & hope that it may be of use to yourself, & I will carefully return the M.S. at a future time with my own M.S.

You may well say that it would be a mystery if I could explain such facts; I expressly bring these antarctic cases forward as the most difficult to account for on any theory whatever;[3] not excepting, as I imagine, even continental extension, but on this point I shall enter a note in my note-book to hear what you have to say, in the autumn, when we meet.— I believe much in Lyell's iceberg action in bringing seeds, & now that I have put a good many facts together, it seems to me perfectly extraordinary if plants have not been sometimes thus transported;[4] but I know that you do not believe in this.

The particular point in regard to Tristan which I wanted to know was whether there were more representative species of American *parentage* in Tristan than in Kerguelen,[5] but of that we will talk— if that were to prove so, it would harmonise well with my notions; but I have not yet studied your M.S. which I am sorry to see includes a good many plants unknown to you.—

Farewell | My dear Hooker | Ever yours | C. Darwin

I have read half your Review & like it very much. D.C. ought to be very much pleased;[6] but I suppose the sugar is at the top & the sour at the bottom.

DAR 114.3: 174

[1] Dated by the relationship to the letter from J. D. Hooker, 10 July 1856, in which Hooker's list of Tristan d'Acunha plants was also referred to, and by the reference to [J. D. Hooker] 1856 (see n. 6, below).

[2] See letter to J. D. Hooker, 5 [July 1856], n. 4. For CD's difficulties in interpreting the flora of Tristan d'Acunha, see also letter to J. D. Hooker, 5 July [1856].

[3] See *Natural selection*, pp. 560-5.

[4] See *Natural selection*, pp. 561-2.

[5] In CD's discussion in his species book of extraordinary cases of distribution, he reported that twelve of the Tristan d'Acunha species were common to South America and seven Kerguelen Land plants common to Fuegia. 'Hence', CD wrote, 'Tristan d'Acunha, like Kerguelen Land, is botanically more nearly related to Fuegia (from which it is almost 2300 miles distant) than to any other country' (*Natural selection*, p. 561).

[6] Hooker's review of A. de Candolle 1855 ([J. D. Hooker] 1856).

To W. B. Tegetmeier   14 August [1856][1]

Down Bromley Kent
Aug[t] 14th

My dear Sir

When at Anerly Show the end of Catalogue which I got, was not completed;[2] & I sh[d] be very much obliged if you would take the trouble (if you can remember) to tell me whose some very much elongated, Blue Runts, with black bars, belonged to. They were in the last pen but one in a row. I want to find out, if I can, something of their history:[3] they struck me as very singular birds.—

The two young Scanderoons are coming on well & becoming feathered, & in about 3 weeks will be ready for you.—[4]

I hope you are recovered from your gigantic labours at the Show.—[5]

Yours very sincerely | C. Darwin

These Blue Runts stood next to a pair of black mottled Runts.—

New York Botanical Garden Library (Charles Finney Cox collection)

[1] Dated by CD's reference to the Anerley poultry show (see n. 2, below).

[2] The second annual poultry and pigeon show held at Anerley Gardens took place from 29 July to 1 August 1856.

[3] CD discussed these and similarly marked pigeons in *Variation* 1: 195-6. He reported that blue birds of any race almost invariably showed black wing-bars. These markings characterised wild *Columba livia* and CD believed they indicated the origin of the domestic races.

[4] CD had promised Tegetmeier a pair of Scanderoons in letter to W. B. Tegetmeier, 24 June [1856].

[5] Tegetmeier was called in at the last minute to supervise the Anerley show (*Cottage Gardener* 16 (1856): 355-6).

To W. B. Tegetmeier   [15-22 August 1856][1]

Down Bromley Kent

My dear Sir

I write one line merely to thank you for your most clear chart of the Runts.— With respect to your very kind offer of Runts, I will hereafter write to you: if, as I imagine, you allude to the Victoria or Smyrna Runts, though **extremely** pretty, I do not think they differ in proportions & that is my *sole* object.—[2] There would be no end to my collection if I allowed myself to be influenced by beauty.—

Far from thinking that you treated me "cavalierly" at Anerly, I was struck by your kindness, as on all other occasions, considering what a weight of work you had on your shoulders—[3]

In Haste | Yours very sincerely | Ch. Darwin

New York Botanical Garden Library (Charles Finney Cox collection)

[1] Dated by the relationship to the preceding letter and to the letter to W. B. Tegetmeier, 23 August [1856].
[2] The characteristic features of runts are discussed in *Variation* 1: 142–4.
[3] See preceding letter.

To Frances Mackintosh Wedgwood   18 [August 1856 – January 1858][1]

Down
Monday 18th—.

My dear Fanny.

I am much pleased & flattered at Mr Gregs proposal,[2] I should have liked to have tryed my hand at Reviewing but I have so many years work in prospect in my present book on species & varieties, that I am not willing to give up my time to any other occupation— I should think Reviewing if one did it heart & soul,[3] would be particularly interesting but I have the greatest doubt whether I should succeed, but for the future I shall vote myself quite capable & worth £15 per sheet; already I stand wonderfully higher in all the children's estimation—

You will be very sorry to hear how much illness we have here at present.[4]

Yours very affectionately. | C. Darwin.

Copy
DAR 148

[1] CD's reference to having 'so many years work in prospect' indicates that he had only recently begun work on his species book (*Natural selection*). He began writing in May 1856. 'Monday 18th—.' occurred in August 1856, May 1857, and January 1858. The next possible Monday fell in October 1858, but it is unlikely that the letter was written then because CD was at that time working on *Origin*.
[2] The copy reads "Grey", but was corrected to "Greg" by Francis Darwin. William Rathbone Greg, a prominent contributor to the *Edinburgh Review* and the *Quarterly Review*, carried considerable influence with the editors of both periodicals. In 1855 he had also briefly been editor of a new Unitarian periodical, the *National Review*, but he resigned before publication began in 1856 (*Wellesley index* 3: 135). Hensleigh and Fanny Wedgwood were active in London's Unitarian circles; Hensleigh Wedgwood also occasionally reviewed books for the *Quarterly Review*.
[3] Francis Darwin corrected the copyist's 'sould' to read 'soul'.
[4] It has not been possible to identify the specific illnesses to which CD refers. In August 1856 Henrietta Darwin was ill, and Emma Darwin recorded in her diary on 25 August that CD took Henrietta to consult Benjamin Collins Brodie.

To E. W. V. Harcourt[1] 19 August [1856][2]

Down Farnborough Kent
Aug[t] 19th

My dear Sir

I think you told me long ago that you had Brehm's Book on German ornithology.[3] If this be so, & you can spare it, will you be so kind as to lend it me for a fortnight.— In this case will you put on enclosed address & send it by Parcels Deliv. Co[y]—

If you have it not, or cannot spare it, tear up the address & I shall understand why it does not come.—[4]

I want to see how far Brehm goes in splitting species.—[5]

I have been thinking over the many curious facts which you told me,[6] & wish more than ever that you would publish a paper on the subject.—[7]

I took some of the little finches from Madeira to Brit Mus. & found the Goldfinch, Linnet & greenfinch rather smaller than the British specimens; but *not* the Blackbird,—nor, as you saw the Swift.—

I wish I had taken more for comparison.—

My dear Sir | Yours very sincerely | Ch. Darwin

Houghton Library, Harvard University

[1] The recipient is identified by CD's reference to Madeiran birds (see n. 7, below).
[2] Dated from the reference to Brehm 1831. CD recorded having read this book on 30 August 1856 (*Correspondence* vol. 4, Appendix IV, 128: 20). CD's notes on the volume are in DAR 71: 162–3.
[3] Brehm 1831. CD had skimmed Christian Ludwig Brehm's work ten years earlier and was aware of his tendency to split species (*Correspondence* vol. 3, letter to Leonard Jenyns, 17 October [1846]).
[4] See letter to E. W. V. Harcourt, 23 August [1856].
[5] CD referred to Brehm 1831 in relation to taxonomic practice in *Natural selection*, p. 114. Brehm was cited as having added 576 species of birds to the German avifauna.
[6] See the concluding paragraphs of the letter from E. W. V. Harvourt, 31 May 1856.
[7] CD probably refers to Harcourt's remarks about the peculiarities of distribution in the letter from E. W. V. Harcourt, 31 May 1856. Harcourt had already published two works on Madeiran ornithology, both of which CD read. CD's notes on Harcourt 1851 are in DAR 71: 87–8, and an annotated copy of Harcourt 1855 is in DAR 196.4.

To Thomas Campbell Eyton 21 August [1856][1]

Down Farnborough Kent
Aug[t] 21st

Dear Eyton

I have just been using your capital facts on the skeletons of Pigs in Zoolog. Proc[s][2] & I want very much to beg a favour of you,—it is to know whether the offspring of the African pig & common were fertile.[3] If you do not know, would it be asking too great a favour to beg you to enquire of Lord Hill;[4] & let me publish the answer on your authority; for this would complete the evidence in regard to fertility.[5]

Also can you tell me whether Ld Hill's African pigs appeared domesticated? Do you know what part of Africa they came from?

I am getting on with my collection of Pigeon skeletons & have every breed alive. I have not yet compared carefully the skeletons; but when I do I shall probably have occasion to beg your assistance; for it would *greatly* add to value of any few remarks which I might make, if I could say that you had seen them & thought my remarks accurate.

I am working away very hard in compiling my Book on Variation, but hardly know when I shall be ready to go to press, for I find it very slow work.—

I hope Mrs Eyton[6] is better than when I last heard of her, now sometime ago.—

How I wish that we lived nearer each other & could sometimes meet. Believe me | Dear Eyton | Yours very sincerely | Ch. Darwin

American Philosophical Society (135)

[1] Dated by CD's reference to writing his chapter on variation (see n. 3, below).

[2] Eyton 1837a. CD's annotated copy of the *Proceedings of the Zoological Society of London* in which the article appeared is in the Darwin Library–CUL.

[3] CD had previously noted Eyton's argument that although the bone structure and number of vertebrae in African and English pigs were very different, they could interbreed (see Notebook B, p. 162 (*Notebooks*)). During the summer of 1856, CD was busy writing up his material on variation for his species book; by October, he had finished the section on 'variation under domestication', which formed the first two chapters of the manuscript of *Natural selection* ('Journal'; Appendix II). From the table of contents CD made for these two chapters, it appears that CD had discussed the variation of the domestic pig (*Natural selection*, p. 25).

[4] Rowland Hill was a Shropshire neighbour of Eyton's.

[5] CD wrote on Eyton's authority that 'cross-bred animals from the African and English races were found by Lord Hill to be perfectly fertile' (*Variation* 1: 74).

[6] Elizabeth Frances Eyton.

To E. W. V. Harcourt 23 August [1856][1]

Down Bromley Kent
Augt 23^{d}

My dear Sir

I have received Brehm all safe & am *very* much obliged for it, & will return it in 2 or 3 weeks time;[2] I would not have kept it so long, but I am going out for a week.—

Yours very sincerely | Ch. Darwin

Wellcome Institute Library

[1] Dated by the relationship to the letter to E. W. V. Harcourt, 19 August [1856].

[2] CD had asked to borrow Brehm 1831 in his letter to E. W. V. Harcourt, 19 August [1856].

To W. B. Tegetmeier 23 August [1856][1]

Down Bromley Kent
Aug 23^{d}

My dear Sir

I shall be **particularly** glad of the Victoria Runt for skeleton & measurement.:[2] will you send it exactly thus addressed per Coach

> Ch. Darwin Esq^{e}
> care of Mr Acton
> Post office
> Bromley
> Kent

Will you enclose a bit of paper telling me where I could send in London in a fortnight or 3 weeks time the 2 Scanderoons, which seem to me very fine Birds.[3] Our Carrier leaves every Wednesday night, & w^{d} deliver pretty early the Birds anywhere in London on the Thursday morning.— I will not send them off till quite strong, & w^{d} advise you of their going.—

About what age is your dead Runt?

yours very sincerely | Ch. Darwin

New York Botanical Garden Library (Charles Finney Cox collection)

[1] Dated by the relationship to the letters to W. B. Tegetmeier, 14 August [1856] and [15–22 August 1856].
[2] Tegetmeier's letter has not been found. He had apparently told CD that the proportions of the Victoria runt would be of significance to him. See letter to W. B. Tegetmeier, [15–22 August 1856].
[3] See letter to W. B. Tegetmeier, 24 June [1856].

To Asa Gray 24 August [1856]

Down Bromley Kent
Augt 24th

My dear Dr Gray

I am much obliged for your letter,[1] which has been *very* interesting to me.— Your "indefinite" answers are perhaps not the least valuable part; for Botany has been followed in so much more a philosophical spirit than Zoology, that I scarcely ever like to trust any general remark in Zoology, without I find that Botanists concur. Thus with respect to intermediate varieties being rare, I found it put, as I suspected, much too strongly (without the limitations & doubts which you point out) by a very good naturalist, Mr Wollaston, in regard to insects;[2] & if it could be established as true it w^{d}, I think, be a curious point.—[3] Your answer in regard to the introduced plants *not* being particularly variable, agrees with an answer, which Mr H. C. Watson has sent me in regard to British Agragrian plants,[4] or such (whether or no naturalised) are now found only in cultivated land. It seems to me very odd, without any theoretical notions of any kind, that such plants should not be variable; but the evidence seems against it.—

Very sincere thanks for your kind invitation to the U. States: in truth there is *nothing* which I should enjoy more; but my health is not, & will, I suppose, never be strong enough, except for the quietest routine life in the country. I shall be particularly glad of the sheets of your paper on Geograph. Distrib; but it really is unlikely in the highest degree that I could make any suggestions.—

With respect to my remark that I supposed that there were but few plants common to Europe & U. States, *not* ranging to the Arctic Regions; it was founded on vague grounds, & partly on range of animals. But I took H. C. Watsons Remarks (1835) & in the table at the end I found that out of 499 plants believed to be common to the Old & new worlds, only 110, did not range on neither side of the Atlantic up to Arctic region.[5] And on writing to M[r] Watson, to ask whether he knew of any plants *not* ranging northward of Britain (say 55°) which were in common,[6] he writes to me that he imagines there are very few; with M[r] Syme's assistance he found some 20–25 species thus circumstanced, but many of them, from on⟨e⟩ cause or other, he considered doubtful.[7] As ⟨exa⟩mples, he specifies to me, with doubt, Chrysopleniu⟨m⟩ oppositifolium; Isnardia palustris; Astragalus Hypoglottis; ⟨Thl⟩aspi alpestre; Arenaria verna; Lythrum hyssopifolium.—

I hope that you will be inclined to work out for your next Paper, what number of your 321 in common, do *not* range to Arctic Regions.[8] Such plants seem exposed to such much greater difficul⟨ ⟩ in diffusion.—

Very many thanks for all your kindness & answers to my questions, & believe me. Yours very sincerely & obliged | Ch. Darwin

If anything sh[d] occur to you on variability of naturalised or agragrian plants; I hope that you will be so kind as to let me hear, as it is a point, which interests me greatly.—

Postmark: AU 26 1856
Gray Herbarium of Harvard University

[1] Letter from Asa Gray, [early August 1856], which CD had marked 'Received Aug 20th /1856/'.
[2] Thomas Vernon Wollaston. See letter to Asa Gray, 14 July [1856].
[3] CD discussed the point in *Natural selection*, p. 268, giving Hewett Cottrell Watson, Gray, and Wollaston as his sources.
[4] Letter from H. C. Watson, 5 June 1856.
[5] Watson 1835.
[6] Letter to H. C. Watson, [after 10 June 1856].
[7] Letter from H. C. Watson, 20 June 1856. Watson's information was used in *Natural selection*, p. 539, although CD later replaced the figures with information supplied by Gray (see letter from Asa Gray, 4 November [1856]).
[8] The number 321 refers to the number of species found in both North America and Europe, as given in a list in A. Gray 1856a, p. xxviii. This page had been forwarded to CD enclosed in the letter from Asa Gray, [early August 1856]. Gray addressed CD's question in the second part of his paper on the statistics of the flora of the United States (A. Gray 1856–7).

To T. C. Eyton 27 [August 1856]¹

Down Bromley Kent
27th

Dear Eyton

Will you forgive me troubling you once again?— I believe that you have attended much to Herefordshire cattle.² I have somewhere seen an account of two strains of this cattle differing slightly in colour,—I think in more white on face. Now what I want to know, is whether in these two strains there is any other difference *whatever*, so that you or a good judge, could generally distinguish these breeds by any slight character, independently of the one of colour.³ Will you be so very kind as to enlighten me, & not abuse me much as being abominably troublesome.

Your's most truly | Ch. Darwin

American Philosophical Society (136)

¹ Dated by the relationship to the letters to T. C. Eyton, 21 August [1856] and 31 August [1856].
² Eyton had published *The herd book of Hereford cattle* (Eyton 1846[–53]).
³ See letter to T. C. Eyton, 31 August [1856].

To W. B. Tegetmeier 30 August [1856]¹

Down Bromley Kent
Augst 30th—

My dear Sir

I received the Victoria Runt all safe & fresh, & very much obliged I am for it.—² I did not write to thank you for it, as I knew I sh^d have to write soon about the Scanderoons. They left the nest a few days ago, & I will send them off on Wednesday night the 10th, so that they will arrive *about* 10 oclock on Thursday 11th—

Many thanks for your offer of dead Turbits & Trumpeters, but I have both.— If you want any other kinds, (alive) please tell me that I may see whether I can supply you.— But I may be prevented as I am crossing all my kinds to see whether crosses are fertile & for the fun of seeing what sort of creatures appear.— I much doubt, indeed disbelieve, that there can be any general law about colours &c for different species, even if there be for varieties of same species. A Mr Orton has lately published some lectures with old theory of externals following males & internals female.—³

I fear (but do not know) that both Laughers are males: The one has shed very many of its dark feathers & is *much* whiter now.— That is a curious remark of yours about pencilling appearing in cross-bred fowls.— How gets on your paper on Fowl's skulls?⁴ I ask because this morning I had a letter from Mr Eyton of Eyton, who has grand collection of skeletons of Birds (& who has published on pig's skeletons) & he says he has been making skeletons of Hamburgh & Dorking

& "will send me some notes". I mention this that you may not be forestalled. I did not know that he was at work on Poultry; though I knew he was on dogs. I wrote to Mr Adkins[5] & had a very obliging answer but no accurate information.—

My dear Sir | Yours very sincerely | C. Darwin

Do you ever see Mr Gulliver;[6] if so I wish you w^d. say how glad to sh^d be to have one of his gigantic Runts dead.—

New York Botanical Garden Library (Charles Finney Cox collection)

[1] Dated by CD's reference to having received a letter from Thomas Campbell Eyton, presumably the same one as referred to in the following letter, and by the relationship to the letter to W. B. Tegetmeier, 23 August [1856].
[2] See letter to W. B. Tegetmeier, 23 August [1856].
[3] Reginald Orton's *Physiology of breeding* (Orton 1855) is cited in *Variation* 1: 404 n. 138. CD's annotated copy is in the Darwin Pamphlet Collection–CUL.
[4] Tegetmeier 1856.
[5] Possibly the breeder who had exhibited blue pigeons with bars on their wings, about whom CD had inquired in his letter to W. B. Tegetmeier, 14 August [1856].
[6] See letter to W. B. Tegetmeier, 24 June [1856].

To T. C. Eyton 31 August [1856][1]

Down Bromley Kent
Aug. 31st

Dear Eyton

I thank you heartily for your note & for your promise of more information on Pigs, about which I am very curious.— [2] By the way Bechstein asserts that the number of incisors varies greatly in domestic pigs:[3] I am myself going to collect Pigs jaws (no other part) to see whether he is to be trusted. Have you ever noticed this? I sh^d like to confirm Bechstein on your authority.[4]

I had no idea that your Stud Book was so well illustrated;[5] I suppose you mean me to return the Plates, so I do. But I have been very glad to see them. Just after writing to you,[6] I found what my memory told me of: viz an article by some apparently well informed person in Quarterly Review (1849 p. 392) who says there is a split of unknown origin in this breed,—one strain having white face tawny sides & upward directed horns; the other a speckled face, generally white line down back, *shorter legs* & more horizontal horns.[7] In my Book on Variation which is progressing (but Heaven knows what it will turn out) I sh^d. like to give this case *trifling* as it is; for it is so rare to trace commencement of even a sub-breed of a sub-breed;[8] & I sh^d. like to quote your remarks in your note, & will append Author of the "Stud Book on Herefordshire Cattle" Is this correct title? If you can give me any other information about these two families of Herefordshire I sh^d. be very *grateful*: I suppose you do not believe about *short-legs*; but just bear this point in mind.—

What a wonderful collection of Birds you have! I had not the least idea of your richness.[9]

I remember well your case of the geese & shall have to quote it.[10] What became of the grandchildren geese? did you continue to breed from them? I have lately been making enquiries in India, where there are flocks of the half-bred-geese habitually kept.—[11]

One of the subjects which gives me most trouble for my work, is means of distribution in the case of species found on distant islands; I have lately been trying the powers of resistance of seeds to sea-water,—their powers of floating— the number of living seeds in earth & mud &c &c.— Would you render me a little assistance in this line? My walking days are over, never to return. I want to know whether on a wet muddy day, whether birds feet are dirty:[12] I am going to send my servant out with some keeper & he shall wash all the partridges feet & save the dirty water!![13]

But I want especially to know whether herons or any waders (we have no ponds hereabouts) or water-birds when suddenly sprung have *ever* dirty feet or beaks? I found in 2 large table-spoon full of mud from a little pond from beneath the water 53 plants germinated.—[14]

Do you know when owl or Hawk eats a little bird, how soon it throws up pellet? Can it throw up pellet whilst on wing? How I sh^d like to get a collection of pellets & see whether they contained any seeds capable of germination. Could your gamekeepers find a roosting place, & collect a lot for me?—

Lastly (if you are not sick of my enquiries) have you ever examined the stomachs of dace & other white fish? Do they ever eat seeds; I know it is good to bait a place with grains. For like the house which Jack built, a heron might eat a fish with seed of water plant & then fly to another pond.

I have been trying for a year with no success to get some dace &c. Have you any & could you catch some in net. & order your kitchen maid to clean them, & you c^d send me the whole stomach & I would sow the contents on burnt earth with every proper precaution. If ever your goodnature sh^d lead you to send me any such rubbish; it might be put in bladder or tin foil & sent by Post, & if you will not think me very impertinent I could repay you the shilling or two for postage; as the rubbish w^d thus come much quicker & cheaper to me.

Do you mean to collect cats' skeletons: Sir C. Lyell has odd Persian & I have *heard* of another odd cat & I w^d request their carcases to be sent to you, if you cared about them. But I fancy cats are much mixed beings.—

Well I have put your words, that you like hearing from old naturalist friends, to a severe test. So forgive me & believe me, | Your's most truly | Ch. Darwin

American Philosophical Society (137)

[1] Dated by CD's reference to Bechstein [1789–95] (see n. 3, below) and to his experiments on seeds carried by the muddy feet of wading birds (see n. 13, below).

[2] See letter to T. C. Eyton, 21 August [1856].

[3] Johann Matthäus Bechstein's book on the natural history of Germany was frequently cited by CD in *Natural selection*. Although CD had previously read Bechstein [1789–95] in 1842, he reread the first volume in February 1856 (see *Correspondence* vol. 4, Appendix IV, 119: 12a; 128: 16). CD owned a mixed set of the four-volume work, an annotated copy of which is in the Darwin Library–CUL.

[4] Bechstein is not cited on this point in either *Natural selection* or *Variation*.

[5] Eyton 1846[–53].

[6] Letter to T. C. Eyton, 27 [August 1856].

[7] An anonymous review of Henry Stephens's *The book of the farm* (Edinburgh and London, [1849]) appeared in the *Quarterly Review* 84 (1848–9): 389–424. The author was Thomas Gisborne, an agricultural writer and MP for Nottingham (*Wellesley index* 1: 732). The review dealt mainly with the breeding of fat stock for the meat market. The particular point mentioned by CD appeared on p. 393 n.

[8] Discussed in *Variation* 2: 214.

[9] Eyton, who kept a large number of animals as well as a museum collection on his estate at Eyton, Shropshire, may have sent CD the first part of his privately printed *Catalogue of the species of birds in his possession* (Eyton 1856), a copy of which is in the Darwin Pamphlet Collection–CUL.

[10] Eyton 1840 was referred to by CD in *Natural selection*, pp. 431 and 439.

[11] In *Natural selection*, pp. 439–40, CD described how hybrids between the Chinese and common goose, as discussed in Eyton 1840, were found by Edward Blyth and Thomas Hutton to be completely fertile in India. CD refers to his correspondence with Blyth: the point is discussed in *Correspondence* vol. 5, letter from Edward Blyth, 8 December 1855.

[12] CD's discussion of birds as 'highly effective agents in the transportation of seeds' is in *Origin*, pp. 361–3.

[13] CD's experiments on the transport of seeds on the feet of birds are listed in his Experimental book, p. 15 (DAR 157a).

[14] Experimental book, pp. 5v. and 6 (DAR 157a).

From E. L. Layard[1] [September–October 1856][2]

Mongrel he is, & will be;—but I have met with them all shapes & colors & yet wearing the—the—I don't know how to describe it,—undescribable mark of Mongrelism.—[3] A dog may look like a Spaniel, or Terrier, or any thing else, but there is mongrel, (& generally coward,) written in his face. by the way, I think the above resemblance may be traced to "*impression*" Fry & I had a very curious confab: respecting this, & he named many very interesting instances; I will mention one He was taken to see a Horse, which was said to be full Blood, & imported here for breeding— After looking at it, Fry said, "that Horse's Mother's first foal was got by a Jackass.—this horse wears the impression then received"— he was much laughed at for his pains, but held to his opinion— Some time after he was taken to see another horse, also imported from England, and he made *the same remark*. this excited the curiosity of parties here, & strick inquiries were made in England, it then turned out that these two horses were *Brother & Sister*, & their Mother had accidentally be covered by a Jackass, & got her first foal to him.[4]

I have known myself one or two instances of Bitches being lined for the *first time*, by dogs of a different kind & ever after throwing one, or more, in each litter, like the old lover, tho' far removed from him.

Fry also mentioned a curious thing regarding *Mules*.— A Stallion he says will never cover a Mare, if he can get to Mule Mares,—nor will a Jackass cover a She ass, if he can get to a Mule Mare,—nor will a Drake tread a Duck of the pure breed, if he can get to a hybrid Muscovy;—so also with the Muscovy drake, & this is so well known among the farmers here, that they take the most extraordinary precautions to prevent the Stallions from scenting the Mules, if their services are required for Mares.

Fry never was accustomed to distinguish between the Pigeons, till I pointed out to him the Differences between them, & therefore cannot remember if the Ascension birds had black bars on the Wings, & white rumps, he calls them "Blue rocks".[5]

Respecting the Greyhound with "*very* short and much curled tail", Fry says, he has seen a dog of the greyhound kind among the caffres, with a curled tail *slightly bushed*. I shall try & make inquiries about these dogs, but I hope you will not think me negligent if I do not succeed;—this country is not like England & for all purposes of this kind, you are nearer to Kafraria than I am!!! however if God Spares my life, I hope in a year or so, to take a trip thro' this country. I have offered to take the place of one of the Judges Secretarys & Registrars, & go on circuit with the court, this will enable me to judge for myself, & ask many questions.[6] No one can tell my craving for travel, & ten minutes of Humbolt, upsets me for a week—[7] I literally dare not read a book of Travels,—I cannot

Incomplete
DAR 83: 185–6

CD ANNOTATIONS
1.1 Mongrel he is,] *underl pencil*
1.1 Mongrel . . . *& Sister*, 1.13] *crossed pencil*
1.14 accidentally . . . to him.] *underl pencil*
2.1 I have . . . from him. 2.3] *crossed pencil*
2.1 lined] '?' *added pencil*
3.1 Fry also . . . *Mules*,—] *double scored pencil*
4.1 Fry never . . . questions. 5.9] *crossed pencil*
5.9 ten minutes . . . cannot 5.10] *cross added pencil*
Top of first page: 'E. Layard' *pencil*; '(Hybridity)' *brown crayon*

[1] The correspondent is identified by CD's annotation.

[2] Dated on the assumption that this letter is a reply to CD's letter to E. L. Layard, 8 June [1856], which would have arrived in Cape Town in late August or early September.

[3] See letter to E. L. Layard, 8 June [1856].

[4] Although CD did not use the information given by Layard in *Natural selection* or *Variation*, he was greatly interested in the general phenomenon of telegony. The most famous case, which was cited by CD in *Natural selection* pp. 330–1, was of Lord Morton's mare which, after successfully breeding with a quagga, subsequently produced offspring resembling the quagga, even though the actual father was a black Arabian horse. In *Variation* 1: 404, CD stated that: 'many similar and well authenticated facts have been published, and others have been communicated to me, plainly showing the influence of the first male on the progeny subsequently borne by the mother to other males.'

5 See letter to E. L. Layard, 8 June [1856], n. 7. CD used this information in *Variation* 1: 190 and n. 18, stating: 'For Ascension I rely on MS. information given me by Mr. Layard.'

6 Layard worked for the judicial civil service in the Cape Colony. Kaffraria was the name given to the south-east part of the Cape province. In 1847, it had become the Crown colony of British Kaffraria (*EB*).

7 A reference to Alexander von Humboldt's classic natural history travelogue, translated as *Personal narrative* (Humboldt 1814–29).

To John Lubbock 5 September [1856][1]

[Down]
Sept 5

Dear Lubbock

In a wretched Brochure by Gerard[2] I find the following, but he is not to be trusted & he does not give authority "Le Lycus miniatus, Lepidopt. des parties boréales de l'Europe, se trouve sur le Cantal,[3] et l'on a decouvert en Suisse le Prionus depsacarius de la Suède. On retrouve sous notre climat a une élévation de 12 a 1,500 metres l'Apollon qui est commun dans les montagnes de Suède."[4]

Ever yours | C. D.

Please do not forget acreage of Larch Wood

Georgy w^d delight to see your insects, but do not send them, without you can quite safely.—[5]

I forgot to ask after the Railway,[6] my invariable question.—

Down House

1 The year is provided by CD's reference to the geographical distribution of alpine butterflies, which was relevant for his work on chapter 11 of his species book (see n. 4, below).

2 CD refers to Frédéric Gérard's pamphlet, an extract from the *Dictionnaire universel d'histoire naturelle*, on 'Géographie zoologique' (Gérard 1845). CD first read the work in 1845 (*Correspondence* vol. 4, Appendix IV, 119: 16a). His copy of the pamphlet is in the Darwin Pamphlet Collection–CUL.

3 A region of south central France.

4 Quoted, with some minor errors, from Gérard 1845, p. 136 (p. 27 of CD's reprint). The passage is marked in CD's copy. CD was interested in these beetles and butterflies as evidence that northern species had migrated south during the cold period and remained behind at higher altitudes when the ice retreated toward the pole. The information from Gérard 1845 is cited in *Natural selection*, p. 535 n. 2, without attribution.

5 George Howard Darwin, aged 11, was an ardent Lepidopterist.

6 John Lubbock's father, John William Lubbock, was chairman of the railway company set up to build an extension to Beckenham (see *Correspondence* vol. 5, letter to W. E. Darwin, [25 April 1855]). CD was a stock holder in the venture.

From J. D. Dana 8 September 1856

New Haven,
Sept. 8, 1856.

My dear Sir:—

I received your most welcome letter a few days before the meeting of our Scientific Association:[1] and as I should meet Prof. Agassiz there, who could best answer your queries respecting the Mammoth Cave Animals, I concluded to defer my answer till my return. Here I am, back again, at last and I seat myself for a few words with you, socially and Scientifically.—

First as to the Mammoth Cave.— Professor Agassiz told me that the family to which the Fishes belong—the Cyprinodonts—was rather strikingly American.[2] With regard to the Insects, Dr John L. LeConte an Excellent Entomologist says that the genera of beetles are not American, but the same that occur in Caverns in Europe & elsewhere.[3] The genus of fly Anthomyia is common in Europe.[4] The Crustacean, *Astacus pellucidus*, belongs to that subdivision of the genus, (Cambarus, as it has been called), which is peculiarly American. Cambarus is made a distinct genus by some writers: the only difference is in the number of branchiæ: Cambarus has 17 on a side or one less than Astacus.— The Crustacean genus *Triura*, has not been found any where except at the Mammoth Cave. You may have seen some notice of the species of the Cave in the Amer. Jour. Sci., vol. xi, p. 127 (1851).—[5] Of the spiders I cannot speak definitely.— I would add respecting the genus Cambarus, that its Species are very numerous and widely spread over North America. Agassiz has collected a large amount of information on the peculiarities of the North American Fauna, but he has not yet embodied them in any work or article. One of the most interesting of our peculiar tribes, as you undoubtedly know, is that of the Gar-pikes, of which there are several genera & near two dozen known species occurring over the Continent between Cuba & the northern Lakes—and not represented elsewhere over the globe.— It is not to the point in view, yet I may mention here a fact of geological interest brought out by Agassiz at our Assoc. meeting a fortnight since. There were some young individuals, alive, shown, which had the tail of the Ancient Ganoids— That is, the vertebræ were actually continued to the extremity of the upper lobe—[6] This upper lobe, as here drawn, drops off as the animal grows & the fish then is of the modern type of form.

AL incomplete
DAR 205.3 (Letters)

CD ANNOTATIONS
0.1 New Haven . . . Cave.— 2.1] *crossed pencil*
2.2 Cyprinodonts] *underl pencil*
2.2 rather strikingly American] *underl pencil*
2.7 peculiarly American] *underl pencil*
Top of first page: 'Dana' *pencil*; '19'[7] *brown crayon*

[1] Letter to J. D. Dana, 14 July [1856]. The 1856 meeting of the American Association for the Advancement of Science was held in Albany, New York from 20 to 28 August.
[2] See letter to J. D. Dana, 14 July [1856]. This information had already been given in Agassiz 1851.

[3] John Lawrence LeConte was recognized at home and abroad as the leading American entomologist (*DAB*).

[4] This sentence was added in the margin.

[5] Agassiz 1851.

[6] This fact was eventually published in Agassiz 1857.

[7] The number of CD's portfolio of notes on geographical distribution of animals.

To J. D. Hooker 8 September [1856][1]

Down Bromley Kent
Sept. 8th.

My dear Hooker

I got your letter of the 1st. this morning;[2] & a real good man you have been to write. Of all the things I ever heard, M[rs] Hooker's pedestrian feats beats them. My Brother is quite right in his comparison of "as strong as a woman",—as a type of strength.— Your letter, after what you have seen in Himalaya &c, gives me a wonderful idea of the beauty of the Alps. How I wish I was one half or one quarter as strong as M[rs] Hooker: but that is a vain hope. You must have had some very interesting work with glaciers &c.[3] When will the glacier structure & motion ever be settled! When reading Tyndall's paper it seemed to me that movement in the particles must come into play in his own doctrine of pressure;[4] for he expressly states that if there be pressure on all side, there is no lamination: I suppose I cannot have understood him, for I sh[d]. have inferred from this that there must have been movement parallel to planes of pressure. Sorby read paper to Brit. Assoc. & he comes to conclusion that Gneiss &c may be metamorphosed *cleavage or strata*; & I think he admits much chemical segregation along the planes of division.[5] I quite subscribe to this view, & sh[d]. have been sorry to have been so utterly wrong, as I sh[d]. have been, if foliation was identical with stratification.[6]

I have been no where & seen no one & really have no news of any kind to tell you.— I have been working away as usual (floating plants in salt-water inter alia & confound them, they all sink pretty soon, but at *very* different rates) working hard at Pigeons &c &c By the way I have been astonished at differences in skeletons of domestic Rabbits:[7] I showed some of the points to Waterhouse & asked him whether he could pretend that they were not as great as between species, & he answered "they are a great deal more".—[8] How very odd it is that no zoologist sh[d]. ever have thought it worth while to look to the real structure of varieties.—

I most earnestly hope that at Vienna you will make particular enquiries about the *pure* Laburnum, which one year bore the hybrid flowers & on one sprig the C. purpureus.—[9] D[r] Reissik(?) is name of man I think.— Bentham[10] will not believe that it was a **pure** Laburnum, & it does seem quite incredible, notwithstanding the clear statements in the Flora.[11] Please enquire particularly whether the hybrid or purple or pure bears seeds: I have just got the seeds of a yellow branch from the sterile hybrid to sow & see what will come up.— Really this case ought to be investigated,[12] & if you, the King of Sceptics believe, all others may.—

My poor wife keeps very uncomfortable, but rather better than she was.[13]

With our kindest regards to Mrs Hooker. | Believe me | My dear Hooker | Ever yours | C. Darwin

DAR 114.3: 176

[1] Dated by the reference to Hooker's holiday in Switzerland (see n. 3, below).

[2] The letter has not been found.

[3] The Hookers were on a walking holiday in Switzerland, where they by chance met Thomas Henry Huxley and John Tyndall who had gone to the Alps to investigate James David Forbes's theory of glacial movement (L. Huxley ed. 1900, 1: 145, and Eve and Creasey 1945, p. 64).

[4] Tyndall 1856. See letter to J. D. Hooker, 5 August [1856].

[5] Sorby 1856.

[6] Henry Clifton Sorby proposed that cleavage took place when metamorphosed stratified rock that had been foliated was subjected to contortion and pressure. Thus slaty cleavage could not, as CD had thought, be partially developed foliation. The problem of the origin of cleavage and foliation had been the subject of much of CD's correspondence with Daniel Sharpe and Charles Lyell in 1854 and 1855 (see *Correspondence* vol. 5).

[7] Rabbits and their skeletal variations are discussed in chapter 4 of *Variation*.

[8] George Robert Waterhouse had classified wild rabbits in Waterhouse 1846–8.

[9] See letter to J. D. Hooker, 22 June [1856].

[10] George Bentham.

[11] CD refers to the brief account of Siegfried Reissek's anomalous laburnum given in Hornschuch 1848, pp. 25–7. See letter to J. D. Hooker, 22 June [1856], n. 2.

[12] Hooker wrote the following note at the end of the letter: 'Only that one time produced red &c flowers impossible that it was a graft The same thing observed in Hungary & Bohemia in many gardens also at Schoenberg'.

[13] Emma Darwin was in her sixth month of pregnancy. Her last child, Charles Waring Darwin, was born on 6 December 1856.

To George Varenne Reed   8 September [1856][1]

Down Kent
Sept. 8th.—

My dear Sir

As you met George at the Crystal Palace, you know that he has been at home with us for a couple of days.[2] I have been ascertaining how he gets on & I find that he is not nearly so low in the school as I anticipated. & I find that in all those parts of his lessons, which shows that he has been well grounded he keeps at the *top* of his Class, & as I owe this to your kind labours, I cannot resist the great pleasure of again returning to you my cordial thanks[3]

My dear Sir | Yours sincerely & obliged | Ch. Darwin

Pray do not trouble yourself to acknowledge this note.—

Buckinghamshire Record Office (D22/39/1)

[1] The year of the letter is established by an entry in Emma Darwin's diary for 5 September 1856 that reads: 'G. came from school the boys meet him at the Crystal Palace'.

[2] George Howard Darwin came home on 5 September and returned to Clapham Grammar School on 8 September (Emma Darwin's diary).

[3] George Varenne Reed had been George Darwin's tutor before he entered Clapham School. CD later sent Francis, Leonard, and Horace to Reed (see J. R. Moore 1977).

To Joseph Augustin Hubert de Bosquet 9 September [1856][1]

Down Farnborough | Kent
Sept. 9th.—

My dear Sir

I am much obliged for your very kind & interesting letter.[2] I am astonished & delighted at your discovery of a Cretacean Chthamalus. It seems to me a very curious discovery. The fact seems to me eminently worth publishing,[3] with a careful & full description & enlarged drawing of the shell. Generally I have not the smallest faith in *negative* Geological evidence;[4] but in the case of sessile cirripedes, the evidence did appear (vide my remarks in Introduct. p. 5 to Fossil Lepadidæ) of some value;[5] & now you show that the evidence is worth nothing.

Do insist strongly on the caution requisite in our *ever* presuming to say *when* a new group first has appeared on the earth. Are you familiar with structure of recent Chthamalus: I think the woodcut at p. 39 of recent Balanidæ shows well the arrangement of the comparments. See my remark p 172 on the odd fact of no extinct Chthamalinæ having been discovered.[6]

The drawings you have sent me are beautiful; but I have been so hard at work for two years at other subjects that cirripedes are gone rather out of my head, which could never boast of a good memory. In truth I have not the least doubt that you are **far** more capable of forming a correct judgment than I am on every point of the subject.

Nevertheless I will just give my *impression* on one or two points.— I should hesitate *greatly* about admitting a specimen to be Lithotrya, without I could see the serration on the upper scales of Peduncle (see Pl. VIII fig 3d. Balanidæ)[7] for they are of such importance for burrowing: in your fig 1 the tergal margin of scutum seems too simple for Lithotrya: can fig. 2. be a carina; if I understand the drawing

AL incomplete
American Philosophical Society (138)

[1] Dated by the reference to Bosquet's identification of a new species of fossil cirripede, which was described in Bosquet 1857. Bosquet's letter notifying CD of the discovery seems to have been written well in advance of publication (see n. 3, below).

[2] Bosquet's letter has not been found. CD had corresponded with Bosquet between 1852 and 1854 about fossil sessile Cirripedia. He was the author of a work on fossil Crustacea (Bosquet 1854).

[3] Bosquet 1857. In the published version, Bosquet named the new cirripede *Chthamalus darwinii* in CD's honour. There are two copies of Bosquet 1857 in the Darwin Library–Down, one of which is a presentation copy from Bosquet.

[4] See Rudwick 1974 for a description of CD's mishaps with 'negative evidence' in geological theorising.

[5] The passage in *Fossil Cirripedia* (1851), p. 5, reads:

> No true Sessile Cirripede has hitherto been found in any Secondary formation; considering that at the present time many species are attached to oceanic floating objects, that many others live in deep water in congregated masses, that their shells are not subject to decay, and that they are not likely to be overlooked when fossilized, this seems one of the cases in which negative evidence is of considerable value.

[6] The references are to *Living Cirripedia* (1854). CD had presented Bosquet with copies of his Cirripedia monographs (see *Correspondence* vol. 5).

[7] A slip for 'Lepadidae'. *Lithotrya* is figured on pl. 8 of *Living Cirripedia* (1851).

From Peter Wallace   10 September 1856

Island of Ascension | Green Mountain
September 10th. 1856

Dear Sir—

I received your note dated June 22nd. 1856 on the 4th. of this Month,[1] and in answer beg to state, that I will do everything in my power to forward your wishes,

Your informant relative to the Domesticated Fowl, and Pigeon, being found here in a wild state, must have laboured under some mistake, as nothing of the kind exist here at this time,=[2] There are four pigeons living in the neighbourhood of N.E Cottage, in the cinder cliffs of the Sheep-penn Ravine, in a half wild state, but have boxes in which they roost at night, they have been there for upwards of four years, without encreasing in number why I cannot well explain, unless their young are killed by cats or their eggs destroyed by rats, during the daytime they pass their time in Black Rock Ravine, flying in the evening to roost in the Sheep-penn-Cliffs, they have a wild appearance, fly very fast, are Brown and white in colour, within twenty yards of where they roost is a fowl house belonging to the Marine who lives in N.E. Cottage but I have never seen or heard of them feeding with the fowls, These four Pigeons are the only ones existing in anything like a wild state,

Two varieties of Doves are plentiful on the Island living generally about the base of the Mountain, and in Cricket Valley, neither of which kind is the Ring Dove although they Coo in the same manner, one is light and dove coloured, the other is dark Brown, I cannot ascertain how long it is since they were introduced, and think there are about 200 on the Island,

I have heard there were some years ago a kind of Malay or Indian-fowl, which ran wild on the Island, but were all shot about 10 years since=

Guinea Fowl are tolerablely numerous,[3] being now about 400 on the Island and just the end of a four months shooting season, in which as near as I can tell about 300 have been shot, they fly very fast, generally running some distance before they rise, unless suddenly approached, when they take wing at once, four birds are considered a fair days sport, they are very wary and difficult of approach, I have shot eight Guinea Fowl and a Goat in one day=being the best days sport I have heard of since I have been here, Guinea Fowl feed on the Black Cricket and a large kind of Grasshopper or Locust, both being very numerous here, also on "Woodlice" and the seeds of a kind of "Shepherds Purse" very common here after heavy rains, in every part of the Island within a Mile of the base of the Mountain=which may be said to be the boundary of the Guinea Fowl districts,,

Formerly the Guinea Fowl used to frequent the higher parts of the Mountain, but latterly have become shy and rarely asscend above the level of the Mountain Hospital, about 1800 feet above the level of the sea,

In just about the Same range Rabbits are found not very numerous, and likewise very wary= they are very fine in size and flavour, and weigh when full grown about 3 and 3½ lbs, Rabbits are found from the before named districts to the very top of the Peak=

I beleive both Guinea Fowl and Rabbits are much persecuted when young by Cats and "Land Crabs" by the latter more especially, I remember one day disturbing (when out among the Sheep) a young brood of Guinea Fowl which ran among some stones for protection, hearing some of them cry as if hurt, I went to the place and found a Monstrous Crab, with one in each claw I killed the Crab and set the young Birds at liberty, they were too much hurt to recover, I imagine the young Rabbits fall a prey to the Crabs in a similar way=or they would be more numerous than they are, Cats not frequenting very much, the vicinity of the Game, they confine themselves more to the localities of the "Fairs," a name given here to places where the sea Birds, breed, I omitted to mention that the Guinea Fowl here is a strong and fast flier, particularly when once up= they do not rise well against a steep,, The other kinds of Game here, are Pheasants and Red-legged Partriges, the former being the least numerous may number about 100,, and Partriges about double that quantity, both frequent the higher parts of the Mountain, and exist on (Blackberries Common Bramble) and small insects, they destroy Sweet Potatoes, and Indian Corn crops for food, but in general do little damage, Many attempts were made to introduce Pheasants and Partriges but failed, untill about 15 years ago the Common Bramble was introduced from St Helena and spread with such marvelous rappidity from about a dozen plants, that at this time about 200 acres of the higher part of the Mountain are overrun with it Sometime after the 'Bramble' was introduced Pheasants and Partriges were sent and have done well unfortunately the parts of the Mountain they frequent is overrun with Rats (two kinds Black and Brown) which I believe destroy both eggs and young Birds= The Birds are never fed, but left to breed and flourish (au naturel)

The Cock Pheasant is a very handsome bird having a white ring under his throat and generally brighter plumage, than they have at home, (I may here mention I have never seen a greater variety of domestic fowls, nor so rich in plumage as they are here, I will take some of handsomest home with me when I go,) the Pheasant and Partriges have much the same habits here as in England, except the Phesant seldom roosts in Trees, owing I presume to the want of high trees, in their district by this means they fall an easy prey to Cats, I have counted as many as fifteen in a brood of young Pheasants but seldom saw more than 4 or 5 reared=owing probablely as much to not having proper drinking places as anything else=

Cats are very numerous in the Lowlands and attain a large size, I have seen them killed weigh in 15 lbs and have heard of them weighing 20 lbs= I have shot a fine black She Cat for you and have her preserved in a jar, in a solution of "Corosive Sublimate" as soon as I have a collection worth sending I will do so=

Goats are very fine, and first rate sport to shoot them, they were very numerous when I came here, but have been greatly reduced, owing I think, to poachers, I have a skin and a pair of horns of one I shot, which I intend for Dr Lindley[4] being the only ones I have saved, but if an opportunity occurs of getting others I will do so for you at present Goat shooting owing to their number being so much reduced, I believe there are about 200 on the Island now 4 years ago there were 600= they are of various colours, some white some red, and fawn colour, but chiefly black they are difficult to stalk, and bound over the Rocks with wonderful fleetness,

I am sorry I am not skillfull at skinning and curing skins, there in most case I must send them whole, but I find they will preserve well in Corrosive Sublimate solution—

I have many interesting and curious stories relative to the Wild Cat, and Goats, which time will not admit of my telling now=

The plants which have spread most widely over the Island, I mean naturally is the Bramble, the Guava, and the (Pride of India *Melia* azerderachta) and Aloe Vulgaris=also a plant known here as the (Madagascar Rose, Vinca rosea, and the Prickley-pear are very numerous=

Being forced to conclude in a hurry to catch a mail (things very uncertain here) I beg you will excuse any errors I have made

I beg leave to remain | dear Sir | Your very obedient Servant | Peter Wallace

To Mr C. Darwin | *Down, Bromley—Kent*

P.S. I have just heard that a colony of Pigeons is found on a detached rock near English Bay my informant stating that they fed on fish, it the first time I have heard of them, and will go down at once to ascertain the Truth of the statement which I have every reason to doubt=

I omitted to state that about 9 years ago I received some Starlings from the Admiralty which I turned out, they have done well and fly about in flocks of about 40, being in all about 150—reared from about a dozen Birds which survived out of twenty Starlings sent by the Admiralty[5]

Peter Wallace

CD ANNOTATIONS

1.1 I received . . . state, 2.13] *crossed pencil*
3.1 Two varieties . . . Island, 3.5] 'Doves feral' *added pencil*
5.1 Guinea Fowl . . . season, 5.2] *double scored pencil*
7.1 In just . . . wary= 7.2] *double scored pencil*; 'Rabbits' *added pencil*
8.9 they confine . . . breed, 8.10] *double scored pencil*
8.12 The other . . . insects, 8.15] *scored brown crayon*
8.12 Pheasants . . . Partriges 8.13] *underl pencil*
8.17 but failed . . . rappidity 8.19] *double scored brown crayon*
10.1 Cats . . . so= 10.4] *scored brown crayon*
11.5 being . . . now 11.6] *scored brown crayon*
19.1 9 years . . . Admiralty 19.4] *scored brown crayon*
Top of first page: 'Naturalised Animals' *ink*; '18'⁶ *brown crayon, circled brown crayon*; 'no wild Pigeons or Fowls' *pencil*

[1] The letter has not been found. CD wrote to several overseas naturalists in June 1856 (see letters to E. L. Layard, 8 June [1856], and to Robert Everest, 18 June [1856]). CD had included Wallace's name in an earlier list of people to contact for pigeon and poultry skins (see *Correspondence* vol. 5, CD memorandum, [December 1855]).
[2] CD's 'informant' was probably Edgar Leopold Layard, whose acquaintance 'Mr Fry' had resided on Ascension. See letter to E. L. Layard, 8 June [1856].
[3] CD himself saw wild guinea-fowl on Ascension during the *Beagle* voyage (*Journal of researches*, p. 587). In *Variation* 1: 190, he stated that the guinea-fowl 'has become perfectly wild at Ascension'.
[4] John Lindley.
[5] The island of Ascension was run by the British Admiralty from 1815. It was governed by a naval captain borne on the books of the flagship of the admiral stationed in Gibraltar.
[6] The number of CD's portfolio of notes on the means of geographical dispersal of animals and plants.

To Laurence Edmondston 11 September [1856][1]

Down Bromley Kent
Sept. 11th

My dear Sir

I have lately been drawing up descriptions of Pigeons, & you can have no idea how valuable I have found the Shetland Specimen.[2] But there were some points which I omitted to examine, for instance the eye-lid, which I find in the Barb, twice as long as in some other breeds.— Again the exact shape of crop I neglected to observe.— You will guess that this is a preface to beg you once again, if you will be so very kind, as to send me per post, (*allowing* me to pay postage) a wild Rock: though it would be best in early winter as keeping better, & the bird would be more sure to be adult.— Will you be so kind as thus far to aid me?—

Have you succeeded in finding out about the attempts at domestication mentioned by Mr. Macgillivray?[3] Are the wild Birds ever chequered with black on their wing coverts? I am interested in this for I find in India, Madeira & the Gambia the quite or half-wild all present this strong variation in plumage.—

Is the Rabbit wild in the Shetlands? I have just lately been comparing my collection of skeletons of domestic & wild Rabbits,[4] & I have been very much surprised to find how much some important points vary, as shape of foramen ovale of the atlas vertebra &c &c.— A Shetland specimen put in a jar with lots of salt w^d be a treasure to me;^{5} the more so to compare with a specimen, I have received from the little isl^d of P. Santo.—^{6} I presume such c^d be sent by ship to London or some Port.? That is if in your power to oblige me.— The Rabbit beyond everything sh^d not be killed by blow on head.

I fear that you will think that you have fallen on a **most** troublesome petitioner.—

I was not aware till I received your letter some two months ago, that you were the Father of the Naturalist, whose fate, I assure you, I most sincerely deplored at the time.—^{7} I can well believe, for I am a father, how this loss must have damped all your zeal for Natural History.—

Pray believe me with sincere respect | Dear Sir | Yours truly obliged | Ch. Darwin

D. J. McL. Edmondston

^{1} Dated by the relationship to the letter to Laurence Edmondston, 3 May [1856].
^{2} See letter to Laurence Edmondston, 3 May [1856], n. 5.
^{3} See letter to Laurence Edmondston, 3 May [1856] and n. 3.
^{4} See letter to J. D. Hooker, 8 September [1856].
^{5} Measurements of a wild rabbit from the Shetland Islands are given in the table of comparisons with other breeds, including feral Porto Santo rabbits, in *Variation* 1: 127.
^{6} The specimen was given to CD by Richard Thomas Lowe via Thomas Vernon Wollaston (see letter from T. V. Wollaston, [February 1856]).
^{7} Thomas Edmondston was accidentally shot dead in 1846 during the voyage of HMS *Herald*. Before Edmondston sailed, CD prepared a memorandum for him listing a number of subjects he recommended for investigation during the voyage (see *Correspondence* vol. 3, letter to Edward Forbes, 13 May [1845]).

To W. B. Tegetmeier [18 September 1856]^1

Leith Hill Place | Dorking Surrey
Thursday

My dear Sir

Your letter has been forwarded here.— I am very glad that you approve of the Scanderoons.—

Many thanks for your offers about dead Pigeons.— I return home tomorrow & will then look over my Catalogue & gratefully tell you what I want.^{2}

I have got to care very much about Rabbits & Ducks, as the evidence is *better* of single origin than in most cases, & therefore differences in their structure the more concerns me. Should any odd Duck (domestic) or Rabbit be ever offered not very dear at Steven's I sh^d be *very* glad to purchase.—

I was most sincerely sorry to hear of the death of our old & excellent friend,— Yarrell.—^{3}

Your's very sincerely | C. Darwin

Henrietta Emma Darwin.

(Courtesy of Mrs Sophie Gurney.)

Down House as it was in the 1870s. On the left is the new dining-room (later the drawing-room) built in 1857; the verandah was added in 1872.

(By permission of the Syndics of the Cambridge University Library.)

New York Botanical Garden Library (Charles Finney Cox collection)

[1] Dated by the relationship to the letter to W. B. Tegetmeier, 30 August [1856], and by CD's reference to being away from home. CD recorded that he visited his sister Caroline and her husband, Josiah Wedgwood III, at their home, Leith Hill Place, from 13 to 19 September 1856 ('Journal'; Appendix II).

[2] CD's Catalogue of Down Specimens (Down House MS) lists his specimens of experimental animals, including the pigeons.

[3] William Yarrell died on 1 September 1856. CD had consulted him before sailing in the *Beagle* in 1831 and had maintained a correspondence with him after his return. It was Yarrell who had first introduced CD to Tegetmeier (see *Correspondence* vol. 5, letter to W. B. Tegetmeier, 31 August [1855], n. 2).

From Victor de Robillard[1]   20 September 1856

Port Louis, Ile Maurice,
le 20 Septembre 1856.

A Monsieur Charles Darwin.

Monsieur,

Mˍ Ch. T. Beke[2] a communiqué ici à la Société d'histoire naturelle dont je suis l'un des membres, une lettre qu'il a reçue de vous, sur différens points qui sont l'objet de vos recherches. Comme depuis de longues années je m'occupe de former une collection de coquilles, ce qui m'a donné l'occasion d'aller souvent sur les bords de mer, je vais vous communiquer mes remarques, quoique le sujet qui vous occupe, n'attirait pas particulièrement mon attention.

1ˍᵉ Je n'ai jamais vu d'arbres rejetés par la mer sur le rivage; après les ouragans les débordemens de rivières en charroient quelquefois à leur embouchure, et alors la mer les rapporte sur le rivage, mais on reconnait alors que ce sont des arbres de l'Ile même— Après les ras-de-marée j'ai vu souvent beaucoup de graines sur le rivage, mais elles appartenaient aux plantes marines— Je me rapelle avoir vu cependant des graines d'un assez gros volume, mais je ne les ai pas assez examinées pour savoir si elles appartenaient à des plantes marines ou à des arbres; si elles étaient la graine d'arbres, il pourrait se faire qu'elles arrivent à la mer avec les eaux qui débordent des rivières & qu'elles seraient venues de l'intérieur de l'Ile, dans les parties où les rivières traversent les bois & forêts— Je ne puis donc rien préciser sur ce point.

2ˍᵉ Il n'est pas à ma connaissance qu'il y ait eu ici des troupes d'oiseaux qui émigrent et arrivent sur l'île, tous ceux qui vivent ici, ont été introduits de différens pays— Je n'ai jamais su non plus que des espèces nouvelles d'oiseaux aient été trouvées, étant arrivées sur l'île, après s'être égarées en émigrant d'une terre à une autre.

3ˍᵉ Pour les animaux domestiques & autres, on a introduit beaucoup d'espèces qui se sont acclimatées, venant de diverses contrées— les chevaux nous sont venus du Cap, de l'australie, des Iles Timors, du Pégou,[3] d'Angleterre & de France, de Buenos-Ayres—des mules de Buenos-Ayres, de France, de la Mer Rouge, du Cap, du golfe Persique. les bœufs et Vaches nous viennent du Cap, de l'australie, de

Madagascar, de l'Inde—les moutons & chevres, de l'Inde, de l'arabie, de l'australie, du Cap, de l'abyssinie; les poules, de l'Inde, de Madagascar, quelques espèces d'Europe et de la cote Est d'afrique. des Pigeons de France, de l'Inde; des chiens, de France, d'angleterre, du Cap—des porcs, de France, d'angleterre, de l'Inde, des Iles Malaises, de chine, de Siam—des Pintades de Madégascar. des Perdrix, de l'Inde—des Canards & des oies, du cap, ⟨de⟩ l'Inde, de Madagascar—

Voilà, Monsieur, avant de faire des recherches

AL incomplete
DAR 205.3 (Letters)

CD ANNOTATIONS
1.1 M.ʳ . . . marines— 2.5] 'No trees— ['seeds' *del*] except endemic— Has seen seeds, but they may be endemic' *added pencil*
3.1 2.ᶜ . . . autre. 3.5] 'No Birds, [strays] important' *added pencil*
Top of first page: '19'[4] *brown crayon*; 'From M. V. de Bobillard' *pencil*; 'As there are many land shells, probably Birds do not introduce shells.—' *ink*

[1] For a translation of this letter, see Appendix I. Robillard, a resident of Mauritius, was an active member of the natural history society of the island. He published several papers on Mauritian shells in the *Transactions of the Royal Society of Arts and Sciences of Mauritius*.
[2] Charles Tilstone Beke, an explorer of Abyssinia, had in 1853 become a partner in a Mauritius mercantile house with the intention of opening up trade routes between Britain and Abyssinia. His second wife, Emily Alston, was from Mauritius. CD evidently contacted him in London, although Beke's name is not on CD's list of individuals who could provide him with overseas contacts (see *Correspondence* vol. 5, CD memorandum, [December 1855]).
[3] A district of Burma.
[4] The number of CD's portfolio of notes on the geographical distribution of animals.

To W. B. Tegetmeier 21 September [1856][1]

Down Bromley Kent
Sept. 21ˢᵗ

My dear Sir

In accordance with your kind offer, I write to say that what I want far most of all is one or two good Carriers, & secondly a Runt of largest size & thirdly a first-rate Fan-tail. I hope to get latter from M.ʳ H. Weir[2] or Wicking,[3] but when Heaven knows only. I have asked M Corker[4] for a Carrier & M.ʳ Gulliver for a Runt[5] but strangers very naturally forget or will not take the trouble to send.—

I am become as much interested in Rabbits as in Pigeons & sh.ᵈ be very glad of any assistance in that line. The Angora has made a capital skeleton & is in some *small* respects rather peculiar. I want much the Great Hare Rabbit. Ducks, also, are very valuable to me & I want a black Buenos Ayrean & Rouen very much. I w.ᵈ buy old Drakes at a moderate rate gladly.—

Would you object at any time to put in Cottage Gardener, a query whether any one in England has Finnikin or Turner Pigeons?[6]

At Leith Hill[7] I noticed some very fine rather dark-coloured Dorkings with 5 toes, & all the quarter-grown chickens had hardly any tail at all, but which comes subsequently. Did you know that this was characteristic of any breed: I went to a Farm House whence my Brother got this breed & the good woman assured me that this was their general, but not quite invariable characteristic. The old Birds seemed all alike & true, so that I can hardly suppose the tailless condition to be owing to cross with Cochin, but so it may be.

I have only got as yet 1st nor of the Poultry Book,[8] so do not know yet how you are getting on.

My dear Sir | Yours sincerely | C. Darwin

New York Botanical Garden Library (Charles Finney Cox collection)

[1] Dated by the reference to Leith Hill Place, the home of CD's sister Caroline Sarah Wedgwood. CD stayed there from 13 to 19 September 1856 ('Journal'; Appendix II).
[2] Harrison Weir.
[3] Matthew Wicking.
[4] E. L. Corker Esq. of 11 Queen Street, Cheapside was mentioned as having won the prize for the best carrier pigeons at the Anerley poultry show, 29 July – 1 August 1856 (*Cottage Gardener* (1856) 16: 340). In *Variation* 1: 132 n. 2, CD recorded that: 'Mr. Haynes and Mr. Corker have given me specimens of their magnificent Carriers.'
[5] See letter to W. B. Tegetmeier, 24 June [1856].
[6] Tegetmeier was a regular contributor to the *Cottage Gardener*, which included the *Poultry Chronicle*. Tegetmeier complied with CD's request (see letter to W. B. Tegetmeier, 3 November [1856]). In *Variation* 1: 156, finnikins and turners were mentioned as having existed 'until recently' in England and possibly in France and Germany.
[7] See n. 1, above.
[8] Tegetmeier 1856–7. The first number was issued in May 1856. By September, three further numbers had been issued. Owing to an error by his bookseller, CD did not actually acquire a copy of the *Poultry book* edited by Tegetmeier until the summer of 1857 (see letter to W. B. Tegetmeier, 18 May [1857]). Eleven numbers, all that were published, are in the Darwin Library–CUL.

To Philip Henry Gosse 22 September [1856][1]

Down Bromley Kent
Sept. 22nd

My dear Sir

I want much to beg a little information from you.— I am working hard at the general question of variation, & paying for this end special attention to domestic Pigeons. This leads me to search out how many species are truly *rock* Pigeons, i.e. do *not* roost or willingly perch or nest in trees. Temminck puts C. leucocephala (your Bald-Pate) under this Category.[2] Can this be the case? Is the loud Coo to which you refer in your interesting "Sojourn" like that of domestic Pigeon.[3]

I see in this same work you speak of Rabbits run wild: I am paying much attention to them, & am making a large collection of their skeletons. Do you think you could get any of your zealous & excellent correspondents to send me an adult (neck *not* broken) feral specimen:[4] it would be of *great* value to me. It might be sent, I sh^d. think in jar with profusion of salt & split in abdomen.—[5] I sh^d., also, be very glad to have one of the wild Canary Birds for same object:[6] I have specimen in spirits from Madeira.—

Do you think you could aid me in this & shall you be inclined to forgive so very troublesome a request? As I have found the goodnature of fellow Naturalists almost unbounded, I will venture further to state that the body of any domestic or Fancy Pigeon, which has been for some generations in the W. Indies, would be of extreme interest, as I am collecting specimens from all quarters of the world.

Trusting to your forgiveness, I remain, My dear Sir | Your's sincerely | Ch. Darwin

Harry Ransom Humanities Research Center, University of Texas at Austin

[1] Dated by CD's interest in 1856 in collecting pigeon and rabbit carcasses and by his reading of the books mentioned in the letter (see n. 2, below).

[2] Probably Temminck and Knip 1811, which CD recorded as having read in May 1856 (see *Correspondence* vol. 4, Appendix IV, 128: 18). *Columba leucocephala* was described in Temminck and Knip 1811, 1: 11–12 and 37, and figured on pl. 13. This pigeon was described by Gosse as arboreal in P. H. Gosse 1847, pp. 299–304. CD recorded having read P. H. Gosse 1847 on 11 May 1856 (*Correspondence* vol. 4, Appendix IV, 128: 18). See also letter to P. H. Gosse, 28 September 1856.

[3] P. H. Gosse 1851, p. 173. Gosse apparently kept one as a pet (see letter to P. H. Gosse, 28 September [1856]). CD had read P. H. Gosse 1851 in 1852 (*Correspondence* vol. 4, Appendix IV, 128: 2).

[4] Gosse had lived and collected in Jamaica in 1844 and 1845.

[5] No specimen seems to have been sent to CD. In *Variation* 1: 112 and n. 21, the description of the feral rabbit of Jamaica is taken from P. H. Gosse 1851, p. 441.

[6] No mention is made of the wild canaries of Jamaica in *Variation*.

From Asa Gray 23 September 1856

Cambridge, Mass.
Sept. 23^d 1856.

My Dear Mr. Darwin

Dr Engelmann,[1] of St. Louis, Missouri—who knew European botany well before he came here, and has been an acute observer generally for 20 years or more in this country, in reply to your question I put to him, promptly said that introduced plants are not particularly variable—are not so variable as the indigenous plants generally, perhaps.

The difficulty of answering your question, as to whether there are any plants *social* here, which are not so in the old world,—is, that I know so little about European plants in nature.[2] The following is all I have to contribute. Lately I took Engelmann and Agassiz on a botanical excursion over half a dozen miles of one of our sea-board counties; when they both remarked that they never saw in Europe altogether half so much *Barberry* as in that trip. Through all this district B. vulgaris may be said to have become a truly social plant, in neglected fields and copses, and even penetrating into rather close old woods. I always supposed that birds diffused the seeds. But I am not clear that many of them touch the berries. At least these hang on the bushes over winter in the greatest abundance. Perhaps the barberry belongs to a warmer country than N. of Europe, and finds itself more at home in our sunny summers. Yet out of New England it seems not to spread at all.

Maruta Cotula,[3] fide Engelmann, is a scattered & rather scarce plant in Germany. Here, from Boston to St. Louis, it covers the road-sides, and is one of our most social plants. But this plant is doubtless a native of a hotter country than N. Germany.

St. Johns-wort (H. perforatum) is an intrusive weed in all hilly pastures, &c.— and may fairly be called a social plant. In Germany it is not so found, fide Engelmann.

Verbascum Thapsus is diffused over all the country,—is vastly more common here than in Germany, fide Engelmann.

I suppose *Erodium cicutarium* was brought to America with cattle from Spain; it seems to be widely spread over S. America out of tropics—&c In Atlantic U.S. it is very scarce and local. But it fills California and the interior of Oregon, quite back to the west slope of the Rocky Mts.— Fremont mentions it as the first spring food for his Cattle when he reached the Western side of the Rocky Mts.—[4] And hardly any body will believe me when I declare it an introduced plant. I dare say it is equally abundant in Spain. I doubt if it is more so.

Engelmann and I have been noting the species truly indigenous here which, becoming *ruderal* or *campestral*, are *increasing* in the number of individuals instead of diminishing, as the country becomes more settled & forests removed.— The list of our wild plants which have become true weeds is larger than I had supposed, and these have probably all of them increased their geographical range—at least, have multiplied in numbers in the Northern States, since settlements.—

Some time ago I sent a copy of the 1st part of my little essay on the Statistics of our N. States plants to Trübner & Co. 12, Pater Noster Row, to be thence posted to you.[5] It may have been delayed or failed, so I post another from here. I regret that I can[not] prepay on U.S. postage on pamphlets.— So I suppose it will cost you 6d—at least, and be an unnecessary expence.

This is only a beginning. Range of species in latitude must next be tabulated,— *disjoined* species catalogued, (i.e. those occurring in remote and entirely separated areas,—e.g. Phryma, Monotropa uniflora &c.—N. American & Himalayan.)— then some of the curious questions you have suggested.—the *degree* of consanguinity between the related species of our country & other countries, &c and the comparative range of species in large & small genera, &c &c &c So is it worth while to go on at this length of detail?— There is no knowing how much space it may cover. Yet after all facts in all their fullness is what is wanted—and those not gathered to support (or even to test) any foregone conclusions. It will be prosy; but it may be useful.

Then I have no time properly to revise *mss.* and correct oversights.— To my vexation, in my short list of our alpine species I have left out, in some unaccountable manner, two of the most characteristic, viz. *Cassiope hypnoides* & *Loiseleuria procumbens*. Please add them, on p. 28.—

There is much to be said about our introduced plants—

But now, and for some time to come, I must be thinking of quite different matters.— I mean to continue this essay in the January no.—for which my mss. must be ready about the 1st of November.—

I have not yet attempted to count them up; but of course I am prepared to believe that fully *three fourths* of our species common to Europe will found to range northward to the arctic regions, I merely meant that I had in mind a number that do not; I think the number will not be *very* small: and I thought you were under the impression that *very few* absolutely did not so extend northward.[6]

The most striking case I know is that of *Convallaria majalis*, in the Mts. Virginia & N. Carolina—and not northward— I believe I mentioned this to you before—

By the way Guyot, who is reliable has newly measured the height of Black Mts, of North Carolina, and makes it higher than I supposed. He had satisfactory data, he says (barometric), and makes it summit 6700 feet![7] It should have alpine vegetation, one would think, but I know not a single truly alpine phænogamous plant there.

Yours most faithfully Asa Gray

I suppose the sea is not harmful to you. As to quietness if you will come over for a holiday, you shall be as quiet here as in Kent, and could travel about the country very leisurely.

DAR 165: 94

CD ANNOTATIONS
1.3 said . . . perhaps. 1.5] *scored brown crayon*
2.5 remarked . . . trip. 2.6] *double scored brown crayon*
3.2 Here . . . plants. 3.3] *double scored brown crayon*
3.3 social] *underl brown crayon*
4.2 and may . . . Engelmann. 4.3] *scored brown crayon*
4.2 social] *underl brown crayon*
6.5 And hardly . . . plant. 6.6] *scored brown crayon*
8.1 Some time . . . November.— 12.3] *crossed pencil*
9.6 So . . . detail?— 9.7] *double scored pencil*
10.3 the most . . . *procumbens*. 10.4] *double scored brown crayon*
13.1 but of . . . Europe 13.2] *double scored brown crayon*
15.1 By the . . . leisurely. 17.3] *crossed pencil*
Top of first page: 'Naturalised Plants variable' *pencil*; '(C)' *brown crayon*; 'Social in America' *brown crayon*;
    '& not in Europe. Ch. 5.'[8] *pencil*

---

[1] George Engelmann, who had emigrated to America from Germany, was a botanist and meteorologist (*DAB*).
[2] See letter to Asa Gray, 2 May [1856].

[3] 'Maruta = Anthemis' was written in pencil in an unknown hand above 'Maruta Cotula'.

[4] John Charles Frémont had led several exploring expeditions across the Rocky Mountains to Oregon and California. In 1856 he was the Republican nominee in the American presidential elections.

[5] A. Gray 1856–7, the first part of which appeared in September 1856. CD's annotated copy is in DAR 135 (3).

[6] See letter to Asa Gray, 12 October [1856], for CD's response to this information.

[7] Arnold Henri Guyot, who had emigrated from Switzerland to Cambridge, Massachusetts in 1848, undertook to measure many thousands of heights in order to construct a new topographic map of the Appalachian Mountains. The map was eventually published under the direction of the Smithsonian Institution.

[8] Chapter 5 of *Natural selection*, 'The struggle for existence as bearing on natural selection', includes a discussion of social plants (*Natural selection*, pp. 203–5).

To John Lubbock 23 September [1856][1]

Downe
Sept. 23^{d}

Dear Lubbock

As I know you are interested in Owen's Parthenogenesis[2] et id genus omne, I send you the enclosed Review by Quatrefages:[3] the 1st half seems to me very dull, but the second half very good. & very true.

I, also, enclose some of Huxley's Lectures.—[4]

I hope Mrs Lubbock is pretty well again; some time ago I heard but an indifferent account of her.—[5]

What wonderfully grand doings you had at the marriage;—I sh^{d} think you can hardly be all rested yet. We made Etty give us a very full account of all she saw.—[6]

Your very sincerely | C. Darwin

P.S. I had almost forgotten what made me think chiefly of writing, viz a letter from Dana,[7] whom I had told of your marriage, in which he says "I was, also pleased to hear of Mr Lubbocks happiness. A long life to them full of joys. Please thank him from me for the pamphlets, which he has sent me. He is doing excellent work among the Entomostraca, & *dear* little things they are".—[8]

Please respeak for me to have any Bantam which sh^{d} chance to die.[9]

Down House

[1] Dated by the reference to the marriage of Lubbock's sister (see n. 6, below) and to the letter from J. D. Dana, 8 September 1856.

[2] Owen 1849. Lubbock was at this time working on the two kinds of reproduction—one being parthenogenic—of *Daphnia* (see letter to John Lubbock, 27 October [1856]). CD's annotated copy of Owen 1849 is in the Darwin Library–CUL.

[3] Quatrefages de Bréau 1855–6, in which Jean Louis Armand de Quatrefages de Bréau reviewed current work on sexual and asexual generation. Owen 1849 was discussed in the fifth part ((1856), pp. 60–4). When Lubbock came to write up his researches, he cited Quatrefages's paper (Lubbock 1857, p. 80).

[4] Huxley's lectures on general natural history given at the School of Mines in 1856 and published in the *Medical Times & Gazette* (T. H. Huxley 1856–7).

[5] Ellen Frances Lubbock was pregnant. The Lubbocks' first child, a daughter, was born on 15 March 1857 (*Burke's peerage* 1970).

[6] Lubbock's sister, Diana Hotham Lubbock, was married to William Powell Rodney on 11 September 1856 (*The Times*, 12 September 1856). The wedding took place in Down church. CD's daughter, Henrietta Emma, aged 13, apparently attended.

[7] Letter from J. D. Dana, 8 September 1856. The passage quoted was probably in the section of the letter that is now missing.

[8] Lubbock had published his researches on Entomostraca in Lubbock 1855.

[9] See letter to John Lubbock, [14 January 1856].

From Louis Sulpice Bouton to Charles Tilstone Beke[1] 24 September [1856][2]

Mon cher Mr Beke

Je vais faire des recherches à *tête reposée* dans le Museum & les Transactions pouvoir vous envoyer plus tard des renseignements plus précis—[3]

T à vs | L. Bouton

24 Sep

Extrait d'une lettre de M Autard de Bragard— Savane—[4] 17 Sept— 1856

En 1835—36—ou 37, j'ai envoyé à la Société d'histoire naturelle un Merle huppé de Bourbon[5] "Turdus cafer" pris au Morne Brabant peu de temps après un coup de vent— aussi un autre oiseau tué à la Savane reconnu appartenir à une espèce de Madagascar—

Vous trouverez ces individus au Museum et ce qui les concerne inséré dans les transactions de la Société—

DAR 205.3 (Letters)

CD ANNOTATIONS

Top of first page: 'Case of Madagascan Bird blown to Mauritius & one from Bourbon.—' *pencil*; '19'[6] *brown crayon*

[1] For a translation of this letter, see Appendix I.

[2] The year is given by the extract included in the letter.

[3] See letter from Victor de Robillard, 20 September 1856, in which it is mentioned that Charles Tilstone Beke had communicated a request from CD to the Société d'Histoire Naturelle of Mauritius. Bouton was secretary of the society until his death in 1878.

[4] Savanne is a district of south-west Mauritius.

[5] Bourbon was the former name of Réunion Island, 130 miles south-west of Mauritius. The bird was possibly the now extinct Réunion huppe (see letter from Victor de Robillard, 26 February 1857).

[6] The number of CD's portfolio of notes on the geographical distribution of animals.

To P. H. Gosse 28 September 1856

Down, Bromley, Kent,
September 28, 1856.

My dear Sir,

I thank you warmly for your extremely kind letter,[1] and for your information about the bald-pate, which is quite sufficient.[2] When we meet next I shall beg to hear the actual coo!

I will by this very post write to Mr. Hill,[3] and will venture to use your name as an introduction, which I am sure will avail me much; so you need take no trouble on the subject, as using your name will be all that I should require.

With my sincere thanks, Yours truly, | Ch. Darwin.

I am very anxious to get all cases of the transport of plants or animals to distant islands. I have been trying the effects of salt water on the vitality of seeds—their powers of floatation—whether earth sticks to birds' feet or base of beak, and I am experimenting whether small seeds are ever enclosed in such earth, etc. Can you remember any facts? But of all cases whatever, the means of transport (and such I must think exist) of land mollusca utterly puzzle me most.[4] I should be very grateful for any light.

E. Gosse 1890, pp. 267–8

[1] The letter has not been found, but see n. 2, below.

[2] See CD's letter to P. H. Gosse, 22 September [1856]. In *Variation* 1: 182 n. 8, CD discussed Coenraad Jacob Temminck's assertion that *Columba leucocephala* was a true rock pigeon and noted that 'I am informed by Mr. Gosse that this is an error.'

[3] Richard Hill had assisted Gosse with his books about Jamaica, particularly P. H. Gosse 1847 on the birds of the island. See letters from Richard Hill, 10 January 1857 and 12 March 1857.

[4] CD gave the results of his investigations on the possible means of dispersal of land molluscs in *Origin*, p. 397. The problem was a 'puzzle' because the eggs could not withstand sea-water and yet oceanic islands were always well stocked with land shells. CD eventually concluded that the opercular membrane provided a water-tight seal over the opening of the shell and that hibernating molluscs could be floated across the ocean.

To J. D. Hooker 28 September [1856][1]

Down Bromley Kent
Sept. 28th

My dear Hooker

I suppose you will soon be home: at first you will be very busy, but after a little time, I sh^d glad, *if you could by giving orders to any assistant* send me some dozen or so capsules with *some* stalk & leaves (ripe or nearly ripe) of any water plants, named in pencil, to try floating powers in sea-water.[2]

I have been trying in vain to get Callitriche verna & Montia fontana, which w^d be very good being such wide rangers.

If you can send me any, please send them any day, addressed

C. Darwin
care of Mr Acton
Post Office
Bromley
Kent
(Per Coach.)

In the course of some weeks, you unfortunate wretch, you will have my M.S. on one point of Geograph. Distribution—³ I will, however, never ask such a favour again; but in regard to this one piece of M.S. it is of infinite importance to me for you to see it; for never in my life have I felt such difficulty what to do, & I heartily wish I could slur the whole subject over.—

I hope your tour has answered in every way. I wrote to you at Vienna.⁴ With kindest remembrances to Mrs Hooker, Ever yours | C. Darwin

We shall meet at Phil. Club on 16th.—⁵

DAR 114.3: 177

¹ Dated by the reference to CD's manuscript on geographical distribution, which was completed in October 1856 ('Journal'; Appendix II).
² The list has not been preserved. The plants required were probably those mentioned in letter to J. D. Hooker, 26 [July 1856].
³ The forty pages of manuscript relating to geographical distribution (*Natural selection*, chapter 11) that CD wanted Hooker to read (see letter to J. D. Hooker, 13 July [1856]).
⁴ Letter to J. D. Hooker, 8 September [1856].
⁵ Both Hooker and CD attended the 16 October meeting of the Philosophical Club (Royal Society Philosophical Club minutes).

To W. B. Tegetmeier 28 [September 1856]¹

Down Bromley Kent
28th.—

My dear Sir

Very many thanks for the Catalogue just received. If the Black Hen Carrier sh^d be a good Bird (& if you do not know yourself, perhaps someone there can inform you), & I mean for my purpose with long beak & narrow flat head, I sh^d be glad of one, for I might pick up a cock afterwards. The Roman Runt is, I presume the same as Runts of Anerly, of which breed I have pair from Mr Baily.² By the way I doubt whether Mr Baker's³ Scanderoons differ much from Runts, they have rather longer necks, & hardly any other difference. They have not the red seam down the breast.—

The Rabbits being in Hutches will go dear; but if any old one went cheap & appeared good, I sh^d be very glad of it alive or strangled, as I have become almost more interested in Rabbits than anything else. In poultry I am getting less interested, though I am going on, & this very morning have sent four good Hens to be skeletonised.—⁴

For live Birds, I find the following the quickest address.

C. Darwin
George Inn
Farnborough
☞ Kent⁵
(To be forwarded)

You know so well what I want & are so very kind that I will leave to your judgment. Carriers are the only distinct Pigeons, which I have not had dead or alive.

Your's very truly obliged | C. Darwin

What a splendid development of skull your young Polands have.[6]

New York Botanical Garden Library (Charles Finney Cox collection)

[1] Dated by the relationship to the letter to W. B. Tegetmeier, 21 September [1856].

[2] John Bailey Jr of Mount Street and his father were expert breeders and dealers in ornamental birds, from whom CD purchased a number of different breeds.

[3] Charles N. or Samuel C. Baker.

[4] According to his Account book (Down House MS), CD paid for these skeletons on 7 October 1856.

[5] A carrier left the George Inn, Borough High Street, and called at Farnborough, Kent every Monday, Wednesday, Friday, and Saturday (*Post Office London directory* 1856).

[6] See letters to W. B. Tegetmeier, 20 March [1856], n. 8 and 24 June [1856].

To J. D. Dana 29 September [1856][1]

Down Bromley Kent
Sept. 29th

My dear Sir

I thank you warmly for your letter,[2] which like every one received from you, was most kind & interesting.— Your information on the points, on which I asked, is *much fuller* than I anticipated; I am glad to find that I deduced nearly or quite accurate results from your work with respect to the Crustacea,[3] but I am always afraid of falling into error when coming to conclusions on subjects on which I am almost ignorant. Very many thanks, also, for facts about the Caves, which to me is a wonderfully interesting subject: one point I forgot & if there be a specimen of the Cave Rat at New Haven & if you have anyone there au fait at Rodents, I sh[d]. *very much* wish to know whether its teeth are on the New or Old world type as pointed out by Waterhouse in regard to the great genus Mus.[4] But you must **not** take *much* trouble on this head, as I have too often given you trouble, & I look at your time as very precious.—

I shall be very curious to see your *Embryogeny of N. America!*[5] What a striking case of vertebræ in tail of young Gar-Pike; I wish with all my heart that Agassiz would publish in detail on his theory of parallelism of geological & embryological development;[6] I *wish* to believe, but have not seen nearly enough as yet to make me a disciple.

I am working very hard at my subject of the variation & origin of species, & am getting M.S. ready for press, but when I shall publish, Heaven only knows, not I fear for a couple of years but whenever I do the first copy shall be sent to you.— I have now been for 19 years with this subject before me; but it is too great for me, especially as my memory is not good. I have of late been chiefly at work on domestic animals, & have now got a considerable collection of skeletons: I am surprised how little this subject has been attended to: I find very grave differences in the skeletons for instance of domestic rabbits, which I think have all certainly descended from one parent wild stock. But Pigeons offer the most wonderful case of variation, & as it seems to me conclusive evidence can be offered that they are all descended from C. livia.

In the case of Pigeons, we have (& in no other case) we have much *old* literature & the changes in the varieties can be traced. I have now a grand collection of living & dead Pigeons; & I am hand & glove with all sorts of Fanciers, Spital-field weavers & all sorts of odd specimens of the Human species, who fancy Pigeons.—

I know that you are not a believer in the doctrine of single points of creation, in which doctrine I am strongly inclined to believe, from *general* arguments; but when one goes into detail there are certainly **frightful** difficulties.[7] No facts seem to me so difficult as those connected with the dispersal of Land Mollusca.[8] If you ever think of, or hear of, any odd means of dispersal of any organisms I sh^d be *infinitely* obliged for any information; as no one subject gives me such trouble as to account for the presence of the same species of terrestrial productions on oceanic islands; for I cannot swallow the prevalent fashion in England of believing that all islands within recent times have been connected with some continent.—

You will be rather indignant at hearing that I am becoming, indeed I sh^d say have become, sceptical on the permanent immutability of species: I groan when I make such a confession, for I shall have little sympathy from those, whose sympathy I alone value.— But anyhow I feel sure that you will give me credit for not having come to so heterodox a conclusion, without much deliberation. How (I think) species become changed I shall explain in my Book, but my views are very different from those of that clever but shallow book, the Vestiges.—[9]

It is my intention to give fully all the facts in favour of the eternal immutability of species & I have taken as much pains to collect them, as I possibly could do. But what my work will turn out, I know not; but I do know that I have worked hard & honestly at my subject.

Agassiz, if he ever honours me by reading my work, will throw a boulder at me, & many others will pelt me; but magna est veritas &c, & those who write against the truth often, I think, do as much service as those who have divined the truth; so that if I am wrong I must comfort myself with this reflection. It may sound presumptious, but I think I have to a certain extent staggered even Lyell.—[10]

But I am scribbling (in a very bad handwriting moreover) in a shameful manner all about myself; so I will stop with cordial good wishes for yourself & family, & pray believe me, my dear Sir | Your sincere & heteredox friend | Ch. Darwin

We are all here now much interested in American politics—[11] You will think us **very impertinent**, when I say how fervently we wish you in the North to be free.—

P.S. | I have long thought that geologists not having found this or that form in this or that formation was *very poor* evidence of such forms not having then existed;

in this respect differing, as wide as the poles, from the great Agassiz, who seems to me to retreat a step & take up a new position with a front so bold as to be admirable in a soldier.— Well, in case of cirripedes I thought, as stated in Preface in my Fossil Lepadidæ, that the evidence was so good, that I did believe that no *Sessile* cirripede existed before the Tertiary period. But yesterday I received from M. Bosquet of Maestricht a beautiful drawing of **perfect** Chthamalus from the Chalk!!—[12]

Never again will I put any trust in negative geological evidence.—

Yale University Library (Manuscripts and archives, Dana papers)

[1] Dated by the relationship to the letter from J. D. Dana, 8 September 1856.

[2] Letter from J. D. Dana, 8 September 1856.

[3] The answer to CD's question was probably in the missing portion of the letter from J. D. Dana, 8 September 1856. In the manuscript of his species book, CD stated that Dana doubted that southern species of Crustacea could have passed through the torrid zone to the northern hemisphere (*Natural selection*, p. 556).

[4] In the *Zoology of the Beagle*, George Robert Waterhouse stated that the North and South American species of *Mus* agreed in their dentition and in this respect differed from Old World species (*Mammalia*, pp. 74–8).

[5] A reference to Dana 1855, in which the geological history of North America was discussed.

[6] Louis Agassiz had published only short statements of his view that the geological succession of extinct forms was in some degree parallel to the embryological development of living forms. See Agassiz 1849 and 1850.

[7] In the concluding pages of Dana 1853, Dana reviewed the geographical distribution of Crustacea with the intention of determining whether species that were located in different, widely separated regions could have migrated there from a single location. He concluded that this was unlikely and indicated that his data favoured the idea of multiple creation (Dana 1853, pp. 1587–91). CD's annotated copy of the work is in the Darwin Library–CUL.

[8] See also letter to P. H. Gosse, 28 September 1856.

[9] [Chambers] 1844.

[10] Lyell's prolonged examination of CD's theories and his fears about the implications of a transmutationist view of nature are recorded in his scientific journals (Wilson ed. 1970).

[11] The 1856 presidential elections of the United States were about to be held. The Democratic nominee was James Buchanan (minister to Great Britain, 1853–6), who wrote regular letters to *The Times* covering the election. Buchanan upheld the position of states' rights with regard to the slavery question; his Republican opponent, the explorer John Charles Frémont, campaigned as an anti-slavery, pro-union candidate. Buchanan was elected.

[12] See letter to J. A. H. de Bosquet, 9 September [1856].

To W. D. Fox 3 October [1856][1]

Down Bromley Kent
Oct. 3^{d}

My dear old Friend

I am very sorry to hear how much you have been ailing. I had not heard of this affection of the leg. I do most sincerely hope that the water-cure will complete the good work which it has begun: the loss of locomotion to a man so active &

energetic as yourself would be grievous. No one can wish more truly for your recovery than I do.—

Thank you for telling me about our poor dear child's grave.[2] The thought of that time is yet most painful to me. Poor dear happy little thing. I will show your letter tonight to Emma.— About a month ago I felt overdone with my work, & had almost made up my mind to go for a fortnight to Malvern; but I got to feel that old thoughts would revive so vividly that it would not have answered; but I have often wished to see the grave, & I thank you for telling me about it.[3] I sh^d not be surprised if I went for a fortnight to Moor Park Hydropathic establishment for a fortnight, some time;[4] for I have great faith in treatment, & no faith whatever in ordinary Doctoring. It was *very* kind in Dr Gully to speak so of me: if you go there again, pray remember me *most* kindly to him, & say that never (or almost never) the vomiting returns, but that I am a good way from being a strong man. Do not mention Moor Park; but sh^d I ever go there, I sh^d *certainly* inform him.—

My poor wife has been having a bad summer; for poor soul, we shall have another Baby this autumn! Our second Boy, George (an enthusiastic *Herald*! & Entomologist)[5] has been for the last six weeks, at his first School; the Revd. C. Pritchard at Clapham,—a school which has been much patronised by scientific men, Herschel, Airy, Grove & Gassiott:[6] I like it very well, & we can have him home monthly for a day.—

You know that our Aunt M^rs S Wedgwood lives here: a month ago she slipped on the road & fractured the head of thigh-bone & is utterly crippled,— —a grievous accident for her with all her peculiarities,—far worse than death.—[7]

I remember you protested against Lyells advice of writing a *sketch* of my species doctrines; well when I begun, I found it such unsatisfactory work that I have desisted & am now drawing up my work as perfect as my materials of 19 years collecting suffice, but do not intend to stop to perfect any line of investigation, beyond current work. Thus far & no farther I shall follow Lyell's urgent advice.— Your remarks weighed with me considerably. I find to my sorrow it will run to quite a big Book.—

I have found my careful work at Pigeons really invaluable, as enlightening me on many points on variation under domestication. The copious old literature, by which I can trace the gradual changes in the Breeds of Pigeons has been extraordinarily useful to me.— I have just had Pigeons & Fowls *alive* from the Gambia! Rabbits & Ducks I am attending pretty carefully, but less so than Pigeons. I find most remarkable differences in skeletons of Rabbits— Have you ever kept any odd breeds of Rabbits & can you give me any details? Your Call Drake is quite hearty: I have not watched it much, but have not noticed its loquacity; the beak seems short & breast very protuberant. If at any time you could spare time, I sh^d very much like to hear any particulars about habits of Call Ducks.— Do they show any migratory restleness in Autumn?—

One other question, you used to keep Hawks, do you at all know, after eating a Bird, how soon after they throw up the pellet? No subject gives me so much trouble & doubt & difficulty, as means of dispersal of the same species of terrestrial productions on to oceanic islands.— Land Mollusca drive me mad, & I cannot anyhow get their eggs to experimentise on their power of floating & resistance to injurious action of salt-water.—

I will not apologise for writing so much about my own doings, as I believe you will like to hear.— Do sometime, I beg you, let me hear how you get on in health; & *if so inclined* let me have some words on Call Ducks.

My dear Fox | Yours affectionately | Ch. Darwin

Christ's College Library, Cambridge (Fox 100)

[1] Dated by the reference to CD's aunt, Sarah Elizabeth (Sarah) Wedgwood, who suffered a bad fall in September 1856 (see n. 7, below).

[2] Anne Elizabeth Darwin had died at James Manby Gully's hydropathic establishment in Malvern, Worcestershire, on 23 April 1851 (see *Correspondence* vol. 5).

[3] CD had left Malvern before Anne's burial, which was arranged by Frances Mackintosh Wedgwood.

[4] Moor Park, in Surrey, was run by Edward Wickstead Lane. CD paid his first visit there in April 1857 ('Journal'; Appendix II).

[5] George Howard Darwin, aged 11, had developed an interest in collecting moths and butterflies, as well as continuing his earlier love of drawing soldiers and coats-of-arms. CD preserved several of his detailed coloured drawings (DAR 210.7).

[6] John Frederick William Herschel, George Biddell Airy, William Robert Grove, and John Peter Gassiot. Charles Pritchard, astronomer and educational reformer, had founded the Clapham Grammar School and instituted a curriculum that included the sciences. He later became Savilian professor of astronomy at Oxford. For the care taken by CD in the education of his sons, see J. R. Moore 1977.

[7] Sarah Wedgwood lived at Petleys, near Down House. On 3 September 1856, Emma Darwin recorded in her diary: 'Aunt S. fall'.

To T. C. Eyton 5 October [1856][1]

Downe Bromley Kent
Oct. 5th

Dear Eyton

I have had sent me by Dr Daniell,[2] long resident at Sierra Leone the skin of a Dog, with skull in it & feet. He says "The Dog skin is an excellent specimen of the country Dog & forms the standard type of most here in W. Africa." Should you not like to have this? If so will you tell me how I shall send it you. Have you any House of Call in London, so far I could send it carriage free. If you like to have it to add to your collection, perhaps you would just tell me whether the skull offers any marked peculiarities,—any further details I could hereafter read in any account which you might hereafter publish on Dogs.—

After writing to you my former long troublesome letter,[3] I bethought me that I could experimentise on Hawks in Zoolog. Gardens, how long it is after eating a bird they throw up a pellet; & I will put seeds in crop & try if they are ejected & will grow.—[4] Some friends in this country are observing the Partridge feet: I found on one the other day 11 grains of dry earth.[5] Nevertheless if you can help me with any information, I sh^d be very grateful.—

I hope sometime to hear from you in regard to Bechstein's statement on the incisors of Pigs, & on Lord Hill's crossed African Pig, & whence it came.—[6]

I suppose you did not continue the cross of the Geese; I have just been consulting your two Papers.—[7]

I sh^d be much pleased if the Dog skin & skull sh^d prove of any interest to you.—

Yours very sincerely | Ch. Darwin

American Philosophical Society (139)

[1] Dated by the relationship to the letter from W. F. Daniell, 8 October – 7 November 1856.

[2] William Freeman Daniell had been an army surgeon in various postings in Africa before settling on the Isle of Wight (see letter from W. F. Daniell, 8 October – 7 November 1856).

[3] Letter to T. C. Eyton, 31 August [1856].

[4] See letter to J. D. Hooker, [19 October 1856].

[5] An entry in CD's Experimental book, p. 15 (DAR 157a), dated 23 September 1856, reads: '1 Partridge shot by Mr Innes, had 10 grams of dry [*interl*] earth chiefly under claws; very firmly attached (Oct. 19th nothing grew) | Sept 25. Parslow shot 2 Partridges, after heavy rain: very little dirt, but yet some; nail of one clogged with fibrous matter, ['fea' *del*] plumose feather?—planted.'

[6] See letters to T. C. Eyton, 21 August [1856] and 31 August [1856].

[7] Eyton 1840 and either Eyton 1837a or 1837b.

To J. D. Hooker 5 October [1856][1]

Downe Bromley Kent
Oct 5th

My dear Hooker

I write merely to ask you to send the seeds, for which I am in no hurry, & for which I am *extremely* much obliged,[2] addressed (not like water plants) to.

> C. Darwin
> care of G. Snow
> Nag's Head
> Borough.

This is safer, course, though not prompt enough for water pods.—

Also hearty thanks for enquiries about Cytisus;[3] I quite agree evidence not sufficient for so marvellous a statement; & how extraordinary that any Botanist, believing that he had witnessed such a phenomenon, & yet that it should not for ever after have impressed his mind! Such obtusity alone would make me doubt his evidence.—

Adios | C. Darwin

You will think it odd but I am glad to disbelieve the Cytisus story. Species would change too quickly, if such stories were true!

DAR 114.3: 178

[1] Dated by the relationship to the letter to J. D. Hooker, 8 September [1856].
[2] It is not clear from the correspondence what seeds CD had requested.
[3] See letters to J. D. Hooker, 22 June [1856] and 8 September [1856].

From William Freeman Daniell 8 October – 7 November 1856

Marine View. Ventnor | Isle of Wight
Oct 8th 1856

My dear Sir

I have just received your letter[1] and am much pleased that any cursory observation of mine may prove of utility to you, and as I am now just out of bed after a weeks *very* severe sickness, I think there is no better opportunity than the present when the mind is purified by this physical tornado. Any observations contained in my letters are always at your service for publication if you think them sufficiently worthy. The pigeons although procured from different houses evidently belong to the same breed, in fact the only domesticated one in Sierra Leone. With regard to the wild fowls I am of your opinion that they originally came from a domesticated breed, and were set free by the ravages of civil commotions in the neighbour where they abound.[2] It is a country which has always been famous for intestinal wars, and even now is a kind of "*debateable land*". I think it may therefore be assumed with some confidence that their descent may be claimed from the African domestic fowl.

1st With regard to your first qn. whether any tendency to temporary infertility or sterility exists in any European Animal &c With regard to dogs there is no difference in West Africa as in England.[3] They breed almost immediately, cats ditto, with regard to fowls it is very doubtful: I have no data to give European women also become frequently pregnant in S. Leo on the whole I should think there might be a temporary infertility but only for a brief period, or until the animals were fully acclimated—

2. With regard to your 2d qn. whether differences in constitution with reference to light or dark complexions in the European resisting the influences of an African climate, I am distinctly of opinion based on the results of a vast experience in human suffering, that a sanguineous or choleric, or light complexioned man stands the African climate *twice* as well and *as long again* as the melancholic or dark complexioned man— I am of a light complexion myself, and have suffered from yellow bilic remittent fevers, dysentery, ulcers &c and in fact most of the tropical diseases to which Europeans are subject and yet am still alive— I will give you an anedote which will prove at least that your qy. has been solved some 18 years ago by an African potentate. When I was a boy, I went with a party to visit the King of Warré, who resided on an island situated on a communicating stream between the rivers Rio Formosa and Niger in the Bight of Benin. The King after alluding

among other topics to the mortality that occurred so frequently among his European friends who resided at the mouth of the former river, and particularly to some recent deaths that had taken place turned round and looked me fully in the face, at the same time inquiring what age I was. "Ah! said his sable majesty! it is the right age to bring white men to Africa, the younger the better, and he is a true child of the sun, his fire (light) hair will save him from many bad diseases; he and others like him, *will* live"!!

London. Oct 20th. The physiological explanation of the sanguineous or choleric temperament enjoying better health than the melancholic, may be chiefly attributed to the greater vascular organization of the skin of the former, by which from perspiration being more easily excited, they are enabled to throw off the febrile paroxysms, and relieve the congested states of the internal organs. It is true, they suffer severely while the disease exists, but they throw it off much sooner than the melancholic temperament. In the latter, disease is much slower in its progress, the cutaneous surface is more difficult to act upon, and the patient suffers greatly from despondency and permanent debility. I do not know whether you can understand me sufficiently, but I perhaps could explain my self better verbally. I may observe that I have always had less difficulty in curing light than dark complexioned men.[4]

The only notices you can find of the Mammalia of the Isles of Anno Bon Principe and St Thomas, will be in some Portuguese works published on the subject, a general view of the animals (most of which have been imported from the main-land, and blended with European species) will be found in Barbot Astleys or Churchills collection of voyages—[5] I have been to all these islands, and see no difference in the live stock from the main land. The voyagers (D. J. Santarem and Don Juan Escobar) were the first Portuguese that discovered and visited the island of Princes' &c I published a work some years since with the information relative to the earlier Portuguese voyagers, but it is unfortunately out of print or you should have had a copy with pleasure.[6] I will however look out the information you require.

With regard to the soundings between Fernando Po, and the mainland, they vary from 24 to 36 or 40 fathoms, clayey mud, black or dark green, nearer the lowlands or alluvial flats— The soundings between Anno Bon and the continent are so deep that they have not been recorded. The water is blue [having] frequently passed down in that direction in sailing vessels.

Nov. 7. I hope you will excuse this imperfect account, the greater part of which has been written while laboring under sickness— With the exception of a slight enlargement of the spleen, I am now quite recoved, and have my usual John Bull looks—

Trusting to have the pleasure of seeing you in town soon | I remain | Yours ever sincerely | W. F. Daniell

C. Darwin Esqr. | Down—

DAR 205.2 (Letters)

CD ANNOTATIONS
1.6 The pigeons . . . Leone. 1.8] *double scored pencil*
1.8 With . . . abound. 1.10] *double scored pencil*
3.3 I am . . . subject 3.8] *double scored pencil*

CD note:

Abstract of Letter.[7]

1 Thinks Fowls of Africa, feral.
(2) No infertility in Dogs or women first arriving in W. Africa
d[itt]o the African Cocks & Pigeons when introduced here.—
(3) Light or Sanguineous men withstand climate better than melancholic men: so, also, African King thought
(4) O. J. Santarem & Don J. Escobar first visited Annobon. St. Thomas & Principe— *Ocean* between them & mainland.
(5) Between Fernando Po & mainland vary from 24 to 36 or 40 fathoms—

[1] This letter has not been located, but it was presumably written after CD received the pigeons and fowls Daniell sent from Sierra Leone (see n. 2, below).
[2] CD received live fowls and pigeons sent by Daniell from Sierra Leone (*Natural selection*, p. 80). CD later gave some of them to William Bernhard Tegetmeier (see letter to W. B. Tegetmeier, 19 October [1856]). Information from this letter was repeated in *Variation* 2: 161.
[3] This information was cited by CD in *Natural selection*, p. 80, and in *Variation* 2: 161.
[4] Daniell's information was used by CD in *Descent* 1: 244–5.
[5] John Barbot's description of his travels in western Africa was published in several collections of voyages. Barbot 1732 is volume 5 of the collection published by Awnsham Churchill. Astley 1745–7 extensively cites Barbot's work and that of his brother James Barbot, who travelled in the same region.
[6] Daniell 1849.
[7] This abstract is preserved with the letter in DAR 205.2 (Letters). CD marked it '18' in brown crayon, the number of his portfolio of notes on the means of dispersal of plants and animals.

To ? 9 October [1856][1]

Down Bromley | Kent
Oct. 9th

My dear Sir

I am very much obliged for your kind note & for all the very great trouble which you have so kindly taken for me. Next summer would not be at all too late, & if you can remember it, I sh^d be *extremely glad* to get some for my experiments.[2]

I have been myself keeping Helix Pomatia in confinement all summer, but they have not laid a single egg, so that I have not at all profited by my scheme.[3]

With many thanks, & hopes that your correspondent may succeed for me next summer. I remain | Dear Sir | Yours very faithfully | Chas. Darwin

Houghton Library, Harvard University

[1] The year is given by CD's interest in the possible means of dispersal of land molluscs, as expressed in the letters to P. H. Gosse, 28 September 1856, and to J. D. Dana, 29 September [1856].

[2] CD had probably requested eggs of land molluscs (see letter to W. D. Fox, 3 October [1856]). His experiments on the possible means of transport of snails' eggs and on the soaking of Mollusca in salt water, begun in October 1856, are recorded in his Experimental book (DAR 157a).

[3] See *Correspondence* vol. 5, letter to John Lubbock, 14 [July 1855], for CD's first consignment of *Helix pomatia*.

To J. D. Hooker 9 October [1856][1]

Down Bromley Kent
Oct. 9th

My dear Hooker

I do not remember sending any list, except one very long ago; but anyhow that does not signify.— The only seeds **in pods** (with some stalk) which I want are water or marsh plants, *named*. The widest rangers would be the most interesting to me, as Montia fontana & Callitriche verna.— We have no water here.— I have been trying some 60 or 70 pods, heads capsules &c to see about their floatation, & I shd. very much like to try some water plants, & likewise I shd like to try whether the vitality these same seeds can resist salt-water.—[2] I shall come to London on 15th & stay till Saturday 18th at noon.[3]

Supposing it proved convenient these seeds (& Vilmorins) cd. be sent to 57 Queen Anne St,[4] by that time per Deliverance Coy.—or as soon *after* the 18th, as you please as before requested.—

With many thanks | Ever yours | C. Darwin

I read long ago your Review on Decandolle with very great interest.[5] He will be disgusted at your estimation of Botanical Geography; though I shd. think, pleased with all the first part.—

DAR 114.3: 180

[1] Dated by CD's reference to a trip to London from 15 to 18 October (see n. 3, below).

[2] Documented in CD's Experimental book, pp. 11–14 (DAR 157a). CD gave the results in *Origin*, p. 359. The species mentioned by CD in the letter were not, however, listed in his experimental records.

[3] On 15 October 1856, Emma Darwin recorded in her diary: 'Ch went to London'.

[4] The residence of CD's brother Erasmus Alvey Darwin, with whom CD usually stayed when in London.

[5] [J. D. Hooker] 1856. See letters to J. D. Hooker, 5 August [1856] and 7 August [1856].

To Asa Gray 12 October [1856][1]

Down Bromley Kent
Oct. 12th

My dear Dr. Gray

I received yesterday your most kind letter of the 23d2 & your "Statistics" & two days previously another copy.[3] I thank you cordially for them. Botanists write, of course, for Botanists; but as far as the opinion of an "outsider" goes, I think your paper *admirable*. I have read carefully a good many papers & works on Geograph. Distribution, & I know of only one Essay (viz Hooker's N. Zealand)[4] that makes any approach to the clearness with which your paper makes a non-Botanist appreciate the character of the Flora of a country. It is wonderfully condensed (what labour it must have required!): you ask whether such details are worth giving, in my opinion there is literally not *one* word too much.

I thank you sincerely for the information about "social" & "varying plants"; & likewise for giving me some idea *about* the proportion (ie $\frac{1}{4}$) of European plants, which you think do not range to the extreme north: this proportion is *very much* greater than I had anticipated from what I picked up in conversation &c.—[5]

To return to your Statistics: I daresay you will give how many genera (& orders) your 260 introduced plants belong to: I see they include 113 genera *non* indigenous: as you have probably a list of the introduced plants, would it be asking too great a favour to send me per Hooker or otherwise just the *total* number of genera & orders to which the introduced plants belong:[6] I am much interested on this, & have found De Candolles remarks on this subject very instructive.[7]

Nothing has surprised me more than the greater generic & specific affinity with E. Asia than with W. America. Can you tell me (& I will promise to inflict no other question) whether climate explains this greater affinity? or it is one of the many utterly inexplicable problems in Bot. Geography? Is E. Asia nearly as well known as West America? so that does the state of knowledge allow a pretty fair comparison?[8]

I presume it would be impossible, but I think it would make in one point your tables of generic ranges more clear (admirably clear as they seem to me) if you could show, even roughly, what proportion of the genera in Common to Europe (ie nearly half) are very general or mundane rangers; as your results now stand at the first glance the affinity seems so very strong to Europe, owing, *as I presume* to the (nearly) half of the genera including very many genera common to the world or large portions of it. Europe is thus unfairly exalted.— Is this not so? If we had the number of genera strictly or nearly strictly European, one could compare better with Asia & southern America &c. But I daresay this is a Utopian wish owing to difficulty of saying what genera to call mundane. Nor have I my ideas at all clear on subject, & I have expressed them even less clearly than I have them.

I am so very glad that you intend to work out N. range of the 321 Europæan species; for it seems to me the by far most important element in their distribution.—

And I am equally glad that you intend to work out range of species in regard to size of genera ie number of species in genus.— I have been attempting to do this in a very few cases; but it is folly for any one but Botanist to attempt it: I must think that De Candolle has fallen into error in attempting to do this for Orders instead of for genera,—for reasons with which I will not trouble you.—[9]

In second column Heading p. 27 (or p. 229) there is misprint "and," for "not", which might seriously mislead an *idle* reader who only looked at general totals.

Many of our Societies always page their *separate* copies of papers with the *proper* pages for reference. Is not this good scheme & worth Prof. Silliman attending to? by a reference in body of your own paper I have corrected your paging.[10]

Hooker has lately returned from his continental trip & I am going to see him on Friday. Have you seen his Review on Decandolle: I cannot but think he is rather too severe on want of originality, & I hope much too severe on whole great & noble subject of Bot. Geograph.—[11]

With most sincere & hearty thanks for all your great kindness. Your's very truly | C. Darwin

Gray Herbarium of Harvard University

[1] Dated by the reference to A. Gray 1856–7 and to the letter from Asa Gray, 23 September 1856.
[2] Letter from Asa Gray, 23 September 1856.
[3] A. Gray 1856–7. CD's annotated copy is in DAR 135 (3).
[4] J. D. Hooker 1853–5.
[5] CD had written 'extremely few' in the manuscript of *Natural selection* (p. 539). Later he added the note: 'Asa Gray thinks there are not a few plants common to U.S. & Europe, which do *not* range to Arctic regions.' To this Hooker added, 'Certainly J.D.H.'. See also letter from Asa Gray, 4 November 1856.
[6] Gray did not give the desired figure in the second part of A. Gray 1856–7. In A. Gray 1856a, pp. xxv–xxviii, he had listed the number of introduced species (giving the total of 260 species, as mentioned by CD in the letter), but these had only been allocated to their taxonomic orders, not genera. The same list was repeated, with additional information but still excluding the number of genera, in A. Gray 1856–7, pp. 208–11. In CD's copy of A. Gray 1856a there is a manuscript list, in the hand of an amanuensis, giving the names and genera of these introduced species. It is not clear whether CD had the list drawn up at Down House or whether it was sent to him by Gray at a later date. The information was eventually used in *Natural selection*, p. 232 n. 3.
[7] A. de Candolle 1855. CD cited pages 745, 759, and 803 on the subject of naturalised plants. Alphonse de Candolle's statistics are compared with Gray's in *Natural selection*, p. 232.
[8] See letter from Asa Gray, 4 November 1856.
[9] See *Correspondence* vol. 5, letter to J. D. Hooker, 8 [November 1855], n. 3. CD thought the statistical relationships Candolle had discerned were probably due only to 'parentage' and common descent when applied to large groups like families and orders.
[10] In CD's copy of A. Gray 1856–7, he added the correct page numbers in pencil to to the pages of his independently paginated reprint.
[11] See preceding letter.

To W. B. Tegetmeier 15 October [1856]

Down Bromley Kent
Oct— 15th

My dear Sir

Many thanks for the Runt received this morning.[1]

I have had some live fowls sent me from some way in interior of Sierra Leone, so that they are genuine Africans.—[2] Would you like a Cockrel to look at & then kill if it so pleases you. I could send it to Carstang's carriage free.[3] But I must tell you that M.ͬ Brent[4] called here the other day & looked at them & says they have no marked character, having a good deal of the Game in them, with perhaps a dash of the Malay, especially a Hen, which I shall keep for skeleton. I doubt whether the Cock is worth your having, but it is most entirely at your service.—

By the way I sh.ᵈ be very glad of a Malay Cock, if you can ever get one dead for me.—

With many thanks | Your's very sincerely | C. Darwin

Endorsement: 'Oct 15ᵗʰ/56'
New York Botanical Garden Library (Charles Finney Cox collection)

[1] See letter to W. B. Tegetmeier, 21 September [1856].
[2] See letter from W. F. Daniell, 8 October – 7 November 1856.
[3] Carstang is listed in CD's Address book (Down House MS) under 'pigeons'. His address was Ship Tavern Passage, Leadenhall Market, an area of London associated with meat and poultry markets.
[4] Bernard Philip Brent.

To J. D. Hooker[1] [16 October 1856][2]

D.ͬ Hooker

Please read this, first, I want, especially to know whether Botanical facts are *fairly* accurate. 2.ᵈ any general or special criticisms: please observe if you will mark margin with pencil, if your criticisms run to any length, I would gladly & gratefully come to Kew, to save you writing.

I *really* hope no other chapter in my book will be so bad; how *atrociously* bad it is, I know not; but I plainly see it is too long, & dull, & hypothetical.

Do not be *too* severe, yet not too indulgent: remember that it will be *extra* dull to you, for it will be a compilation with hardly anything new to you.—

It is only fragment of chapter, & assumes some points as true, which will require *much* explanation,—as to close relation of plants to plants rather than to conditions: again I am unfortunately forced not to admit continental extensions as you know.—

Glance at the notes, at back of Pages.—

In truth you are doing me a very **great** kindness in reading it, for I am sorely perplexed what to do & how much to strike out.—

AL
DAR 50 (ser. 5): 9

[1] This note accompanied the fair copy of CD's manuscript on geographical distribution for his book on species (see letters to J. D. Hooker, 13 July [1856] and 28 September [1856]). The fair copy is now in DAR 14.
[2] The date of the letter is that of the meeting of the Philosophical Club of the Royal Society at which CD intended to meet Hooker (see letter to J. D. Hooker, 28 September [1856]). Although it is

possible that CD sent the manuscript by post, it seems more likely that CD gave it to Hooker on this occasion, along with the covering note printed here. The manuscript had been completed on 13 October 1856 ('Journal'; Appendix II) and was read by Hooker by 9 November (see letter from J. D. Hooker, 9 November 1856).

To J. D. Hooker [19 October 1856][1]

Down Bromley Kent
Sunday

My dear Hooker

The seeds are come all safe, many thanks for them.

I was very sorry to run away so soon & miss any part of my *most* pleasant evening; & I ran away like a goth & vandal without wishing Mrs Hooker good bye; but I was only just in time, as I got on the platform the train had arrived.

I was particularly glad of our discussion after dinner; fighting a battle with you always clears my mind wonderfully. I groan to hear that A. Gray agrees with you about the condition of Botanical Geography. All I know is that if you had had to search for light in zoological geography you would by contrast respect your own subject a vast deal more than you now do.— The Hawks have behaved like gentlemen & have cast up pellets with lots of seeds of them;[2] & I have just had a parcel of partridges feet well caked with mud!!![3]

Adios | Your insane & perverse friend | C. Darwin

DAR 114.3: 179

[1] The Sunday after CD's trip to London, 15–18 October (see letter to J. D. Hooker, 9 October [1856]).

[2] The experiment, which was carried out in the gardens of the Zoological Society, was recorded in CD's Experimental book, p. 15 (DAR 157a) on 19 October 1856:

> Killed some sparrow *on 14th [*interl*], one with wheat inside, put in Oats, Canary seed, Tares, Cabbage & Clover—gave *3 Birds [*above del* 'it'] to small S. African Eagle. (Bateleur): bolted them; threw up pellet in 18 hours, *ie on morning of 16th [*interl*] charged with seed: planted these seeds on 19th.—

[3] See letter to T. C. Eyton, 5 October [1856].

To W. B. Tegetmeier 19 October [1856][1]

Downe Bromley Kent
Oct. 19th

My dear Sir

I am glad you think the young Cock worth having: it will do for the Pot, after you have looked at it.— It shall be at Carstangs on Thursday **morning**, which is first day our Carrier goes up.—[2] These Fowls came from interior village of Sierra Leone, & please state that they were sent to me by the kindness of Staff-Surgeon Dr Daniell.[3] He states that they offer a good idea of the pure fowls of that country & for some way north of it. The fowls further S. on the west Coast, viz at Congo

&c are smaller, not *much* larger than Bantams & are very beautiful of red, white & black colours. But *since* I wrote, I have on making further enquiry heard that he has some fears that another cock had access to the Hen, & this may account for one of my two Hens having more of the Malay aspect. He says these Cockrels quite resemble in plumage the pure mother.— I observe that they are *very pugnacious* even the hens chasing my other fowls.— It is odd that they do not seem to have felt the prodigious change of climate, & from the first have been very amorous! I notice they eat the leaves of the Dandelion,—a plant they could never have seen before arriving here.

I cannot think of anything else to tell you about them.—

If you kill the cockrel, you had better keep the head.—

I am *truly* pleased that the Scanderoons have been worth your receiving.—[4]

I sh^d like to have a Catalogue of poor excellent Mr Yarrells Library &c—[5]

Yours in Haste | C. Darwin

P.S. I sh^d be extremely glad of a *few* pure Malay Eggs.

New York Botanical Garden Library (Charles Finney Cox collection)

[1] Dated by the reference to the sale of William Yarrell's books (see n. 5, below).
[2] See letter to W. B. Tegetmeier, 15 October [1856].
[3] See letters from W. F. Daniell, 8 October – 7 November 1856 and 14 November 1856.
[4] See letters to W. B. Tegetmeier, 23 August [1856] and 30 August [1856].
[5] William Yarrell had died on 1 September 1856. His collection was sold by John Crace Stevens, natural history auctioneer, 4–6 December 1856. The catalogue was nineteen pages long (Chalmers-Hunt 1976, p. 96).

To W. D. Fox 20 October [1856][1]

Down Bromley Kent
Oct. 20

My dear Fox

I am so sorry to hear of all your suffering: it is most grievous that you should so soon fall back after Malvern, & how patiently you seem to bear it.[2] I suppose the pain is extremely severe: very oddly I had yesterday morning just a touch of it; my back feeling locked & rigid! I never felt such a thing before.—

I should very much like to have ever such brief a note before very long to hear how you are getting on.— I had no idea that lumbago ever became so long-continued & severe as in your case.

I fear I sh^d not possibly be able to join you, at Malvern; for I could hardly leave my wife now for the next two or three months.[3]

You ask Georgy's age; he is just turned eleven: perhaps a little, & but very little earlier would have been better for his going; as he is thrown back by his entire ignorance of Greek: he is very forward in everything except that & Arithmetick, of which much account is made.[4] Georgy was at first put under care of Penfold, & as he says that he (or his brother I forget which) is your God Child, he must be son of P. of Christ Coll:.—⁵

Many thanks for such answers as you could give to my diverse queries: I have since taken birds with seeds in crops to Eagles & Owls at Zoolog. Soc., & the pellets with all seeds apparently perfect were thrown up in 18 & 16 hours; but the Keepers thought this *much* sooner than often happens. We have thus an effective means of distribution of any seed eaten by any Birds, for Hawks & owls are often blown far out to sea: I am trying whether the seeds will germinate.—⁶

If *when you are better* you do not dislike writing to your nephew in Jamaica, it would be really an *important gain* to me, to know whether the eggs of Lizards, (or snakes) & Shells will float in **sea-water** say for a week (or 10 days); & if they will float for that time, if he would wash them in pure water & keep them to see whether they would hatch, it would be still greater Service. If many could be obtained it would be well to try whether some would hatch after 5 days floating & some after 10 days. But if they sink, I shᵈ care less.—

I have given up in despair trying to get lizards eggs in England.—⁷

But I fear your nephew wᵈ think the experiment too foolish.

My dear old friend, no one can more sincerely wish for your recovery than I do.— With our very kind remembrances to Mʳˢ Fox | Your's most truly | Ch. Darwin

Christ's College Library, Cambridge (Fox 99)

¹ Dated by the reference to Emma Darwin's forthcoming confinement (see n. 3, below) and to George Howard Darwin being 11 years old.
² See letter to W. D. Fox, 3 October [1856].
³ Emma Darwin was seven months pregnant. Charles Waring Darwin was born on 6 December 1856.
⁴ George had just entered the Clapham Grammar School. See also letter to W. D. Fox, 3 October [1856].
⁵ James Penfold had been a contemporary of CD's and Fox's at Christ's College, Cambridge, 1825–33 (*Alum. Cantab.*).
⁶ See letter to J. D. Hooker, [19 October 1856]. The description of the experiment with owls' pellets is given in CD's Experimental book, p. 15 (DAR 157a): 'Oct 19ᵗʰ Planted pellet from Snowy owl, charged with seeds. planted pellet whole— 16¼ in stomach. Such seeds wᵈ be well manured— Birds wᵈ float if died in sea.—'.
⁷ See *Correspondence* vol. 5, letters to W. D. Fox, 17 May [1855] and 23 May [1855].

To John Lubbock 27 October [1856]¹

Down.
Oct. 27ᵗʰ

My dear Lubbock.

I received this morning your paper & have read it attentively.² It is to me *decidedly* interesting, & as a whole very clear. But without some special object (& trusting not much to my own judgment) I shᵈ have almost thought it wᵈ have been better to have waited a little longer before publishing. For you discuss (& in a *very interesting* manner) such high points towards the close, that the premises ought to be extra certain:[3] in the beginning you put very modestly & candidly the deficiences in your evidence about not having traced every step in the formation of the ephippial eggs, & in regard to impregnation.[4] Certainly if these points could have been more thoroughily cleared up, your paper would have been much more valuable. (V. Back of Page)— But do not think that I wish to underrate the novelty viz about the spermatozoa, male organs, structure of case of ephippium, & the stages, as far as you have traced them, of formation of ephippium.— The two sorts of eggs from a *"female"* is a new & striking point. To go into a few details.

In p. 1. I sh^d have thought the expression that "at once evident" that Daphnia was a case of "Lucina sine C." was rather strong;[5] for why sh^d not a priori the Ephippium have been produced without impregnation as well as the "ordinary eggs".— it may be very probable that Daphnia is case in point.—

p. 1. you use word "former"; in my opinion every author who uses "former & latter" ought to be executed; & this w^d clear the world of all authors except Maccaulay.—[6]

p. 10 is not very clear in parts, owing, I think, to your varying your terms "cells" "eggs" "darkened" "brown"[7]

p. 18. Surely ought you not to give your own facts pretty full (& references to others) about ova being produced by females for successive times without males.— When I met this page, I turned back, thinking that I had overlooked some whole page.—

I sh^d have doubted whether it was worth while to have given such long extracts from Baird & M. Edwards, as you state that Straus more accurate.[8]

I have appended a few pencil marks to some sentences, which required twice reading over, which no sentence ought to do.

Do not mistake my first remarks, & suppose for one moment that I do not think your present materials worth publishing: only I sh^d have liked to have seen them still more perfect;—but it is quite likely that I carry this notion to an extreme.— Trust more to Huxley's opinion than to mine, if you can get him to read the M.S.— — I do most honestly admire your powers of observation & zeal; & you will do, much in Nat. History, notwithstanding your terrible case of "pursuit of knowledge under riches", as I said the other day.—

Farewell— I am sorry that you are poorly— I hope & fully expect that I shall be well enough to see you on Wednesday if you come

Adios | C. Darwin

The evidence in regard to fecundation & the 2 sorts of eggs is thus, is it not?—

Ordinary eggs are produced for 2 or 3 successive times in same individual without males; but not I suppose for successive generations.— I suppose existence of a spermatheca is very improbable.—[9]

Ephippial eggs are never produced without presence of males, (& you have a good long experience in this?)[10] but the presence of males does not necessarily in your experience, induce ephippial eggs.— ie Whilst males are present & even seen attached to females, "ordinary" eggs are produced.— Is this not so? Certainly evidence seems pretty strong for your view: but yet, if I have put the case right, still further evidence or still longer experience w^d be desirable.—

W^d it not be adviseable to give some summing up of evidence

Down House

[1] Dated by the reference to the manuscript of Lubbock 1857.

[2] Lubbock sent CD the manuscript of his paper on reproduction in *Daphnia* (Lubbock 1857). CD communicated the paper to the Royal Society on 22 December 1856 (see *Proceedings of the Royal Society of London* 8 (1856–7): 352–4). It was published in full in the *Philosophical Transactions* of the society.

[3] In the closing pages of the published version (Lubbock 1857, pp. 95–9), Lubbock summarised recent work on parthenogenesis in the Articulata, including Crustacea and insects. He drew a parallel between 'agamic' reproduction in *Daphnia* and in plants that apparently produced agamic seeds, and suggested that there was no fundamental difference between the two kinds of eggs produced by *Daphnia*: the eggs formed 'parts of one and the same series' (p. 99).

[4] Lubbock distinguished the common, parthenogenic mode of reproduction in *Daphnia* from the less frequent and less understood mode of sexual reproduction. Although he stated that had not been able to prove that the ephippial eggs (so named because they were found in a specialised compartment of the animal's carapace known as the ephippium) were the result of sexual reproduction, he presented his reasons for believing that this was the case. He also claimed to have identified the male sexual organs and the animal's spermatozoa, hitherto unobserved.

[5] In the final version of the paper, Lubbock did not use the expression 'at once evident' (Lubbock 1857, p. 79).

[6] Thomas Babington Macaulay.

[7] CD refers to the section in Lubbock's paper in which Lubbock described the appearance and subsequent development of agamic eggs after their deposition (Lubbock 1857, p. 83).

[8] Lubbock cited William Baird (Baird 1850) and Henri Milne-Edwards (Milne-Edwards 1834–40) in Lubbock 1857, pp. 84–5, before stating that Hercule Eugène Gregoire Straus-Durckheim's description of the anatomy of the ephippium was 'very accurate'. Lubbock refers to Straus-Durckheim 1819.

[9] In the published paper, Lubbock stated his belief that there was no spermatheca in which the female could retain live sperm to fertilise successive broods (Lubbock 1857, p. 88).

[10] Lubbock stated: 'I have not succeeded in . . . obtaining ephippial eggs from isolated specimens' (Lubbock 1857, p. 87). However, he went on to describe experiments in which isolated females developed ephippia and concluded that 'these experiments prove that ephippia *can* be produced without male influence. I only, however met with seven instances, though I have had at least 400 broods of agamic eggs produced by females kept separate from males' (p. 88).

From T. V. Wollaston [early November 1856][1]

Scarites (a genus of the *Carabidæ*), there also the mandibles are greatly developed (*in proportion*, almost as much as in the *Lucanidæ*), & *in size* eminently variable.[2]

And, as I suppose the principle is the same whether the organs be above *or below* the medium standard of development, we might perhaps cite the wing-cases (elytra) of *Meloë* as a case in point. These are unusually reduced in dimensions (for the Coleopterous type), scarcely covering two-thirds of the abdomen; & they are, in some of the species, *very inconstant in size.*—

The connateness of elytra (as having merely a *character*, & not *an organ*) will not perhaps suit you.— otherwise I might mention that the only *Harpalus* (I believe) on record in which the wing-cases are ever joined is the *H. vividus* of the Madeiran Islands (opus diab. p.p. 56, 57);[3] but that character (anomalous as it is) does not *always* occur in that species,—the elytra being sometimes connected, sometimes sub-connected, & occasionally almost (if not entirely) free.—[4]

Such are a few facts which strike me primâ facie, from very ordinary & commonplace material. With the extravagancies of Nature (such as the tropics may produce) I have nothing to do; but, if this principle be an universal one, an examination of the Leaf-insects & (for instance) those remarkable *Homopterous* creatures in which the thoracic projections take every conceivable form, might perhaps throw some additional light upon it.—

The basal joint of the feet of some of the Madeiran *Tarphii* is **wonderfully** developed into an elongated spine; but I have not yet observed any variability in this,—unless indeed (as is not likely) I have been mistaken in regarding it as sexual.

I must however cease, for Time fails, & moreover I am not in great condition for either thought or work. Good luck to the *Helices*: may *they live,*—tho' *not*, I trust, in salt-water.[5]

Yours very sincerely | T V Wollaston.

Incomplete
DAR 181

CD ANNOTATIONS

1.1 *Scarites* . . . variable. 1.2] *crossed pencil*; 'Q'[6] *added pencil, circled pencil*
2.1 And . . . principle] '10'[7] *added brown crayon*
2.1 And . . . point 2.3] 'whole genus. rudimentary organ' *added pencil*
3.1 The connateness . . . connected, 3.5] *crossed pencil*; 'Q'[8] *added pencil, circled pencil*
3.4 but . . . species, 3.5] *double scored pencil*
4.1 Such . . . Wollaston. 7.1] *crossed pencil*

[1] Dated by the reference at the end of the letter to the land snails that Wollaston had given to CD. CD recorded his first experiments on Wollaston's Porto Santo snails in his Experimental book, p. 17 (DAR 157a) on 30 November 1856.
[2] CD had apparently requested further details concerning cases mentioned in chapter 4, 'Organs and characters of variation', of Wollaston 1856.
[3] The 'opus diab[olicus]' is Wollaston 1854.
[4] The tendency of the wing-cases in *Harpalus* to be joined was discussed in Wollaston 1856, pp. 96–7, and *Natural selection*, pp. 124–5, 313.
[5] Wollaston had given CD a bag of live molluscs he had collected on Porto Santo, Madeira to be used in CD's experiments to test their ability to withstand the effects of sea-water. Land molluscs presented a considerable challenge, through their ubiquitous presence on oceanic islands, to CD's views on the dispersal of seeds and ova by sea transport. Wollaston was a primary advocate of the land-bridge theory of geographical distribution. See letter from T. V. Wollaston, [11 or 18 December 1856].

[6] The 'Q' stands for quoted. CD used this information from Wollaston in *Natural selection*, p. 313.

[7] The number of CD's portfolio on abortive organs.

[8] This information was quoted in *Natural selection*, p. 313.

To John Lubbock [1 November 1856][1]

Down
Sunday

Dear Lubbock

I have written to Hooker to enquire. The case has been described in Linn: Transactions.—[2] Many of Spallanzanis experiments seemed to show that plants have been produced without pollen;[3] but the subject has been discussed backwards & forwards of late years,—& largely by Gærtner in his Beitrage zur Kenntniss &c.[4] But I cannot *possibly* spare this Book for about a fortnight. for I am using it daily.[5]

I think your explanation of the inner saddle perfectly satisfactory; it is in fact *Continuous* with inner membrane of carapace, like the outer saddle is with the outer membrane of carapace.—

But I demur vehemently to your calling the under membrane of carapace "Corium"[6]

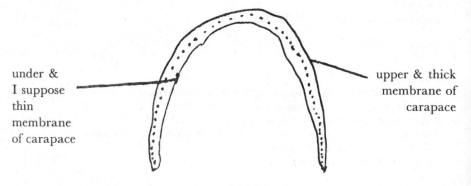

under &
I suppose
thin
membrane
of carapace

upper & thick
membrane of
carapace

The "corium", I apprehend, w^d be represented by the dotted line between these two membranes, & at each moult, would form a new upper & a new under membrane. Corium or true skin will never, I imagine, be freely exposed to water or air; its very nature, I apprehend, is to be covered with epidermis or chitine membrane, which it generates at each fresh moult.—

Your's very sincerely | C. Darwin

No doubt the corium itself is double or a fold

Would not **longitudinal section** of posterior end, *together* with tranverse section make things clearer?.—[7]

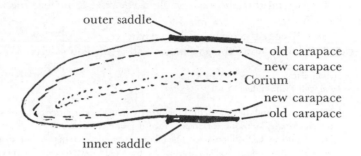

This would show how both inner & outer saddle were moulted together. You could give true outline longitudinally of carapace.

Down House

[1] Dated by the discussion of Lubbock's work on *Daphnia* (Lubbock 1857), which CD communicated to the Royal Society on 22 December 1856 (see letter to John Lubbock, 27 October [1856], n. 2). The first Sunday after the letter to John Lubbock, 27 October [1856], was 1 November 1856.

[2] In Lubbock 1857, pp. 96–7, Lubbock described all the known cases in which agamic seeds had been found in plants, including that reported by John Smith, curator of the Royal Botanic Gardens at Kew, of the germination of seeds of *Coelebogyne ilicifolia* apparently without the presence of pollen (J. Smith 1841).

[3] Spallanzani 1769. CD had read this work in 1839 (*Correspondence* vol. 4, Appendix IV, 119: 4a).

[4] Gärtner 1844. Lubbock stated that Karl Friedrich von Gärtner doubted the truth of such observations (Lubbock 1857, p. 97).

[5] CD was working on his chapter on crossing (*Natural selection*, pp. 35–91), in which Gärtner 1844 was frequently cited. He finished the chapter on 16 December 1856 ('Journal'; Appendix II).

[6] Lubbock's observations on the internal structure of the ephippium led him to believe that during the process of moulting the new carapace of *Daphnia* was formed between the upper and lower surfaces of the old carapace and that the new carapace was formed around a thin membrane, the corium (Lubbock 1857, p. 87).

[7] Lubbock included both longitudinal and transverse sections in his plate of the carapace (Lubbock 1857, Pl. VII, figs. 4 and 5).

To J. A. H. de Bosquet [before 3 November 1856][1]

great rogues & not to be trusted.— The skins (though I fear there is little chance of your hearing of these breeds) could be sent through Bookseller & I could somehow manage to repay you.

I sh^d very much like also to have one of your good Carriers (old Cock bird) skinned.—

Could you find out for me *authentically* at what rate Carriers have flown long distances such as 200, or 300 or 400 miles; telling me the names of places whence flown & where to.—

Forgive all this *immense* trouble if you can, & believe me with every feeling of friendliness,

Your's very sincerely | Ch. Darwin

I sincerely hope the specimens will reach you safely.

Incomplete
American Philosophical Society (138a)

[1] The recipient and the date of the letter are suggested by the letter being preserved in the collection of the American Philosophical Society together with the letter to J. A. H. de Bosquet, 9 September [1856], and by CD's reference in the following letter to having written to a geologist in Maastricht for information about finnikin pigeons. Bosquet and CD had corresponded when CD was preparing his monographs on cirripedes (see *Correspondence* vol. 5). Bosquet had recently sent CD drawings of sessile cirripedes he had discovered in Secondary formations (see letter to J. A. H. de Bosquet, 9 September [1856]).

To W. B. Tegetmeier   3 November [1856][1]

Down Bromley Kent
Novr 3^d

My dear Sir

Many thanks for your kind note: I ought to have mentioned, that the two other African cocks, have a treble comb; the two external parallel crests being small: otherwise the 3 birds are hardly distinguishable.

I have Mr Eaton's curious Book.—[2]

The Malay Eggs would be much better in Spring; but if ever you have opportunity I sh^d be very glad to buy an old Malay Cock for skeleton.—

Since writing to you I have received from the Honble Walter Elliot of Madras, *skins* of following poultry,

2. Game Cock & Hen (of great size)
2. Black-boned, hairy Cock & Hen
2. Caffir Cock & Hen with curled feathers.
1 Cock of *doubtful* origin ticketed Rangoon
2 Common or Pariah Cock & Hen, such as are
   commonly kept by the country People.

Now if you would like to examine & describe these 9 skins, I shall be delighted to send them you, ie if you *yourself* wish it.— They would require to be sent in large Box, but not very heavy & I could send them on Thursday week. to Carstangs or elsewhere, carriage free.— But I sh^d require them back in a few weeks, & you could then return me the head of the wild Jungle Fowl.—

I expect soon to receive some Persian Fowls, which are at your service, if you like, but please to understand that I do not wish to give you trouble & I make these offers only on the chance that you yourself would like them; for if you did, it would give me very sincere pleasure to repay you in **small** part, your kindnesses towards me.—[3]

Many thanks for your enquiries about Finnikins.[4] I have written to Geologist in Maestricht about these same birds, who declares he will take any trouble for me.—[5]

I have received to day catalogue from M[r] Bult of Steven's next sale,[6] but there seems nothing that concerns me.—

Please in any notice about African Fowls, do not forget to introduce about their being brought over by the kindness of D[r] Daniell. Also if you wish to see & notice the Madras Fowls say that they were sent to me by great kindness of the Hon[ble] Walter Elliot.—

Would it be of any use, & sh[d] you object to put in enquiry in Cottage Gardener, whether any one has crossed Call, or Hook-billed, or Penguin Ducks with common Ducks, & whether *Hybrids were fertile with other Ducks or inter se* If there w[d] be any *chance* of getting information, would you enquire for me?[7]

My dear Sir | Yours sincerely | C. Darwin

New York Botanical Garden Library (Charles Finney Cox collection)

[1] Dated by the relationship to the letter to W. B. Tegetmeier, 19 October [1856].
[2] Eaton 1852. CD's annotated copy is in the Darwin Library–CUL.
[3] See letter to W. B. Tegetmeier, 19 November [1856].
[4] See letter to W. B. Tegetmeier, 21 September [1856]. In the 11 November issue of the *Cottage Gardener* 17 (1856–7): 103, Tegetmeier reported on the 14 October meeting of the Philoperisteron Society and asked if any readers could inform him about the present existence of the old finnikin or turner pigeon.
[5] See preceding letter.
[6] Samuel Bult was a leading pigeon fancier and, according to CD, 'the most successful breeder of Pouters in the world' (*Variation* 1: 208). John Crace Stevens's monthly poultry sale was to take place on Tuesday, 4 November 1856 (*Gardeners' Chronicle and Agricultural Gazette*, 25 October 1856, p. 719). Catalogues for the sale were available from his offices at 38 King Street, Covent Garden.
[7] No notice relating to CD's query has been found in *Cottage Gardener*.

From Asa Gray 4 November 1856

Of our say 320 Phænogamous species common to Europe—there are *230*, or 72 per cent, which have not been detected here within the arctic circle.[1] About a *dozen* of these are among our alpine and subalpine species! Excluding these last (alpine & strictly subalpine)

218	species do not cross the arctic circle.
155	" do not much if any pass lat. 60°
113	" " " " " " " " 55°
56	" " " " " " " " 50°
20	" " " " " " " " 45°
5	" " " " " " " " 40°

But of these 5 only one is really a reliable case in point, and that is *Convallaria majalis*.

Cambridge 4th Nov. 1856.

Dear Mr. Darwin

Your welcome favor of Oct. 12th. came in the nick of time, and encouraged me to go on with Statistics.[2] I am now working on a portion. The difficulty is that I cannot fix my attention on such subjects long enough to get into the spirit, and do any thing of any moment—except to arrange a few facts, which you can shape and use.

Above is the *upshot* as to our species common to Europe ranging N. I have gone into this matter with no small pains.— Of course further observation all tends one way—i.e to carry the range further north—; but so it is according to our present knowledge— I think you will be a little astonished, at the result.

As to *introduced species*, I am saving them up to the last, as I wish to discuss them somewhat particularly— You will find the orders they belong to in 1st table (p. 208) of my Statistics—[3]

I meant to have copied out for you a list of the genera & species; not time to-night. But get the Botany of N. States, at *Trubner's 12 Paternoster Row*, and you can easily gather what you want, as the introduced species are in a different type.[4]

I cannot tell you whether "climate will explain the greater affinity with E. Asia than W. America"—but it stands in *direct relation* to climate; ours & that of E. Asia being extreme climates; & Oregon the contrary.

Of course a great many of our genera common to Europe are *mundane* or nearly.— Thank you for the hint. I am going on to treat of the relations of our flora to European more particularly— It will be neither "Utopian" nor difficult to exclude the mundane genera—and then discuss the real points of likeness.— Also, to exclude the identical species, and investigate the various degrees of resemblance between W. European & E. N. American plants. Only I fear I shall not get time to do it well, before my mss. must go to the printer.

I think I clearly see what you want, and it seems not difficult to do— But continue your questions & suggestions—if you please. The only way to get anything out of me is to set me my work and show the way.

Thank you for correcting misprint. There are some bad clerical errors— I shall reprint list of alpine species, &c which are very faulty, I find.

The pages of *Journal* itself ought to be kept in extras, even when there is separate paging. I neglected to give proper directions—thinking little of the extra-copies.

I have read with much instruction Hooker upon De Candolle's book—think he is too hard at the end, both upon DC, & upon the subject, and getting dreadfully paradoxical to contend that Coniferæ are the *highest style of plants*.[5]

A considerable part of our alpine plants (more than our subalpine) are not known in our *arctic continental regions*, but are connected with Scandinavia through Labrador & Greenland alone—

A few (such as *Spiræa Aruncus*) seem as if they had come to us from Central Europe via Siberia—

Excuse such an epistle as this. I am much pressed for time.—

Yours ever | Asa Gray

Kindly post enclosures—putting that to Hooker in an envelope.[6]

DAR 165: 95

CD ANNOTATIONS

1.1 say] *parentheses added pencil*
1.6 155] *pencil line drawn below*
1.10 5] *pencil line drawn below;* '194' *added pencil*
2.1 5] *comma added pencil*
5.1 As to . . . type. 6.3] *crossed pencil*
5.2 You will . . . Statistics—5.3] *double scored pencil*
8.1 Of course . . . *plants.* 12.3] *crossed pencil*
13.1 plants . . . alone— 13.3] *double scored pencil;* 'very hostile | No. *Labrador* is part of *Continent*, but shows how much extinction there must have been inland | On White mountain only 5 not found in Asia' *added pencil*
13.3 Labrador] *underl pencil*
14.1 A few . . . time.— 15.1] *scored pencil*
15.1 Excuse . . . time.—] *crossed pencil*
17.1 Kindly . . . envelope.] *crossed pencil*

[1] In his letter to Asa Gray, 24 August [1856], CD had indicated that he considered the northern range of the 321 species common to Europe and North America to be an important element in studying their distribution. He had been surprised to hear from Gray that few of these species had been found to range northwards (see letter to Asa Gray, 12 October [1856]). When he received the figures given by Gray in this letter, CD decided to use them in *Natural selection*, replacing information given by Hewett Cottrell Watson (see *Natural selection*, p. 539). To his account of Watson's views on the northern range of plants, he added the comment: '(**No.** give Asa Gray's facts, far more accurate)' and wrote in the margin, 'dele'.

[2] Letter to Asa Gray, 12 October [1856], in which CD referred to the first part of A. Gray 1856–7.

[3] CD had asked Gray for some indication of the genera (not orders) to which the introduced species belonged (see letter to Asa Gray, 12 October [1856]).

[4] A. Gray 1856a. See letter to Asa Gray, 12 October [1856], n. 6.

[5] Gray refers to [J. D. Hooker] 1856, a review of A. de Candolle 1855.

[6] CD forwarded Gray's letter to Joseph Dalton Hooker with his own letter to J. D. Hooker, 18 November [1856].

From J. D. Hooker 9 November 1856

Kew
Nov 9th/56

Dear Darwin

I have finished the reading of your mss. & have been very much delighted & instructed.[1] Your case is a most strong one & gives me a much higher idea of *change* than I had previously entertained; &, though, as you know, never very stubborn about unalterability of specific type, I never felt so shaky about species before. The first half you will be able to put more clearly when you polish up, I have in several cases made pencil alterations in details as to words &c, to enable myself to follow better—some of it is rather stiff reading.[2] I have a page or two of notes for discussion, many of which were answered as I got further on with the mss., more or less fully.[3]

Your doctrine of the cooling of the tropics is a startling one, when carried to the length of supporting plants of cold-temperate regions, & I must confess that, much as I should like it, I can hardly stomach keeping the Tropical genera alive in so very cool a greenhouse. Still I must confess that all your arguments pro may be much stronger put than you have

I am more reconciled to Iceberg transport than I was also, the more especially as I will give you any length of time to keep vitality in ice, & more than that will let you transport roots that way also— Many of these subjects which I never myself studied for myself, I wanted put in the systematic form you have put them, for proper appreciation—

I think that you might support your cause by making more use of gulf streams & oblique lines of transport—you appear to dwell too much upon meridional lines of migration This mode of handling at once suggested the Query are the Arctic & Antarctic American genera more allied than the Tasmanian & Siberian—the former offering every possible facility in continuous land,—the latter none.— It also makes you appear to shirk the question of transport from E to W. or vice versa.— you offer no explanation of the vegetation (not littoral) of Abyssinia & Indian Penins. being so similar; of the Carnatic, Ava, & N.W. Australia being in so many points alike—of the curious parallels or representatives between Madagasca, Ceylon & the Sunda Islands. In short Meridional migration alone occupies you. Nor do I like putting Iceland, Ferroe, & Spitzbergen out of the Category of the glacially peopled countries, & leaving Shetlands Orkneys Scotland in it. This is however a trifle.

Ch. Martins arguments seem to apply no more to these islands than to any other area Continental or insular—[4] If they presented any anomalies as the presence of Lapland plants or Greenland ones I might then believe them to be peopled by accidental migration, but if Icebergs are to be so powerful why did they bring no Greenland, American or other plants to these Islands, which are so well situated for the purpose.

Thanks for your note received this morning— We shall hope (if not look) for you on Wednesday to meet Lindley & Henslow.—or on Friday to meet Tyndall & Henslow.[5]

Owen I hear committed a cutting telling & flaying alive assault on Huxleys adaptation views at the Geolog. Soc. & read it with the cool deliberation & emphasis & pointed tone & look of an implacable foe.—& H. I fear did not defend himself well (though with temper) & perhaps had not a popular champion in Carpenter who barbed him— These embroglios are very bad indeed & must insensibly have a bad effect upon Huxley—the best natures insensibly deteriorate under such trials.[6]

I shall bring your mss to the Club in the Club box so you need not come on purpose.—[7] I shall be **really** glad of more.

Ever most sincerely yrs | Jos D Hooker

P.S. Was it not with you? that I argued that Lyell had assumed as demonstrable a change in the temp of the *whole globe* to be producable by altered condition of surface see Principles 9th Ed. 103, 104[8]

[Memorandum][9]

Note I.[10] Would Forbes suppose that the presence of the South Shetland Aira Antarctica on the Falklands was due to Iceberg transportation North?[11] Is it not more natural to suppose that A. Ant. was produced by creation or variation on the American continent & thence either transported South to S. Shetland or that it inhabited an intermediate sunk area. I am against making arctic regions centres of creation either by variation or by specific creation.

I think it would facilitate our researches much not to look beyond the epoch of the existence of those continents having the required climate for the existence of the scattered productions whose migrations we seek to account for. It is enough to admit a glacial land & sea over central Europe & do not let us speculate on the origin of its species. Never wander further back into Geological time than is necessary—it bewilders.

On the whole then I would perhaps confine this part of the discussion to the migration North & vertical ascent of species inhabiting a cold country.

Note B.[12] Might not much of this difficulty be got over by supposing the E & W. parts of the glacial continent differently heated, & that currents flowed East & West or NE & NW.

Thus the connecting land of Europe & America might be much warmer than those parts of either continent in the same latitude where the mountains were

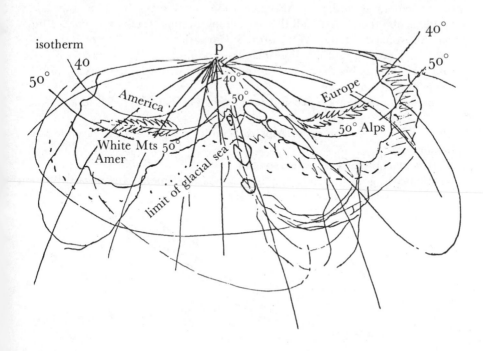

Note C.[13] I cannot see why the colonization of Iceland Ferroe & Spitzbergen should come under a different category from other lands—this is most unphilosophical Surely a theoretical inflexion of the isothermals should not be wholly lost sight of, during the glacial epoch, as it manifestly is after it. The gradual accession of the Gulf Stream influence would warm all that part of the glacial sea coast or chain of Islands that included Iceland Ferroe &c before any other part of the glacial region & induce migration along that line, however cold the preexisting arctic desert in which they were situated may be assumed to be

Note D.[14] I cannot understand this. Why do the Gentians not go North? these not being more Alpine than the Arctic species— Why should they have spread over the intervening country?

Note E.[15] Then why no peculiar species or varieties in Iceland, Spitzbergen &c.

Note F[16] The same argument must hold for the Arctic & Antarctic representative Crustacea—on which Ross was always insisting & swearing that some were identical with what he had described in Capt. Parry Voy &c[17]

Note G[18] In fact the Flora of analogous elevations of Ceylon, Nilghiri, Khasia & Himal is to great extent specifically the same

Note H.[19] After which why did not any ascend the Himalaya?

An argument in favor of alteration induced by isolation afforded by fact that so many well known species when found isolated have as much differences as to deceive botanists & then when dried lose all distinguishing characters

Change of Tropical Climate demanded is far too great Where were many tropical genera & orders— Also migration not always N. & S. but across continents obliquely— Also all this leaves longitudinal distribution unaccounted for as Abyssinia & India—W. Austral & Carnatic.

Ordinary laws of reproduction includes modif. of specific forms.

But it is improbable that similar forms be generated from specifically different parents in different places

Hence will propagation account for presence of identical forms in all parts of globe

Plants, insects—common to Alps & Scandinavia Steinbock, variable Hare, Chamois.

Forbes glacial epoch accounts for this

Help may be got by introducing humidity as an element—quote very different levels on Himal. Khasia & Ceylon for same species.[20]

DAR 100: 105–8; 109–10

CD ANNOTATIONS

1.4 The first . . . reading. 1.7] *scored pencil*
1.4 The first . . . polish up, 1.5] *double scored pencil*
2.3 I can . . . greenhouse. 2.4] *double scored pencil*; 'Not so very cool, but northern ones c^d range further South, if not opposed.' *added pencil*
2.5 arguments pro . . . have 2.6] *double scored pencil*
4.3 suggested . . . land,— 4.5] *double scored ink*
4.11 Nor . . . Scotland 4.12] *scored pencil*; 'I do not understand' *added pencil*
5.1 Ch. Martins . . . purpose. 5.6] 'Not icebergs, because from S. to North & currents from N. to S.' *added pencil*
8.2 glad of more] *double scored pencil*
End of letter: 'Ask Falconer, Royle what proportion of Brit [*interl*] plants could live in Calcutta Bot. Gardens.'

[Memorandum]
6.1 Note C . . . after it. 6.4] 'Make clearer' *added pencil*
7.1 Note D . . . country? 7.3] 'Make clearer' *added pencil*

[1] Hooker refers to the manuscript on geographical distribution (a section of chapter 11 of *Natural selection*), which CD gave to him for his comments in October (see letter to J. D. Hooker, [16 October 1856]). The manuscript is now in DAR 14.

[2] Hooker's comments are on the fair copy in DAR 14. They have also been transcribed as footnotes in the published version of *Natural selection* .

[3] See the memorandum transcribed following the letter.

[4] Martins 1849, in which Charles Frédéric Martins claimed that the number of European plants decreased on northern islands as the distance from Europe increased. See *Natural selection*, p. 541.

[5] CD visited London on 13 November 1856 to attend a meeting of the Philosophical Club (Royal Society Philosophical Club minutes) but did not dine with Hooker.

[6] Richard Owen had read a paper on a much-debated fossil mammal, named by him *Stereognathus ooliticus*, at a meeting of the Geological Society on 5 November 1856. In the paper (Owen 1857a), Owen made it clear that he was using the case as an example of how a single fossil tooth could legitimately lead to the determination of affinities and organisation, a point sharply criticised by Thomas Henry Huxley in T. H. Huxley 1856b.

[7] The Philosophical Club of the Royal Society (see n. 5, above).

[8] In C. Lyell 1853, pp. 103–4, Charles Lyell used information on soundings given to him by James Clark Ross and Hooker after the voyage of the *Erebus* and *Terror* (1839–43) to establish the point that the movement of large segments of the earth's crust, up and down relative to the sea, could produce great changes in climate.

[9] Hooker's notes on CD's manuscript on geographical distribution were probably given to CD at a later date, but they have been transcribed here for clarity. They had been received by CD by 15 November 1856 (see letter to J. D. Hooker, 15 November [1856]). The notes are in DAR 100: 109–10.

[10] Note A was presumably intended. On the manuscript, Hooker wrote: 'It is difficult to believe that during glacial epoch the northern land was warm enough for any plants at all. See Note A.' He refers to CD's point that the seeds of northern plants were transported via icebergs to the mountains of southern Europe during a former cold period (see *Natural selection*, p. 536).

[11] *Aira antarctica* was described in J. D. Hooker 1844–7, 2: 377. He called it an 'elegant grass'.

[12] On the manuscript, Hooker wrote: 'Certainly J. D. H. Note B'. He refers to CD's proposal that there was a connection between Arctic regions before the former cold period began (see *Natural selection*, p. 538).

[13] Hooker refers to CD's discussion of the colonisation of northern islands such as Iceland and Greenland. CD believed that they were covered by snow and ice during a former cold period and their flora and fauna were introduced at a later time. He cited Martins 1849 in his discussion (see n. 4, above). Hooker also made reference to Martins's work in his letter.

[14] CD explained that exclusively alpine species, such as gentians, would have extended their range in the glacial period and 'on the returning warmth . . . would together with the arctic species have reascended the mountains' (*Natural selection*, p. 542). He did not explain why they would not also have migrated northwards with the Arctic plants.

[15] CD's manuscript reads: 'we might have expected that there would have been many representative species & strongly marked varieties, on the several alpine summits of Europe, when compared one with another & with the arctic regions' (*Natural selection*, p. 542). Beside this passage, CD wrote: 'Dr. Hooker: I wish I knew whether this was so: Forbes thought so, but I do not know whether he is to be trusted.' Hooker added the note: 'Certainly true J. D. H', but indicated the further query in note E.

[16] Note F refers to CD's discussion of the existence of species in the two hemispheres that 'represented' each other in the sense that they were very closely similar. Hooker's point relates to CD's concluding sentence: 'Some of the fish, also, from Madeira, as I am informed by the Rev: R. B. Lowe represent those of Japan.—' (*Natural selection*, p. 543).

[17] James Clark Ross in Ross 1826, Appendix, pp. 116–20. Hooker had accompanied Ross on the Antarctic expedition of 1839–43.

[18] Hooker refers to CD's statement: 'Dr. Hooker believes abundance of plants on the Nilghiri are common with those on the mountains of Ceylon & the Himalaya.' (*Natural selection*, pp. 545–6).

[19] Hooker refers to CD's description of the flora and fauna of the Far East and India during a former cold period (*Natural selection*, p. 552). He wondered why the 'woolly-covered Rhinoceros tichorinus and Elephas primigenius' did not ascend the mountain heights with the plants at the end of the cold period.

[20] CD subsequently responded to Hooker's criticisms in a note now in DAR 50 (ser. 5): 40 that reads:

> The way to put the question is,—cool Tropics, & imagine all plants killed, then w^d they migrate over *bare land*, if so they w^d have some chance [*illeg del*] with distressed Tropical productions. If come to this how much do you attribute S. range to other productions & how much to temperature.
>
> Your objection to *meridional* migration is because whole discussion an offset of *Alpine* distribution & so connected with cold.—

Following this, CD added in pencil: 'Not more temperate plants S. than almost arctic | Effects of damp on Khasia.—' He then deleted 'Effects . . . Khasia.—' in pencil.

To Charles Lyell   10 November [1856][1]

Down Bromley Kent
Nov. 10th

My dear Lyell

I am writing to you in order to answer Lady Lyells note to Emma,—as she find writing or indeed doing anything whatever a considerable exertion.

We have been most sincerely grieved to hear such very indifferent accounts of Mrs Horner:[2] paradoxical as it may appear, I think, illness after so many years of good health, seems all the more to be deplored.— I had hoped to have come to London this week, & I had calculated on the very great pleasure of seeing you &

Lady Lyell; but several combined circumstances will stop me; & chiefly Emma's state. I do not suppose I shall see you till January.—

Last week my Aunt, M[rs] Wedgwood, expired here quite suddenly & easily.;—a great relief to her, as her life had become a heavy burthen to her.—[3]

I wish I could see you sooner than I shall, for I sh[d] like to hear what you have been about.— I suppose the Madeira paper will soon be sent in.—[4]

I am working very steadily at my big Book;—I have found it quite impossible to publish any preliminary essay or sketch; but am doing my work as complete as my present materials allow, without waiting to perfect them. And this much acceleration I owe to you.[5]

I know you like all cases of negative geological evidence being upset. I fancied that I was a most unwilling believer in negative evidence; but yet such negative evidence did seem to me so strong that in my Fossil Lepadidæ I have stated, giving reasons, that I did not believe there could have existed any Sessile Cirripedes during the Secondary ages. Now the other day Bosquet of Maestricht sends me a perfect drawing of a perfect Chthamalus, (a recent genus) from the Chalk! Indeed it is stretching a point to make it specifically distinct from our living British Species.— It is a genus not hitherto found in any Tertiary bed.[6]

Farewell | Yours most truly | C. Darwin

American Philosophical Society (140)

[1] Dated by the reference to the death of Sarah Elizabeth (Sarah) Wedgwood (see n. 3, below).

[2] Anne Susan Horner, Charles Lyell's mother-in-law.

[3] Sarah Wedgwood died on 6 November 1856 (*Emma Darwin* 2: 161). She had broken her hip in an accident in September (see letter to W. D. Fox, 3 October [1856]).

[4] Lyell had been working for several years on an analysis of the volcanic geology of Madeira and intended to publish a paper in collaboration with Georg Hartung (see K. M. Lyell ed. 1881, 2: 232, 282). The paper was eventually abandoned, but Hartung later published most of their results (Hartung 1864).

[5] Lyell had urged CD to publish a short sketch of his theory in order to establish his priority (letter from Charles Lyell, 1–2 May 1856).

[6] See letter to J. A. H. de Bosquet, 9 September [1856].

From H. C. Watson 10 November 1856

Thames Ditton
Nov 10. 1856

My dear Sir

With great interest I have just been reading your experiments on the germination of seeds after lengthened immersion in salt water.—[1]

Botanists appear to have taken for granted or assumed too carelessly, that seeds must lose their vitality by immersion in salt water for a few weeks or days. It is an important matter to demonstrate a power of resistance to the supposed noxious influence, of one, two, or three months' time.

Perhaps you in turn allow too much weight to the objections against migration over salt water, founded on the tendency of seeds to sink, & of plants to decay & sink.

Your experiments & observations are made in *still* water, of *small depth*, and small bulk.

1. Would not plants resist putrefaction much longer in agitated water, than in still water,—especially on the agitated surface of a sea, as compared with the still surface of a tub?

2. Take a glass of muddy water, from a turbid stream during flood. Speedily the fine particles of earth, &c. which make it muddy, will settle to the bottom. It would be contrary to fact, to assume from this, that streams, currents, tides, cannot carry muddy particles a long distance.

Before you can positively say, that the sinking of seeds in still water to a few inches, or few feet of depth, will prevent their crossing a sea, you must be convinced that they will sink so deep as to fall below motion in the sea. Not an easy matter to establish; altho' the deeper they go, the less likely to get ashore again,— if not at the bottom.

Millions upon millions of seeds are carried to the sea yearly. Vast numbers of these must become entangled or resting among Algæ. Might not a sea weed occasionally float a land seed over a great extent of marine surface?— This seems to me as likely as timber floats, or whole plants from the land.

Sincerely yours | Hewett C. Watson

DAR 205.3 (Letters)

CD ANNOTATIONS
0.3 My . . . tub? 5.3] *crossed pencil*
7.1 Before . . . bottom. 7.5] *crossed pencil*
8.2 Might . . . surface?— 8.3] *double scored pencil*
Top of first page: '18'[2] *brown crayon*
Bottom of last page: 'Limosella' *pencil, del pencil*; '*Hawks* /Birds feet/ Earth in roots/' *pencil, del pencil*

[1] CD's paper reporting the results of his and Miles Joseph Berkeley's experiments 'On the action of sea-water on the germination of seeds' was read at the Linnean Society, 6 May 1856 (*Collected papers* 1: 264–73).
[2] The number of CD's portfolio of notes on the means of geographical dispersal of plants and animals.

To J. D. Hooker   11–12 November [1856][1]

Down Bromley Kent
Nov. 11th.—

My dear Hooker

I thank you more *cordially* that you will think probable, for your note.[2] Your verdict has been a great relief.— On my honour I had no idea whether or not you would say it was (& I knew you would say it very kindly) so bad, that you would have begged me to have burnt the whole. To my own mind my M.S relieved me of some few difficulties, & the difficulties seemed to me pretty fairly stated, but I had become so bewildered with conflicting facts, evidence, reasoning & opinions, that I felt to myself that I had lost all judgment.— Your general verdict is *incomparably* more favourable than I had anticipated.

Very many thanks for your invitation: I had made up my mind on my poor wifes account not to come up to next Phil. Club; but I am so much tempted by your invitation, & my poor dear wife is so goodnatured about it, that I think I shall not resist, ie if she does not get worse.— I w^d come to dinner at about same time as before, if that w^d suit you & I do not hear to contrary, & w^d go away by the early train ie about 9 olock.— I find my present work tries me a good deal & sets my heart palpitating, so I must be careful.— But I sh^d so much like to see Henslow, & likewise meet Lindley if the fates will permit.[3] You will see, whether there will be time for any criticism in detail on my M.S. before dinner. Not that I am in the *least* hurry, for it will be months before I come again to Geograph. Distrib.; only I am afraid of your forgetting any remarks.—

I do not know whether my very trifling observations on means of distribution are worth your reading, but it amuses me to tell them.

The seeds which the Eagle had in stomach for 18 hours looked so fresh that I would have bet 5 to 1 they would all have grown; but some kinds were **all** killed & 2 oats 1 Canary seed, 1 Clover & 1 Beet alone came up! now I sh^d have not cared swearing that the Beet w^d not have been killed, & I sh^d have fully expected that the Clover would have been.— These seeds, however, were kept for 3 days in moist pellets damp with gastric juice after being ejected which would have helped to have injured them.—[4]

Lately I have been looking during few walks at excrement of small birds; I have found 6 kinds of seeds, which is more than I expected. Lastly I have had a partride with 22 grains of dry earth on *one* foot, & to my surprise a pebble as big as a tare seed; & I now understand how this is possible for the bird scartches itself, & little plumose feathers make a sort of very tenacious plaister. Think of the millions of migratory quails, & it w^d be strange if some plants have not been transported across good arms of the sea.—[5]

Talking of this, I have just read your *curious* Raoul Is^d paper:[6] this looks more like a case of continuous land, or perhaps of several intervening, now lost, islands, than any, (according to my heteredox notions) I have yet seen; the concordance of the vegetation seems so complete with New Zealand & with that land alone.

I have read Salters paper, & can hardly stomach it: I wonder whether the lighters were ever used to carry grain & Hay to ships?—[7]

Adios, my dear Hooker, I thank you most honestly for your assistance,— assistance by the way now spread over some dozen years.—

Farewell | C. Darwin

P.S. Wednesday

I see from my wife's expression that she does not really much like my going, & therefore I must give up of course this pleasure.— If you sh^d have anything to discuss about my M.S. I see that I c^d get to you by about 12, & then c^d return by the 2° 19' olock train & be home by 5½ oclock, & thus I sh^d get 2 hours talk.— But it would be a considerable exertion for me, & I would not undertake it for mere pleasure sake, but would **very gladly** for my Book's sake.—

DAR 114.3: 181

[1] Dated by the relationship to the letter from J. D. Hooker, 9 November 1856.

[2] Letter from J. D. Hooker, 9 November 1856.

[3] See letter from J. D. Hooker, 9 November 1856, in which Hooker invited CD to dinner on Wednesday, 12 November, to meet John Lindley and John Stevens Henslow, or on Friday 14 November, to meet John Tyndall and Henslow. CD attended neither dinner but did go up to London on 13 November (see following letter).

[4] See letter to J. D. Hooker, [19 October 1856] and n. 2.

[5] CD recorded this case on 19 October 1856 in his Experimental book, p. 15 (DAR 157a). Following the entry, CD added: '(Nov. 13th. Nothing came up.)'.

[6] J. D. Hooker 1857.

[7] James Salter had reported that mud scraped from the bottom of Poole harbour in 1843 and deposited on the shore eventually gave rise to a vegetation different from that of the surrounding area (J. Salter 1857).

To George Howard Darwin and W. E. Darwin[1] 13 [November 1856][2]

Down
Thursday 13th

My dear Willy & Georgy.

I have thought that you would like to hear about poor Aunt Sarah's funeral.[3] Aunt Elizabeth, & Uncles, Jos, Harry, Frank, Hensleigh, & Allen all attended, so that we had the house quite full.[4] The Funeral was at 3 olock, & Mr Lewis managed it all.[5] We walked down to Petleys & there all put on black Cloaks & crape to our hats, & followed the Hearse, which was carried by six men; another six men changing half way.— At the Church Door Mr Innes came out to meet the Hearse.[6] Then it was carried into the Church & a short service was read. Then we all went out, & stood uncovered round the grave whilst the Coffin was lowered, & then M^r. Innes finished the service, but he did not read this very impressive service well. Hemmings, M^rs. Moray & Martha attended & seemed to cry a good deal.—[7] Then we all marched back to the House, Mr Lewis & his two sons carrying a sort of black standard before us; & we then went into the House & read Aunt Sarah's will aloud. She desired her Funeral to be as quiet as possible, & that no tablet should be erected to her. She has left a great deal of money to very many Charities.—[8]

All the Uncles are gone away today. And I am going to London today, so cannot write anymore.[9] Do you Georgy send this letter soon on to Willy.

Hemmings & the maids will stay here about a month more I sh^d think; so that you Georgy will see them again, but I fear Willy will not at present; but no doubt he will some time when visiting Barlaston.— [10] They have behaved admirably toward Aunt Sarah; & she has left them a little money.—

My very dear Boys, Your affect. Father | C. Darwin

DAR 210.6

[1] The letter indicates that CD sent it first to George Darwin, at school in Clapham, who was instructed to forward it to William Darwin, then at Rugby School.

[2] Dated by the reference to the funeral of Sarah Elizabeth (Sarah) Wedgwood, who died on 6 November 1856.

[3] Sarah Wedgwood, CD's and Emma's aunt, was the unmarried sister of Josiah Wedgwood II. She had moved to Petleys, a house in Down village, in 1847 (see *Correspondence* vol. 3, letter to Emma Darwin, [24 June 1846] and *Emma Darwin* 2: 105). The funeral had taken place on 12 November 1856 (Emma Darwin's diary).

[4] These uncles were Emma's brothers and her cousin, John Allen Wedgwood. Aunt Elizabeth was Emma's older sister, Sarah Elizabeth (Elizabeth) Wedgwood.

[5] John Lewis was the carpenter and undertaker in Down village.

[6] John Innes, perpetual curate of Down.

[7] Sarah Wedgwood's servants, of whom the Darwin children were very fond. Henrietta Litchfield spelled 'Moray' as 'Morrey' when she recalled: 'Mrs Morrey's gingerbread was like no other we have ever tasted before or since, and Martha would sing us songs which only gained by repetition.' (*Emma Darwin* 2: 106).

[8] Henrietta Litchfield later recalled that Sarah Wedgwood 'lived in her books, and the administration of her charities, and her only society was that of my mother and a few old friends and relations. She had no gift for intercourse with her neighbours, rich or poor, and I do not believe ever visited in the village.' (*Emma Darwin* 2: 105).

[9] Although CD had told Joseph Dalton Hooker that he would not be able to dine with him (see preceding letter), CD did go to London for the day on 13 November, as proposed in the postscript in the preceding letter. His name appears in the Royal Society Philosophical Club minutes as having attended the 13 November meeting, and the expenses for the trip were recorded in CD's Account book (Down House MS) on 15 November 1856.

[10] The home of Francis Wedgwood near the family pottery works at Etruria, Staffordshire.

From W. F. Daniell 14 November 1856

> 25. Great Russel Street | Bloomsbury.
> Nov. 14. 1856.

My dear Sir.

I have just received your letter and regret I did not give you my town address however, my general Agents, McGregors. *17 Charles St. St James Square*, will always give it you when I am either in England or abroad.[1] I write this letter, without any delay, as I think I can afford you some information about the *mamalia*, of St Thomas, & Princes— I have also to state the pleasing news, that I have received a letter from Africa, from H.M.S. Scourge,[2] stating that Mr Gabriel of the mixed commiss. on coast,[3] is making an extensive collection of fowls &c for you— I suppose ere this, you will have received information of this collection.[4]

In looking over some old notes, extracted from a Portuguese history of these islands, I find that numerous monkeys & civit cats (the same probably as those of Fernando Po and the main land)—belong to these localities. The following are the precise words in Portuguese— *"A unica especie do genero Mamalia, que se achou nestas Ilhas ao tempo do descobrimento eram macacos de differentes castas, e muitos ratos assas daninhos. Os Portuguezes alli introduziram logo gado, vaccum, lanigero cabrum, e cavallar, o qual propagou sufficientemente, e mais que tudo as cabras"* — *"Varias viverras se acoutam tambem nessas matas entre ellas uma especie de gato dalgalia ou viverra civetta. Lagartos, lagartixas sapos acham-se por toda a parte, e dos amphibios a rãa, e o cágado—e nas praias destas Ilhas sahem muitas tartarugas, de que a casca se aproveita para o commercio por ser da melhor qualidade"*[5]

I hope this long quotation may afford you a slight insight I will however keep you memo in my sight. It is most difficult to gain any information on these subjects. Fernando was inhabited by a black race of men, supposed originally to have passed over from [Camarões][6] river, as there are several words in both languages the same— It is not unlikely I may go abroad again early in next year, and if I can give you any information or make inquiries into any particular subject for you, I will do all I can in these respects.

I remain my dear Sir | ever yours sincerely | W. F. Daniell

C. Darwin Esq— | &c &c

The island of St. Thomas' was discovered in 1470 on Dec 21. by Joān de Santarem and Pedro de Escobar. It was colonized in September 1485

Navegação de Lisboa a Ilha de St Thomes e Principe por um Piloto Portuguese about 1500— published

DAR 205.3 (Letters)

CD ANNOTATIONS
Top of first page: '19'[7] *brown crayon*
After valediction: 'Is it volcanic? Wooded?' *ink*; 'Are there such animals now?' *ink, circled ink*; 'Was St. Thomas inhabited? | [Sandrige] off coast far out' *ink*; 'Is it certain that F. Po was not meant.' *ink*; 'St. T. about as big as Madeira; about 160 miles from Mainland' *ink*
After postscript: 'This is the authority for J. de Limas'

CD note:

Discovered about [*interl*] 1470 by Santarem, believed St Thomas uninhabited. when discovered. ([*added ink*] 32 × 8 *miles in size [*interl ink*]— 66 miles S.W. by W of Princes | J. de Lima.— *Statistical work [*added ink*] *in 1550 [*ink over pencil*] exported [*interl ink*] *much sugar [*ink over pencil*] exported.' *pencil*; 'I must give this as hostile case, saying species unknown.— & remarking how easily isl^{ds} are colonised.—[8] *ink*

---

[1] CD's letter, probably written after he had received the letter from W. F. Daniell, 8 October – 7 November 1856, has not been found. Charles Roderic and Walter McGrigor were army agents at 17 Charles Street, St James's Square (*Post Office London directory* 1857).
[2] HMS *Scourge* was serving off the west coast of Africa in 1856 (*Navy list* 1856).

[3] Edmund Gabriel was an anti-slave-trade commissioner in Luanda, Angola. See also *Correspondence* vol. 5, CD memorandum, [December 1855].

[4] See letters to W. B. Tegetmeier, 19 November [1856] and 29 November [1856].

[5] 'The only mammalian species found on these islands when they were discovered were monkeys of various types and many rather noxious rats. The Portuguese immediately introduced there herds of cattle, wool-bearing goats, and horses, which multiplied adequately, especially the goats' — 'Various *Viverra* also find shelter in the forests there, among them a type of civet-cat or *Viverra civetta*. Large and small lizards and toads are found everywhere, and of the amphibians the frog and the fresh-water turtle—and on the shores of these islands appear sea turtles whose shells are of use commercially, being of the best quality'. The editors thank Mario di Gregorio for this translation.

[6] Presumably the Cameroons estuary, originally named in Portuguese 'Rio dos Camarões'.

[7] The number of CD's portfolio of notes on the geographical distribution of animals.

[8] CD had inquired about the mammals on the islands of Principe and São Tomé, some 200 miles off the coast of Gabon in the Gulf of Guinea, and about the depth of the sea separating the islands from the mainland (see letter from W. F. Daniell, 8 October – 7 November 1856). The statistical work referred to is Lopes de Lima 1844–62. The ability of monkeys, reptiles, and frogs, for example, to cross open seas (see n. 5, above) was difficult to explain by natural means. See also following letter.

To J. D. Hooker 15 November [1856][1]

Down Bromley Kent
Nov. 15th—

My dear Hooker

I shall not consider all your notes on my M.S. for some weeks, till I have done with crossing; but I have not been able to stop myself meditating on your powerful objection to mundane cold period, viz that **many** fold more of the warm-temperate species ought to have crossed the Tropics that of the sub-arctic forms.—[2] I really think that to those who deny modification of species, this would *absolutely* disprove my theory. But according to the notions which I am testing, viz that species do become changed & that time is a *most* important element (which I think I shall be able to show very clearly is the case) in such change, I think the result would be as follows. Some of the warm-temperate forms would penetrate the Tropics long before the subartic, & some might get across the Equator long before the sub-arctic forms could do so, (ie always supposing that the cold came on slowly) & therefore they must have been exposed to new associates & new conditions much longer than the sub-arctic. Hence I sh^d infer that we ought to have in warm temperate S. hemisphere more representative or modified forms, & fewer identical species than in comparing the colder regions of the N. & S.— I have expressed this very obscurely, but you will understand, I think, what I mean.— It is a parallel case, (but with greater difference) to the species of the Mountains of S. Europe compared with the artic plants. The S. European alpine species having been isolated for a longer period than on the arctic islands. Whether there are *many* tolerably close species in the warmer temperate lands of the S. & N. I know not; as in La Plata, C. of Good Hope, & S. Australia compared to the North, I know not.— I presume it would be very difficult to test this; but perhaps you will keep it a little before your mind. For your argument strikes me as *by far* the most serious difficulty which has occurred to me.— All your critisms & approvals are in simple truth *invaluable* to me.—

I fancy I am right in speaking in this note of the species in common to N. & S as being rather sub-arctic than arctic.—

This letter does not require any answer: I have written it to ease myself & to get you just to bear your argument under the modification point of view in mind.—

I have had this morning a most cruel stab in the side on my notion of distribution of Mammals in relation to soundings.[3]

My dear Hooker | Yours most truly | C. Darwin

DAR 114.3: 182

[1] Dated by the relationship to the letter from J. D. Hooker, 9 November 1856.
[2] A point presumably made by Hooker in person since it is not addressed in either the letter from J. D. Hooker, 9 November 1856, or the memorandum transcribed after that letter itemising Hooker's queries about CD's manuscript on geographical distribution.
[3] See preceding letter.

From J. D. Hooker [16 November 1856][1]

Dear Darwin

I write only to say that I entirely appreciate your answer to my objection on the score of the comparative rarity of Northern warm-temperate forms in the Southern Hemisphere.[2] You certainly have wriggled out of it by getting them more time to change, but as you must admit that the distance traversed is not so great as the Arctics have to travel & the extremes of modifying cause not so great as the Arctics undergoe, the result should be considerably modified thereby.

Thus

The Sub Arctics have 1) to travel twice as far, 2) taking twice the time, 3) undergoing manyfold more disturbing influences.—

All this you have to meet by giving the North temp. forms simply more time—I think this will hardly hold water.

Ever Yrs | Jos D Hooker

Kew Sunday

DAR 100: 162

CD note:[3]

In answer to this show from similarity of American & European & Alp. Arctic plants, that they have travelled enormously without any change.[4]

As *sub-arctic*, temperate & Tropical are all slowly marching towards the equator, the Tropical will be first checked & ['then' *del*] distressed, then *some stray [*interl*] the temperate will invade, [*illeg*] to height or [*even*], *then come the struggle for life & death legion after legion in the long line of march from the far North [*added pencil*] after ['wards' *del*] the temperate can advance or do not wish to advance further, the arctics will be checked & will invade— The temperate will hence be far longer in Tropics than sub-arctic— The subartic will be first here to cross temperate & then Tropics.—

They w^d penetrate amongst strangers just like the many naturalised plants brought by man, from some unknown advantage— But more, for nearly [*interl*] *all* have Chance of doing so—

[1] The Sunday after the letter to J. D. Hooker, 15 November [1856].

[2] See preceding letter.

[3] This note follows the letter in DAR 100: 163.

[4] In *Natural selection*, pp. 548–9, CD maintained that floras generally migrated in a body and did not experience different selective pressures until encountering a new of mixture of plants and animals. See letter to J. D. Hooker, 18 November [1856], and J. Browne 1983.

To John Murray 17 November [1856–7][1]

Down Bromley Kent
Nov. 17th.—

Dear Sir

I find that I have for little civilities in Natural History often occasion to give away copies of my Journal.[2] I hope that you will not think it an unreasonable favour to beg to be supplied with 4 copies in cloth at the price charged to wholesale Booksellers.— If you can so far oblige me, would you be so kind as to have the parcel sent, anytime before Thursday morning.[3] (with account enclosed) addressed thus

> C. Darwin Esq^e
> care of Mr G. Snow
> Nag's Head
> Borough.—

I do not know whether you keep any account of *total* sales of Books, but if so I sh^d. very much like to hear how many copies of my Journal have been sold altogether, if you would be so kind as to direct one of your clerks to put the number on a slip of paper, & enclose it with Books, if you are willing to let me have them; apologising for this trouble | I remain Dear Sir | Your's very faithfully | Charles Darwin

John Murray Esq^e

John Murray Archive

[1] See letter to John Murray, 20 November [1856–7], n. 1, for the basis of this date.

[2] The second edition of *Journal of researches* was published by John Murray (see *Correspondence* vol. 3).

[3] Thursday was the day that the carrier left London for Down village.

To J. D. Hooker 18 November [1856][1]

Down Bromley Kent
Nov. 18th

My dear Hooker

I send enclosed, received this morning.—[2] I send my own,, also, as you might like to see it; please be sure return it.—[3] As the facts about N. range are quite invaluable for me for my theory of transport to America. If your letter is *Botanical & has nothing private*, I sh^d like to see it. I do not know whether I ought to send to you his to me; as you will see there is a little rap for you. But, as I know full well, you are not thin-skinned & can stand a blow (& by Jove return it) as well as any man, I send it.—

Many thanks for your note received this morning, & now for another "*wriggle*"[4] According to my notions, the sub-arctic species would advance in a body, advancing so as to keep climate nearly the same, & as long as they did this, I do not believe there would be any tendency to change, but only when the few got amongst foreign associates. When the tropical species retreated as far as they could to the equator, they would halt, & then the confusion would spread back in the line of march from the far north & the strongest would struggle forward &c &c (But I am getting quite poetical in my wriggles) In short *I think* the warm temperate would be exposed very much longer to those causes which I believe are *alone* efficient in producing change than the sub-arctic; but I must think more over this, & have a good wriggle I cannot quite agree with your proposition that because the sub-arctic have to travel twice as far, they w^d be more liable to change. Look at the two Journeys which the Arctics have had from N. to S. & S. to North, *with no change*, as may be inferred, if my doctrine is correct, from similarity of Arctic species in America & Europe & in the Alps.— But I will not weary you; but I really & truly think your last objection is not so strong as it looks at first. You never make an objection without doing me *much* good.—

Hurrah a seed has just germinated after 21½ hours in Owls stomach. This according to ornithologists calculation w^d carry it, God knows how many miles; but I think an owl really might go in storm in this time 400 or 500 miles.—[5]

Adios | C. Darwin
Owls & Hawks have often been seen in mid Atlantic.

DAR 114.3: 183

[1] Dated by the relationship to the letter from Asa Gray, 4 November 1856 (see n. 2, below).
[2] A letter to Hooker from Asa Gray, sent by Gray to CD enclosed in the letter from Asa Gray, 4 November 1856.
[3] Letter from Asa Gray, 4 November 1856.
[4] See letter from J. D. Hooker, [16 November 1856].
[5] This experiment was recorded in CD's Experimental book, p. 17 (DAR 157a).

To J. W. Lubbock 18 November [1856][1]

Down.
Nov. 18th

Dear Sir John Lubbock

I am much obliged for your note, which I will forward to Mr Wedgwood. I will speak about the Cards of Admission.—[2]

I see it was an entire blunder of A. Woods about your wanting the Furniture.[3]
Pray believe me | Yours sincerely | Ch. Darwin

Royal Society (LUB: D24)

[1] Dated on the basis that the letter refers to the auction of Sarah Elizabeth (Sarah) Wedgwood's effects (see n. 2, below).
[2] This is probably a reference to admission to the auction of Sarah Wedgwood's house and its contents. The auction was to be held on 9 December (see letter to W. E. Darwin, 25 [November 1856]).
[3] A. Wood was probably overseeing the sale of Sarah Wedgwood's house and effects. CD's Account book (Down House MS) records a payment of £4 to him, on behalf of Josiah Wedgwood III, on 20 November 1856.

To John Higgins 19 November [1856][1]

Down Bromley Kent
Nov. 19th

My dear Sir

I am very glad to hear of the increased Rent, which will make my Farm a very good investment; & I truly thank you for your kind attention to my interests.[2]

With respect to the £105, I sh[d] prefer your second plan of a Lien for 15 years; but if you think the other plan of payment with interest to myself at 5 per cent, the fairer plan to M[r] Hardy (who seems so excellent a tenant) I have the money & could transmit it immediately; but for myself, as I have said, I sh[d] prefer the lien.—[3]

With my very sincere thanks, pray believe me my dear Sir | Your's very faithfully | Ch. Darwin

To | John Higgins Esq[e]

Lincolnshire Archives Office

[1] Dated by the reference to the rent increase on the Beesby farm (see n. 2, below).
[2] John Higgins was the agent for CD's farm near Beesby, Lincolnshire. According to his Investment book (Down House MS), CD received £215 6s. 5d. rent on his farm on 12 June 1856 and a payment of £238 18s. 8d. on 19 December 1856. Subsequent payments continued to be made at around the higher level.
[3] Francis Hardy was CD's tenant. A lien was the right to retain possession of another person's property until the owner paid a debt.

To W. B. Tegetmeier 19 November [1856][1]

Down Bromley Kent
Nov. 19th

My dear Sir

I sh[d] very much like to hear your paper on Dec[r] 9th,[2] but at present owing to the state of M[rs] Darwin's health it is impossible for me to leave home. This has prevented my attendance at the Philo-perist. last time & it will next time.

What a very odd & curious fact about the colours & absence of down—it is quite new to me: I fear that I have no birds of these colours matched to observe it in. It really strikes me as very curious, & I shall be curious to ask M[r] H. Weir how often he has observed it, & in what breeds; but this I can do at some future Philo-peristeron.—[3]

Your confirmation seems a strong one. Unfortunately I killed but the other day two yellow Tumblers.— Many thanks for your *kind* offer of Brunswicker, but I do not think the Breed distinct enough to be worth crossing.—

No news yet of the Persian Pigeon;[4] but I have just heard of large collection of skins for me from further S. on W. coast of Africa.[5]

Yours very sincerely | Ch. Darwin

I hope that you are getting on well with your Poultry Book; I have seen only the 1st no[r], as my Bookseller has neglected to send them; but I must touch him up.—[6]

New York Botanical Garden Library (Charles Finney Cox collection)

[1] Dated by the reference to Emma Darwin's expected confinement and to the letter from W. F. Daniell, 14 November 1856.

[2] The paper has not been identified, although it was almost certainly the same as Tegetmeier 1856, an account of the abnormal development of the skull of Polish fowls, which was delivered at the Zoological Society on 25 November 1856. It was probably read at a meeting of the Philoperisteron Society of pigeon fanciers. Tegetmeier later published an abstract of Tegetmeier 1856 in *Poultry Chronicle* (see *Cottage Gardener* 17 (1856–7): 284).

[3] Harrison Weir was a pigeon fancier and painter of animals. He did the illustrations for Tegetmeier's edition of the *Poultry book* (see n. 6, below). CD asked Tegetmeier for this information again in letter to W. B. Tegetmeier, [18 June 1857].

[4] These pigeons were also mentioned in letter to W. B. Tegetmeier, 3 November [1856]. See letter to W. B. Tegetmeier, 29 November [1856], in which CD reported their arrival.

[5] See letter from W. F. Daniell, 14 November 1856.

[6] Tegetmeier ed. 1856–7.

From H. C. Watson 19 November 1856

Thames Ditton
Nov. 19. 1856

My dear Sir

I had occasionally noticed common earth completely embedded within roots, & on one occasion had grounds *almost* beyond a doubt, to suppose that embedded seeds had germinated on exposure,— still, the idea of this, as a mode of navigation for seeds, never occurred to me until reading your letter,[1] yet it is fairly within possibility, & perhaps even probability as a rare event.

In alluding to still water as hastening decay, I had in thought your remarks on the upper part of page 135,—not the germinating power of the seeds.[2] It seems likely that plants would float longer, & thus be carried farther, on moving water, than might be inferred from their early decay & sinking in still water.—

As to the London Catalogue.—[3] For the convenience of sorting duplicates, labelled accordingly, the Nos. prefixed to the specific names in the first edition have been kept on; the changes being usually interpolated by the Numeral repeated with an *.

I have run over the last, 4th, edition & get these results.—

Species by Numerals 1428
Numbers interposed 132
      1560
Numbers dropped out 17
Actual Number 1543

On reading over these 1543 specific names, I think about 200 of them represent 'species', which have either been treated expressly as varieties, or have not been distinguished from their nearest allied species, by Authors publishing more lately than the age of Linneus. Nearly 40 (36, it appears) of these belong to the one very unsettled genus Rubus.

You will recollect that all the names of varieties in the 4th edition of the Catalogue, with Comparatively few exceptions (chiefly in Salix & Rubus), also represent 'species', so considered by some Authors.

In the lump, we may say of the Catalogue, that it includes over 1800 names of plants, which are (or recently have been) deemed species by one or more botanical writers of some authority. And that among these 1800 names of so supposed species, there are over 450 which represent what other botanical writers call varieties, not species.[4]

You will better see what I would convey by the alternative description, 'treated expressly as varieties or not distinguished from near species', if you will turn up the genus Ranunculus, in the first column of the Catalogue. Less than a century ago, all the species above Ficaria were held two only:—

R. aquatilis.
R. hederaceus.

From the first of these, many years ago, were separated

R. circinatus.
R. fluitans.

Opinions being much divided & long wavering, whether these two latter should be held species or varieties. The great majority, but not yet all botanical systematists, now admit them as species. D[r] J. D. Hooker apparently holds them varieties.

A few years ago, two others were separated from R. hederaceus:—

R. tripartitus.
R. cœnosus.

The former of these two is seemingly very local, & may not have ever been noticed apart from hederaceus (or even aquatilis, for it stands between) in England, until taken up as a species, & identified with a Continental one.— But R. cœnosus is widely & plentifully spread through England, & must have been often & constantly seen, & passed by *as* R. hederaceus, scarcely a variety, less than 20 years ago.

Still more lately, R. confusus has been cut off from R. aquatilis. And now, there are two or three others suggested, as similar "splits".

You will see from this, that no clear separation can be made between "species" treated as varieties, & "species" not distinguished formerly.

I believe such instances as this may be found; the dates fanciful for the moment:—

　　1700 a. species
　　　　　b. species
　　1750 a. species
　　　　　b. variety
　　1800 ab species (sole)
　　1850 a. species
　　　　　b. variety
　　1856 a. species.
　　　　　b. species.

The short truth is, that we have no real proof or test of a species in botany. We may indeed occasionally **dis**prove an alleged species by seeing its descendants become another such species,—or we may **unite** two by finding a full series of intermediate links. Many botanists assume all describable forms, if not (or until) so disproved or united, to be distinct species, ab initio ad finem.

I am making a sad scrawl, & just as the thoughts come of themselves, evidencing it to be a "rough draft".—

Yours very truly | Hewett C. Watson

C. Darwin | Esq

DAR 98: 7–10

CD ANNOTATIONS
8.4 species] '7' *added pencil*
Top of letter: 'All used' *pencil, circled pencil*
Whole letter crossed pencil

[1] CD's letter has not been found. It was probably a reply to the letter from H. C. Watson, 10 November 1856.
[2] Watson refers to CD's paper 'On the action of sea-water on the germination of seeds' published in the *Journal of the Proceedings of the Linnean Society (Botany)* (see *Collected papers* 1: 268).

³ The *London catalogue of British plants* (Watson and Syme eds. 1853). CD had corresponded with Watson in 1855 about this work (see *Correspondence* vol. 5).
⁴ Watson's figures from Watson and Syme eds. 1853 are cited in *Natural selection*, pp. 112–13.

To John Murray 20 November [1856–7][1]

Down Bromley Kent
Nov. 20

My dear Sir

I am extremely much obliged for your very kind present & note.—[2] I assure you that I had no intention of begging, but as I have given away during the last year or two a dozen copies, I thought the difference of price for the future was worth saving.— Your present is very kind & most acceptable.—

I hope that you are pretty well satisfied with the sale: it is a good deal more than I had anticipated.

Pray believe me | My dear Sir | Yours truly obliged | Charles Darwin

P.S. | I apologise for having given you the trouble to return me a note put in the wrong cover: this note has been returned to me from Devonshire. It was a very stupid blunder.—

John Murray Archive

¹ The date of this letter and the letter to John Murray, 17 November [1856–7], is based on the 'Down Bromley Kent' address, which was in use in the second half of 1853 and from October 1855. If Murray had sent copies of CD's *Journal of researches* 2d ed. on the Thursday carrier, as requested in letter to John Murray, 17 November [1856–7], they would have been received on Thursday, 20 November 1856 or Thursday, 19 November 1857. Either possibility would fit well with both letters.
² See letter to John Murray, 17 November [1856–7].

From Charles Cardale Babington 22 November 1856

St John's College | Cambridge
22 Nov. 1856

Dear Darwin

I do not now remember to have ever gathered the *Subularia* otherwise than totally submersed.[1] Koch (Syn. Fl. Germ.) describes the flowers as they appear when open above the water & so I think that we may fairly believe that it does flower above the water—at least sometimes.[2] On thinking again, I strongly suspect that I have seen it out of water at the edges of Welsh lakes when they had run low. I cannot be quite sure either that I have or have not seen the aerial flowers.

I have not been able to find any anonymous book upon Pigeons in the University Library.[3] The word is in the Catalogue and refers to a class that has been "broken up many years since" and no trace of the book is to be found. The officials think, after consulting all the probable records in their possession that the book is not now in the library. The Catalogue does not describe the book.

Yours truly | Charles C. Babington—

I have very seldom seen *Limosella* growing, but believe that it is an aerial flowerer.

DAR 207 (Letters)

[1] *Subularia aquatica*, a hermaphroditic water-plant, was thought to flower under water with the corolla closed, which would make crossing with another individual impossible. CD, who was at this time writing his chapter on the crossing of animals and plants (*Natural selection*, pp. 35–91), doubted that such 'eternal and absolute' hermaphrodites existed and was investigating all instances of alleged cases that he could find.

[2] Koch 1843–4. The work is cited in *Natural selection*, p. 63, and in a note CD further commented: 'I am indebted for this reference to Mr Babington & to Mr H. C. Watson'. See also letter from H. C. Watson, 26 November 1856, and letter to J. D. Hooker, [early December 1856].

[3] CD had probably asked Babington to look for [J. Moore] 1765 in the Cambridge University Library.

From J. D. Hooker   22 November 1856

Nov 22/56 Saty

Dear Darwin

Thanks for your letter & its enclosure from A. Gray[1] which contains nil Botanices & is a reminder that I owe a collector of his for plants purchased by Thomson at Calcutta—[2] I do expect a Botanical letter ere long & will send it you if it contains any interest.

Your arguments have certainly some force, but the interferences of space, time & temperature are becoming so complicated, that really it makes me almost giddy. There is one point that appears to me a fundamental one, which is, that in transporting the Sub Arctic sp. across the tropics you expose them to more extreme conditions than any other plants can ever be subjected to— they suffer more change of climate, & of association than any others, & they ought to be proportionally the most altered; I must confess that however much you may modify the effects of such operations, I do not see how you can subvert the first obvious deduction from such facts. I would almost rather allow them an expansive power, a sort of absolute amount of specific character that however much masked & so to speak *latent* in the tropical part of their course, becomes apparent again in them as they reach subantarctic regions— this is horrible I know to you. Much depends however upon preliminary points not yet settled— What are your causes *alone* efficient?—

You cannot but allow of Time + altered conditions + altered associations—& that all these are more or less convertible terms.— You know that I think there are species now existing in all regions from subarctic to subantarctic, but where Trop. specimens are not recognized as identical specifically with the subarctic & subantarctic—which latter *are* recognized as identical.

Do send your notice (if only 2 lines) of owl seed germinating to Linnæan— the L. Journal should be the deposit for all such isolated facts.[3]

Thanks for A Gray's letter. I do rub my hands & chuckle (like Lyell) at the happy idea of my being caught in a *Paradox*—[4] I know the human soul loves paradox, even to miracle, & that this love of it is one of the curses of science, but Lord bless you my dear Darwin it is the greatest paradox in the world to think of Conifers as any thing but very high in the Vegetable Kingdom.— Like[5]

The Seeds you sent were Raspberry.—[6]

AL incomplete
DAR 100: 111–12

CD ANNOTATIONS
0.2 Dear . . . giddy. 2.3] *crossed pencil*
2.3 There] *after* '[' *added pencil*
2.5 they suffer . . . altered; 2.7] *double scored pencil*
2.12 I know] *del pencil*; 'I know' *added pencil*
3.2 all these . . . terms.—] *scored pencil*; 'No' *added pencil*
4.1 Do send . . . Raspberry.— 5.1] *crossed pencil*; 'I agree about D. C. not having given you credit.—' *added pencil*

[1] See letter to J. D. Hooker, 18 November [1856] and n. 2.
[2] Thomas Thomson was superintendent of the botanical garden in Calcutta.
[3] See letter to J. D. Hooker, 18 November [1856]. CD did not communicate this information (see following letter).
[4] See letter from Asa Gray, 4 November 1856, which CD had sent to Hooker to read.
[5] The remainder of the letter is missing. The concluding sentence about raspberry seeds was written above the salutation.
[6] CD recorded an experiment in his Experimental book, p. 15 (DAR 157a), in which he planted the seeds found in birds' dung. The seeds were identified by CD as 'plenty of Thorn, *(Briony I am almost sure) [*interl*] Yew, Laurel, & 3 other kinds.' Above '3 other kinds' he has added the comment: 'Raspberry'.

To J. D. Hooker 23 November [1856][1]

Down Bromley Kent
Nov. 23^d

My dear Hooker

I fear I shall weary you with letters; but do not answer this, for in truth & without flattery, I so value your letters, that after a heavy batch, as of late, I feel that I have been extravagant & have drawn too much money & shall therefore have to stint myself on another occasion.—

When I sent my M.S. I felt strongly that some preliminary questions on causes of variation ought to have been sent you. Whether I am right or wrong in these points is quite a separate question, but the conclusion which I have come to, quite independently of geographical distribution, is that external conditions (to which naturalists so often appeal) do by themselves *very little*. How much they do is the point of all others on which I feel myself very weak.— I judge from facts of variation under domestication, & I may yet get more light. But at present, after drawing up a rough copy on this subject, my conclusion is that external conditions do *extremely* little, except in causing mere variability. This mere variability, (causing the child *not* closely to resemble its parent) I look at as *very* different from the formation of a marked variety or new species.— (No doubt the variability is governed by laws, some of which I am endeavouring very obscurely to trace).— The formation of a strong variety or species, I look at as almost wholly due to the selection of what may be incorrectly called *chance* variations or variability. This power of selection stands in the most direct relation to time, & in state of nature can be only excessively slow.— Again the slight differences selected, by which a race or species is at last formed, stands, as I think can be shown (even with plants & obviously with animals) in far more important relation to its associates than to external conditions.— Therefore, according to my principles, whether right or wrong, I cannot agree with your proposition that Time + altered conditions + altered associates are "convertible terms". I look at first & last as **far** more important;—time being important only so far as giving scope to selection.— God knows, whether you will perceive at what I am driving.— I shall have to discuss & think *more* about your difficulty of the temperate & sub-arctic forms in S. hemisphere, than I have yet done.— But I am inclined to think I am right (if my general principles are right) that there would be little tendency to the formation of a new species, during the period of migration, whether shorter or longer; though considerable variability may have supervened.—

I will think whether to send notice to Linn. Journal about Owl seed; I almost doubt whether worth it.— Thanks for answer about Raspberry seeds.— [2] Where the Deuce can the Birds **now** get so many from? I have now found 9 kinds in Dung.—

I always thought from beginning to end, that D. C. had not done you justice; & independently of justice, that he was a goose not more to have used your materials[3]

Adios | My dear Hooker | Your's | C. Darwin

DAR 114.3: 184

[1] Dated by CD's reference to his manuscript on geographical distribution (see letter from J. D. Hooker, 9 November 1856).
[2] See preceding letter, n. 6.
[3] CD refers to A. de Candolle 1855.

From John Obadiah Westwood 23 November 1856

My dear Sir

The descriptions given by Tellkampf of the insects found by him in the

Kentucky Cave are not sufficiently precise to enable us to determine the points in question with precision[1]

His adelops is however evidently identical generically with the European species of the same Genus ten of which have been already described as cave insects & several others not cave insects but found in damp dark places— We have one if not two of this Genus in England.— I do not know if the genus Adelops has been found elsewhere in America Neither can I tell whether the american Anophthalmus has been found out of the Caves or whether there is any other American Species of the same genus as I have not the last American Catalogue of Coleoptera at hand (Lacordaire only gives one American Anophthalmus. & no American Pristonychus.)[2]

I cannot do better than refer you to Dr Wallich's Translation of Schiodtes remarkable paper published in the Trans. Ent.^l Society New Ser. Vol. 1.—which I think from your letter you cannot have seen—[3]

As there have only been three or 4 American Cave insects descri^d it follows that, as at present known, many of the European cave genera have no American representatives All we can say is that Adelops & Anophthalmus occur especially in caverns & in both America & Europe.

My impression is that old coprophagous or at least fossorial insects blunt the points of the spines or teeth of their fore tibiæ with working—but I do not think they break off their tarsi ordinarily. There is in fact generally a place of lodgement for the tarsus on the inside of the extremity of the tibiæ

Believe me | Yours very sincerely | Jno O Westwood

Hammersmith | 23 Novr 1856

[Enclosure][4]

Lepdoderus Hohenwarti (Coleopt? new Fam.)[5]
—— angustatus
—— sericeus
Pristonychus[6] elegans *var.* Schreibersi is this blind?[7]
Blothrus Schiödtii (scorpion)[8]
Titanethes albus (oniscus like)
Polydesmus subteraneus
Niphargus stygius (allied to Polydesmus)[9]
Anurophurus stillicidii (do?)[10]
Phalangopsis caricola. Grasshopper
Cavicularia anophthalma (spider)
Phalangium cancroides (do)[11]
Adelops[12] Khavenhulleri[13] (Coleopt.)[14]
Anophtalmus stomoides (Coleopt.)

——— Bilimeki[15]
——— Scopolii[16]
F. J. Schmidt | "Laibacher Zeitung" | No^r 146. 4 Aug 1852.

DAR 205.3 (Letters)

CD ANNOTATIONS
2.3 not cave . . . places] *double scored pencil*
Top of first page: 'Cave Insects' *pencil*; '19'[17] *brown crayon*

[1] Tellkampf 1844a and 1844b. CD had previously asked James Dwight Dana for information on the fauna of the Kentucky caves, as described in Agassiz 1851 and Silliman 1851. See letter to J. D. Dana, 14 July [1856], and letter from J. D. Dana, 8 September 1856.

[2] Lacordaire 1854–75.

[3] Schiödte [1849] was translated and read as a paper by Nathaniel Wallich at a meeting of the Entomological Society of London on 6 January 1851 (Wallich 1851). The paper is cited in CD's discussion of cave animals in *Origin*, p. 138.

[4] The enclosure is in CD's hand and comprises a list of cave insects taken from a paper by Ferdinand Joseph Schmidt, as indicated at the bottom of the list. It has not been possible to trace CD's source. The list was evidently sent to Westwood at some earlier stage with a covering letter asking him to answer the questions and return it to CD. The list was annotated by Westwood (see nn. 5–16, below).

[5] Westwood deleted 'Coleopt? new Fam.' and bracketing the three species wrote in pencil: 'A genus of beetles forming a separate family quite distinct'.

[6] Westwood wrote in ink below this name: 'A British genus'.

[7] In reply to CD's query, Westwood wrote in pencil: 'not quite'.

[8] Westwood deleted 'scorpion' and wrote in ink: 'Q^y Blothrus spelæus Schiodte Cheliferidæ'.

[9] Westwood deleted '(allied to Polydesmus)' and added in pencil: 'an Amphipod'.

[10] Westwood deleted '(do?)' and added in pencil, partially overwritten in ink: 'Poduridæ'.

[11] Westwood circled this entry and added in ink: 'Q^y Chelifer caneroides?— Q^y Blothrus spelæus'.

[12] Westwood here interlined 'Tellkampf' in pencil.

[13] Under this name, Westwood wrote '= G. Bathyscia Schiodte—' in pencil.

[14] Westwood added in pencil: '10 Species now known'.

[15] Westwood added in pencil: 'Carabidæ'.

[16] Following this list, Westwood wrote in pencil: 'A considerable N°. of new additional cave insects have since been published in the Stettin Zeitung & the Proceedings of the Zool–Bot Verein of Vienna.' He refers to the journals *Entomologische Zeitung . . . zu Stettin* and *Verhandlungen des Zoologisch-Botanischen Vereins in Wien*.

[17] The number of CD's portfolio of notes on the geographical distribution of animals.

To Asa Gray 24 November [1856][1]

Down Bromley Kent
Nov. 24th

My dear Dr Gray

Although I have nothing particular to say, yet I cannot resist thanking you warmly for your letter of Nov. 4th.—[2] Your facts on northern range have indeed astonished me; they will be **preeminently** useful for my especial purpose.[3]

I forwarded your enclosures the same day of their receipt.—[4]

I am delighted to hear that you intend to attack the naturalised plants, with especial care.— I have ordered (indeed I ordered it many months ago) your new Edit. & hope soon to get it.—[5]

The last sentence in your letter at first surprised me & troubled me to a degree which would have made you laugh, had you seen me, viz "that a considerable part of our Alpine plants are not known in our Arctic continental regions". I did not perceive that you had added but are connected with Scandinavia through *Labrador* &c.— And this made me happy again. But looking at *the Globe* is it not rather a forced expression to exclude Labrador from your "Arctic continental regions"?— Pray think of this, or you may confound some one else as you did me.

When you say that the only method to make you work is "to show you the way"—it convinces me of one thing alone,—that, as the very best workman sees blemishes in his work which other & poorer workmen cannot even perceive, so you cannot appreciate your own work in the generalising line.— Good Heavens if I had written a paper half as good as yours, how conceited I should have been![6]

With my hearty thanks for all your kindness,—& I have just been looking over some of your old letters— Believe me | Your's most truly | C. Darwin

I cannot get over my surprise at your naturalised & agragrian plants not being *variable* (I do not mean by this presenting marked races)—[7] Pray keep this point in mind.

Gray Herbarium of Harvard University

[1] Dated by the reference to the letter from Asa Gray, 4 November 1856.
[2] Letter from Asa Gray, 4 November 1856.
[3] See letter to Asa Gray, 12 October [1856] and n. 5.
[4] See letter to J. D. Hooker, 18 November [1856].
[5] A. Gray 1856a. There is a copy in the Darwin Library–CUL.
[6] CD refers to the first part of A. Gray 1856–7. See letter to Asa Gray, 12 October [1856].
[7] See letter from Asa Gray, 23 September 1856.

To W. E. Darwin 25 [November 1856][1]

Down.
25th

My dear old Willy

First in regard to your clothes you had better get a warm waist-coat & Trowsers, black & white,—something that will do for half mourning & yet will do afterwards—[2] If your dress Coat & waistcoat will not do, you must get others; but I thought they were in good state: you must judge for yourself.— I will write on separate paper to show M[r] Mayor.[3]

The House has been in a wretched way lately; Mamma very bad; Etty, from catching cold has quite gone back; & I have had a bad cold. Georgy came home on Saturday & on Sunday he became bad & is not gone to school yet; but I suppose he will be fit tomorrow morning.[4] Georgy's Holidays are on 22ᵈ: are yours fixed for 18ᵗʰ? *Answer this.*— On Saturday *all* the children went down for a farewell tea-party at Aunt Sarah's, & Georgy staid till 10½ very jolly.— How you will miss that house.—[5] The sale by Auction is to be this day fortnight, & about that day Hemmings & the maids will leave.—

The grey mare has been growing thin & we find she will not stand being turned out; & now she proves rather frisky, & whether she will not be too frisky for you I do not know; if so I will sell her again, & I daresay I shall not lose much money: she wants regular daily work, & then she is not at all too frisky.[6]

It is very jolly to think how soon you will be at home now.— Aunt Elizabeth will come here in a few days, & that will aid in making the House more cheerful.[7] The new Railway to Beckenham will be open, we hear, on January 1ˢᵗ.—[8] Now that we shall not have the use of[9]

AL incomplete
DAR 210.6

[1] Dated by CD's reference to George Howard Darwin having returned home from school (see n. 4, below).

[2] Sarah Elizabeth (Sarah) Wedgwood, CD's aunt, had died on 6 November.

[3] Robert Bickersteth Mayor was the mathematics master and William's housemaster at Rugby School (*Rugby School register*).

[4] Emma Darwin recorded in her diary that 'George came from school' on 22 November; 'G. very poorly' on 25 November; and that George returned to school on 27 November.

[5] See letter to G. H. Darwin and W. E. Darwin, 13 [November 1856], n. 7.

[6] CD had bought the grey mare in October. A payment of £30 was recorded in his Account book (Down House MS) on 6 October 1856.

[7] Sarah Elizabeth (Elizabeth) Wedgwood, Emma Darwin's older sister and CD's cousin. Emma Darwin recorded in her diary that 'Eliz came' on 27 November 1856.

[8] The Mid-Kent Railway. Beckenham was two miles closer to Down than the nearest station on the London, Brighton, and South Coast Railway.

[9] Probably a reference to the use of Sarah Wedgwood's phaeton, which would now cease (see *Emma Darwin* 2: 105).

To George Bentham   26 November [1856][1]

Down Bromley Kent
Nov. 26ᵗʰ

My dear Mʳ Bentham

I venture a beg a favour of you. I have rather a wild bit of speculation afloat on the crossing of plants; & the Leguminosæ are my determined enemies, & worst of all are any forest-trees of this order.[2] Now the only book which I have as any sort of guide is Loudon Encyclop. of Plants;[3] & as far as I can most imperfectly judge; the Leguminosæ in the two sub-orders Cæsalpinieæ & Mimoseæ are mostly Trees or bushes— is this so? But what concerns me most is to know whether there are many timber trees with papilionaceous flowers, ie with a keel enclosing the stamens & pistil, so as to shut them up as in a common Pea.— In Loudon to my joy the little woodcuts seem to show that the trees Dipterix, Parivoa & Erythrina have the stamens protruding, unlike common papilionaceous flowers:— on the other hand, to my sorrow the trees Dalbergia Pongamia, Pterocarpus, Butea & Piscidia have flowers shut up just like a Pea.— Do you know these trees, & is my inference right? And can you say, whether the papilionaceous division of the Leguminosæ have as many trees as the two other divisions; or (but this is a very loose question) as many trees as most orders which have any trees?

(I have forgotten Robinia, but this I can myself watch to see if the keel opens next summer.)

Will you humour me in giving me a little information on this head, which I am very curious about, though the notion which I am testing is very wild.— I ought to be ashamed of myself to ask you to take so much trouble, but a brief answer, if you will **kindly** give me one, will suffice.—

Pray believe me, | Your's very sincerely | Ch. Darwin

P.S. | I have been comparing all the evidence which I can collect on the natural crossing of the *varieties* of cultivated Leguminosæ; & it is most conflicting; but preponderates against crossing ever taking place.[4] Do you happen to know of any facts throwing light on this question?—

Are many of the Cæsalpinieæ polygamous like many of the Mimoseæ?

Royal Botanic Gardens, Kew (Bentham letters: 684)

[1] The year is based on the subject of the letter, which was discussed in CD's chapter 'On the possibility of all organic beings occasionally crossing' (*Natural selection*, pp. 35–91). The chapter was completed on 16 December 1856 ('Journal'; Appendix II).

[2] CD described how the flowers of Leguminosae enclose the plant's reproductive organs as tightly as 'a bivalve shell' which, to many naturalists, indicated the impossibility of any cross-fertilisation (*Natural selection*, p. 68). CD believed that insects were a prime factor in the transport of pollen from one flower to another in this family, but he came to the conclusion that Leguminosae only rarely cross (*Natural selection*, p. 71). Trees were a problem because they had so many flowers opening at the same time that any crossing would take place with another flower on the same tree (see *Natural selection*, pp. 61–2, 71). Bentham was an expert on Leguminosae.

[3] Loudon 1842 was cited in *Natural selection*, p. 61.

[4] See n. 2, above, and letter to *Gardeners' Chronicle*, [before 6 December 1856].

To John Thompson?[1]   26 November [1856][2]

Down Bromley Kent
Nov. 26th.—

My dear Sir

I thank you much for your promise of the rabbit: whenever the misfortune happens to you will be time enough for the good fortune for me.—[3]

I am much obliged to you for telling me about the Rabbit at Zoolog. Gardens; & I have written to the Secretary to bespeak the carcase.—⁴ I think you spoke as if you knew M.ʳ Vivian well; should you have any communication with him, I sh.ᵈ be extremely much obliged, if you would state to him how great a favour he would confer if he would let me have the carcase of one of his Creve-Coeur, *old* **Cocks**, whenever one should die.⁵ This Breed is mentioned as very peculiar in the Poultry Chronicle, as having 2 horn-li⟨ke⟩ combs, & when the comb is much developed, I find to my surprise the skull is modified to support it.—⁶ You might, if so inclined & intimate with Mʳ Vivian, say that I sh.ᵈ be most grateful for the bodies of any *old* birds of *curious* breeds of Rabbits, Poultry, Pigeon, or Ducks, as I mean to work at the skeletons of all these, & horrid work it will be.— You see I have taken you at your word of kindly offering me assistance with a vengeance.

Pray do not trouble yourself to acknowledge this note; only forgive me sending it. & believe me, my dear Sir | Yours very sincerely | Charles Darwin

The best address for parcels (I mention it for *chance* of your being able to assist me) is

> C. Darwin Esqᵉ
> Care of M.ʳ Acton
> Post Office
> Bromley
> Kent.—

Cambridge University Library (Add 4251: 337)

¹ The recipient is conjectured from an entry in CD's Account book (Down House MS), dated 22 January 1857, recording a payment of 12s. 6d. to 'Thompson' for a silver grey and Himalayan rabbit. John Thompson was superintendent of the Zoological gardens in London from 1852 to 1859 (Scherren 1905, p. 104).

² The year is given by the relationship to the letter to W. D. Fox, 14 June [1856], in which CD refers to having asked for the carcass of a Himalayan rabbit from the gardens of the Zoological Society. The letter also precedes a payment for rabbits made in January 1857 (see n. 1, above).

³ Probably the silver grey rabbit paid for in January 1857 (see n. 1, above). CD's interest in silver grey rabbits later led him to study the origin of their peculiar colouring and their possible contribution to the ancestry of other well-known breeds, in particular the Himalayan rabbit (see *Variation* 1: 109–11). CD was assisted in his investigation by Abraham Dee Bartlett, Thompson's successor as superintendent of the Zoological gardens.

⁴ The secretary of the Zoological Society was David William Mitchell. A previous letter to William Darwin Fox, 16 June [1856], indicates that CD had requested a carcase of a Himalayan rabbit.

⁵ Probably Edward Vivian, who was a founder of the Torquay Natural History Society and an exhibitor of ducks and poultry (*Cottage Gardener* 10 (1853): 386).

⁶ William Bernhard Tegetmeier had described the French fowl known as crève-coeur in *Cottage Gardener* 16 (1856): 265–6. In *Cottage Gardener* 18 (1857): 336–7, he reported the results of cross-breeding experiments with his own stock of crève-coeur fowl. CD had possibly acquired a carcass from him after his first notice. CD described this breed briefly in *Variation* 1: 229.

From H. C. Watson 26 November 1856

Thames Ditton
Nov.ʳ. 26. 1856

My dear Sir

It is some score of years since I gathered the Subularia, & once only. The specimens are in advanced bud, but not quite to the flowering (impregnating) point. From recollection, too distant & unclear for reliance, the plants were under water, & where they would still be under water when (or, possibly, **if**) the pollen would be perfected.— This does not meet your question.[1] There is discrepancy or contradiction in English writers, Smith, Hooker, Lightfoot.[2] But the following passage seems simple & clear on the point, & is from an author in general very reliable:—

"S. aquatica"— "In piscinis sub aqua, et, aqua æstate exsiccata, ad margines piscinarum"—

"Sub aqua clandestine floret; extra aquam flores parvi albi explicantur"— Koch Synopsis Floræ Germanicæ et Helveticæ, ed. 2 p. 73.[3]

Of Limosella aquatica I have probably seen thousands of plants in flower. It grows in hollows, & sides of ponds liable to fluctuations in the level of the water,— so that it is frequently left uncovered, & certainly flowers out of water. From the situations in which it grows, a day or two of heavy rain in autumn will submerge it while in flower, & I have seen it *apparently* flowering under water, under such circumstances. I do not recall ever having seen it in flower under water in places where it *must have remained under water during* the growth of its buds up to the actual flowering. In the wet summer of 1843 I found two plants of it, one nearly in flower, the other in flower & past flower, on a spot where they never had been *under* water, but must have grown on the ground, out of water,—very proby. carried to the spot in a gardener's watering-can, either as seeds or as very young plants, from a pond some 60 yards distant; a public road, & thick hedge intervening.—

This is a longer answer than you required, I hope not too long. D.ʳ Dickie, Prof. of Bot.ʸ in Queen's College, Belfast, could perhaps answer your question as to the Subularia.[4] The specimens in my herbarium, the appearance of which most suggests the idea of flowering out of water, are from him, gathered in his former vicinity of Aberdeen.—

Sincerely Yours | Hewett C. Watson

DAR 207 (Letters)

[1] See letter from C. C. Babington, 22 November 1856, n. 1.

[2] J. E. Smith 1824–36, W. J. Hooker 1830b, and Lightfoot 1777 were standard reference works on the British flora.

[3] Koch 1843–4. See letter from C. C. Babington, 22 November 1856, n. 2.

[4] See letter from George Dickie, 1 December 1856. George Dickie was professor of natural history at Belfast University, but was particularly noted for his botanical studies of the area around Aberdeen, his birthplace (R. Desmond 1977).

To W. B. Tegetmeier   29 November [1856][1]

Down Bromley Kent
Nov. 29th

My dear Sir

I have got skins of common black (handsome) Persian Cock & Hen, & Cock & Hen of larger breed, called Lâri from S. Persia, sent me by Honble Ch. Murray.—[2] I hear some Poultry skins are on their voyage from Natal for me; & that a large collection is making for me by Mr Gabriel (Her. M. Commissioner) on the W. & S. coast of Africa.—[3] Would you like the Persians to be sent to you at once or wait till I receive others? If I do *not* hear I shall understand they may wait.—

I really now think I shall have materials to judge of Poultry of World, for Mr Wallace is collecting in the Malay Archipelago;[4] but the carriage is costing me a fortune! The Persian Box alone cost £4 · s4 · d10.—

Though I had originally intended to describe these breeds myself; I am convinced it will be better for science that you sh^d do it, & everything which I can get, (if you think it worth while to give up your valuable time to it) shall be placed at your disposal; & I think it will at least in the eyes of scientific men, be a new & valuable feature in your Book, & will be quoted years after we are both in our graves.— But you must not for science sake, & still less for my sake, give up time to this, if you do not think it worth your while.

Your's very sincerely | C. Darwin

New York Botanical Garden Library (Charles Finney Cox collection)

[1] Dated by the relationship to the letter from W. F. Daniell, 14 November 1856.

[2] CD had asked Charles Augustus Murray for specimens of domestic birds and animals in 1855 (*Correspondence* vol. 5, letter to C. A. Murray, 24 December 1855). These pigeons were mentioned in letters to W. B. Tegetmeier, 3 November [1856] and 19 November [1856]. On 25 November 1856, CD recorded a payment in his Account book (Down House MS): 'Wheatley & Co. Persian Pigeons *All Repaid by Brit Mus & Wheatley [interl]*'.

[3] Edmund Gabriel. See letter from W. F. Daniell, 14 November 1856.

[4] Alfred Russel Wallace had left England to collect in the Malay Archipelago in 1854, and his name was included in CD's list of individuals to ask for specimens (see *Correspondence* vol. 5, CD memorandum, [December 1855]). In 1856 Wallace visited the islands of Bali and Lombok on his way to Celebes. From there, in a letter dated 21 August 1856 (Cambridge University Library (Add. MS. 7339/234)), he told his agent Samuel Stevens that his latest shipment of specimens included items for CD: 'The domestic duck var. is for Mr. Darwin & he would perhaps also like the jungle cock, which is often domesticated here & is doubtless one of the originals of the domestic breed of poultry.' Wallace evidently also wrote directly to CD, for CD referred to a letter from Wallace, dated 10 October 1856, when he wrote to Wallace in May 1857 (see letter to A. R. Wallace, 1 May 1857). The letter of 10 October has not been found.

To George Bentham 30 November [1856][1]

Down Bromley Kent
Nov. 30th

My dear Sir

I am extremely much obliged to you for answering my loose questions so fully & clearly.— I see I was quite wrong (as might have been expected from so vague a guide as Loudon's Encycl.) in regard to few trees having papilionaceous flowers.—[2]

I was not in the least aware that those Leguminosæ, which have apetalous flowers, "were almost without anthers": you once told me before, about the apetalous Leguminosæ, & I think I wrote down other names besides Ononis, Lespediza & Clitoria: I will, if I do not hear to the contrary quote this fact on your authority, viz the apetalous condition of "many" Leguminosæ, which in this condition are almost without anthers & yet produce more seed than the ordinary flowers.[3] I wish to Heaven this case was more general, it would be splendid for me, as it would seem that pollen must be brought from some other flowers, without indeed it be a case of agamic seeding.— I have for many years watched insects on flowers,—them carrying pollen &c &c, & I have some curious *little* facts; however, loose & wild the speculations are, which they tend to lead me to. In my enemies the Leguminosæ, or rather in some of them, I cannot conceive how it is that the Bees do not carry pollen from one variety or individual to another, & yet in some cases, as in Sweet Peas, I have good evidence that the varieties never cross.—[4] How strange it is that the Cruciferæ sh[d] cross to such an extent as they do! I suppose there is no common English Leguminous plant, which has apetalous flowers; I sh[d] so like to watch it & insects.—

I thank you *warmly* for your very kind offer of giving me further information; but I will not be exorbitant, yet sometime I daresay I may trouble you again

Your's very sincerely | C. Darwin

Royal Botanic Gardens, Kew (Bentham letters: 685)

[1] Dated by the relationship to the letter to George Bentham, 26 November [1856].
[2] See letter to George Bentham, 26 November [1856].
[3] No quotation from Bentham on this subject appears in *Natural selection* or in *Origin*.
[4] CD summarised his views in *Natural selection*, pp. 68–71. His conclusions were repeated in a paper on 'Bees and fertilization of Kidney beans' published in 1857 (see letter to *Gardeners' Chronicle*, 18 October [1857]).

To J. D. Hooker [early December 1856][1]

[Down]

My dear Hooker.

Will you return this to me. with any remarks?—[2]

A more curious case is offered by Podostemon, which D[r] Hooker informs me flowers with its corolla closed in the rocky beds of the rapid torrents of the

Himalaya. From the habits of the Family to which it belongs, it probably never flowers in the open air; & as the corolla is closed it seems impossible that there should ever be a cross between two individuals.[3]

But Lindley (Veg. Kingdom p. 482)[4] says the flowers of the species of the Family of Podostemaceæ are usually hermaphrodite, & as he says all the species are submersed it would appear (without there be some error) that there must be some means in the mono- or diœcious species, of the pollen being carried under water from flower to flower.—[5]

Can you illuminate me? For this in my present state of ignorance seems the strongest case of "Darwin, an eternal & necessary hermaphrodite".—[6]

I send directed envelope to give as little trouble as possible—

Ever yours | C. Darwin

DAR 205.5 (Letters)

[1] The conjectured date is suggested by CD's interest in the possible crossing of plants that flower under water. See letters from C. C. Babington, 22 November 1856, and from H. C. Watson, 26 November 1856. CD referred to the crossing of these aquatic plants in *Natural selection*, p. 63. He recorded having completed the chapter on crossing on 16 December 1856 ('Journal'; Appendix II).

[2] The following paragraph was evidently intended by CD to form part of his chapter on crossing (see *Natural selection*, p. 63). The text was corrected by Hooker (see nn. 3 and 5, below) and presumably returned to CD. Further information was given in a letter from Hooker that may have accompanied the returned manuscript (see following letter).

[3] CD put three reversed interrogation marks next to this sentence in the margin. Hooker deleted them when he made his revisions (see n. 5, below).

[4] Lindley 1846, p. 482.

[5] The statement, as revised by Hooker, reads:

> A more curious case is offered by Podostemon, some species of which D.r Hooker informs me flowers *with their corolla closed*, in the rocky beds of the rapid torrents of the Khasia mountains in Eastern Bengal. The species referred to are annual, & only appear in the the rainy season when the torrents they inhabit are swollen for several months & they appear never to flower in the open air; & *as the corolla is closed* it seems impossible that there should ever be a cross between two individuals.
>
> The flowers of the species of the Family of Podostemaceæ of which there are many & several genera are almost invariably are hermaphrodite, but all the species are not *submersed* at the time of flowering some caulescent species raising their flower stalks above the water at that period: the Indian species however to which D.r Hooker refers grow appressed to the rocks & below a considerable depth of water It is not known whether the mono- or diœcious species admit of the pollen being carried under water from flower to flower or whether they are fecundated above its surface.

For the final form of this paragraph, see *Natural selection*, p. 63.

[6] See letter to T. H. Huxley, 1 July [1856] and n. 2.

From J. D. Hooker [early December 1856][1]

Dear Darwin

A M. Tulasne has published lately a splendid monograph of *Podostemaceæ* &

says that some species certainly fecundate under water, though others may not do so & some *caulescent* ones raise their flower stems above water to flower.[2]

My species were gathered at the end of the rainy season when the streams were very swollen, they grew **appressed** like Lichens to rocks, only under water & only away from the rocks of the bank, they were covered with minute flowers that are almost sessile—& grow in the most rapid water.

No species grow in the Himalayan rivers, which I consider very curious & suppose to be owing to the great vicissitudes of their temperature. They are found in the Khasia, Mountains of the Peninsula & Ceylon besides America, Madagascar & various other countries. (see Himal. Journals. II. 314 for footnote.)[3]

Furthermore some species actually cover the whole bottoms of the streams that *are never dry* & these flower & fruit abundantly

In Kerguelens Land I most particularly attended to Limosella flowering in 2 feet water; with a bubble of air in each corolla & I assume it is the same thing with Podostemon but I have a bottle full in spirits & will examine.

I do not see why the occasional partial drying of a stream might not let some flowers of Podostemon be exposed & be left for you to cross by any means you chuse.[4]

Ever Yrs | Jos D Hooker

DAR 100: 149

[1] The letter is dated by its relationship to the preceding letter. It seems to supplement the information given by Hooker in his notes written on the letter (see preceding letter, n. 5) and may have been sent to CD at the same time.
[2] Tulasne 1852.
[3] J. D. Hooker 1854, 2: 314 n. The passage is marked in CD's copy, now in the Darwin Library–CUL.
[4] CD repeated this point in *Natural selection*, p. 63: 'Dr. Hooker has never seen them flowering in the open air; but he will not assert that this may not sometimes occur, when the torrents sink.'

From George Dickie 1 December 1856

Belfast
Dec.ʳ 1ˢᵗ | 1856

Dear Sir,

I have much pleasure in stating my own observations on Subularia.[1] A few miles W from Aberdeen—my native place—it is abundant, & I have repeatedly, year after year, visited the locality with my students.

I always found it *submerged*, but never in deep water; generally round the edges of the Loch.

We generally found flower buds, but never saw them expanded. On being opened I have often found fully developed Anthers &c, and have never failed to get ripe seeds in August & September.[2]

Koch may have seen the plant growing and flowering in the air, I can only state my own experience as to it's habits. The question can be settled by cultivation &

careful regulation of the supply of water. I generally visit Aberdeen every summer & should feel glad to send you a supply of living plants by post. I believe we have similar habits in the Elatines, one of which I have often gathered, growing as it does at Aberdeen in company with Subularia.

faithfully yours | G Dickie

P.S. I shall always be glad to reply to any similar communications | GD.

DAR 207 (Letters)

CD ANNOTATIONS
Top of first page: 'Subularia | Limosella | Menyanthes | Podostemon | Zostera & Co.' *pencil*

[1] See the two preceding letters and the letters from C. C. Babington, 22 November 1856, and from H. C. Watson, 26 November 1856.
[2] Dickie's information was cited by CD in *Natural selection*, p. 62.

To J. D. Hooker   1 December [1856][1]

Down Bromley Kent
Dec[r] 1st

My dear Hooker

I am become a good deal interested in crossing speculations, though I can come to no certain conclusion.— With respect to Trees, which from number of flowers on same individual offer, considerable difficulty to the probability of crossing, I want to know whether the following would cost you much trouble, & whether you feel the least interest on the point.[2] I have had copied in good hand a list in order, divided into Families, of the 700–800 New Zealand plants.[3] Now I have no idea how far you have in your mind the memory of the N.Z plants: if distinctly it would not take you half an hour to go over the list (as many Families might be skipped) & mark with pencil cross the Trees & double Pencil cross those which were also mono- or diœcious or polygamous.[4] I tried to do it, but broke down in not being able to decide between bushes & trees, & this must be arbitary.— I have done the few Trees of England, taking Loudon as rule what to call trees & what bushes, & you will see result.[5] If N. Zealand with so different a Flora gave at all same result, it would appear *probable* that there was some connection between trees & the separation of sexes (which w[d] favour crossing).—[6] Loudon calls the Viburnum, Box, Arbutus & Juniper Bushes.—

I think in your N. Zealand Flora you have given the number of plants of all kinds with separated sexes, which would have to be considered in the result.—

Now it will wholly depend on the state of your memory whether this would be worth doing, as it would by no means be worth your having to consult your own Flora.— What say you? Shall I send the M.S. list? Can you tell me *genus* of enclosed seed, which was in Birds Dung & has since germinated.—[7]

Ever yours | C. Darwin

DAR 114.3: 185

[1] Dated by the reference to the crossing of trees, also discussed in the letters to George Bentham, 26 November [1856] and 30 November [1856].

[2] See letters to George Bentham, 26 November [1856] and 30 November [1856]. CD addressed the question of the crossing of trees in *Natural selection*, pp. 61–2.

[3] CD's list, in the hand of an amanuensis, is now located in the back of CD's copy of the introductory essay of J. D. Hooker 1853–5, in the Darwin Library–CUL.

[4] See letter from J. D. Hooker, 7 December 1856.

[5] CD's results are given in *Natural selection*, pp. 61–2. He refers to Loudon 1842.

[6] CD's point was that a hermaphrodite tree would have so many flowers that any crossing that took place would be with a flower from the same tree. If the flowers were all of one sex on a tree, crossing could only take place between flowers of different trees (see *Natural selection*, p. 61).

[7] Hooker apparently lost the seed (see letter to J. D. Hooker, 24 December [1856]). Hooker had previously identified other seeds from birds' dung for CD (see letter from J. D. Hooker, 22 November 1856).

To George Bentham 3 December [1856][1]

Down Bromley Kent
Dec. 3ᵈ

My dear Sir

I write merely to thank you for your information on the apetalous flowers.—[2] The case seems too exceptional to help me; & the whole order will remain my detestable enemies. I will get some Russian violets & will watch them next summer.— I have attended a little to Heartease, & have observed after the *rare* visits of Bees, that numbers of flowers suddenly became in a few days fertilised.[3]

Your's truly obliged | Ch. Darwin

Royal Botanic Gardens, Kew (Bentham letters: 687)

[1] Dated by the relationship to the letter to George Bentham, 30 November [1856].

[2] Bentham's letter has not been found. It was probably a reply to the letter to George Bentham, 30 November [1856].

[3] CD's observations from 1841 and 1846 are in DAR 46.2 (ser. 3): 15. He described this case in *Natural selection*, p. 57.

To W. B. Tegetmeier 4 December [1856][1]

Down Bromley Kent
Dec. 4ᵗʰ

My dear Sir

I have no details about the Madras Fowls; except that the Pariah one is the common Form; & that the history of the Rangoon one was stated to be unknown.—[2]

Do not trouble yourself to write me descriptions of Fowls, as I shall see them in time in Poultry Book.—[3]

Since writing I have heard from the Rajah, Sir J. Brook that he has sent me some Fowls from the interior of Borneo,[4] & I think I mentioned that some from Natal are on the road, so I will wait till I get some more before sending the Persians; ie without I hear that from any cause you w[d] like them sooner. The expence of sending them to London is nothing.—

I am very sorry to hear that you have been a sufferer & from so painful as Rheumatism.—

I am *extremely* glad that you are willing to describe all the Poultry, which I can get, in the Poultry Book, as I am sure it will be well done; & my only object is to know what amount of difference there is in the Fowls of different parts of world—

In Haste. | Yours very sincerely | C. Darwin

New York Botanical Garden Library (Charles Finney Cox collection)

[1] Dated by the relationship to the letter to W. B. Tegetmeier, 29 November [1856].
[2] See letter to W. B. Tegetmeier, 3 November [1856].
[3] A reference to Tegetmeier's work describing the various breeds of poultry, issued in parts during 1856 and 1857 but never completed (Tegetmeier ed. 1856–7).
[4] The letter from James Brooke, raja of Saráwak, has not been found. CD had included him in his list of people to ask for specimens of foreign poultry and domestic animals in 1855 (see *Correspondence* vol. 5, CD memorandum, [December 1855]).

To *Gardeners' Chronicle* [before 6 December 1856][1]

I have been lately collecting all the evidence which I can get from the observation of others and my own, on the natural crossing of varieties of plants. The evidence in regard to Leguminous plants is curiously conflicting, but preponderates against their ever crossing without artificial aid.[2] I should esteem it a singular favour if any of your correspondents would give in your paper or send me any evidence showing either that Leguminous crops, when grown close together, do sometimes cross; or, on the other hand, that they may invariably be grown close together without any chance of deterioration.[3] *Charles Darwin, Down, Bromley, Kent.*

Gardeners' Chronicle and Agricultural Gazette, 6 December 1856, p. 806.

[1] Dated from the issue of the *Gardeners' Chronicle and Agricultural Gazette* in which the letter appeared.
[2] CD summarised contemporary views of this point in *Natural selection*, pp. 69–71. He was principally concerned with the conflicting opinions of Arend Friedrich August Wiegmann in Wiegmann 1828 and Karl Friedrich von Gärtner in Gärtner 1849.
[3] No response has been located in *Gardeners' Chronicle*.

To J. S. Henslow [after 6 December 1856][1]

[Down]

You can tear off the first page & send it to Fisher & you will have no more trouble on subject.—[2]

I was very glad to have your letter & hear a little news of you.— Your success with your village girls strikes me as nothing less than marvellous.[3] I am delighted to hear how well your son is going on,[4] & that the by me truly honoured name of Henslow will have a Botanical successor.— We are all pretty well, & my Boys are *now* d.[6]—[5] Is it not awful? I am working away steadily & very hard at my work on Variation; & I find the whole subject deeply interesting, but horribly perplexed.—

My dear Henslow | Your affectionate old Pupil | C. Darwin

Incomplete
DAR 93: 115

[1] The letter could have been written at any time after the birth of Charles Waring Darwin, but CD's reference implies that this is the first that Henslow knew of the baby (see n. 5, below).
[2] Francis Fisher was a fellow of Jesus College, Cambridge (see *Correspondence* vol. 5, letter from J. S. Henslow, 29 June 1855).
[3] Henslow offered a course in botany for the school children of his parish in Hitcham. In 1856 the *Gardeners' Chronicle and Agricultural Gazette* carried a series of his 'Practical lessons for beginners in botany'. (See also *Correspondence* vol. 5, letter to J. S. Henslow, 28 July [1855]).
[4] George Henslow, aged 21, was an undergraduate at Christ's College, Cambridge.
[5] Henslow referred to CD's children as little 'd's' (see *Correspondence* vol. 2, letter from J. S. Henslow, 9 October 1843). Charles Waring Darwin, born 6 December, was CD's sixth son.

From J. D. Hooker 7 December 1856

Kew
Decr 7th/56

Dear Darwin

I have roughly done your N.Z. list & am glad to have had my attention drawn to the subject I have roughly marked your list & have made a separate calculation with some slight modifications, due, partly to swaying to & fro between what to call monœcious & what diœcious, & partly to a few additions.[1] The Upshot is that I find on adding up my data, 756 species. (some of this 26 over your list is due to unnumbered species & perhaps some to over calculation on my part, I suppose 6 or 8 errors of the latter kind will not influence result.

I make *now* 108 trees in all; whereas in N. Zeald Flora Introd Essay XXVIII I say there are 113—a near enough hit considering how vague the terms are, which it would be easy to cook into exact accordance or further discordance.[2] I say N.Z. Fl. l.c. 156 shrubs & plants with woody stems I now make 150—which is near enough also (far nearer both than I anticipated; for I did not at all remember even approximately what I had done before, & made my calculations before looking to the Introd Essay.[3]

I also say l.c. that "considerably more than 200 have unisexual, or incomplete flowers as far as reproduct: organs go. I now make the number 234.

I was a great deal too wise in my generation to say in N.Z. Flora how many are monoeecious, dioecious & polygamous for I find that I cannot do this with any approach to accuracy— Perhaps very few are really everlastingly & unalterably dioecious: & I do not think that my diœcious figure is too small.— On the other hand no doubt

1. sundry of my hermaphrodite plants may be virtually though abnormally monœcous
2. Many of my monœcious may bear many perfect flowers, as *Clematis*
3. Many of my hermaphrodite have normally unisexual flowers mixed with the ♀
4 Lastly various dioecious trees often bear a few flowers of the other sex.

On the opposite side you have the best approximation I can give.

	flowers		
	Hermaph.	Monœcious	Dioecious
Herbs	379	102*	19
Shrubs	88	30	31
Trees	56	40.	12.

* including 55 Compos.

The results will be satisfactory to your theory.[4]

I will do the V.D.L. Flora for you far better in a couple of months.[5]

I dare say that you will not be so surprized as I was at Lyell's asking my wife if I was the author of the Review (very laudatory of self) in Ed. Review of ADC. + Fl. Ind. + N.Z. Essay—to me the idea is monstrous & revolting—he must think of me all the same as of Owen. by way of mending the matter, when my wife told him I was not, he added that he did not think it very well done![6] This latter bit tickled me amazingly, I can afford that, but not the first imputation. Poor Lady L.[7] who was present, seemed much annoyed, my wife laughed at it all, happily— had it been my mother—the Lord have mercy on poor Lyell's bones— Pray do not let this go further than your own house. I would not that poor Lyell heard of it again.

Ever Y^rs | Jos D Hooker

Do tell us how Mrs Darwin is when you next have occasion to write—

Clematis, dioecious but some ♂ flowers have a few ripening pistilla

Viola. hermaphrodite but the winter flowers which have scarcely any stamina appear to be the most productive (? is this so in the N.Z. sp.)

Compositæ. I class with monœcious.

Grasses various called ♀ have also ♂ flowers.

DAR 100: 113–14

CD ANNOTATIONS
1.5 756] *underl pencil*; '*say about 750*' *added pencil*
2.1 108] *underl pencil*
Paragraph five:
On the table CD has added in pencil an extra row of figures totalling the figures for herbs and shrubs '467 132 50' *and an extra column of figures (including an entry for his extra row) headed* 'Mono & Dioec' *totalling the figures for these classes* '121 61 182 52'.

379] $\dfrac{'121}{500}$ *added under, pencil*

In margin of table: '108
        149
        500
        757' *added pencil*

        '88
        30
        31
149' *added pencil*

Top of first page: 'allude to cause of doubt—' *pencil*

[1] See letter to J. D. Hooker, 1 December [1856]. The list, in the back of CD's copy of the introductory essay of J. D. Hooker 1853–5 in the Darwin Library–CUL, has been annotated in pencil by Hooker.
[2] See J. D. Hooker 1853–5, 1: xxviii.
[3] See J. D. Hooker 1853–5, 1: xxviii. Hooker's abbreviation 'l.c.' stands for '*loc. cit.*' CD used the information given in this letter in *Natural selection*, p. 62.
[4] CD's theory was that trees tended to have separate sexes (see *Natural selection*, pp. 61–2).
[5] Hooker was working on his flora of Tasmania (J. D. Hooker 1855[–60]).
[6] Hooker refers to the review of A. de Candolle 1855, J. D. Hooker and Thomson 1855, the introductory essay of J. D. Hooker 1853–5, and T. Moore 1855 that was published in *Edinburgh Review* 104 (1856): 490–518. The author was George Bentham (*Wellesley index* 1: 506).
[7] Mary Elizabeth Lyell.

From J. D. Dana 8 December 1856

New Haven,
Dec. 8, 1856

My dear Mr Darwin—

I have received from Professor Agassiz, thro' Dr Gray, the reply that the mice or rats of the Mammoth Cave are *American* in type, as to teeth &c.—[1]

You say you doubt the principle that the progress of life on the globe is parallel with the development in different tribes. I would not accept of it to the same extent as Agassiz. Yet although far from universal, it appears now & then to be exhibited, as in the case of the Ganoids mentioned in my former note.—[2] The principle is subordinate to a more general law of progress—a law which involves the expression of a type-idea in forms or groups of increasing diversity, and generally of higher elevation; always resulting in a purer & fuller exhibition of the type. There may be an expansion to lower as well as higher groups, as the development of an embryo brings out the *[illeg]* of the surface as well as the brain & viscera. But in all cases, it is a more & more complete exhibition or unfolding of the type-idea, until its maximum display is made.

It is the comprehensive before a multiplicity of specialities, and under this restriction it is the simple before the Complex.— I cannot but read this much at least from our knowledge of the life of the globe, and see *so far* a parallelism with the law of development in the embryo, which brings multiplied details, beginning with those of most comprehensive signification, from the memberless unit. Under such a principle, it is natural that the history of an individual in its particulars should sometimes run parallel with that of the palæontological history of the tribe to which it pertains. But the fundamental parallelism is in the grand law alluded to, not in these occasional lines under it. Looking at the progress of life from this point of view, I should find no principle upset by the discovery of barnacles far lower in the series of rocks than they now are known to occur.[3] The details under the law are yet to be fully made out; and when so, I believe the law will appear in far bolder Characters that now are manifest.

Excuse my homily on this Subject, and believe me always glad to aid you in any way within my power; for I believe there is real truth in the results of your labors, and the best of foundations for general laws or principles.

very truly yours | James D. Dana

DAR 205.9 (Letters)

CD ANNOTATIONS
1.1 Professor . . . teeth &c.— 1.2] *double scored pencil*
1.8 lower] 'lower' *added pencil for clarity*
Top of first page: '19 & 22'[4] *brown crayon*; 'See Back'[5] *pencil*
Verso of last page: 'This note contains fact of Cavern Rat being American Form | & Dana's belief that in Embryonic changes & geological expression there is a certain parallelism from the unfolding of the type idea to its full display— It is comprehensive before a *multiplicity of [interl] specialities' ink
Whole letter crossed pencil

[1] CD had requested this information in his letter to J. D. Dana, 29 September [1856]. Louis Agassiz had discussed the cave fauna in Agassiz 1851. Agassiz and Asa Gray were both professors at Harvard University.
[2] See letter from J. D. Dana, 8 September 1856.
[3] CD had written to Dana about the discovery of a sessile cirripede (*Chthamalus darwinii*) in a formation that was much older than he had anticipated (see letter to J. D. Dana, 29 September [1856]).
[4] The numbers of CD's portfolios on the geographical distribution of animals and on palaeontology and extinction, respectively.
[5] This note refers to CD's annotation on the verso of the last page of the letter.

To Harriet Lubbock [8 December 1856][1]

> Down.
> Monday

Dear Lady Lubbock

I thank you cordially for your very great kindness in thinking of us, & being willing to take so much trouble to keep the house quiet. But Mrs Darwin is going on perfectly well, & as Etty[2] is not well, & Miss Thorley[3] not with us, we think we had had better keep our little Boys at home, & we shall be able to get on perfectly well with them in the house.—

My wife desires me to give you her kindest thanks.— We have now half-a dozen Boys, which seems perfectly awful to me, but will seem less dreadful to you.—[4] Permit me again to say how very sincerely I thank you for your great kindness, & believe me dear Lady Lubbock | Your Ladyship's | Most truly obliged | Charles Darwin

American Philosophical Society (141)

[1] The first Monday after the birth of Charles Waring Darwin.
[2] Henrietta Emma Darwin.
[3] Miss Thorley was the governess of the Darwin children until 6 January 1857 (Emma Darwin's diary). Miss Pugh took her place in April 1857.
[4] The Lubbocks had eight sons and three daughters.

To T. H. Huxley 9 December [1856][1]

> Down Bromley Kent
> Dec 9th—

My dear Huxley

I am infinitely obliged for your note with Siebolds most wonderful facts, which stretch one's faith to the utmost in that admirable observer.—[2] Males always coming from unimpregnated eggs beats everything one ever heard!—[3] I will send your letter to Lubbock.—[4] Now that I hear of vitality of spermatozoa, I begin to think a few must have got into the ovarium of Ld Moreton's Mare.—[5] But *incidentally* I am *extremely* curious about hybridising of Bees; for as you speak about "pure races"; it seems to imply that there are varieties of the Domestic Bee, about which I feel an especial interest; for Bees offer in one respect by far my greatest theoretical difficulty.[6]

I am delighted to hear of your work on Crustacea, on the gland of antennæ.[7] The conviction of an author, who has published goes for nothing; but from the many specimens which I have dissected, I **cannot** get it out of my head that the cement receptacle, as called by me, is continuous with & opens into the ovarian tubes; but that at the earliest age, before the ovarian tubes are developed, the cement-duct leads to those glands, which I *formerly* called the "true ovaria".[8] Before you publish I do so wish you could get a good specimen of Otion (=

Conchoderma aurita) for in this the cement receptacles are visible within the sack,[9] on the sides low down & you could *easily* dissect the connection between cement-receptacle, & the ovarian tubes.— This would alone convince me of my error.[10]

I shall not be at Club,[11] for my wife has just been confined of our sixth Boy! Adios, I thank you heartily for your letter.

Ever yours | C. Darwin

Many thanks for your Lectures, which seem to me most valuable[12]

Sometime when you can spare Siebold's Book for 2 or 3 weeks (for I am slow at German) I sh^d very much like to borrow it— Will it not be translated?[13]

By Jove I can hardly believe it— Oh if neuters were agamic offspring, what a load it w^d remove from me.—

If ever you stumble on reference to Hicks on Sexual organ in Flustra, let me have it.[14]

Imperial College of Science, Technology and Medicine Archives (Huxley 5: 42, 374)

[1] Dated by CD's reference to the birth of Charles Waring Darwin on 6 December 1856 ('Journal'; Appendix II).

[2] Huxley spoke about Karl Theodor Ernst von Siebold's *Wahre Parthenogenesis bei Schmetterlingen und Bienen* (Siebold 1856) at a meeting of the Philosophical Club of the Royal Society on 11 December 1856 (Bonney 1919, p. 134). He called attention to Siebold's

> strong evidence to show that the females of certain *Pychidae* [Psychidae] and of *Bombyx mori* produce fertile ova without previous fecundation, and with the former the process may be repeated for several generations. With the bees, not only can the unimpregnated queen-bee lay fertile eggs, but she also appears to fertilize, after impregnation, only those ova which are laid in neuter or female cells, those laid in drone cells remaining unfertilized.

CD's annotated copy of the English translation of Siebold's book (Siebold 1857) is in the Darwin Library–CUL. He discussed this work in *Natural selection*, p. 365.

[3] CD cited the fact 'that in the Queen Hive Bee it is exclusively the unimpregnated eggs which produce males' in *Natural selection*, p. 365 n. 3.

[4] John Lubbock was preparing a paper on the parthenogenesis of *Daphnia* (see letter to John Lubbock, 27 October [1856]).

[5] See letter from E. L. Layard, [September–October 1856], n. 4.

[6] For CD's discussion of the varieties of the common hive-bee, see *Natural selection*, p. 372 n. 5. The difficulties posed by neuters and sterile forms among social insects are discussed at length in *Natural selection*, pp. 364–8.

[7] Huxley's lectures were published in the *Medical Times & Gazette* beginning in May 1856 (T. H. Huxley 1856–7). In the fourth part of the tenth lecture, published in April 1857 and treating the internal anatomy of *Astacus*, Huxley discussed the antennae and the secreting glands connected to these, which he related to glands found by CD in cirripedes. In T. H. Huxley 1856–7, 14: 354, Huxley stated that:

> the similarity in the position, structure, and contents of the antennary "gland" in the decapods with those of the "gut-formed organ," described as the "true ovary" by Mr. Darwin, in the Cirripedes, leaves no doubt on my mind that they are homologous organs, and that the sac is the homologue of the "cement duct" in the Cirripedes.

[8] In *Living Cirripedia* (1851) and (1854), CD had advanced these observations to support his view that the ovarian system of cirripedes had been transformed to fulfil the new function of cementing the organism to a substrate (see *Correspondence* vol. 4, Appendix II, pp. 397–9). CD had previously asked

Huxley to investigate this point independently (see *Correspondence* vol. 5, letter to T. H. Huxley, 29 [September 1855]).

9 This species was singled out by CD because his observations had convinced him that it clearly displayed the cementing gland as a part of a modified ovarian tube (*Living Cirripedia* (1851): 34).

10 Huxley devoted part of his twelfth lecture to the Cirripedia, and there he discussed this matter. Huxley confirmed CD's observation that the gut-formed glands were part of a continuous organ with the cement ducts and the peduncular tubules. But he differed with CD over the interpretation of the functions of the glands and tubules, believing that the latter and not the former were the 'true ovaria' (T. H. Huxley 1856–7, 15: 239). CD's (and Huxley's) error was later pointed out by August David Krohn in Krohn 1859.

11 See n. 2, above.

12 The most recent part of T. H. Huxley 1856–7, the second part of lecture nine, had appeared on 29 November 1856.

13 Siebold 1856 was translated by William Sweetland Dallas (Siebold 1857). CD read the translation early in 1857.

14 See letter to T. H. Huxley, 8 July [1856], in which Thomas Hincks's work on Polyzoa was mentioned.

To W. E. Darwin 10 [December 1856][1]

Down.
10th

My dear old Willy

Mamma is going on perfectly well (as is your beautiful little Brother) & I hope she will make a quicker recovery than usual.— Georgy's holidays are on 22[d] & he will go up to Cumberland Terrace & come back with you;[2] as Erny & H. Hemming[3] will be in London, we think you had both better stay till Friday 26th or Saturday if Aunt Fanny will ask you.[4] And we have written to ask her.— How do you like this plan? This House will be rather dull, as Mamman cannot be perfectly recovered by that time.—[5]

I thought your skates would not prove of any use, but the dear old Mammy would send them off.—

Parslow rode the grey mare the other day & found her **very** pleasant & quiet. Today & yesterday were the auction days at At. Sarahs.— The things have sold very well. The Brougham 10½ guineas. The 3 Vases sold for £7 · s5.— An American clock which cost when new 8·6 sold for 15[s]— The arm chair fetched £6 · s12 · d6.—[6]

Good Bye my old man. I heartily hope that you will get on well in your examination

Your affect. | C. Darwin

What a nice letter that was of Georgys.—

DAR 210.6

1 Dated by the reference to the birth of Charles Waring Darwin.

[2] Emma Darwin noted in her diary that after his school-term ended, George Howard Darwin went directly to the home of Hensleigh and Frances Mackintosh Wedgwood in Cumberland Terrace, London.

[3] Ernest Hensleigh Wedgwood was the son of Fanny and Hensleigh Wedgwood and a fellow student of William Erasmus Darwin's at Rugby. Henry Hemmings, a servant of the recently deceased Sarah Elizabeth (Sarah) Wedgwood, was probably on his way to Barlaston (see letter to G. H. Darwin and W. E. Darwin, 13 [November 1856]).

[4] According to Emma Darwin's diary, the Darwin boys stayed with Fanny Mackintosh Wedgwood until Saturday, 27 December, which was William Darwin's birthday.

[5] Emma Darwin recorded in her diary on 26 December, 'came down to breakfast'.

[6] CD recorded spending £5 at the sale of Sarah Wedgwood's goods. An entry in his Account book (Down House MS) on 11 December 1856 lists 'Furniture At Sarahs Sale. Crockery'.

To J. D. Hooker 10 December [1856][1]

> Down Bromley Kent
> Dec 10th

My dear Hooker

I thank you sincerely about the trees: it must have cost you much more trouble than I anticipated. Now that the result is what it is in N. Zealand (I find trees occur in 38 Families, & in exactly half of these the trees have separated sexes; so that it is not an attribute related to any particular systematic structure) & with the much feebler case of England, & with the great number of trees in Persoon's Synopsis[2] in the mono- & dioicous classes (though I have here no accurate standard of comparison), I think it pretty certain there is a relation between trees & separation of sexes; whether or not my explanation is correct.—[3] I have had lately some hard blows against my crossing theory, & hardly know what to think.[4] I shall be very glad to hear about Tasmanian trees, & perhaps you will find the result worth a paragraph in your introduction.— It is a most tiresome drawback to my satisfaction in writing, that though I leave out a good deal & try to condense, every chapter runs to such an inordinate length: my present chapter on causes of fertility & sterility & on natural crossing has actually run out to 100 pages M.S., & yet I do not think I have put in anything superfluous.—[5]

My wife is going on capitally in every respect.— Give my best thanks to Mrs Hooker for her kind note.—

What a funny story about Lyell & the Review (which I have not yet seen); but really I think the most probable solution is that he forgot at moment that the Review was laudatory of yourself; you are the very last man whom he would suspect of praising yourself.—[6]

I have for last 15 months been tormented & haunted by land mollusca, which occur on *every* oceanic island; & I thought that the double creationists or continental extensionists had here a complete victory. The few eggs which I have tried both sink & are killed. No one doubts that salt-water wd be eminently destructive to them; & I was really in despair, when I thought I would try them when torpid; & this day I have taken a lot out of sea-water, after exactly 7 days immersion. Some *sink*, & some swim; & in both cases I have had (as yet) one come to life again, which has quite astonished & delighted me. I feel as if a thousand pound weight was taken off my back.[7]

Adios my dear kind friend | C.D.

P.S. | I must tell you another of my *profound* experiments! Franky said to me, "why sh^d not a bird be killed (by hawk, lightning, apoplexy, hail &c) with seeds in crop, & it would swim." No sooner said, than done: a pigeon has floated for 30 days in salt water with seeds in crop & they have grown splendidly & to my great surprise even tares (Leguminosæ, so generally killed by sea-water) which the Bird had naturally eaten have grown *well*.— You will say gulls & dog-fish &c w^d eat up the carcase, & so they w^d. 999 out of a thousand, but one might escape: I have seen dead land bird in sea-drift.—[8]

Keep in mind some difficulty about distribution of F. W. Plants, which you alluded to at Kew when I was there; & I had not time to ask you about.— I do not mean write it, but for when we meet.—

I wonder whether Dr Harvey would think it a bore to illuminate me about separation of sexes, how far general, in Algæ?[9] What do you think?

DAR 114.3: 186, 186a

[1] Dated by the relationship to the letter from J. D. Hooker, 7 December 1856.
[2] Persoon 1805–7.
[3] CD discussed this point in *Natural selection*, pp. 61–2.
[4] See letter to George Bentham, 26 November [1856], letter from H. C. Watson, 26 November 1856, and letter to George Bentham, 30 November [1856]. The difficulties that CD felt in relation to the theory that all organic beings occasionally cross are given in *Natural selection*, pp. 63–71.
[5] CD was preparing chapter 3 of his species book, 'On the possibility of all organic beings occasionally crossing, & on the remarkable susceptibility of the reproductive system to external agencies' (*Natural selection*, pp. 35–91). It was completed on 16 December 1856 ('Journal'; Appendix II).
[6] See letter from J. D. Hooker, 7 December 1856.
[7] See letter to P. H. Gosse, 28 September 1856, n. 4, and the following letter. CD's experiment, dated 3 December 1856, is recorded in his Experimental book, p. 17 (DAR 157a).
[8] CD's experiment was entered in his Experimental book, pp. 16–17 (DAR 157a). CD mentioned Francis Darwin's suggestion in *Origin*, p. 361, where he proposed the floating carcases of birds as one of a number of 'occasional' means of dispersal.
[9] William Henry Harvey was an expert on Algae. CD had sent him Algae specimens from the *Beagle* voyage (*Correspondence* vol. 4, letter to W. H. Harvey, [7 April 1847]).

From T. V. Wollaston [11 or 18 December 1856][1]

10. Hereford S^t | Park Lane.
Thursday evening.

My dear Darwin,

Those Helices ought to be ashamed of themselves for surviving in salt water; I s^d have given them credit, I confess, for better behaviour.[2] However, in the meantime I must congratulate you on the addition of another ♂ to your vivarium;[3] &,

although I sympathise with you in your disappointment, the result of this infamous experiment will compensate for the difference between a ♂ and ♀, I am sure.—

Those dishonest Mollusks were collected, in Porto Santo, during the *2ⁿᵈ. week of May*, 1855.[4] It is possible, perhaps, that, like seasoned casks, which are proof against the vicissitudes of this nether world, they may become more tenacious of life in proportion *to their age*: so that if my thesis be true that some of them have lived since miocene times, the patriarchs whose opercula have become thickened into solid rock might survive for ten years (instead of days) in aquis valde diabolicis maritimis.—

I have no others, I think, except more of the Porto Santans; & those not in such good condition as the bag-full wʰ. you took,—being chiefly (if not altogether) defunct.—

Of shell-less Mollusks, there are in Insulis Maderensibus a few slimy *Limaces* (all the same as European species), 2 *Testacellæ* (D°.), & 3 *Vitrinæ*,—wʰ. are supposed by Lowe (who however shaves things very finely) to be peculiar to those islands.[5] They are however the exact "representatives" of the 3 European *Vitrinæ*; hence *I*, who am less morbidly sensitive about *specific* differences (though *believing* in "species" more & more every day),[6] sᵈ. be inclined to regard them as mere insular modifications. The others *we suppose* have been introduced by man,—but if you choose to substitute "mare" instead of "homo", I will not refuse my assent.— I am taking a week's interlude, between the paroxisms of Madeira work, to revise a British genus of Coleopterous atoms,—the 25 species of wʰ. might be put into the compass of a hollow pea.—

Vivat Darwinianus masculus sextus.—

Your's very sincerely | T V Wollaston.

DAR 205.3 (Letters)

CD ANNOTATIONS
0.3 My . . . defunct. 3.3] *crossed pencil*
4.9 Madeira . . . sextus.— 5.1] *crossed pencil*
First page: '19'[7] *brown crayon*

[1] The two most likely Thursdays after the birth of Charles Waring Darwin and the conclusion of CD's experiments on Mollusca (see n. 2, below).

[2] In his Experimental book, p. 17 (DAR 157a), CD recorded that he had been given several hundred snails from the island of Porto Santo by Wollaston. Half of these, plus some snails collected around Down House, were put to float in sea-water by CD on 3 December 1856. CD's results were given in his letter to J. D. Hooker, 10 December [1856], and in *Origin*, p. 397. In his Experimental book, p. 17 (DAR 157a), he recorded:

> Nov. 30ᵗʰ Put small ½ of Mʳ Wollaston P. Santo Land shells to soak *These P. Santo shells were collected middle of May 55, so that now 18 months old. [*added*] & about. 5. *All Helix pulvinata [*added*] came quite to life & crawled about*— 17 other partly protruded their bodies, but hardly or not at all moved & soon died.— *There were several hundred.— [*added*]

Dec. 3ᵈ 9½ A.M. Put the other larger ½ of Mᵣ Wollaston; & my own Down Mollusca to float, in real Sea-Water. Only one *Helix pulvinata [*interl*] came to life **& moved**; but soon died. The weather during the exact week they were in the sea water was unfortunately very warm.—

3 Wollaston refers to the birth of Charles Waring Darwin on 6 December. He goes on to allude to CD's preference for daughters (see *Correspondence* vol. 5, letter to W. D. Fox, 7 March [1852]).

4 Wollaston had spent March to November 1855 in Madeira (see *Correspondence* vol. 5, letters from T. V. Wollaston, 2 March [1855], and from J. D. Hooker, 14 November [1855]).

5 Richard Thomas Lowe was an expert on Madeiran flora and fauna.

6 See letter from Charles Lyell, 1–2 May 1856, n. 7.

7 The number of CD's portfolio of notes on the geographical distribution of animals.

To T. H. Huxley 13 [December 1856]¹

Down.
13ᵗʰ

My dear Huxley

I must just thank you for your note. For quickness sake, I think I shall read Translation, at least if Book is big.—² I fear Entomologist will not let the Italian Bee pass as a variety: it is reckoned good species.³

I am much pleased at what you say about relations of cement-gland & organs in higher Crust.—⁴ I shall be quite content to be moderately right on subject; & I do very much hope you will dissect the receptacle in Conchoderma. Remember, that the *so-called* "true ovaria" yet act, as I saw pellets of yellow stuff on one or two occasions in transitu in the unbranched part between the "true ovaria" & ovarian tubes or cæca.⁵

I hardly myself remember, *at present*, what I asked you (thanks for Hincks),⁶ but I will put down 2 or 3 points on next page, for chance of your coming across good examples; then enter them in your note Book, but do *not* take trouble to write.—

With cordial thanks for all your kindness | My dear Huxley | Ever yours | C. Darwin

I hope that Mᵣˢ— Huxley is pretty well.—⁷

Can "Darwin" be an eternal & necessary hermaphrodite?⁸

Cases of organs in which there is no apparent passage or transition from other organ: or still better, if such transition can be shown in an unexpected manner.⁹ E.G. Electrical organs in Fish, *seem* to be really new organ & not any other changed. Some think poison-gland of Snakes are *not* salivary gland modified. I require passages, but I always give all the facts which I can collect, *hostile* to my notions.

Cases of odd & inexplicable connection, between different parts of structure, so that if one changes the other changes. E.G. *All* cats with blue eyes are deaf.— or Hairless dogs are nearly toothless, which latter we can understand.—

Imperial College of Science, Technology and Medicine Archives (Huxley 5: 44)

1 Dated by the relationship to the letter to T. H. Huxley, 9 December [1856].

[2] Siebold 1857, the English translation of Siebold 1856.

[3] CD referred to the Italian dark bee in *Natural selection*, p. 372 n. 5.

[4] See letter to T. H. Huxley, 9 December [1856], n. 7.

[5] See *Correspondence* vol. 5, letter to T. H. Huxley, 29 [September 1855].

[6] Hincks 1852 was discussed in the letter to T. H. Huxley, 8 July [1856].

[7] Henrietta Anne Huxley was pregnant. The Huxleys' first child, a son, was born 31 December 1856 (L. Huxley ed. 1900, 1: 151).

[8] See letter to T. H. Huxley, 1 July [1856], n. 2.

[9] CD was gathering such cases for chapter 8 of his species book, 'Difficulties on the theory of natural selection in relation to passages from form to form' (*Natural selection*, pp. 339–86). CD mentioned a case provided by Huxley (*ibid.*, p. 357 and n. 1).

To Thomas William St Clair Davidson 23 December [1856][1]

Down Bromley Kent

Dec. 23[d]

Dear Sir

I do not know whether you will forgive me, a stranger to you personally though not to your works, taking the liberty of begging a favour of you.—[2] Namely to ask for a piece of information, which you are more likely to be able to give than anyone, though it is doubtful whether you will be able.— I am particularly anxious to know & be permitted to quote (if I find it desirable) any fact showing, that a variable species is or is not equally variable at all times & places.—[3] Some facts seem to show that a species may vary far more in one area than in another; & some other, & perhaps more numerous, facts seem to indicate that a variable species is always equally variable.

With your profound knowledge of Brachiopoda, you may be able to give me examples & inform me how this is, with the species which have had a long existence or wide range.— Of course in order to judge, specimens must have been examined nearly equal in number at the two times or places.— Variability in close connexion with or caused by attachment to various substances would probably be alike at all times; so that there are many difficulties in coming to any conclusion.[4]

I formerly talked on this subject with the late E. Forbes, & more recently with M[r] Woodward of the British Museum, but could get no definite light.—[5] If you are willing to give me information on this head, I sh[d] esteem it a great kindness, which I have no right to expect, except as a fellow student in Natural History.

I beg to remain | Dear Sir | Yours faithfully | Ch. Darwin

American Philosophical Society (142)

[1] Dated by the relationship to the letter from T. W. St C. Davidson, 29 December 1856.

[2] Davidson was a fellow of the Geological Society of London and the author of works on Brachiopoda, particularly a monograph on fossil brachiopods (Davidson 1851–86).

[3] CD had begun writing chapter 4 of his species book, 'Variation under nature' (*Natural selection*, pp. 95–171). In his discussion of individual variations, CD stated that: 'I have applied, also to Mr.

Davidson, whose vast experience in Brachiopodous shells, makes his opinion of the highest value'
(*Natural selection*, p. 106).

4 See letter from T. W. St C. Davidson, 29 December 1856.

5 In *Natural selection*, p. 106, CD reported that: 'innumerable examples could be given of the foregoing cases [of variability] & this was all that I could learn on this subject from the late Prof. E. Forbes & from Mr. Woodward.' See letters from S. P. Woodward, 2 May 1856 and 15 July 1856.

To J. D. Hooker 24 December [1856][1]

Down Bromley Kent
Decr 24th

My dear Hooker

I am particularly glad of the reference about Leguminosæ, as I am at a dead lock with regard to them.[2] Botanists seem to differ so much about Campanulaceæ, that I have given them up almost in despair, & cannot but think old C. C. Sprengel was right.—[3]

What a letter of Col. Portlock;[4] it beats hollow all the many proofs which I have seen how little men, who are not naturalists, understand of Nat. History.—

I shall write to Harvey as it will cost him only a brief answer.—[5]

How I do wish I lived near you to discuss matters with.— I have just been comparing definitions of species, & stating briefly how systematic naturalists work out their subject:—Aquilegia in F. Indica was a capital example for me.—[6] It is really laughable to see what different ideas are prominent in various naturalists minds, when they speak of "species" in some resemblance is everything & descent of little weight—in some resemblance seems to go for nothing & Creation the reigning idea—in some descent the key—in some sterility an unfailing test, with others not worth a farthing. It all comes, I believe, from trying to define the undefinable.

I suppose you have lost the odd black seed from Birds Dung, which germinated—anyhow it is not worth taking trouble over— I have now got about a dozen seeds out of small Birds dung.—[7]

Adios— | My dear Hooker | Ever yours | C. Darwin

DAR 114.3: 187

1 Dated by CD's interest in Leguminosae, which extended through the autumn and winter of 1856 (see letter to George Bentham, 26 November [1856]).

2 CD had tried to ascertain the probability of cross-fertilisation in the Leguminosae (see letter to George Bentham, 26 November [1856], and letter to George Bentham, 30 November [1856]). No reference is made to Hooker in the discussion of Leguminosae in *Natural selection*, pp. 68–71.

3 Christian Konrad Sprengel maintained that fertilisation of Campanulaceae takes place after the flower is opened (Sprengel 1793, p. 117). An annotated copy of Sprengel 1793 is in the Darwin Library–CUL.

4 Joseph Ellison Portlock was inspector of studies at Woolwich. It is not known to what letter CD refers.

5 See letter from W. H. Harvey, 3 January 1857.

6 J. D. Hooker and Thomson 1855, p. 44. CD had earlier discussed the case of *Aquilegia* with Hooker (see *Correspondence* vol. 5, letter to J. D. Hooker, 14 [August 1855]).

[7] CD's attempts to germinate seeds extracted from birds' dung are recorded in his Experimental book, p. 15 (DAR 157a). Hooker had previously identified some of the seeds for him (see letter from J. D. Hooker, 22 November 1856).

From H. C. Watson [28 December 1856][1]

The comparative rarity of the subspecies or simple varieties, cut off from alleged true species, might be shown by numerous examples.[2] Indeed, this greater rarity would seem to be matter of course; for otherwise the subspecies or variety would have been itself described for the type species in most instances. In the following short list of names, the first of each pair is that of a common British plant, generally received as a true species; while the second name belongs to a subspecies, closely allied to the first, and much more scarce or local:—

Caltha palustris L. C. radicans, Forst.
Ranunculus Flammula, L. R. reptans, Lightf.
Draba verna, L. D. præcox, Reich.
Ulex europæus, L. U. strictus, Mack.
Genista tinctoria, L. G humifusa, Dicks.
Agrimonia Eupatoria, L. A. odorata, Ait.
Apargia autumnalis, L. A. Taraxaci, Sm.
Centaurea nigra, L. C. nigrescens, Aut.
Gentiana germanica, L. G. germanica, W.
Erythræa Centaurium, Pers. E. latifolia, Sm.
Veronica officinalis, L. V. hirsuta, Hopk.
Veronica serpyllifolia, L. V. humifusa, Dicks.
Taxus baccata, L. T. fastigiata, Lindl.
Asplenium Adiantum negriem L— A. acutum, Bory.

The comparative rarity of intermediates is not very readily proved by examples, without the aid of explanatory descriptions that would be tedious. The difficulty lies partly in the fact that they are often without special names; partly also because they are usually assigned as varieties of the species to which they approximate most, and are thus not viewed as intermediates by many botanists. The grounds for designating them intermediates are shortly these. 1. Some of them form a series of varieties passing gradually into both species. 2. In other instances, both species vary, and the varieties approximate. 3. In more numerous instances, perhaps, one of the species varies; its varieties assuming some of the characters of the other species, either positive or negative, but without passing into that other species.[3]

Examples of No 1.

Geum urbanum (L). and G. rivale (L). are so connected by a series of intermediate forms, that botanists hesitate to which species the intermediates should be assigned; some aggregating them into a third species (G. intermedium Ehrh.); others pronouncing them hybrid forms.

Saxifraga umbrosa (L.) & S. Geum (L). are also connected by endless intermediates; among which the S. elegans (Mack) and S. hirsuta (L.) are often deemed to be real species.

Primula veris & P. vulgaris (of English authors) are closely connected by a full series of intermediates, always now held varieties or hybrids.

Pyrus Aucuparia (Gaertn.) and P. Aria (Sm.) are imperfectly connected by the P. pinnatifida (Sm.), which seems to be usually called a species because botanists are at a loss to choose its real ally between the other two; its alledged distinguishing characters are variable, & perhaps irregular growths.

Examples of No. 2.

Arenaria rubra (L.) and A. marina (Oed.) both vary into forms that resemble each other more closely than the type species; and probably the Arenaria media (Aut.) may occasionally include varieties of each.

Viola tricolor (L.) and V. lutea (Huds.) have numerous varieties; some of which almost touch or pass into each other, as the V. Curtisii (Forst.) and V. sabulosa (Boreau); the former supposed a variety of V. lutea, the latter a variety of V. tricolor.

Linaria vulgaris (Mill.) and L. repens (Ait.) both seem to vary; the varieties approximating. The most intermediate form is the L. sepium (Allm.) conjectured to be a hybrid.

Circæa Lutetiana (L.) and C. alpina (L.) both vary; and it is occasionally difficult to assign their varieties correctly to the species, so closely do they approximate. The intermediates are more frequent than C. alpina, less so than C. lutetiana.

Examples of No. 3

Erica Tetralix (L.) and E. ciliaris (L.) are very distinct in their typical forms; but a series of varieties occurs, perhaps all of them forms of E. Tetralix, which show a gradual passage from Tetralix towards ciliaris. One of these forms is the E. Mackaii (Hook.) known only in single localities in Ireland and Spain. The other intermediates (Watsoni & Borreri) occur in Cornwall, & only near Truro.

Stachys ambigua (Sm.) is by some authors held a species, by others a variety of S. palustris (L.) In so far as it differs from the latter, it may be said to shade off towards S. sylvatica (L.)

Luzula Borreri (Bromf.) lately described for a species, seems a variety of L. pilosa (W.). Where it differs from the latter, it approaches to L. Forsteri (DC.).

More briefly put it may be said that the following intermediate subspecies &c. are more rare than the species.

Geum intermedium (Ehrh)
    Series of vars. between G. rivale & G. urbanum
Saxifraga hirsuta (L) &c.
    Subspecies between S. umbrosa and S. Geum.
Primula intermedia (var.) &c.
    A series of varieties between P. veris and P. vulgaris.

Pyrus pinnatifida (Sm.)

A variable subspecies between P. Aucuparia and P. Aria.

Linaria sepium (Allm.), &c.

A series between L. vulgaris and L. repens, including varieties of each probably.

Arenaria media (Aut.) &c.

Varieties of A. rubra and A. marina, closely approximating.

Viola Curtisii (Forst.) & V. sabulosa (Bor, &c.)

Varieties of V. tricolor and V. lutea, closely approximating.

Erica Mackaii (Hook.) &c.

Varieties of E. Tetralix passing towards E. ciliaris.

Stachys ambigua (Sm.)

A variety, itself rather variable, of S. palustris passing towards S. sylvatica, but not closely so.

Potentilla procumbens (Sibth.)

A variety of P. Tormentilla passing rather slightly towards P. reptans.

Luzula Borreri (Bromf.)

A subspecies mingling the characters of L. pilosa and L. Forsteri.

Juncus diffusus (Hoppe)

A doubtful species between J. effusus and J. glaucus in general characters.

A memorandum
DAR 98: 15–18

CD ANNOTATIONS
1.1 The . . . instances. 1.4] 'Q' *added pencil, circled pencil*
2.1 The . . . botanists. 2.5] 'Q' *added pencil, circled pencil*
14.20 Stachys . . . closely so. 14.22] 'Q' *added pencil, circled pencil*
Top of first page: 'On Rarity of intermediate forms.' *pencil*; 'Dec. 28 1856. (8th letter)' *pencil*
End of last page: '12' *pencil*

[1] The date is given by CD's annotation.
[2] CD used information from this memorandum on the rareness of intermediate forms and on the occurrence of varieties connecting distinct species in *Natural selection*, pp. 102, 268.

From T. W. St C. Davidson 29 December 1856

48 Park Crescent | Brighton.
29. Dec. 1856.

Dear Sir

I beg to thank you for your kind note of the 23ᵈ which I received only yesterday morning, & happy would I feel were it in my power (which I regret it is not) to transmit the information required.[1] The subject is one to which I have myself paid much attention without being able to arrive at a satisfactory conclusion; for no shells can be more variable than the Brachiopoda, so much so that it seems very often difficult if not at times impossible to know where to draw a line of demarcation or to know what are the true limits of variation peculiar to individuals of a same species— The vast number of Brachiopoda that have passed through my hands or that have been examined by myself both in situ and in collections at home & abroad, have amply confirmed the opinion you have expressed that some forms even genera are more variable than others both in time and space. While preparing my monographs I have had **much more difficulty** in dealing with *some* than with other Species, and I have found that a variable Species is so in all localities however distant from which I have been able to examine Specimens and vice versa a form less variable is also so both in time & space. Thus the *Spirifera Rostrata* of Schlotheim has been found far spread both in Europe & America & every where has it presented that great variabili⟨ty⟩ of shape & character visible in any particular locality, being elongated or transverse more or less inflated smooth or ribbed & with or without a defined mesial fold & sinus—thus tempting naturalists to propose a number of names for what in reality constitutes but *one single species* all these extreme shapes being connected by every intermediate shade or passage to such an extent that it would be really impossible to trace a line of stoppage without violating nature which has granted to its forms a large range of variation. Ter. *resupinata* T. *punctata* and many hundred examples could be adduced in proof of this nature of things. (it is understood I only allude to full grown individuals) Now on the contrary certain other forms such as *Rh. wrightii.* and in particular many species of *Lingula* vary but little both in time & space—[2]

What causes these variations is difficult to explain but they are no doubt due to various causes depending on the same law which holds good with all the animal kingdom. That is to say that no two animals of a same species are exactly alike or cast in the same mould and t⟨hat⟩ the differences we perceive in our species are equally common to all types of life although our eyes or minds may not be able to weigh or appreciate them to an equal extent. Therefore variability is common to all species in a more or less degree dependent often on local circumstances such as habitat, climate, food etc—which forms races & true varieties. Thus for example some shores & bottoms seem to have been more favourable than others for the regular growth of individuals of a same species & here we have the cause why individuals of a same species are all more different in two such localities. Thus the *Terebratella Menardi* at Mans in France attained there its largest dimensions & most elegant shapes its ribs are sharp its shell thin and delicate, while at Farringdon in England where the same form occurs all the examples are stunted in growth the ribs more blunted & the shell more thickened & if we examine the bottoms on which they lived we find at once the cause to account for this remarkable difference. All these facts are as well or better known to yourself & to all naturalists & I repeat them merely to show that I cannot produce any ne⟨w⟩ explanation than that whi⟨ch⟩ has been already advocated, nor in repeated conversations I have had with several distinguished Zoologists have I been able to obtain any definite light on the subject of your note—

M.ʳ *Bouchard* Director of the Museum of Boulogne-sur-Mer³ has commissioned me to purchase if possible for him a memoir of yours, *A Monograph of the Subclass Cirripeda* 1854⁴ but he does not inform me where it is published, and as I have never seen the work would feel greatly obliged if you would kindly inform me where it can be had in town, as M.ʳ Bouchard is also writing a paper on the same subject and is most anxious to possess the work for reference etc—⁵

I am at present preparing a monograph of British Permian & Carboniferous species which will be illustrated by about 50 quarto plates⁶

I beg to remain Dear Sir Yours faithfully | Thoˢ Davidson

DAR 162: 116

CD ANNOTATIONS
1.13 in dealing . . . Specimens 1.15] *double scored pencil*

¹ See letter to T. W. St C. Davidson, 23 December [1856].
² Davidson's information was cited by CD in *Natural selection*, p. 106.
³ Nicholas Robert Bouchard-Chantereaux.
⁴ The two volumes of *Living Cirripedia* (1851) and (1854) were published under the title of *Monograph of the subclass Cirripedia* by the Ray Society, London, and were generally available only to members who subscribed to the society. *Fossil Cirripedia* (1851) and (1854) were also published by subscription by the Palaeontographical Society.
⁵ Bouchard-Chantereaux seems to have published works on Mollusca only.
⁶ The second volume of Davidson's *Monograph of the British fossil Brachiopoda* (Davidson 1851–86), published by the Palaeontographical Society.

To Asa Gray 1 January [1857]¹

Down Bromley Kent
Jan. 1ˢᵗ

My dear D.ʳ Gray

I have received the 2.ᵈ part of your paper,² & though I have nothing particular to say, I must send you my thanks & hearty admiration. The whole paper strikes *me* as quite exhausting the subject, & I quite fancy & flatter myself I now appreciate the character of your Flora. What a difference in regard to Europe your remarks in relation to the genera makes! I have been eminently glad to see your conclusion in regard to species of large genera widely ranging: it is in strict conformity with the results I have worked out in several ways.³ It is of great importance to my notions.⁴ By the way you have paid me a *great* compliment; to be *simply* mentioned even in such a paper, I consider a very great honour.—⁵

One of your conclusions makes me groan, viz that the line of connection of the strictly Alpine plants is through Greenland: I sh.ᵈ *extremely* like to see your reasons published in detail, for it "riles" me (this is a proper expression; is it not?)

dreadfully.—[6] Lyell told me, that Agassiz having a theory about when Saurians were first created, on hearing some careful observations opposed to this, said he did not believe it, "for Nature never lied"— I am just in this predicament & repeat to you that "Nature never lies"; ergo, theorisers are always right.—

In reading your paper, one point struck me as well worth working out, *if it could be done,* viz a comparison of the principal zone of habitation in the U. States of the 320 European plants in comparison with your 115 representative species + the 15 strongly marked varieties = 130 species; & there again rudely with the (4th & 5) classes of Strictly Congeneric & perhaps divergent Congeneric species.—[7] I should be astonished if you do not get a very curious & harmonious result, ie on the great principle that Nature never lies.—

Overworked as you are, I daresay you will say that I am an odious plague;— but here is another suggestion! I was led by one of my wild speculations to conclude (though it has nothing to do with geograph. distribution, yet it has with your Statistics) that Trees would have strong tendency to have flowers with dioicous, monoicous or polygamous structure. Seeing that this seemed so in Persoon, I took our little British Flora, & discriminating trees from Bushes according to Loudon, I have found that the result was in species, genera & Families, as I anticipated.[8] So I sent my notions to Hooker to ask him to tabulate N. Zealand Flora for this end, & he thought my result sufficiently curious, to do so; & the accordance with Britain is very striking, & the more so, as he made 3 classes of Trees, Bushes, & herbaceous plants.—[9] (He says further he shall work the Tasmanian Flora on same principle.) The Bushes hold intermediate position between the other two classes.— It seems to me a curious relation in itself, & is very much so, if my theory & explanation are correct.

With hearty thanks | Your' most troublesome friend | Ch. Darwin

Pray do not forget *variability* of Naturalised plants.—[10]

P.S. | You might give me a valuable piece of information, with very little trouble to yourself.— I have been comparing, as far as I can, Protean genera, & have left off in a maze of perplexity. By Protean genera, I mean such as hardly two Botanists agree in about the species,—what to call species & what varieties. Now what I want to know is, whether such genera as Salix, Rubus, Rosa, Mentha, Saxifraga, Hieracium, Myosotis, &c have equally Protean species in U. States; even if they have only one, but more especially if they have many. I think you have no Rosa, & I forget how it is with some of the other genera.— The converse case w^d be equally valuable to me if you would think over your half-dozen or dozen worst genera which have any European species, & then I could find out whether such are very troublesome in Europe.— I think Hooker told me that in Himalaya, Rubus & Salix, though large genera, were not troublesome to make out.— I think Protean genera of shells are troublesome at all geological times & in all places.

Gray Herbarium of Harvard University

[1] Dated by the relationship to the letter from Asa Gray, 16 February 1857.

[2] A. Gray 1856–7.

[3] In his letter to Asa Gray, 2 May [1856], CD had asked Gray to examine the ranges of species in large and small genera of North American plants. CD expected large genera to include the most successful, and hence most common and most widely ranging, forms. CD had previously found some support for his views in A. de Candolle 1855, but Alphonse de Candolle's evidence was not clear-cut because, CD believed, his comparison had been based on families rather than on genera. Gray had compared the species common to North America and Europe, first using families and then genera, and had concluded: 'To be of any value, at least upon our limited scale, the comparison should be made with genera, as Mr. Darwin suggests' (A. Gray 1856–7, p. 77).

[4] See *Natural selection*, pp. 140–67, in which CD discussed 'the relation of the commonness and diffusion of species to the size of the orders and genera in which they are included'. See also J. Browne 1980.

[5] In mentioning CD's views, Gray stated that 'from some investigations of his own, this sagacious naturalist inclines to think that species in large genera range over a wider area than the species of small genera do.' (A. Gray 1856–7, p. 77).

[6] Comparing the northward range of species common to North America and Europe, Gray concluded that 'it seems almost certain that the interchange of alpine species between us and Europe must have taken place in the direction of Newfoundland, Labrador and Greenland, rather than through the polar regions' (A. Gray 1856–7, p. 73). CD, on the other hand, believed that a circumpolar flora had travelled only in a north–south direction to colonise lands in lower latitudes (*Natural selection*, pp. 539–42). See also letter from Asa Gray, 16 February 1857.

[7] CD refers to the section entitled 'Comparison of the flora of the northern United States with that of Europe in respect to the similar or related species' (A. Gray 1856–7, pp. 78–84), in which Gray described five different categories of representative species: geographical varieties, very close representative species, strictly representative species, strictly congeneric species, and divergent congeneric species.

[8] Persoon 1805–7 and Loudon 1842 (see *Natural selection*, pp. 61–2). For Gray's response to this query, see letter from Asa Gray, 1 June 1857.

[9] See letters to J. D. Hooker, 1 December [1856] and 10 December [1856], and letter from J. D. Hooker, 7 December 1856.

[10] CD had continued to remind Gray about this topic after having first mentioned it in his letter to Asa Gray, 2 May [1856]. See also letters to Asa Gray, 24 August [1856] and 24 November [1856], and letter from Asa Gray, 4 November 1856.

From William Henry Harvey 3 January 1857

Trin. Coll. Dublin
Jan. 3^{d} 1857.

My dear Sir

Your letter arrived during my absence in the Country at Xmas, and I only received it on my return to town this evening.[1] In reply to your query respecting the sexes of Algæ—the great majority with whose male & female organs I am acquainted are *Monoecious* in the Linnæan sense—but the different sexes are frequently placed in juxta position, within the walls of a common receptacle,— much in the same way as those of *Euphorbia* are brought within an involucre. In most of the *Fucoideæ* both *spores* & *antheridia* are found in each little cyst of the receptacle

— But in some *Cystoseiræ* & *Sargassa* they occur in different cysts.— And other genera are truly Diœcious, such as Splachnidium, Durvillœa, Sarcophycus, Hormosira, & some others—& many Sargassa.— *Cutleria multifida* is a good example of a clearly diœcious plant.

I do not remember any invariably Diœcious Rhodosperm (or Floridea)—except perhaps some of the Callithamnia may be so. They are at least *polygamous*. You are, of course, aware of the double female system of the Florideæ—which is invariably *dioecious*—but I think the *male* is common to either kind of female. It is perhaps more common to find antheridia on individuals bearing "*tetraspores*", than on those that bear "spores".—

I am not acquainted with any instance of true hermaphroditism, such as occurs in flowers. The nearest approach to such is found in the Fuci, & in certain Dictyoteæ & Chordarieæ—

No doubt you are familiar with Thurets admirable papers on the Sexes of the Algæ, in various Vols. of An. des Sc. Nat.[2] If not, you should look to them.

I am obliged by your kind congratulations on my return home.[3] I hope we may see you at the Dublin meeting of Brit. Ass[n]. this year, & that Hooker will be here too.[4]

Believe me | my dear Sir | very truly yours | W. H. Harvey.

DAR 166: 115

CD ANNOTATIONS
1.1 Your . . . evening. 1.2] *crossed pencil*
5.1 I am . . . yours 6.1] *crossed pencil*
Top of first page: 'Ch. 3.'[5] *pencil*

[1] CD's letter has not been found, but see letter to J. D. Hooker, 10 December [1856], in which CD mentioned his intention to write to Harvey concerning the separation of sexes in Algae.

[2] CD had earlier asked Joseph Dalton Hooker for a copy of Gustave Thuret's famous paper (Thuret 1854–5) on the sexual reproduction of Algae (letter to J. D. Hooker, 21 [May 1856]).

[3] Harvey had recently returned from a three-year visit to Ceylon and Australia, during which he had made extensive collections of Algae and other botanical and marine invertebrate specimens (*Memoir of W. H. Harvey . . . with selections from his journal and correspondence* (London, 1869), pp. 244–312).

[4] Neither CD nor Hooker attended the Dublin meeting of the British Association for the Advancement of Science, 26 August – 2 September 1857.

[5] A reference to chapter 3, 'On the possibility of all organic beings occasionally crossing', of CD's species book (*Natural selection*, pp. 35–91).

To T. H. Huxley   4 January [1857][1]

Down Bromley Kent
Jan. 4th

My dear Huxley

I congratulate you heartily on M[rs] Huxley's troubles being over.—[2] It is a horrid affair at the best of times; & I fear a repitition will not make you feel younger.—

The Balanus is positively B. balanoides, & you may positively state that the surface to which it adhered was not beneath level of lowest tide[3]

I will bring specimen when next I come to London.—

Ever yours | C. Darwin

Imperial College of Science, Technology and Medicine Archives (Huxley 5: 48)

[1] The year is given by CD's reference to the birth of Noel Huxley (see n. 2, below).
[2] Noel Huxley was born 31 December 1856 (L. Huxley ed. 1900, 1: 150–1).
[3] CD had described the tidal locations in which individuals of *Balanus balanoides* were found in *Living Cirripedia* (1854): 272.

From Richard Hill   10 January 1857

Spanish Town Jamaica
10th January 1857.

My dear Sir

I duly received your communication by the early December packet,[1] and I shall feel pleasure in attending to all and any subject here, in which you are desirous of information.

Mr Shakspear of whom I spoke as a needful assistant, in getting specimens of the Cithagra, has just returned from England.[2] I communicated what you wished. He has promised me his assistance. He told me that M[r] Leyden of Black-river, who is now in England, attempted to breed the Cithagra Braziliensis with the Canary, but they refused to pair, and he found the Brazilian Birds so impatient of restraint that they seldom endured the cage long. Another friend who has offered me his ready aid, tells me that in another part of the island, the true Canary has been naturalized, and that it is a ready singer.— I suspect he has mistaken the Canary Warbler of America for it,—the Sylvicola petechia (æstiva). He is however a good Naturalist, and will not err on investigation.

We have a well ascertained collection of the Land Mollusca of Jamaica in the British Museum presented by Mr Edward Chitty one of our late Chairmen of Quarter Sessions now in England.—[3] He will give you with pleasure information on the limited localities of species, or their special diffusion. He has made that branch of our Natural History a subject of minute study. In very many instances with us particular species have no extension beyond a particular Valley. Mr Edward Chitty will be found at his Brothers, M[r] Thomas Chitty, King's Bench Walk, in the Temple London.

I should ascribe considerable influence to Hurricanes, in the work of diffusion in our West Indian Islands. New Continental birds are occasionally driven before these winds to us, and so doubtless are insects.— Water spouts do something also in this way. There is a record of frogs having been brought to Port au prince Haiti in a whirlwind. In April last year we suffered exceedingly in the vicinity of Kingston by Tornado rains. I was at the East end of the Island, and saw from the Hills the morning after the deluging rains that fell, a spout passing along the coast, and three had been approaching the island the day previous. We owe the diffusion of crustacea in our mountain streams to migratory Ducks, through whom the eggs pass undigested.

Instances have been recorded of some of the large Boas reaching the West Indian Islands on trees pushed out into our Seas, by freshets from the great Cotinental rivers. There is a particular instance recorded of the Island of Saint Vincent receiving such a Visitor.

Seeds are certainty spread by the prevailing Currents. The Hibiscus populneus a Malabar tree mallow,—called by us the Gamboge Mallow, has been naturalized at Port-Royal, where it is very common now. It has been carried out to the out keys and Islets, by the land breeze and will become common enough upon them.

I take the liberty of sending you a little brochure I printed, entitled "a Week at Port Royal",—not that it contains anything of any kind of service to you, but because it is a contribution to our local Natural History.[4]

I have a finished paper rather long on migratory birds, which I will copy out on an early opportunity and send you.[5]

Any-thing and every-thing I have are at your Service to be used when and how you please.[6] So long as they are accepted as information my object is accomplished.

With respectful assurances, believe me, very faithfully, | Your obedient Servant, | Richard Hill

Charles Darwin Esqre

DAR 205.2 (Letters)

CD ANNOTATIONS

0.1 Spanish Town . . . (æstiva). 2.9] *crossed pencil*
3.5 In very . . . Valley. 3.6] *double scored pencil*
4.2 New . . . insects.— 4.3] *triple scored pencil*
4.8 We owe . . . undigested. 4.10] *double scored pencil*; 'Ask' *added pencil*
5.1 Instances . . . Richard Hill 10.2] *crossed pencil*
7.1 I take . . . send you. 8.2] *scored pencil*
Top of first page: '18'[7] *brown crayon*
Bottom of last page: 'Tropical Fish eat seed?' *pencil, del pencil*

1 CD's letter has not been found. CD had written to Hill at the suggestion of Philip Henry Gosse, from whom he had requested information concerning the domesticated animals found in Jamaica. See letters to P. H. Gosse, 22 September [1856] and 28 September 1856.

2 In the description of *Crithagra braziliensis*, the golden crowned canary, in P. H. Gosse 1847, pp. 245–7, Gosse stated:

> These birds are believed in Jamaica to be the descendants of some pairs of the common Canary turned out. "A gentleman of the colony named Shakspeare," observes Mr. Hill, "many years ago, touching at Madeira on his voyage to this island, is said to have procured several male and female Canaries, which he set at large in the fields about the rectory at Black River, where they have multiplied, and have become wild birds of the country. . . ."

3 Edward Chitty had lived in Jamaica since 1840. In 1854 he presented a large collection of land and freshwater shells of Jamaica to the British Museum (British Museum (Natural History) 1904–6, 2: 707).

4 Hill 1855. This paper is in the Darwin Pamphlet Collection–CUL.

5 This paper has not been located in the Darwin archive, but another one on the fishes of Jamaica, published in the *Transactions of the Jamaica Society of Arts* 2 (1855): 115–16, is in the Darwin Pamphlet Collection–CUL.

6 In *Variation* 1: 294 nn. 43 and 44, CD referred to information on the guinea-fowl of Jamaica that Hill provided in a subsequent letter, now missing.

7 The number of CD's portfolio of notes on the means of geographical dispersal of plants and animals.

From Charles Lyell [16 January 1857][1]

		Mammalia in the whole world[2]
Eocene	Thanet Sands	0
	hiatus	0
Secondary	Maestricht beds	0
	White Chalk	0
	Chalk Marl	0
	Upp. G.S.	0
	Gault	0
	Lower G.S.	0
	Weald Clay	0
	Hastings beds	0
	Upper Purbeck	0
	Middle d°.	12
	Lower d°.	0
	Portland Ool.	0
	Kimmeridge	0
	Calc. Grit	0
	Oxf^d Clay.	0
	Lower Ool. (Stonesfield)	4
	Lias.	0
	Trias (upper) Stuttgardt.	1

I have casts & beautiful drawings made at my expence when last at Stuttgard of the Microlestes & am much pleased at knowing what that oldest of yet found mammals was.[3]

So the "Nochnichtgefundenseyn" (a capital specimen of a German substantive) of the angiospermous plants in rocks older than the chalk offers no reason to anticipate the rarity of warm blooded quadrupeds—[4]

I hope you are all well— Your brother gives us news now &

AL incomplete
DAR 205.9 (Letters)

CD ANNOTATIONS
Top of first page: 'Jan 16/57/' *ink*; '22'[5] *brown crayon*

[1] Dated by CD's annotation.
[2] This list of fossil mammals probably resulted from Lyell's interest in the fossils recently discovered by Samuel Beckles. Lyell encouraged Beckles to excavate the Purbeck beds in a quarry in Swanage, Dorset, and had passed the specimens on to Hugh Falconer and Richard Owen. They were identified as new species of insectivorous mammals. Lyell announced the discovery at a meeting of the Philosophical Club of the Royal Society on 18 January 1857 (Bonney 1919, p. 134). He also mentioned the significance of these fossils for his views in a letter to Charles James Fox Bunbury, 13 January 1857 (K. M. Lyell ed. 1881, 2: 238–9). See also following letter.
[3] The fossil molar teeth of this insectivorous mammal had been found in 1847 by Wilhelm Plieninger in a bed of Triassic rock at Württemberg (Wilson ed. 1970, p. xxxv). Lyell regarded the appearance of mammals in such early strata as powerful evidence against the theory of progressive development of living forms.
[4] Lyell's point was that the scarcity of advanced plants in geological formations older than the Chalk did not necessarily mean that mammals would also be scarce. He believed that mammals existed elsewhere during these periods and that fossil remains would be found as more formations were explored, although they would be rare because of the special conditions required for their preservation. See Wilson ed. 1970, pp. xxxi–xxxvii.
[5] The number of CD's portfolio of notes on palaeontology and extinction.

To J. D. Hooker   17 January [1857][1]

Down Bromley Kent
Jan. 17th

My dear Hooker

I saw Dr Daniell on Friday morning & told him there were general difficulties & he agreed to be guided by my advice as far as Royal Soc. was concerned.[2] And I pledged myself to act for him, as I would for myself, ie. give him the best advice I could whether or not to apply to Royal Soc. for grant &c. He expressed great zeal, & did not pretend that he had ever tryed to collect anything besides plants of economical value.—[3] I then went to Benett[4] & *in confidence* told him what you thought, & he agreed with your opinion to a large extent, but yet seemed to think that Daniell would work pretty well.— It seems R. Brown[5] had at once made same suggestion as you about sending out a collector with him; but Bennett agreed with me that it w^d. probably be death to him.[6] D^r. D. said he had not smallest objection to collector being sent out & said he w^d. nurse & look after him, but it would be in his opinion death to him.— I *most entirely* agree with you that if you decide that D^r. D. would not do a fair amount of work, it w^d. be very wrong in us to send him out in this way.— On other hand it seems to me no reason not to send him out, under the very special circumstance, because he is not a first rate man.—[7] I shall be guided by you, after you have consulted others.

Yours affect^y | C. Darwin

Endorsement: '1857'
DAR 114.4: 188

[1] It is probable that some of the endorsements recording the years of CD's letters were made by Hooker long after the date of receipt (see J. Browne 1978). The editors provide corroborative evidence for dates when possible. This letter is dated by the reference to the meeting of the Philosophical Club of the Royal Society that CD attended (see n. 2, below).

[2] At a meeting of the Philosophical Club of the Royal Society held on 15 January 1857, CD had asked whether a sum from the government grant to the Royal Society could be given to William Freeman Daniell for exploration of the mountains of Cameroun (Royal Society Philosophical Club minutes).

[3] Daniell had, since the 1840s, regularly contributed notices of African plants of potential economic or medicinal use to various London journals. His specimens were given to the Royal Botanic Gardens, Kew.

[4] John Joseph Bennett was secretary of the Linnean Society and assistant to Robert Brown in the botany department of the British Museum. Bennett had described several of the specimens that Daniell had collected in Sierra Leone.

[5] Robert Brown.

[6] West Africa was the deadliest malarial environment in the world (see Brockway 1979, pp. 127–33).

[7] CD refers to the private opinion of the Hookers that Daniell was not a particularly good collector, borne out by the difficulty of a later expedition to the River Niger to procure specimens because 'Dr. Daniel's localities did not afford us the plants he attributed to them.' (*Hooker's Journal of Botany and Kew Garden Miscellany* 9 (1857): 250).

To T. H. Huxley 17 January [1857][1]

Down Bromley Kent
Jan. 17th

My dear Huxley

I was most deeply interested by the lecture at the Royal,[2] & was much vexed at being forced to retreat from my head aching so, & a pretty night I had afterwards! It is a horrid shame to trouble you, but I can hardly get subject out of my head, & am so very curious on one point, that I must beg for an answer on a point which I daresay came into the tail of the paper.— The point is, do you suppose if the fragments of ice did *not* after their innumerable fractures become united together by freezing;[3] would the mass flow on? I suppose not, judging from the high angle of a talus of broken rocks though of course rock is far less brittle than ice. I presume the downward pressure could not be communicated through a fragmentary mass, just as you cannot push out an inch of sand in a gun barrel, though you could easily push out a piece of loose sandstone—or again, I suppose Tyndall would have found it far more difficult with his screw to have pushed out granular salt out of his semi-ring-model, than the piece of solid ice[4]—or again I presume a scaffold pole would slide lengthways down a slope, down which a pile of fragments of wood would not slide.— Do tell me whether this is correct, for it seems to me quite beautiful if the freezing of the brittle ice, accounts for its filling all inequalities, its apparent solidity, & its flowing motion.—

By the way Tyndall ought to explain for ignoramuses how he knows that the ice freezes together & not merely coheres, like two pieces of wet glass— I do not understand how the non-coherence of snow or ice under a freezing temperature explains this; for it would be dry & so would not cohere; nor could I understand how he distinguishes the pillars or pinnacles (if I understand rightly) by which the pieces of Wenham ice were united, from preexisting ice thawed away.[5] But I daresay all this is made clear in paper. The point which I want to know is the previous one, whether or not the freezing of the fragments together & the consequent easy transmission of force in one direction, is the cause of the flowing or sliding movement of the glaciers.—[6]

I never heard a more interesting paper than the part, which I did hear.— Forgive me for being so troublesome

Ever yours | C. Darwin

Down House (MS 8: 1)

[1] Dated by CD's reference to Tyndall and Huxley 1857 (see n. 2, below).

[2] Huxley and John Tyndall's paper 'On the structure and motion of glaciers' (Tyndall and Huxley 1857) had been read to the Royal Society of London on 15 January 1857. The paper, which challenged the viscous theory of glaciers proposed in 1843 by James David Forbes, contained observations made by the two during a trip to Switzerland in August 1856 and the results of subsequent experiments conducted by Tyndall.

[3] Assuming ice to be a hard, brittle substance rather than a viscous one, Tyndall suggested that under pressure the ice would break into fragments and that these would subsequently re-unite when they became moist through melting, a process he called 'regelation' on the suggestion of Joseph Dalton Hooker (see Rowlinson 1971, p. 192). Regelation explained how the form of the glacier could be moulded under pressure such that it appeared to flow as if it were viscous.

[4] Tyndall had used a hydraulic press in his demonstrations to show how ice could be moulded into any shape (Tyndall and Huxley 1857, p. 329–31).

[5] Tyndall had observed that ice blocks in a hot shop-window froze together despite the fact that the blocks themselves were melting (Tyndall and Huxley 1857, p. 329). Wenham ice was imported, mainly from Norway, by the Wenham Ice Company of the Strand, London.

[6] Tyndall and Huxley were not so much interested in the 'cause' of glacial flow, which, like Forbes, they ascribed to the weight of the glacier: 'the glacial valley is a mould through which the ice is pressed by its own gravity' (Tyndall and Huxley 1857, pp. 332–3). Instead, they were concerned with how the pressure was communicated through the body of the glacier, whether by virtue of its

true viscosity, as Forbes believed, or by the process of regelation in the fragmentary mass giving the appearance of viscous flow (p. 333). See also letters to T. H. Huxley, 3 February [1857], and to John Tyndall, 4 February [1857].

To J. D. Hooker 20 January [1857][1]

Down Bromley Kent
Jan. 20th

My dear Hooker

Very many thanks for your note.— with the opinions of yourself, Sir William & Bentham against D[r] D. I shall do nothing more.[2] I have written to him to say that under present circumstances &c I strongly advise him not to apply to Royal Soc.[y] for aid, & that will be the end of the affair I hope.— It is a thousand pities that the man is what you say.—

I was quite knocked up on Thursday at R. Soc[y] but all that I heard was most interesting. I have become so curious on subject, that I have written to Huxley to beg for a little light on two points.—[3] I really think that they will beat Forbes.— I sh[d] like to have talked over subject with you.—

Ever yours | C. Darwin

I have written to Sharpey & marking letter "Private" have told him what you say of D[r] D. & told him that I shall not stir in matter.[4]

P.S. I find Fish will greedily eat seeds of aquatic grasses, & that millet seed put into Fish & given to Stork & then voided will germinate.[5]

So this is the nursery rhyme of this is the stick that beat the pig &c &c—[6]

D[r] D. say Cameroons is accessible.—

Endorsement: '/57'
DAR 114.4: 189

[1] Dated by the relationship to the two preceding letters.
[2] William Jackson Hooker, George Bentham, and William Freeman Daniell. See letter to J. D. Hooker, 17 January [1857].
[3] See preceding letter.
[4] William Sharpey was a secretary of the Royal Society. The letter has not been found.
[5] Recorded in CD's Experimental book, p. 19 (DAR 157a).
[6] CD refers to the children's story 'The old woman and the pig'. For CD's earlier experiments to investigate fish as a means of distributing plant seeds, see *Correspondence* vol. 5, letters to W. D. Fox, 7 May [1855], and to J. D. Hooker, 15 [May 1855].

To J. D. Hooker [after 20 January 1857][1]

Down Bromley Kent

My dear Hooker

Very many thanks for your note, all of which always give me pleasure. D[r] Daniel gives up assistance from Royal Soc. very pleasantly, & I am glad I am out of that scrape.—[2]

I have been glad to see A. Grays letter:[3] there is always something in them, which shows that he is a very loveable man. He seems to come quite round to your view of A. de C.[4]

One must judge by one's own light, however imperfect, & as I have found no other Book so useful to me, I am bound to feel grateful: no doubt it is in main part owing to the concentrated light of the *noble art of compilation*. I was aware that he was not first who had insisted on range of Monocots. (Was not R. Brown in Flinders)[5] & I fancy I only used expression "strongly insisted on"—but it is quite unimportant.— If you & I had time to waste, I sh^d like to go over his Book & point out the several subjects, in which I *fancy* he is original.— His remarks on the relations of naturalised plants will be very useful to me—on the ranges of *large* Families seemed to me good, though I believe he has made great blunder in taking Families instead of smaller groups; as I have been delighted to find in A. Gray's last paper.[6] But it is no use going on.—

I do so wish I could understand clearly why you do not at all believe in accidental means of dispersion of plants.[7] The strongest argument, which I can remember at this instant is A. de C. that very widely ranging plants are found as commonly on islands as over continents. It is really provoking to me that the immense contrast in proportions of plants in N. Zealand & Australia seems to me strong argument for *non*-continuous land; & this does not seem to weigh in the least with you.[8] I wish I could put myself in your Frame of mind. In Madeira I find in Wollastons Books a parallel case with your N. Zealand case, viz the striking absence of whole genera & orders *now* common in Europe & (as I have just been hunting out) common in Europe in miocene periods.[9] Of course I can offer no explanation why this or that group is absent; but if the means of introduction have been accidental, then one might expect odd proportions & absences.— When we meet, do try & make me see more clearly than I do, your reasons.—

I should, indeed, most heartily enjoy a dinner such as you describe at the Wellington.

Will you look at enclosed Memorandum & give me any light if you can. I do not want details only your opinion.—

My dear Hooker | Most truly yours | C. Darwin

What a time it takes to make out the truth as we see in the Glaciers![10]

I am glad the Irish orchid is dished.—[11]

[Enclosure]

Memorandum

You know how I work subjects, namely if I stumble on any general remark, & if I find it confirmed in any other very distinct class, then I try to find out whether it is true, if it has any bearing on my work.—

The following *perhaps* may be important to me. D.ʳ Wight remarks that Cucurbitacea is a very *isolated* Family, & has *very diverging* affinities:[12] I find strongly put & illustrated the very same remark in genera of Hymenoptera.— Now it is not to me at first apparent why a very distinct & isolated group, sh.ᵈ be apt to have more divergent affinities than a less isolated group. I am aware that most genera have more affinities than in two ways, which latter perhaps is commonest case. I see how infinitely vague all this is. But I sh.ᵈ very much like to know what you & M.ʳ Bentham (if he will read this) who have attended so much to the principles of classification think of this. Perhaps the best way would be to think of half-a-dozen *most isolated* groups of plants, & *then* consider whether the affinities point in an *unusual number of directions*.

C. Darwin

Very likely you may think the whole question too vague to be worth considering.—

DAR 114.4: 190

[1] Dated by the relationship to the letters to J. D. Hooker, 17 January [1857], and 20 January [1857].

[2] See letters to J. D. Hooker, 17 January [1857] and 20 January [1857].

[3] CD refers to a letter from Asa Gray addressed to both William Jackson Hooker and Joseph Dalton Hooker dated 5 January 1857 (Asa Gray, Kew Correspondence 1839/73 (137/8), Royal Botanic Gardens, Kew).

[4] CD refers to Hooker's criticism of Alphonse de Candolle's *Géographie botanique raisonnée* (A. de Candolle 1855) in his review of the work ([J. D. Hooker] 1856). The review had been discussed by CD and Hooker in 1856 (see letter to J. D. Hooker, 9 October [1856]). In his letter to the Hookers (see n. 3, above), Gray had conceded that J. D. Hooker was correct 'in saying that he [Candolle] has conceived no problem which had not already been put forward, and in that sense has not *advanced* the science at all.' Gray went on to stress that Candolle 'has none of your originality of mind, nor high intellectual activity, but is a mere *worker*—a very good one in his way.' (Asa Gray, Kew Correspondence 1839/73 (137/8)).

[5] R. Brown 1814. Robert Brown had been naturalist to the expedition to survey the Australian coast, 1801–5, under the command of Matthew Flinders.

[6] A. Gray 1856–7. See letter to Asa Gray, 1 January [1857].

[7] See letter to J. D. Hooker, 15 March [1857], for an explanation of what is meant by 'accidental'.

[8] Hooker believed that geographical distribution of plants on the major land-masses in the southern Pacific was best explained by the assumption of a former land-bridge connecting the areas, an idea he had put forth in J. D. Hooker 1844–7 and 1853–5. See the earlier correspondence on this question (*Correspondence* vol. 5, letters to J. D. Hooker, 7 March [1855] and 5 June [1855], and letter from J. D. Hooker, [6–9 June 1855]).

[9] Wollaston 1854 and 1856. As Thomas Vernon Wollaston remarked in the introduction of *Insecta Maderensia*, 'the total absence of numerous genera (and even of whole families) which are looked upon as all but universal, constitutes one of the most striking features of our entomological fauna.' (Wollaston 1854, p. x). See also letter from T. V. Wollaston, [12 April 1857].

[10] See letter to T. H. Huxley, 17 January [1857].

[11] 'American-' has been interlined in pencil before 'Irish', possibly by Hooker.

[12] Wight 1841.

To William Sharpey, Secretary, Royal Society 24 January [1857][1]

Down Bromley Kent
Jan. 24th—

Dear Sharpey

Having never especially attended to the Natural History of the Region in question I really am quite unable to offer any special points of research.[2] And as far as general suggestions go, I cannot add to the Instructions published in the Admiralty Manual.—[3] As it seems that there will be a Geologist attached to the Expedition it seems superfluous to remark, that a collection of the Carboniferous plants from the Coal of that Region would preeminently possess high interest.[4] So again with Glacial action, more especially in regard to sea-borne erratic boulders, it would be highly desirable to ascertain their extension southward, inland, & to what elevation on the land.—[5]

I wish sincerely I could aid in any suggestions, but it is really not in my power.—

Pray believe me, Dear Sharpey | Yours very sincerely | Ch. Darwin

Royal Society (MC 17: 336)

[1] Dated by the reference to the North American exploring expedition (see n. 2, below).
[2] A committee of the Royal Society had been formed to prepare instructions for the guidance of members of a North American exploring expedition, 1857–9, under the direction of John Palliser. The purpose of the mission was to survey the territory between Lake Superior and Vancouver Island to determine the best available route connecting the eastern and western regions of Canada. William Jackson Hooker and Joseph Dalton Hooker prepared instructions for the botanist to the voyage (*Hooker's Journal of Botany and Kew Gardens Miscellany* 9 (1857): 214–19).
[3] The Admiralty *Manual of scientific enquiry* (Herschel ed. 1849), to which CD had contributed a chapter on geology. There had been a second edition in 1851.
[4] For CD's earlier interest in this question, see *Correspondence*, vol. 3, letter to J. D. Hooker, [December 1846].
[5] The subject of erratic boulders had long been of interest to CD. He believed such boulders had been transported by floating icebergs. See *Collected papers* 1: 218–27.

From Henry Doubleday 26 January 1857

Epping
Jany 26th 1857

My dear Sir,

With this you will receive a few specimens of *Tortrix* of which I beg your acceptance— they will illustrate the extraordinary variation of markings in two or three species—[1]

No 1. *Peronea Hastiana.* (*Tortrix Hastiana* of Linné)—the description in the Fauna Suecica applies to the variety marked thus + —and the specimen is still in the Cabinet—[2] This species feeds in the larva state upon the sallow and most of the varieties may be obtained from one brood of larvae—but they seem almost

endless and have been considered by most of the older writers as distinct species—
— I believe I was the first person in this country who felt convinced that the host
of nominal species of these Buttons" as they are called, would sink into two—
Hastiana and *Cristana*—and this ⟨   ⟩ by everyone—[3]

No. 2. *Peronea cristana*— this species is equally as variable as the last but all the
varieties have a raised tuft or button on the upper varying much in size and
colour—

No 3. *Pœdisca sordidana*— this species varies very much as you will see by the two
specimens sent—

In every family of *Lepidoptera* there seem to be two or three species extremely
prone to vary—and it is certain that in some localities they vary more than they
do in others. In rearing *Lepidoptera* from the eggs I have frequently found that if
you have a brood from a typical female of a varying species the moths will all be
like the parents—

Some years ago I made some experiments with the varieties of *Harpalyce
Russata*— Haworth & Stephens described several of these varieties as distinct
species—[4] I felt convinced that they were all one and to prove it I caught several
fem⟨ales⟩ and obtained a numb⟨er⟩ of eggs— I had bro⟨ods⟩ from three
typica⟨l⟩ ⟨   ⟩ of Haworth and when the moths appeared they were all exactly
alike— after this I caught a female of the yellow-banded variety—the *Comma-
notata* of Haworth—and from the eggs of this female I reared all the varieties—
With the central fascia yellow—white—black—and all intermediate shades
constituting the *Centum-notata, Comma-notata, perfuscata* &c of Stephens— —

It is singular that some species should vary so much while others closely allied
should scarcely ever vary at all— species too, really distinct—often approach so
closely as to be separated with difficulty in the perfect state— for instance *Acronycta
Psi, tridens* and *Cuspis*— yet the larvae of these three species are totally different
from each other—

Locality seems to affect the shade of colour of many species— Scotch specimens
of *Lepidoptera* are generally much darker than those of the south of England— this
is particularly the case with *Aplecta occulta, Hadena adusta, Xylophasia polyodon*, and
Aplecta tructa.

I hope you will write again if you want any other information that it is in my
power to give[5] and with best wishes believe me | My dear Sir | Yours most sincerely
| Henry Doubleday

Charles Darwin Esq

I do not want the box returned as it is of no value | H Dy

DAR 162: 235

CD ANNOTATIONS
5.1 In every . . . others. 5.3] *double scored pencil*
5.2 localities . . . others. 5.3] *triple scored pencil and brown crayon*

5.3 frequently . . . parents— 5.5] *scored brown crayon*
6.2 Haworth . . . species— 6.3] *scored brown crayon*
Top of first page: 'Ch. 4'[6] *brown crayon*; '1ˢᵗ letter' *pencil*

[1] CD had just completed chapter 4 on 'Variation under nature' of his species book (*Natural selection*, pp. 95–171). See 'Journal' (Appendix II).

[2] Linnaeus 1761, p. 346. Linnaeus's collections, including his cabinet of insect specimens, had been purchased in 1784 by James Edward Smith, founder of the Linnean Society. The specimen is listed in the catalogue of the Linnean collection (Jackson 1913, p. 32).

[3] Doubleday's chief interest was in standardising the system of nomenclature used in England with that followed by continental naturalists. Parts of his *Synonymic list of British butterflies and moths* appeared between 1847 and 1850, and a much more complete edition was brought out in 1859. Further supplements appeared in 1865 and 1873 (*DNB*).

[4] Haworth 1803–28 and Stephens 1828–46.

[5] See letter to Henry Doubleday, [before 5 February 1857].

[6] A reference to chapter 4 of CD's species book. See n. 1, above.

To T. H. Huxley 3 February [1857][1]

Down Bromley Kent
Feb 3ᵈ

My dear Huxley

Knowing how busy you are, it was a shame in me to trouble you; but you can form no idea how anxious I was about the flowing or sliding part; & I did not know that it was so difficult.—[2] Many thanks for telling me what you can.[3] Will not Tyndall experimentise upon broken ice in this horrid frost, & explain how two pieces of ice can freeze together,—[4] I hope & daresay he has.—

I am sorry to hear of the "jolly row" with Owen;[5] though I do not want to doubt that you are, as you once called yourself, as meek as a Dove—

With many thanks for your note, & most humble apologies for having bothered you | believe me | My dear Huxley | Ever yours | C. Darwin

Imperial College of Science, Technology and Medicine Archives (Huxley 5: 104)

[1] Dated by the relationship to the letter to T. H. Huxley, 17 January [1857].

[2] See letter to T. H. Huxley, 17 January [1857], in which CD posed several questions relating to the joint paper by Huxley and John Tyndall on the structure and flow of glaciers (Tyndall and Huxley 1857).

[3] Huxley had contributed little to the researches described in Tyndall and Huxley 1857. He claimed to have done nothing more than suggest that Tyndall might profitably apply his views on slaty cleavage to glacial phenomena. He also accompanied him to Switzerland to make first-hand observations of glaciers. As Huxley wrote to Joseph Dalton Hooker, 'Tyndall fairly *made* me put my name to that paper, and would have had it first if I would have let him, but if people go on ascribing to me any share in his admirable work I shall have to make a public protest.' (L. Huxley ed. 1900, 1: 144).

[4] Tyndall continued to experiment with ice, in order to explain the veined or laminated structure in some glaciers and to show that ice was a good conductor of heat (Eve and Creasey 1945, pp. 301–8). He did not, however, specifically address CD's question about whether regelation could occur when the temperature was below freezing (see letter to T. H. Huxley, 17 January [1857], and following letter).

[5] Richard Owen had announced a course of lectures 'on the osteology and palæontology, or the framework and fossils of the class Mammalia' to be delivered in the Museum of Practical Geology, Jermyn Street, beginning 26 February 1857 (*Athenæum*, 14 February 1857, p. 197) and also assumed the title of professor of palaeontology at the School of Mines. Huxley, professor of natural history at the School of Mines, whose own lectures included palaeontology, considered Owen's action was intended to undermine his position and broke off all personal relations with him (see L. Huxley ed. 1900, 1: 142, and A. Desmond 1982, p. 38).

To John Tyndall 4 February [1857][1]

Down Bromley Kent
Feb. 4th

Dear Tyndall

I am *very* much obliged to you for your note. My only excuse for having troubled Huxley, was my very great curiosity to hear something more of your views.—[2]

I am as ignorant of mechanics as a pig as you will have perceived; but Glaciers for years & years have interested me greatly.[3]

I am so very glad to hear that you are continuing your experiments on ice; & I hope to hear that you will explain about the freezing together of ice under the freezing point.—

I can fancy a man so ignorant of nat. History as to advise Owen to compare a skull with a vertebra;[4] on exactly same principle, I hope that you will squeeze together pieces of ice quite dry as far as water is concerned, but wetted with something which will not freeze. There is a valuable suggestion for you!![5]

I wish you all sorts of good fortune in your most interesting investigations; & the Lord have mercy on you, when Forbes answers you is my prayer[6]

Most truly yours | C. Darwin

It is beautiful your having given cleavage to ice.[7]

Down House (MS 8: 2)

[1] Dated by the relationship to the preceding letter.

[2] See letter to T. H. Huxley, 17 January [1857], and preceding letter.

[3] In *Volcanic islands*, pp. 70–1, CD compared the lamination he had observed in volcanic rocks to the zoned structure James David Forbes described in glaciers. See also *Correspondence* vol. 3, letters to J. D. Forbes, 11 October [1844] and [November? 1844].

[4] Probably an allusion to Thomas Henry Huxley's intention to attack Richard Owen's theory of the vertebral origin of the skull as put forward in Owen 1846. Huxley attempted to demolish Owen's vertebral theory in his Croonian lecture of 1858 (T. H. Huxley 1859), by which time the 'jolly row' mentioned in the preceding letter was well advanced (see A. Desmond 1982 and di Gregorio 1984).

[5] This was presumably a suggestion as to how to investigate the question, posed in CD's letter to T. H. Huxley, 17 January [1857], of how Tyndall knew 'that the ice freezes together & not merely coheres, like two pieces of wet glass—'.

[6] Forbes had previously engaged in a controversy with William Hopkins over the viscous theory of ice. Hopkins, like Tyndall, had challenged the physical principles upon which the viscous

theory rested. See C. Smith 1989. For the ensuing controversy between Tyndall and Forbes, see Rowlinson 1971.

[7] Tyndall had been drawn to the study of glaciers through his investigation of the role of pressure in producing slaty cleavage. Accepting the mechanical theory of cleavage put forth by Daniel Sharpe (Sharpe 1855), Tyndall held that the lamination observed in glaciers was, like the cleavage of slate, the result of pressure and that bedding and structure were different in nature and origin (Tyndall and Huxley 1857, pp. 339–46). CD, who had arrived at this conclusion through his consideration of cleavage and foliation in clay-slate, had been actively interested in the question for over a decade (*Correspondence* vols. 3, 4, and 5).

To Henry Doubleday [before 5 February 1857][1]

[Down]

have bred *all* these varieties from the same set of eggs, so that there can be no doubt that they are all the same species?[2] I trust & think that your kindness will make you forgive my writing so long & troublesome a letter.— I can assure you, that in so far as giving me knowledge, which I value, your note will not have been thrown away upon me.

Pray believe me | My dear Sir | Your's truly obliged | Charles Darwin

I must express my admiration at the skill with which your Box was packed up; & must confess, until I read your letter, I was somewhat dismayed to think how I sh^{d} ever return it half so neatly!

I think there is no difference in the Peronias except in the wondrous differences of colour, & somewhat in size.— I presume that there is no rule of any of the coloured vars. tending to be smaller or larger.[3]

Incomplete
American Philosophical Society (Getz 2032)

[1] Dated by the relationship to the following letter.
[2] See letter from Henry Doubleday, 26 January 1857. CD refers to Doubleday's description of having reared eggs of what had been claimed to be two distinct species of *Harpalyce* and finding, as he expected, that the resulting moths were identical.
[3] See following letter.

From Henry Doubleday 5 February 1857

Epping
Feby 5th 1857.

My dear Sir,

I am truly glad to find that the few insects which I sent were acceptable.— you are most heartily welcome to them and I regret that you did not write before my stock of duplicates was so much reduced— I could then have sent you varieties of several other species.

If all is well this year, I shall be most happy to send any specimens that I may obtain likely to interest you.

The variations of colour and markings in the *Peronea* are really astonishing—Haworth and the late J. F. Stephens did not at all understand the two species—*Cristana* and *Hastiana*— they created thirty or forty species founded upon the variations in colour and markings.[1] Very soon after I had paid any attention to them I felt convinced that all these reputed species would sink into two. I believe at that time every Entomologist in London thought that my ⟨o⟩pinion was ⟨err⟩oneous—but they now all admit that I was right.

In my last letter I alluded to a *Geometra*—*Harpalyce Russata*— there is an allied species—*immanata* of Haworth—typical specimens of which so much resemble those of *Russata* that no one but an Entomologist could separate them.— This species runs through similar variations to those of *Russata*—with the exception of the yellow-banded variety—the *Comma-notata* of Haworth.— The central fascia in *immanata* varies from white through all shades of grey to black, *but is never yellow.*

In answer to your query as to whether I can give any good instances of species varying in one locality and not in another I may mention the above-named insect *H Russata*— M[r] Logan[2] tells me that at Edinburgh he only takes the type—that it does not vary at all.

There is another Geometra—*Angerona prunaria* the typical specimens of which are pure orange with slight dusky irrorations— in the south of England a clouded variety with the wings broadly margined with deep brown is almost as common as the type— I beli⟨eve⟩ this variety ⟨ne⟩ver occurs in the north of England— Many species of Lepidoptera are always double-brooded—that is there are two distinct broods in each year— in many species there is little or no difference between the individuals of the two broods—but in others the differences in size and colour are constant and striking— *Selenia illustraria* and *S. illunaria* may be mentioned as instances— the spring broods appear in April—and the specimens are large and highly-coloured— from the eggs of these the second brood of moths comes forth in July— these are always much smaller than the vernal specimens and differently coloured— from these small summer specimens we of course have the large specimens the following spring—[3]

In the common White butterflies—*Pieris brassicæ* and *P. Rapæ* the reverse is the case— the vernal specimens are always smaller and more faintly marked than the summer ones.

I enclose a portion of my list with the varieties of the two species of *Peronea*[4]—all the varieties of *Hastiana*, I and others have reared in numbers from larvæ found upon sallows at the same time and which were exactly alike—and the perfect insects copulate indiscriminately—so that there is not a shadow of doubt that they all belong to one species The larva of *Peristera* is unknown but there is not the slightest doubt of all the varities belonging to one species as they appear simultaneously and every intermediate shade may be found. They differ from *Hastiana* in always having a tuft of Scales on the wing— All the varities of *Peronea* are liable, like most other species, to slight variations in size— I may just add that it extremely difficult to rear larvae of the minute moths from the eggs—but three or four years I confined two females of *P. Hastiana* under gauze upon a young sallow in our garden and upon which I had never seen a larva of this species before and in August I took a number from the curled up leaves and bred the greater part of the varieties— I have thus stated a few facts but can throw little or no light upon this interesting subject— There was nothing particular about the packing of the box—⁵ I always send parcels to the post done up in a similar manner—

I shall always be most happy to hear from and glad to serve you in any way in my power and with best wishes believe me | My dear Sir | Yours very Sincerely | Henry Doubleday

C Darwin Esq

DAR 162: 236–236/2

CD ANNOTATIONS
1.1 I am . . . species. 1.4] *crossed pencil*
4.3 this species . . . *Russata*— 4.4] 'parallel variation' *added pencil*
6.5 Many species . . . striking 6.8] 'Differences in double broods mentioned by Brown also' *added pencil*
6.8 *Silenia* . . . spring— 6.13] 'Q'⁶ *added pencil, circled pencil*
Top of first page: '2ᵈ letter' *pencil*

CD note:⁷

Feb 8ᵗʰ 57/
I may say the same genus (let us see what A. Gray says) may have variable (different) species in one country & fixed species in another.— In same country of closely allied species some are variable some are not— And lastly the same identical species may be variable in one district & not in another.—

¹ Haworth 1803–28 and Stephens 1828–46. See letter from Henry Doubleday, 26 January 1857.
² Robert Francis Logan.
³ CD cited this case in *Natural selection*, p. 358.
⁴ This list is preserved in DAR 162: 236/1.
⁵ See preceding letter.
⁶ The 'Q' stands for 'quoted' (see n. 3, above).
⁷ Now bound following Doubleday's letter in DAR 162: 236/2.

To W. B. Tegetmeier   6 February [1857]¹

Down Bromley Kent
Feb. 6ᵗʰ

My dear Sir

Many thanks for all the many *valuable* kindnesses in your letter.

I certainly shall be very much obliged for an egg or two of any Rumpless Fowl.— I am anxious just to see how early in our domestic breeds a rudimentary organ is rudimentary.²

If you hatch any for yourself, it would save me trouble, & not cost you any more to send me per Post, an egg *within* (ie before) about 24 hours of the period of hatching; & for this I shᵈ be very much obliged, though it is unreasonable to rob you of your valuable Birds. If you do not intend hatching,, I sh^d be very glad of an egg to hatch.—

Very many thanks, also, for Black Drake, for which I cannot send till next Thursday. The carrier shall ask whether he can pay Carstang 6^s · 6^d for it, as it will save my paying you, for it.—³ I have not as yet noticed much difference in skeletons of Ducks, but I shall now have all the principal Breeds.⁴

I am very glad to hear of Mr Gulliver's Runt; I am now well off for skeletons of all the Breeds; as I have splendid Carrier from Mr Hayne.⁵

Your letter is full of good news for me; as I was the other day wishing there had ever been any account of crosses between Pheasants & *different* breeds of Poultry.—⁶

With very sincere thanks for your great kindness to me, pray believe me, my dear Sir | Yours sincerely | Ch. Darwin

P.S. | By oddest chance this very morning I have heard of a German Pouter on sale! so will not trouble you.—⁷

New York Botanical Garden Library (Charles Finney Cox collection)

¹ Dated by the relationship to the letter to W. B. Tegetmeier, 11 February [1857].
² CD later discussed this topic in *Variation* 2: 315–18, where he explained the significance of the relatively greater development of rudimentary organs early in life. Rumpless fowls, however, are not mentioned.
³ Mr Carstang of Ship Tavern Passage, Leadenhall Market, London (CD's Address book (Down House MS)).
⁴ The osteological differences of ducks are discussed in *Variation* 1: 282–6.
⁵ Watson W. Hayne.
⁶ Tegetmeier described a cross between a male of the common wild pheasant and a hen of the domestic Hamburgh fowl, reported to him by Edward Hewitt, in part 10 of Tegetmeier 1856–7, which had appeared in January 1857. CD's copy of the work is in the Darwin Library–CUL. This passage is annotated (*ibid.*, pp. 123–4).
⁷ In a deleted postscript to this letter, CD had requested a common Dutch or German pouter (see Manuscript alterations and comments).

To W. D. Fox 8 February [1857]¹

Down Bromley Kent
Feb. 8th

My dear Fox

I was very glad to get your note; but it was really too bad of you not to say one single word about your own health. Do you think I do not care to hear?—

It was a complete oversight that I did not write to tell you that Emma produced under blessed Chloroform our sixth Boy almost two months ago.² I daresay you will think only half-a-dozen Boys a mere joke;³ but there is a rotundity in the half-dozen which is tremendously serious to me.— Good Heavens to think of all the sendings to School & the Professions afterwards: it is dreadful.—

I am very sorry to hear of your ¾ child!⁴

We shall be most heartily rejoiced to see you here at any time: we have now R^y to Beckenham which cuts of 2 miles & gladly will we send you both ways at any time.[5]

But the other morning I was telling my Boys about some of our ancient entomological expeditions to Whittlesea meer &c;[6] & how we two used to drink our tea & Coffee together daily. We had not then $20\frac{3}{4}$ children between us; & I had no stomach.

I do not think I shall have courage for Water Cure again: I am now trying mineral Acids, with, I think, good effect.[7] I am not so well as I was a year or two ago.

I am working very hard at my Book, perhaps too hard. It will be very big & I am become most deeply interested in the way facts fall into groups. I am like Crœsus overwhelmed with my riches in facts. & I mean to make my Book as perfect as ever I can. I shall not go to press at soonest for a couple of years.

Thanks about W. Indies. I have just had a Helix pomatia withstand 14 days well in Salt-water; to my very great surprise.[8]

I work all my friends: Are there any Mormodes at Oulton Hothouses[9] or any of those Orchideæ which eject their pollen-masses when irritated: if so will you examine & see what would be effect of Humble-Bee visiting flower: w^d pollen-mass ever adhere to Bee, or w^d it always hit direct the stigmatic surface?—

You ask about Pigeons: I keep at work & skins are now flocking in from all parts of world.—

You ask about Erasmus & my sisters: the latter have been tolerable; but Eras. not so well with more frequent fever fits & a good deal debilitated: Charlotte Langton has been very ill with Asthma & Bronchitis; but I hope is recovering.

Farewell, my dear old Friend. | Yours affect^y | C. Darwin

Are castrated Deer larger than ordinary Bucks? Do you know?

Christ's College Library, Cambridge (Fox 110)

[1] The letter is dated by the reference to the birth of Charles Waring Darwin (see n. 2, below).

[2] Charles Waring Darwin was born on 6 December 1856. CD had administered chloroform to Emma Darwin during previous labours (see *Correspondence* vol. 4, letters to W. D. Fox, [17 January 1850], and to J. D. Hooker, 3 February [1850]).

[3] Fox had twelve children in all, five by his first marriage and seven by his second. Of these, four were boys.

[4] An allusion to the advanced stage of Ellen Sophia Fox's pregnancy. Edith Darwin Fox was born on 13 February 1857 (*Darwin pedigree*).

[5] In April 1855, a plan was drawn up for an extension to the South Eastern Railway to run from Lewisham to Beckenham. John William Lubbock was the chairman of the committee; CD was a shareholder (*Correspondence* vol. 5, letter to W. E. Darwin, [25 April 1855], n. 4). The Mid-Kent Railway was opened on 1 January 1857 (*The Times*, 9 February 1857, p. 7).

[6] Whittlesea, a village near Cambridge, was the site of entomological excursions during CD's and Fox's undergraduate days. See *Correspondence* vol. 1, letter to W. D. Fox, May 1832, and letter from W. D. Fox, 30 June 1832.

[7] 'Mineral acids', according to Colp 1977, p. 57, 'probably meant a mixture of muriatic (hydrochloric) acid and nitric acid'. It was believed that some cases of dyspepsia were caused by a lack of acid secreted by the stomach and could be treated by administering replacement acids. For CD's precise prescription, transcribed from the 'Receipts' book (the medical prescription book used by the family), see Colp 1977, p. 157.

[8] See letter to J. D. Hooker, 10 December [1856], and letter from T. V. Wollaston, [11 or 18 December 1856]. On 22 January, after one of the *Helices* provided by Thomas Vernon Wollaston had survived the effects of immersion in salt water, CD began a new experiment with *Helix pomatia* and *H. aspersa*. In his Experimental book, p. 16 (DAR 157a), CD recorded on 5 February 1857 that a snail 'moved distinctly after 14 days in salt-water—'.

[9] Probably Oulton Park, Cheshire, the home of CD's and Fox's friend, Philip de Malpas Grey-Egerton.

From Alfred Christy to W. B. Tegetmeier[1] 11 February 1857

Surrey Square[2]
11 Feby 1857

My dear Sir,

I have to apologize for not answering your note before but I have been down to Stockport. I was in hopes I could have found some letters I had in reply to mine respecting the pigeons I had to let fly for a Society at Brussells but I cannot find them so I must have lent them to someone to read and they have never returned them

I let fly 64 Pigeons in London and the first bird returned to Brussells in $7\frac{1}{2}$ hours and several in a few minutes after

I have had a pigeon come from Stockport to Surrey Square in 7 hours 180 miles and have had many come from Goodward[3] in 1 hour and 8 and 10 minutes that is about 68 to 70 miles. a good pigeon flys about 1 mile a minute, my Pigeons come from Aperfield to Surrey Square in about 17 to 18 minutes and from Dovor in about 1 hour and 20 to 30 minutes, it depends very much on the weather and a fine clear day and if a pigeon is in good practice from a place and the wind in the right direction it should blow very little, and the same way the pigeon travels, if it does blow at all but the stiller the day the better and not too hot the heat affects them much

I trust this will be a little information for your friend

I am | Yours very truly | Alfred Christy

DAR 205.2 (Letters)

CD ANNOTATIONS
2.1 to Brussells] 'about 140 miles' *added pencil*
Top of first page: '18'[4] *brown crayon*

CD note:[5]
Feb. 13th/57/
    It seems from Mr Christy facts that for distances under 60 miles Pigeons Fly for 30 to 40 *geographical* miles.[6] Therefore I think estimates of ornithologist exaggerated—[*yet Major Carter*].—
    For longer distances only about 20 geograph. miles.— I suspect that they rest on road.—

But how far this will apply to bird blown by gales of wind, I do not know; & such I sh^d think was origin of those which arrive at Madeira, many of which are not regular migrators.

I think 30 miles per hour in gale w^d not be too much, when one think of rate which a Bird will fly down a gale.

Hurricane blow Birds to Mauritius & Jamaica[7]

---

[1] The recipient is identified on the basis that CD frequently called upon William Bernhard Tegetmeier for information from various pigeon-fanciers with whom he was acquainted. CD had previously inquired about the distances carrier pigeons could fly (see letter to J. A. H. de Bosquet, [before 3 November 1856]).

[2] In the *Post Office London directory* 1857, Alfred Christy is listed in the court directory as residing in Surrey Square, Old Kent Road.

[3] Christy probably refers to Goodwood in Surrey.

[4] The number of CD's portfolio of notes on the means of dispersal of plants and animals.

[5] The note is with the letter in DAR 205.2 (Letters).

[6] A geographical or nautical mile is one minute of longitude measured along the equator.

[7] The last sentence of this note was written in pencil. Associated with this letter is another page of notes by CD (preserved in DAR 205.2 (Letters)) in which he noted the times and distances of pigeon flights as given in various sources on pigeons.

To Charles Lyell 11 February [1857]

Down Bromley Kent
Feb. 11

My dear Lyell

I was glad to see in the newspapers about the Austrian Expedition:[1] I have nothing to add geologically to my notes in the Manual.—[2] I do not know whether the Expedition is tied down to call at only fixed spots. But if there be any choice or power in the scientific men to influence the places, this w^d be very desirable; it is my most delibreate conviction that nothing would aid more Natural History, than careful collecting & investigating *all the productions* of the most insulated islands, especially of the southern hemisphere.— Except Tristan d'Acunha & Kerguelen Land, they are very imperfectly known; & even at Kerguelen land, how much there is to make out about the lignite beds, & whether there are signs of old Glacial action— Every sea-shell & insects & plant is of value from such spots.

Someone in Expedition *especially* ought to have Hookers N. Zealand Essay.[3] What grand work to explore Rodriguez with its fossil birds & little known productions of every kind—[4]

Again the Seychelles, which with the Cocos de mar,[5] must be a remnant of some older land.— The outer isl^d of Juan Fernandez is little known.— The investigation of these little spots by a band of naturalists would be grand.— St. Pauls & Amsterdam would be glorious botanically & geologically.— Can you not recommend them to get my Journal (& Volcanic islands) *on account of Galapagos*.[6] If they come from North, it will be a shame & sin if they do not call at Cocos islet, N. of the Galapagos.— I always regretted that I was not able to examine the great craters on Albemarle Is^d, one of the Galapagos. In New Zealand urge on them to look out for erratic Boulders, & marks of old Glaciers.—

Urge the use of the Dredge in Tropics; how little or nothing we know of limit of life downwards in the hot seas.—

My present work leads me to perceive how much the domestic animals have been neglected in out-of-the way countries.—

The Revilligago isl^d off Mexico, I believe, have never been trodden by foot of naturalist

If the expedition sticks to such places as Rio, C. Good Hope, Ceylon & Australia &c, it will not do much.—[7]

Ever yours most truly | C. Darwin

I have just had Helix Pomatia quite alive & hearty after *20* days under sea-water; & this same individual about six-weeks ago had a bath of 7 days.—[8]

P.S. I have really nothing to suggest to Mr Forbes—[9]

I am delighted to hear about the Coal Plant & Purbeck Fossils.—[10]

Endorsement: 'Feb. 13. 1857'
American Philosophical Society (145)

[1] The first Austrian scientific expedition to circumnavigate the globe was announced in the *Athenæum*, 10 January 1857, p. 53. The leader of the expedition, Karl von Scherzer, requested advice to render the voyage 'efficient and fruitful in valuable results.'

[2] CD refers to his contribution to Herschel ed. 1849 (see also letter to William Sharpey, 24 January [1857]).

[3] J. D. Hooker 1853–5.

[4] The recently extinct, dodo-like solitaire (*Pezophaps solitarius*) was discussed in detail in Strickland and Melville 1848. Rodriguez Island had been a British possession since 1809.

[5] The coco-de-mer, or double coconut, of the Seychelles palm, *Lodoicea sechelarum*. Edward Blyth had earlier commented to CD on the large number of plants and animals peculiar to the Seychelles, noting in particular the coco-de-mer, which he believed to thrive on only two of the islands. See *Correspondence* vol. 5, letter from Edward Blyth, [1–8 October 1855].

[6] *Journal of researches* and *Volcanic islands*.

[7] In the event, the expedition visited, among the places CD mentioned, only the islands of St Paul and Amsterdam in the South Indian Ocean. CD's recommendations, along with those of Roderick Impey Murchison, William Jackson Hooker, Charles Lyell, and others, were acknowledged in the *Narrative* of the voyage (Scherzer 1861–3, 1: 3–4). The personal thanks of Archduke Maximilian of Austria were read out at a meeting of the Philosophical Club of the Royal Society on 11 June 1857 (Bonney 1919, p. 136).

[8] See letter to W. D. Fox, 8 February [1857], n. 8.

[9] Probably David Forbes, who was preparing to depart for South America in search of nickel and cobalt ores (*DNB*).

[10] See letter from Charles Lyell, [16 January 1857] and n. 2.

To W. B. Tegetmeier   11 February [1857][1]

Down Bromley Kent
Feb. 11th.

My dear Sir

The Bearer shall pay Carstang for the Drake & very many thanks to you for having got it for me.—

By all means show the Poultry;[2] & I will this evening send to Carstang the 2 pair of Persian Fowls, sent to me by Hon^e C. Murray.—[3]

I shall be **very glad** of the young Silk fowl, if you will kindly send it me, *about* 12 hours, before it breaks out of shells.— Also many thanks for your kind promise of the rumpless Chick.

I will *not* kill the Scanderoons; at least without giving you notice.—

The German Pouters which Mr Wicking gave me are not the old fashioned German Pouters, but fancy birds, crossed I imagine for they do not breed true.—[4]

I am glad to hear that you are succeeding so well with your Pigeon Breeding. With thanks, in Haste | Believe me | Yours very sincerely | C. Darwin

I hope you may prove right that Mr B.' affairs are not so bad, as I supposed.

New York Botanical Garden Library (Charles Finney Cox collection)

[1] Dated by the reference to Tegetmeier's wish to exhibit some of CD's poultry (see n. 2, below).
[2] The account in the *Athenæum* (7 March 1857, p. 313) of the 24 February meeting of the Zoological Society of London stated that 'Mr. Tegetmeier exhibited a portion of the collection of Asiatic poultry skins which has been entrusted to him by Mr. C. Darwin, with the view of illustrating the variations which take place in the domestic fowl. The collection contained some curious birds from Persia, India and Singapore, the peculiarities of which were successively pointed out by the exhibitor.'
[3] Charles Augustus Murray was minister plenipotentiary to the court of Persia. The description of these fowl given in the *Proceedings of the Zoological Society of London* 25 (1857): 46, stated: 'The interior of Persia furnished a very beautiful steel-black variety, perfectly distinct from any known in this country, and which was stated to be the common Fowl of the district.'
[4] See the postscript of CD's letter to W. B. Tegetmeier, 6 February [1857]. CD obtained several different pigeon breeds from Matthew Wicking.

From Asa Gray 16 February 1857

Cambridge, Mass.
Feb. 16th, 1857

My Dear Mr. Darwin

I meant to have replied to your interesting letter of the 1st Jany long before this time, and also that of Nov. 24, which I doubt if I have ever acknowledged.— But after getting my School-book or Lessons in Botany off my hands[1]—it taking up time far beyond what its size would seem to warrant—I had to fall hard at work upon a collection of small size from *Japan*—mostly N. Japan,—which I am only just done with.[2] As I expected the number of species common to N. America is considerably increased in this collection, as also the number of closely representative species in the two, and a pretty considerable number of European species too.— I have packed off my mss. (tho' I hardly know what will become of it) or I would refer you to some illustrations.[3] The greater part of the identical species (of Japan & N. Amer.) are of those extending to or belonging to N.W. Coast of America; but there are several peculiar to Japan & E. U. States, E.g. our *Viburnum lantanoides* is one of Thunberg's species.[4] De Candolle's remarkable case of *Phryma*, which he so dwells upon,—turns out, as Dr Hooker said it would, to be only one out of a great many cases of the same sort.[5] (Hooker brought Monotropa uniflora, you know from the Himalaya's; and now, by the way, I have it from almost as far south, i.e. from St. Fé de Bogota, New Granada).—

There is another Japanese collection I shall have a chance to look over, presently, that made by Charles Wright.[6] When that comes to be studied, it will be worth while to compare the Japanese & N. American Floras rather critically.[7]

Your anecdote of Agassiz, "Nature never lies" is most characteristic.[8] Instead of learning caution from experience A. goes on *faster* than ever, in drawing positive conclusions from imperfect or conjectural data, confident that he reads Nature through and through, and without the least apparent misgiving that anything will turn up that he cannot explain away.

—Well, I never meant to draw any conclusions at all, and am very sorry, that the only one I was beguiled into should "rile" you, as you say it does:—that on p. 73 of my 2.^d article: for if it troubles you it is not likely to be sound.[9] Of course I had no idea of laying any great stress upon the fact (at first view so unexpected to me) that one third of our Alpine species common to Europe do not reach the arctic circle; but the remark which I put down was an off hand inference from what you geologists seem to have settled, viz. that the northern regions must have been a deal colder than they are now,—the northern limit of vegetation therefore much lower than now,—about the epoch when it would seem probable that the existing species of our plants were created. At any rate, during the glacial period, there could have been no phenogamous plants on our continent anywhere near the polar regions; and it seems a good rule to look in the first place for the cause or reason of what *now is* in that which immediately preceded. I don't see that Greenland could help us much; but if there was any interchange of species between N. America & N. Europe in *those times*, was not the communication more likely to be in lower latitudes than over the pole?

If, however, you say—as you may have very good reasons for saying that the existing species got their present diffusion before the glacial epoch, I should have no answer. I suppose you must needs assume very great antiquity for species of plants in order to account for their present dispersion, so long as we cling—as one cannot but do, to the idea of the single birth place of species.

I am curious to see whether, as you suggest, there would be found a harmony or close similarity between the geographical range in this country of the species common to Europe and those strictly representative or strictly congeneric with European species. If I get a little time I will look up the facts,—though as Dr Hooker rightly tells me, I have no business to be running after side game of any sort, while there is so much I have to do—much more than I shall ever do, probably, to finish undertakings I have long ago begun.

— I wish you would tell me *why* you were led to conclude *a priori*, that trees would have a stronger tendency to be mono-diœcious than herbs.[10]

As to your P.S.— If you have time to send me a longer list of your Protean genera, I will say if they seem to be Protean here.— [11] Of those you mention—

Salix, I really know nothing about.

Rubus: the N. American species, with one exception, are very clearly marked indeed.

Mentha. We have only one wild species! that has two pretty well-marked forms, which have been taken for species: one smooth, the other hairy.

Saxifraga, gives no trouble here.

Myosotis: only one or two species here, and those very well-marked.

Hieracium: few-species; but pretty well-marked.

Rosa: putting down a set of nominal species, leaves us 4: two of them polymorphous, but easy to distinguish.

Our genera which I take to be most Protean (restricting myself to those of Bot. N. U. States) are—Ranunculus, Viola, Lechea, **Vitis**, *Ceanothus*, Polygala, *Amorpha*, Lespedeza, *Lathyrus*, **Cratægus**, **Amelanchier**, Calycanthus, *Œnothera*, *Ribes*, *Zizia*, Viburnum, Galium, *Oldenlandia*, VERNONIA, *Liatris*, *Eupatorium*, ASTER, Erigeron, *Solidago*, Silphium, *Xanthium*, Echinacea, *Helianthus*, Coreopsis, Bidens, *Artemisia*, SENECIO, *Cirsium*, **Nabalus**, *Mulgedium*, Lobelia, **Vaccinium**, *Azalea*, *Pyrola*, *Bumelia*, *Plantago*, **Dodecatheon**, *Lysimachia*, **Gerardia**, *Dipteracanthus*, Verbena, *Lycopus*, **Pycnanthemum**, **Monarda**, *Scutellaria*, *Stachys*, **Phlox**, Gentiana, *Apocynum*, **Fraxinus**, Polygonum, *Euphorbia*, Acalypha, *Celtis*, Carya, **Quercus**, (*Salix* & *Populus*, because not well known) ABIES, *Sparganium*, POTOMAGETON, **Sagittaria**, *Spiranthes*, Iris, *Smilax*, *Lilium*, Juncus, *Commelyna*, *Tradescantia*, **Xyris**, CYPERUS, *Scirpus*, *Eriophorum*, Rynchospora, *Scleria*, **Carex**, Agrostis, **Panicum**, *Andropogon*.— I have given you both *large* & *small* genera— and have marked the worst by underscoring, according to their degree of badness.[12]

It is not so easy to answer the question in your last P.S.— How I distinguish the introduced & aboriginal stocks of the same species in this country. We have yet a great extent of country in a state of nature,—especially woods. When a plant grows there widely there is no doubt.— While our *weeds* in cult. lands &c are so generally introduced plants that there is a strong tendency to view them all as such. We have, however, a considerable number of indigenous species becoming weeds. I mean to catalogue them: also some time or other the question will be reared whether they were indigenous to this country. *Œnothera biennis*, Erigeron annuum & strigosum & Canadense (if really indigenous at the north). Antennaria margaritacea, Asclepias Cornuti, &c—&c

Generally the wild and the introduced stocks *look different*, more or less, e.g. Triticum repens & caninum, etc—

Sometimes it is mere guess-work, but generally I feel sure, tho' I could not always tell perfectly well why.

A great many plants came with seed-grain—with cattle, &c—just as they have been carried to various parts of the world— Now these (take Dandelion &

Agrostis vulgaris for examples)—these were none the less likely to come here in this way *because* they were already a part of the indigenous flora!

But I will not ask you to read more of my blind handwriting now. I will not leave your welcome letters so long unnoticed again, if I can help it.

Ever Yours | Asa Gray

DAR 165: 96–7

CD ANNOTATIONS
0.1 Cambridge . . . warrant— 1.4] *crossed pencil*
1.10 The greater . . . States, 1.12] *double scored pencil*
1.14 to be . . . south, 1.17] *double scored pencil*; 'south' *added pencil*
2.1 There is . . . away. 3.5] *crossed pencil*
4.3 had no idea . . . settled, 4.7] *double scored pencil*
6.1 I am . . . herbs. 7.2] *crossed pencil*
18.1 Generally . . . etc.— 18.2] *double scored pencil*
Top of first page: 'A' *brown crayon, circled brown crayon*
Top of last sheet: 'Please return to me'[13] *ink*

[1] A. Gray 1857a was a text-book designed for school students.

[2] Gray refers to the collection made by James Morrow during Matthew Galbraith Perry's expedition to Japan, 1852–4 (see Dupree 1959, pp. 208–9).

[3] A. Gray 1857b. Although Perry's report was already in the hands of the printers, he instructed them to include Gray's botanical descriptions, but the illustrations were not included (see Perry 1856–7, 2: 299–301).

[4] Thunberg 1784 and 1794. Carl Peter Thunberg had collected plants near Nagasaki and Tokyo. Gray concluded his description of this *Viburnum* species by stating: 'This adds another to the interesting list of species peculiar to Eastern North America and to the Chino-Japanese region.' (A. Gray 1857b, p. 313).

[5] Alphonse de Candolle felt that the uniqueness of the distribution of *Phryma leptostachya* could not be explained by continental extension and appeared to favour the multiple centres of creation hypothesis (A. de Candolle 1855, 2: 1328).

[6] Charles Wright, who had previously collected plants in Texas and Mexico for Gray, was the botanist to the United States North Pacific expedition, 1853–6, which visited the ports of Shimoda and Hakodate in Japan (Dupree 1959, pp. 209–10).

[7] Gray's paper on the relationship between the Japanese and North American floras (A. Gray 1859) was described by Gray's biographer as 'the most important paper he ever wrote' (Dupree 1959, p. 210). In it Gray discussed *Phryma leptostachya*, stating that so many species were known to be common to eastern North America and eastern and northern Asia 'that De Candolle would now explain these cases in accordance with the general views of distribution adopted by him, under which they fall,— so abandoning the notion of a separate creation.' (A. Gray 1859, pp. 444–5).

[8] See letter to Asa Gray, 1 January [1857].

[9] See letter to Asa Gray, 1 January [1857] and n. 6.

[10] See letters to J. D. Hooker, 1 December [1856] and 10 December [1856].

[11] See letter to Asa Gray, 1 January [1857].

[12] Gray's underscoring has been typographically reproduced as italics for one underline, bold type for two underlines, and small capitals for three underlines.

[13] The annotation refers to a later occasion when CD forwarded the last sheet of Gray's letter (paragraph sixteen onwards) to Hewett Cottrell Watson for his comments (see letter from H. C. Watson, 10 March 1857).

To John Innes [after 16 February 1857][1]

[Down]
Wednesday

Dear Innes

I strongly recommend you to read enclosed beginning at p. 38 to 91.—[2] Von Siebold is about the most careful & profound naturalist in Europe.[3] I have found in this Book some facts, such as I was enquiring from you in regard to Bees.—[4]
    Ever yours | C. Darwin

American Philosophical Society (149)

[1] Dated from the reference to Karl Theodor Ernst von Siebold's work on parthenogenesis (Siebold 1856), an English translation of which (Siebold 1857) was published between 31 January and 16 February 1857 (*Publishers' Circular*, 16 February 1857, p. 94). CD's copy of Siebold 1857 is in the Darwin Library–CUL.
[2] The page numbers refer to the chapter in Siebold 1857 entitled 'True parthenogenesis in the honey bee'.
[3] Siebold had been appointed *Ordinarius* professor of zoology and comparative anatomy at the University of Munich in 1853. He was noted for his detailed microscopical researches on invertebrates.
[4] CD cited Siebold 1857 in his discussion of the two varieties of the common hive-bee in *Natural selection*, p. 372 n. 5.

To W. E. Darwin [17 February 1857][1]

[Down]
Tuesday Night.

My dear Willy

I am very glad indeed to hear that you are in the sixth;[2] & I do not care how difficult you find the work: am I not a kind Father? I am even almost as glad to hear of the Debating Society, for it will stir you up to read.— Do send me as soon as you can the subjects; & I will do my very best to give you hints; & mamma will try also.— But I fear, as the subjects will generally be historical or political, that I shall not be of much use.— By thinking at odds & ends of times on any subject, especially if you read a little about it, you will form some opinion & find something to say; & in truth the habit of speaking will be of greatest importance to you. Uncle Harry was here this morning,[3] & we were telling him that we had settled for you to be a Barrister (he was one) & his first question was, "has he the gift of the gab"? But then he added, he has got industry, & that is by far the most important of all.— Mamma desires that you will read the Chapters *very well*;[4] & the dear old Mammy must be obeyed. Her lip is plaistered up, so we cannot tell yet how she will look.—[5]

Parslow[6] has looked for the guard of the Razor, & it is not in your room; & he remembers putting it in paper, & he thinks it was probably thrown on one side with the paper, so you had better enquire.—

Lenny, Franky & Co.Y were rather awe-struck to hear that you had bought a cane to whip the Boys.—[7]

Be sure tell me about the Optics—& how you get on with the Reading in Chapel. Read slow & read the chapter two or three times over to yourself first; *that will make a great difference*. When I was Secretary to the Geolog. Soc, I had to read aloud to Meeting M.S. papers; but I always read them over carefully first; yet I was so nervous at first, I somehow could see nothing all around me, but the paper, & I felt as if my body was gone, & only my head left.—[8]

Snow came here today, in the carriage which took Harry back, & this, between ourselves, is rather a bore.[9] On Saturday all the Josselinas come.— [10]

When you write tell me how long the Boys make their speeches, & whether many get up & answer.—

Good night, my dear old fellow & future Lord Chancellor of all England.— Your's' most affectionately | Ch. Darwin

DAR 210.6

[1] Dated by the reference to the guests in the house (see n. 9, below) and to Emma Darwin's surgery (see n. 5, below).

[2] CD is referring to the sixth form at Rugby School.

[3] Henry Allen (Harry) Wedgwood was Emma Darwin's brother.

[4] The daily reading in the school chapel of a short lesson or passage from the Bible was assigned to the sixth form pupils.

[5] Emma Darwin, who had been in London from 11 to 16 February, recorded in her diary on 13 February 1857: 'Mr. Paget. lip done'. James Paget was assistant surgeon at St Bartholomew's Hospital, London.

[6] Joseph Parslow was the Darwin family butler.

[7] Leonard Darwin (aged 7) and Francis Darwin (aged 9).

[8] CD had been secretary of the Geological Society of London from 1838 until 1841. See *Correspondence* vol. 2.

[9] Frances Julia Wedgwood, or 'Snow', aged 24, the daughter of Hensleigh and Frances Mackintosh Wedgwood, was 'a young woman of extreme passions and fastidious principles' (B. Wedgwood and H. Wedgwood 1980, p. 259). Emma Darwin recorded that 'Snow came' on 17 February 1857 (Emma Darwin's diary).

[10] 'Josselinas' was evidently a family nickname for the daughters of Caroline and Josiah Wedgwood III. Sophy Wedgwood was 15 years old; Margaret, 14; and Lucy, 11. According to Emma Darwin's diary, the relatives from Leith Hill Place, the home of Caroline and Josiah Wedgwood III, did not arrive until 27 February and stayed until 5 March.

To W. B. Tegetmeier   18 February [1857][1]

Down Bromley Kent
Feb. 18th

My dear Sir

I have just received some Fowls from the Rajah Sir James Brooke,[2] which are not of much interest, but as I have thought you would like to display all you cd at Zoolog. Soc.[3] I will this night send them to Carstang's. They consist of white silk

Fowls originally from China, one crested, & some frizzled Fowls of unknown origin procured at Singapore. They are labelled.—

My dear Sir | Yours very sincerely | C. Darwin

New York Botanical Garden Library (Charles Finney Cox collection)

[1] Dated by the relationship to the letter to W. B. Tegetmeier, 11 February [1857].
[2] James Brooke, raja of Saráwak, was one of the correspondents to whom CD had written in December 1855 (see *Correspondence* vol. 5, CD memorandum, [December 1855]).
[3] See letter to W. B. Tegetmeier, 11 February [1857] and n. 2.

To Syms Covington 22 February 1857

Down Bromley, Kent,
February 22, 1857.

Dear Covington,—[1]

I received a short time since your letter of September 14, and was glad to hear how you were getting on, though the account of your affairs was not quite so prosperous as in some former letters, owing, as I understand, chiefly to the expense of your new house. But with your good sense and steadiness I have great hopes that you will tide over the time of difficulty. You must console yourself with thinking what a position you would be in here with six boys and two girls (which is now exactly my number). I never meet any one who is not perplexed what to do with their children. My eldest boy is almost a young man, and has just got into the head form at the great school of Rugby, and is very steady and good. We think of making him a Barrister, though it is a bad trade.

Captain Sulivan has been very lucky and has got a high place, of £1000 a year, I believe, and has beaten two Admirals and Captain Fitz Roy, who tried for the same place.[2] By the way, Captain F. with Mrs. F. are coming to-morrow to lunch with me on their road home from the Marquis of Camden.[3] Poor Captain F. has lately lost his only daughter, a beautiful and charming girl of about 16 or 17 years old.[4]

I lately dined with one of your great Australian potentates, Sir W. Macarthur, and heard a good deal of news of Australia, and drank some admirable Australian wine.[5] Yours is a fine country, and your children will see it a very great one. With every good wish for your health and prosperity, I am, dear Covington, yours sincerely, | CH. DARWIN.

Sydney Mail, 9 August 1884

[1] Misprinted as 'Corington' in the *Sydney Mail.*
[2] Bartholomew James Sulivan had been appointed the naval officer of the marine department of the Board of Trade in December 1856 (*DNB*).
[3] George Charles Pratt, whose country residence was the 'Wildernesse', near Sevenoaks, Kent.
[4] Robert FitzRoy, at that time chief of the meteorological department of the Board of Trade, had several children by his first wife, Mary Henrietta. He had married Maria Smith in 1854.

[5] William Macarthur had been knighted and made an officer in the Legion of Honour at the close of the Paris Exhibition of 1855, which he attended as commissioner for New South Wales. A son of John Macarthur, who introduced the vine to Australia, William Macarthur encouraged winemaking, bringing out German vignerons to New South Wales (*Aust. dict. biog.*).

To W. D. Fox 22 February [1857][1]

Down Bromley Kent
Feb. 22^{d}

My dear Fox

I am much obliged for your various enclosures, viz (1st) about yourself, & most heartily glad I am that Dr Gully has done you some good.—[2]

Emma desires to be most kindly remembered to Mrs Fox & we are very glad that she & the little girl are both well.—[3]

I hope that your nephew may succeed in finding some lizard eggs;[4] for it seems that he will try his best to ascertain the point in question.— By the way I have just had Helix Pomatia quite healthy after 20 days submersion in salt-water.—

Thanks about Pea case: it is a very great puzzle to me; for if I could trust to my observations on Bees, I cannot see how they can avoid being crossed; but the evidence certainly preponderates on your side, & most heavily in case of Sweet Peas.—[5] I suppose the Queen Pea flowered at same time with adjoining Peas: are you sure of this?

With respect to Clapham School: I think favourably of it: the Boys are not so exclusively kept to Classics: arithmetic is made much of: all are taught drawing, & some modern languages.—[6] I was rather frightened by having heard that it was rather a rough school; but young Herschel did not agree to this;[7] & Georgy is rather a soft Boy & I cannot find out that he has anything to complain of, though of a very home-sick, disposition. I will at any time answer any queries in detail: I do not know, but could find out, whether Clergymen's sons are charged less.—

My wife agrees very heartily with your preachment against overwork, & wishes to go to Malvern; but I doubt: yet I suppose I shall take a little holiday sometime; perhaps to Tenby: though how I can leave all my experiments, I know not.—[8]

I am got most deeply interested in my subject; though I wish I could set less value on the bauble fame, either present or posthumous, than I do, but not, I think, to any extreme degree; yet, if I know myself, I would work just as hard, though with less gusto, if I knew that my Book w^{d} be published for ever anonymously

Farewell, my dear Fox | Ever yours | C. Darwin

Christ's College Library, Cambridge (Fox 101–2)

[1] Dated by the reference to the birth of Fox's daughter (see n. 3, below).

[2] James Manby Gully ran a hydropathic establishment in Malvern. For CD's and Fox's previous correspondence about Gully and hydropathy, see *Correspondence* vol. 4.

[3] Ellen Sophia Fox had given birth to a daughter, Edith Darwin Fox, on 13 February 1857 (*Darwin pedigree*).

[4] See letter to W. D. Fox, 20 October [1856]. CD had first requested Fox's assistance in helping him procure lizard eggs in 1855 (*Correspondence* vol. 5, letters to W. D. Fox, 17 May [1855] and 23 May [1855]). CD wished to see whether they could withstand prolonged exposure to salt water to help explain the pattern of dispersal of lizards.

[5] Of all the Leguminosae, the sweetpea (*Lathyrus odoratus*) presented CD with the greatest challenge to his hypothesis that all organic beings must occasionally cross-fertilise. See *Natural selection*, p. 70, for a discussion of the evidence for and against the possibility of cross-fertilisation in sweetpeas.

[6] George Howard Darwin was attending Clapham Grammar School. See J. R. Moore 1977.

[7] Alexander Stewart Herschel, son of John Frederick William Herschel, was also at Clapham Grammar School.

[8] CD eventually went to Moor Park, a hydropathic establishment in Surrey, from 22 April until 6 May. CD and Emma's daughter, Anne Elizabeth Darwin, had died at Gully's hydropathic establishment in Malvern in 1851 (see *Correspondence* vol. 5).

To Richard Kippist 23 February [1857]

Down Bromley Kent
Feb. 23[d]

My dear Sir

I beg leave to enclose to you my Subscription[1]
Yours very sincerely | C. Darwin

R Kippist Es[q]

Endorsement: 'Ackn[d] Febr[y] 23/57. R. K.'
Linnean Society

[1] Kippist was the clerk, librarian, and housekeeper of the Linnean Society. The society had opened a subscription in June 1856 to meet the cost of moving from Somerset House to its new rooms in Burlington House. Kippist noted on the letter, '(Cheque for £20 on Union B[k] of London)'. The society held its first general meeting in Burlington House on 25 May 1857. See Gage 1938, pp. 44, 51. CD recorded this payment on 22 February 1857 in his Account book (Down House MS).

From Victor de Robillard 26 February 1857[1]

Port Louis, | Ile Maurice,
le 26 Fév[r] /57.

Mon cher Monsieur,

J'ai reçu votre lettre du 7 D[bre] dernier & vous remercie des démarches que vous avez faites près de Mr Cuming pour les coquilles;[2] je vois par la réponse écrite qu'il vous a faite qu'il désire plutôt vendre qu'échanger; je lui ai écrit alors et lui envoie une liste des coquilles que je désire, afin d'en connaître les prix.

Je vais maintenant vous donner les renseignemens que vous me demandez, dans le sens de mes idées:

Vous dites que dans quelques cas, les mêmes mollusques se trouvent habiter différentes iles, quoique séparées par la mer. Dans les iles qui nous environnent, il est bien rare de voir les mêmes coquilles terrestres sur des iles différentes— Maurice a beaucoup plus de terrestres que Bourbon,[3] et ce ne sont pas les mêmes espèces, à l'exception d'une agathine commune & d'un petit cyclostome des bois— toutes les coquilles terrestres des Iles Seychelles n'existent pas à Maurice ni celles de Madagascar, ce qui confirmerait l'opinion de ceux qui pensent que les iles se sont élevées isolément du sein de l'océan & n'ont pas été séparées par des révolutions volcaniques et des feux sous-marins, comme quelques uns le prétendent—

Je ne pense pas que des oeufs puissent être transportés d'une île à une autre, par la mer; ils rencontreraient trop d'ennemis pour pouvoir arrivés sur un autre rivage, où les chocs dans tous les cas pourraient aussi détruire le principe vital—

Il y a plusieurs années on a transporté ici une belle agathine à bouche rouge de Madagascar, qui s'est beaucoup répandue & s'est de suite acclimatée.

Quant aux mollusques d'eau douce, il y a deux à trois espèces de Maurice qu'on rencontre à Bourbon; je les crois indigènes aux deux Iles, sans qu'on les ait portés de l'une à l'autre. Le voyage de leurs oeufs, devant descendre à la mer par les débordemens de Rivière, ne doit cependant pas se faire d'une ile à l'autre, devant être détruits de la même manière que ceux des mollusques terrestres—

Maintenant je crois qu'il est très facile de porter des espèces d'une ile à l'autre, soit par les mollusques eux-mêmes soit par leurs oeufs, il ne suffirait pour cela que la volonté de l'homme.

Quant aux plus récentes découvertes d'oiseaux de Maurice ou de Bourbon, on n'a rien rencontré de nouveaux, depuis le Dronte & le Solitaire—[4]

à Bourbon il y a un oiseau qui vit dans les solitudes des bois des montagnes, qui y devient très rare, il est connu sous le nom de Huppe;[5] il n'a jamais été rencontré à Maurice—

Parmis les insectes, je remarque depuis peu d'années ici, un papillon bleu, qu'on dit exister à Madagascar & un autre blanc, qui serait de la côte d'afrique— Comment sont-ils arrivés ici? on présume que c'est par des plantes sous lesquelles les oeufs se seraient trouvées attachées lorsque l'insecte a pondu.

Vous me demandez si pendant les ouragans des objects légers pourraient être transportés de Bourbon à Maurice? d'après la théorie admise pour la formation de ce grand ébranlement de la nature & de la manière dont ils se forment & se déplacent, il est un fait certain acquis par nombre d'observations, c'est

AL incomplete
DAR 205.3 (Letters)

CD ANNOTATIONS
0.3 Mon cher . . . idées: 2.2] *crossed pencil*
3.3 bien rare] *underl pencil*
3.5 l'exception . . . Madagascar 3.7] *scored pencil*
4.1 Je ne pense . . . vital— 4.3] *crossed pencil*
5.2 sans] 'sans' *added pencil*
5.3 Le voyage . . . le solitaire— 7.2] *crossed pencil*
8.1 à Bourbon . . . montagnes,] *double scored pencil*
9.1 Parmis . . . c'est 10.4] *crossed pencil*
10.1 Vous . . . c'est 10.4] 'Hurricanes circular so not good to transport' *added pencil*
Top of first page: '19'[6] *brown crayon*

[1] For a translation of this letter, see Appendix I.

[2] Hugh Cuming had assembled one of the largest shell collections in Britain.

[3] Bourbon, now called Réunion, was a French colony in the Indian Ocean, 130 miles south-west of Mauritius.

[4] The dodo of Mauritius and the solitaire of Bourbon were extinct (Strickland and Melville 1848, pp. 27, 60).

[5] The Réunion huppe is now extinct (Staub 1976).

[6] The number of CD's portfolio of notes on the geographical distribution of animals.

To *Gardeners' Chronicle* [after 28 February 1857][1]

As you have noticed hybrid Dianths,[2] you may like to hear that the summer before last I fertilised a poor single pale red Carnation with the pollen of a crimson Spanish Pink; and likewise a Spanish Pink with the pollen of the same Carnation.[3] I got seed from both crosses in fair number; namely, 77 seed from two pods of the Spanish Pink, and raised plenty of seedlings. In the eyes of a florist they would be, I presume, quite worthless from their straggling habit; but they were showy, and like most hybrids produced during a long time an extraordinary abundance of flowers. They varied somewhat in colour, but in no other respect; and one variety was of a really beautiful pale crimson. Taken in a mass there was no difference between the reciprocal crosses. Not one plant of either lot set a single seed. One plant came up identical with the Spanish Pink; no doubt owing to a few grains of the pollen of the Spanish Pink not having been removed; for Gærtner has shown that this is sometimes the result when a flower is fertilised with mixed pollen.[4] I may add that Gærtner raised many hybrids between various species of Dianthus. *C. Darwin, Down, Bromley, Kent.*

Gardeners' Chronicle and Agricultural Gazette, 7 March 1857, p. 155

[1] Dated on the basis that the letter was written in response to an item published in the *Gardeners' Chronicle*, 28 February 1857, p. 132 (see n. 2, below).

[2] *Gardeners' Chronicle*, 28 February 1857, p. 132, described two hybrid *Dianthus* plants with quite different flowers that were believed to have grown from seeds from the same seed-capsule.

[3] CD also carried out hybridising experiments on wild *Dianthus caryophyllus* in the summer of 1855 (see *Correspondence* vol. 5, letter to J. S. Henslow, 11 July [1855]).

[4] Gärtner 1849.

To John Lubbock [6 March 1857][1]

57 Queen Anne St., | Cavendish Square
Friday

Dear Lubbock

It is true that it is against rules to be proposed under 18 as I found out to my sorrow; for Willy was going to be elected as one of the 8, being mistaken for me!![2] and as there was some fuss about it, I thought it incumbent to write to Committee, stating my utter ignorance of rules and withdraw his name. On other hand I have reason to believe that annually the Rule is often broken. As I did not know the rule when I seconded you, I have no scruple.[3] But it depends on whether you have any reason to suppose that your case will be noticed, or whether of late the Regulation has been attended to, whether you had better withdraw, and on that I really have no means of forming opinion. I have spoken to some to vote, and from all that I can hear from old members, my Brother and others, I should say that your election was *certain*.

I leave this place on Saturday at about 2 o'clock; so please let me hear either here (if time allows) or at Down, that I may know as soon as I can, whether I shall have to come up on Monday, which I shall of course do to canvass for you, though your name by itself is really quite sufficient. My movements, as you know, are uncertain from health, but I pledge my word to come if able; for if not decently well in truth I could not stand any excitement so late in evening. I think you would hereafter regret if you for no reason gave up being elected.

Ever yours | C. Darwin

Copy
DAR 146

[1] Dated by the reference to the forthcoming Athenæum Club elections and to Lubbock's candidature (see n. 3, below). CD was in London from 4 to 7 March (Emma Darwin's diary).
[2] See letter to W. E. Darwin, [26 February 1856].
[3] Apparently Lubbock had been proposed for membership of the Athenæum Club before his eighteenth birthday in 1852. His name had been put forward by Lord Hotham and seconded by CD. He was elected on 9 March 1857 (see Hutchinson 1914, 2: 166).

To Hugh Falconer 8 March [1857][1]

Down
March 8th

My dear Falconer

I have written down what I gathered from you on Thibet Dogs;[2] and if at any time you could add a few details, the case probably would be a very valuable and interesting one for me, as I know of nothing parallel to it.

Yours most sincerely | Ch. Darwin

I enjoy so much a chat with you, that you will be sure to see me, when next in London.

Copy
DAR 144

[1] The year is the earliest that CD could have written to Falconer in London. Falconer, formerly superintendent of the Calcutta botanic garden, had retired in 1855. After his return to England late in 1855, he travelled on the Continent to pursue his palaeontological researches and did not settle in London until early 1857 (*DNB*).
[2] CD cited Falconer's case of 'the Thibet mastiff and goat, [which] when brought down from the Himalaya to Kashmir, lose their fine wool' as support for his claim that climate influences the hairy covering of animals (*Variation* 2: 278). He gave, however, a later article by Falconer (Falconer 1863) as his source.

To Robert Patterson 10 March [1857][1]

DOWN, BROMLEY, KENT
March 10th. [1857]

MY DEAR SIR

I am going to beg a great and troublesome favor of you,—I have been collecting skeletons of all varieties of Rabbits, & I want very much a real Irish Rabbit,[2] the L. veomicule of our poor friend Thompson—[3] Would you have the great kindness to take the trouble to procure me one. The only care requisite to be to get one not very severely shot, but especially not struck on the back of the head to kill it, as that part is easily injured & is very characteristic.

I enclose an address; and you will know whether to steamer to Liverpool & then per Railway, will be the cheapest and quickest route— I am fearful you will think me rather unreasonable in begging this favor.

Truly believe me | My dear Sir | Yours sincerely | CH. DARWIN

Praeger 1935, p. 714

[1] The date as given by Praeger 1935.
[2] In a note dated 7 March 1857 (DAR 45: 76), CD recorded that: '*Irish Rabbit* in Brit. Mus is named Lepus vermicula of Thompson.— so marked var.' In his continuation of William Thompson's work on Irish natural history (see n. 3, below), Patterson classified this rabbit as *Lepus cuniculus* (Thompson 1849–56, 4: 30).
[3] Thompson, who had died in 1851, had specified in his will that his manuscript and notes for the remaining volume of his *Natural history of Ireland* were to be given to Patterson and James R. Garrett to prepare for publication. After Garrett's death in 1855, Patterson assumed the task and published volume four in 1856.

From H. C. Watson 10 March 1857

Thames Ditton
March 10th/57

My dear Sir

I am not quite sure of understanding your question about "variable genera".[1] To explain my uncertainty, I will endeavour to define or state the differences for choice.

1. Genera, of which the species are close, & difficult to distinguish by reason of their similarity;—but the species themselves not remarkably variable. Ex: Carex & Ranunculus (excluding Batrachium)

2. Genera, of which the quasi species are so close that it becomes highly difficult to say whether the genus is composed of a comparatively few extremely variable species, or of many very close species. Ex: Rubus & Hieracium.

3. Genera, the species of which are themselves so variable, & approximating, that it becomes difficult to say where one species ends & the next begins. Ex: Viola & Saxifraga, at least in certain sections or subgenera—

It seems to me that Dr. A. Gray may have inclined to the first, while you perhaps yes certainly intend 2 or 3. I will copy the three categories, that you maybe

AL incomplete
DAR 181

CD ANNOTATIONS
5.2 I will . . . maybe 5.3] *crossed pencil*
Top of first page: 'Please return to me' *ink*; '1)' *ink*[2]

[1] CD had asked Watson to comment on the list of 'protean' genera (in which the species present a great amount of variation) included in the letter from Asa Gray, 16 February 1857. CD discussed protean genera in *Origin*, p. 46.
[2] The annotations relate to a later occasion when CD forwarded Watson's letter to Asa Gray. See letter to Asa Gray, [after 15 March 1857].

From Richard Hill   12 March 1857

Spanish Town Jamaica
12th March 1857.

My dear Sir,
I am unwilling to let the packet depart without acknowledging the receipt of your letter.[1]

I am fearful I cannot promise you any thing more precise respecting Ducks being instrumental in the dispersion of Crustacea through Mountain streams, except common report. One can scarcely conceive the ova resisting the grinding and crushing of so powerful an instrument as a Duck's gizzard. However I will attend to all precise facts on the subject and communicate them to you.

With regard to instances in which Hurricanes have brought us Continental birds, not known to us as visitors ordinarily, I can mention the acquisition of a specimen of the Anas maxima, in the year 1848. I must first direct your attention to the bird described under the name of the Green-backed Mallard in Gosse's Birds of Jamaica at page 399.[2] He assigns it the name of Anas maxima, and on the suspicion of M[r] Gray,[3] sets it down as a possible hybrid, between Anas boschas and Cairina moschata— He tells you where and when it came to his hand, and the prevalent knowledge of it as a distinct species. It is a rare visitor. In the October of 1848 we were smitten by a Hurricane, which spun its rotary course with such violence in the gulf of Mexico, that the impeded waters of the gulf stream, created a prodigious tidal rise about us, if I may so speak, and submerged the Islands of the Caymanas— The inhabitants of those Solitary Islets were obliged to betake themselves to the few hills they have, to escape drowning. I mention this fact to shew the intensity of the Tornado between the coasts of Florida north, and Yuccatan south. The westerly whirl of the breeze brought to us multitudinous flocks of Ducks. They were picked up by persons as they dropped exhausted in and

Philip Henry Gosse. Photograph by Maull & Polyblank, 1855.
(Courtesy of the National Portrait Gallery, London.)

Alphonse de Candolle. After a painting by Antoine Maurin.
(Courtesy of the Bibliothèque publique et universitaire de Genève.)

about our towns;—among the number a sister of mine, had brought to her, by the servants, picked up in our yard at Montego bay, a specimen of the Anas maxima. She kept it some fifteen or sixteen months. The sexual instinct was so strong that it laid infertile eggs. In one season they were laid without any apparent injury, but in the second the bird died under the passionate influence;—a disease in the anal passage, being the consequence of the unsatisfied desire for a mate. I mention this case distinctively because the Anas maxima is altogether a new bird to naturalists.

Our winter just past,—for we are now swept by vernal equinoctial breezes, was exceedingly cold. The thermometer under the influence of strong north westerly gales sunk to 43° Fahrenheit on our mountains near here. In Cuba they report that the thermometer was as low as barely above Réaumur's zero. In the midst of this weather, the gunners from our marshes, brought in two specimens of the Anas maxima. I received this information from a friend who has for some three years past been busy collecting birds for the Royal Society of Dublin of which Society he is an Honorary Member. I have mentioned this second fact in connexion with the first, to shew, that stormy breezes in both instances brought the birds to us, one being a hurricane, and the other boisterous winds from the North-West.

On the port royal beach, the Cassia obovata—Alexandrine Senna, is growing. It grows on the sand beaches of the further side of Port royal Harbour, and it is found on the sands at Salt-river in Old Harbour Bay Westward, but no where else. It is supposed to have been planted at port royal by the Spaniards, and to have drifted over to the other shores. We have in our salt pond Savannas, a very beautiful lofty and shady tree, bearing a sweet gummy pulp very much sought after by cattle, that collect under it, and wait the fall of the seed-pods. It was brought hither by Spanish Stock from Caraccas, and has been dispersed by them in their dung. Though not of fifty years introduction, it is fast being carried in the same way into the interior. It is seen now upon the banks of mountain streams. It is tracked upward not downward;—not dispersed by the rivers, but carried up by the stock feeding on the banks of rivers. The prosopis juliflora,—the tree known with us as Cashaw and with the French of Haiti as Bahiahond, a hard timbered Acacia, was introduced in a similar way. It now forms miles of lowland forest in both Jamaica and St Domingo.

Port Royal is unfitted for breeding and rearing poultry from the absence of all Fresh water, save that that is brought into it by the sailing-tanks of Government;—and the incapability of the land to sustain any succulent vegetation whatever. There is a scattered growth of the Tribulus,—a yellow blossom that covers the Savannas of the Leguanca and St Catherine's plains, like buttercups in English meadows, on whose petals the running poultry fatten, but this is not constant, and is at best only *an aid* to poultry feeding.

We have two distinctive river Mullets the Mountain Mullet, and the Hog-nose Mullet and both feed on the seeds of a Laurel called the Timber Sweet-wood, and wild figs. They take the fruit as they float down the stream, and can be caught with a line, if it be baited with these fruits.

The Whistling Duck, Dendrocygna arborea, breeds in captivity.— The pinioned Ducks take to nest making in the underwood about the ponds, in which they feed in association with the Muscovy Duck, Cairina moschata, and the Common domesticated Mallard, Anas Boschas. They perform the part of excellent decoys when bred and not pinioned for their own species in the Guinea Corn Season, when the wild flocks scour our Savannas at nightfall.

No attempt has been made to domesticate and breed the palemadeas occasionally brought hither from the Spanish Main— D'Azara speaks of the Indians rearing them, but says nothing distinctly implying that they breed them.[4] When he says, he saw some brought up, among the domestic poultry at Country houses, and that they were as tame as fowls, and that they lived faithfully with their mates, we should judge he *had seen them pairing* in a state of domestication.

I have enquired and I find no one who has ever known the parrot in its submission to a life in human homes,—copulating— It seems of all birds the most completely to abandon its instincts;—for without cutting the wing, it is reconciled to relinquish flight, and to go climbing about.

Our Quails the Ortyx virginiana will breed in captivity:—but the enclosure must be spacious, for it is restless, and always running, and if the barred space in which it is kept is not close-boarded for about two feet up, it gets blind by the alternating effect of light and shadow from the bars upon the eye.

Our pimenta, Myrtus pimenta, is propagated entirely by birds. Germination seems only to be insured by the exciting influence of a passage through the birds intestines. This spice is peculiar to Jamaica among the West-Indian Islands. It is known only at Yuccatan on the Continent. The intestinal process of germination seems to be essential for the guava, psidium pyriferum; the forest hogs disperse it.

I will not forget my promises of other matters for you; but I find every body slow;—I wish they would be sure also;—tardy only to guarantee success.

With all respect I subscribe myself | My dear Sir | Your obedient Servant | Richard Hill

Charles Darwin Esq[re]

DAR 205.2 (Letters)

CD ANNOTATIONS

1.1 I am . . . to you. 2.5] *crossed pencil*; 'Ducks carrying eggs of Crustacea not known' *added pencil*
3.1 With . . . boschas 3.6] 'Birds Hurricanes A Duck' *added pencil*
3.7 He . . . Tornado 3.14] *crossed pencil*
3.15 The westerly . . . towns;— 3.17] *double scored pencil*
3.15 breeze] '[*illeg*]' *added pencil*
3.19 She kept . . . maxima. 4.6] *crossed pencil*
4.8 I have . . . North-West. 4.10] *scored pencil*
4.9 breezes] 'N' *added pencil*
5.1 On the . . . where else. 5.3] *crossed pencil*
5.5 We have . . . rivers. 5.12] 'Trees brought with mammals from Caraccas' *added pencil*

5.12 The prosopis . . . St Domingo. 5.15] *scored pencil*; 'd[itt]o.' *added pencil*
6.1 Port Royal . . . feeding. 6.7] *crossed pencil*
7.1 We have . . . fruits. 7.4] 'Fish eating seeds.' *added pencil*
8.1 The Whistling . . . nightfall 8.6] 'Dendrocygna arborea bred in Captivity' *added pencil*
9.1 No attempt . . . domestication. 9.6] *crossed pencil*
10.1 I have . . . about. 10.4] 'Parrots not breeding' *added pencil*
11.1 Our Quails . . . eye. 11.4] 'Quails breeding if pretty large space' *added pencil*
12.1 Our pimenta, . . . Islands. 12.3] 'Myrtus by Birds' *added pencil*
12.4 The . . . it. 12.5] 'Guava by forest Hogs' *added pencil*
13.1 I will . . . myself 14.1] *crossed pencil*
Top of first page: '18'⁵ *brown crayon*

[1] CD's letter has not been found, but it was evidently a response to the letter from Richard Hill, 10 January 1857.
[2] P. H. Gosse 1847, to which Hill had contributed.
[3] George Robert Gray.
[4] Azara 1801, 2: 323–4. CD's annotated copy of this work is in the Darwin Library–CUL. The section referring to black-skinned Paraguayan fowl or *Palamedea* is marked with a 'Q'. CD cited this passage in *Variation* 2: 209. See letter from Edward Blyth, 8 January [1856] and n. 12.
[5] The number of CD's portfolio of notes on the means of geographical dispersal of animals and plants.

From H. C. Watson to Asa Gray[1] 13 March 1857

13 March 1857

First, with regard to a point in Mr. D's last letter.[2] The categories 2 & 3 are different chiefly by the manner of viewing & grouping the facts, rather than by the natural facts of themselves;—& they even pass into 1 from the like cause; altho' 1 is essentially more different. Suppose, that 200 forms under one genus can be recognized by the eye, & described by words. Suppose, that one botanist groups these into 60 species; and that another groups them into 20 species;—in the latter case the 20 species will seem to be, on an average, thrice as variable as the 60 species of the former case. Such differences, & even greater differences, are found among botanical systematists; and they bear importantly on investigations such as Mr. D's. Thus, in arranging a collection of British Hieracia into one dozen species (Hudson, 12)[3] we readily distinguish the type examples from each other; but then there still remain many other examples so aberrant from these types, & more or less intermediate, as to render their assignment among the twelve species very difficult. In arranging the same collection into nearly three dozen species (Backhouse, 33),[4] the difficulty of assignment is found more in the closeness of the presumed species, than in the number of aberrant & intermediate examples. There is left less space for varieties (so to write) between the type forms of the more numerous species, than between those of the less numerous & less similar species. Not only does the same genus thus vary in regard to species in the works of different Authors;—but further, by a sort of conventional consent of the let-alone kind, some genera are habitually less divided into species than others. It is only of late years that Rubus & Hieracium have been so greatly subdivided into species.

On the contrary, Rosa and Mentha are now less so subdivided than formerly. Or, to take an example from Species merely:—Ranunculus aquatilis & Polygonum aviculare, two common British plants, were held single species by Linneus & his early successors. The former is now divided into a dozen or upwards; the latter left entire. Both are proteiform; & as far as external form goes, either might be made into a dozen species; the subdivisions of Polygonum aviculare quite as describably so divided, as are those of Ranunculus aquatilis. Now, Mr. D. wants the facts of Nature for his investigations, but is thus forced on conventional interpretations & arrangements instead. I find the difference between these two things enormous impediments.

European genera, including several variable species, or some few very variable species; the species more or less gliding into each other by the varieties.—
SALIX — RUBUS — ROSA — MENTHA — **Helianthemum** — **Cirsium** — **Hieracium** — **Viola** — *Festuca* — *Poa* — Triticum — *Saxifraga* — **Potamogeton** — **Atriplex** — Chenopodium — Verbascum — Polygala — BATRACHIUM (section of Ranunculus) Ulmus — Lastrea — *Galium* — Epilobium — Taraxacum — Fragaria — Polygonum — Potentilla — Campanula — Betula — Sedum — Erythræa — Euphrasia — Lamium. *Ribes* — Plantago — Cerastium — Phagnalon — Teucrium. Stachys — Galeopsis — Cochlearia — Ophrys — Daucus.

The above names are set down as they came into recollection. In general, the earlier names may be considered to indicate genera in which the species most run together by their varieties. The underscoring is an attempt to imitate Dr. Gray, but without perfectly understanding his signs.[5] There are other large genera, with close species, having one or few variables among them,—but not entered above, because the variables are *proportionably* few. Thus, the species of Silene are very numerous, many closely alike, but not variable as a characteristic, although among them one or more species have been subdivided (wrongly) into several; Silene anglica, gallica, lusitanica, & 5-vulnera, for example, being forms likely of a single species.

I wrote down the above names without having the list from Dr. A. Gray before me. I find the following are in his list also,—Viola, Polygala, Ribes, Galium Cirsium, *Plantago*, *Stachys*, Polygonum, POTAMOGETON, *Salix*,—only one of them strongly underlined.

Remarks on the European genera in the list by Dr Gray.— *Ranunculus*:—The species are numerous, & close; but not remarkably variable in Europe after deducting the subgenus Batrachium. R. Flammula however is very proteiform. *Lathyrus*:—Numerous Europeans, not variable, unless in colour, & breadth of leaves. *Ribes* has few European species; the Gooseberry is now held a single species, though deemed 3 species by Linneus; the red Currant has been divided into 4, but is perhaps only 1 species; the alpine Currant has had two sub-species carved out of it. These changes show a considerable variability. Galium is perhaps an unsettled & imperfectly understood genus, & thus its species may seem more variable than should be. *Xanthium*:—doubtful whether 3, 4, or 5 species in Europe. *Artemisia*:—

species numerous & close, rather than variable if taken singly; but A. camphorata is a group of species according to some botanists, a varying species according to others. *Senecio*:—same remark, three or four confused or variable species. *Cirsium*:—many intermediates, supposed hybrids. *Potamogeton* is the most strongly underlined, & its species differ a good deal in Europe, running into varieties as the water is deep or shallow, stagnant, still, or running, &c. *Quercus*:—variable in the form of leaves, & position of fruit; the English forms being assigned to 1, or 2, or 3 species by different botanists. *Juncus*:—some of the species in very close pairs (conglomeratus & effusus—compressus & cœnosus—&c.) but the species separately not very variable. *Carex*:—species numerous, many of them close; but not much variable, unless we adopt the views of Læstadius that several alpine Carices are reduced states of the lowland species. *Scirpus*:—several pairs & trios are very close, or some species sport into strong varieties, undecided which, tho' botanists in general call them species, not varieties. Several of the other genera mentioned in Dr. Gray's list are chiefly or exclusively American;—& on the other side, several of those in my European list are chiefly European, with few American representatives.

On the whole, I fear that nothing satisfactory can be got out of these lengthy & rather vague notes.—

My **impression** is, that some species tend more to vary than others, apart from external influences,—though such influences may call out or augment the tendency.[6]

But let me repeat the remark that species may be made to appear more or less variable, according as a genus is divided in books into few or many species. How very variable, for instance, is the Rubus fruticosus (of Linneus) when including about 50 modern species! Or, the Helianthemum variabile (Spach),[7] formed by the re-union of a dozen species, & several subspecies; many of them long supposed, or still supposed, quite distinct.

H. C. W.

A memorandum S
DAR 181

CD ANNOTATIONS
2.3 SALIX] *cross added pencil*
2.3 RUBUS] *cross added pencil*
2.3 ROSA] *cross added pencil*
2.3 MENTHA] *cross added pencil*
2.3 SALIX . . . *Saxifraga* 2.4] 'Those with pencil cross have been mentioned in your former letter'[8] *added pencil*
2.4 **Hieracium**] *cross added pencil*
2.4 *Saxifraga*] *cross added pencil*
2.10 Daucus.] '42 genera' *added pencil*
7.1 But . . . species. 7.2] *triple scored pencil*
Top of first page: 'Please return to me' *ink*; '2' *ink*[9]

[1] The manuscript of the letter indicates that Watson began by writing a letter to CD; he subsequently changed 'you' to 'Mr. D.' throughout, evidently with the intention that CD could send it to Asa Gray. The letter provides information on variable genera in Britain that would enable Gray to make comparable remarks about the same genera in the United States (see letter to Asa Gray, [after 15 March 1857]).

[2] CD's letter has not been found, but he had evidently questioned the categories listed by Watson in the letter from H. C. Watson, 10 March 1857.

[3] Hudson 1762.

[4] Backhouse 1856.

[5] See letter from Asa Gray, 16 February 1857. The underscoring has been printed as italics for one underline, bold for two underlines, and small capitals for three underlines.

[6] CD discussed this point in *Natural selection*, pp. 105–8.

[7] Edouard Spach specialised in the taxonomy of Cistaceae, to which *Helianthemum* belonged.

[8] The pencil crosses and written remarks refer to a later time when CD forwarded this letter to Asa Gray (see letter to Asa Gray, [after 15 March 1857]).

[9] These notes are for Asa Gray (see n. 8, above).

To J. D. Hooker 15 March [1857][1]

Down Bromley Kent

My dear Hooker

I have thought you would like to see enclosed *which please return*.[2] & I sh^d have sent it earlier, but I sent the last page to H. C. Watson for advice.—[3] The pencil scores mean nothing.—

I asked A. Gray whether he c^d tell me about Trees in U. States; & I told him that I had expected they w^d have sexes tending to be separate from theoretical notions, & I told him result for Britain & N. Zealand from you.—[4]

I have been thinking over your casual remarks at the Club, versus "accidental" dispersal, in contradistinction to dispersal over land *more or less* continuous;[5] & your remarks do not quite come up to my wishes; for I want to hear whether plants offer any *positive* testimony in favour of continuous land.— Your remarks were that the dispersal & more especially *non*-dispersal could not be accounted for by "accidental" means; which of course I must agree to & can say only that we are quite ignorant of means of *trans-oceanic* transport. But then all these arguments seem to me to tell equally against "continuous more or less" land; & you must say that some were created since separation on mainlands, & some extinct since on island.— Between these excuses on both sides, there seems not much to choose, but I prefer my answer to yours.—[6]

The same remark, seems to me applicable to your observation on the commonest species not having been transported; for it seems bold hypothesis to suppose that the commonest have been generally last created on the mainland or soonest extinguished on the island.— But I sh^d like to hear whether you are prepared on reflexion to uphold this doctrine of the commonest being least widely disseminated on outlying isl^{ds}.— I know it holds in New Zealand & feebly owing to distance in Tristan d'Acunha., but generally I sh^d have taken from De Candolle a different impression:—[7] I am referring only to *identical* species in these remarks.—

What I sh^d call positive evidence would be if proportions of Families had been exactly same on island with mainland.— If all plants were common to some mainland & island (as in your Raoul Is^d.)[8] more especially if some *other* main-land was nearer.— If soundings concurred with any great predominance of species from any country—or any other such argument of which I know nothing.

Do not answer me, without you feel inclined, but keep this part of subject before your mind for some future essay. I have written at this length that you may see, what I for one sh^d like to see discussed. But I will stop for I could go on prosing for another hour.—

I hope to get a feeble ray of light on Protean genera from A. Gray: how infinitely kind he has been to me.

I have just heard from H. C. Watson with whom I have been corresponding on protean genera; & I find I shall have to send back to Gray the latter half—of his letter,[9] which I daresay you w^d not care to see.

So adios | Ever yours | C. Darwin

March 15th—

Endorsement: 'March 15/57.'
DAR 114.4: 193

[1] Dated by the references to Hewett Cottrell Watson's letters about protean genera (see n. 3, below).
[2] Letter from Asa Gray, 16 February 1857.
[3] CD had sent Watson the last page of Asa Gray's letter, which contained a list Gray had prepared of protean genera. See letter from H. C. Watson, 10 March 1857 and letter from H. C. Watson to Asa Gray, 13 March 1857.
[4] See letter from J. D. Hooker, 7 December 1856.
[5] CD and Hooker had discussed means of geographical dispersal when they met at a meeting of the Philosophical Club in February (see letter to J. D. Hooker, [after 20 January 1857]).
[6] CD summarised the botanical relations that led Hooker to believe in a former land-mass connecting New Zealand, Kerguelen Land, Tristan d'Acunha, and Tierra del Fuego in *Natural selection*, pp. 560-5. There he also discussed his own view of the agencies that may have accounted for these particular relationships.
[7] A. de Candolle 1855.
[8] In a paper on the botany of Raoul Island, Hooker stated: 'The most interesting circumstance connected with the vegetation of Raoul Island is the identity of most of the flowering plants, and all but one of the ferns, that have been collected upon it, with those of New Zealand.' (J. D. Hooker 1857, p. 126). Hooker found this difficult to explain by assuming accidental transportation, for the island lies 450 miles north-west of New Zealand.
[9] See n. 2, above.

To Asa Gray [after 15 March 1857][1]

Down Bromley Kent

(I have divided my letter to save Postage)
My dear Dr Gray

Your last letter, like all its predecessors has been very valuable to me; & every word in it has interested me.[2] When I said that your remarks on your alpine plants "riled" me; I did not mean to doubt them, except in the Agassian sense that they went against some theoretic notions of mine. These notions are too long to give & indeed not worth giving, as far as America is concerned, & I can see from your letter that we sh^d take very much the same view. I am very glad to hear that you think of discussing the relative ranges of the identical & allied U. States & European species, *when you have time*. Now this leads me to make a very audacious remark in opposition to what I imagine Hooker has been writing & to your own scientific conscience. I presume he has been urging you to finish your great Flora, before you do anything else. Now I would say it is your duty to generalise as far as you safely can from your as yet completed work. Undoubtedly careful discrimin-ation of species is the foundation of all good work; but I must look at such papers as yours in Silliman as the *fruit*.[3] As careful observation is far harder work than generalisation & still harder than speculation; do you not think it very possible that it may be overvalued? It ought never to be forgotten that the observer can generalise his own observations incomparably better than anyone else.

How many astronomers have laboured their whole lives on observations & have not drawn a single conclusion; I think it is Herschel who has remarked how much better it would be, if they had paused in their devoted work & seen what they could have deduced from their work.[4] So do pray look at this side of question, & let us have another paper or two like the last admirable ones. There, am I not an audacious dog!

You ask about my doctrine which led me to expect that Trees would tend to have separate sexes.[5] I am *inclined* to believe that no organic being exists which *perpetually* self-fertilises itself. This will appear very wild, but I can venture to say that if you were to read all my observations on this subject, you would agree it is not so wild as it will at first appear to you, from flowers said to be always fertilised in bud &c &c &c.— It is a long subject to which I have attended to for 18 years! Now it occurred to me that in a large tree with hermaphrodite flowers, we will say it w^d be ten to one that it would be fertilised by the pollen of its own flower, & a thousand or ten-thousand to one that if crossed, it would be crossed only with pollen from another flower of same tree, which would be opposed to my doctrine. Therefore on the great principle of "Nature not lying"[6] I fully expected that trees would be apt to be dioicous or monoicous (which as pollen has to be carried from flower to flower everytime, would *favour* a cross from another individual of the same species) & so it seems to be in Britain & N. Zealand. Nor can the fact be explained by certain families having this structure & chancing to be trees, for the rule seems to hold both in genera & families, as well as in species.

I give you full permission to laugh your fill at this wild speculation; & I do not pretend but what it may be chance which, in this case, has led me apparently right. But I repeat that I feel sure that my doctrine has more probability, than at first it appears to have. If you had not asked, I sh^d not have written at such length, though I cannot give any of my reasons.

The Leguminosæ are my greatest opposers; yet if I were to trust to observations on insects made during many years, I sh^d fully expect crosses to take place in them; but I cannot find that our garden varieties ever cross each other. I do **not** ask you to take *any* trouble about it, but if you should by chance come across any intelligent nurserymen, I wish you w^d enquire whether they take any pains in raising the vars. of papilionaceous plants apart to prevent crossing. (I have seen statement of naturally formed crossed Phaseoli near N. York) The worst is that nurserymen are apt to attribute all variations to crossing.—

Finally I *incline* to believe that every living being requires an *occasional* cross with distinct individual; & as trees from mere multitude of flowers offer obstacle to this, I suspect this obstacle is counteracted by tendency to have sexes separated. But I have forgotten to say that my maximum difficulty is trees having papilionaceous flowers: some of these, I know, have their Keel-petals expanded when ready for fertilisation; but Bentham does not believe that this is general:[7] nevertheless on principe of nature not lying I suspect that this will turn out so, or that they are eminently sought by Bees dusted with pollen. Again I do **not** ask you to take trouble, but if strolling under your Robinias when in full flower, just look at stamens & pistils whether protruded & whether Bees visit them.— I must just mention a fact mentioned to me the other day by Sir W. Macarthur, a clever Australian gardener, viz how odd it was that his Erythrinas in N. S. Wales would not set a seed, without he imitated the movements of the petals which Bees cause.—[8] Well, as long as you live, you will never after this fearfully long note ask me why I believe this or that.

I am *particularly* obliged for your information about Protean genera; as this is one of the greatest of my many great puzzles; viz to know or conjecture whether the great variability of such genera is due to their conditions of existence, or whether it is apt to be innate in them at all times & places (I am aware that this cannot be strictly predicated of any genus, for all have some fixed species).— Now I have thought that you would not object to my sending the latter half of your note with list of such American genera to Mr H. C. Watson, of whose great clearness of mind & acuteness I have from long correspondence the *highest* opinion.— I have sent his note (1)[9] & Essay (2)[10] for the *chance* of your liking to see them; but if more busy even than usual, do *not* read them; & I daresay in course of half year, you will be able to return them *together with your own list*; but **pray** do not lose them. There is only one point, to which for myself, I wish to call your attention, viz whether you rightly understood that my question did not refer to genera having very close species, but to genera having very variable species. Watson thought that you might not have understood me.— If you do read Watsons papers & have anything to remark on subject, I need not say, how *very much* I sh^d like to hear it.[11] I am sure I do not know whether I have acted wisely in sending Watsons letters, but I repeat again pray do not read them, without you feel inclined: as, I suppose, in course of few months you would be sending some parcel to England, it will not cost you trouble to return them.

With hearty thanks for all your kindness, & begging forgiveness for length of this letter, which is chiefly your own fault (as far as trees are concerned), believe me, Your's most sincerely | Ch. Darwin

Gray Herbarium of Harvard University

[1] The date is based on CD's remark that he is enclosing with this letter some notes and a letter from Hewett Cottrell Watson (see letters from H. C. Watson, 10 March 1857, and from H. C. Watson to Asa Gray, 13 March 1857). The letter from Watson to Gray was received by CD on or around 15 March 1857, since he mentioned having just received it in his letter to J. D. Hooker, 15 March [1857].

[2] Letter from Asa Gray, 16 February 1857.

[3] A. Gray 1856–7 was published in 'Silliman's Journal', the *American Journal of Science and Arts*.

[4] John Frederick William Herschel. The exact passage has not been located in Herschel 1831, but see p. 266.

[5] CD discussed this subject in *Natural selection*, pp. 61–2, and *Origin*, pp. 99–100. See also letters to J. D. Hooker, 1 December [1856] and 10 December [1856].

[6] CD refers to his anecdote about Louis Agassiz related in the letter to Asa Gray, 1 January [1857].

[7] See letters to George Bentham, 26 November [1856] and 30 November [1856].

[8] See letter to Syms Covington, 22 February 1857. William Macarthur had strong interests in horticulture and gardening and played an active role in the governance of the Royal Botanic Gardens of Sydney (Gilbert 1986, pp. 58, 72–3).

[9] Letter from H. C. Watson, 10 March 1857.

[10] Letter from H. C. Watson to Asa Gray, 13 March 1857.

[11] See letter from Asa Gray, [*c.* 24 May 1857].

To Edward Sabine[1]   16 March [1857][2]

Down Bromley Kent
March 16

My dear Sir

By some accident I received your note only this morning, for which I am much obliged; as it w^d be very inconvenient to me to attend I would much rather not be on the Committee. Indeed it w^d be superfluous, as I know not much of the Nat. History of N. America, & as Sir Roderick Murchison & Dr Hooker are on it.[3] Sir John Richardson would be the man for Zoological suggestions.—

As the extension in Lat. & Long. & all the phenomena of Glacial action & erratic boulders will, no doubt, be one chief object of attention to the geologist of the Expedition, I may make one suggestion, viz to attend most carefully to the state of the rocks in those rivers, in which annually large quantities of ice are carried down with great force. Sir John Richardson many years ago, told me that they were beautifully polished.—[4] Are they scored also? & is the scoring on the upper side. &c &c.— are stones & mud embedded in river ice? It is a great desideratum in geology to be able to distinguish between rocks polished & scored by glaciers, & by floating ice.—[5]

I suppose the expedition will not visit any arctic shore; but it may fall across some ancient line of cliff, with beds of shingle at its base, formed during the Glacial epoch, & I think a most minute examination of the character of the shingle on arctic shores would be very desirable, for comparison with the sub-angular drift of the southern counties of England. The tertiary strata with fossil plants & lignite would be a very interesting point for examination; but is quite obvious.— If I sh^d think of any point worth noticing, I will write.

Pray believe me, my dear Sir | Yours sincerely | Ch. Darwin

Royal Society (MM 4: 39)

[1] Edward Sabine was chairman of the Royal Society North American exploring expedition committee (Royal Society committee minute book 42).

[2] The minutes of the expedition committee of 16 March 1857 record that Sabine had written to CD (Royal Society committee minute book 42).

[3] See letter to William Sharpey, 24 January [1857]. Both Roderick Impey Murchison and Joseph Dalton Hooker were members of the committee.

[4] CD had consulted John Richardson, author of *Fauna Boreali-Americana* (J. Richardson 1829–37), in 1838 when he was writing the addenda to *Journal of researches* in order to explain the distribution of erratic boulders by iceberg transport (pp. 619–20).

[5] CD had long wished to differentiate the action of glaciers from that of floating ice and was especially concerned to demonstrate the role of icebergs in the distribution of erratic blocks (see 'On the distribution of the erratic boulders . . . of South America', *Collected papers* 1: 145–163, and 'Notes on the effects produced by the ancient glaciers of Caernarvonshire', *Collected papers* 1: 163–171).

To J. D. Hooker [21 March 1857]

[Down]

in regard to species, & then all is horrid fog.— You told me that you could lend me Drege's work or list on the distribution of the Cape Plants[1] (not the paper in the Flora,[2] which I know) & I sh^d be *particularly* obliged for it within a week & I will keep it not more than a fortnight. Will you send it per Post, & let me pay postage. I want to see whether there are materials to work out range of the species in large genera contrasted with those in small genera.— A. De C. has done it for families,[3] but as these will include small & large genera, I think this is not the right way. Asa Gray took *genera* as I asked him & the result was as it sh^d be, for as Agassiz says, nature never lies.[4]

I am amusing myself with several little experiments; I have now got a little weed garden & am marking each seedling as it appears, to see at what time of life they suffer most.—[5]

I congratulate you on having done so much of the Indian Flora, & am astounded how you possibly could have made out 7000 species & ticketed 15000 species.[6] I would not have done such work for a guinea a specimen! Such materials will give some splendid general results.— I envy your power of work & noble zeal.—

Some time ago you told me of two reputed species of Thistle, which in the Himalaya, *alone*, were blended by a perfect series of intermediate forms: some time I sh^d. be very glad to have particulars *briefly*, & be allowed to quote; I sh^d. like to know whether both the distinct forms grew mingled with the intermediates.

My dear Hooker | Your's affectionately | C. Darwin

Incomplete
Endorsement: 'March 21/57'
DAR 114.4: 192C

[1] Probably Drège 1837–39, 1840, or [1847], which described plants collected by Jean François Drège in South Africa. The Kew library held all three catalogues (Royal Gardens, Kew 1899).
[2] Drège 1843. See letter from J. D. Hooker, 7 May 1856 .
[3] A. de Candolle 1855.
[4] See letter to Asa Gray, 1 January [1857].
[5] CD recorded this experiment under the heading 'Weed Garden' in his Experimental book, p. 25 (DAR 157a). Selecting a small plot of land in the orchard protected from large animals, he cleared it of all perennials in January 1857. His next entry reads: 'Early in March seeds began to spring up: marked each daily.' He continued to monitor the plot, marking new plants, counting the ones that had perished, and suggesting possible causes of death, until 1 August. Out of 357 plants he had marked, he found 62 had survived.
[6] Hooker continued to work on his Indian plants, despite being frustrated in his attempts to obtain financial support to continue publishing the *Flora Indica*, only one volume of which had appeared (J. D. Hooker and Thomson 1855). On 3 March 1857, he and Thomas Thomson had contributed a paper to the Linnean Society that was intended to provide a 'temporary substitute' by way of analysing several of the natural orders of Indian plants (J. D. Hooker and Thomson 1858).

From James Tenant[1] 31 March 1857

March 31st/1857

Dr Sir

I now send the result of the experiments of the Seeds[2] which I have been trying ever since you was at the Gardens

I have succeded in getting Several of them to take them by letting them go a day or two without food the Minnows take the millet very and so does the Gold fish one minnow took 5 Seeds this day and the Tench and Common Carp and Barbel take the Wheat after it is Soaked well in Water for 12 Hours

As an Illustration of the fact that Barbel will take Wheat Dr Crisp—a Fellow of the Society who is a great Angler told me, that he has taken Barbel and dissected them And found a quantity of Wheat in them and he says he has caught them near the water mills where the wheat has been spilt into the river but never could get them to take it as a bait[3] I should be most happy to make any Experiment you might suggest and

I remain your | obedient Servt | Ja^s Tenant

C Darwin Esq^e

DAR 205.2 (Letters)

CD ANNOTATIONS
Top of first page: '18'⁴ *brown crayon*

[1] Tenant was keeper of the aquarium at the Zoological Society's gardens.
[2] CD was repeating experiments first carried out in 1855 (see *Correspondence* vol. 5, letter to W. D. Fox, 7 May [1855]). Initially, he had hoped to find that seeds swallowed by fish could be transported and subsequently germinate. By 1857, he was investigating the possibility that birds might eat fish or other animals that had seeds in their guts (see letter to W. D. Fox, 20 October [1856]). CD also persuaded his nephew Edmund Langton to perform experiments similar to the ones described here. Langton wrote to CD's son Francis on 21 February [1856] (DAR 205.2 (Letters)):

> Will you tell your papa that I have tried the experiments with all the seeds but the minnows only took a very little Dutch clover and spit it out again, and the Prussian carp took one anthoxanthum seed and spit it out again but it was a rather cold day so I will try again.

The results of Langton's experiments are recorded in CD's Experimental book, p. 8 (DAR 157a).
[3] See next letter.
[4] The number of CD's portfolio of notes on the means of geographical dispersal of plants and animals.

From Edwards Crisp[1] 4 April 1857

21 Parliament St
April 4 1857

Dear Sir

I caught the Barbel (several of them) last autumn near a mill-dam and was surprised to find wheat in their stomachs—wondered how the fish could get it— but the mystery was solved by the miller, who told me "that a large quantity of wheat passed into the river from the mill" the largest Barbel (about 2 lbs) had about 200 grains of wheat in its stomach: the greater part of it was entire. If taken soon after it was swallowed, it would probably vegetate, but (as you know) the digestion of a fish is so rapid, that most seeds I imagine would soon be destroyed? Should I meet with any I will not fail to send them—

You are probably aware that a kernel of green wheat is a good bait for roach and probably some other fish— To return to the Barbel an angler of my acquaintance used to eat these fish, and contrary to the general taste he thought them excellent. On one occasion however he caught a big fish, and when he was removing the hook, a large quantity of cow[x]-dung escaped from its mouth; he has never eaten Barbel since!

Ys very faithfully | Edwards Crisp

[x]The cow often deposits its excrement in the water as you know.

C Darwin Esqre

DAR 205.2 (Letters)

CD ANNOTATIONS
2.1 green wheat is a good bait] *underl pencil*
Top of first page: '18'² *brown crayon*

[1] CD had been directed to Crisp, a fellow of the Zoological Society, by James Tenant (see previous letter).

[2] The number of CD's portfolio of notes on the means of geographical dispersal of animals and plants.

To J. D. Dana 5 April [1857][1]

> Down Bromley Kent
> April 5th

My dear Sir

You were so kind as to say that I might trouble you occasionally for information. There is now a *point* on which I am very curious, & which I think I could make out from your Memoir,[2] but, as I once said before, it is incomparably safer not to infer but to quote direct opinion of author.— Sir J. Richardson says the Fish of the cooler temperate parts of the S. Hemisphere present a much stronger analogy to the fish of the same latitudes in the North, than do the strictly *Arctic* forms to the *Antarctic*.[3]

Now I sh[d] very much like to know how this is with Crustaceans.—[4] I have quoted your remarks on the relation of the N. Zealand Crust. to those at these Antipodes;[5] but can you tell me how it is with those further south. I fear that there are hardly materials.— Cape Horn may throw some light, but it is hardly far enough south: I think, as far as I can remember, very few Crust. are known from the S. Shetland or icy regions.— But if you will give me a sentence on Crustacea, in relation to Sir J. Richardson's remark, I sh[d] be particularly obliged.—[6]

When I shall publish my Book, Heaven only knows, for it daily grows on me; but I do some work every day; but my day's work, from ill-health is ridiculously short.— I am sorry that I have no scientific news to communicate, for I have left home very seldom of late owing to my health having been worse than usual. The most interesting discovery, I think, made for some years, has been the astonishing find of Mammalian remains in the Purbeck beds;[7] I have seen at D[r] Falconer's many of the specimens—[8] They give one an astonishing idea of the richness of the Fauna at that period. Lyell, as you may suppose is delighted. I never saw anything more curious than the manner in which the Plagyoulax (or some such name) connects the living Hypsiprimnus & the Triassic mammifers, about which the doubts formerly held must now be given up; for it must assuredly have been a Mammifer.—[9]

Some small & very highly organised Lizards in same bed are, I think, even more interesting than the Mammalian remains.— This discovery has made a deep impression on some of our geologists, as Prestwich, who have been strongly inclined to trust in negative evidence. Lyell will very soon publish a little supplement to his Elements & will give an outline of these new facts & many others which he picked up on the continent last autumn.—[10] He visited with Barrande his celebrated region; & will discuss B.'s colonies, which always troubled me as great anomaly.—[11]

Owen has lately published a new Classification of mammals, taken from structures of Brain;[12] so great an authority ought to be right, but I cannot help always having doubts on a classification founded on one character, however important.— I have had of late a good deal of correspondence with Asa Gray, who has been infinitely kind in giving me valuable information.— I sometimes hope that my Book will be useful as comparing the results which different authors from different data have arrived at; however erroneous my general conclusions may prove.—

Whenever you have time to write, tell me a little what you are about, & believe me, My dear Sir | Your's very sincerely | Ch. Darwin

Yale University Library (Manuscripts and archives, Dana papers)

[1] The year is provided by the references to the recent publication of Owen 1857b and to the forthcoming publication of C. Lyell 1857a.

[2] Dana 1853. Dana had published the section from Dana 1852[–3] on the classification and geographical distribution of Crustacea separately in 1853. CD's presentation copy is in the Darwin Library–CUL.

[3] In *Natural selection*, p. 555, CD cited J. Richardson 1845, p. 189, on this point. See also letter from John Richardson, 17 July 1856.

[4] CD was gathering cases from the zoological realm to test his hypothesis of the migration of plants and animals 'from north to south during the glacial epoch' (*Natural selection*, p. 554). He was seeking examples of northern temperate species, or closely allied species, that were also found in the southern hemisphere.

[5] 'Well does Prof. Dana remark that "it is certainly a wonderful fact that New Zealand should have a closer resemblance in its Crustacea to Great Britain, its antipode, than to any other part of the world," ' (*Natural selection*, p. 557). CD refers to Dana 1853, p. 1587.

[6] See letter from J. D. Dana, 27 April 1857.

[7] See letter from Charles Lyell, [16 January 1857], n. 2.

[8] Charles Lyell had asked Hugh Falconer to examine the fossils from the Purbeck beds (Wilson ed. 1970, p. lii). Falconer published a description of two species of a new mammalian genus *Plagiaulax* later in the year (Falconer 1857).

[9] According to Falconer, the *Plagiaulax* was decidedly 'a marsupial form of rodent, constituting a peculiar type of the family to which *Hypsiprymnus* [the kangaroo-rat of Australia] belongs' (Falconer 1857, p. 274). Descriptions were also incorporated into a supplement to Lyell's *Elements of geology* (see n. 10, below).

[10] C. Lyell 1857a. Copies of this and the revised edition, C. Lyell 1857b, are in the Darwin Library–CUL. Both contain annotations.

[11] Joseph Barrande, a French geologist living in exile in Prague, had found a 'colony' of Upper Silurian fossils in the Lower Silurian strata, which Lyell, in a letter of 23 August 1856 to Leonard Horner, called 'the most singular and, at first sight at least, anomalous fact I ever remember to have verified in paleontological geology.' (K. M. Lyell ed. 1881, 2: 223). In his supplement to the *Manual of elementary geology* (C. Lyell 1857a), Lyell reported Barrande's conclusion that each period of geological time was characterised by a fully diversified flora and fauna, rather than a uniform assemblage spread out over the globe. To Barrande, the 'colony' was merely a pocket of animals that then increased greatly in numbers during the following geological epoch, so giving the appearance of an anomaly in the Lower Silurian. See C. Lyell 1857a, p. 31–4.

[12] Owen 1857b.

To J. D. Hooker 8 April [1857][1]

Down Bromley Kent
Ap— 8th

My dear Hooker

Drege contains no materials for seeing range of genera, so I return it by Deliverance Coy, & I hope you will get it safe by Thursday night or Friday morning.—[2]

I now want to ask your opinion & for facts on a point; & as I shall often want to do this during next year or two; so let me say once for all, that you must not take trouble out of *mere* goodnature (of which towards me you have a most abundant stock) but you must consider, in regard to trouble any question may take, whether you think it worth while, (*as all loss of time so far lessens your original work*) to give me facts to be quoted on your authority in my work. Do not think I shall be disappointed if you cannot spare time; for already I have profited enormously from your judgment & knowledge.— I earnestly beg you to act as I suggest, & not take trouble *solely* out of goodnature.—

My point is as follows—Harvey gives cases of Fucus varying remarkably, & yet in same way under *most* different conditions. D. Don makes same remark in regard to Juncus bufonius in England & India.— Polygala vulgaris has white red & blue flowers in Faröe, England, & I think Herbert says in Zante.[3]

Now such cases seem to me very striking, as showing how little relation some variations have to climatal conditions.

Do you think there are *many* such cases? Does Oxalis corniculata present *exactly same* varieties under **very different climates**?

How is it with any other British plants in N. Zealand, or at foot of Himalaya?— Will you think over this & let me hear result.—[4]

One other question; do you remember, whether the *introduced* Sonchus in N. Zealand, was less, equally, or more common than the aboriginal stock of same species, where both occurred together: I forget whether there is any other case parallel with this curious one of the Sonchus.—

My wife starts with Etty on Thursday for Hastings: she is no better.—[5]

I have been making good, though slow, progress with my Book, for facts have been falling nicely into groups, enlightening each other.—

My dear Hooker | Farewell | C. Darwin

Endorsement: '1857'
DAR 114.4: 191

[1] The year is given by the reference to Emma and Henrietta Emma Darwin's trip to Hastings (see n. 5, below).
[2] See letter to J. D. Hooker, [21 March 1857].
[3] Harvey 1849, Don 1841, and Herbert 1846 are all cited in *Natural selection*, p. 284.
[4] See following letter. In *Natural selection*, p. 284, CD stated that while he believed these instances were not common, 'Dr. Hooker thinks a good many could be collected'.

5 Henrietta Darwin's health had begun to fail in 1856. In August 1856, CD had taken her to London to consult Benjamin Collins Brodie (Emma Darwin's diary). In March 1857 Henrietta's condition worsened, and on 9 April Emma took her to the seaside resort of Hastings, Sussex, where Henrietta remained until 12 May (Emma Darwin's diary).

From J. D. Hooker [11 April 1857][1]

Hastings
Saturday.

Dear Darwin

I came down here on Thursday, when also my wife came from Brighton & yours from the North— the latter met on the Railway station. Mrs. D. seems well & strong, Etty is thin & pale but not looking worn or anxious, I do hope the change will do her good. they have taken lodgings, which look well, close by here & go in today.[2]

If you knew how grateful the turning from the drudgery of my "professional Botany" to your "philosophical Botany" was, you would not fear bothering me with questions— the truth in its positive nakedness is, that I really look for & count upon such questions, as the best means of keeping alive a due interest in these subjects. I indulge vague hopes of treating of them some day, but days & years fly over my head & all I do is done in correspondence to you, but for which I should soon loose sight of the whole matter.

Harveys observation on Fucus varying much & yet in same way under *most* different conditions goes with me for a great deal & I would endorse it.[3] D. Don's on Juncus bufonius in England & India I would not put in the same scale but is good By the oddest chance I was, on the day of the arrival of your letter, doing Indian Juncus bufonius, several hundred specimens from 8 or 10 different localities, from plains of Panjab to elev. 9000 feet in Sikkim. Now I find the best marked English varieties (& these are very wide) amongst the Indian ones, just as Don did, & as none of the conditions except that of Sikkim (which maybe compared rudely to W of Scotland) are at all like Britain, we may I suppose assume it to be a good case in point. The Polygala case is as good in its way, & you may add Anagallis arvensis which varies Red, blue & white in N. W India, as in England.

Cardamine hirsuta, presents in Fuegia & N. Zealand, most (if not all) its European phases, besides many more.

There are I think heaps such cases, they have so often struck me, that one of my sketched out methods of treating of the Indian plants common to W. Europe & India is by dividing them into

1. Identical unvarying species
2. Identical variable species
   a variations equal & similar in both countries
   β. variations unequal or dissimilar or both

Oxalis corniculata I believe to be as good a case, but it is as well to avoid a genus upon which Botanists are so notoriously divided in opinion

I am sorry that I cannot at all answer the *Sonchus* question; but as the wild *Sonchus* is not very common I suppose the introduced will soon be by far the most so.

The Taraxacum dens-Leonis must be a parallel case; as also *Alsine media*, & *Cardamine hirsuta*, though they have not yet forced themselves on the notice of Botanists.

I shall return to Kew on Monday & send you any further notices of plants varying simila⟨r⟩l⟨y⟩ under widely different conditions.

Ever Yrs | J D Hooker

DAR 104: 198–200

CD ANNOTATIONS
0.1 Hastings . . . matter. 2.7] *crossed pencil*
10.1 I shall . . . simila⟨r⟩l⟨y⟩ 10.2] *crossed pencil*

CD note:[4]

Ch 4[5] [*added brown crayon*]
    I must introduce a discussion under Polymorphism, **after it**—viz *cases of [*interl*] ['in' *del*] some degree of *definite* variation under most different climates— see what I have said in beginning part of Chapt. 7. & strike that out & *there* (in Ch. 7) only *allude* to it.[6]
    This is a very perplexing subject—shows how preponderant the organization of the being is, & how little external conditions has to do with it, & how little selection has to do.—. But such variations are probably not correlation to other beings.

[1] Dated by the reference to Emma and Henrietta Darwin's visit to Hastings (see preceding letter & n. 5) and by the relationship to the preceding and following letters.
[2] See preceding letter.
[3] See preceding letter.
[4] This note is in DAR 104: 201.
[5] CD refers to the *Natural selection* chapter 'Variation under nature'.
[6] CD refers to chapter 7, 'Laws of variation', of his species book (*Natural selection*, pp. 279–338), which he was writing at this time. The chapter was completed on 5 July 1857 ('Journal'; Appendix II). In *Natural selection*, p. 284, CD mentioned the cases of *Fucus* and *Juncus* and concluded: 'These cases . . . lead us back to the perplexing facts of polymorphous species & genera, discussed in the fourth Chapter; they show us how ignorant we are on the subject of variation'. See also the following letter.

To J. D. Hooker 12 April [1857][1]

Down Bromley Kent
Ap. 12th

My dear Hooker
    Your letter has pleased me much, for I never can get it out of my head, that I take unfair advantage of your kindness, as I receive all & give nothing. What a

splendid discussion you could write on whole subject of variation! The cases discussed in your last note are valuable to me, (though odious & damnable) as showing how profoundly ignorant we are on causes of variation.—

I shall just allude to these cases, as a sort of sub-division of polymorphism—a little more definite I fancy than the variation of for instance the Rubi, & equally or more perplexing.—[2]

I have just been putting my notes together on variations *apparently* due to the immediate & direct action of external causes;[3] & I have been struck with one result. The most firm stickers for independent creation admit, that the fur of *same* species is thinner towards south of range of same species than to north—that *same* shells are brighter coloured to S. than N.; that same is paler-coloured in deep water—that insects are smaller & darker on mountains—more lurid & testaceous near sea—that plants are smaller & more hairy & with brighter flowers on mountains: now in all such (& other cases) cases, distinct species in the two zones follow the same rule, which seems to me to be most simply explained by species, being only strongly marked varieties, & therefore following same laws as recognised & admitted varieties. I mention all this on account of variation of plants in ascending mountains; I have quoted the foregoing remark only generally with no examples, for I add there is so much doubt & dispute what to call varieties; but yet I have stumbled on so many casual remarks on *varieties* of plants on mountains being so characterised, that I presume there is some truth in it. What think you? do you believe there is *any* tendency in *varieties*, as *generally* so called, of plants to become more hairy & with proportionally larger & brighter coloured flowers in ascending a mountain.—

I have been interested in my "weed garden" of 3 × 2 feet square:[4] I mark each seedling as it appears, & I am astonished at number that come up. & still more at number killed by slugs &c.— Already 59 have been so killed; I expected a good many, but I had fancied that this was a less potent check than it seems to be; & I attributed almost exclusively to mere choking the destruction of seedlings.— Grass-seedlings seem to suffer much less than exogens.—

I have *almost* finished my floating experiments on salt-water:[5] $\frac{72}{94}$ sunk under 10 days—seven plants, however, floated *on average* 67 days each.— I then dried all these (with in each case, with pods, bits of twig & few leaves) & $\frac{62}{94}$ sunk under 10 days, so that generally the drying had no great effect, but here comes the odd part sometimes it had great effect, thus

	Not dried floated	*Dryed* floated
Asparagus	22–23	85–86 (germinated excellently)
Lychnis dioica	21–22	44–45
Honysuckle	3–4	21–22
Helosciadium	1–2	21–22 & seeds above 90 days & germinated splendidly

	Not Dryed	Dryed
Barbery	20–21	40–41
Viscaria oculata	2–3	30–31
Dianthus	1–2	28–29
Sweet Briar	2–3	21–22
Juniper	12–13	38–39
Nuts	0–1	60–70
	&c &c	&c &c

I *think* it will turn out on average from my very few experiments, of **very** little value, but better than mere conjecture, that about $\frac{1}{10}$ of all plants of a country will float when dryed 30 days & *the seeds then germinate*; & this on *average* current of 33 miles per day will carry them a good way.[6] I would wager that the pods of the Acacia(?) scandens which get to the Azores had been dried first.—[7] I suppose the oriental species does not fruit at Kew: if it did, I sh^d. like to try.—

Are there any **hardy** garden plants or shrubs endemic to the Canaries, Madeira or Azores.— I sh^d. like to know; as it shows that the constitution of an endemic plant is not absolutely fitted to its home, perhaps in more striking manner than the hardiness or naturalisation of a plant from a continent.— The Chiococca racemosa, the suckers of which I cannot weed out of my garden, is a very striking case, as it is, I believe, confined to W. Indies.[8]

Farewell my dear Hooker, everything which I write about or think of, I long to talk over with you, as I have shown in this note.

Farewell | C. Darwin

P.S Strictly according to my experiments a little above $\frac{1}{7}$ (.140) of the plants of any country could be transported 924 miles & *would then germinate*! for $\frac{18}{94}$ have floated above 28 days & $\frac{64}{87}$ is proportion of seeds which germinate after 28 days immersion.— & average of current in Atlantic is 33 miles per diem.—

I have just had a letter from Emma & she speaks with much pleasure at having seen you & Mrs Hooker, whose state (you are as bad as I am) is to be pitied.[9]

P.S. Can any general character be predicated of water-plants; if so, & again if any plant has a variety growing in damp ground, does it take in ever so slight a degree the characteristic features of aquatic plants.— It w^d. be another case to the many which I have collected.

A plant *abruptly* having two forms like the aquatic Ranunculus seems something different & very unpleasant to me.

Endorsement: '1857'
DAR 114.4: 192

[1] The year is established by the reference to Emma Darwin meeting Hooker and his wife in Hastings (see n. 9, below).

[2] See the CD note transcribed with the preceding letter.

[3] See *Natural selection*, pp. 281–5.

[4] See letter to J. D. Hooker, [21 March 1857].

⁵ The notes on these experiments, begun on 3 December 1856 and headed 'Dryed Seeds & Fruits in Salt-Water' and '1857 Dryed Seeds', are in CD's Experimental book, pp. 10v.–14 (DAR 157a).

⁶ These results are summarised in *Origin*, pp. 359–60.

⁷ See *Correspondence* vol. 5, letter from J. R. Crowe, 27 September 1855, and letter to J. R. Crowe, 9 November 1855, for CD's attempts to float seeds of *Acacia scandens*.

⁸ See *Natural selection*, pp. 285–6.

⁹ Frances Harriet Hooker was expecting the Hookers' fourth child. Marie Elizabeth Hooker was born on 10 August 1857. There is an entry on 11 April 1857 in Emma Darwin's diary, written during her stay in Hastings: 'drank tea with Hookers'.

From T. V. Wollaston [12 April 1857]¹

10 Hereford Street | Park Lane
Easter Sunday (evening).

My dear Darwin,

Many thanks for your note rec⁴ late last night (just before going to bed), & which kept me awake (thinking) almost until "daylight doth appear": & as I foresee several questions to answer, I may as well get out one of my *larger* note-sheets, to scribble upon.—

Imprimis, however, I am very sorry to hear that you (like me) have been at a low ebb of late. Our "par" standard is I think about equal; & we may be pronounced therefore as "Arcades ambo", in our way.² We have however one advantage over the rest of Christendom,—viz. that it make us *contented* where other people would swear; a very small amount of prosperity raising us into the seventh heaven. For my own part, I have been seedy all through the spring,—not however suffering from chest (wʰ has been tolerably free), but from a complication of miseries, having been rejoicing latterly (inter alia diabolica diversa) under an attack of the *mumps*! However I am beginning to "pick-up" again now, & may hope to be once more in statu normali some of these fine days. Where are you going to for your Hydropathic settler? Why not try Malvern,—taking your books & papers with you for writing.³ The oxygen of those glorious hills wᵈ do you good, & perhaps give you some new ideas zoologically. I have been grinding *slowly on* (though as quick as the fates wᵈ, under the circumstances, permit) at my new Catalogue;⁴ but really so much accuracy is required (where one works **conscientiously**, so as to satisfy **one's own** individual ideas of what is right, & not merely those of Dʳ Gray)⁵ even in so small a matter, that I scarcely expect to finish it before the end of next month. However I am arranging my collection *simultaneously* (which is more than half the battle, in tediousness), replenishing & correcting the Museum drawers, & overhauling Mason's bottles⁶ & my own duplicates,⁷—all of which involves so much *manual* labour, that it is impossible (even if one was bursting with plethora & health) to go beyond a certain (Helix) pace. Unfortunately a B.M. "blue-book" does not admit, I find, of anything beyond the dryest detail, so that I can have no hope of generalisations, or separate lists, or anything else of interest; & I must therefore content myself with merely *settling my premises* (w^h is after all, however, the main point), & leave all after deductions & considerations for a separate paper elsewhere.— This however can be easily (& quickly) done, if desirable, for I always find it a simple matter to draw general conclusions *when once one's data are fixed*, & the difficult questions of "species", &c have been fairly digested in one's own mind. Indeed, *until this is done*, I always find myself in a maze whenever I attempt to decypher my material *as a whole*; & therefore I do not regret the amount of time and labour which I am now bestowing on these preliminary points (dry & monotonous tho' they be),—feeling convinced that it will pay in the end.—

And now, as regards your questions about the *absent* Madeiran groups, I will take them *seriatim*,—merely premising that I think you are right in supposing that they belong *par excellence* to the winged families.—

Cicindelidæ. There is not one of this family; & the species are the most active fliers in the Coleoptera.

Buprestidæ. This family was undetected when I published my volume; but my last expedition produced **one single specimen** of an *Agrilus*. (which I have lately described, by the way, as "Agrilus Darwinii",—thinking it a more worthy steed for you to ride down to posterity upon, than a mere "devil's coach-horse", stercoraceous tho' he be).[8] However even this one *species* would appear to be so rare, that the family may be practically considered as absent; and, next to the Cicindelas, the *Buprestidæ* are I suppose amongst the most active of all Coleopterous flyers.—

Pselaphidæ. *One* **specimen**, also, I have detected, of this family since my little novel made its debut; but the *Pselaphidæ* are not usually **very** active with their wings, though I believe that nearly all of them *are flyers*.

Carabus (genus). None in Madeira; but rare flyers.

Nebria (genus). None in Madeira; but (also) rare flyers.

Silpha and **Necrophorus**. *Great & powerful flyers*; but none in Madeira.

Cetoniadæ (& indeed **all** the "Thalerophagous Lamellicorns"). *Wonderful & active flyers*; & only one species in Madeira,—**if** *indeed "one"* (for the Chasmatopterus nigrocinctus **is still unique**, & from the coll. of Dr Heinecken; so that it **may** have been accidentally imported into the island).—[9]

Telephoridæ. Quite as great flyers (& universally so) as the *Cicindelas*, yet there is only one small, & insignificant species (the Malthodes Kiesenwetteri) in these islands.

Tentyria, **Pimelia**, **Akis** and **Asida**. *All* still absent from the Madeiran Group,—**but all apterous elsewhere**. This however does not tell against your theory, as there is **a reason for** their non-appearance,—viz. that they are are all *exactly* (& certainly) *represented* by the genera *Helops*, *Hegeter* and *Hadrus*,—w^h are equally apterous, & which do the work of the others.—

Otiorhynchus. Totally absent, & generally apterous everywhere. this genus however **is exactly** represented by *Laparocerus* (peculiar I believe to Madeira, though reported by entomological liars to be "European") and *Atlantis*,—both of which some naturalists regard as mere modifications of Otiorhynchus. Both are apterous.

Elateridæ. An enormous (indeed ponderous), universal and mundane family. It may be said to be absent from the Madeiran group, since up to the present time only *one, minute, anomalous* & **apterous** species has been detected (*& that one only in Porto Santo*). They are generally great flyers, though not so much so as the *Cicindelidæ* & *Telephoridæ*.—

Œdemeridæ. A large family & monstrous flyers; but only one species (that one however *a great flyer*) in Madeira.—

So much for your queries. It is needless to add that all the great groups of the island *in w*ʰ *the species are endemic* are apterous. Thus the *18 Tarphii* are apterous: all the *Helops* are apterous: all the truly indigenous *Ptini* are apterous; & nearly all the *Curculionidæ* are apterous.—

I think you give me too much credit for the mere observation of a self-evident fact,—the *apterous tendency* of Madeiran (and perhaps of all insular) influences.[10] I believe that no living (British) entomologist has made a single remark on any one of the instances wʰ I enumerated of wingless insects: they do not interest them; & all suchlike remarks fall as dull & ditchwater on the surface of our London society. There are not, alas, many Darwins in the world; & there is not (certainly) one single *entomological* D. in all Beetledom: hence everything relating to development or non-development does not so much as ruffle the waters of our Slough of Despond in Bedford Row.—[11]

I am not aware that insects will ever "fight for their females". I know however, from painful experience, that they will fight for *themselves* (having been bitten by a *Scarites* very severely some months ago); & I believe that "Nº 1" is their motto,—caring little for *N*º *2* (or their mates), except now & then to *eat* them.—

Nor can I say (in my ignorance) why the tarsi of coprophagous beetles sᵈ be so often missing,—unless it be that they **use** them so constantly & violently, in burrowing, that they perhaps are apt (like the tails of the Monkeys, when rubbed) to come off.—[12]

I think (without going afresh into the statistics) that you **may** say that "$\frac{23}{29}$ endemic genera have *all* their species either apterous, or incapable of flight".—[13]

Have you seen the last Punch? They have made Bright & Cobden *cherubs*, with wings coming out of their heads; & so, having given them wings, they have (on the law of compensation) taken away their bodies.[14] I must however conclude this note, so Believe me | my dear Darwin | Yours very sincerely | T V Wollaston.

DAR 181

CD ANNOTATIONS
0.3 My dear . . . end.— 2.34] *crossed pencil*
5.1 I think . . . Wollaston. 9.4] *crossed pencil*

[1] Dated by the heading 'Easter Sunday', which fell on 12 April in 1857, and by the reference to Wollaston 1857 (see n. 4, below).

[2] Virgil, *Eclogues*, 7. 4: 'Arcadians both, and each as ready as the other to lead off with a song, or to give an apt response.'

[3] CD had undergone treatment at James Manby Gully's hydropathic establishment in Malvern in 1849 (see *Correspondence* vol. 4).

[4] Wollaston was preparing the *Catalogue of the coleopterous insects of Madeira in the collection of the British Museum* (Wollaston 1857). The copy received by the British Museum is dated October 1857 (Entomology department, British Museum (Natural History)).

[5] John Edward Gray was keeper of the zoological collections of the British Museum.

[6] Philip Brookes Mason had made available to Wollaston a collection of over 20,000 Coleoptera from Madeira (Wollaston 1857, p. xiv n.).

[7] The British Museum had purchased Wollaston's collection of 4000 Coleoptera from Madeira and the Salvages in 1855 (British Museum (Natural History) 1904–6, 2: 562).

[8] *Agrilus darwinii* is described by Wollaston as a dung-eating beetle (Wollaston 1857, p. 82). He wrote: 'I have dedicated the species to Charles Darwin, Esq., M.A., V.P.R.S., whose inquiries into the obscurer phenomena of geographical zoology have contributed more than those of any other man living to our knowledge, in the general questions of animal distribution.'

[9] The specimen was described in Wollaston 1854, p. 236, where Wollaston stated that it had been 'communicated to me by the Rev. R. T. Lowe from the collection of the late Dr. Heinecken, by whom it was captured, many years ago, near Funchal.'

[10] See *Correspondence* vol. 5, letter from T. V. Wollaston, 2 March [1855]. Wollaston had discussed the apterous tendency among Madeiran Coleoptera in Wollaston 1856, pp. 82–7. He agreed with CD that wings would be disadvantageous to insects on these windy islands (pp. 86–7).

[11] Wollaston is equating the Entomological Society of London, housed at 12 Bedford Row, with the spiritual bog in John Bunyan's *Pilgrim's progress*.

[12] See *Natural selection*, pp. 293–4, and *Origin*, p. 135.

[13] CD used this information, and further information contained in Wollaston 1856, pp. 82–7, in *Natural selection*, pp. 291–3, and *Origin*, pp. 135–6.

[14] See *Punch, or the London Charivari*, 11 April 1857, p. 147. Richard Cobden's motion in the House of Commons in March 1857, condemning the declaration of war on China, led to the collapse of Henry John Temple Palmerston's government and a general election on 26 March in which both Cobden and John Bright lost their seats.

To Charles Lyell   13 April [1857][1]

Down Bromley Kent
Ap. 13th

My dear Lyell

I have been particularly glad to see Wollaston's letter.[2] The news did not require any breaking to me; for though as a general rule I am much opposed to the Forbesian continental extensions, I have no objection whatever to its being proved in some cases. Not that I can admit that W. has by any means proved it; nor, I think, can anyone else, till we know something of the means of distribution of insects.—[3] But the close similarity or identity of the two Faunas is certainly very interesting.— I am extremely glad to hear that your Madeira paper is making progress; & I shall be most curious to see. I sh^{d} be infinitely obliged for a separate copy, whenever printed.—[4]

My health has been very poor of late, & I am going in a week's time for a fortnight of hydropathy & rest.—[5] My everlasting species-Book quite overwhelms me with work— It is beyond my powers, but I hope to live to finish it.—

Farewell | My dear Lyell | Ever yours | C. Darwin

Edinburgh University Library (Lyell 1)

[1] Dated by the reference to CD's intention to visit a hydropathic establishment 'in a week's time' (see n. 5, below).

[2] The letter from Thomas Vernon Wollaston to Lyell has not been found, but its subject matter may be inferred from a passage in the introduction to Wollaston 1857. Having stated that the Coleoptera of Porto Santo and the Dezertas had more species in common than they did with Madeira as a whole, Wollaston wrote (Wollaston 1857, p. xv):

> And, without attempting to solve a geological problem, upon which Sir Charles Lyell will probably be able in a short time to throw considerable light, or to add any real evidence either in favour or against the existence of an ancient connective land; it does certainly appear to me, judging simply from Coleopterous data, as if the insect-population had possessed wonderful facilities, at some remote period, of migrating to and fro (as though along a slightly elevated mountain-ridge) between Porto Santo and the Dezertas, and in like manner . . . between the latter rocks and the eastern extremity of Madeira.

[3] Both Lyell and Wollaston supported the idea of specific centres of creation and the land-bridge doctrine as an explanation for geographical distribution. See letters to Charles Lyell, 16 [June 1856] and 25 June [1856], and letters from Charles Lyell, 17 June 1856, and from T. V. Wollaston, [27 June 1856].

[4] See letter to Charles Lyell, 10 November [1856], n. 4.

[5] CD left for Edward Wickstead Lane's hydropathic establishment at Moor Park, Surrey, on 22 April and returned to Down on 6 May 1857 ('Journal'; Appendix II).

From George Robert Waterhouse 14 April 1857

British Museum
April 14. 57

My dear Darwin

I have had scarcely any time at my command during the last two or three days—that is to say, whilst *in* the Museum, the only time that I could get to see one of the books referred to by you— I have now looked into them & find nothing connected with bee's cells in either paper— one of the papers—on the habits of *Osmia atricapilla*—threatens to give, upon a future occasion, an account of the mode of constructing the cell of the insect and some other bees, but I cannot find that it was done—certainly not in the same book.[1]

Faithfully Yours Geo R Waterhouse

I have always laboured under the impression that I *did* publish an account of the Osmia atricapilla, & that in the account I alluded to the structure of the cell, but now that I have found a paper on the insect with the cell question being disputed, I strongly suspect that I have never done it— I think I have told about the cell of this bee (*Osmia atricapilla*)—a little oval mud cell, with the inner side smooth & the outer side rough, & with a lid, the exact counterpart of the first commencement of the cell—rough on the *inner* side & with a *smooth* concavity at the top or outer side— the insect worked out the inner side & *smoothed* it with her jaws—

At the last meeting of the Entomological Society there was some discussion upon the subject of the Hive-bee's cell—introduced by Mr Westwood from whose brief observations upon the subject I conclude that his views are the same as mine—² I was made to say something, & in that say I alluded to a small piece of a Hornet's comb which I possess & which is composed of three cells only— presenting a section like this—

Mr Tegetmier said he had got a small piece of honey-comb like it—i.e with cells cylindrical externally & with two straight sides where they came in contact with the other cells—& that he would exhibit the same on a future occasion.³

DAR 181

CD ANNOTATIONS
Top of first page: 'Valuable Note' *ink*

¹ CD was gathering information on the cell-making powers of bees for a section of his chapter on 'Mental powers and the instincts of animals' (*Natural selection*, pp. 513–16). His notes and diagrams relating to the construction of the cells of bees are in DAR 48. In *Origin*, p. 225, CD stated: 'I was led to investigate this subject by Mr. Waterhouse, who has shown that the form of the cell stands in close relation to the presence of adjoining cells'.
² John Obadiah Westwood spoke about hive bees at a meeting of the Entomological Society of London on 6 April 1857, although there is no record of any discussion on bees' cells (*Transactions of the Entomological Society of London* n.s. 4 (1856–8), *Journal of Proceedings*, p. 67).
³ William Bernhard Tegetmeier, primarily known as an authority on pigeon breeding and on poultry, was also interested in the behaviour of bees (E. W. Richardson 1916, pp. 42–50). He delivered a paper 'On the formation of the cells of bees' at the 1858 meeting of the British Association for the Advancement of Science (Tegetmeier 1858), in which he described his experiments to show that the original form of a bee's cell is cylindrical and that the hexagonal shape results from the pressure exerted on it by six contiguous cells surrounding it.

To Laurence Edmondston   19 April [1857]¹

Down Bromley Kent
April 19th

My dear Sir
    I ought to have written sooner to have thanked you for the very fine Pigeons received per post, a few days ago;² but as there was a scrap, inside saying that you

intended writing, so I delayed, but I will delay no longer thanking you cordially for all the very kind trouble which you have taken to oblige me.— The specimen was very fine & very valuable to me, for there were several little points, which I had omitted to observe in the former specimen.

I shall certainly not want to trouble for any other specimen of Pigeon.— I see you most kindly note that the Rabbit is not forgotten. I sh^d be very glad to know whether there is any tradition of the introduction of the Rabbit. If you ever have any information on the domestication of wild Rock Pigeons in the Shetland Isl^d. I sh^d be very glad to hear.[3] Likewise whether birds chequered with black marks all over wing coverts & back are ever met with in wild state.— Graba states that this is the case in Färoe.—[4] And Col. King near Hythe near Hythe has stock of Dovecots which he informs me, are all descended from *wild* young procured in the Hebrides, & these are all chequered; but he cannot remember whether they were so at first now more than 20 years ago.[5]

With my very sincere thanks for all your kindness, I remain, my dear Sir | Yours sincerely | Ch. Darwin

D. J. McL. Edmondston

[1] Dated by the relationship to the letters to Laurence Edmondston, 11 September [1856] and 2 August [1857], both of which discuss specimens from the Shetland Islands.

[2] CD had asked Edmondston to send him a rock pigeon (*Columba livia*) and, if possible, a rabbit from the Shetland Islands in his letter to Laurence Edmondston, 11 September [1856].

[3] See letter to Laurence Edmondston, 2 August [1857].

[4] Graba 1830, which CD recorded having read on 2 April 1856 (*Correspondence* vol. 4, Appendix IV, 128: 18).

[5] CD included this information in *Variation* 1: 184, where he stated: 'Colonel King, of Hythe, stocked his dovecot with young wild birds which he himself procured from nests at the Orkney Islands; and several specimens, kindly sent to me by him, were all plainly chequered.'

From James Tenant 23 April 1857

London
April 23rd/57

Dr Sir

I have tried the Fish with Oats and the two sorts of Seeds that you sent[1] The Tench and Cm Carp and Gold Fish take Both Sorts equally well but I cannot get them to take Oats but if I ever succeed will let you know the result

Your obedient | Servant | J Tenant

C Darwin Esq^e

DAR 205.2 (Letters)

CD ANNOTATIONS
Top of page: '18'[2] *brown crayon*

Bottom of page: 'over' *ink*
Verso: 'Yellow Water Lily & Potamogeton Natans'[3] *ink*

[1] See letter from James Tenant, 31 March 1857.
[2] The number of CD's portfolio of notes on the means of dispersal of animals and plants.
[3] CD's annotations refer to an entry in his Experimental book, p. 18 (DAR 157a), that reads: 'Fish ate
   Yellow Water Lily & Potamogeton seeds— 3 Water Lily seeds passed through a Stork, but did not
   germinate'.

From J. D. Dana 27 April 1857

New Haven—
Ap. 27. 1857.

My dear Sir:—

I had intended to have replied to your last letter by the last mail,[1] and commenced my sheet. But a heavy cold in my head compelled me to lay it aside for the time. Your former letter also remains unanswered.[2] I assure you I take pleasure in contributing in any way I can to your labors, and wish I could do more than is now possible. The world needs a much more thorough search before the principles that flow from the Geog. Distribution of Species will be satisfactorily known.—

On the point on which you enquire, I can only say we need facts.[3] We know almost nothing of the species of the Antarctic. Fuegia is the most Southern point— & this is in latitude 50 to 56° as you know. The South Shetlands have furnished a gigantic Idotæa (called by Dr Eights Glyptonotus Antarcticus)[4]—and the Arctic has analogous large species, possibly—I cannot say certainly,—the same subgenus. This exhausts our Antarctic Catalogue, excepting a few Oceanic Amphipods, and the Serolis an Antarctic type.

With regard to the Fuegian species, There are the genera *Serolis* & *Halicarcinus* (Brachyman) peculiar Southern types, unrepresented, as far as now known, at the north. Halicarcinus has a representative genus, (Hymenicus) at New Zealand (Bay of Ids. lat 35½°) and another (Hymenosoma) at Cape of Good Hope; so that the subfamily (*Hymenicinæ*, as I call it) is cold temperate rather than Polar.

Another Characteristic Fuegian genus is *Lithodes*— But this has species in the cold waters of the North Pacific & North Atlantic, and one species extends as far South as Puget's Sound on the Columbia river. *Eurypodius* occurs from Cape Horn to Valparaiso,—being a cold temperate genus; and although unknown to the North, there is the closely related genus *Oregonia*, found in Puget's Sound.—

In the Northern Seas, the genera Hippolyte & Crangon abound in species, while we know thus far of but one Hippolyte from the colder Southern latitudes (Hip., ignobilis of Dr Kinahan, Jour. Roy. Dublin Soc. Oct. 1856),[5] and no Crangon. More investigation will probably bring some to light; yet these will still stand, we may presume, as northern genera. Hippolyte however has tropical as well as cold water species. The Crustacea genera of the cold temperate waters have fewer species and are more numerous, especially among the Maioids, or triangular Crabs, and the Amphipods, than those of the tropics. The cold-water species of Crustacea are more apt to be spinous Species.— I do not know of any species of the cold temperate South, identical with those of the cold temperate north.

You have no doubt observed what I have written, on p. 1579 of my Report, on the Crustacea of the colder zones.[6] It is my impression that the species of the regions South of Valparaiso (cold temperate) differ more in genera from those north of San Francisco, that those of the warm temperate South from those of the warm temp. north.[7] But without more observations, it would be difficult to prove this by tables for comparison. The Galapagos afforded Dr Bell a great variety of Maioids,[8] and they would counterbalance any diversity among the maioids of more extratropical regions. The Maioids, you remember are prominently a cold water group, and the cold southern waters that sweep over or around the Galapagos account for the abundance there of these Crustacea.

The paper of Dr Kinahan, to which I alluded, is not quite satisfactory in some respects. In making a comparison between Dublin Bay & Port Philip Australia, he is not careful to mention only representative genera or species. He says (p. 133) that *Ozius serratifrons* of Port Philip is replaced by *Carcinus mænas* & *Pilumnus hirtellus* of Dublin Bay. So, *Cyclograpsus 8-dentatus* is spoken of as representing a *Portunus* in Dublin Bay. It is true that in all cases, the species of a region are those that replace the species of another region: and a Catalogue of the two, exhibits the contrast. But these genera are not representatives in any proper sense. His other comparative statements on the same page are of this kind.

I have just learned that Prof. Agassiz & a naturalist of Boston have a collector now in the Pacific, who is to collect every thing at the Islands, commencing with the Sandwich Isds.[9] I wish some person could spend a few months between valparaiso and the Antarctic in as diligent search, with dredges &c. It would bring much to light. The U.S. North Pacific expedition that returned home last year had an excellent Naturalist on board by the name of Wm. Stimpson, who is now at the Smithsonian Institution Washington City, studying and describing his specimens collected between the East Indies and Behrings Straits & along the American Coast down to San Francisco. He is an *Invertebrate* man in his pursuits and you will gather some idea of his Collections from the fact that he has 900 species of Crustacea Here is a very large addition to the Cold-water Fauna. He is now upon the Crustacea, but has not yet any results to bring out, as his genera species are not named.[10]

The Purbeck discovery is a grand one.[11] If you take hold of geology by its deeper fundamental principles of progress, you find all such facts only illustrations of those principles. But if we look only to the more superficial conclusions such as those relating to the range of special groups,—which in fact Can hardly be called principles,—we find our notions often upset. I have just received a copy of Lyell's Supplement, and believe I have to thank you for it.— [12]

If my letter does not satisfy you, and you think I can aid you farther, I shall expect other enquiries. I always value your letters, and shall ever remain | with warm esteem | Sincerely yours | James D. Dana

C. Darwin, Esq.—

DAR 162: 39

[1] See letter to J. D. Dana, 5 April [1857].

[2] The 'former' letter to which Dana refers has not been located. Presumably it was a reply to the letter from J. D. Dana, 8 December 1856.

[3] In his letter to J. D. Dana, 5 April [1857], CD had asked whether the Crustacea of the temperate northern seas bore a much stronger analogy to those of the temperate southern seas than those of the Antarctic did to the Arctic forms.

[4] Eights 1852.

[5] Kinahan 1856.

[6] Dana 1853 discusses the geographical distribution of Crustacea. CD's copy in the Darwin Library–CUL is heavily annotated.

[7] In the last part of this sentence, the terms 'South' and 'north' were transposed, presumably by Dana.

[8] Bell 1841.

[9] It has not been possible to identify the collector.

[10] William Stimpson, naturalist to the North Pacific exploring expedition, published a long description of new species of Crustacea he had collected (*Boston Journal of Natural History* 6 (1850–7): 444–532). The monograph he had been preparing since 1856 describing these and other specimens was destroyed, along with his and Dana's type specimens and other valuable manuscripts, in the great Chicago fire of 1871 (*DSB*). In 1907, the Smithsonian Institute published what materials remained.

[11] See letter to J. D. Dana, 5 April [1857], and letter from Charles Lyell [16 January 1857], n. 2.

[12] C. Lyell 1857a.

To P. H. Gosse   27 April [1857][1]

Moor Park | Farnham | Surrey
April 27th

My dear Sir

I have thought that perhaps in course of summer you would have an opportunity & would be so very kind as to try a *little* experiment for me.—[2] I think I can tell best what I want, by telling what I have done. The wide distribution of same species of F. Water Molluscs has long been a great perplexity to me: I have just lately hatched a lot & it occurred to me that when first born they might perhaps have not acquired phytophagous habits, & might perhaps like nibbling at a Ducks-foot.— Whether this is so I do not know, & indeed do not believe it is so, but I found when there were many *very* young Molluscs in a *small* vessel with aquatic plants, amongst which I placed a dried Ducks foot, that the little barely visible shells **often** crawled over it, & that they *adhered* so firmly that they c^d not be shaken off, & that the foot being kept out of water in a damp atmosphere, the little

Molluscs survived well 10, 12 & 15 hours & a *few* even 24 hours.—[3] And thus, I believe, it must be that Fr. W. shells get from pond to pond & even to islands out at sea. A Heron fishing for instance, & then startled might well on a rainy day carry a young mollusc for a long distance.—

Now you will remember that E. Forbes argues chiefly from the difficulty of imagining how *littoral* sea-molluscs could cross tracts of open ocean, that islands, such as Madeira must have been joined by continuous land to Europe:[4] which seems to me, for many reasons, very rash reasoning.— Now what I want to beg of you, is, that you would try an analogous experiment with some sea-molluscs, especially any strictly littoral species,—hatching them in numbers in a smallish vessel & seeing whether, either *in larval* or *young shell state* they can *adhere* to a birds foot & survive say 10 hours in *damp* atmosphere out of water. It may seem a trifling experiment, but seeing what enormous conclusions poor Forbes drew from his belief that he knew all means of distribution of sea-animals, it seems to me worth trying.—[5]

My health has lately been very indifferent, & I have come here for a fortnight's water-cure.—[6]

I owe to using your name a most kind & most valuable correspondent, in Mr Hill of Spanish-Town.—[7]

I hope you will forgive my troubling you on the above point & believe me, | My dear Sir | Your's very sincerely | Ch. Darwin

P.S. | Can you tell me, you who have so watched all sea-creatures, whether male Crustaceans ever fight for the females: is the female sex in the sea, like on the land, "teterrima belli causa"?[8]

I beg you not to answer this letter, without you can & will be so kind as to tell about Crustacean Battles, if such there be.

Leeds University Library (Brotherton collection)

[1] The year is given by CD's visit to Moor Park (see n. 6, below) and by the reference to the correspondence between CD and Richard Hill (see n. 7, below).

[2] Gosse and his family usually spent the winter in London and the summer at the seaside, where he pursued his researches on marine organisms.

[3] This experiment is recorded in CD's Experimental book, pp. 21–2 (DAR 157a). CD included this case in the discussion of the geographical distribution of freshwater organisms in *Origin*, p. 385.

[4] Edward Forbes discussed at length the geographical distribution of Mollusca, and particularly those inhabiting shallow tidal zones (E. Forbes 1846, pp. 352–90), and concluded that 'there must have been either a connexion or such a proximity of land as would account for the transmission of a non-migratory terrestrial, and a littoral marine fauna.' (p. 383).

[5] According to Edmond Gosse, his father sent CD 'ample notes' on this subject (E. Gosse 1890, p. 269).

[6] CD was at Moor Park, Surrey, from 22 April to 6 May 1857 ('Journal'; Appendix II).

[7] See letters from Richard Hill, 10 January 1857 and 12 March 1857.

[8] A reference to Horace, *Satires*, 1. 3. 107, in which women were deemed 'the most foul cause of war'. CD had asked T. V. Wollaston the same question (see letter from T. V. Wollaston, [12 April 1857]).

To J. D. Hooker [29 April 1857][1]

Moor Park | Farnham | Surrey
Wednesday

My dear Hooker.

Your letter has been forwarded to me here, where I am undergoing Hydropathy for a fortnight, having been here for a week, & having already received an amount of good, which is quite incredible to myself & quite unaccountable.— I can walk & eat like a hearty Christian; & even my nights are good.— I cannot in the least understand how hydropathy can act as it certainly does on me. It dulls one's brain splendidly, I have not thought about a single species of any kind, since leaving home.

Your note has taken me aback; I thought the hairiness &c of alpine *species* was generally admitted; I am sure I have seen it alluded to a score of times.[2] Falconer was haranging on it the other day to me.[3] Meyen or Gay or some such fellow (whom you would despise) I remember makes same remark on Chilian Cordillera plants.—[4] Wimmer has written a little Book on same line, & on *varieties* being so characterised in Alps.—[5] But after writing to you, I confess I was staggered by finding one man (Moquin Tandon I think) saying that alpine flowers are strongly inclined to be white;[6] & Linnæus saying that cold makes plants *apelatous*; even the same species![7] Are arctic plants often apetalous?

My general belief from my *compiling work*, is quite to agree with what you say about little direct influence of climate; & I have just alluded to the hairiness of alpine plants as an *exception*. The odoriferousness would be a good case for me, if I knew of *varieties* being more odoriferous in dry habitats.—

I fear that I have looked at hairiness of Alpine plants as so generally acknowledged that I have not marked passages, so as at all to see what kind of evidence authors advance.— I must confess the other day, when I asked Falconer, whether he knew of *individual* plant losing or acquiring hairiness when transported, he did not.— But now *this second*, my memory flashes on me & I am certain I have somewhere got marked case of hairy plants from Pyrenees, losing hairs when cultivated at Montpellier.—[8]

Shall you think me very impudent if I tell you, that I have sometimes thought that (quite independently of the present case) you are a little too hard on bad observers—that a remark made by a bad observer *cannot* be right—an observer who deserves to be damned, you would utterly damn— I feel entire deference to any remark you make out of your own head; but when in opposition to some poor devil, I somehow involuntarily feel not quite so much, but yet much deference for your opinion.

I do not know in the least whether there is any truth in this my criticism against you, but I have often thought I would tell you it.—

I am really very much obliged for your letter, for though I intended to put only one sentence & that vaguely, I sh^d probably have put that much too strongly.

Ever, my dear Hooker | Your most truly | C. Darwin

This note, as you will see, has not anything requiring an answer.—

The distribution of F.W. Molluscs, has been a horrid incubus to me, but I think I know way now; when first hatched they are very active, & I have had 30 or 40 crawl on a dead Duck's foot; & they cannot be jerked off, & will live 15 & even 24 hours out of water.—[9]

DAR 114.4: 194

[1] Dated by CD's reference to having been at Moor Park for a week. He arrived there on Wednesday, 22 April ('Journal'; Appendix II).

[2] See letter to J. D. Hooker, 12 April [1857].

[3] CD had visited Hugh Falconer at his London home (see letter to J. D. Dana, 5 April [1857]).

[4] Franz Julius Ferdinand Meyen and Claude Gay had both prepared works describing the natural history of Chile (Meyen 1834[–5] and Gay 1845[–53]).

[5] Probably Wimmer 1840.

[6] Moquin-Tandon 1841, p. 42. CD scored this passage in pencil and wrote 'Q' beside it in his copy (DAR Library–CUL). The 'Q' indicated he had quoted it. See *Natural selection*, p. 283.

[7] Linnaeus 1783, p. 79. In the back of his copy of this work, CD wrote: 'p. 79 Flowers apetalous from cold'.

[8] CD mentioned this case in *Natural selection*, p. 283, citing Moquin-Tandon 1841, p. 62 (see letter to J. D. Hooker, [2 May 1857], n. 2). Next to this passage in his copy of Moquin-Tandon 1841 (Darwin Library–CUL), CD wrote 'Ch 7' in pencil, a reference to the 'Laws of variation' chapter in his species book (*Natural selection*, pp. 279–338).

[9] See preceding letter.

To W. D. Fox [30 April 1857][1]

Moor Park | Farnham | Surrey
Thursday

My dear Fox.

I have now been here for exactly one week, & intend to stay one week more.— I had got very much below par at home, & it is really quite astonishing & utterly unaccountable the good this one week has done me.— I like Dr Lane & his wife & her mother, who are the proprietors of this establishment very much.— Dr L. is too young,—that is his only fault—but he is a gentleman & very well read man.[2] And in one respect I like him better than Dr Gully, viz that he does not believe in all the rubbish which Dr G. does; nor does he pretend to explain much, which neither he or any doctor can explain.—[3] I enclose a paper for the strange chance of your ever knowing anyone in the S. in want of Hydropathy.— I really think I shall make a point of coming here for a fortnight occasionally, as the country is very pleasant for walking.—[4]

But I, also, think it highly probable that we *all* shall move to Malvern this summer, not for my sake, but for Etty's, who has now been out of health for some six or 8 months. I hardly know yet when we shall go, if we do go; but I very much wish that we might meet you there. Etty is now & has been for some time at

Hastings.[5] I am well convinced that the only thing for Chronic cases is the water-cure.—[6] Write to me either here or after Wednesday next to Down, & tell me how the world goes on with you, & how, especially, the Sciatica has been, if it was sciatica, which caused you so much suffering.—[7]

I believe that I worked too hard at home on my species-Book, which progresses, but very slowly. Whenever you write, as you are, I know, very learned in Pigs, pray tell me, whether any breed, known to have originated or to have been greatly modified, by a cross with the Chinese or Neapolitan Pig, whether any such crossed-breed, breeds true or nearly true.—[8] I am pretty sure I have read of some breed known to have been formed by a cross with one of the above Breeds, but I cannot remember particulars.— I must now take a sitz Bath, my treatment being,—daily Shallow, Douche, & Sitz.[9]

Farewell, my dear Fox | Yours most truly | C. Darwin

Postmark: 1 MY 1857
Christ's College Library, Cambridge (Fox 103)

[1] The date is provided by the postmark and by CD's reference in the letter to having been at Moor Park 'for exactly one week' ('Journal'; Appendix II).

[2] Edward Wickstead Lane was 34 years old. He took over the lease of Moor Park, a large country house with associated parkland, from Thomas Smethurst. Smethurst had turned the house into a hydropathic establishment in or around 1850; by 1855, Lane had become the proprietor (*Post Office directory of the six home counties* 1851 and 1855). The patients at Moor Park were accommodated in the same building as Lane and his family. It was Lane's belief that this arrangement was greatly beneficial to the patient (Lane 1857, p. 79).

[3] James Manby Gully advocated homoeopathy, and clairvoyance in some cases, in his water-cure establishment in Malvern. For CD's opinion of these remedies, see *Correspondence* vol. 4, letters to Susan Darwin, [19 March 1849], and to W. D. Fox, 4 September [1850].

[4] Lane believed that a change of scenery was an essential constituent of therapy, the site of Moor Park having been chosen for its location in a 'picturesque district abounding in pleasant and varied walks, with a dry soil under-foot and the fresh breezes of health playing about . . . over-head from morning till night' (Lane 1857, p. 43).

[5] Emma Darwin had taken her daughter Henrietta Emma to Hastings on 9 April 1857 to see whether her health would improve at the seaside. Henrietta returned home on 12 May. The family did not go to Malvern for the summer; instead, Emma took Henrietta to Moor Park on 29 May where she remained until 7 August (Emma Darwin's diary). CD returned to Moor Park for two weeks in June ('Journal'; Appendix II).

[6] For CD's belief that his condition was much relieved by the water-cure, see *Correspondence* vol. 4, letter to W. D. Fox, 7 [July 1849].

[7] Fox had recently visited Gully's hydropathic establishment in Malvern for treatment. See letter to W. D. Fox, 3 October [1856].

[8] CD discussed this subject in *Variation* 1: 78–9, but Fox's name is not mentioned.

[9] See *Correspondence* vol. 4, letter to Susan Darwin, [19 March 1849], for a description of CD's previous treatment at Gully's establishment in Malvern. Lane described his method of treatment in a handbook on hydropathy issued in 1857. He attributed most disorders to imperfect digestion (Lane 1857, pp. 51–78).

To Alfred Russel Wallace 1 May 1857

Down Bromley Kent[1]
May 1.— 1857

My dear Sir

I am much obliged for your letter of Oct. 10th from Celebes received a few days ago:[2] in a laborious undertaking sympathy is a valuable & real encouragement. By your letter & even still more by your paper in Annals, a year or more ago,[3] I can plainly see that we have thought much alike & to a certain extent have come to similar conclusions. In regard to the Paper in Annals, I agree to the truth of almost every word of your paper; & I daresay that you will agree with me that it is very rare to find oneself agreeing pretty closely with any theoretical paper; for it is lamentable how each man draws his own different conclusions from the very same fact.—

This summer will make the 20th year (!) since I opened my first-note-book, on the question how & in what way do species & varieties differ from each other.— I am now preparing my work for publication, but I find the subject so very large, that though I have written many chapters, I do not suppose I shall go to press for two years.—[4]

I have never heard how long you intend staying in the Malay archipelago; I wish I might profit by the publication of your Travels there before my work appears, for no doubt you will reap a large harvest of facts.—[5] I have acted already in accordance with your advice of keeping domestic varieties & those appearing in a state of nature, distinct; but I have sometimes doubted of the wisdom of this, & therefore I am glad to be backed by your opinion.— I must confess, however, I rather doubt the truth of the now very prevalent doctrine of all our domestic animals having descended from several wild stocks; though I do not doubt that it is so in some cases.— I think there is rather better evidence on the sterility of Hybrid animals that you seem to admit: & in regard to Plants the collection of carefully recorded facts by Kölreuter & Gærtner, (& Herbert) is *enormous.*—[6]

I most entirely agree with you on the little effects of "climatal conditions", which one sees referred to ad nauseam in all Books; I suppose some very little effect must be attributed to such influences, but I fully believe that they are very slight.—[7] It is really *impossible* to explain my views in the compass of a letter on the causes & means of variation in a state of nature; but I have slowly adopted a distinct & tangible idea.— Whether true or false others must judge; for the firmest conviction of the truth of a doctrine by its author, seems, alas, not to be slightest guarantee of truth.—

I have been rather disappointed at my results in the Poultry line; but if you sh[d] after receiving this stumble on any curious domestic breed, I sh[d] be very glad to have it;[8] but I can plainly see that this result will not be at all worth the trouble which I have taken.— The case is different with the domestic Pigeons; from its study I have learned much.— The Rajah has sent me some of his Pigeons & Fowls & **Cats** skins from interior of Borneo, & from Singapore.—⁹

Can you tell me positively that Black Jaguars or Leopards are believed generally or always to pair with Black?¹⁰ I do not think colour of offspring good evidence.— Is the case of parrots fed on fat of fish turning colour, mentioned in your Travels?¹¹ I remember case of Parrot with, (*I think,*) poison from some Toad put into hollow whence primaries had been removed.

One of the subjects on which I have been experimentising & which cost me much trouble, is the means of distribution of all organic beings found on oceanic islands; & any facts on this subject would be most gratefully received: Land-Molluscs are a great perplexity to me.—

This is a very dull letter, but I am a good deal out of health; & am writing this, not from my home, as dated, but from a water-cure establishment.

With most sincere good wishes for your success in every way I remain | My dear Sir | Yours sincerely | Ch. Darwin

British Library (Add 46434)

¹ Although CD put his home address on the letter, he actually wrote it at Moor Park, Farnham, Surrey, as he states in the letter.

² Wallace's letter of 10 October 1856 has not been found. None of their earlier correspondence has been preserved, but see letter to W. B. Tegetmeier, 29 November [1856] and n. 4.

³ Wallace 1855. For CD's notes on the paper, see *Correspondence* vol. 5, letter from Edward Blyth, 8 December 1855, n. 1.

⁴ In the original, the following lines are underlined, presumably by Wallace: 'I agree . . . & I'; 'the 20ᵗʰ year (!)'; 'way do species & varieties differ from each other.—'.

⁵ Wallace did not publish an account of his travels in the Malay Archipelago until 1869 (Wallace 1869).

⁶ CD had carefully studied the works on hybridism of Joseph Gottlieb Kölreuter, Karl Friedrich von Gärtner, and William Herbert. See *Correspondence* vol. 5.

⁷ For CD's views on the direct effect of the environment in causing variations, see *Natural selection*, pp. 281–9, and *Origin*, pp. 131–4 and 139–43.

⁸ CD had first approached Wallace in 1855 or early 1856 to request specimens of domestic and wild fowl from Malaysia (see *Correspondence* vol. 5, CD memorandum, [December 1855]).

⁹ James Brooke, raja of Saráwak, Borneo (see letter to W. B. Tegetmeier, 18 February [1857]).

¹⁰ See letter from A. R. Wallace, [27 September 1857].

¹¹ In Wallace 1853, p. 321, Wallace had recounted how one of his two tame parrots 'was a most omnivorous feeder, eating rice, farinha, every kind of fruit, fish, meat, and vegetable, and drinking coffee too'. He did not mention any subsequent change of colour in the parrot.

To J. D. Hooker [2 May 1857]

Moor Park
Saturday

My dear Hooker

You have shaved the hair off the alpine plants pretty effectually.—¹ The case of the Anthyllis will make a "tie" with the believed case of Pyrenees plants becoming glabrous at low levels.—[2] If I *do* find that I have marked such facts, I will lay the evidence before you.— I wonder how the belief c^d have originated: was it through final causes, to keep the plants warm! Falconer in talk coupled the two facts of woolly alpine plants & mammals.—

How candidly & meekly you took my Jeremiad on your severity to second class men.[3] After I had sent it off, an ugly little voice asked me once or twice how much of my noble defence of the poor in spirit & in fact, was owing to your having not seldom smashed favourite notions of my own.— I silenced the ugly little voice with contempt, but it would whisper again & again.— I sometimes despise myself as a poor compiler, as heartily as you could do, though I do *not* despise my whole work, as I think there is enough known to lay a foundation for the discussion on origin of species.— I have been led to despise & laugh at myself as compiler, for having put down that "alpine plants have large flowers," & now perhaps I may write over these very words "alpine plants have small or apelatous flowers"!

The most striking case, which I have stumbled on, on *apparent* but false relation of structure of plants to climate, seems to be Meyers & Dreges remark that there is not one single even moderately sized Family at the Cape of Good Hope which has not one or several species with Heath-like foliage;[4] & when we consider this together with the number of true Heaths; anyone would have been justified, had it not been for our own British Heaths, in saying that Heath-like foliage must stand in direct relation to a dry & moderately warm climate: Does this not strike you as a good case of false relation?

I am so pleased with this place & the people here, that I am greatly tempted to bring Etty here; for she has not on *whole*, derived any benefit from Hastings.—[5]

With thanks for your never failing assistance to me— Ever yours | My dear Hooker | C. Darwin

I return home, thanks be to God, on Wednesday.—

I remember that you were surprised at number of seeds germinating in pond mud: I tried a 4th Pond, & took about as much mud, (rather more than in former cases) as would fill a very large breakfast cup, & before I had left home 118 plants had come up; how many more will be up on my return I know not.—[6] This bears on chance of Birds by their muddy feet transporting F.W. plants.—

It w^d not be a bad dodge for a collector in country, when plants were *not* in seed, to collect & dry mud from Ponds.

Endorsement: 'May 2^d/57'
DAR 114.4: 195

[1] See letters to J. D. Hooker, 12 April [1857] and [29 April 1857].
[2] CD did indeed 'tie' the two cases together in *Natural selection*, where he wrote (p. 283):

> It has often been asserted that the same plant is more woolly when growing on mountains than on lowlands, & Moquin Tandon asserts that this change occurred with several species from Pyrenees when placed in the Botanic Garden at Toulouse: but Dr Hooker informs me that the Anthyllis vulneraria is glabrous in the Alps & woolly on hot dry banks: moreover Dr Hooker

after tabulating some Alpine floras does not find that in truly alpine species the proportion of woolly plants to be large. He is inclined to believe that dryness has a stronger tendency to produce hairs on plants.

[3] See letter to J. D. Hooker, [29 April 1857].

[4] Drège 1843, p. 26. CD reported this case in *Natural selection*, p. 209.

[5] See letter to W. D. Fox, [30 April 1857].

[6] CD recorded these experiments in his Experimental book, pp. 20–1 (DAR 157a). Having first collected mud on 11 January from a pond near Down and 'on road to Crystal Palace' and finding that a large number of seeds sprouted, he then took samples on 10 February from different parts of a pond on the road to Westerham. He recorded the number of plants that germinated in a table; an entry on 21 April shows a total of 104 'Dicots & some Monocots' and 14 'Grasses'. The last entry of 1 August 1857 gives a total of 537 plants. CD reported this experiment in *Origin*, pp. 386–7, in the chapter on the geographical distribution of freshwater animals and plants.

To J. D. Hooker [3 May 1857]

> Moor Park
> Sunday morning

My dear Hooker

I have just received your final & *most conclusive* destruction of woolly-alpine doctrine.[1] I sh^d. hope that these materials (now returned) would be well worth publishing in something:—

Would they not come in well in your Arctic memoir;[2] so that I do not apologise for having been the means of stirring you up to prove that your old suspicions were thoroughily well grounded.

It has given my nervous system, as a compiler, a severe shock; I am sure to publish many small generalisations as false as you have shown this to be.— Whenevever you turn to Arctic work, bear in mind Linnæus & Adamson's remark about flowers *becoming* apelatous in cold, & therefore Arctic species *ought* to be normally apetalous in many cases.—[3]

I am not sure whether you meant me to return you your note *accompanying* your Statistics, if so, let me hear & I will return it.[4]

I return Asa Gray's list of Alpines.—

I suppose you will soon have back M^rs. Hooker.— Her presence has been a great thing for Etty, & pray give her my best thanks for all her kindness to Etty.

My dear Hooker | Most truly yours | Ch. Darwin

Endorsement: 'May 3/57'
DAR 114.4: 196

[1] See preceding letter, n. 2.

[2] J. D. Hooker 1862.

[3] See letter to J. D. Hooker, [29 April 1857], n. 7.

[4] CD refers to a manuscript list, drawn up by Hooker, tabulating alpine plants according to their 'woolliness' (see preceding letter).

To Asa Gray 9 May [1857][1]

Down Bromley Kent
May 9th

My dear Dr Gray

I must thank you for your new part of Statistics:[2] if feeling the most lively interest in reading it can make me worthy of receiving it, assuredly I am worthy. I will not trouble you by specifying the many points which have particularly struck me; but the note at p. 387 I must allude to, as I want to ask a question which I think it cannot take 5 minutes to answer, namely to how many genera the 49 species belong: as there are six species of Carex there cannot be more than 44: I tried to go through list, but I c^d not feel sure in separating the 3^d head from 1st & 2^d. Heads.—[3] I want to know to see more clearly in proportion to your whole Flora how large the proportion of monotypic genera is in the disjoined species. This subject interests me very much: I began to try to work out this point in all the cases of much disjoined species which I met with; but I failed from want of knowledge: I tried also to make out whether the disjoined species would not on average belong to small Families, but here again I failed from want of knowledge; though the cases in which I could find out something, confirmed my **very strong** expectation that species having disjoined ranges would belong to *small* genera;[4] so you may imagine how much interest I felt in coming on your note on this very subject.—

Your list of the Trees made my mouth rather water to know what proportion had sexes in some degree separated,—on which subject I wrote you a ridiculously long letter some weeks ago.[5]

I am so glad that you are going to attack your introduced plants in the next number:[6] I may mention that two or three years ago I compared the proportions of the British introduced species, to the native Flora & it was in several cases ridiculously close: I then took your first Edition[7] & did the same, but the proportions here were very different; but I think this point w^d be just worth looking to, for *chance* of some result.

I have just looked at my old useless notes, & I see I made out in your Manual 206 introduced plants & of these Compositæ form $\frac{1}{8}$ & so do (as I thought) your indigenous compositæ.—[8]

In Britain from H. Watson's Cybele[9]

	Introduced	*Indigenous*
Compositæ	$\frac{1}{10}$	$\frac{1}{9}$
Umbelliferæ	$\frac{1}{22}$	$\frac{1}{23}$
Labiatæ	$\frac{1}{28}$	$\frac{1}{30}$
Leguminosæ	$\frac{1}{20}$	$\frac{1}{18}$

I happened to stumble on these results *first*, & was inclined to think something of them; but I suppose all was chance or errors. The standard proportion ought to be, I sh^d think, that for world in same latitude, & not the standard of the individual country. Though why I sh^d trouble you with an old exploded notion of mine, I know not.

With my sincere thanks & good wishes.—

Believe me | Your's most sincerely | Ch. Darwin

Gray Herbarium of Harvard University

[1] The year is given by CD's acknowledgment of the receipt of the third part of A. Gray 1856–7 (see n. 2, below).

[2] The third and final part of A. Gray 1856–7 was published in the May 1857 issue of the *American Journal of Science and Arts*. CD's copy is in DAR 135 (3).

[3] Discussing species 'of widely sundered habitation', or the so-called 'disjoined' species of Alphonse de Candolle, Gray enumerated those species that 're-appear in Japan, the Himalaya, or some part of North Asia, but are not European' (A. Gray 1856–7, p. 387). Dividing this group into two subclasses, the first being those common to north-west America and Japan and the second being those common to the Himalayas, Europe, and North America, Gray added the following note (A. Gray 1856–7, p. 387 n.), to which CD refers:

> Out of 49 species belonging to these first and second heads, as many as 10 belong to monotypic genera, and 21 to genera of less than ten good species;—six of the species belong to the vast genus Carex;—on the whole rather militating against the idea that the geographical extension of species bears some proportion to the size of the genus they belong to.

Part of CD's confusion is a result of an error Gray made in his note (see letter from Asa Gray, 1 June 1857).

[4] CD explained why he expected species with disjoined ranges to be found in small genera in letter to Asa Gray, 18 June [1857].

[5] A. Gray 1856–7, p. 400. See letter to Asa Gray, [after 15 March 1857], and letter from Asa Gray, 1 June 1857.

[6] In the event, Gray was not able to prepare this paper. See letter from Asa Gray, 1 June 1857.

[7] A. Gray 1848.

[8] These notes are in DAR 46.2 (ser. 2): 38–41. CD discussed the naturalised plants of North America in *Natural selection*, p. 232 and n. 3, and in *Origin*, p. 115, but having retabulated his results using the second edition of Gray's *Manual* (A. Gray 1856a).

[9] Watson 1847–59. The calculations are in DAR 46.2 (ser. 2): 42–58. See *Correspondence* vol. 5, letter to J. D. Hooker, 13 April [1855], for CD's earlier comments on these statistics.

To W. B. Tegetmeier   12 [May 1857][1]

Down Bromley Kent
12th

My dear Sir

Many thanks for your most kind offer of the Eggs of the Rumpless Polands.—[2] Will you be so kind as to send them by enclosed address as soon as you conveniently can, as we have a Hen ready to sit: it is really a shame accepting so many as a dozen, but I shall by having so many be able to see whether the Breed comes *true* as Rumpless.—[3] *Have you found them true*??

With respect to the Scanderoons my pair has never laid an egg since those which produced your pair.[4] I am ashamed to say that I do not know Cock from

Hen.—[5] Would you like to have the pair to try to breed from; I sh^d think the change might very likely make them breed. All that I now want of them is a carcass for skeleton & measurement in fresh state. If you like to have them, I must send them off on a Wednesday *night*, so that you would have somehow to get them home before late on the Thursday.[6]

I have now got some Fowls from the Dyaks of the interior of Borneo sent me by the Rajah:[7] shall I send them to you at same time?

Will you tell me what *colour* your turbits are which have thrown the dark tails: you are quite right it is just the class of facts which interest me:[8]—in Mayors' time, I see that the red & yellow Turbits had white tails.—[9]

I know I have your note, (but I have mislaid it) about the colour of Pigeons & their down.— Will you kindly tell me briefly the fact again?[10] By the way I wonder if I have any Pigeons which you want: do you want black Barbs? Just mention what you want, & then I can see whether I have anything to suit.—

I have not yet got your Poultry Book;[11] though I presume that it is lying at my Brother's (for I have not been for a long time in London; my health having been of late very indifferent), & therefore I do not know whether you describe the plumage of chickens in their down: Dixon describes most of them;[12] but the chicks of Gold & Silver *Pencilled* & *spangled* Hamburghs are not described, & I sh^d like to know them.

I wonder whether Fowls when crossed throw odd & unexpected colours like Pigeons do.— Do you know of any such facts? For instance if you were to cross black Spanish with Black or Silver Polands, do you suppose ever red or other marked new colour would appear?—[13]

Let me have a line about the Scanderoons; if you do not choose to have the pair to breed from; I then will try & make out which is Hen, & send her, & I think I will kill the Cock.— Also say shall I send Borneo Fowls?

Pray believe me | My dear Sir | Your's very sincerely | Ch. Darwin

P.S. I have just been looking at Scanderoons, but I really fear the Hen is ill, for I think I know which is which.— The servant in charge reported 3 weeks ago (I have been from home last fortnight) that one of them looked moping, & so she certainly does with wings a little drooping; & that may account for the pair not having gone to nest this spring.— but I will observe more carefully during this coming week

New York Botanical Garden Library (Charles Finney Cox collection)

---

[1] The date is suggested by the relationship to the letter to W. B. Tegetmeier, 18 May [1857], which it precedes, and by CD's reference to having been away from home a fortnight. CD returned to Down from Moor Park on 6 May 1857 ('Journal'; Appendix II).

[2] Tegetmeier had earlier promised CD a rumpless chick (see letter to W. B. Tegetmeier, 11 February [1857]). CD was collecting specimens of the young of various breeds of fowl to note the first appearance of the typical characters of each.

3 See letters to W. B. Tegetmeier, 23 June [1857] and [19 July 1857]. CD found that the rumpless Polish fowl did not breed true.

4 CD had sent Tegetmeier a pair of Scanderoons in August 1856 (see letter to W. B. Tegetmeier, 30 August [1856]).

5 According to Eaton 1852, it was not easy to tell Scanderoon cocks from hens: 'in this point the best and oldest Fanciers have been sometimes deceived'.

6 See letter to W. B. Tegetmeier, 18 May [1857].

7 James Brooke, raja of Saráwak, had sent CD several specimens of domestic fowls kept in Borneo.

8 CD collected such facts as instances of reversion to what he considered was the progenitor of the races of domestic pigeons, *Columba livia*. See *Variation* 1: 195–203.

9 [J. Moore] 1765, pp. 127–8. This book is an anonymously edited, revised, and enlarged edition of a treatise on domestic pigeons originally prepared by John Moore in 1735 (J. Moore 1735). The later work was inscribed to John Mayor, by whose name it was sometimes known.

10 CD cited Tegetmeier in *Variation* 1: 170 concerning 'a curious and inexplicable case of correlation, namely, that young pigeons of all breeds, which when mature become white, yellow, silver (*i.e.* extremely pale blue), or dun-coloured, are born almost naked; whereas other coloured pigeons are born well clothed with down'.

11 Tegetmeier ed. 1856–7, issued in parts.

12 Dixon 1848.

13 See letter to W. B. Tegetmeier, 18 May [1857].

To W. E. Darwin 13 May [1857][1]

Down
May 13th

My dear Willy

We have got your Sunday & Tuesday letters.— Will you tell Mrs Kynnersley[2] that I am very sorry to say that really have not *one* single aquaintance left in S. America: it is now 24–25 years since I was there, & the English population is a floating & changeable one.—

Mamma has used you very ill, & quite forgot about your Flute, but I have found it, & it shall go off tomorrow morning by the Post together with this note.— So you may "tootle" to your heart's content. What a dining out gentleman you have become!—

With respect to allowance we will settle it this summer Holidays: £40 sounds a good lot, nearly as much as Parslow's wages;[3] but I daresay it is not too much. Try & find out more particulars, & whether any other Boys have an allowance.—

Mamma, Miss Thorley & Etty returned yesterday:[4] I grieve to say Etty is not one bit better: everything must bend to her health, & Mamma will take her to Moor Park in a fortnight's time, stay a fortnight, & then I shall go there again for a bit.—[5] What we shall do in summer, it is really impossible to say: if we do not go to Malvern, we will try our best to think of some lark or amusement for all of you.—[6] Mamma thinks if Water Cure does suit Etty, we had better go to Malvern to complete it, but I shd. prefer leaving her longer at Moor Park.—[7] It is very disheartening for me, that all the wonderful good which Moor Park did me at the time, has gone all away like a flash of lightening, now that I am at work again. And eheu, eheu, I have left off Snuff for nothing!—[8]

On Friday Aunt Susan is coming, on Saturday Uncle Harry & Aunt Jessie, on Monday Fanny Frank, & on tomorrow week Aunt Catherine, so we have lots of visitors expected,—a good lot too many, now poor dear Etty is so indifferent.[9]

What a *capital* Hand you have directed your two last notes in; it was quite Ducal.—

Ever my dear Son | Your' affect Father | C. Darwin

DAR 210.6

[1] The year is established by the references to family trips and visitors to Down House (see nn. 4, 5, and 9, below).

[2] Mrs Clements John Kynnersley, née Mary Sneyd. The Sneyds of Staffordshire were old friends of the Darwin family.

[3] According to CD's Classed accounts book (Down House MS), Joseph Parslow received £44 per year for his services as butler at Down House.

[4] Miss Thorley was the governess of the Darwin children until January 1857, when she was replaced by Miss Pugh. According to Emma Darwin's diary, she had cared for Henrietta Emma Darwin at Hastings from 30 April until 11 May 1857. Emma and Henrietta Darwin returned to Down on 12 May.

[5] Emma Darwin took Henrietta to Moor Park, where she was to undergo the water-cure under Edward Wickstead Lane's care, on 29 May 1857 (Emma Darwin's diary). CD went there on 16 June, returning home on 30 June ('Journal'; Appendix II).

[6] According to her diary, Emma Darwin took the children to Barlaston and then to Shrewsbury from 23 June to 6 July 1857.

[7] Henrietta remained at Moor Park until 7 August 1857 but returned there for further treatment on 22 August, staying until 31 October (Emma Darwin's diary).

[8] Like James Manby Gully, Lane insisted that CD stop taking snuff while undergoing hydropathy (see Lane 1857, p. 82).

[9] Susan Elizabeth Darwin arrived at Down on 15 May 1857; Henry Allen (Harry) Wedgwood and his wife Jessie and daughters Louisa and Caroline (Carry) came on 16 May; and Frances Mosley Wedgwood, wife of Francis (Frank) Wedgwood, and her daughter Cecily joined them on 18 May (Emma Darwin's diary). Emily Catherine Darwin's visit was not recorded.

To J. D. Hooker 16 [May 1857]

Down Bromley Kent
16th

My dear Hooker

You said, I hope honestly, that you did not dislike my asking questions on general points,—you of course answering, or not, as time & inclination might serve.

I find in animal kingdom that the proposition that any part or organ developed (normally (ie not monstrosity)) in a species in any *high* or *unusual* degree, compared with the same part or organ in allied species, tends to be *highly variable*.[1]

I cannot doubt this, from my mass of collected facts.— To give instance, the Cross-Bill is very abnormal in structure of Bill compared with other allied Fringillidæ, & the Beak is *eminently variable.*—[2] The Himantopus remarkable from wonderful length of legs, is *very* variable in length in length of legs.—[3] I c^d give **many** most striking & curious illustrations in all classes;—; so many that I think it cannot be chance. But I have *none* in vegetable kingdom, owing, as I believe, to my ignorance.— If Nepenthes consisted of *one* or two species in group with pitcher developed, then I sh^d have expected it to have been very variable; but I do not consider Nepenthes case in point, for when a whole genus or group has an organ, however anomalous, I do not expect it to be variable,—it is only when one or few species differ greatly in some one part or organ from the forms *closely allied* to it in all other respects, that I believe such part or organ to be highly variable.—[4] Will you turn this in your mind: it is important apparent *law*(!) for me.—[5]

Ever your's | C. Darwin

P.S. I do not know how far you will care to hear, but I find Moquin-Tandon treats in his Teratologie on villosity of plants & seems to attribute more to dryness than altitude;[6] but seems to think that it must be admitted that mountain plants are villose, & that this villosity is only in part explained by De Candolles remark, that the dwarfed condition of mountain plants, would condense the hairs, & so give them the *appearance* of being more hairy. He quotes Senebier Phys. Veg. as authority, I suppose first authority, for mountain plants being hairy.—[7]

If I could show positively that the endemic species were more hairy in dry district, then the case of the vars. becoming more hairy in dry ground, would be fact for me.—

Endorsement: 'May | 57'
DAR 114.4: 197

[1] CD had previously asked Hooker, Hewett Cottrell Watson, and Asa Gray about this particular phenomenon (*Correspondence* vol. 5). CD was at this time working on his chapter on the 'Laws of variation' (*Natural selection*, pp. 279–338), which was completed on 5 July 1857 ('Journal'; Appendix II).

[2] In his discussion of this apparent rule in *Natural selection*, pp. 307–18, CD mentioned the case of the crossbill and cited Macgillivray 1837–52, 1: 423, on the great variability 'in length, curvature & the degree of elongation of lower mandible' (*Natural selection*, p. 310).

[3] 'The long-legged Plover or Himantopus forms a small genus with closely allied species, quite remarkable from their extraordinary length of the legs compared with their nearest allies.' (*Natural selection*, p. 310).

[4] Hooker had offered *Nepenthes* as a case in point in response to CD's earlier query (see *Correspondence* vol. 5, letter to J. D. Hooker, [April 1852], n. 10).

[5] See letter to J. D. Hooker, 3 June [1857].

[6] Moquin-Tandon 1841, pp. 65–7. See also *Natural selection*, p. 283.

[7] Horace Bénédict Alfred Moquin-Tandon cited Senebier [1800] in his discussion of this point (Moquin-Tandon 1841, p. 66 n. 2).

To W. B. Tegetmeier 18 May [1857]¹

Down Bromley Kent
May 18th

My dear Sir

I am extremely much obliged for your to me very interesting letter, which I will go through seriatim.²

The Scanderoons shall be sent (I have seen since I wrote some tendency to go to nest) to the address carriage paid on *Thursday* morning before 12 oclock. The Borneo Fowls shall be sent at same time; for as coming from interior of Borneo (sent by the Rajah Sir J. Brooke) I think they are worth notice, even if not presenting any peculiar character.—

I doubt whether the young Barbs are old enough to travel; but they shall be saved for you.— If Scanderoon dies please remember I shd wish carcase sent per coach by enclosed address *as soon as possible to arrive fresh.*—

Mr Gulliver never sent me the fine Bird to which you *formerly* alluded, as dying;³ I shd be very glad of one: if ever you see him pray remind him.— I have long wished for one of the long-winged & long-tailed Runts for examination & Skeleton; When you have done breeding with it, I wish you would sell it me, & tell me at what you value it: I have received so many favours of you, that you must let me, **I beg you**, purchase it from you, without indeed it is too valuable for my purpose of seeing it alive for a few weeks & then slaughtering it.

Please sometime inform me; & likewise tell me whether I can purchase the Poultry Book which you edit, & who are publishers:⁴ the deuce is in my Booksellers, why they have not got it me yet: I must give them a rowing: I think I told you that they had sent me an old Edition, not edited, apparently, by you.—⁵ Is your Edition completed yet?⁶

Thanks for information about Turbits & for your kind offer of your Birds, but they wd. be of no use to me: the Meave is German: Mr Wicking has the Breed.—

Also, thanks for offer of Fowls eggs for me to see the chickens, but the simple truth is that I have so many objects to observe & experimentise on, that I *literally* have no time to watch anything else, & as it is, I work too hard.—

Your sitting of the Rumpless, Polands which you have so kindly sent me, are put under a Hen today.—

I shd be **very glad** to hear what colour your Pigeons from Magpie & Helmet, with the down turns into.—

Lastly thanks for information about crossed Fowls: I am *surprised* that red does not appear in some crosses.

With my very sincere thanks. My dear Sir | Your's very truly | Ch. Darwin

New York Botanical Garden Library (Charles Finney Cox collection)

¹ The year of the letter is suggested by the relationship to the letter to W. B. Tegetmeier, 12 [May 1857].
² See letter to W. B. Tegetmeier, 12 [May 1857].

³ CD had been trying for some time to acquire a runt from Mr Gulliver (see letter to W. B. Tegetmeier, 21 September [1856]).

⁴ Tegetmeier ed. 1856–7 was published in parts, beginning in May 1856, by William S. Orr and Co., London.

⁵ Tegetmeier's *Poultry book* was a rearranged and enlarged edition of an earlier work of the same title published in 1853 by William Wriothesley Wingfield and George William Johnson. CD had first ordered the work from his booksellers in the spring of 1856 (see letter to W. B. Tegetmeier, 25 April [1856]).

⁶ Publication of Tegetmeier's *Poultry book* ceased with part 11, which appeared in July 1857. A new and complete edition (Tegetmeier 1867) was published ten years later, with parts issued regularly in 1866 and 1867.

From Asa Gray [*c.* 24 May 1857][1]

Now about Protean genera, and Mr. Watson's very proper discriminations.[2] For want of these discriminations, my hasty list of "variable genera" can be of little use.

As to Mr. Watson's three Categories,[3] the 2^d & 3^d differ only in degree, and I see not how to draw any clear line of distinction between them. Where "species are so close that it is highly difficult to say whether . . . etc."—there it must "become difficult to say where one species begins and another ends". This, certainly i.e. Categories 2 & 3,—is what I had in view, with also an eye to the Category no 1.— For, see you, Mr. Watson cites *Carex* & *Ranunculus* under this head. Now *Carex*, viewed according to D.ʳ Boott, is a good case in point.[4] The species very difficult to distinguish by reason of their similarity, but not remarkably variable; because he regards almost every *definable* form as a separate species. But if D.ʳ Hooker were to elaborate the genus, how would it be? Would it not fall at once into no. 3? Potamogeton, would be placed by Mr. Tuckerman under no 1.[5]—by me under no. 3. etc—

As to Ranunculus—what I call *R. repens* (though, I wish I were surer it is the European sp.) figures as a dozen in books; and where I limit it, I am not clear about its boundaries, on one side. Several of the other species vary a good deal, but none are so very proteiform. D.ʳ Hooker, who refers our R. abortivus (I doubt if rightly) to R. auricomus, would rank Ranunculus under no. 2 or no. 3, incontinently.

So you must take my list, especially the underscored names, in the rough, as including those genera that I find most difficulty with here, in the complete limitation of the species,—or some of them, either, because of the great variability of a certain species, or the very close approximation of a greater number of species.—

I should say the greater part of those underscored in my list were of the former sort.—excluding, Liatris, Eupatorium, Solidago? Salix, & Populus,—& Carex.

I have not time just now to put my mind on the subject, however, and must close my rambling letter without reading it over.

Kindly give me any remarks that strike you, on reading over my last article,— which I sent you Journal sheets of.[6] Your letters are always most instructive to me, and I only regret I have not the opportunity to think and write more upon the topics they bring up.—

Kindly post the 3 enclosed letters, and believe me to remain

Ever Yours | A. Gray

Incomplete
DAR 165: 97

CD ANNOTATIONS
7.1 Kindly . . . Gray 9.1] *crossed pencil*
Top of first page: 'Ch. 4'[7] *pencil*; 'June 9th—1857' *ink*

[1] The date of the letter is conjectured from Gray's reference in letter from Asa Gray, 1 June 1857, to having '*despatched*' a letter to CD 'last week' and from CD's endorsement, which probably records the date of receipt (see CD's annotations).

[2] CD had returned Gray's list of protean genera (letter from Asa Gray, 16 February 1857); he had also sent Hewett Cottrell Watson's comments on this list to Gray (see letters from H. C. Watson, 10 March 1857 and from H. C. Watson to Asa Gray, 13 March 1857).

[3] See letter from H. C. Watson, 10 March 1857.

[4] Francis Boott was a specialist on the genus *Carex*.

[5] Edward Tuckerman had worked on *Potamogeton* in 1849 (Tuckerman 1849).

[6] Gray had sent CD the third part of A. Gray 1856–7 (see letter to Asa Gray, 9 May [1857]).

[7] CD refers to the chapter of his species book on 'Variation under nature' (*Natural selection*, pp. 92–171).

To J. D. Dana 25 May [1857][1]

Down Bromley Kent.
May 25th

My dear Sir

Although I have nothing particular to say I must thank you for all the trouble you have so kindly taken in answering my question on the relation of Arctic & Antarctic Crustacea as fully as the present state of knowledge permits.[2] I was very curious on the point, otherwise I would not have troubled you. And indeed I have been many times very troublesome to you, & you have invariably received my letters of enquiry in the *kindest* manner.

How the U. States are going ahead in Natural History! I had not heard of the late expedition, though I had heard of Mr Stimpson before: indeed he formerly sent me some cirripedes.[3] Prof. Huxley has lately been working on the homologies of Crustacea, & has come to some important differences with Milne Edwards: he has published an outline in some Lectures in a Medical Journal, but I suppose will soon publish in extenso.—[4]

My neighbour J. Lubbock is working on the anatomy of the larvæ of Diptera, & has made most minute & beautiful drawings of their muscular system:[5] I wish he had more time & he would do good work. As I sometimes tell him, he is a case of the "pursuit of knowledge under riches", which seems as great a drawback as poverty.[6] I am glad Lyell has sent you his supplement, for it strikes me as full of remarkable facts.[7]

Farewell—Floreat Scientia—with very sincere thanks for all your kindness. Yours very truly | C. Darwin

Yale University Library (Manuscripts and archives, Dana papers)

[1] The year is given by the relationship to the letter from J. D. Dana, 27 April 1857.

[2] See letter to J. D. Dana, 5 April [1857], and letter from J. D. Dana, 27 April 1857.

[3] In his last letter to CD, 27 April 1857, Dana had mentioned that Louis Agassiz had a collector who was to explore Pacific islands. He also told CD of William Stimpson's progress in describing the Crustacea collected by the North Pacific Exploring Expedition.

[4] Thomas Henry Huxley devoted the final three lectures of his course on natural history to the Crustacea. These lectures, delivered at the School of Mines, were published in a series in the *Medical Times & Gazette* (T. H. Huxley 1856–7). In lecture eleven, published in the issue of 23 May 1857, Huxley put forward a new interpretation of the division of the segments of the archetypal crustacean originally proposed by Henri Milne-Edwards.

[5] John Lubbock was studying variation in larval musculature. In Lubbock 1859, he presented his observations on the muscles of the larvae of the moth *Pygaera bucephala*. In DAR 45: 105, there is a note in CD's hand which reads: 'Lubbock's Muscles. Wonderful variation & attachment.'

[6] Lubbock had joined his father's bank, Lubbock, Forster & Co., as a partner in 1848 at the age of 15 (Hutchinson 1914, 1: 22).

[7] C. Lyell 1857a included a discussion of the fossils recently discovered in the Purbeck beds near Swanage (see letter from Charles Lyell, [16 January 1857]). CD had mentioned the work in his previous letter to Dana (letter to J. D. Dana, 5 April [1857]).

To J. D. Hooker[1] [June 1857][2]

Any account of Flora of Marianne Isl[ds] in N.W. Pacific. or of Bonin Isl[ds]?[3]
Any account of Greenland Flora some Danish Man.—[4]

In Cnestis & Connaris, Brown[5] says differ in the species of, having one or more ovaria, in existence & absence of albumen, in æstivation of Calyx. Are any of these points slightly variable in the individuals of the same species.[6]

AL memorandum
DAR 114.4: 222b

[1] These queries are bound together with another list of questions for Hooker. In the letter to J. D. Hooker, 20 October [1857], CD stated that he had been collecting questions to ask Hooker during some future visit to Down: these questions may be those referred to, or some of them. The second set of questions is transcribed following the letter to J. D. Hooker, 20 October [1857].

[2] The final question in this memorandum is related to a point discussed by CD in the letters to J. D. Hooker, 16 [May 1857] and 3 June [1857]. The chapter of CD's species book in which the topic is discussed was completed on 5 July 1857 ('Journal'; Appendix II).

[3] The Marianne and Bonin Islands, off the east coast of Japan, had been visited in 1853 by Matthew Calbraith Perry. Hooker, who had worked on the flora of a neighbouring Pacific island in 1857

(J. D. Hooker 1857), would presumably have known whether any botanical studies of the area had been recently published. CD did not, however, refer to any such work in *Natural selection*.

4 Probably Lange 1857.

5 Robert Brown.

6 CD had previously asked Hooker to comment on the proposition that a characteristic that is particularly developed in any species is apt to be highly variable (letter to J. D. Hooker, 16 [May 1857]). See also letters to J. D. Hooker, 3 June [1857] and 5 June [1857], in which CD expressed his surprise that examples could not be found in plants. The subject is discussed in *Natural selection*, pp. 307–11. CD cited Hooker's opinion that plants were altogether too variable for the phenomenon to be apparent (*ibid.*, p. 308).

From Asa Gray 1 June 1857

Cambridge, Mass. U.S.A.
June 1st, 1857

My Dear Darwin

Yours of the 9th came last week.

I do not wonder you were somewhat puzzled to make out just the 49 species spoken of in my note p. 387.[1] It was a clear mistake my speaking of 6 species of Carex as belonging to 1st & 2d heads—as evidently there are only *three* of the 1st head and none of the 2d.

I mail you a fresh copy of the article, with the 49 species I must have had in view marked with a — in pencil.—

The 49 or rather 50, species belong to 46 genera,—which is as you would have it.—

I did not know at all that you suspected *disjoined species* to belong to small genera & small orders, as a general thing.

The monotypic genera of these 50 species are—

> Brasenia, Hippuris, Cryptotænia, Crantzia, Phryma, Monotropa (in the restricted sense) Anacharis(?) Hemicarpha(?), Zannichellia(?), Camptosorus.

The only *good-sized* genera are Anemone, Silene (S. Antirrhina is diffused as a weed & by the agency of man?) Cerastium, Potentilla, Plantago, Primula, Veronica, *Carex*, *Poa*, Festuca, Adiantum.

My 76 disjoined species belong to 34 families,—and I cannot see that they incline to belong to small families.

15 are Gramineæ which form $\frac{1}{13}$ of our Flora.

8 " Cyperaceæ " $\frac{1}{10}$ "

The 1 Leguminosa & 1 Composita are as you would like; but that is because these orders are remarkable for their species being of narrow range.

3 are Rosaceæ

2 " Scrophulariaceæ (the 1 orchid is to be erased)

3 " Ranunculaceæ. &c &c

6 " Umbelliferæ

As to our trees, what proportion have flowers more or less separated. Number the orders on p. 400— 1. Magnoliaceæ, and so on.[2] And append

p. = polygamous more or less.

m = monœcious

d = diœcious.

		separated flowers
1.	Magnoliaceæ	o
2	——	o
3	——	o
4	——	o
5	——	1 p
6	——	8 p.
7	——	2 d
8	——	o
9	——	1 m
10	——	1 p
11	——	2 p. d.
12		o
13		o
14		1-p
15		1 p
16		o
17	——	7 p. d.
18	——	2 p
19	——	8 p. d.
20	——	1 m
21	——	9 m
22	——	21 m.
23	——	5 m.
24		7 d
25	Coniferæ——	18 m. d.

Out of 132 trees, those with separated flowers more or less—are 95.—and for the greater part very decidedly separated.

I must think it by chance—that your introduced plants are in so near the proportion by families that the indigenous species are.[3]

		Indigenous		Introduced
Our	Compositæ	$\frac{1}{8}$	——	nearly $\frac{1}{10}$.
"	Cyperaceæ	$\frac{1}{10}$	——	$[\frac{1}{60}]$
"	Gramineæ	$\frac{1}{13}$	——	$\frac{1}{8}$
"	Leguminosæ	$\frac{1}{24}$	——	$\frac{1}{18}$
"	Rosaceæ	$\frac{1}{29}$	——	$\frac{1}{52}$. &
"	Orchidaceæ	—		o
"	Ranunculaceæ	$\frac{1}{43}$	——	$\frac{1}{43}$
but "	Labiatæ—	$\frac{1}{43}$	——	$\frac{1}{11}$!

I am very glad if my published notes or my jottings are of any use to you.

This is my season of greatest and most distracting occupation. I shall have no article in the July no. of Sill. Journal—nor in the *Sept.* either, I fear.

I wrote—or rather *despatched* a letter to you last week——[4] Watson's memoranda will be sent back to you a week or two hence—[5]

Ever Yours | A. Gray

DAR 8: 47bA

CD ANNOTATIONS
5.1 I did . . . thing. 5.2] *crossed ink*
10.1 As to . . . so on. 10.2] *crossed ink*
10.7 1. . . . 25 10.31] 'These *no[s] of [*interl*] Families apply to A. Grays Paper p. 400'[6] *added ink*;
    '95/132 = .72' *added ink, circled ink*; '17/25 Families' *added ink, circled ink*
10.31 m. d.] '95' *added below pencil*
12.1 I must . . . A. Gray. 16.1] *crossed ink*
Top of first page: '**Trees** & disjoined species' *pencil*; 'p. 387' *pencil*

[1] See letter to Asa Gray, 9 May [1857].
[2] See letters to Asa Gray, 1 January [1857] and [after 15 March 1857]. Gray refers to his tabulation of the trees of the northern United States in A. Gray 1856–7, p. 400.
[3] See letter to Asa Gray, 9 May [1857].
[4] See letter from Asa Gray, [*c.* 24 May 1857].
[5] See letter to Asa Gray, [after 15 March 1857], in which CD enclosed some notes and a letter from Hewett Cottrell Watson.
[6] A. Gray 1856–7. See n. 2, above.

To J. D. Hooker 2 June [1857][1]

Down Bromley Kent
June 2[d]

My dear Hooker

I will write about science &c another day, but I must just say how *very very* kind you always are in helping me.—

Now about medals:[2] I thought that you had quite fixed not to propose Lindley except for Copley:[3] for a Royal I sh[d] never *dream* of putting Hancock & Prestwich in opposition to Lindley.[4] I will write to Sharpey,[5] for I really have no opening to write to Sabine;[6] & indeed it puts me in an awkward position to express any opinion whatever on Lindley, as I am not a Botanist, but I will do my best, & will make some, I fear it must be egotistical apology.

I will, however, (subject to my manifest ignorance) express strongly my opinion: the letter shall go in day or two.— I am rather taken aback by hearing that my letter to Sharpey was read aloud, as I meant it only for him & fear I expressed myself dogmatically: I am rather annoyed at this, for I cannot remember exactly what I said.—[7]

I am exceedingly glad to hear that Lyell has chance of Copley sometime.—[8]

Now for your remarks on some of the other men: I must think that you underrate Hancock. Do pray talk with Huxley. I think he will be surprised at your speaking of him as a "local *cork*" sic.— (No No I see it is local *work*!!!)[9] I sh^d. say he was second as comparative anatomist only to Owen in this country. His work on the circulation of the Mollusca I have always looked at as grand.[10] His papers I believe on the Brachiopoda & Bryozoa are first rate:[11] good anatomical & physiological work, in *Zoology* ranks, in my opinion, far far above descriptions of species.— His paper on the boring powers of Sponges & Mollusca is admirable:[12] do pray talk with Huxley, for I have for so many years respected Hancock, that I am astounded at your manner of speaking of him.— How difficult it is to judge of Scientific claims! *Private* If I were put on my oath, though I sh^d. say that Huxley was a *far* cleverer man than Pretwich, I sh^d. say that the latter had done *far more* for Science than Huxley![13] So how we differ, & God knows which is right.— With respect to older names, I had not thought about those you mention, in chief part from thinking that the Royal medals, from restriction of date, were especially intended for younger men. But I rather think that we take distinct views on medals; I would try to disregard general reputation, which no doubt the elder men mentioned by you have, & would ask what *especial* points in physical science Ross,[14] Beaufort[15] &c have done.— I am far from wishing to hint that they have not done good work, but in the *Natural* Sciences, which we have to consider, I do not know what.— Although we are very apt, I have observed, at the first approach of a subject, to take different views we generally come to a near approach after a talk, & I daresay we sh^d. in this case; but anyhow my opinion does not signify as I am not on Council.

My only fear is that Lindley might not think the Medal any honour after ourselves have had it, before him.—

I am heartily sorry you will not be at Club.[16] I had *quite* calculated on meeting you there.

My wife & Etty have just started for Moor Park: she will stay a fortnight, & then I shall relieve guard for another fortnight.—[17] It is most provoking that a cold on leaving Moor Park suddenly turned into my old vomiting, & I have been *almost* as bad since my return home as before, notwithstanding the really surprising state of health I was in there. I fear that my head will stand *no* thought, but I would sooner be the wretched contemptible invalid, which I am, than live the life of an idle squire.

Yours affect^y | C. Darwin

Have you got settled your Household troubles?

P.S. | In rereading your letter I see I have misunderstood one part, viz that Prestwich was in a quite inferior class to Huxley— I see you do not think so.—

I have written my letter to Sharpey,[18] and as I could not remember any previous expressions of yours, I have given my independent impression, which I have picked up in general reading.— I sincerely hope that you may succeed.— I am glad that Richardson is the last preceder to Lindley, as no one c^d. dislike following him—[19]

I hope that my letter to Sharpey will satisfy you, but reflect how absurd it must strike anyone in my expressing an opinion.

Endorsement: '57'
DAR 114.4: 199

1 The year is given by the discussion of nominations for medals of the Royal Society in 1857 and by CD's reference to visits to Moor Park (see n. 17, below).

2 Hooker was on the council of the Royal Society of London in 1857 and 1858. CD had retired from the council in November 1856. Balloting for the society's awards took place early in June.

3 Hooker had been attempting to secure the Copley Medal, the Royal Society's highest honour, for John Lindley for a number of years. In 1856, Hooker had suggested Lindley's name to CD for this medal, but CD apparently convinced Hooker that Lindley had little chance of being selected that year (see letter to J. D. Hooker, 8 April [1856]).

4 The previous year, CD had supported Joseph Prestwich and Albany Hancock as 'eminently well qualified for the Royal Medal' (letter to Edward Sabine, 23 April [1856]).

5 See following letter. William Sharpey was one of the secretaries of the Royal Society.

6 Edward Sabine was treasurer of the Royal Society.

7 This letter to Sharpey has not been found, but see the following letter in which CD refers to this incident again.

8 Charles Lyell received the society's Copley Medal in 1858.

9 Perhaps a reference to the fact that Hancock lived and worked in Newcastle upon Tyne, far from the scientific circles of London.

10 Albany Hancock's insistence on the existence of a circulatory system in Mollusca had been recently confirmed, after years of controversy with Jean Louis Armand de Quatrefages de Bréau and Henri Milne-Edwards, by an independent commission set up by the Société de Biologie of Paris. Hancock's views, first put forward in Alder and Hancock 1844, were given in detail in Alder and Hancock 1845–55, pt 7. The work is in the Darwin Library–CUL.

11 A. Hancock 1850 and 1858.

12 A. Hancock 1848 was a study of the boring powers of Mollusca; Hancock 1849 was concerned with the excavating powers of sponges.

13 Prestwich, a specialist on the Tertiary geology of England and Europe, was the author of numerous scientific memoirs.

14 James Clark Ross, naval officer and explorer, had commanded the Antarctic expedition (1839–43) on which Hooker had served as assistant surgeon and botanist. He subsequently commanded an expedition to find Sir John Franklin and was generally regarded as the first authority on matters relating to Arctic navigation (*DNB*).

15 Francis Beaufort, a former naval officer, had retired as hydrographer to the Admiralty in 1855. He was 83 years old and died at the end of 1857.

16 The Philosophical Club of the Royal Society was to meet on 11 June 1857 (Bonney 1919, p. 136).

17 Emma Darwin took Henrietta Emma Darwin to Moor Park for hydropathy on 29 May 1857 (Emma Darwin's diary). See also letter from Henrietta Emma Darwin, [2 August 1857].

18 See following letter.

19 John Richardson had received one of the Royal Medals in 1856. CD had supported his nomination (see letter to Edward Sabine, 23 April [1856]).

To William Sharpey 2 June [1857][1]

Down Bromley Kent
June 2ᵈ

Dear Sharpey

I have heard from Hooker that he has proposed Lindley for the Royal medal; & as I expressed my opinion to you on Hancock & Prestwich, I hope you will forgive me troubling you with a few lines.[2] You may with truth think it absurd in a man not a Botanist, expressing an opinion on a Botanist, but my work for several years has led me to read a good deal on Botany, especially in foreign Journals &c., & it has been of consequence to me to form the best opinion I could, how much to trust the remarks & generalisations of various authors. This being so, I may state that I have been led to form a very high opinion of Lindley's work, so that in my opinion neither Hancock or Prestwich could for a minute be placed in competition with him for one of the R. medals. His claim under many points of view seems to me very strong. To make a first rate monograph on any one small department seems not to be very difficult,—still less so to gain a wide but superficial knowledge; but to have so profound a knowledge as to discuss on sound principles the classification of the whole vegetable kingdom, as Lindley has done, shows extraordinary talents & knowledge:[3] I have observed frequently that foreign authors on whatever class or family they are treating, seem to consider Lindley's opinions as deserving serious consideration. If I were on the Council, I sh[d] give without doubt & with the highest satisfaction, my vote for Lindley.[4]

Permit me to say one other word: Hooker tells me that my letter to you was read to the Council:[5] I had thought that you had asked merely for my private opinion,—all that I said in that letter expressed my *deliberate* conviction, whatever that may be worth, but I am rather alarmed that I expressed myself dogmatically under the impression that I was writing to a private friend & not to a Body like the Council of the Royal Soc.y.— I most sincerely hope that this may not have been, as I fear it was.—

Believe me dear Sharpey | Your's sincerely | Ch. Darwin

Houghton Library, Harvard University

[1] Dated by the relationship to the preceding letter.
[2] See preceding letter. CD's previous note to Sharpey supporting Joseph Prestwich and Albany Hancock has not been found.
[3] John Lindley was the author of *The vegetable kingdom; or, the structure, classification, and uses of plants, illustrated upon the natural system* (Lindley 1846; 2d ed., 1853).
[4] Lindley was awarded a Royal Medal in 1857. Hancock received one of the two Royal Medals in 1858, and Joseph Prestwich was a recipient in 1865.
[5] See preceding letter.

To J. D. Hooker   3 June [1857][1]

Down Bromley Kent
June 3[d]

My dear Hooker

I am going to enjoy myself by having a prose on my own subjects to you, & this is a greater enjoyment to me than you will readily understand; as I for months together do not open my mouth on Nat. History.— Your letter is of great value to me & staggers me in regard to my proposition. I daresay the absence of Bot. facts may in part be accounted for by the difficulty of measuring slight variations.[2] Indeed after writing this occurred to me; for I have Crucianella stylosa coming into flower & the pistil ought to be very variable in length, & thinking of this I at once felt how could one judge whether it was variable in any high degree. How different, for instance, from Beak of Bird!— But I am not satisfied with this explanation, & am staggered. Yet I think there is something in law; I have had so many instances as following: I wrote to Wollaston to ask him to run through Madeira Beetles & tell me whether any one presented anything very anomalous in relation to its allies: he gave me a unique case of enormous head in female & then I found in his Book already stated that the size of head was *astonishingly* variable.—[3] Part of difference with plants may be accounted for by many of my cases being secondary male or *female* characters, but then I have striking cases with hermaphrodite Cirripedes.—[4] The cases seem to me far too numerous for accidental coincidences of great variability & abnormal development. I presume you will not object to my putting a note saying that you had reflected over case & though one or two cases seemed to support, quite as many or more seemed wholly contradictory. This want of evidence is the more surprising to me, as generally I find any proposition more easily tested by observations in Bot. works, which I have picked up, than in Zoological works.— I never dreamed that you had kept the subject at all before your mind.[5] Altogether the case is one more of my *many* horrid puzzles.—

My observations, though on so infinitely a small scale, on the struggle for existence, begin to make me see a little clearer how the fight goes on: out of 16 kinds of seed sown on my meadow, 15 have germinated, but now they are perishing at such a rate that I doubt whether more than one will flower.[6] Here we have choking, which has taken place likewise on great scale with plant not seedlings in a bit of my lawn allowed to grow up. On other hand in a bit of ground 2 × 3 feet, I have daily marked each seedling weed as it has appeared during March, April & May, and 357 have come up, & of these 277 have *already* been killed chiefly by slugs.—[7] By the way at Moor Park, I saw rather pretty case of effect of animals on vegetation: there are enormous commons with clumps of old Scotch firs on hills, & about 8–10 years ago some of these commons were enclosed & all round the clumps nice young trees are springing up by the millions, looking exactly as if planted so many are of same age. In other part of common, not yet enclosed, I looked for miles & not *one* young tree cᵈ be seen; I then went near (within ¼ of mile of the clumps & looked closely in the heather, & there I found tens of thousands of young scotch-firs (30 in one square yard) with their tops nibbled off by the few cattle which occasionally roam over these wretched Heaths. One little tree 3 inches high, by the rings appeared to be 26 years old with a short stem about as thick as stick of sealing wax.— What a wondrous problem it is,— what a play of forces, determining the kinds & proportions of each plant in a square yard of turf! It is to my mind truly wonderful. And yet we are pleased to wonder when some animal or plant becomes extinct.—[8]

I am so sorry that you will not be at Club. I see Mrs Hooker is going to Yarmouth; I trust that the health of your children is not motive.—

Good Bye | My dear Hooker | Ever yours | Ch. Darwin

P.S. | You must not forget *sometime* to let me have the Himalayan thistle case in short abstract.

I believe you are afraid to send me a ripe Edwardsia pod for fear I sh^d float it from N. Zealand to Chile!!!

Endorsement: '57'
DAR 114.4: 200

[1] The year is given by CD's reference to experiments carried out in 1857 (see nn. 6 and 7, below) and by dated observations made while he was at Moor Park (see n. 8, below).

[2] See letter to J. D. Hooker, 16 [May 1857]. In the discussion of this subject in *Natural selection*, p. 308, CD wrote:

> I applied to Dr. Hooker on this subject, who after careful consideration, informs me that though some facts seem to countenance the rule, yet quite as many or more are opposed to it. In plants one large class of cases, namely secondary sexual characters are not present. Moreover, as Dr. Hooker has remarked to me, in all plants there is so much variability, that it becomes very difficult to form a judgment on the degrees of variability: . . .

[3] See letter from T. V. Wollaston, [early November 1856].

[4] In *Natural selection*, p. 308, CD wrote: 'Moreover from Cirripedes being hermaphrodite, the cases are the more valuable, as clearly showing that the law holds good without any relation to sexual distinctions.'

[5] CD had first mentioned this topic to Hooker in 1852 (see *Correspondence* vol. 5, letter to J. D. Hooker, [April 1852]).

[6] In an entry in his Experimental book, p. 27 (DAR 157a), headed '1857. April 8th,', CD recorded '16 K. Garden seeds sown on Grass field.—'. Commenting several weeks later on the fate of the young seedlings, CD wrote: 'I sh^d think most germinated, & then came battle for life with slugs & insects & other plants'. (DAR 157a, p. 28).

[7] This experiment, labelled 'Weed Garden', is recorded in CD's Experimental book, pp. 24, 25 (DAR 157a). An entry of 1 June 1857 tallies the results to June and gives the figures that CD cited to Hooker.

[8] These observations are described in notes dated 3 May 1857 in DAR 46.1: 38–9. They are transcribed as an appendix to *Natural selection*, pp. 570–1.

To J. D. Hooker 5 June [1857][1]

Down Bromley Kent
June 5th

My dear Hooker

I honour your conscientious care about the medals.[2] Thank God I am only an amateur (but a much interested one) on subject.—

It is an old notion of mine that more good is done by giving medals to younger men in the early part of career, than as a mere reward to men whose scientific career is nearly finished.— Whether medals ever do any good is question which does not concern us, as there the medals are.— I am almost inclined to think I would rather lower standard & give medal to young workers than to old ones, with no *especial* claims. With regard to *especial* claims, I think it just deserves your attention, that if general claims are once admitted, it opens the door to great laxity in giving them.— Think of case of very rich man who aided *solely* with his money but to a grand extent—or to such an inconceivable prodigy as a minister of Crown who really cared for science. Would you give such men medals—perhaps medals could not be better applied than *exclusively* to such men! I confess at present I incline to stick to especial claims, which can be put down on paper.— I do not quite agree with your estimate of Richardson's merits.[3] Do, I beg you, (whenever you quietly see) to talk with Lyell on Pretwich: if he agrees with Hopkins[4] I am silenced; but as yet I must look at the correlation of the Tertiaries as one of the highest & *most* frightfully *difficult* tasks a man could set himself, & *excellent* work, as I believe, P. has done. I confess I do not value Hopkins' opinion on such a point. I confess I have never thought, as you show ought to be done, on the *future*.— I quite agree under all circumstances with propriety of Lindley.—[5] How strange no new geologists are coming forward!— Are there not lots of good young chemits & astronomers or physcicists? Fitton is the only old geologist left,[6] who has done good work, except *Sedgwick*.— Have you thought of him? He w^d. be brilliant companion for Lindley. Only it w^d never do to give Lyell a Copley, & Sedgwick a Royal on same year. It seems wrong that there sh^d be *three* Natural Science medals on the same year.— Lindley, Sedgwick & Bunsen[7] sounds well.— And Lyell next year for Copley???—[8] You will see that I am speculating as mere idle amateur.—

I am much confounded by your showing that there are not obvious instances of my (or rather Waterhouse's) Law of abnormal developments being highly variable.—[9] I have been thinking more of your remark about difficulty of judging or comparing variability in plants from great general variability of parts.— I should look at the law as more completely smashed, if you would turn in your mind for a little while for cases of great variability of an organ, & tell me whether it is moderately easy to pick out such cases, *for if they can be picked out*, & notwithstanding do not coincide with great or abnormal development, it w^d be a complete smasher: it is only beginning in your mind at the variability end of the question instead of at the abnormality end. *Perhaps* cases in which a part is highly variable in all the species of a group should be excluded, as possibly being something distinct, & connected with the perplexing subject of polymorphism.— Will you perfect your assistance by further considering for a little the subject this way?—

I have been so much interested this morning in comparing all my notes on the variation of the several species of genus Equus & the results of their crossing: Taking most strictly analogous facts amongst the blessed Pigeons for my guide, I believe I can plainly see the colouring & marks of the grandfather of the Ass, Horse, Quagga, Hemionus & Zebra, some millions of generations ago! Should not I sneer at any one who made such a remark to me a few years ago!—but my evidence seems to me so good, that I shall publish my vision at end of my little discussion on this genus.[10]

I have of late inundated you with my notions, you best of friends & philosophers.—

Adios | C. Darwin

I can most truly say that anyhow I am worthy of your letters, if being interested by them makes me worthy.—

I am extremely glad to hear my letter to Sharpey did not appear dogmatic or presumptuous: I had quite annoyed myself by trying to remember exactly what I had said.

Endorsement: '57'
DAR 114.4: 201

[1] The year is given by the reference to topics first mentioned in the letter to J. D. Hooker, 2 June [1857].

[2] See letter to J. D. Hooker, 2 June [1857].

[3] CD had supported the nomination of John Richardson for a Royal Medal in 1856 (see letter to Edward Sabine, 23 April [1856]).

[4] William Hopkins.

[5] Hooker had nominated John Lindley for one of the Royal Medals (Royal Society council minutes).

[6] William Henry Fitton was best known for work carried out between 1824 and 1836 on the succession of strata between the Oolite and the Chalk (*DNB*).

[7] The German chemist Robert Wilhelm Eberhard Bunsen was awarded the Copley Medal in 1860. In 1857, the recipient was the French chemist Michel Eugène Chevreul.

[8] Charles Lyell did receive the Copley Medal in 1858. In 1857, the Royal Medals were given to John Lindley and to the chemist Edward Frankland.

[9] See letter to J. D. Hooker, 3 June [1857]. George Robert Waterhouse had first formulated the generalisation concerning variation in his *Natural history of the Mammalia* (Waterhouse 1846–8, 2: 452 n. 1).

[10] CD included this material at the end of his chapter on 'Laws of variation' (*Natural selection*, pp. 328–32).

To T. C. Eyton   9 June [1857][1]

Down Bromley Kent
June 9th

Dear Eyton

I must thank you for the sheets completing your Catalogue received this morning.[2] What a superb collection you have: I am quite astounded at it!

I fear that you have not been able to find out whether the half bred African Pigs of Lord Hills' were fertile, which I regret as it would have made your capital case on Pigs perfect.[3] How go on the skeletons of dogs &c &c?—[4]

By the way I mentioned in a former letter that I had a skin with skull in & legs of a pure W. African domestic dog; but I presume you do not care about it.—[5] Some day I will send it to Brit. Mus. if of no use to you.—

Believe me dear Eyton | Yours very sincerely | C. Darwin

P.S. Do you breed horses? If you did & would make a few observations for me merely on the colouring of colts of different colours, it would be of *much* use to me, & I do not think troublesome;[6] but I daresay you have your hands quite full enough with your own work.—

Can you tell me whether the convolutions in the Trachea of the *male* of the *same* species of Bird ever varies much?—

American Philosophical Society (146)

[1] Dated by reference to other letters in the Eyton correspondence (see nn. 5 and 6, below).

[2] CD refers to Eyton's *Catalogue of the species of birds in his possession* (Eyton 1856), which was issued in parts. A copy of the work is in the Darwin Pamphlet Collection–CUL.

[3] In Eyton 1837a, Eyton had mentioned the case of a litter of pigs from a cross between Rowland Hill's African boar and a female common pig, stating 'time will show whether they will again be fruitful' (Eyton 1837a, p. 23). CD cited the case in *Variation* 1: 74.

[4] See *Correspondence* vol. 5, letters to T. C. Eyton, 25 October [1855] and 3 December [1855].

[5] Letter to T. C. Eyton, 5 October [1856].

[6] For CD's interest in the colouring of horses, see the preceding letter and letter to T. C. Eyton, 26 [June 1857].

To *Gardeners' Chronicle* [before 13 June 1857][1]

Will any reader of the Gardeners' Chronicle be so kind as to take the trouble to inform me how Dun or Mouse-coloured Ponys with a dark stripe down their backs are bred? This breed is common in Norway on the banks of the Indus & in the Malayan archipelago, & is in some respects very interesting in relation to the origin of the domestic Horse.[2] Is this peculiar colour thrown from Ponys of any other colour, or must one or both parents be Dun?

Occasionally ponys of this colour have a cross stripe on the shoulder, like that on the Ass, & likewise bars on the legs. If anyone who has bred ponys of this colour, would inform me whether their stripes are more distinct in the colt than in aged ponys, I sh$^{\underline{d}}$ be greatly obliged. The transverse bars sometimes seen on the legs of the Ass are said to be plainest during growth.

Ch. Darwin

Down, Bromley Kent

Gloucestershire Record Office (D1021/8/4. T. C. Morton deposit)

[1] This letter appears to be the original of the notice published in the *Gardeners' Chronicle and Agricultural Gazette*, 13 June 1857, p. 427. There are discrepancies between the manuscript and the published text, presumably resulting from changes made by the editor of the *Gardeners' Chronicle*.

[2] For CD's interest in this subject, see the preceding letter and letters to J. D. Hooker, 5 June [1857], and to T. C. Eyton, 26 [June 1857]. The issue was further discussed in *Natural selection*, pp. 328–32.

From H. C. Watson   14 June [1857][1]

My dear Sir

Looking for something else, I stumbled on the passage extracted on preceding page. It seems to meet a query you some time ago addressed to me respecting the *Limosella* & *Subularia*, to which I was able to reply from personal observation only in respect to the former.—[2]

Very truly | Hewett C. Watson

Thames Ditton June 14th

To C. Darwin | Esqe

[Enclosure]

9th February 1843

Professor Graham read an account of a Botl Excursion in Ross-shire during August, 1842.[3]

The party left Edinburgh on 21st August, &c - - - - -

"The season having been remarkably dry, all the lakes were far below their usual level, & in consequence such plants as *Lobelia Dortmanna*, *Subularia aquatica*, &c. were seen in flower and fruit on dry ground"— Trans. Bot. Soc. Edinb. vol. 1, p. 201.[4]

DAR 207 (Letters)

[1] Dated by the reference to Watson's previous letter on *Limosella* and *Subularia* (see n. 2, below).
[2] See letter from H. C. Watson, 26 November 1856.
[3] Robert Graham was professor of botany at Edinburgh University, 1820–45.
[4] This information was not used by CD in his discussion of the possible cross-fertilisation of water plants (*Natural selection*, pp. 62–3).

To Asa Gray   18 June [1857][1]

Down Bromley Kent
June 18th—

My dear Dr Gray

I must thank you for your two very valuable letters.[2] It is extremely kind of you to say that my letters have not bored you very much, & it is almost *incredible* to me, for I am quite conscious that my speculations run quite beyond the bounds of true science. One chief object of this note is to say that I have **not** received the last part of your Silliman Papers:[3] Hooker has, & he says he will lend it me, if, as is very likely, you have not another copy. But it may come with the Watson correspondence.—[4] Your remarks on that head will be of *real* use to me, when I return to the subject, for a man must be blind not to see how cautious a reasoner you are.

Thank you much for your remarks on disjoined species:[5] I daresay I may be quite in error: I saw so much difficulty even *theoretically* & so much impossibility practically from my ignorance, that I had given up notion till I read your note to your Article. I had only just copied out a few striking cases out of Hooker's Him: Journal[6] & turned to Steudel[7] to see what the genera were. The notion was grounded on the belief that disjoined species had suffered much local extinction & therefore (*conversely* with the case of genera with many species having species with wide ranges.) I inferred that genera & Families with very few species (ie from Extinction) would be apt (not necessarily always) to have narrow ranges & disjoined ranges. You will not perceive, perhaps, what I am driving at & it is not worth enlarging on,—but I look at Extinction as common cause of *small* genera & *disjoined* ranges & therefore they ought, *if they behaved properly* & as nature does not lie to go together!—[8]

I have not the least doubt that the proportions of British naturalised plants were due to simple chance; but I thought it was just worth mentioning to you: I had from your former Edition of Manual quite given up idea.—

It has been *extremely* kind of you telling me about the trees: now with your facts, & those from Britain, N. Zealand, & Tasmania, I shall have fair materials for judging:[9] I am writing this away from home,[10] but I think your fraction of $\frac{95}{132}$ is as large as in other cases, & is at least a striking coincidence.—

I thank you much for your remarks about my crossing notions, to which I may add, I was led by exactly the same idea as yours, viz that crossing must be *one* means of eliminating variation, & then I wished to make out how far in animals & vegetables this was *possible*.— Papilionaceous flowers are almost dead floorers to me, & I cannot experimentise as castration alone often produces sterility. I am surprised at what you say about Compositæ & Gramineæ. From what I have seen of latter they seemed to me (& I have watched Wheat owing to what L. Deslongchamps has said on their fertilisation in bud)[11] favourable for crossing; & from Cassini's observations[12] & Kölreuters[13] on the adhesive pollen & C. C. Sprengels',[14] I had concluded that the Compositæ were eminently likely (I am aware of the pistil brushing out pollen.) to be crossed. If in *some months* time you can find time to tell me whether you have made any observations on the early fertilisation of plants in these two orders, I sh^d be *very glad* to hear, as it w^d save me from great blunder. In several published remarks on this subject in various genera it has seemed to me that the early fertilisation has been inferred from the early shedding of the pollen, which I think is clearly false inference. Another cause, I sh^d think, of the belief of fertilisation in the bud, is the not-rare **abnormal** early maturity of the pistil, as described by Gærtner.—[15] I have hitherto failed in meeting with detailed account of regular & normal impregnation in the bud.— Podostemon & Subularia under water (& Leguminosæ) seem & are strongest cases against me, as far as I as yet know.

I am so sorry that you are so overwhelmed with work; it makes your **very great** kindness to me the more striking. Believe me, Your's gratefully | C. Darwin

It is really pretty to see how effectual insects are: a short time ago I found a *female* Holly 60 measured yards from any other Holly & I cut off some twigs & took by chance 20 stigmas, cut off their tops & put them under microscope: there was pollen *on every one* & in *profusion* on most! Weather cloudy & stormy & unfavourable, wind in wrong direction to have brought any.[16]

Gray Herbarium of Harvard University

[1] The year is given by CD's reference to recent observations of holly plants, his notes on which are dated 25 May 1857 (see n. 16, below).

[2] Letters from Asa Gray, [*c.* 24 May 1857] and 1 June 1857.

[3] Gray had promised to send CD a new and corrected copy of A. Gray 1856–7, the third part of his 'Statistics of the flora of the northern United States' (see letter from Asa Gray, 1 June 1857). In letter from Asa Gray, 7 July 1857, Gray informed CD that he had dispatched the memoir to him.

[4] In his letter of 1 June 1857, Gray had also mentioned that he would send back to CD 'in a week or two' the letters from Hewett Cottrell Watson.

[5] See letter to Asa Gray, 9 May [1857], and letter from Asa Gray, 1 June 1857.

[6] J. D. Hooker 1854.

[7] Steudel 1840–1.

[8] See letter to Asa Gray, I January [1857], in which CD first mentioned this aphorism.

[9] See *Natural selection*, p. 62 n. 1, and *Origin*, pp. 99–100.

[10] CD was writing from Edward Wickstead Lane's hydropathic establishment at Moor Park, where he had arrived on 16 June 1857 ('Journal'; Appendix II).

[11] Loiseleur Deslongchamps 1842–3. The first part of this work is in the Darwin Library–CUL.

[12] In *Natural selection*, p. 48 and n. 1, CD cited T. Smith 1822, p. 595, as the reference for Henri Cassini's observations. Thomas Smith discussed Cassini's discovery that in dioecious species of Compositae the different times of flowering prevent self-fertilisation, citing Cassini 1813 as the source (T. Smith 1822, pp. 600–1).

[13] Kölreuter 1761–6, cited in *Natural selection*, p. 55.

[14] Sprengel 1793.

[15] Gärtner 1849.

[16] These observations are described in a note in DAR 49: 45 dated 'May 25. 57/'.

To W. B. Tegetmeier [18 June 1857][1]

D.̲ Lane's Hydropathic Estabmt | Moor Park | Farnham | Surrey
Thursday

My dear Sir

I write merely to thank you for your offer crossed yellow Magpie & Helmet, but it will be of no use to me.[2]

I am very glad you are investigating the tail question, & I hope that you will work out the down & colour point.—[3] I am really delighted that the Borneo fowls turn out in the least interesting.—[4]

With respect to the Runt, I fancied by your having only one that it was not valuable.— Have you any interest with M.̲ Bridge to **supplicate** him for a dead body sh^d one die.⁵ I am much interested in Runts, as they seem to vary or differ more than other Breeds.⁶ It is a pity to take your young Gulliver Runt, for I sh^d hope I c^d get one some time dead by chance from Mr Gulliver.—

My dear Sir | Yours very sincerely | C. Darwin

P.S. I presume you do not care for any of the Pigeons mentioned in my last note.

New York Botanical Garden Library (Charles Finney Cox collection)

¹ Dated by CD's arrival at Moor Park on Tuesday, 16 June 1857 ('Journal'; Appendix II), and by the relationship to the letters to W. B. Tegetmeier, 18 May [1857] and 23 June [1857].
² See letter to W. B. Tegetmeier, 18 May [1857].
³ See letter to W. B. Tegetmeier, 12 [May 1857].
⁴ See letter to W. B. Tegetmeier, 18 May [1857].
⁵ Mr Bridge has not been identified.
⁶ CD introduced the description of this breed in *Variation* 1: 142 with the remark: 'Inextricable confusion reigns in the classification, affinities, and naming of Runts.'

To W. B. Tegetmeier 23 June [1857]¹

Moor Park | Farnham | Surrey
June 23^d

My dear Sir

I quite forgot to say one thing in my last note.² Of the eggs of the rumpless Fowls only 3 hatched—2 jet black & one a little banded. Now would you believe it, owing to my fault in not having given clear enough instructions these three were hatched two or three days before I heard of it!— So that after all your kindness I have not got a chicken about 24 hours before Birth.—³ Shall you hatch any more? What would you say to exchange two of my chickens when nearly full grown, for an egg with young chick in it, about 24 hours before birth.— You could have two nearly full-grown instead of one not born.—

In case you had eggs hatching soon, I may state that I shall be here till the 30th morning, & after that at Down.— Forgive, if you can, my stupidity: I beg you not to take the trouble to answer this merely to acknowledge it.—

Your's very sincerely | C. Darwin

New York Botanical Garden Library (Charles Finney Cox collection)

¹ The year is established by CD's reference to remaining at Moor Park until 30 June. In his 'Journal' for 1857, CD recorded returning to Down from Moor Park on this day (Appendix II).
² Preceding letter.
³ CD wished to know how early in development rudimentary organs became rudimentary (see letter to W. B. Tegetmeier, 6 February [1857]).

To J. D. Hooker 25 June [1857][1]

Moor Park, Farnham | Surrey
June 25th

My dear Hooker.

This requires no answer, but I will ask you whenever we meet.— Look at enclosed seedling gorzes, especially at one with top knocked off.— The leaves succeeding cotyledons being almost Clover-like in shape, seem to me feebly analogous to embryonic resemblances in young animals,—as, for instance the young Lion being striped.— I shall ask you whether this is so.—

Etty is gaining strength steadily but slowly: her pulse has certainly improved, & she is very happy here.[2] The owners, D[r] Lane & wife & mother-in-law Lady Drysdale are some of the nicest people, I have ever met.

I return home on 30th.—

GoodBye— My dear Hooker | Ever yours | C. Darwin

Endorsement: '1857.'
DAR 114.4: 205

[1] The year is given by CD's stay at Moor Park hydropathic establishment at the same time as Henrietta Emma Darwin (see n. 2, below).
[2] Henrietta Darwin had been at Moor Park since 29 May 1857 undergoing hydropathic treatment under the care of Edward Wickstead Lane (Emma Darwin's diary). See also letter from H. E. Darwin, [2 August 1857].

To W. B. Tegetmeier 25 [June 1857][1]

[Moor Park]
Thursday 25th

My dear Sir

One egg will amply suffice, if you will be so kind as to send it to Down as soon as one chips the shell.— I presume if the egg were put in cold water it would easily & soon kill the Chick: it w[d] be as well just to see that there was a chick inside. To save trouble I enclose some stamps which I presume will cover the weight.—

With many thanks | Yours very sincerely | C. Darwin

At last just before leaving home I got all the published numbers of the Poultry Book & shall soon study them.—[2] At some *future* time I shall be glad of skull of the G. Bankiva again.—[3]

Do you think it w[d]. be any use my writing to M[r] Bridges to ask for Carcase of the long-winged Runt? If so when you send egg will you tell me address.—

New York Botanical Garden Library (Charles Finney Cox collection)

[1] Dated by the relationship to the letter to W. B. Tegetmeier, 23 June [1857].
[2] Tegetmeier ed. 1856–7. See letter to W. B. Tegetmeier, 18 May [1857].
[3] See *Correspondence* vol. 5, letter to W. B. Tegetmeier, 6 December [1855] and n. 4.

Henri Milne-Edwards.
(Courtesy of the New York Botanical Garden Library.)

Thomas Henry Huxley. Photograph by Maull & Polyblank, 1857.

To T. C. Eyton 26 [June 1857][1]

Moor Park, Farnham | Surrey
26th

Dear Eyton

Your letter has been forwarded to me here, where I am staying at a Hydropathic establishment as my health has lately been very indifferent.— I thank you for the reference on Hybrid & for your very kind note.—[2] I am delighted to hear what fine Natural History prospect you have before you.—

It is a new & interesting line on the growth of Bones of Birds: I was not at all aware that there was anything peculiar in that subject.[3] My confounded health interferes terribly with all my work, but yet I am making steady progress in my Book on Variation of Species. & on domestic varieties.

As you are a great man for Horses, you might aid me, if not irksome to you, by looking out in Ireland or elswhere for any cases of Horses or Ponies with transverse bars on legs like those of zebra, or on shoulder & along the back, as with the ass. These stripes sometimes occur in duns, & mouse-coloured & chesnut horses; & I sh^d be infinitely obliged for any cases of this kind.—[4]

I sh^d most thoroughily enjoy visiting you at Eyton; but I am not likely to be in Shropshire soon.—

Pray believe me | Dear Eyton | Yours very truly | C. Darwin

American Philosophical Society (147)

[1] The month and year are given by CD's stay at Moor Park (16–30 June 1857).
[2] Eyton's letter has not been found. It was evidently a reply to CD's letter to T. C. Eyton, 9 June [1857].
[3] Eyton was at work on his *Osteologia avium* (Eyton 1867). A copy is in the Darwin Library–Down.
[4] CD had previously asked Eyton for this information in his letter to T. C. Eyton, 9 June [1857]. See also letter to *Gardeners' Chronicle*, [before 13 June 1857].

From J. D. Hooker [27] June 1857[1]

7. Brighton Terr. | Gt. Yarmouth
Saturday June 28/57.

Dear Darwin

Many thanks for the Embryonic Furze.[2] The subject is one of great interest. A quite identical case is that of the Phyllodineous Acacias, which have pinnate leaves very early: & Eucalypti which have all opposite leaves when young—also many simple leaved Barberries have compound at very early state. A great stumbling block in developement to me has been the very great differences between the Cotyledonary leaves of plants—even of the same Nat. Order.— Leguminosæ for instance— this has always prevented me from understanding the embryonic developement in plants being so good an evidence of affinity as in animals. Comparative developement would appear to begin with the post

Cotyledonary leaves, & the Cotyledonary may be regarded as placenta? Amnios? &c which vary? in allied animals. Is this not a shadow of a generalization?— I have often recommended germination & first formed leaves as the most interesting enquiry a young Botanist could take up, & particularly urged it upon G. Henslow—[3] The latter two days ago sent me a letter that has given me great concern— he writes curtly to say that

AL incomplete
DAR 100: 115

CD ANNOTATIONS
1.7 the embryonic . . . animals. 1.9] 'Can you not account for difference by shape of seed.' *added pencil*
Top of first page: '12'[4] *brown crayon*

[1] Hooker made a slip in writing the date: Saturday was 27 June in 1857.
[2] See letter to J. D. Hooker, 25 June [1857].
[3] George Henslow, the youngest son of John Stevens Henslow, was studying natural sciences at Cambridge University. See letter to J. S. Henslow, [after 6 December 1856].
[4] The number of CD's portfolio of notes on embryology.

To J. D. Hooker 1 July [1857][1]

Down Bromley Kent
July 1st

My dear Hooker

I write to say how truly I think you the best of men for proposing to pay us a visit this summer, I shall most truly & **heartily** enjoy it. In a few weeks time we shall know our plans better, as Etty & our Boys holidays make everything doubtful I am alone at present Emma visiting some relations with a tail of six children![2] I am very sorry to hear about G. Henslow: an attack of religion is a most serious thing: the whole affair must be a great disappointment to you.—[3]

Thanks for your interesting note about embryonic leaves: after I had sent it, I began to think about cotyledons, & marvelled that I could not remember having ever read any discussion on their resemblances & dissimilarities in allied plants. How curious that the subject sh^d never have been taken up! I do not even know whether functions of the cotyledons are same as leaves, or whether they serve, also, as receptacle of nutriment: I have noticed in my weed-garden that their destruction *seems always* to kill the plant.— I was speculating in my ignorance that the form of cotyledon was probably related to the shape of seed & its embryo & radicle; & if this were so, as seeds are adapted to various contingencies, we might expect the cotyledons to differ; but probably this is not the case, as it w^d have occurred to you. I am not learned enough in animal embryology to compare cotyledons with amnios &c &c.—

If cotyledons have a relation to the external conditions of existence distinct or partially different from what the leaves of the mature plant have to the conditions,

I think the differences of the cotyledons in the same Family, would be explicable in the same way as in some Diptera & some Neuroptera, there is a *wonderful* amount of difference in the larvæ:— no doubt these larvæ have much in common, & so I presume the cotyledons have much in common.

I have had another letter from Asa Gray: *most* kindly he has worked out trees of U. States & finds out of 132, 95 have sexes more or less separate (= .72); so that the Rule seems here to hold good whether or no my explanation is correct.[4]

Farewell | My dear Hooker | Ever yours | C. Darwin

Endorsement: '57'
DAR 114.4: 198

[1] The year is established by the subjects discussed and by CD's reference to Emma Darwin and the children being away from Down (see n. 2, below).
[2] Emma and the children were visiting Francis (Frank) Wedgwood and his family in Barlaston, Staffordshire. They later travelled to Shrewsbury, the children returning to Down on 4 July 1857 and Emma on 6 July (Emma Darwin's diary).
[3] The missing portion of the preceding letter may have discussed this point.
[4] See letter from Asa Gray, 1 June 1857.

To J. D. Hooker 5 July [1857][1]

Down Bromley Kent
July 5th

My dear Hooker

Am I right in supposing that in Linnean Journal p. 5. your words "some *states* of the Wahlenbergia saxicola" means a variation of this species; & does it mean that this variation *alone* presents "the unequal inclined anthers &c".[2]

I ask because this wd be a *valuable* case to add to my list of variations of one species assuming the character of a distinct, but allied species or genus.—[3]

I am particularly interested on this point.—

The briefest answer will suffice. If you had been a stranger I shd have quoted the sentence, but it is far safer to ask.—

Ever yours | C. Darwin

P.S. I have just had a seed case which has surprised me: Fresh Nuts sink in seawater; but old Spanish nuts float for between 75 & upwards of 90 days— after this prolonged bath, I opened one & it looked so fresh, that I planted 2 others & one has germinated well—[4] I had fancied a seed with so much organized matter wd have assuredly gone bad when after having been long dry, & then kept in water for 3 months.—

What a capital number of the Linnean Journal! Owen's is a grand Paper; but I cannot swallow Man making a division as distinct from a Chimpanzee, as an ornithorhynchus from a Horse: I wonder what a Chimpanzee wd say to this?[5]

Endorsement: '/57'
DAR 114.4: 203

[1] The year is established by CD's reference to experiments testing the viability of nuts, which were recorded in his Experimental book (see n. 4, below).

[2] The passage referred to discusses the problem with using the anthers and corolla as characters for classifying the Lobeliaceae (J. D. Hooker and Thomson 1858, p. 5).

[3] CD included this case in his chapter on 'Laws of variation' (*Natural selection*, p. 327).

[4] Under the heading '*1857* Dryed Seeds' in CD's Experimental book, pp. 12–13 (DAR 157a), there is a list of the various nuts that he floated in salt water. In the Experimental book, p. 12, an entry dated 16 May 1857 reads: '3^{d} nut still floating after 90 days'. On 2 July CD recorded: 'one of these two germinated'.

[5] In a paper on the classification of the Mammalia, Richard Owen proposed to use the brain to delimit the 'most natural primary divisions' of the class (Owen 1857b, p. 14). Applying this criterion to man, he concluded: 'I am led to regard the genus *Homo*, as not merely a representative of a distinct order, but of a distinct subclass of the Mammalia' (*ibid.*, p. 20).

To T. H. Huxley　5 July [1857][1]

Down Bromley Kent
July 5th

My dear Huxley

Will you be so kind as to read the two enclosed pages as you said you would & consider the little point therein referred to.—[2] I have not thought it worth troubling you with how far & in which way the case concerns my work,—the point being how far there is any truth in M.M. Brullé & Barneoud.[3] My plan of work is just to compare partial generalisations of various authors & see how far they corroborate each other. Especially I want your opinion how far you think I am right in bringing in Milne Edwards' view of classification.[4] I was long ago much struck with the principle referred to; but I could then see no rational explanation why affinities sh^{d} go with the *more or less early* branching off from a common embryonic form. But if MM. Brullé & Barneoud are true, it seems to me we get some light on Milne Edwards' views of classification; & this particularly interests me.[5] I wish I could anyhow test M. Brullé's doctrine: as in Vertebrata the head consists of greatly altered vertebræ, according to this rule, in an early part of the embryonic development of a Vertebrate animal, the head ought to have arrived more nearly to its perfect state, than a dorsal or cervical vertebra to its perfect state: How is this?[6]

I have been reading Goodsir, but have found no light on my particular point.[7] The paper impresses me with a high idea of his judgment & knowledge, though, of course, I can form no independent judgment of the truth of his doctrines. But by Jove it w^{d} require a wonderful amount of evidence to make one believe that the head of an Elephant or Tapir had more vertebræ in it, than the head of a Horse or Ox.[8]

Many thanks for your last Lecture. How curious the development of Mysis![9]

Yours very sincerely | Ch. Darwin

Do you know whether the embryology of a Bat has ever been worked out?

Imperial College of Science, Technology and Medicine Archives (Huxley 5: 67)

[1] Dated by the relationship to the letter from T. H. Huxley, 7 July 1857.

[2] CD sent Huxley a fair copy of folios 41–4 of his chapter on 'Laws of variation: varieties and species compared' (*Natural selection*, pp. 303–4). Huxley returned them to CD with his letter of 7 July 1857, and both the letter and these pages are interleaved with the folios of the original manuscript in DAR 11.1: 41–5. They were the concluding pages to the section entitled 'Correlation of growth' and discuss the connection between embryonic development and genealogical relationship.

[3] CD refers to the memoirs by Gaspard Auguste Brullé on the development of appendages in the Articulata (Brullé 1844) and that by François Marius Barnéoud on the development of ovules, embryos, and anomalous corollas in plants (Barnéoud 1846). Brullé proposed that it was a general rule in zoology that the most complex or highly developed organs of the adult animal were the first to appear in the embryo. Barnéoud noted that abnormally developed flower parts tended to appear early and develop quickly.

[4] Milne-Edwards 1844, which CD read in December 1846 (*Correspondence* vol. 4, Appendix IV, 119: 17a). CD's notes on the paper are in DAR 72: 117–22. See also *Correspondence* vol. 4, Appendix II, pp. 392–3, and Ospovat 1981.

[5] CD hoped that if Brullé's law held it would help explain Henri Milne-Edwards's generalisation that the greater the differences between adults, the earlier in development their embryos diverged from a common path. For a transcription of portions of CD's notes on Milne-Edwards 1844 and on Brullé 1844, see *Correspondence* vol. 4, Appendix II, pp. 392–4. CD also noted in his abstract of Brullé 1844: 'This is most remarkable & accords with M. Edwards . . . statement respect to whole animals in same great class, ie the more changed from each other, earlier they separate in embryonic characters.—' (DAR 72: 123). In a note keyed to this passage, CD further commented: 'states that in Crust, antennæ & parts of mouth appear before legs— the mouth-parts appearing before antennæ— The former have assumed their form, when the legs begin to appear.' Later, CD added in pencil: 'V. Rathke's paper & M. Edwards Treatise on Crustaceæ— It w^d be important to show that in diff. families, parts did appear in different order, even if no rule c^d be established. (Ask Huxley)' (DAR 72: 123v.).

[6] See letter from T. H. Huxley, 7 July 1857.

[7] Goodsir 1857. Huxley praised this paper in one of his lectures on Crustacea (see n. 9, below).

[8] In Goodsir 1857, John Goodsir challenged the view 'that the number of segments in the vertebrate head is the same in all its forms'. Goodsir maintained that in most fish, amphibia, reptiles, and birds there were six 'sclerotomes', while there were seven in all mammals excluding the proboscidians, which 'present indications of a great number' (Goodsir 1857, p. 136).

[9] Huxley discussed the development of the crustacean *Mysis* in his eleventh lecture in the *Medical Times & Gazette* (T. H. Huxley 1856–7). In *Mysis*, 'the larva is inactive, and its changes are undergone within the incubating pouch of the parent.' (T. H. Huxley 1856–7, 14: 639).

To Francis Galton 7 July [1857][1]

Down Bromley Kent
July 7th

My dear Galton

I return the enclosed signed with very great pleasure.—[2]

Many thanks for information about Dr Barths work which I will read.[3]

I continue much interested about all domestic animals of all savage nations, though I shall not take up cattle in detail.[4] If on reading I sh.[d] have anything to ask I will accept your kind offer & ask.— Anything about savages taking any the least pain in breeding & crossing their domestic animals is of particular interest to me.

With kind remembrances to Mrs Galton,[5] pray believe me | Yours very sincerely | Ch. Darwin.

University College London (Galton 190)

[1] Dated by the reference to 'D.[r] Barths work' (see n. 3, below).
[2] The item has not been identified, although it may have related to the Royal Geographical Society, of which Francis Galton was honorary secretary. Galton had been elected to the position in May 1857. CD had been a fellow since 1838.
[3] The German geographer and traveller Heinrich Barth, a member of the British expedition to explore North and Central Africa, 1849–55, returned to England in September 1855 and began writing up an account of his African travels. The first three volumes of Barth 1857–8 were published between 15 and 30 May 1857 (*Publishers' Circular*, 1 June 1857, p. 230).
[4] CD discussed the origins and variations of domestic cattle in *Variation* 1: 79–93.
[5] Francis Galton had married Louisa Jane Butler in 1853.

From Asa Gray 7 July 1857

Cambridge, Massachusetts,
July 7.th 1857.

My Dear Mr. Darwin

Your letter of June 18th came last week. Surely I need not say your letters are always most welcome. Among the great amount and variety of letters I receive yours and D.[r] Hooker's are almost the only ones that set me thinking. I am sorry I can make only such insignificant returns.

By a private hand, a passenger by sailing ship to Liverpool I last week sent addressed to you (ready stamped for the Post. at Liverpool) an envelope containing, your Watson correspondence:[1]—also a copy of my pamphlet from Silliman's Journal—. the same you have sheets of already.—[2] and a copy of my little *hornbook* for beginners, the First Lessons in Botany.[3]

I have not the least remembrance of what I wrote last to you. I am interested in what you say about *extinction* as giving the key to many of your ingenious views, & the clue to the probable explanation of a good many otherwise inexplicable things.[4]

I accept it as best explaining *disjoined species*. I see that the same cause must have reduced many species of great range to small, and that it *may* have reduced large genera to small, and so of families. But why is it not just as likely that there were as many small genera (nearly) at first as now, and as great a disproportion in the number of their species? There is *Sarraceniaceæ* of three genera—belonging to E. U. States, California, & Guiana, two of them of one known species each, and all of narrow ranges Is it philosophical, is it quite allowable, to assume (without evidence from fossil plants) that the family or any of the genera was once larger and wide spread? and occupied a continuous area? The tree of narrowest range in U.S. is *Cladrastis*: must I suppose it ever had a wide range, or even a wider range than it has now? That its area was once *somewhat* larger is extremely probable. But, granting you lots of species which have been greatly reduced by extinction, must you not admit that there are and were many more that never succeeded in getting a wide range? Would it not be pretty sure to be so of the later-comers? And there must have been late comers, comparatively. What chance would a new species, of average powers, have had of getting a foot-hold, still less of ever getting a wide range, if introduced locally into the forest of N. America while yet untouched by man?

Of any given local species, it seems to me *a priori* quite *as* likely (to say the least) that it never had a wide dominion, as that it had, and is on the road to extinction:— That is of species of continents.— As to islands, when you know their geological history, the presumption may be quite the other way, no doubt.

That is the way the question strikes an *outsider* like me,—a cautious and *slow* one,—just at first view. If I had ever thought over the matter, and investigated it as you have, very likely I should think quite differently. But it is just such sort of people as I that you have to satisfy and convince, and I am a very good subject for you to operate on, as I have no prejudice nor preposessions in favor of any theory at all.

I never yet saw any good reason for concluding that the several species of a genus must ever have had a common or continuous area. Convince me of that, or show me any good grounds for it (beyond the mere fact that it is *generally* the case),— i.e. show me why it *ought* to be so, and I think you would carry me a good way with you—as I dare say you will, when I understand it.

About crossing of individuals of a species, I have as you know a strong inclination to suppose it so, and that it plays an important part in *repressing variation*, and perhaps even in ensuring the *perpetuation* of species.[5] That it is **a** *vera causa* I surely believe; but how much operation it has is the question. Perhaps you may overestimate it. As you say, Papilionaceæ are dead against you. There the blossoms are self-fertilized—no doubt. I will give you a stronger case still, in *Fumariaceæ*! Here it cannot be otherwise.[6] But are Fumariaceæ sportive? *Fumaria off.* seems to be. *Corydalis aurea* runs to variation. All the rest of ours are remarkably *steady*; and that this is, on the whole, *decidedly so in the family*, Scc *Hook. & Thom. Fl. Ind. p. 259.*[7]

As to *Gramineæ* I must have taken for granted (in my last letter) that what has of late been so positively asserted was true. But I have just been looking at the principal Grasses around my door. Festuca elatior, Triticum repens, Dactylis, Arrhenatherum, Phleum, Alopecurus, Panicum clandestinum, Phalaris: not a soul of them shed a grain of pollen till the anthers and the stigmas hang out:—and the conspicuous way they hang out in anthers should have prevented anybody from talking of fertilization in the bud till they had observed it.[8] So I can confirm your own observations here.

As to Compositæ my hasty statement may need equal correction. It was based on the fact that the anther-cells are empty of pollen as soon as the corolla opens, or very soon after, the style as it elongates pushing out the pollen. But as the branches do not diverge until after protrusion, I will not swear that any pollen reaches the stigmatic surface while in the bud.—probably *not* generally.—[9] But I think I know some cases where the fertilization occurs in the bud. I have not time now to look at some common plants.— What you say about the pollen carried to a female *Holly* so abundantly, and to such a distance is very remarkable.

Ever most cordially Yours | Asa Gray

DAR 205.9 (Letters) and DAR 165: 98

CD ANNOTATIONS
0.3 My Dear . . . species? 4.5] *crossed pencil*
4.5 There . . . Guiana. 4.6] *double scored pencil*
4.11 But . . . doubt. 5.4] *crossed pencil*
6.1 That . . . overestimate it. 8.5] *crossed pencil*
6.3 But . . . at all. 6.6] *double scored pencil*
7.4 show me . . . understand it. 7.5] *double scored pencil*
9.3 Festuca . . . hang out: 9.5] *double scored ink*
Top of first page: '22'[10] *brown crayon*
Top of last page: 'Ch. 3'[11] *pencil; circled pencil*

[1] See letter from H. C. Watson, 10 March 1857, letter from H. C. Watson to Asa Gray, 13 March 1857, and letter to Asa Gray, [after 15 March 1857].
[2] A duplicate copy of the third part of A. Gray 1856–7 (see letter to Asa Gray, 9 May [1857], and letter from Asa Gray, 1 June 1857).
[3] A. Gray 1857a.
[4] See letter to Asa Gray, 18 June [1857].
[5] Gray refers to CD's belief in the possibility of all organic beings occasionally cross-fertilised (see letters to Asa Gray, [after 15 March 1857] and 18 June [1857]).
[6] See letter to Asa Gray, 20 July [1857].
[7] J. D. Hooker and Thomson 1855.
[8] Jean Louis Auguste Loiseleur Deslongchamps had made such a claim for wheat in Loiseleur Deslongchamps 1842–3 (see *Natural selection*, p. 59).
[9] CD had made the same observation (see *Natural selection*, p. 48).
[10] The number of CD's portfolio of notes on palaeontology and extinction.
[11] This letter was divided into two parts by CD. The first page was placed in portfolio 22 (see n. 10, above) and the second page (paragraph six onwards) was placed with his notes 'On the possibility of all organic beings occasionally crossing' for chapter 3 of his species book (*Natural selection*, pp. 35–91).

From T. H. Huxley 7 July 1857

14 Waverly Place,
July 7th 1857

My dear Darwin—

I have been looking into Brullé's paper, and all the evidence I can find for his generalization (adduced by himself) is contained in the extract which I inclose—[1]

Let us dispose of this first— Paragraph N°. 1. is true but does not necessarily either support or weaken his view, which rests on paragraph N°. 2.— Now this paragraph is a mass of Errors.— You will find in my account of the development of *Mysis* that the antennæ appear before the gnathites are any of them discoverable[2]— & Rathke states the same thing with regard to *Astacus*[3]—and I believe it to be true of *Crustacea* in general.[4]

The second statement, that the legs do not appear until the buccal appendages have taken on their adult form is equally opposed to my own observations & to those of all who have worked in this field.

It would have been very wonderful to me to find Brullé resting such a generalization on such a basis, even had his two affirmations as to matter of fact, been correct But as they are both *wrong*—one can only stand on one's head in the spirit—

Next as to the converse proposition marked 3) It is equally untrue— From the Antennules backwards the appendages in *Mysis* & in *Astacus* appear in regular order from before backwards wholly without respect to their future simplicity or complexity—and, what is still worse for M. Brullé, the opthalmic peduncles, which as you know well are the most rudimentary & simple of all the appendages in the adult make their appearance at the most very little later than the mandibles & increase in size at first out of all proportion to the other appendages

M. Brullé bases his whole generalization upon what he supposes to occur in the *Crustacea*—whereas the development of both *Astacus* & *Mysis*—affords the most striking refutation of his views Tant pis pour Brullé!

And now having *bruler'd* Brullé (couldn't help the pun) I must say that I can find no support for his generalization elsewhere— There are two organs in the *Vertebrata* whose developmental history is especially well qualified to test it—the Heart & the nervous system—both presenting the greatest possible amount of variation in their degree of perfection in different members of the vertebrate series— The heart of a Fish is very simple as compared with that of a Mammal & a like relation obtains between the brains of the two— If Brullés doctrine were correct therefore the Heart & Brain of of the Fish should appear at a later period relatively to the other organs than those of the Mammal— I do not know that there is the least evidence of anything of the kind— On the contrary the history of development in the Fish & in the Mammal shews that in both the relative time of appearance of these organs is the same or at any rate the difference if such exist is so insignificant as to have escaped notice—

With regard to Milne Edwards views—I do not think they at all involve or bear out Brullé's.[5] Milne Edwards says nothing, as far as I am aware about the relative time of appearance of more or less complex organs— I should not understand Milne Edwards doctrine as you put it in the passage I have marked:[6] he seems to me to say that, not the *most highly complex*, but the most *characteristic* organs are those first developed—[7] Thus the chorda dorsalis of vertebrates—a structure characteristic of the group but which is & remains excessively simple, is one of the earliest developed— The animal body is built up like a House—when the judicious builder begins with putting together the simple rafters— According to Brullés notion of Nature's operations he would begin with the cornices, cupboards, & grand piano—

It is quite true that "the more widely two animals differ from one another the earlier does their embryonic resemblance cease"[8] but you must remember that the differentiation which takes place is the result not so much of the development of new parts as of the *modification of parts already existing and common to both of the divergent types*—.

I should be quite inclined to believe that a more complex part requires a longer time for its development than a simple one—: but it does not at all follow that it should appear *relatively* earlier than the simple part. The Brain, I doubt not, requires a longer time for its development than the spinal cord— Nevertheless they both appear together—as a continuous whole, the Brain continuing to change after the spinal cord has attained its perfect form

The period at which an organ appears therefore, seems to me not to furnish the least indication as to the time which is required for that organ to become perfect

You see my verdict would be that Brullés doctrine is quite unsupported—nay is contradicted by development—so far as animals are concerned & I suspect a Botanist would give you the same opinion with regard to plants—

Ever yours faithfully | T. H. Huxley

[Enclosure][9]

1) En suivant, comme on l'a fait dans ces derniers temps les phases du developpement des Crustacés, on voit que les pièces de la bouche et des antennes se manifestent avant les pattes; celles ci ne se montrent que par suite des developpements ultérieurs—

2) De leur côté, les antennes sont encore fort peu développées que les pièces de la bouche le sont déjà plus; enfin c'est lorsque les appendices buccaux ont revêtu la forme qu'ils doivent conserver que les pattes commencent à paraître. Il en résulte donc cette conséquence remarquable *que les appendices se montrent d'autant plus tôt que leur structure doit être plus complexe.* On trouve, en outre, dans ces développements divers une nouvelle preuve de l'analogie des appendices. Ainsi les pattes n'ont pas de transformation à subir elles ne se montrent que quand les autres appendices ont déjà revêtu la forme de mâchoires ou d'antennes.

3) Donc dans un animal articulé *les appendices se montrent d'autant plus tard qu'ils ont moins de transformations à subir*: c'est le complement de la loi précédente

On peut par conséquent juger du degré d'importance et de complication d'un appendice par l'époque même a laquelle il commence à se manifester

pp. 282. 283—

CD ANNOTATIONS

0.3 My dear . . . notice— 6.13] *crossed pencil*

6.6 The heart . . . Mammal] 'good' *added pencil*

7.2 Milne Edwards says nothing] 'See to this' *added pencil*; '?' *added pencil*

7.6 Thus the . . . developed— 7.8] *double scored pencil*

9.4 Nevertheless . . . form 9.6] *double scored pencil*

11.1 You . . . plants— 11.3] *double scored pencil*

[1] See the enclosure transcribed following the letter.

[2] T. H. Huxley 1856–7, 14: 639.

[3] Rathke 1840.

[4] Huxley's observations would, therefore, refute Gaspard Auguste Brullé's rule that the more essential and differentiated organs of an animal develop prior to those of less functional importance.

[5] See letter to T. H. Huxley, 5 July [1857], n. 5. In the manuscript CD had sent to Huxley, he cited both Milne-Edwards 1844 and Milne-Edwards 1845 (see *Natural selection*, p. 303).

[6] CD had stated (*Natural selection*, p. 303): 'he [Milne-Edwards] seems to think that according as the organs in question are most developed in any class, the earlier they appear in the embryo in that class'. Huxley marked the passage in the fair copy that CD had sent him (see letter to T. H. Huxley, 5 July [1857], n. 2).

[7] The passage in question (Milne-Edwards 1845, p. 176) reads as follows:

> Chez les Vertébrés, où l'appareil circulatoire doit acquérir une perfection très grande, et doit remplir un des rôles les plus importants, le cœur et les vaisseaux sanguins se forment, dès l'une des premières périodes de la vie embryonnaire, longtemps avant que le tube alimentaire se soit constitué, ou que le petit être en voie de formation ait acquis aucun des caractères propre aux animaux de sa classe.

> For CD's response to Huxley's point, see the following letter.

[8] The quotation is taken from CD's manuscript (*Natural selection*, p. 303).

[9] Huxley copied out the text, with a few alterations and errors, from Brullé 1844, pp. 282–3.

To T. H. Huxley 9 July [1857][1]

Down Bromley Kent
July 9th

My dear Huxley

I am *extremely* much obliged to you for having so fully entered on my point,[2] I knew I was on unsafe ground, but it proves *far* unsafer than I had thought. I had thought that Brulle had a wider basis for his generalisation; for I made the extract several years ago, & I presume (I state it as some excuse for myself) that I doubted it, for differently from my general habit, I have not extracted his grounds.—[3] It was meeting with Barneouds paper which made me think there might be truth in the doctrine.[4] Your instance of Heart & Brain of Fish seems to me very good.—

It was a very stupid blunder on my part, not thinking of the posterior part of the time of development. I shall, of course not allude to the subject, which I rather grieve about, as I wished it to be true;[5] but alas a scientific man ought to have no wishes, no affections,—a mere heart of stone.—

There is only one point in your letter which at present I cannot quite follow you in: **supposing** that Barneoud's (I do *not* say Brulle's) remark were true & universal, ie that the petal which have to undergo the greatest amount of development & modification begins to change the soonest from the simple & common embryonic form of the petal,[6]—if this were a true law, then I cannot but think that it would throw light on Milne Edwards' proposition that the wider apart the classes of animals are, the sooner do they diverge from the common embryonic plan,—which common embryonic[7] may be compared with the similar petals in the early bud—the several petals in *one* flower being compared to the distinct but similar embryos of the different classes.—

I much wish, that you w^d so far keep this in mind, that whenever we meet, I might hear how far you differ or concur in this.— I have always looked at Barneoud's & Brulle's proposition as only in some degree analogous.—

With hearty thanks for your very kind assistance. | Your's most truly | C. Darwin

P.S. | I see in my abstract of M. Edwards paper he speaks of "the most perfect & important organs" as being first developed,[8] & I sh^d have thought that this was usually synononymous with the most developed or modified.—

P.S. | Allman's account of the fertilisation of the ova in his F.W. Polyzoa seems dreadfully opposed to "Darwin not an eternal hermaphrodite."[9]

Imperial College of Science, Technology and Medicine Archives (Huxley 5: 50)

[1] Dated by the relationship to the preceding letter.
[2] See preceding letter.
[3] CD's abstract of Brullé 1844 is in DAR 72: 123–4. See also letter to T. H. Huxley, 5 July [1857] and n. 5.
[4] Barnéoud 1846.
[5] Because Huxley had refuted Gaspard Auguste Brullé's 'law', CD decided to omit this discussion from his chapter on 'Laws of variation', renumbering folio 45 as '40 to 45' (see DAR 11.1: 45 and *Natural selection*, p. 303 n. 1).
[6] Barnéoud 1846, pp. 287–8.
[7] In a draft of this letter preserved in DAR 11.1: 41a, this phrase reads: 'which common embryonic ['plan' *del*; 'type' *del*] plan'.
[8] See letter to T. H. Huxley, 5 July [1857], n. 4.
[9] CD had asked Huxley to provide him with cases that might oppose his view that all organic beings crossed, even if only occasionally (see letters to T. H. Huxley, 1 July [1856], 8 July [1856], and 13 [December 1856]). Such a case seemed to be presented by the freshwater Polyzoa described by George James Allman (Allman 1850). Allman stated his grounds for believing that the polyzoans he had dissected were hermaphrodites and that they must also, from lacking 'some orifice through which ova may escape from the cells', be perpetually self-fertilised (Allman 1850, p. 321–5).

To J. D. Hooker   14 July [1857][1]

Down Bromley Kent
July 14th

Please Read last part. soon.

My dear Hooker

The enclosed specimens do *not* concern me, but *perhaps* you would like to see them: they are seedling Lathyrus nissolia. In the 3 or 4 I have looked at lower scale-leaf is

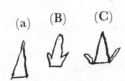

a); the second alternate (B); & the third with lateral stipules (C)

As these seedlings are about 12 days old, the cotyledons seem never to come out, which if true, tells me, what I did not know how subordinately their function is that of a leaf.[2]

Many thanks for your last note & answer to my query.[3]

(N.B I once saw a laburnum tree with terminal flower in each raceme regular or peloric)[4]

I quite agree with what you say on my nut case:[5] the *utmost* I try to show is that the means of distribution are very imperfectly known, & therefore that anyone must be cautious in arguing for double Creations of insular plants (& in extending continents!).—

I write now to supplicate most earnestly a favour, viz the loan of Boreau Flore du Centre de la France either 1st or 2d Edit, last best.—[6] Also "Flora Ratisbonensis by Dr. Furnrohr in Naturhist. Topographie von Regenberg 1839.".[7] If you can *possibly* spare them will you send them at once by enclosed address.— If you have not them, will you send one line by Return of Post. As I must try whether Kippist can anyhow find them, which I fear will be nearly impossible in Linnean Library, in which I know they are.—[8]

I have been making some calculations about varieties &c. & talking yesterday with Lubbock, he has pointed out to me the grossest blunder which I have made in principle, & which entails 2 or 3 weeks lost work; & I am at a dead lock till I have these Books to go over again, & see what the result of calculation on right principle is.—[9] I am the most miserable, bemuddled, stupid Dog in all England, & am ready to cry at vexation at my blindness & presumption.

Ever yours | Most Miserably | C. Darwin

DAR 114.4: 204

[1] The year is established by the reference to John Lubbock (see following letter).

[2] See letter to J. D. Hooker, 1 July [1857].

[3] The letter has not been located. For CD's query, see letter to J. D. Hooker, 5 July [1857].

[4] CD had mentioned this case of pelorism to Hooker in 1847 (*Correspondence* vol. 4, letter to J. D. Hooker, [12 June 1847]).

[5] See letter to J. D. Hooker, 5 July [1857].

[6] Boreau 1840.

[7] Fürnrohr 1839.

[8] Richard Kippist was the librarian of the Linnean Society. During 1857, the society moved its premises into Burlington House, Piccadilly (Gage 1938).

[9] See following letter.

To John Lubbock 14 [July 1857][1]

Down.—
14th

My dear Lubbock

You have done me the greatest possible service in helping me to clarify my Brains. If I am as muzzy on all subjects as I am on proportions & chance,—what a Book I shall produce!—

I have divided N. Zealand Flora as you suggested. There are 339 species in genera of 4 & upwards & 323 in genera of 3 & less. The 339 species have 51 species presenting one or more varieties—[2] The 323 species have only 37: proportionally (as 339:323 :: 51.:48.5) they ought to have had $48\frac{1}{2}$ species presenting vars.— So that the case goes as I want it, but not strong enough, without it be general, for me to have much confidence in.

I am quite convinced yours is the right way; I had thought of it, but sh[d] never have done it, had it not been for my most fortunate conversation with you.

I am quite shocked to find how easily I am muddled, for I had before thought over the subject much, & concluded my way was fair. It is dreadfully erroneous. What a disgraceful blunder you have saved me from. I heartily thank you—[3]

Ever yours | C. Darwin

It is enough to make me tear up all my M.S. & give up in despair.—

It will take me several weeks to go over all my materials. But oh if you knew how thankful I am to you.—

Down House

[1] Dated by the relationship to the letter to J. D. Hooker, 1 August [1857].

[2] CD's calculations on J. D. Hooker 1853–5 are in DAR 16.2: 239a–46.

[3] Since 1855, CD had been engaged in recording the incidence of varieties in various botanical and zoological catalogues. He wished to show that varieties were most frequent in genera that also contained a large number of species; further calculations aimed to demonstrate that these large genera were also geographically widespread and their component species individually abundant. His calculations consistently gave him the correlations he hoped for, but he failed to notice that his results were the entirely artificial consequence of a faulty method that would always have indicated an apparent positive correlation between genus size and any chosen characteristic. CD's manuscripts (DAR 15.2, 16.1, and 16.2) show that in July 1857 he changed his method of calculation. Under Lubbock's guidance, he began to compare the proportion of variable species in large genera with the proportion in small genera using a valid method like the one described in the letter. See J. Browne 1980.

To W. B. Tegetmeier [19 July 1857][1]

Down Bromley Kent
Sunday

My dear Sir

I am very sorry to hear of the anxiety about your child which you have undergone, but now over.—

Mr Gilbert has sent me as a present two young Runts one a fine young Cock of apparently the long-winged kind,[2] but not so long as some I have seen; so I sh^d not want the one you purchased for me, & which in any case you had a right to keep.— I will not accept your young Runt, for it is a pity to have so fine a Bird to kill: I will this very day write to Mr Gulliver & beg him to keep me in mind. for a dead Bird.[3]

Thanks for information on Rumpless—but I am sorry to say that the *one* mottled chicken of the 3 which I hatched has got a very decided tail.[4]

On Thursday morning I will send you the following Birds.

2 young Jacobins bred from (2^d Prize) Mr H. Weirs Birds:[5] they are yellow, but their parents red: I know not whether they will become dark with age.—

2 young Barbs—one red (I had thought both were black): the parents are black & good Birds.

2 white Turbits;—I know not sexes.—

Whether I can send Trumpeters, I am not sure, on account of their having young. Nuns I find I have none at present, & there is only one young in nest at present, but I will not forget.— I have not begun yet to read Poultry Book,[6] but I assuredly will some day, but I have my hands very full of work

Yours very sincerely | C. Darwin

P.S. | I wish I could get chicken on point of hatching of a Breed of Fowls positively & certainly rumpless.—

If ever you fall across a man who keeps Penguin Ducks, I wish you w^d remember that I much want to ask some questions about them.

I am sorry to find I have 4 Cock Trumpeters & only 1 Hen! & all are crossed at present, so I cannot send them. I thought I had a white Cock & Hen.—

New York Botanical Garden Library (Charles Finney Cox collection)

[1] Dated by the reference to CD's continuing attempts to acquire a specimen of Mr Gulliver's runts (see n. 3, below) and the relationship to the letter to W. B. Tegetmeier, 23 June [1857].
[2] Henry Gilbert was a dentist and pigeon-fancier.
[3] CD had been trying to acquire one of Mr Gulliver's runts for some time (see letters to W. B. Tegetmeier, 21 September [1856], 6 February [1857], and [18 June 1857]).
[4] See letter to W. B. Tegetmeier, 23 June [1857].
[5] Harrison William Weir was a noted breeder of fancy pigeons.
[6] Tegetmeier ed. 1856–7. See letter to W. B. Tegetmeier, 25 [June 1857].

To Asa Gray 20 July [1857][1]

Down Bromley Kent
July 20th

My dear Dr Gray

What you say about extinction, in regard to *small* genera & local disjunction, being hypothetical seems very just.[2] Something direct, however, could be advanced on this head from fossil shells; but hypothetical such notions must remain. It is not a little egotistical, but I sh^d like to tell you, (& I do not *think* I have) how I view my work. Nineteen years (!) ago it occurred to me that whilst otherwise employed on Nat. Hist, I might perhaps do good if I noted any sort of facts bearing on the question of the origin of species; & this I have since been doing. Either species have been independently created, or they have descended from other species, like varieties from one species. I think it can be shown to be probable that man gets his most distinct varieties by preserving such as arise best worth keeping & destroying the others,—but I sh^d fill a quire if I were to go on. To be brief I *assume* that species arise like our domestic varieties with *much* extinction; & then test this hypothesis by comparison with as many general & pretty well established propositions as I can find made out,—in geograph. distribution, geological history—affinities &c &c &c.. And it seems to me, that **supposing** that such hypothesis were to explain general propositions, we ought, in accordance with common way of following all sciences, to admit it, till some better hypothesis be found out. For to my mind to say that species were created so & so is no scientific explanation only a reverent way of saying it is so & so. But it is nonsensical trying to show how I try to proceed in compass of a note. But as an honest man I must tell you that I have come to the heteredox conclusion that there are no such things as independently created species—that species are only strongly defined varieties. I know that this will make you despise me.— I do not much underrate the many *huge* difficulties on this view, but yet it seems to me to explain too much, otherwise inexplicable, to be false. Just to allude to one point in your last note, viz about species of the same genus *generally* having a common or continuous area; if they are actual lineal descendants of one species, this of course would be the case; & the sadly too many exceptions (for me) have to be explained by climatal & geological changes. A fortiori on this view (but on exactly same grounds) all the individuals of the same species sh^d have a continuous distribution. On this latter branch of the subject I have put a chapter together, & Hooker kindly read it over:[3] I thought the exceptions & difficulties were so great that on the whole the balance weighed against my notions, but I was much pleased to find that it seemed to have considerable weight with Hooker, who said he had never been so much staggered about the permanence of species.—[4]

I must say one word more in justification (for I feel sure that your tendency will be to despise me & my crotchets) that all my notion about *how* species change are derived from long-continued study of the works of (& converse with) agriculturists & horticulturists; & I believe I see my way pretty clearly on the means used by nature to change her species & *adapt* them to the wondrous & exquisitely beautiful contingencies to which every living being is exposed.

Thank you *much* for what you say about *possibility* of crossing of the grasses: I have been often astounded at what Botanists say on fertilisation in the bud: I have seen *Cruciferæ* mentioned as instance, which every gardener knows how difficult it is to protect from crossing!

What you say on Papilionaceous flowers is very true; & I have no facts to show that varieties are crossed; but yet (& the same remark is applicable in a beautiful way to Fumaria & Diclytra as I noticed many years ago) I must believe that the flowers are constructed partly in direct relation to the visits of insects; & how insects can avoid bringing pollen from other individuals I cannot understand.[5] It is really pretty to watch the action of a *Humble*-Bee on the scarlet Kidney Bean, & in this genus (& in Lathyrus grandifloris) the honey is so placed that the Bee invariably alight on that *one* side of the flower towards which the spiral pistil is protruded (bringing out with it pollen) & by the depression of the wing-petal is forced against the Bee's side all dusted with pollen. N.B if you will look at bed of scarlet Kidney Bean you will find that the wing-petals on the *left*-side alone are all scratched by the tarsi of the Bees.[6] In the Broom the pistil is rubbed on centre of back of Bee &c. &c.— I suspect there is something to be made out about the Leguminosæ which will bring the case within **our** theory;[7] though I have failed to do so.

Our theory will explain why in vegetable & animal kingdom the act of fertilisation even in hermaphrodites usually takes place sub-jove, though thus exposed to the *great* injury from damp & rain.[8] In animals in which the *semen* cannot, like pollen, be *occasionally* carried by insects or wind; there is *no case* of *land*-animals being hermaphrodite without the concourse of two individuals.

But my letter has been horribly egotistical: but your letters always so greatly interest me; & what is more they have in simple truth been of the *utmost* value to me.

Your's most sincerely & gratefully | C. Darwin

Gray Herbarium of Harvard University

[1] Dated on the basis of a note about kidney beans (see n. 6, below), the substance of which is repeated in the letter.

[2] See letter from Asa Gray, 7 July 1857.

[3] CD had asked Joseph Dalton Hooker to read and comment on the draft of his chapter on geographical distribution (see letters to J. D. Hooker, 30 July [1856] and [16 October 1856]).

[4] See letter from J. D. Hooker, 9 November 1856.

[5] CD alluded to the cross-pollination of *Fumaria* by insects in his species book (*Natural selection*, p. 53).

[6] This sentence was added in the margin. In DAR 49: 47, there is a note dated 'July 19/57' that reads: 'The left-wing-petal flower *(to you [*interl*] *facing* *it [*above del* 'you']) of Kidney Bean are all scratched & disfigured by the tarsi of the Bees.—'.

[7] That is, the theory that all organic beings must occasionally cross-fertilise.

[8] Literally, under Jupiter (the sky-god), that is, in the open air. CD maintained that although fertilisation in the bud did occur, it was never normal and regular (see letter to Asa Gray, 18 June [1857]).

To W. E. Darwin 21 [July 1857][1]

Down.
21st.

My dear old Willy or William

I am delighted that you went to Manchester, & had so prosperous an expedition.[2] You seem to have worked capitally & seen it well. We are amused at your adoration of the haughty Lady.[3] I quite agree with your admiration of Gainsborough's portraits: one of the Pictures which has ever most struck me is a portrait by him in the Dulwich Gallery.— By the way how stupid it has been in us never to have suggested your riding to Dulwich & seeing the capital publick Gallery there.[4] Then, again, there are some few good pictures at Knole.—[5]

You want a jobation about your handwriting—dreadfully bad & not a stop from beginning to end! After severe labour in deciphering we *rather think* that your outlay was, 1 · 12 · 0 & accordingly I send that, but I hope it is too little to punish you for such a scrawl. I am glad that you were tipped but that makes no difference in my repaying your outlay.—

By the way have you no paper, so that you cross your letter, or do you think your handwriting is too clear? You want pitching into severely.—

I have had a letter from M[r] Mayor[6] (about his Bankers mistake) in which he says he has heard so grand an account of your future master's, M[r]. Temple, attainments, that he wants to persuade me to leave you at Rugby till October.—[7] M[r] Mayor says he shall *very much* miss you.— Think over this well & deliberately, & do not be guided by fleeting motives. You shall settle for yourself, whatever you think will be really best, not pleasantest, shall be done— The school would be least trouble & expence; but I cannot help *rather* thinking that the Tutor would get you on best.[8] Let me hear before very long. And if you still think of tutor, talk to Mayor, & then I will write to the Tutor to know whether he can take you. When ought you to go? & what summer vacations?

Your most affect[te] | C. Darwin

PS. | Though I have abused you so for your bad writing, I must say that you were *very* good to send us such a *capital* account of all you did & saw.—

DAR 210.6

[1] The date is provided by an entry in CD's Account book (Down House MS), added to an earlier one of 10 July 1857, recording expenses for 'Willy to Manchester'. On 22 July, there was a further entry of a payment to 'Mayor' for William's school fees (see nn. 6 and 8, below).

[2] The Art-Treasures Exhibition in Manchester, opened by Prince Albert on 5 May 1857 and visited by Queen Victoria on 29 June, contained over 5,000 paintings and drawings, over 1,000 of which were old masters. By the time the exhibition closed on 17 October, over a million people had attended. See *The Art-Treasures examiner: a pictorial, critical, and historical record of the Art-Treasures Exhibition, at Manchester, in 1857* (Manchester, 1857).

[3] Thomas Gainsborough's 'Portrait of the Hon. Mrs Graham' was exhibited in the gallery of modern paintings at the Manchester Art-Treasures Exhibition. The reference to the 'haughty Lady' may have come from a description in one of the guide-books to the exhibition: 'Who is there will bid

farewell to our glorious Exhibition without taking a last fond look at pretty Mistress Graham . . . She looks petted and spoiled, made haughty by flattery and adulation.' (*The Art-Treasures examiner*, p. 296).

4 The gallery at Dulwich College, built by John Soane, was the first public art gallery in England. CD had visited the picture gallery there in September 1847 (see *Correspondence* vol. 4, letter to J. D. Hooker, [12 September 1847]). Of the Gainsborough portraits at Dulwich, the most famous was the double portrait of Elizabeth Ann and Maria Linley.

5 Knole House, near Sevenoaks, Kent.

6 Robert Bickersteth Mayor was William's housemaster at Rugby School.

7 Frederick Temple, who had obtained a double first class in mathematics and classics at Oxford and was a fellow of Balliol College, became headmaster of Rugby School on 12 November 1857.

8 CD refers to the forthcoming school year that ran from August 1857 to August 1858. He was undecided about whether to keep William at Rugby until 1858 or whether to take him away from school at the end of 1857 and send him to a private tutor for a few months prior to entering Cambridge University in October 1858. If the autumn term was to be William's last at Rugby, his housemaster would need advance notice. However, CD was still undecided in October (see letter to W. E. Darwin, 29 [October 1857]).

To T. C. Eyton   22 [July 1857][1]

Down Bromley Kent
22^{d}

Dear Eyton

I have been rather amiss in not before sending you the W. African Dog-skin; but it shall be sent off this very night.— I was assured by Dr Daniell that it was a very characteristic specimen of the *pure* native dog of Sierra Leone &c.[2]

Dear Eyton | Yours very sincerely | C. Darwin

American Philosophical Society (148)

1 The date is conjectured by the relationship to the letter to T. C. Eyton, 9 June [1857], in which CD offered to send Eyton the skin of a domestic dog of West Africa. Since CD states that he will send the skin 'this very night', the letter must have been written on a Wednesday, before the weekly carrier departed for London at 2 AM on Thursday morning. Wednesday, 22 July, is the most probable date.

2 William Freeman Daniell provided CD with several specimens of domestic animals and fowls from West Africa (see letter to T. C. Eyton, 5 October [1856]).

To *Gardeners' Chronicle*   [before 25 July 1857][1]

[Down]

The subject of Deep Wells has been sometimes discussed in your columns.[2] I have a well 325 feet deep, and the 12-gallon bucket actually weighs 40 lbs. For many years I used a chain weighing 232 lbs; this, with the water, itself 96 lbs., amounts to 481 lbs. I have made an enormous saving of labour by using for the last half year Newall's patent wire rope. Now, will any one have the charity to say from experience whether there could not be a great saving in the weight of the bucket. Would zinc, or gutta percha, or leather serve? The bucket must be strong enough to withstand being occasionally dashed against the side of the well. Or must I stick to my old substantial oaken friend?

    C. D.

Gardeners' Chronicle and Agricultural Gazette, 25 July 1857, p. 518

[1] The original letter has not been found. It was published in the 25 July 1857 issue of *Gardeners' Chronicle and Agricultural Gazette*.

[2] CD had contributed to this earlier discussion about wells (see *Correspondence* vol. 5, letter to *Gardeners' Chronicle*, [before 10 January 1852]).

To W. B. Tegetmeier   27 July [1857][1]

Down Bromley Kent

July 27th

My dear Sir

    I have to thank you for two notes.— Pray never speak of obligation to me, as I am *certain* the balance lies quite the other way.[2]

    I am glad you like the Barb: I shall most probably have more to offer before long.— I am sorry to hear the Jacobins are so poor; I never looked at the eyes.— Were the white Turbits or owls (I forget which they are called) good: they came from a Lady who is & ought to be a good Fancier.— Anyhow they are easily killed.—

    Very sincere thanks for the offer of Gulliver's Runt, which as you offer it so very kindly, I will gladly accept.— I sh^d not kill it till next year; if you think the Hen w^d be quite full-grown by that time, she w^d do as well as the Cock, otherwise the Cock w^d be better, but you must not sacrifice the Cock if of use to you.—

    With respect to the Ghondook Hen, if you really do not intend to breed from her, I sh^d certainly be very glad of her for skeleton;[3] & she c^d be sent to Nag's Head,[4] recently killed so as to come sweet.—

    I think you had better send the Birds to the Nag's Head. I suppose you c^d get messenger to go for about 1^s · 6^d & I c^d repay you by stamps as that w^d be safer & as cheap as sending up to Tottenham C^t Road & your sending there likewise.[5]

    My dear Sir | with thanks | Your's very sincerely | Ch. Darwin

New York Botanical Garden Library (Charles Finney Cox collection)

[1] Dated by the relationship to the letter to W. B. Tegetmeier, [19 July 1857].

[2] CD had given Tegetmeier a number of pigeons that he no longer needed for his researches (see letter to W. B. Tegetmeier, [19 July 1857]).

[3] CD described the skull of a Ghoondook specimen in *Variation* 1: 265. The Ghoondooks are a subbreed of the crested Polish fowls originally bred in Turkey.

[4] The Nag's Head was an inn in south London where parcels were collected every Thursday by the Down carrier, George Snow.

[5] Tegetmeier had moved from Wood Green to Muswell Hill in 1856 (E. W. Richardson 1916, p. 47).

From Asa Gray [August 1857][1]

No one can have worked at systematic botany as long as I have, without having many misgivings about the definiteness of species.[2] My notions about *varieties* are I believe just what you would have them See Sill. Jour. Sci. Jan. 1856. p. 136.:[3] i.e. I believe every constitutional variety has a strong tendency to be perpetuated by seed, and the 2^d & 3^d generations a stronger tendency still to transmit their inherited peculiarities. And when we see that every plant man takes in hand developes into varieties with readiness, when favorably circumstanced, we cannot avoid suspecting they may do the same thing—.i.e. sport in some way in the wild state also;—and that there is some law, some power inherent in plants generally prompting them to originate varieties.— —which is just what you want to come to, and I suppose this is your starting point.

Here you begin then with good, tangible facts; and I am greatly interested to see what is to be made out of them. First, can you get at the *law* of variation? or throw any ⟨*section missing*⟩

There *is* a good deal of fertilization in the bud, in various plants.[4]

I must look at Kidney beans in view of what you say.[5]

As to Fumariaceæ, I can't imagine how, in *Adlumia* for instance, insects can get at the pollen very well, and still less how they can take any to the stigma of other blossoms.

But in *most cases* cross-fertilization seems a most likely thing to happen—

Kindly post the enclosed. I write in greatest haste, and am, with the *highest regard*

Ever Yours most faithfully | A. Gray

Incomplete
DAR 165: 101, 100

CD ANNOTATIONS
Top of first page: 'Ch. 4' *ink*
Bottom of last page: 'Ch. 3.' *ink, circled ink*;[6] 'Adlumia' *ink, circled ink*

[1] This letter falls between the letters to Asa Gray, 20 July [1857] and 5 September [1857].
[2] Gray is responding to CD's letter to Asa Gray, 20 July [1857], in which CD revealed his belief in the transmutation of species.
[3] Gray added this reference to A. Gray 1856b in the margin of the letter in pencil.
[4] See the earlier discussion of this point in the letter to Asa Gray, 18 June [1857], and the letter from Asa Gray, 7 July 1857.
[5] See letter to Asa Gray, 20 July [1857].
[6] The chapter numbers refer to chapter 3, 'On the possibility of all organic beings occasionally crossing', and chapter 4, 'Variation under nature', of CD's species book (*Natural selection*, pp. 35–91, 95–171).

To J. D. Hooker 1 August [1857][1]

Down Bromley Kent
Aug 1st

My dear Hooker

It is a horrid bore you cannot come soon, & I reproach myself that I did not write sooner, but I had fancied M[rs]. Hooker's confinement was later.[2] N.B. I found that I could do with giving much less Chloroform last time, for I never gave it till she skriked out for it, & yet she never suffered at all.—[3]

How busy you must be! with such a heap of Botanists at Kew.— Only think I have just had letter from Henslow, saying he will come here between 11th & 15th! Is not that grand?[4] Many thanks about Furnrohr: I must humbly supplicate Kippist to search for it.: he most kindly got Boreau for me.—[5]

I am got **extremely** interested in tabulating according to mere size of genera, the species having any varieties marked by greek letters or otherwise: the result (as far as I have yet gone seems to me one of the most important arguments I have yet met with, that varieties are only small species—or species only strongly marked varieties. The subject is in many ways so very important for me; I wish much you would think of any well-worked Floras with from 1000–2000 species, with the varieties marked. It is good to have hair-splitters & lumpers.—[6] I have done or am doing.

> Babington[7]
> Henslow[8] British Flora
> London Catalogue. H. C. Watson[9] }
>
> Boreau. France
> Miquel.[10] Holland
> Asa Gray.[11] N. U. States
> Hooker.[12] N. Zealand
> ———— Fragment of Indian Flora
> Wollaston[13] Madeira Insects.

Has not Koch published good Germany Flora:[14] does he mark varieties? Could you lend it me? Is there not some grand Russian Flora which perhaps has vars. marked.—[15] The Floras ought to be well known.—

I am in no hurry for a few weeks.— Will you turn this in your head, when, if ever, you have leisure. The subject is very important for my work, though I clearly see *many* causes of error.—

Do not forget that you *must* come before very long here.— Most sincerely do I wish M[rs] Hooker through her troubles

Ever yours | C. Darwin

Endorsement: '57'
DAR 114.4: 206–7

1 The year is given by the reference to Frances Harriet Hooker's imminent confinement (see n. 2, below).

2 The Hookers' fourth child, Marie Elizabeth Hooker, was born on 10 August 1857.

3 For CD's earlier remarks concerning the administration of chloroform during childbirth, see *Correspondence* vol. 4, letters to W. D. Fox, [17 January 1850] and [May 1850], and *Correspondence* vol. 5, letter to J. D. Hooker, 27 [June 1854], and letter from J. D. Hooker, [29 June 1854]. The proper dosage of chloroform had recently been the subject of discussion in various medical journals following an increase in the number of deaths resulting from its use.

4 See letters to J. S. Henslow, 10 August [1857], and to John Lubbock, 11 August [1857] and 12 [August 1857].

5 Fürnrohr 1839 and Boreau 1840. See letter to J. D. Hooker, 14 July [1857].

6 By 'hair splitters' CD means those taxonomists who formed many genera, species, and varieties; 'lumpers' tended to group diverse forms together as one species or genus.

7 Babington 1851.

8 Henslow 1835.

9 Watson and Syme eds. 1853.

10 Miquel 1837.

11 A. Gray 1856a.

12 J. D. Hooker 1853–5 and J. D. Hooker and Thomson 1857–8.

13 Wollaston 1854.

14 Koch 1843–4.

15 Ledebour 1842–53 (see letter to J. D. Hooker, 22 August [1857]).

From Henrietta Emma Darwin [2 August 1857][1]

Sunday

My own dear Papa

I look forward with such pleasure to coming home & seeing you all again.[2] Thank you for your very nice letter. It seems a long time since I was at home, it was 9 weeks on Friday. I am very sorry to hear of Mama's headache.

I don't know whether I told you that Mŗ Crawford is gone & Mŗ Butterworth come back. There is a Mŗ Beh come with a very red face but he is not a patient. I believe the Combes go on Monday. This ought to be the full season but it is very empty, only 8

I have very nearly done the "War Trail"[3] & I've done one of the vols of Miss Strickland[4] so I hope Mama has got plenty of books.

I must go out now so Goodbye My dear Papa | Yours | H E D

I feel very well today, but I can't do much walking.

CUL (Richard Darwin Keynes deposit)

1 Dated by Henrietta Darwin's reference to having been away from home '9 weeks on Friday'. Emma Darwin had taken her to Edward Wickstead Lane's hydropathic establishment at Moor Park on Friday, 29 May 1857, where she remained until 7 August (Emma Darwin's diary). Friday, 31 July, would have been nine weeks exactly.

2 Henrietta, aged 14, had been without either of her parents during much of her time at Moor Park: she had last seen Emma Darwin on 21 July (Emma Darwin's diary) and CD had been with her from 16 to 30 June (see 'Journal'; Appendix II).

[3] An adventure story by Mayne Reid (Reid 1857).

[4] This is perhaps a reference to one of Agnes Strickland's twelve volumes of *Lives of the Queens of England, from the Norman conquest* (London, 1840–8). Some years previously, CD had also read one volume (see *Correspondence* vol. 4, Appendix IV, 119: 21b).

To Laurence Edmondston 2 August [1857][1]

Down Bromley Kent

Aug 2^{d}

My dear Sir

I am very much obliged for your letter of the 23^{d}, & for your information on the wild Pigeons, which is of much value to me.— I enclose the stamps for the Rabbit, for which I heartily thank you, as it will prove interesting in comparison with some other insular individuals.—[2] You are so kind as to offer to take the great trouble of sending me some young Rock Pigeons: if I could have had them ten years ago, they w^{d} have been worth their weight in gold to me, but now, I think, I could have hardly any chance of breeding them for a sufficient number of generations to make it worth while to have them; though this does not make it the less kind of your offering to send them.—[3]

About a month ago I wished extremely to ask you a question but I refrained solely because I thought I had already trespassed to a *quite* unreasonable degree on your kindness; but as you offer with so much goodnature to assist me further, I will ask my question, as I do not think it can cost very much trouble, & it is a point on which I have vainly sought for information.

In most parts of N. Europe, small horses or ponys are common (Eel-backs) of a Dun or Mouse-colour. Do such occur in the Shetlands? These duns generally, (perhaps *always* (?)) have a black stripe along the spine, & sometimes, (as I have seen) transverse dark Zebra-like marks on the legs, & I have been assured on perfect authority a transverse shoulder stripe like that of the Ass. Now any information on ponys of this colour w^{d} be of *extreme* interest to me, as it is a most widely geographically extended & ancient breed.—[4]

Is the spinal stripe universal with the duns? Have you *ever* seen the shoulder asinine stripe? are the transverse leg marks common? But *especially* I want to know whether these marks when they occur are plainer on the very young foal before the first hair is shed. And more than all I want to know whether the Dun or Mouse colour ever appears when neither parent is dun, and *as far as known*, no ancestor has been of this colour; though this must always be very doubtful.

Is the Dun (with spinal stripe & occasional other stripes) a very hereditary colour, ie will *one* parent of this colour generally transmit it, when two horses of different colours, one being dun, are crossed? When duns & other colours are crossed what colour results? I have written to Norway to beg for information;[5] & if you can give me any (& coming from you, I know it will be reliable) it will in truth be of extreme use to me. Something analogous occurs in cream- & roan &

chesnut horses; though in the two latter colours I have heard (& seen) only the spinal stripe. I do not know whether these colours ever appear in the Shetland herds.

I can trust only to your great kindness, so often shown to me, to forgive this long note, & I beg to remain, with hearty thanks, My dear Sir | Your's sincerely | Ch. Darwin

Anything about the stripes on Chesnuts interests me almost as much as in Duns.

Endorsement: 'Darwin | 1857'
D. J. McL. Edmondston

[1] The year is given by CD's request for information on dun-coloured horses (see letter to *Gardeners' Chronicle*, [before 13 June 1857]) and by the reference to the rabbit sent by Edmondston (see n. 2, below).

[2] Edmondston's letter has not been found, but see the letter to Laurence Edmondston, 19 April [1857], in which CD asked Edmondston to provide him with a specimen of the feral rabbit of the Shetland Islands.

[3] CD had already received two dead rock pigeons from Edmondston in April (see letter to Laurence Edmondston, 19 April [1857]), the measurements of which he later used in *Variation* 1: 134 n. 5.

[4] CD discussed the dun-coloured ponies, but without mentioning any from the Shetlands, in his species book (*Natural selection*, pp. 329–32). He believed that they, like the occasional reversion of breeds of domestic pigeons to the coloration and markings of *Columba livia*, might illustrate the 'Laws of variation'.

[5] The letter has not been found, but a note in *Variation* 1: 58 n. 31 indicates that CD had written to John Rice Crowe, British consul-general in Norway, who provided him with information on this from several university professors. Crowe had previously corresponded with CD (see *Correspondence* vol. 5). In his species book (*Natural selection*, p. 329), CD stated that dun-coloured ponies were found 'commonly (as informed by a friend) in Norway'.

To J. S. Henslow   10 August [1857][1]

Down Bromley Kent
Aug. 10th

My dear Henslow

I am delighted at your letter just received.

For fear of a mistake, I write to say the Train leaves London Bridge (N. Kent Division) at 5.20'—(not 5.*30'*) & remember how crowded near London Bridge is.— Our last visitors missed the Train.— We will send you back at any hour on Friday you like.

Ever yours | C. Darwin

DAR 93: 122

[1] Dated by the reference to Henslow's proposed visit to Down. Although this letter and the following two letters give no indication of the year in which they were written, it is clear from the letter to J. D. Hooker, 1 August [1857], that Henslow's visit took place in 1857.

To John Lubbock 11 August [1857][1]

Down.
Aug.[t] 11th

My dear Lubbock

I am going to do a most ungracious thing, viz to ask you not to call here on Thursday, for I heard yesterday that a very old friend is coming to spend the day here,[2] & childish as it must seem to you, this is a *very great* exertion to me, & the last straw breaks the camel's back,—especially such a miserable worthless camel as I am.— I am sure I need not say how much pleasure a talk with you gives me whenever you have time & inclination to call.—

Forgive me | Dear Lubbock| Yours very sincerely | Ch. Darwin

I did not write in answer about the Chinese fowls, as I thought I sh[d] see you before very long. When I do see you & hear about them, I will settle whether the *long journey* to Bromley is worth while.

I went yesterday to Crystal Palace for Poultry show, & returned rather soured in temper, as I saw nothing worth seeing in my line.—[3]

Down House

[1] The dates of this and the following letter are based on John Stevens Henslow's proposed trip to Down (see preceding letter, n. 1).

[2] Henslow was the expected visitor (see preceding letter and letter to J. D. Hooker, 1 August [1857]).

[3] Beginning in 1857, the Crystal Palace was the site of a twice yearly poultry show, the largest in the London area (Secord 1981, p. 171).

To John Lubbock 12 [August 1857][1]

Down.
12th

Dear Lubbock

I should like you very much to meet my best of old friends, Prof. Henslow & I have been thinking that perhaps you would come & dine here (if not engaged) on Thursday at 7¼, for I am sure you will forgive me, if tired, going & sitting by myself for ½ an hour after dinner.[2] It is an intolerable evil my being so easily done up.— you may remember the last time you dined here I was not able to appear; but if you will come & take chance, I think you would like Henslow.—

We had intended asking you to persuade M[rs] Lubbock to come with you, but my wife was quite suddenly called to London by dangerous illness of a relation.—[3] & I fear there is hardly a chance of her returning tomorrow by the Omnibus, but I mean to have dinner late for the chance. Do not trouble yourself to answer this, but come if you can & are so inclined.

Ever yours | C. Darwin

Down House

[1] See the preceding letter, n. 1, for the basis of the date. It is further supported by the reference to Emma Darwin having been called to London because of an illness in the family (see n. 3, below).

[2] See the preceding letter. John Stevens Henslow was to visit Down on Thursday, 13 August 1857.

[3] Emma Darwin recorded in her diary on 11 August 1857: 'I came to London to Eliza'. Sarah Elizabeth (Eliza) Wedgwood, Emma's cousin, died one month later, on 11 September 1857.

To J. D. Hooker 22 August [1857][1]

Down Bromley Kent
Aug. 22^{d}

My dear Hooker

Your hand-writing always rejoices the cockles of my heart; though you have no reason to be "overwhelmed with shame", as I did not expect to hear.— We heartily rejoice that the birth is all safe over.—[2]

I write now chiefly to know whether you can tell me how to write to Hermann Schla*genheit* (is this spelt right?) for I believe he is returned to England,[3] & he has Poultry skins for me from W. Elliot of Madras.[4]

I am very glad to hear that you have been tabulating some Floras about varieties. Will you just tell me roughly the result?— Do you not find it takes much time? I am employing a laboriously careful Schoolmaster, who does the tabulating & dividing part into two great cohorts more carefully than I can.[5] This being so I sh^{d} be very glad some time to have Koch[6]—Webb's Canaries[7]—& Ledebour,[8] & Grisebach, but I do not know even where Rumelia is.[9] I shall work British Flora with 3 separate Floras; & I intend dividing the varieties into 2 classes as Asa Gray & Henslow gives the materials,[10] & further A. Gray & H. C. Watson have marked for me the forms, which they consider real species, but yet are very close to others;[11] & it will be curious to compare results. If it will all hold good it is very important for me; for it explains, as I think, all classification, ie the quasi-branching & sub-branching of forms, as if from one root, *big* genera increasing & splitting up &c &c, as you will perceive.— But then comes, also, in what I call a principle of divergence, which I think I can explain, but which is too long & perhaps you would not care to hear.— As you have been on this subject, you might like to hear what very little is *complete* (for my schoolmaster has had 3 weeks holidays): only 3 cases as yet, I see.

Babington British Flora[12]

593 species in genera of 5 & upwards have in a thousand species presenting vars. $\frac{134}{1000}$	593 (odd chance equal) in genera of 3 & downwards have in a thousand presenting vars $\frac{97}{1000}$

Hooker New Zealand[13]

genera with 4 sp. &	with 3 species &
upwards	downwards
$\frac{150}{1000}$	$\frac{114}{1000}$

Godron. Centre France[14]

| 5 species & upwards | 3 species & downward |
| $\frac{160}{1000}$ | $\frac{105}{1000}$ |

I do not enter in details on omitting introduced plants & very varying genera as Rubus, Salix Rosa. &c.—which w^{d} make result more in favour.—

I enjoyed seeing Henslow extremely, though I was a good way from well at the time.[15]

Farewell, my dear Hooker do not forget your visit here sometime.—

Ever yours | C. Darwin

What sort of man is Swede Anderson?[16] Does he care for generalities & other branches of Nat. History.— Is there any chance that he w^{d} investigate a little point on colouring of Horses in Norway for me, about Zebra-like stripes on them.?[17] If there was I w^{d} write out questions, but it depends on nature of man.—

Endorsement: '57'
DAR 114.4: 208

[1] The year is given by the reference to the birth of the Hookers' child (see n. 2, below) and by CD's wish to write to Hermann Schlagintweit (see n. 3, below).

[2] Marie Elizabeth Hooker was born on 10 August 1857.

[3] Hermann Rudolph Alfred Schlagintweit and his brother Robert had recently returned to England after a three-year scientific expedition to India and Tibet. See letter from Robert Schlagintweit, 25 September 1857.

[4] CD had asked Walter Elliot, a member of the council of the governor of Madras, to send him specimens of pigeons and poultry of India. A consignment had reached CD in November 1856 (letter to W. B. Tegetmeier, 3 November [1856]). A payment for the arrival of a second consignment was recorded on 6 September 1857 in CD's Account book (Down House MS).

[5] Ebenezer Norman, the schoolmaster of the national school in Down, was CD's copyist for many years (*LL* 1: 153). The first payment for copying by Norman was recorded by CD in his Account book (Down House MS) on 17 August 1856. There are numerous subsequent entries.

[6] Koch 1843–4.

[7] Webb and Berthelot 1836–50.

[8] Ledebour 1842–53.

[9] Grisebach 1843–4 is a catalogue of the flora of Rumelia, a Turkish possession in the Balkans that includes present-day Bulgaria.

[10] A. Gray 1856a and Henslow 1835.

[11] See *Correspondence* vol. 5, letter to Asa Gray, 8 June [1855], and letter from H. C. Watson, 13 August 1855.

[12] Babington 1851. CD's calculations are in DAR 16.1: 143, 160 and DAR 16.2: 231a–7.

[13] J. D. Hooker 1853–5. CD's calculations are in DAR 16.1: 148 and DAR 16.2: 239a–46.

14 CD made a mistake in this citation: Alexandre Boreau, not Dominique Alexandre Godron, was the author of *Flore du centre de la France* (Boreau 1840). CD's calculations are in DAR 16.1: 137a, 199–203 and DAR 16.2: 204–6. For CD's later use of these calculations, with some revisions, see the tables in *Natural selection*, pp. 149–52.

15 John Stevens Henslow had visited Down on 13 August (see preceding letter).

16 The Swedish botanist Nils Johan Andersson was then working at Kew (see letter to J. D. Hooker, 11 September [1857]).

17 For CD's interest in the dun-coloured ponies of Norway, see letter to *Gardeners' Chronicle*, [before 13 June 1857].

To Asa Gray 5 September [1857]¹

Down Bromley Kent
Sept. 5ᵗʰ

My dear Gray

I forget the exact words which I used in my former letter, but I daresay I said that I thought you would utterly despise me, when I told you what views I had arrived at, which I did because I thought I was bound as an honest man to do so.²

I shᵈ have been a strange mortal, seeing how much I owe to your quite extraordinary kindness, if in saying this I had meant to attribute the least bad feeling to you. Permit me to tell you, that before I had ever corresponded with you, Hooker had shown me several of your letters (not of a private nature) & these gave me the warmest feeling of respect to you; & I shᵈ indeed be ungrateful if your letters to me & all I have heard of you, had not strongly enhanced this feeling. But I did not feel in the least sure that when you knew whither I was tending, that you might not think me so wild & foolish in my views (God knows arrived at slowly enough, & I hope conscientiously) that you would think me worth no more notice or assistance. To give one example, the last time I saw my dear old friend Falconer,³ he attacked me most vigorously, but quite kindly, & told me "you will do more harm than any ten naturalists will do good"— "I can see that you have already *corrupted* & half-spoiled Hooker"(!!). Now when I see such strong feeling in my oldest friends, you need not wonder that I always expect my views to be received with contempt. But enough & too much of this.—

I thank you most truly for the kind spirit of your last letter.⁴ I agree to every word in it; & think I go as far as almost anyone in seeing the grave difficulties against my doctrine. With respect to the extent to which I go, all arguments in favour of my notions fall *rapidly* away the greater the scope of forms considered. But in animals, embryology leads me to an enormous & frightful range. The facts which kept me longest scientifically orthodox are those of adaptation—the pollen-masses in Asclepias⁵—the misseltoe with its pollen carried by insects & seed by Birds the woodpecker with its feet & tail beak & tongue to climb trees & secure insects. To talk of climate or Lamarckian habit producing such adaptations to other organic beings is futile. This difficulty, I believe I have surmounted. As you seem interested in subject, & as it is an *immense* advantage to me to write to you &

to hear **ever so briefly**, what you think, I will enclose (*copied* so as to save you trouble in reading) the briefest abstract of my notions on the **means** by which nature makes her species. Why I think that species have really changed depends on general facts in the affinities, embryology, rudimentary organs, geological history & geographical distribution of organic beings. In regard to my abstract you must take immensely on trust; each paragraph occupying one or two chapters in my Book. You will, perhaps, think it paltry in me, when I ask you not to mention my doctrine; the reason is, if anyone, like the Author of the Vestiges, were to hear of them, he might easily work them in,[6] & then I sh^{d} have to quote from a work perhaps despised by naturalists & this would greatly injure any chance of my views being received by those alone whose opinion I value.—

I have been lately at work on a point which interests me *much*; namely dividing the species of several Floras into two as nearly as equal cohorts as possible—one with all those forming large genera, & the other with the small genera. Thus in your U. States Flora,[7] I make (with omissions of naturalised & of a few protean genera & Carex from its unusual size) 1005 sp. in genera of 5 & upwards, & 917 in genera with 4 & downwards; & the large genera have $\frac{88}{1000}$ varieties & the small genera only $\frac{50}{1000}$. This rule *seems* to be general. & Hooker is going to work out some Floras on same plan.— But to my disgust your *vars*. marked by big-type are only in proportion $\frac{48}{1000}$ to $\frac{46}{1000}$.[8]

Several things have made me confidently believe that "close" species occurred most frequently in the larger genera; & you may remember that you made me the enclosed list.[9] Now to my utter disgust, I find that the case is somewhat the reverse of what I had so confidently expected, the close species hugging the smaller genera. Hence I have enclosed the list. & beg you kindly to run your eye over it, & see whether, not understanding my motive, you c^{d} have attended more to the small than to the large genera: but I can see that this is not probable.[10] And do not think that I want you to "cook" the results for me.— Are the close species *very generally* geographical representative species: this might make some difference?

Lately I examined *buds* of Kidney Bean with pollen shed, but I was led to believe that the pollen c^{d} *hardly* get on stigma by wind or otherwise, except by Bees visiting & moving the wing petals:[11] hence I included a small bunch of flowers in two Bottles, in everyway treated the same: the flowers in one I daily just momentarily moved as if by a Bee; these set 3 fine pods, the other *not one*.[12] Of course this little experiment must be tried again, & this year in England it is too late, as the flowers seem now seldom to set. If Bees are necessary to this flower's *self*-fertilisation, Bees *must* almost cross them, as their *dusted* right-side of head & right legs constantly touch the stigma.

I have, also, lately been reobserving daily Lobelia fulgens— this in my garden is never visited by insects & never sets seeds, *without pollen be put on stigma*. (whereas the small blue Lobelia is visited by Bees & does set seed); I mention this because these are such beautiful contrivances to prevent the stigma ever getting its own pollen; which seems only explicable on the doctrine of the advantage of crosses.

I forget whether I ever said I had received safely M.ʳ Watson's papers.[13] & your Lesson in Botany,[14] for which very many thanks & which I am now reading. But I have never had the *last* part of your paper on Naturalised Plants.[15] If you have a spare copy (which is not likely) I sh.ᵈ be *very* glad of it: otherwise I will borrow Hooker's. I ought to feel ashamed of the length of this letter, knowing how busy you are.

My dear D.ʳ Gray | Believe me with much | sincerity Your's truly | C. Darwin

I will try if I can anyhow get seed of the Adlumia cirrhosa & observe it next summer. Perhaps they have it at Kew.[16]

[Enclosure][17]

I. It is wonderful what the principle of Selection by Man, that is the picking out of individuals with any desired quality, and breeding from them, and again picking out, can do. Even Breeders have been astonished at their own results. They can act on differences inappreciable to an uneducated eye. Selection has been *methodically* followed in *Europe* for only the last half century. But it has occasionally, and even in some degree methodically, been followed in the most ancient times. There must have been, also, a kind of unconscious selection from the most ancient times,—namely in the preservation of the individual animals (without any thought of their offspring) most useful to each race of man in his particular circumstances. The "rogueing" as nurserymen call the destroying of varieties, which depart from their type, is a kind of selection. I am convinced that intentional and occasional selection has been the main agent in making our domestic races. But, however, this may be, its great power of modification has been indisputably shown in late times. Selection acts only by the accumulation of very slight or greater variations, caused by external conditions, or by the mere fact that in generation the child is not absolutely similar to its parent. Man by this power of accumulating variations adapts living beings to his wants,—he *may be said* to make the wool of one sheep good for carpets and another for cloth &c.—

II. Now suppose there was a being, who did not judge by mere external appearance, but could study the whole internal organization—who never was capricious,—who should go on selecting for one end during millions of generations, who will say what he might not effect! In nature we have some *slight* variations, occasionally in all parts: and I think it can be shown that a change in the conditions of existence is the main cause of the child not exactly resembling its parents; and in nature geology shows us what changes have taken place, and are taking place. We have almost unlimited time: no one but a practical geologist can fully appreciate this: think of the Glacial period, during the whole of which the same species of shells at least have existed; there must have been during this period millions on millions of generations.

III. I think it can be shown that there is such an unerring power at work, or *Natural Selection* (the title of my Book), which selects exclusively for the good of each organic being. The elder De Candolle, W. Herbert, and Lyell have written strongly on the struggle for life; but even they have not written strongly enough. Reflect that every being (even the Elephant) breeds at such a rate, that in a few years, or at most a few centuries or thousands of years, the surface of the earth would not hold the progeny of any one species. I have found it hard constantly to bear in mind that the increase of every single species is checked during some part of its life, or during some shortly recurrent generation. Only a few of those annually born can live to propagate their kind. What a trifling difference must often determine which shall survive and which perish—

IV. Now take the case of a country undergoing some change; this will tend to cause some of its inhabitants to vary slightly; not but what I believe most beings vary at all times enough for selection to act on. Some of its inhabitants will be exterminated, and the remainder will be exposed to the mutual action of a different set of inhabitants, which I believe to be more important to the life of each being than mere climate. Considering the infinitely various ways, beings have to obtain food by struggling with other beings, to escape danger at various times of life, to have their eggs or seeds disseminated &c. &c, I cannot doubt that during millions of generations individuals of a species will be born with some slight variation profitable to some part of its economy; such will have a better chance of surviving, propagating, this variation, which again will be slowly increased by the accumulative action of Natural selection; and the variety thus formed will either coexist with, or more commonly will exterminate its parent form. An organic being like the woodpecker or misletoe may thus come to be adapted to a score of contingencies: natural selection, accumulating those slight variations in all parts of its structure which are in any way useful to it, during any part of its life.

V. Multiform difficulties will occur to everyone on this theory. Most can I think be satisfactorily answered.— "Natura non facit saltum" answers some of the most obvious.— The slowness of the change, and only a very few undergoing change at any one time answers others. The extreme imperfections of our geological records answers others.—

VI. One other principle, which may be called the principle of divergence plays, I believe, an important part in the origin of species. The same spot will support more life if occupied by very diverse forms: we see this in the many generic forms in a square yard of turf (I have counted 20 species belonging to 18 genera),—or in the plants and insects, on any little uniform islet, belonging almost to as many genera and families as to species.— We can understand this with the higher, animals whose habits we best understand. We know that it has been experimentally shown that a plot of land will yield a greater weight, if cropped with several species of grasses than with 2 or 3 species. Now every single organic being, by propagating so rapidly, may be said to be striving its utmost to increase in numbers. So it will be with the offspring of any species after it has broken into varieties or sub-species or true species. And it follows, I think, from the foregoing facts that the varying offspring of each species will try (only few will succeed) to seize on as many and as diverse places in the economy of nature, as possible. Each new variety or species, when formed will generally take the places of and so exterminate its less well-fitted parent. This, I believe, to be the origin of the classification or arrangement of all organic beings at all times. These always **seem** to branch and sub-branch like a tree from a common trunk; the flourishing twigs destroying the less vigorous,—the dead and lost branches rudely representing extinct genera and families.

This sketch is *most* imperfect; but in so short a space I cannot make it better. Your imagination must fill up many wide blanks.— Without some reflexion it will appear all rubbish; perhaps it will appear so after reflexion.— | C. D.

This little abstract touches only on the accumulative power of natural selection, which I look at as by far the most important element in the production of new forms. The laws governing the incipient or primordial variation (unimportant except as to groundwork for selection to act on, in which respect it is all important) I shall discuss under several heads, but I can come, as you may well believe, only to very partial & imperfect conclusions.—

Gray Herbarium of Harvard University

[1] The year is established by the relationship to the letter to Asa Gray, 20 July [1857], and by the reference to CD's experiments on controlling fertilisation in kidney beans (see n. 12, below).

[2] See letter to Asa Gray, 20 July [1857].

[3] Hugh Falconer.

[4] Letter from Asa Gray, [August 1857].

[5] In his chapter on the possibility of all organic beings crossing, CD described the 'beautiful adaptation' of the Asclepiadeae and Orchideae that ensured that the pollen masses adhered to the backs of insects that visited the flowers (*Natural selection*, p. 53).

[6] *Vestiges of the natural history of creation* ([Chambers] 1844) had run to ten editions by 1853. Each one was revised in the light of criticisms and new information.

[7] A. Gray 1856a.

[8] CD refers to the varieties considered by Gray to be doubtful, and distinguished in his catalogue (A. Gray 1856a) by being printed in bold type.

[9] CD had asked Gray to prepare a list of 'close species' in genera of North American plants, which Gray did in August 1855 (see *Correspondence* vol. 5, letters to Asa Gray, 8 June [1855] and 24 August [1855]). The list is now in DAR 165: 92/3.

[10] Gray's response to this query has not been found. In DAR 16.1: 126 there is a sheet headed 'Asa Gray. Close *Species*' that appears to indicate Gray's comments. It reads, in part:

> He marked the close. 160 in number (omitting Salix), but several of these were allied to 2 others, these I have omitted, as they were difficult to compare with any standard & which has injured the result for me. Nor have I much confidence in result of any kind, as Dr G. told me in subsequent letter, not to trust much. Moreover they seem to cling to the smaller genera.

[11] See letter to Asa Gray, 20 July [1857].

[12] A note in DAR 49: 48 headed 'Aug 19th 1857.' describes this experiment. See also letter to *Gardeners' Chronicle*, 18 October [1857].

[13] See letter from Asa Gray, 7 July 1857.

[14] A. Gray 1857a.

[15] Gray had already sent CD the third and final part of his 'Statistics of the flora of the northern United States' (A. Gray 1856–7), but CD continued to think (perhaps because of what Gray had told him in the letter from Asa Gray, 4 November 1856) that he had not received a further part that discussed introduced plants (see letters to Asa Gray, 9 May [1857] and 18 June [1857]).

[16] See letter from Asa Gray, [August 1857]. CD was able to obtain the seeds from Kew (see letter to J. D. Hooker, 14 [November 1857]).

[17] The enclosure up to and including the initials 'C. D.', is in the hand of CD's copyist, Ebenezer Norman. CD corrected the copy and filled in gaps left by the copyist. Only those changes that appear to have been made by him are recorded in the Manuscript alterations and comments section. The draft from which the copy was made is in DAR 6 and is reprinted in *Correspondence* vol. 7, Appendix III. This draft, along with extracts from CD's 1844 species essay (DAR 7; *Foundations*, pp. 57–255), formed the basis of the paper read to the Linnean Society on 1 July 1858 as CD's part of the joint contribution with Alfred Russel Wallace entitled 'On the tendency of species to form varieties; and on the perpetuation of varieties and species by natural means of selection' (Darwin and Wallace 1858).

To J. D. Hooker 6 September [1857][1]

Down Bromley Kent
Sept 6th

My dear Hooker

If at home & you can find time, I sh^d be very glad of the Books, (Koch or Ledebour, or Webb or anything) in which varieties are marked to tabulate them.[2]

Can you at same time lend me the Cybele Brit.[3] as Linn. Soc^y is closed.—

About varieties, I have had one slap in face, viz in Miquel's Holland, but he only gives 46 vars, in whole Flora & does not in the least show what they are.—[4]

I have had, also, a far more serious blow in a list of *close* species sent my by A. Gray, which I expected confidently would follow same *apparent* law with varieties; & I have written to him to ask whether he marked the species in the big genera with equal care with the little.[5]

I sh^d be very glad to hear ever so roughly the result of your tabulation of Indian Flora, as some guide to see how things are tending.[6]

I do not know whether you care about Galls, the enclosed wild Carrot seems funny, each with bright orange larvæ within.

Ever yours | C. Darwin

DAR 114.4: 209

[1] Dated by the relationship to the letter to J. D. Hooker, 11 September [1857].

[2] See letters to J. D. Hooker, 1 August [1857] and 22 August [1857].

[3] Watson 1847–59.

[4] CD's calculations on Miquel 1837 are in DAR 15.2: 32–4, DAR 16.1: 146–6a, and DAR 16.2: 218–222v. They are given in *Natural selection*, p. 150.

[5] See preceding letter.

[6] In the event, Hooker did not make the calculations himself (see letter to J. D. Hooker, 11 September [1857]).

To Secretary, Academia Caesarea Leopoldino-Carolina Naturae Curiosorum
8 September [1857][1]

> Down Farnborough Kent
> Sept. 8th.

Sir

I beg most respectfully to acknowledge the receipt of your letter of Sept. 3d, and to express my sense of the very high honour which the Academia Cæs. Leopold. Car. Nat. Cur. has conferred on me.[2]

In answer to your questions I beg to state that my name is Charles Robert Darwin; that I was born at Shrewsbury on Feby. 12th 1809; that I am Magister Artium Cantabriensis, and Fellow of the Royal, the Geological, Linnean, Royal Geographical, Zoological, and Entomological Societies of London.—

With the highest respect, I have the honour to remain, Sir, | Your obliged & obedient servant | Ch. R. Darwin

To the Honourable Secretary | Acad. Cæs. Leopold. Car. &c &c

Deutsche Staatsbibliotek (Darmstadt 1926.10)

[1] CD was elected a member of the German Academy of Naturalists in 1857 (see n. 2, below).
[2] The German Academy of Naturalists, known as the Leopoldina, was the most prestigious scientific academy in Germany and the first in Europe to elect CD to membership. It was the custom of the academy to give a new member a cognomen, a name celebrated in the branch of science to which he belonged. CD was named after Johann Reinhold Forster (1729–98), the German naturalist who had accompanied James Cook on his circumnavigation of the globe, 1772–5 (*LL* 3: 375 n.). For CD's reaction to his election, see the letter to J. D. Hooker, 11 September [1857].

To W. E. Darwin [before 11 September 1857][1]

> [Down]

know where Long Stratton is.— I have been much pleased at the way Mayor speaks of you.—[2] It seems very odd not having you rushing up & down the House with your Photographs & very dirty hands.—[3] Mamma has been today & yesterday in London to see Aunt Eliza, (who seems very weak & ill),[4] but she comes back tonight. By the way I think, it not amiss that you did not take your Photography, as you wd. have had no time for school work; but it was very kind in Miss Mayor offering a room.— The foundations are laid out for new rooms, & on Monday the Bricklayers begin in earnest.— the room looks jolly & big, & I often feel dreadfully ashamed of my extravagance.[5] How fond Lenny is of metaphors, which he uses so unconsciously, I heard him this morning in sand-walk wood shouting "here is a tree full of nuts—oh such a lot—they are all crushed overboard."[6]

Good Bye my dear old Gulielmus
Your' affect. Father | C. Darwin

Incomplete
DAR 210.6

[1] The date of the letter comes from CD's reference to the illness of 'Aunt Eliza' (see n. 4, below).

[2] Robert Bickersteth Mayor was the mathematics master at Rugby and also William Darwin's housemaster (*Rugby School register*).

[3] There are several entries in CD's Account book (Down House MS) beginning in July 1857 for expenditures on photographic equipment for William Darwin. William had returned to school on 20 August (Emma Darwin's diary).

[4] Sarah Elizabeth (Eliza) Wedgwood, Emma Darwin's cousin, died in London on 11 September 1857.

[5] In CD's Account book (Down House MS), there is an entry on 6 September 1857 for 'Laslett Repairs' and one on 10 September for 'Laslett Advanced for Dining Room'. Isaac Laslett was the general builder and bricklayer in Down. A new dining-room and bedroom were being added to Down House (see letter to W. D. Fox, 30 October [1857]).

[6] Leonard Darwin, then aged 7, was known in the family for his amusing sayings (see *Correspondence* vol. 4, Appendix III).

To J. D. Hooker   11 September [1857][1]

Down.
Sept. 11th

My dear Hooker

The magnificent & awful Box of Books arrived quite safely this morning, & I thank you *heartily* for so valuable assistance.[2]

I shall not, of course, try to do all, but will invest a handsome sum with our Schoolmaster,[3] & will see that the Books are covered & are taken scrupulous care of. But it is slow work, & if I keep the Books too long you must order them home. I have not yet fixed what to begin with. I will, hereafter, do the Indian Flora, as I had intended;[4] I understood that you had tabulated it for same end, but I suppose I quite misunderstood you, & was at the time surprised that you thought it worth so much of your valuable time.

I have to thank you & Dr Seeman (& please give him my particular thanks) for the honour of being elected a member of the Soc. Cæs. Leopold &c. &c.:[5] I confess that I know nothing of this Socy but no doubt it is an honour.

I have just finished Asa Gray's Lessons in Botany:[6] how wonderfully clear they are: so clear & simple that I infer that he must have passed over all difficulties: at least my very ancient remembrance of similar works tells me that Botany was formerly much more complicated.

I will see how the present Books work out about varieties before I think of De Candolle:[7] I am not at all sure that I see my way theoretically when the whole range of a plant is taken in, for then come in geographical representatives, which may *perhaps* form a somewhat different case from a species in the same country splitting up into varieties.

I enclose questions for Anderson, & if he can help me, it will be valuable assistance.—[8]

I have just been writing an audacious little discussion, to show that organic beings are not perfect, only perfect enough to struggle with their competitors; & I have been giving the pollen-case of Coniferæ, which we talked over; & Drone Bees give a good parallel in animals—viz 2000 to fertilise 2 or 3 Queens, & the act causing the inevitable death of the Male![9]

I am so very sorry that you cannot come soon here.

Farewell | My dear Hooker | C. Darwin

If you can remember, will you ask Henslow to enquire whether Water-Fowl have ever been seen in Ransome's great tank, when shells reappear.—[10]

P.S. Speaking of tank & water-fowl reminds me to mention as bearing on water-waders transporting seeds, that in some mud, got on Feb. 10 from 3 or 4 spots under water on edge of small pond (& therefore no *one* nest of seeds)—that out of this mud, which when stiff filled break-fast cup & when dried at ordinary heat, weighed 6¾ oz, I have picked out, during last six months 537 plants! many different kinds of plants appeared.—[11]

[Enclosure]

Dun or Mouse-coloured Horses are said to be common in Norway.

Have they invariably a dark line along the spine? How generally have they the transverse Stripe on the shoulder like that on ass; & how generally transverse zebra-like stripes on legs?

Is this colour (with or without the stripes) particularly inheritable? Must both parents be dun to produce a foal of this colour? Does this colour ever appear in the offspring, when *both* parents are of a different colour?

Especially I want to know, whether the stripes, such as those on the legs, are plainer on the foal, before it has shed its first coat of Hair.— If such be the case, do the leg- or shoulder-stripes ever quite disappear when the horse grows old?

Any other information on Horses of this colour would be very valuable to me.— C. Darwin

Down Bromley Kent | Sept. 11ᵗʰ.—

DAR 114.4: 211, DAR 115.1: 73a

[1] Dated by the reference to CD's election to the German Academy of Naturalists (see letter to the Secretary, Academia Caesarea Leopoldino-Carolina Naturae Curiosorum, 8 September [1857], and n. 5, below).

[2] See letters to J. D. Hooker, 1 August [1857], 22 August [1857], and 6 September [1857].

[3] Ebenezer Norman made extensive tabulations from the books sent by Hooker during the last quarter of 1857. The tabulations are in DAR 15.2, 16.1, and 16.2.

[4] CD's calculations on data in J. D. Hooker and Thomson 1855 are in DAR 16.2: 230. For the final figures, see *Natural selection*, p. 151.

[5] Berthold Carl Seemann and his brother Wilhelm E. G. Seemann edited *Bonplandia*, the botanical journal of the Academia Caesarea Leopoldino-Carolina Naturae Curiosorum. Berthold Seemann, who worked at Kew, was elected an adjunct member of the academy in 1857.

[6] A. Gray 1857a.

[7] Candolle and Candolle 1824–73.

[8] Nils Johan Andersson (see letter to J. D. Hooker, 22 August [1857]).

[9] CD was preparing chapter 8 of *Natural selection*. For the passage referred to, see *Natural selection*, pp. 381–2, and *Origin*, pp. 202–3.

[10] The 'great tank' was probably part of James and Robert Ransome's agricultural works in Ipswich.

[11] The case is reported in CD's Experimental book, pp. 20–1 (DAR 157a).

To T. H. Huxley 15 September [1857][1]

Down Bromley Kent
Sept 15th

My dear Huxley

I must just thank you for your three last Lectures which I have read with much interest (& have forwarded to J. Lubbock), & for your magnificent compliment to me.—[2] I declare you will turn my head right round. You have given, as it seems to me, a capital account of the Cirripedes. I have been glad to read what you say on the value of the Group; & I daresay you are right; but how difficult, not to say impossible it is to classify the higher groups.[3] Take the Crustacea & see what differences in opinion in the half-dozen best judges, without much difference in the facts they go on.

I am, also, particularly obliged for the Lecture on the Nerves: which has struck me as *eminently* curious & interesting; & all new to me.—[4]

I suppose you will soon set off for a Holiday or perhaps are gone,[5] so I have marked this note not to be forwarded.— I hope Mrs Huxley & the wonderful Baby are well.—[6]

My dear Huxley | Yours very truly | C. Darwin

Imperial College of Science, Technology and Medicine Archives (Huxley 5: 137)

[1] Dated by the reference to Huxley's 'Lectures' (see n. 2, below).
[2] CD refers to the published version of a lecture Huxley delivered before the Royal Institution on 15 May 1857 (T. H. Huxley 1857a) and to lectures eleven and twelve from his course on general natural history delivered at the Royal School of Mines (T. H. Huxley 1856–7). The second part of lecture twelve covered the Cirripedia, and in a footnote Huxley praised CD's monograph, stating: 'It is one of the most beautiful and complete anatomical and zoological monographs which has appeared in our time' (T. H. Huxley 1856–7, 15: 238). CD had previously forwarded his copies of successive parts of T. H. Huxley 1856–7 to John Lubbock (see letter to John Lubbock, 23 September [1856]).
[3] Huxley believed that the Cirripedia were closely allied to the Branchiopoda (T. H. Huxley 1856–7, 15: 241). CD believed that the resemblances between cirripedes and higher Crustacea were merely analogical (*Correspondence* vol. 5, letter to T. H. Huxley, 8 March [1855]).
[4] T. H. Huxley 1857a.
[5] Huxley and John Tyndall had just returned to London after a second visit to the Swiss Alps to continue their researches on the motion of glaciers (see L. Huxley ed. 1900, 1: 145–7).
[6] Noel Huxley had been born on 31 December 1856 (L. Huxley ed. 1900, 1: 151).

To J. S. Henslow 25 September [1857][1]

Down Bromley Kent
Sept. 25th—

My dear Henslow

I have this minute writing to my son at Rugby,[2] to tell him of your magnificent present. The Copper will especially delight him; & I write now to give you our hearty thanks.[3]

I hope Hooker will come here this autumn & then he will bring the specimens; & if he does not come I must run over to Kew for a day.

Please remember sometime when at Ipswich to enquire whether water-fowl have ever been seen on the great tank of M.ʳ Ransome's.[4]

Ever my dear Henslow | Your's most truly | C. Darwin

DAR 93: 58–9

[1] The year is established by the reminder to Henslow to find out about birds on the tank at Ipswich, as previously mentioned in the letter to J. D. Hooker, 11 September [1857].

[2] William Erasmus Darwin had returned to Rugby School on 20 August 1857 (Emma Darwin's diary).

[3] Henslow proposed to give CD some butterfly specimens (see letter to J. D. Hooker, [23 October 1857]). 'Copper' is short for the rare 'copper-butterfly', the common name of the genus *Lycaena*, so called from the metallic colouring of their wings (*OED*).

[4] See letter to J. D. Hooker, 11 September [1857].

From Robert Schlagintweit 25 September 1857

192, Jermyn Street, | London,
25.ᵗʰ Sept. | 1857

My dear Sir,

It gave us great pleasure to receive your kind note of the 20ᵗʰ, and as I see that you kindly took interest in the communications, made by us lately in Dublin,[1] I am very happy, to give you, as desired, some more particulars of the offspring of the Yak, the Chooboo, whom we really found fertile as far as to the 7.ᵗʰ generation, a statement which is correctly contained in the Athenæum.[2]

As we were very anxious ourselves to have decided the question about the fertility of the Chooboos, we tried to collect as much information as possible; in all instances we met with fertile Chooboos, moreover not unfrequently even with castrated one's.

I shall be most happy to give you more details, as soon as we shall have packed out & worked up our manuscripts & collections[3]

Believe me to be | dear Sir | Yours truly | Robert Schlagintweit

DAR 177 (fragile)

CD note:[4]

British Assoc. Dublin 1857 | H Schlagintweit say cross from Yak with Indian cow fertile to 7ᵗʰ generation & said to be always fertile, called Chooboos, & these 7ᵗʰ [*above del* 'later'] generation, which he saw, were not much altered from the early one which is I think a proof, that his account is true; enquire.

[1] Robert and Hermann Rudolph Alfred Schlagintweit had made several reports on the results of their expedition to India and Tibet, made with their brother Adolphe, at the British Association for the Advancement of Science meeting in Dublin, 26 August to 6 September 1857.

[2] The *Athenæum*, 12 September 1857, p. 1156, gave a description of Hermann and Robert Schlagintweit's 'Notes on some of the animals of Tibet and India' (Schlagintweit and Schlagintweit 1857). See also CD's note, above.

[3] Hermann and Robert Schlagintweit returned from their expedition with a vast collection of geological, botanical, and zoological specimens and ethnographic materials. The results of their travels were published in H. R. A. Schlagintweit *et al.* 1861–6.

[4] This note is with the letter in DAR 177 (fragile). CD has marked it with '17' in brown crayon. The number is that of his portfolio of notes on hybridism.

To T. H. Huxley 26 September [1857][1]

Down Bromley Kent
Sept. 26th

My dear Huxley

Thanks for your very pleasant note.—[2] It amuses me to see what a bug-bear I have made myself to you; when having written some very pungent & good sentences it must be very disagreeable to have my face rise up like an ugly ghost.— I have always suspected Agassiz of superficiality & wretched reasoning powers; but I think such men do immense good in their way.[3] See how he stirred up all Europe about Glaciers.— By the way Lyell has been at the Glaciers, or rather their effects, & seems to have done good work in testing & judging what others have done.[4]

In regard to Blatta I have looked into Westwood,[5] & see Marcel de Serres have written on anatomy of Orthoptera in Annal. du Museum. Tom. 12, 14, & 17.[6]

Audouin & Brullé in Hist. Nat. Ins. vol 9. 1836. but I do not know what work this is.—[7]

Duthiers in Annal. des Sc. Nat. 3d series. Zoolog. Tom. 12th & 13th, has written good paper on homologies & structure of ovipositor in Orthoptera & Hymenoptera,[8] but I cannot remember whether he attended to Male organs; Audouin, I know, has attended especially to male organs in Hymenoptera.—[9] I do indeed pity the *Hen-Cock-Roach* from your description of the male apparatus.—[10]

I am very glad to hear how hearty you have returned from the Glaciers,[11]

Farewell | Yours very truly | C. Darwin

In regard to Classification, & all the endless disputes about the "Natural System which no two authors define in same way, I believe it ought, in accordance to my heteredox notions, to be simply genealogical.—[12] But as we have no written pedigrees, you will, perhaps, say this will not help much; but I think it ultimately will, whenever heteredoxy becomes orthodoxy, for it will clear away an immense amount of rubbish about the value of characters &—will make the difference between analogy & homology, clear.—[13] The time will come I believe, though I shall not live to see it, when we shall have very fairly true genealogical trees of each great kingdom of nature.—

I see Hummel has written on the development of Blattæ, (whether anatomical I know not) in his Essais Entomologique No.ʳ 1. S.ᵗ Petersburgh 1821.[14]

Imperial College of Science, Technology and Medicine Archives (Huxley 5: 54)

[1] The year is given by the reference to Huxley's recent return 'from the Glaciers' (see n. 11, below) and to his criticisms of Louis Agassiz's investigations of glacial phenomena (see n. 3, below).

[2] The letter has not been found.

[3] Huxley had evidently described to CD his forthcoming account of experiments and observations made on glaciers during his visit to the Swiss Alps in August 1857. In a published letter to John Tyndall (T. H. Huxley 1857b), Huxley criticized Agassiz's earlier investigations of the same glaciers (Agassiz 1847).

[4] Charles and Mary Lyell were making a tour of southern Europe. While in Switzerland, Lyell had studied glacial phenomena and had reported some of his observations in letters to Leonard Horner (K. M. Lyell ed. 1881, 2: 249–72).

[5] Westwood 1839–40.

[6] Serres 1809b and 1811 both appeared in the *Annales du Muséum d'Histoire Naturelle*. Another earlier memoir on Orthoptera (Serres 1809a) was published in the *Journal de Physique, de Chimie, et de l'Histoire Naturelle*.

[7] Volume 9 of Audouin and Brullé 1834–7. This series, *Histoire naturelle des insectes*, was intended to run to twelve volumes. Only four volumes, all by Gaspard Auguste Brullé, were published.

[8] Lacaze-Duthiers 1852a and 1852b.

[9] Audouin 1824.

[10] In a paper on *Aphis*, which he was preparing at that time, Huxley stated that the only other adult insect whose sexual anatomy he had studied with care was the common cockroach, *Blatta orientalis* (T. H. Huxley 1857–8, pp. 230–1).

[11] Huxley had returned from the Swiss Alps by 3 September 1857 (L. Huxley ed. 1900, 1: 146).

[12] See letter to T. H. Huxley, 15 September [1857]. For Huxley's response and CD's further remarks, see letter from T. H. Huxley, [before 3 October 1857], and letter to T. H. Huxley, 3 October [1857].

[13] CD explained his views on classification in chapter 13 of *Origin*, where he stated that 'the natural system is founded on descent with modification; that the characters which naturalists consider as showing true affinity between any two or more species, are those which have been inherited from a common parent, and, in so far, all true classification is genealogical' (*Origin*, p. 420).

[14] Hummel 1821, cited in Westwood 1839–40, 1: 421 (see note 5, above).

From A. R. Wallace [27 September 1857][1]

of May last, that my views on the order of succession of species were in accordance with your own, for I had begun to be a little disappointed that my paper had neither excited discussion nor even elicited opposition.[2] The mere statement & illustration of the theory in that paper is of course but preliminary to an attempt at a detailed proof of it, the plan of which I have arranged, & in part written, but which of course requires much ⟨research in English⟩ libraries & collections, a labour which I look ⟨*missing section of unknown length*⟩

With regard to the black Jaguars always breeding *inter se*,[3] it is of course a point not capable of proof, but the black & the spotted animals are generally confined to separate localities, & among the hundreds & thousands of the skins which are articles of commerce I have never heard of a particoloured one having occurred. I *think* there is a difference of form the black being the more slender & graceful animal.

AL incomplete
DAR 47: 145

CD ANNOTATIONS
1.1 of May . . . look 1.7] *crossed ink*
Bottom of fragment: '(Alfred R Wallace. Letter Sept. 1857.)' *ink*

[1] The date is established by CD's annotation and by his reference in the letter to A. R. Wallace, 22 December 1857, to 'your letter of Sept. 27th.—'.
[2] In his letter to A. R. Wallace, 1 May 1857, CD had praised Wallace's paper 'On the law which has regulated the introduction of new species' (Wallace 1855).
[3] CD had inquired about this point in his letter to A. R. Wallace, 1 May 1857.

From Hensleigh Wedgwood [before 29 September 1857][1]

Dear Charles

I do not see that it is at all important to your argument, or rather illustration that the series connecting the unlike relations should be lost in all the other European languages than that in which they may be found. You might consider that language alone and then Head & chief would afford a good illustration in addition to Bishop & the numerals. These are all admitted by every one. Head, OE. heved, AS heafod, G. haupt Goth. haubith Lat capit (is) It. capo Fr. chef E. chief. If we had only E, It & Fr remaining nobody would have guessed it possible that head & chief could be different forms of the same word.

Perhaps one or two striking instances as this & bishop afford a better illustration than a longer series of less decisive ones—[2]

I have often thought that there is much resemblance between language & geology in another way. We all consider English a very mixed language because we can trace the elements into Latin, German &c. but I see much the same sort of thing in Latin itself & I believe that if we were but acquainted with the previous state of things we should find all languages made up of the debris of former tongues just as every geological formation is the grinding down of former continents.[3]

I am going to Hartfield[4] tomorrow to meet Fanny.[5] Mrs Gaskell cannot have them till the 9th which will allow a tidy visit at H—[6]

Adieu | H. W.

DAR 48: 80–1

CD ANNOTATIONS
Top of first page: 'Ch 8'[7] *pencil, circled pencil*
Bottom of last page: 'Day & ['Journa' *del*] Jour | Dies | Diurnes | Giorno | Journal, Jour' *pencil*; 'Episcopus | Obispo | Bishop' *pencil*; 'vescovo | evesque | eveque' *pencil*

[1] The date is based on CD's use of the information given by Hensleigh Wedgwood in the closing pages of his chapter 8, 'Difficulties on the theory of natural selection in relation to passages from form to form' (*Natural selection*, p. 384), which was completed on 29 September 1857 ('Journal'; Appendix II).

[2] Hensleigh Wedgwood was preparing a dictionary of English etymology (H. Wedgwood 1859–65), the first volume of which included the example given in the letter. CD used the case of 'bishop' and 'évêque' in *Natural selection*, p. 384, in order to show how apparently dissimilar words were derived from a common source. The general comparison between animal and plant ancestry and etymology was made again in *Origin*, pp. 422–3.

[3] In H. Wedgwood 1859–65, Hensleigh Wedgwood argued for etymology to be placed on a solid scientific basis, pointing to geology among the physical sciences as a suitable model.

[4] Two of Hensleigh Wedgwood's sisters lived in Hartfield, Sussex: Sarah Elizabeth (Elizabeth) Wedgwood at 'The Ridge' and Charlotte Langton at 'Hartfield Grove'.

[5] Frances Mackintosh Wedgwood, Hensleigh's wife.

[6] The Wedgwoods were friends and distant relations of Elizabeth Cleghorn Gaskell and her husband William. The Wedgwoods' daughter, Frances Julia, had assisted Mrs Gaskell with her biography of Charlotte Brontë (Gaskell 1857) (B. Wedgwood and H. Wedgwood 1980, p. 258). Throughout the summer of 1857, Mrs Gaskell entertained a series of guests who had come to see the Art-Treasures Exhibition in Manchester (A. B. Hopkins 1952, p. 218).

[7] A reference to chapter 8 of CD's species book (see n. 1, above).

To W. B. Tegetmeier   29 September [1857][1]

Down Bromley Kent
Sept. 29th

My dear Sir

Thank you for so kindly always remembering me. I have skeleton of Himalaya;[2] but if you will leave the Head till thoroughily wasted I sh^d. be glad of it as duplicate.—[3] With respect to Pigeons I shall collect no more, for I think for my object I have done enough. Though I am telling a partial story, for I have just written about some Smiters advertised by a Mr Roe at Salisbury.—[4] I have just lost one of the old Barbs the parent of that sent you, so I shall have no more; but I have a grown male, which I can send to you, & will look in course of week or two, whether I have anything else worth sending.

You can make away with the young Cock Scanderoon. Thanks you for your offer of the Cock Rumpless, but I shall not want it. There is only one sort of Fowl, which I sh^d. be glad to get or buy cheap, viz an old Cock Malay, & if you could help me in this I sh^d. be very glad.— Pray take care of the Head of the wild Jungle, as at *some* time I sh^d. like that Back.—

I have heard of arrival of a set of Burmese Fowl-skins, but they are at Berlin, so I suppose I shall not receive them very soon.—[5]

With many thanks— | Your's very sincerely | C. Darwin

New York Botanical Garden Library (Charles Finney Cox collection)

[1] The year is established by a record of a payment in CD's Account book (Down House MS), dated 12 September 1857, for 'Smiters' (see n. 4, below) and by the relationship to the letter to W. B. Tegetmeier, 21 November [1857].

[2] See letter to John Thompson?, 26 November [1856].

[3] CD had previously asked Tegetmeier to send him rabbit specimens (letter to W. B. Tegetmeier, 21 September [1856]). CD refers to the Himalayan breed of rabbits (see *Variation* 1: 108–9).

[4] Smiters are a breed of pigeons remarkable for their manner of flight (*Variation* 1: 156). See also letter
   to James Buckman, 4 October [1857].
[5] Sent by Walter Elliot (see *Correspondence* vol. 7, letter to W. B. Tegetmeier, 17 January [1858]).

To J. D. Hooker 30 September [1857][1]

Down Bromley Kent
Sept. 30th

My dear Hooker

In looking over my scraps I find one from you with some cases of Hybridism
from M[r] Glover of Manchester:[2] Who is he? is he alive & do you know his address,
as I I sh[d] like to write & ask him some questions on one of his crosses of Cereus?[3] Is
he a man to be trusted?—

I hope you are not getting impatient for your Books back;[4] for I have done only
a few of those which I sh[d]. like to do; for it is very slow work, & our Schoolmaster
has only his evenings to spare.— I have chosen

   Koch[5]
   Webb. & B.[6]
   Visiani[7]
   Grisebach[8]
   & Ledebour.[9]

This last will be a tough job; more especially as he gives splendid materials for
working out range of big and small genera. As I have done Britain & France &
U. States, I shall have worked round the N. Hemisphere.—

Hereafter I think I shall borrow 2 or 3 vols. of Decandolle's Prodromus,[10] as you
suggested; & if possible a Flora of Holland;[11] & then I think I shall have taken
ample materials: *as yet* the results go as I like; & my tables will show some
additional results,—as variability & commonness going together, often stated to
be the case, but very strongly demonstrated by some of my tables.[12]

We have been lately taking a very extravagant step & are building a new dining
room & bedroom over; & so are in the midst of brick & rubbish.—

My health has been very indifferent of late; & I have given up on compulsion
going out anywhere.—

It is a strange thing, & I am sure you will sympathise with us, that for the last
ten days our darling little fellow Lenny's health has failed, *exactly* as three of our
children's have before, namely with extremely irregular & feeble pulse; but he is
so much better today that I cannot help having hopes that, unlike the former
cases, it may be something temporary. But it makes life very bitter.—[13]

I hope you enjoyed Manchester & are all the stronger for your trip.— You
never or seldom tell me what you are working at, which I always very much like to
hear; but I daresay the reason is that I ask such lots of questions & you have so
little time to spare.— How I wish you were as idle or rather as busy a man with
free will to do what you like, as I am.—

Farewell my dear Hooker | Ever yours | C. Darwin

Though I work every day, my last two Chapters of rough M.S. have taken me exactly six months![14] Pleasant prospect!

Endorsement: '57'
DAR 114.4: 210

[1] The year is established by CD's reference to the failing health of his son Leonard Darwin (see n. 13, below) and the building work at Down.

[2] Thomas Glover, of Smedley Hill near Manchester, had met Hooker when he visited Manchester in 1836 (L. Huxley ed. 1918, 1: 30). He was an authority on breeding cactuses.

[3] See letter from Thomas Glover, 26 October 1857.

[4] See letter to J. D. Hooker, 11 September [1857].

[5] Koch 1843–4.

[6] Webb and Berthelot 1836–50.

[7] Visiani 1842–52.

[8] Grisebach 1843–4.

[9] Ledebour 1842–53.

[10] Candolle and Candolle 1824–73.

[11] Miquel 1837.

[12] See *Natural selection*, p. 155, and J. Browne 1980.

[13] CD recorded in his 'Journal' (Appendix II) in October 1857: 'In latter part of Sept. for about week, Lenny had very intermittent pulse; but now Oct. 6th seems quite well'. Leonard Darwin was then aged 7. One of CD's fears was that his children had inherited his ill-health (see *Correspondence* vol. 5, letters to W. D. Fox, 7 March [1852], 24 [October 1852], and 17 July [1853]).

[14] CD recorded in his 'Journal' (Appendix II) on 29 September: 'finished Ch. 7 & 8; but one month lost at Moor Park.' These are the chapters entitled 'Laws of variation: varieties and species compared' and 'Difficulties on the theory of natural selection in relation to passages from form to form'.

From T. H. Huxley [before 3 October 1857][1]

Cuviers definition of the object of Classification seems to me to embody all that is really wanted in Science—it is *to throw the facts of structure into the fewest possible general propositions*— This of course leaves out of view & passes by, all questions of pedigree & possible modifications—dealing with existing animals and plants as faits accomplis[2]

I for one believe that a Scientific & logical Zoology & Botany are not at present possible—for they must be based on sound Morphology—a Science which has as yet to be created out of the old Comparative Anatomy—& the new study of **Development** When the mode of thought & speculation of Oken & Geoffroy S. Hilaire & their servile follower Owen,[3] have been replaced by the principle so long ago inculcated by Caspar Wolff & Von Baer & Rathke[4]—& so completely ignored in this country & in France up to the last ten years—we shall have in the course of a generation a science of Morphology & then a Scientific Zoology & Botany will flow from it as Corollaries—

Your pedigree business is a part of Physiology—a most important and valuable part—and in itself a matter of profound interest—but to my mind it has no more to do with pure Zoology—than human pedigree has with the Census— Zoological classification is a Census of the animal world

Ever yours faithfully | T. H. Huxley

Chas Darwin Esq

Incomplete
DAR 205.5 (Letters)

CD ANNOTATIONS
2.2 for they . . . Morphology] 'why from morphology' *added ink*
Top of first page: 'Huxley on Classification' *ink*
Bottom of first page: '11'^5 *brown crayon*

First CD note:[6]

Huxley says that the cleft & loops of arteries in air-breathing mammals is true, not withstanding Baudement.[7] | (**Keep this**)

Second CD note:

No doubt there is classification on mere aggregate resemblance.— But *most* [*interl*] nat. search for something more either tacitly or overtly for [*del illeg*] plan of creation &c.—& this I believe is genealogical.—

Take races of man, suppose them perfectly known, w^d not genealogical be most perfect, even if it did separate some few races from their nearest like form.— Generally no doubt resemblance w^d go with genealogy.—

[1] Dated by the relationship to the following letter.
[2] Huxley is responding to CD's view, as put forward in the letter to T. H. Huxley, 26 September [1857], that classification should be essentially genealogical.
[3] Huxley associated Richard Owen's methodology with Lorenz Oken's *Naturphilosophie* and with the philosophical anatomy of Étienne Geoffroy Saint-Hilaire (see A. Desmond 1982 and di Gregorio 1984).
[4] The study of the developmental history of organisms (*Entwicklungsgeschichte*) that had begun to flourish in Germany owed its origin to the work in embryology of Caspar Friedrich Wolff, Karl Ernst von Baer, and Martin Heinrich Rathke.
[5] The number of CD's portfolio of notes on classification.
[6] Associated with this letter in DAR 205.5 are two separate notes in CD's hand. The first may pertain to the section of the letter that is now missing.
[7] CD refers to the view that at an early stage of development the embryos of mammals possess gill clefts and branchial arteries, somewhat like fish. Émile Baudement had challenged this view in Baudement 1847.

To T. H. Huxley 3 October [1857][1]

Down Bromley Kent
Oct. 3.^d

My dear Huxley.

I know you have no time for speculative correspondence; & I did not **in the least** expect an answer to my last.[2] But I am very glad to have had it, for in my eclectic work, the opinions of the few good men are of great value to me.—

I knew, of course, of the Cuvierian view of Classification, but I think that most naturalists look for something further, & search for "the natural system",—"for the plan on which the Creator has worked" &c &c.— It is this further element which I believe to be simply genealogical.

But I sh^d be very glad to have your answer (either *when we meet* or by note) to the following case, *taken by itself & not allowing yourself to look any further than to the point in question.*

Grant all races of man descended from one race; *grant* that all structure of each race of man were perfectly known—**grant** that a perfect table of descent of each race was perfectly known.— grant all this, & then do you not think that most would prefer as the best classification, a genealogical one, even if it did occasionally put one race not quite so near to another, as it would have stood, if allocated by structure alone. Generally, we may safely presume, that the resemblance of races & their pedigrees would go together.

I sh^d like to hear what you w^d say on this purely theoretical case.

Ever your's very truly | C. Darwin

It might be asked why is development so all-potent in classification, as I fully admit it is: I believe it is, because it depends on, & best betrays, genealogical descent; but this is too large a point to enter on.[3]

Imperial College of Science, Technology and Medicine Archives (Huxley 5: 139)

[1] Dated by the relationship to earlier correspondence with Huxley (see letters to T. H. Huxley, 15 September [1857] and 26 September [1857]).
[2] See preceding letter, which was a reply to CD's letter to T. H. Huxley, 26 September [1857].
[3] In *Origin*, CD included a long section on embryology, arguing that since 'the embryo is the animal in its less modified state', it therefore 'reveals the structure of its progenitor . . . Thus, community in embryonic structure reveals community of descent.' (*Origin*, p. 449).

To James Buckman 4 October [1857][1]

Down Bromley Kent
Oct. 4th

My dear Sir

I want to beg a favour of you & sh^d be very much obliged if you could kindly grant it.— I am keeping all Breeds of Pigeons on account of the subject of Variation, on which I know that you are much interested.[2] I have just ordered a pair of Smiters from a Mr Roe. of Salisbury,[3] who informs me that he got this very rare Breed from M^r **K**inder or **Tr**inder Junr of Cirencester. Now would you be so very kind as to endeavour to find out the gentleman & beg him to answer the questions on the enclosed paper. You would thus confer a great favour on me.

Supposing that for any reason you do not like to do this; will you let me know, & I will ask my cousin Mr Holland of Dumpleton to make the enquiries,[4] but as he is not on the spot, I have ventured to ask you.—

I have lately seen a short abstract in Athenæum of a communication by you on the variation of Plants by culture read before Brit. Assoc.—[5] I feel the deepest & most lively interest in these researches of yours— Will you tell me whether they will be published *in detail* & soon? For I must get the volume whenever published.— Mr Bentham[6] told me sometime ago that you had already published on this subject;[7] will you be so kind, as to give me references of any papers by you on this subject, as I must carefully study all that you have done on this head.—[8]

I hope that you will forgive my troubling you, & believe me | My dear Sir | Yours very faithfully | Ch. Darwin

[Enclosure]

(1) Whence were the Smiters procured: if from abroad what name did they bear?
(2) What are the peculiarities in their habits?
(3) Will they display their peculiarities in a cage, some 5 or 6 yards long by 4 or 5 broad & ten ft high? Or must they be quite free.—
(4). Are the peculiarities the same in *Cock* & *Hen*? are they equally displayed in all; or are some Birds much better than others? I refer of course, only to pure bred Birds.—
(5) Do they breed true in form & colour?

Private collection

[1] Dated by CD's interest in smiter pigeons (see n. 3, below).
[2] Buckman was professor of geology and botany at the Royal Agricultural College at Cirencester from 1848. He worked particularly on varieties of agricultural plants (see nn. 5, 7, and 8, below).
[3] CD recorded a payment of 10s. for smiters on 12 September 1857 (CD's Account book (Down House MS)). See also letter to W. B. Tegetmeier, 29 September [1857].
[4] Edward Holland, who lived in Dumbleton, north Gloucestershire, was CD's second cousin.
[5] Buckman communicated a 'Report on the experimental plots in the botanical garden of the Royal Agricultural College at Cirencester' to the 1857 meeting of the British Association for the Advancement of Science (Buckman 1857). Buckman's paper was reported in *Athenæum*, 12 September 1857, p. 1157. An offprint of Buckman 1857, inscribed 'With the authors compliments.' and containing annotations by CD, is in the Darwin Pamphlet Collection–CUL.
[6] George Bentham.
[7] Buckman had delivered an earlier report to the British Association (Buckman 1856). In this paper, he had concluded that the common agricultural clover, *Trifolium medium*, had merged into *T. pratense*.
[8] There is a note in DAR 47: 65 that reads:

> Such cases as Cowslip & Primrose— Prof. Buckman experiments on Plants are the most hostile to *that part of [interl] my theory *of selection [del] which attributes so much to selection— (so Orchis case?) though favourable to change of some kind— shows effect direct of excess of food Ch 6.

CD refers to Buckman's belief that many agricultural plants usually thought to be distinct species were really only variants that had arisen as a result of cultivation and the differences in fertility of various soils.

To J. S. Henslow 14 October [1857][1]

Down Bromley Kent
Oct. 14th

My dear Henslow

It is a great shame to trouble you about such trifles; but your Myosotis is beginning to sport under my treatment;[2] & I want to know whether it is not an odd thing some of the calyces, (as the one separate one) having 6 nuts.— Some twigs produce green flowers, with the nuts oddly elongated; of which I send a little twig, I do not send more, as I want to save the monstrous seeds.— The shortest scrap in answer anytime w^d much oblige me.

Most truly yours | C. Darwin

DAR 93: 119

[1] Dated by the relationship to the letter to J. S. Henslow, 18 October [1857].
[2] CD's observations are recorded in his Experimental book, p. 31 (DAR 157a) in an entry dated 14 October 1857. It is not clear to which set of *Myosotis* seeds sent by Henslow he refers.

To *Gardeners' Chronicle* 18 October [1857][1]

Mr. Swayne in the 5th volume of the Horticultural Transactions incidentally speaks of the advantage of artificially fertilising the early Bean.[2] Can you tell me to what sort of Bean he refers? (We presume to the Early Mazagan; but we have no special information.)[3] and who has followed this plan, and how has it been effected? My motive for asking is as follows: every one who has looked at the flower of the Kidney Bean must have noticed in how curious a manner the pistil with its tubular keel-pistil curls like a French horn to the left side—the flower being viewed in front. Bees, owing to the greater ease with which they can reach the copious nectar from the left side, invariably stand on the left wing-petal; their weight and the effort of sucking depresses this petal, which, for its attachment to the keel-petal, causes the pistil to protrude. On the pistil beneath the stigma there is a brush of fine hairs, which when the pistil is moved backwards and forwards, sweeps the pollen already shed out of the tubular and curled keel-petal, and gradually pushes it on to the stigma. I have repeatedly tried this by gently moving the wing petals of a lately expanded flower. Hence the movement of the pistil indirectly caused by the bees would appear to aid in the fertilisation of the flower by its own pollen; but besides this, pollen from the other flowers of the Kidney Bean sometimes adheres to the right side of the head and body of the bees, and this can scarcely fail occasionally to be left on the humid stigma, quite close to which, on the left side, the bees invariably insert their proboscis. Believing that the brush on the pistil, its backward and forward curling movement, its protrusion on the left side, and the constant alighting of the bees on the same side, were not accidental coincidences, but were connected with, perhaps necessary to, the fertilisation of the flower, I examined the flowers just before their expansion. The pollen is then already shed; but from its position just beneath the stigma, and from its coherence, I doubt whether it could get on the stigma, without some movement of the wing petals; and I further doubt whether any movement, which the wind might cause, would suffice. I may add that all which I have here described occurs in a lesser degree with Lathyrus grandiflorus. To test the agency of the bees, I put on three occasions a few flowers within bottles and under gauze: half of these I left quite undisturbed; of the other half I daily moved the left wing-petal, exactly as a bee would have done whilst sucking. Not one of the undisturbed flowers set a pod, whereas the greater number (but not all) of those which I moved, and which were treated in no other respect differently, set fine pods with good seeds.[4] I am aware that this little experiment ought to have been repeated many times; and I may be greatly mistaken, but my belief at present is, that if every bee in Britain were destroyed, we should not again see a pod on our Kidney Beans. These facts make me curious to know the meaning of Mr. Swayne's allusion to the good arising from the artificial fertilisation of early Beans. I am also astonished that the varieties of the Kidney Bean can be raised true when grown near each other. I should have expected that they would have been crossed by the bees bringing pollen from other varieties; and I should be infinitely obliged for any information on this head from any of your correspondents. As I have mentioned bees, a little fact which surprised me may be worth giving:—One day I saw for the first time several large humble-bees visiting my rows of the tall scarlet Kidney Bean; they were not sucking at the mouth of the flower, but cutting holes through the calyx, and thus extracting the nectar. I watched this with some attention, for though it is a common thing in many kinds of flowers to see humble-bees sucking through a hole already made, I have not very often seen them in the act of cutting.[5] As these humble-bees had to cut a hole in almost every flower, it was clear that this was the first day on which they had visited my Kidney Beans. I had previously watched every day for some weeks, and often several times daily, the hive-bees, and had seen them always sucking at the mouth of the flower. And here comes the curious point: the very next day after the humble-bees had cut the holes, every single hive bee, without exception, instead of alighting on the left wing-petal, flew straight to the calyx and sucked through the cut hole; and so they continued to do for many following days. Now how did the hive-bees find out that the holes had been made? Instinct seems to be here out of the question, as the Kidney Bean is an exotic. The holes could scarcely be seen from any point, and not at all from the mouth of the flower, where the hive-bees hitherto had invariably alighted. I doubt whether they were guided by a stronger odour of the nectar escaping through the cut holes; for I have found in the case of the little blue Lobelia, which is a prime favourite of the hive-bee, that cutting off the lower striped petals deceived them; they seem to think the mutilated flowers are withered, and they pass them over unnoticed. Hence I am strongly inclined to believe that the hive-bees saw the humble-bees at work, and well understanding what they were at, rationally took immediate advantage of the shorter path thus made to the nectar.[6] *C. Darwin, Down, Bromley, Kent, Oct.* 18.

Gardeners' Chronicle and Agricultural Gazette, 24 October 1857, p. 725

[1] The year is given by the date of publication in *Gardeners' Chronicle and Agricultural Gazette*.
[2] Swayne 1824.
[3] This sentence was added (in square brackets) by the editor of *Gardeners' Chronicle*.
[4] CD's notes on this experiment are in DAR 49: 48.
[5] CD had previously observed hive-bees boring holes in flowers and had written on this to the *Gardeners' Chronicle* in 1841 (see *Correspondence* vol. 2, letter to *Gardeners' Chronicle*, [16 August 1841]).
[6] CD also discussed this subject in *Natural selection*, pp. 475–6.

To J. S. Henslow 18 October [1857][1]

Down Bromley Kent
Oct. 18th

My dear Henslow

I write to thank you for your note & to say that I sh^d be **very** glad to have your Photograph.— You c^d leave it for me at Royal Soc. or Linn. Soc. or Athenæum Club, wherever most convenient to you.—

The plant with green flowers was this year's seedling; but *apparently* only certain twigs on same plant were thus characterised. I had fancied that the 6 seeds was oddest part; but after sending you the specimen, I found flower with 10 sepals & two pistils & 8 or 10 imperfect seeds—; & other flowers with only 3 seeds.—[2]

I feel pretty sure I could make any flower in some degree monstrous in 4 or 5 generations.[3]

I am very glad to hear of the grand success of the Hitcham Hort. Soc. It must be very pleasant to you.—[4]

I noticed the Death of your Aunt.[5] If you keep your health, God grant you may live as long.

Most truly yours | C. Darwin

Endorsement: '1857'
DAR 93: 45–6

[1] Dated by the reference to the death of Henslow's aunt (see n. 5, below).
[2] See letter to J. S. Henslow, 14 October [1857].
[3] An idea of the kinds of experiments CD was conducting in his attempt to break the constitution of plants comes from a letter Joseph Dalton Hooker wrote at CD's urging to George Bentham in the summer of 1857. Hooker asked Bentham: 'Have you ever made any observations on *inducing varieties* by playing tricks with plants? as by high manuring wild species; plucking all their flowers off for several years; pruning; &c. Darwin wants to know who has done such things.' (L. Huxley ed. 1918, 1: 452).

[4] The Hitcham Labourers' and Mechanics' Horticultural Show, held 16 July 1857, was reported in *Gardeners' Chronicle and Agricultural Gazette*, 3 October 1857, p. 679:

> Amidst our numerous harvest homes and happy children's *fêtes*, this gathering maintains, under Professor Henslow, its instructive and pleasant attractions. The lawn in front of the rectory is made gay with flags and streamers, and the visitor comes immediately upon two tents of excellent construction, one of large dimensions for the show, and the other called the Marquee Museum, containing objects of interest on which the Professor gives instructive little lectures in the course of the day.

See also Russell-Gebbett 1977.

[5] Ann Henslow, 'dau. of the late Sir John Henslow, formerly Surveyor to the Navy, and aunt to the Rev. Professor Henslow, of Hitcham Rectory', died at Bildeston, Suffolk, aged 90, on 9 October 1857 (*Gentleman's Magazine* n.s. 3 (1857) 2: 574).

To J. D. Hooker   20 October [1857][1]

Down Bromley Kent
Oct. 20th

My dear Hooker

I am sorry to trouble you, but will you tell me how I can best send back the large Box.[2] Is it not too heavy for Parcels Delivery? Secondly how address it? How to the Library?— You must let me know what the carriage will come to.— I will return all *immediately* I get your answer, except, perhaps 2d, & except 3d & 4th vols. of Ledebour,[3] if you can spare them, but I fear these will be just the Books you are likely most to want. But if you can spare them I will return *each* vol. as soon as finished.

I shd. like also to keep *Grisebach*.[4] also Cybele;[5] this latter, probably, can best be spared. Grisebach I have prepared for classifying, but do not care much about.— So please give me orders, which of the above Books I may keep: Ledebour is by far most useful on account of ranges: I think ½ Ledebour will be done by time Box goes.—

I shall be very glad to do 2 or 3 vols. of Decandolle;[6] if you will think which wd. be best:— I shd. think, *but will be guided entirely by* you, one with some *one* huge nat. Family, & one with *several* small ones.—

Also I shd. **especially** be thankful for any Flora of Holland with varieties marked.— These can be sent in about 3 weeks time, as by enclosed address.

Thank you much for enquiring after our children: Etty is better but far from strong:[7] Lenny much better, but has attacks of intermittent pulse.[8]

How busy you appear to be!

Pray give my very kind remembrances to Mrs. Hooker; How long it is since I have seen her: I hope she is thoroughily well.

Yours most truly | C. Darwin

I have sent little notice to Gardeners' Chron. on fertilisation of Kidney Beans:[9] I have not *yet* looked to your case of hybrid kidney Beans.—

You talked of coming here for a Sunday: is there any hope of it? I shd. most thoroughily enjoy it & I have a frightful lot of questions to torture you with you unfortunate wretch.—

C^d you lay your hand on paper in which a Fucus is mentioned *not* capable of crossing reciprocally?[10] it is **not** worth a hunt.—

If you cannot come here you must let me pay you a call of an hour or two, & do some heavy questioning.—

I have just looked at my list of queries, but it is not so tremendous, as I had fancied.—

Endorsement: '57/'
DAR 114.4: 212, 222c

CD note:[11]

Adlumia cirrhosa, plant or seed for me[12]
Fucus case of crossing
Himalayan Thistle case
Flora of Holland with varieties marked.
Yak. Campbell[13]
Solanum dulcamara winds indifferently to right or left: do some other species wind to right & some to left.—
Ocymum basilicon, have not one stalk leaves of one kind & one of other: is this normal or monstrous variety
Boreau says Linum catharticum has sometimes alternate leaves:[14] what sort of leaves has other members of same genus..
Climbing Droseras. particulars of.—
Ranunculus. Aug St. Hilaire p. 548. says ovules various in position in different species:[15] do they ever vary in same species in ever so little a degree.

[1] The year is given by CD's reference to his notice on the fertilisation of kidney beans (letter to *Gardeners' Chronicle*, 18 October [1857]).
[2] CD was returning several botanical works lent to him by Hooker (see letter to J. D. Hooker, 30 September [1857]).
[3] Ledebour 1842–53.
[4] Grisebach 1843–4.
[5] Watson 1847–59.
[6] Candolle and Candolle 1824–73.
[7] Henrietta Darwin was at Moor Park, where she was undergoing hydropathic treatment. She returned to Down after ten weeks of therapy on 31 October 1857 (Emma Darwin's diary).
[8] See letter to J. D. Hooker, 30 September [1857].
[9] Letter to *Gardeners' Chronicle*, 18 October [1857].
[10] Thuret 1854–5.
[11] This is CD's 'list of queries' referred to in the letter. It is preserved in DAR 114: 222c.
[12] *Adlumia* had been mentioned in the letter from Asa Gray, [August 1857], as a plant in which insects could hardly be agents of cross-pollination .
[13] CD refers to Andrew Campbell, a British civil servant in India whom Hooker had met and travelled with during his stay in India (see *Correspondence* vol. 4). CD was at this time investigating the breeding of yaks (see letter from Robert Schlagintweit, 25 September 1857).
[14] Boreau 1840. CD discussed this query in a note in DAR 47: 192a: 'Boreau Tom 2. p. 26. Has seen Linum cartharticum with alternate leaves. Is there not division of Linum into 2 sections.' The note was pasted onto another sheet of paper, and CD wrote: 'Ch. 7 | Hooker will enquire'. Hooker answered the query in the letter from J. D. Hooker, [6 December 1857]. CD subsequently added in pencil: '(*not* entered)'.
[15] Saint-Hilaire 1847.

From Bernard Philip Brent 23 October 1857

Castle Farm | Dallington | n.ʳ Hurst Green | Sussex
October 23.ʳᵈ 1857.

Dear Sir,

Yours of yesterday came to hand this morning and now that the children are in bed I take up my pen to answer your inquiries as well as I am able though I fear, I can give but little information;

With respect to breeding Mules from hen Canaries, the only difficulty I have found is to get the cock bird either Goldfinch; Grey Linnet; or Green Linnet; to pair, or associate amicably with the hen canary.[1]

when this is accomplished I have always found the eggs hatch well even better than from two canaries, frequently every eggs prolific, and the young birds seem to me hardier than the canaries,

When in France I had a hen Goldfinch mule paired with a cock Goldfinch they built and the hen mule laid, the eggs were of various sizes one the smallest about like a small pea and round—one the natural size, the third between. Mice destroyed the nest at Bessel's Green,[2] I put up two hen Goldfinch mules, with a cock canary, and they both, built and laid eggs, but they were not properly paired, and there was no chance of produce, one laid natural eggs, the other varied in size, and one laid two nests— one of these hens reared a young one from eggs given

I have had both Goldfinch mule cocks and Green linnet mule cocks paired with, and tread hen canaries, but no produce, from what I hear and my own experience I believe that these unions are occasionally productive, but it is stated that the mule cock must be paired with its own mother, though I am not an eyewitness to the fact, I have only heard it so affirmed; it is also an axiom with fanciers, that if the hen canary is over four years she will never be fertile with a Goldfinch—[3]

Cook-Flower Esq.ʳ a rare old Sportsman with whom I spent many happy days at Calais

AL incomplete
DAR 160.2: 299

CD ANNOTATIONS

0.1 Castle . . . found 2.2] *crossed pencil*
2.2 Goldfinch] *underl pencil*
2.2 Grey Linnet] *underl pencil*
2.2 Green Linnet] *underl pencil*
3.1 I have . . . canaries, 3.3] *scored pencil and brown crayon;* 'Q'[4] *added pencil, circled pencil*
3.3 canaries,] 'All eggs Prolific C. Œne (& reciprocally)' *added pencil*
4.2 were . . . between. 4.3] *scored brown crayon;* 'Q' *added pencil, circled pencil*
5.5 it is . . . Goldfinch— 5.7] *scored brown crayon;* 'Q' *added pencil, circled pencil*
6.1 Cook-Flower . . . Calais 6.2] *crossed pencil*

[1] CD cited this information in his chapter on hybridism (*Natural selection*, p. 429).
[2] Bessels Green, Riverhead, is the first of three addresses recorded for Brent in CD's Address book (Down House MS).
[3] Brent is cited on this point in *Natural selection*, p. 431.
[4] 'Q' stands for quote. See n. 1, above.

To J. D. Hooker [23 October 1857][1]

Down Bromley Kent
Friday Evening

My dear Hooker

I have just sent off the Box to Bromley, addressed to "57 Queen Anne St, Cavendish Sq^e;[2] it will reach there on Saturday Evening", so you can send for it on Monday morning. By keeping the Box till Monday I c^d have returned 2^d vol. of Ledebour, but I thought it best to lose no time.—

I have retained 2^d, 3^d, & 4th vols. of Ledebour, & Grisebach, which shall be despatched by separate parcel per Parcel Delivery when done with, & no time shall be lost.— I have, also, retained the Cybele Brit.

Will you bring with you Henslow's present of Butterflies.[3]

I am delighted you will come: your best plan will be by Beckenham R^y from London Bridge at 5°·20′ (remember there is crush & delay on Saturday) & thence per Bus to our door for late dinner; for we have to send on that day (Oct 31) for Etty who is coming home.[4] Is there any chance of persuading Mrs Hooker to come with Baby;[5] we sh^d be so *very* glad to see her. We can send you back on whatever day you must return

Thank you much for asking me to meet Sulivan:[6] I sh^d like it very much, though I must confess that I sh^d prefer you two by yourselves: but it so great an excertion for me, poor contemptible wretch as I am, that nothing short of a sort of compulsion, ie, seeing you in no other way, would make me.—

Most truly yours | C. Darwin

PS. | Please remember if possible a *Flora* of Holland; for Miquels **list**, with only 47 vars. for whole Flora, works out **slightly** wrong way; the only case as yet.—[7]

DAR 114.4: 214

[1] The date is given by CD's reference to Hooker's intended visit to Down House on the day that Henrietta Darwin was to return home from Moor Park (see n. 4, below). The Friday preceding 31 October 1857 was 23 October.
[2] The address is that of CD's brother, Erasmus Alvey Darwin. The box was full of botanical books (see letter to J. D. Hooker, 20 October [1857]).
[3] See letter to J. S. Henslow, 25 September [1857].
[4] Henrietta Darwin had been at Edward Wickstead Lane's hydropathic establishment at Moor Park since 22 August 1857. She returned home on 31 October (Emma Darwin's diary).
[5] Frances Harriet Hooker had given birth to the Hookers' fourth child, Marie Elizabeth, on 10 August 1857.

[6] Bartholomew James Sulivan.

[7] Miquel 1837, about which CD noted in the manuscript of his species book: 'The list is unsatisfactory for our purpose so few varieties being indicated . . . I should not have given the results from this list, had I not felt bound to do so from honesty, as the result differed from those in all the other Floras, in several respects.' (*Natural selection*, p. 169).

From Thomas Glover 26 October 1857

Smedley Hill Manchester
Oct 26. 1857—

Sir

Your letter of 22^{d} only came to hand this morning, or it should have been sooner replied to—[1]

Of course you know Cactus speciosissimus the beautiful shades of colour in the flowers perhaps surpass all others—but the habit of the plant is bad— I take this to be as true a species as any existing— There is another known by the name of stellatus, closely resembling Ackermanni, perhaps the same, the flowers of which are somewhat *bricky* in colour, but the habit of the plant is good— I dont know its origin but from the not distant resemblance the flowers bear to the one known as Jenkinsoni which is undoubtedly a hybrid, by speciosus on speciosissimus, having myself raised it in quantities, I suspect it may be a hybrid also.

Many years ago, it struck me that *if* a flower resembling speciosissimus could be produced upon a plant of the habit of stellatus, it would be a desirable acquisition & with the view of accomplishing this, I impregnated speciosissimus with stellatus, & stellatus with speciosissimus— I obtained, from each fruit, an abundance of perfectly fertile seeds— It never occurred to me to *count* the number, but my recollection serves me so far as to say *confidently* that the numbers which vegetated were about equal— they amounted in the whole to 3 to 400 & as soon as they could be well handled, were potted off in small pots 5 round the edge of each & carefully marked— The first plant bloomed in 1850 & produced a flower of much more lively colour than stellatus, & had also a tinge of blue in it. I impregnated this bloom with speciosissimus & obtained therefrom about 50 plants, only one of which has bloomed— it has a *strong* shade of blue in it, almost equal to speciosissimus, but the petals are not so even as speciosissimus— The plants that bloomed afterwards were poor in colour except one which was *very* promising, but it was stolen from me & though I offered a reward of £2. I could not recover it. In 1855 the last of the lot bloomed & amongst them were eight of very fine colour, the whole of which I impregnated with speciosissimus & the whole of which have come up as 'thick as mustard'— this is as far as I have gone, but it is sufficient to prove that the hybrids are fertile— I also impregnated speciosissimus with one of the above flowers, & though the seed was to all appearance finer than any of the others fewer plants have come up— I may observe by the way that the difference in appearance of the two first named crosses is so small, that if they were not marked, I could not say which was which but in the second cross I think there are more stems that are angular

—none are so prickly as speciosissimus—

I met with a plant in Guernsey last year but one, having the habit of my second cross & the bloom of a rose colour— I strongly suspect it to be a hybrid, as it came from a person whom I suspect, though entirely unknown to me, has been going on the same tack as myself. This I have also impregnated with speciosissimus & have a most abundant produce.

I have impregnated speciosissimus with flagelliformis & obtained plants almost identical with those known under the names of Palmerii, Mallinsoni &c all of which had bricky red blossoms.[2] I also impregnated stellatus with flagelliformis, & obtained forms similar to Palmerii, but not so prickly but the blooms were rose colour— I never tried them any further but I don't doubt their being fertile. These latter hybrids have all the form of flagelliformis—

There is or was a Mr Lukis & also a Dr Brock in Guernsey or Jersey who have d⟨o⟩ne a good deal in this way.

I have done nothing in Gloxinias beyond impregnating the common sorts one with another, & have obtained no results worth recording, but I have obtained a hybrid from Gloxinia tubiflora (quere, is this not nearer Gesneria?) impregnated with Gesneria Cooperi—the hybrid is barren. I sent them a plant to Kew—

I have also got hybrids from Franciscii confortiflora impregnated with latifolia but the plants are sickly & all variegated in the leaves.

I should have pleasure in sending you any of the above if you deemed them worth the Carriage

I am Yours faithfully | Thomas Glover.

DAR 165: 58

CD ANNOTATIONS
2.1 Of . . . also. 2.8] 'Hybrid impregnated by Speciossissmus very fertile' *added pencil*
2.4 stellatus] *underl pencil*
2.4 Ackermanni] *underl pencil*
4.1 I met . . . produce. 4.5] *crossed pencil*
9.1 I should . . . Carriage 9.2] *scored pencil*
Top of first page: 'Hooker. Are C. spec. & Ackermanni very different forms' *pencil*

---

[1] See letter to J. D. Hooker, 30 September [1857], in which CD asked Joseph Dalton Hooker for Glover's address to inquire about hybrid crosses of species of *Cereus*, a large genus of cactuses.
[2] Hybrid crosses between *Cereus speciosissimus* and *C. flagelliformis* are mentioned in CD's chapter on hybridism, but Glover is not cited as the authority (*Natural selection*, p. 412).

From Walter White 26 October 1857

The Royal Society, | *Burlington House,* | *W.*
Oct 26 *1857*

Dear Sir

At page 688 of the "abominable" catalogue you will find Encyclopaedias.[1]

Do you mention the "Dict. raisonnée universel—1746–93—" as being in our Library?[2] I can't find it in the Catalogue.

We have two works by Seringe, but not the "Flore des Jardins."[3]

We have Hooper's Dictionary of Medicine—;[4] there is a department of "Medicine" in the Encyclopédie Méthodique—[5] we have also "Dictionnaire de Medicine et de Chirurgie &c by Andral & others: 15 Vols.[6]

I can send the other books named in your List: but pardon my saying that I do not understand what this means.

Vol II & IV, if in Library.
Edited by Bomare[7]

Yrs very respectfully | W. White

C. Darwin Esq

Mr Weld is confined to bed by a weakening attack of fever.

Down House (MS 11: 19)

CD ANNOTATIONS
0.1 The . . . Jardins." 3.1] *crossed pencil*
Bottom of last page: 'p 639 see Journal de Medicin' *pencil*
Verso of last page: 'p 668 Osbecks Travels perhaps worth skimming'[8] *pencil*

[1] White was an assistant to Charles Richard Weld, the librarian of the Royal Society of London. The work to which he refers is the *Catalogue of the scientific books in the library of the Royal Society* (Royal Society of London 1839). A copy is in the Darwin Library–CUL.

[2] See n. 7, below

[3] Seringe 1845–9.

[4] Hooper 1831 was the only edition of this medical dictionary in the library of the Royal Society (Royal Society of London 1839).

[5] Vicq d'Azyr 1787–1830.

[6] Andral et al. 1829–35.

[7] White cut out this reference from CD's letter and glued it to his own. It is likely that CD refers to Jacques Christophe Valmont de Bomare's *Dictionnaire raisonné universal d'histoire naturelle*, which appeared in four editions between 1764 and 1800. Valmont de Bomare's work was frequently cited in Lucas 1847–50, which CD recorded having read on 5 September 1856 (*Correspondence* vol. 4, Appendix IV, 128: 20). CD cited Lucas 1847–50 in his species book as the source for Valmont de Bomare's remarks on hybrid peacocks (see *Natural selection*, p. 435 n. 5). An annotated copy of Lucas 1847–50 is in the Darwin Library–CUL.

[8] Osbeck 1771.

To W. E. Darwin [before 29 October 1857][1]

[Down]

We have just begun the horrid mess of plaistering all the new rooms;[2] for *all* the outside plaister had to be stripped off. Lewis found the floor-boards so damp, that he could not lay down the floor in any of the rooms; & the ceilings are so damp, that they cannot be white-washed.—

My dear William | Your affect. Father | C. Darwin

Incomplete
DAR 210.6

[1] The date is suggested by the reported progress of the plastering of new rooms being added to Down House (see n. 2, below). The information appears to precede that included in the following letter to William Darwin.

[2] CD was having a new dining-room and bedroom added to Down House (see the two following letters). In 1858, however, the new downstairs room was put to use as a new drawing-room (Atkins 1974).

To W. E. Darwin 29 [October 1857][1]

Down.
29th

My dear William

I think clearly from what you say, we had better decide, notwithstanding whatever Mr Mayor may say, for you to go after Christmas Holidays to tutor to stay till October.[2] Therefore soon have some talk with Mr Mayor, about what is customary in such cases. I presume I ought to write soon.— What with Christmas & summer vacation, you will not be with Tutor more than six months, I shd. think. Find out all you can, & let me hear & I will write to the Gentleman.—[3]

Your last letter was written splendidly & did one's eyes good to see. Mamma is gone to London today to the Miss Tollets & returns tomorrow:[4] she is gone to buy winter clothing. On Saturday Etty comes home for good & all.[5] I have been so indifferent lately that I have some thoughts of going to Moor Park or Hartfield for a week.[6] But I cant go just yet as Dr Hooker comes here on Saturday next for a day or two.[7]

The Building goes on slowly:[8] the scaffolding is down & outside plaistered & now they are plaistering inside, but the windows are not in yet.

Several of the children are not very brisk & I am poorly myself; so that the House is ready to sing miserere.

Farewell my dear old fellow: I am sorry that you cannot "grind" at present:— progress in life mainly depends on the great art of grinding; there can be no doubt of that.

Your affect | C. Darwin

DAR 210.6

[1] The date is provided by the reference to Joseph Dalton Hooker's anticipated visit and Henrietta Darwin's return to Down House on Saturday, 31 October 1857 (see nn. 5 and 7, below).

[2] CD wanted William to go to a tutor prior to entering Cambridge University, whereas Robert Bickersteth Mayor, William's housemaster at Rugby School, had suggested that William remain at Rugby until October 1858 (see letter to W. E. Darwin, 21 [July 1857]).

[3] See letter to W. E. Darwin, [November 1857].

[4] Emma Darwin recorded a trip to London in her diary on 29 and 30 October 1857. Georgina Tollet and her sisters were some of Emma Darwin's oldest friends, the Tollet family having been neighbours of the Wedgwoods in Staffordshire.

[5] Henrietta Darwin returned from a ten-week residence at Moor Park, her second stay at Edward Wickstead Lane's hydropathic establishment, on 31 October 1857 (Emma Darwin's diary).

[6] CD recorded in his 'Journal' (Appendix II) that he was at Moor Park from 5 to 12 November 1857.

[7] Hooker arrived at Down House on the evening of 31 October (see letter to J. D. Hooker, [23 October 1857]).

[8] See preceding letter.

To W. D. Fox 30 October [1857][1]

Down Bromley Kent
Oct. 30th.—

My dear Fox

I was very glad to get your note with a very fairly good account of yourself.— I cannot say much for myself; I have had a poor summer, & am at last rather come to your theory that my Brains were not made for thinking, for twice I staid for a fortnight at Moor Park,[2] & was so extraordinarily better that I can attribute the difference, (& I fell back into my old state *immediately* I returned) to nothing but to mental work; & I cannot attribute the difference but in a very secondary degree to Hydropathy. Moor Park, I like *much* better *as a place* than Malvern;[3] & I like Dr Lane very much: by the way have you seen his Brochure on Hydropathy;[4] it seem to me very good & worth reading. Unfortunately for me, I believe Dr L. means to look out for some new place.—[5]

We have had Etty there all summer; but she comes home for good next Saturday.—[6] She has received much benefit, I think, from Hydropathy; but can walk very little & is still very feeble. For the last month or two we have had trouble about Lenny, who has been the picture of strength & vigour, & now his pulse has become feeble & often very irregular like three of our other children:[7] it is strange & heart-breaking. A man ought to be a bachelor, & care for no human being to be happy! or not to be wretched.

I make slow progress in my work, which is altogether too much for me; I have done only 2 chapters in rough, first copy during the last six months![8]

I see you ask about Mr Pritchards school: I have *nothing* to say against it; but were it not for the great advantage of having George home on monthly Sundays, & *short* Michaelmas & Easter holidays; I think I sh^d prefer Rugby; but if you ask me why, I declare I c^d give no answer.[9]

You ask about all my Sisters & Eras: all are much as usual: Catherine is thinking of taking a house in London & living there at least during greater part of year, but I do not know how it will answer.[10]

How you have been spinning all about the English world: we have been all fixture, except Moor Park.— We have, however, been recreating ourselves with building a new Dining Room & large bedroom over it; for we found our party, when we had cousins had quite outgrown our old room.— *Your* Pear-trees have born very well this year for the first time; & we had lots on the wall;[11] & pretty well off for Plums; but no other fruit succeeds with us. Louise Bonne & Marie Louise have been splendid. Not only have we been nowhere; but we have hardly had any visitors, except Henslow for 2 or 3 days;[12] & he was, all what he always is,—than which I cannot give higher praise.— I am very glad your children are flourishing.

My dear Fox | Yours affectionately | C. Darwin

Emma desires her very kind remembrances.

Postmark: OC 31 1857
Christ's College Library, Cambridge (Fox 104)

[1] The year is confirmed by the reference to Henrietta Darwin's return from Moor Park (see n. 6, below).

[2] CD had spent a fortnight at Moor Park from 22 April to 6 May 1857 and again from 16 June to 30 June 1857 ('Journal'; Appendix II).

[3] At Fox's suggestion, CD had undergone hydropathic treatment in 1849 and 1850 under the care of James Manby Gully in Malvern, Worcestershire (see *Correspondence* vol. 4 and J. Browne 1990).

[4] Lane 1857.

[5] Edward Wickstead Lane did not move his hydropathic establishment from Moor Park until 1860, when he transferred to Sudbrook Park, near Richmond, Surrey (see Colp 1977, p. 68).

[6] Henrietta Darwin returned to Down House from Moor Park on 31 October 1857 (Emma Darwin's diary).

[7] See letter to J. D. Hooker, 30 September [1857].

[8] CD recorded having finished chapters 7 and 8 of his species book on 29 September 1857 (see Appendixes II and III).

[9] Charles Pritchard was the headmaster of Clapham Grammar School, which George Howard Darwin had entered in August 1856 (see letter to G. V. Reed, 8 September [1856]). CD had sent his oldest son, William Erasmus Darwin, to Rugby School.

[10] See following letter.

[11] For CD's and Fox's earlier discussion of the value of securing fruit-trees to the wall, see *Correspondence* vol. 4, letters to W. D. Fox, 6 February [1849] and 10 October [1850].

[12] John Stevens Henslow visited Down House in August 1857 (see letter to J. S. Henslow, 10 August [1857]).

To W. E. Darwin [November 1857][1]

Thursday Evening

My dear William

I wrote a few days since to Mr Wilson, the Tutor, near Norwich & got an answer this morning saying that his house was full & he could not receive you.[2] I am extremely much provoked that I did not write sooner.— I have written this morning to Mr Mayor to ask if he could advise any one else.[3] I have been a "muff" about the affair.—

Your last note was a very nice one, & written very well, thanks to my jobation to you.— The Grey mare is all right & we have taken her up to get her ready for you.— She has been very seldom in the tax-cart, for she makes such a fuss in starting, that she is hardly safe; & I do not think Parslow much likes riding her,[4] so I am inclined to think we had better sell her after the holidays.— I am glad to hear you are going on with your painting, it so nice & useful an amusement. Aunt Catherine has got a small house in Regents' Park, in York Terrace, on the opposite side to Cumberland Terrace,[5] & she proposes having you for a visitor for a fortnight & giving you some good lessons from some good master.[6]

AL incomplete
DAR 210.6

[1] Dated from the reference to Emily Catherine Darwin taking a house in London (see preceding letter and n. 5, below).

[2] See letter to W. E. Darwin, 29 [October 1857]. William Greive Wilson was the rector of Forncett, Norfolk. Apparently Wilson was eventually engaged, for there is an entry in CD's Account book (Down House MS) on 6 February 1858 for a payment to him.

[3] Robert Bickersteth Mayor was William Darwin's housemaster at Rugby.

[4] See letters to W. E. Darwin, 25 [November 1856] and 10 [December 1856].

[5] Emily Catherine Darwin, CD's younger sister, took a house near the home of Hensleigh and Fanny Mackintosh Wedgwood, who lived at 17 Cumberland Terrace, Regent's Park, London (*Post Office London directory* 1857).

[6] There is an entry in CD's Account book (Down House MS) on 17 July 1857 for 'Willy Clothes Drawing Utensils & present'.

From T. V. Wollaston [November–December 1857][1]

the "ultra-indigenous" ones, but) those which have apparently found their way thither *by natural processes*, & *before* the period of colonisation,—regarding those which have arrived through human agencies (*whether recently* or *not*) as, in a general sense, "introduced".—

Therefore in all generalisations (*except to prove* **certain special** *propositions*) I would, as the safer plan, consider the Fauna (as stated in my foot-note) in its *2-fold light*,—i.e., as composed of asterisked & unasterisked species (or "imported" & "indigenous", *in a* **general** *sense*).[2]

I was not aware that varieties occurred more in large genera than in small ones,—except from the *à priori* certainty, that where there are *more species* **to vary**, there must naturally be more varieties.—[3] I do not know what book to recommend you, about varieties. Stephens is *utterly* worthless (mere waste paper).[4] Perhaps old Gyllenhal's Ins. Suecica would be your best, for he is a most careful & good describer, & notes his varieties well.[5] Some of his "varieties" have perhaps, since his day, been established as species; nevertheless *upon the whole* he is careful & sound, & usually notices varieties, where they are at all well-marked: & I s^d think therefore as the book is *extensive* (4 vols.), any deductions from so large a superficies would stand a chance of being correct.[6]

1.	perhaps T. lineatus, Sch.[7]	370.	— B. —?
31.	— O. fuscatus, Dej.	386.	— C. —?
36.	— O. aterrimus, Hbst.	387.	— H. —?
67.	— B. rufescens, Guèr.	392.	— L. —?
69.	— — tibiale, Meg.	394.	— L. —?
71.	— — callosum, Küst.	395.	— L. —?
81.	— O. punctatus, Steph.	399.	— P. —?
178.	— C. fuscula, Humm.	400.	— P. —?
179.	— — elongata, Schüpp.	404.	— C. minutus, F.
186.	— M. picipes, Payk.	437.	perhaps H. —?
197.	— S. setigera, Ill.	449.	— Blaps clypeata, Germ.
207.	— T. grandicollis, Germ.	468.	— M. murinus, Brandt.
223.	— T. elateroides, Heer	499.	— H. clientula, Er.
225.	— E. hœmorrhoum, Germ.	508.	— O. exoleta, Er.
226.	— M. brevicollis, Payk.	509.	— —. cuniculina, Er.
231.	— D. —?	526.	— T. pilicornis, Gyll.
250.	— A. villosam, Bon.	545.	— P. procerulus, Grav.
258.	— O. glabriculus, Gyll.	546.	— A. depressum, Grav.
261.	— T. —?	560.	— S. cicindeloides, Grav.
319.	— L. Juncii, Dahl		

But I must really conclude this perfectly horrid & disreputable scrawl, for I am starting early tom. morning, & have nearly all my packing &c to do. A line *here* will always find me; but if you have anything to say immediately, "at the Rev.[d] J. F. Dawson's, Woodlands, Bedford"[8] will be my address **till Friday**. Whilst in Ireland I shall not have much opportunity for writing.

Your's very sincerely | T V Wollaston.

Incomplete
DAR 16.2: 223

CD ANNOTATIONS
1.0 the "ultra-indigenous" . . . *sense*). 2.4] *crossed brown crayon*
3.4 Stephens is . . . being correct. 3.10] *double scored brown crayon*
Top of list: "Close Geographical Representatives of European species" *ink*

---

[1] The letter follows the publication of Wollaston 1857, with which it deals. The book was published on 10 October 1857 (Entomology Library, British Museum (Natural History)). Though CD could have requested the information given in this letter at any time from October 1857 to April 1858, when he completed his statistical survey of varieties ('Journal'; Appendix II), it seems likely that the letter was written during the close of 1857. Wollaston visited the Canary Islands from January to June 1858 (Wollaston 1861, pp. 368, 371, and 402).

[2] In the introductory remarks of Wollaston 1857, Wollaston described how he distinguished between those species that had undoubtedly been imported (which he denoted by a double asterisk), the ones

that had arrived on the island through various 'accidental circumstances' since the group was first colonised (marked by a single asterisk), and those that were created in the region ('ultra-indigenous') or had migrated there by natural means before colonisation (Wollaston 1857, pp. viii–ix and ix n.).

[3] CD apparently wrote to Wollaston in connection with his calculations comparing the numbers of variable species in large and small genera. CD had previously written to Wollaston when he was studying aberrant genera (see *Correspondence* vol. 5, letter from T. V. Wollaston, 2 March [1855]).

[4] Stephens 1829.

[5] Gyllenhal 1808–27.

[6] For CD's calculations using Wollaston 1857 and Gyllenhal 1808–27, see *Natural selection*, pp. 151 and 171. His manuscripts relating to the calculations are in DAR 16.1: 138–9 and DAR 16.2: 216–17, 272–4.

[7] In the list, the numbers on the left refer to species descriptions in Wollaston 1857 of Coleoptera found only in Madeira. The rest of the list gives the names of non-Madeiran species that Wollaston considered were closely related to the numbered ones. At the top of the list CD has written: 'Close geographical Representatives of European species' (see CD's annotations). In his copy of Wollaston 1857, the numbered species are marked with a cross so that Ebenezer Norman, CD's copyist, could calculate the ratio of species to genera as instructed by CD (see DAR 16.2: 217a).

[8] John Frederick Dawson, who resided at Woodlands, Bedford, was the vicar of All Saints' Church in Toynton, Lincolnshire.

To T. C. Eyton 2 November [1857][1]

Down Bromley Kent
November 2^{d}

Dear Eyton

Will you forgive me troubling you with a question, which you can answer by "yes" or "no". Did you observe that your hybrids between the Chinese & common form, were at all wilder or less tame than *both* parents?[2] I have been informed that this is sometimes the case with hybrids, as with those from common & Musk Duck.[3]

Yours most truly | C. Darwin

University of Birmingham Library (Special collections, MS Eyton 42)

[1] The information given in this letter was used by CD in chapter 10 of his species book, completed in March 1858 ('Journal'; Appendix II).

[2] Eyton had described these hybrid geese in Eyton 1837b and 1840. The question posed in the letter was evidently answered by Eyton, for CD reported that the offspring were not wilder than their parents (*Natural selection*, p. 486).

[3] See *Natural selection*, p. 486, where CD wrote: 'Mr. Garnett of Clitheroe in a letter to me states that his hybrids from the musk & common Duck "evinced a *singular* tendency to wildness."'

From Henry Coe 4 November 1857

Asylum. Knowle. | N^{r} Fareham. Hants.[1]
4th Novr 1857.

Sir,

In an article headed "Bees and Fertilisation of Kidney Beans" published in the

Gardeners' Chronicle of the 24th ult: I read "I am also astonished that the varieties of the Kidney Bean can be raised true when grown near each other. I should have expected that they would have been crossed by the bees bringing pollen from other varieties; and I should be infinitely obliged for any information on this head from any of your correspondents".—²

I enclose a few beans they are from the "Black Negro".— They were planted between the Dwarf Haricot, a "*White*" Bean, and a Dwarf entirely "Brown", thus

about 10 rows	Haricot	"White"
-"- 4 -"-	Negro	"Black"
then	Dwarf.	"Brown".

Such a mixture, I never before met with, and the result was altogether unlooked for or expected— The Haricots (White) are very little touched with the exception of here and there one— The "Brown" not at all.—

I have the honor to be, | Sir, | Your obed! serv! | Henry Coe. | Gardener

M! C. Darwin— | &c. &c.

DAR 161: 192

CD ANNOTATIONS
Top of first page: 'done' *pencil, circled pencil*
Bottom of last page: 'not 1/5 part of the crossed Beans were pure, all shades betwen Brown & black & one mottled with white— All the pure (if pure) were the black'³ *ink*

¹ The County Lunatic Asylum, built in 1852, was about three miles north-west of Fareham, Hampshire. According to his valediction, Henry Coe was a gardener at the asylum.
² See letter to *Gardeners' Chronicle*, 18 October [1857].
³ See letter from Henry Coe, 14 November 1857.

From Frederick Smith   10 November 1857

British Museum.
10 Nov! 1857

My dear Sir

Some time ago you asked me to furnish you with remarkable instances of desparity in form etc—in workers of Insects living in community—¹ As one is apt to forget these things at the moment they are asked for I send you one that is a truly remarkable instance— In my Monograph on the Genus Cryptocerus I figured & described a species as C. discocephalus—² some time subsequent I received a letter from M! H. W. Bates from Brazil—³ he said—"I have met with your curious species C. discocephalus— the creature figured is only the large size of the worker of the species— I send you both the workers taken from several nests constructed in dead branches of shrubs—"

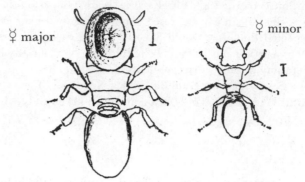

☿ major   ☿ minor

Cryptocerus discocephalus

I send you tracings of the creatures in relative proportion— dont trouble to reply to this but tell me what you think of it when you are next time at the Museum⁴ and believe me

Yours very truly | Fredᵏ Smith

Chaˢ Darwin Esqʳᵉ

DAR 11.2: 65a

CD ANNOTATIONS
Top of letter: 'Ch 8.'⁵ *brown crayon*

¹ Smith, an assistant in the zoological department of the British Museum, was an authority on social insects. In a discussion of the origin of neuter insects in his species book, CD stated in a note: 'I am greatly indebted to Mr. F. Smith of the British Museum, one of the highest authorities on Hymenoptera, for much valuable information on all points in the following discussion.' (*Natural selection*, p. 364 n.1).

² F. Smith 1854.

³ Henry Walter Bates had accompanied Alfred Russel Wallace to the Amazon in 1849 and remained in Brazil to continue his researches and the collection of specimens (*DNB*).

⁴ This letter was attached by CD to a page in the manuscript of his species book relating to neuter insects, part of chapter 8 on 'Difficulties on the theory of natural selection' (see *Natural selection*, pp. 368 and 574). In this chapter, CD attempted to explain how natural selection could bring about changes in the structure of neuter (or worker) forms of ants and bees, which cannot propagate their kind. Certain ants, like those of the genus *Cryptocerus*, which, as Smith's letter indicates, have two kinds of workers, were to CD 'the acme of difficulty on our principle of natural selection' (*Natural selection*, p. 371). For CD's explanation of the possible advantages accruing to parent-ants if they produce some sterile offspring, see *Natural selection*, pp. 370–4.

⁵ CD refers to chapter 8 of his species book (see n. 4, above).

To *Gardeners' Chronicle* [before 12 November 1857]¹

Will the writer of the highly remarkable article on weeds in your last Number have the kindness to state why he supposes that "there is too much reason to believe that foreign seed of an indigenous species is often more prolific than that grown at home?"[2] Is it meant that the plant produced from the foreign seed actually produces more seed, or merely that the introduced stock is more vigorous than the native stock? I have no doubt that so acute an observer has some good reason for his belief. The point seems to me of considerable interest in regard to the great battle for life which is perpetually going on all around us. The great American botanist, Dr. Asa Gray, believes that in the United States there are several plants now naturalised in abundance from imported seed, which are likewise indigenous; and my impression is (but writing from home I cannot refer to his letter to me)[3] that the imported stock prevails over the aboriginal. So again, Dr. Hooker in his admirable Flora of New Zealand has told us that the common Sonchus has spread extensively from imported seed, whilst the same species is likewise an aboriginal; the natives in this instance being able from trifling differences to distinguish the two stocks.[4] Might I further ask whether it is now some years since the seed of Sinapis nigra was accidentally introduced on the farm described; and if so, whether the common Charlock still remains in lessened numbers owing to the presence of the invader, and without, as far as known, fresh seed of the invading S. nigra having been introduced?—whether, in short, it was a fair fight between the two species, ending in the victory of the Black Mustard?[5] Would it be trespassing too much on the kindness of the writer of the article to ask whether he knows of any other analogous cases of a weed introduced from other land beating out, to a greater or lesser extent, a weed previously common in any particular field or farm?[6] *C. Darwin, Down, Bromley, Kent.*

Gardeners' Chronicle and Agricultural Gazette, 14 November 1857, p. 779

[1] The letter was published in *Gardeners' Chronicle and Agricultural Gazette* on 14 November 1857. It was written during CD's stay at Moor Park (see n. 3, below). CD returned to Down on 12 November 1857 ('Journal'; Appendix II).

[2] The passage to which CD refers, in the agricultural section of the *Gardeners' Chronicle and Agricultural Gazette,* 7 November 1857, p. 762, reads:

> The manner in which weeds are spread over some farms may be observed in the increase of exotic species from the use of foreign seeds, a circumstance which accounts for the increase of plants in our English Flora within the last few years. However these, as being wholly foreigners, seldom make rapid progress, whilst there is too much reason to believe that foreign seed of an indigenous species is often more prolific than that grown at home.

[3] CD wrote the letter during his stay at Moor Park. CD underwent hydropathy at the Moor Park establishment from 5 to 12 November 1857 ('Journal'; Appendix II). The reference is to the letter from Asa Gray, 16 February 1857.

[4] J. D. Hooker 1853–5, 1: 152–3.

[5] The reference is to one of the cases given in the *Gardeners' Chronicle and Agricultural Gazette* (see n. 2, above) illustrative of how seeds of unwanted species could be introduced into a field with manure. The article described how a large quantity of black mustard, *Sinapis nigra,* was unwittingly introduced into a crop of flax. In various ways the weed seeds got into the manure heaps,

and the black mustard soon became widespread, 'having nearly expelled the former common Sinapis arvensis—a circumstance which we think partly accounted for by the greater fecundity of the former, when compared with the latter' (*Gardeners' Chronicle and Agricultural Gazette*, 7 November 1857, p. 762).

6 CD received a letter from James Buckman in response to the printed version of this letter. Although, Buckman's letter has not been found, a printed response appeared in *Gardeners' Chronicle and Agricultural Gazette*, 2 January 1858, p. 11, and there is an abstract of it, in CD's hand, in DAR 46.1: 41–2. The abstract describes Buckman's view that many agrarian weeds are the product of cultivation and are not found in the wild.

To T. H. Huxley [before 12 November 1857][1]

Moor Park, Farnham | Surrey

My dear Huxley

Your letter has been forwarded to me here, where I am profiting by a weeks rest & hydropathy.[2] Your letter has interested & amused me much.— I am extremely glad you have taken up the Aphis question,[3] but for Heaven sake do not come the mild Hindoo to Owen (whatever he may be):[4] your Father confessor trembles for you.— I fancy Owen thinks much of this doctrine of his: I never from the first believed it; & I cannot but think that the same power is concerned in producing Aphides without fertilisation, & producing, for instance, nails on the amputated stump of a man's fingers, or the new tail of a Lizard.—

By the way I saw somewhere during the last week or so a statement, of a man rearing from the *same set* of eggs winged & wingless Aphides, which seemed new to me. Does not some Yankee say that the American viviparous Aphides are winged?[5] I am particularly glad that you are ruminating on the act of fertilisation: it has long seemed to me the most wonderful & curious of physiological problems. I have often & often speculated for amusement on the subject, but quite fruitlessly. Do you not think that the conjugation of the Diatomaceæ will ultimately throw light on subject?[6] But the other day I came to the conclusion that some day we shall have cases of young being produced from spermatozoa or pollen without an ovule. Approaching the subject from the side which attracts me most, viz inheritance, I have lately been inclined to speculate very crudely & *indistinctly*, that propagation by true fertilisation, will turn out to be a sort of *mixture* & not true *fusion*, of two distinct individuals, or rather of innumerable individuals, as each parent has its parents & ancestors:— I can understand on no other view the way in which crossed forms go back to so large an extent to ancestral forms.— But all this, of course, is infinitely crude.[7]

I hope to be in London in course of this month, & there are two or three points, which, for my own sake, I want to discuss briefly with you.—[8]

Ever my dear Huxley | Yours very truly | C. Darwin

There is a couple of very clever men here with a taste for natural Science I have just made them roar with laughter at your last Page.

Imperial College of Science, Technology and Medicine Archives (Huxley 5: 58)

[1] Dated by the references to CD's stay at Moor Park (see n. 2, below) and to T. H. Huxley 1857–8 (see n. 3, below).

[2] CD travelled to Moor Park, near Farnham, Surrey, on 5 November 1857 and returned to Down on 12 November ('Journal'; Appendix II).

[3] T. H. Huxley 1857–8, the first part of which Huxley delivered before the Linnean Society on 5 November 1857. Huxley had first discussed parthenogenesis in the aphid in T. H. Huxley 1856–7, 12: 482, in which he had strongly criticised Richard Owen's interpretation of the phenomenon.

[4] CD's admonition refers to the increasingly public feud between Huxley and Richard Owen. In T. H. Huxley 1857–8, pp. 212–18, Huxley attacked Owen's explanation in Owen 1849 of the phenomenon of parthenogenesis as resulting from the retention of a 'spermatic force' in subsequent generations of females, commenting that: 'reference to an undefined "force," of questionable existence, is simply "ignorance writ large" ' (T. H. Huxley 1857–8, p. 216).

[5] Burnett 1854, p. 63.

[6] George Henry Kendrick Thwaites had first reported observing conjugation in the Diatomaceae at the 1847 meeting in Oxford of the British Association for the Advancement of Science. CD attended the meeting and conversed with Thwaites (see letter to G. H. K. Thwaites, 8 March 1856, and *Correspondence* vol. 5, letter to G. H. K. Thwaites, 10 December 1855).

[7] CD developed his theory of inheritance, 'pangenesis', and set it out in *Variation* 2: 357–404.

[8] CD was in London from 17 to 20 November 1857 (Emma Darwin's diary). On 19 November, he attended a meeting of the Philosophical Club of the Royal Society, at which Huxley was also present (Royal Society Philosophical Club minutes).

To Robert Patterson   12 November [1857][1]

Down, Bromley, Kent
Nov. 12 [1857]

My dear Mr. Patterson

The rabbits arrived safely last night after their long journey; & most sincerely do I thank you for the very great trouble you have taken to oblige me.[2] Externally they seem to differ extremely little except perhaps in fulness of head, from the rabbit of this neighbourhood. But they shall be skeletonized.

I have now rabbits from Shetland, Madeira & Ireland and hope to receive one soon from Jamaica, so I shall have good means of comparison for ⟨   ⟩ to several domestic breeds[3]

If you remember whenever you see Lord Massarene I hope you will present my thanks for his great kindness.—[4] When I have done with the Rabbit Skeletons for my own purpose I shall present ⟨   ⟩ to the Brit ⟨   ⟩ been thrown away & I well know that you work for Natural History from a pure love of Science

With my very sincere thanks pray believe me.

Praeger 1935, p. 714

[1] The date is as given in Praeger 1935, p. 714. It is confirmed by the relationship to the letter to Robert Patterson, 10 March [1857].

[2] In his letter to Robert Patterson, 10 March [1857], CD had asked for a specimen of the Irish rabbit.

[3] In *Variation* 1: 127, CD gave a table of osteological comparisons between the wild rabbits mentioned in the letter and various breeds of domestic rabbits.

[4] John Skeffington, Viscount Massereene, of Antrim Castle, Northern Ireland.

From Henry Coe 14 November 1857

Asylum. Knowle— | n.° Fareham. Hants.
14.ᵗʰ Nov.° 1857.

Sir,

I beg most respectfully to acknowledge the receipt of your favour of the 13.ᵗʰ.—
You very kindly thank me, for having forwarded the Sample of Beans; I can only
assure you, that I felt a pleasure in so doing—[1]

The case is indeed an extremely interesting one.— Being short of the Black
Negro Bean, the four rows mentioned in my last as having been planted, were
reserved for Seed, I have therefore several quarts of the *varieties*, which I shall
assort (as near as possible) and plant separately next year, (of course carefully
watching the result)

I will now proceed to answer your questions.—

1.— The pods were gathered by myself from the Negro Bean. There was no
possibility of any mistake— The three sorts (with the add.ⁿ of the Scarlet runner)
being all the sorts that were planted in the grounds—

2.— Having several bundles of the Plants from which the Beans are not yet
removed, I have to day examined several of them, and do not find any difference
in the same actual pod.

3.— I feel a great pleasure in complying with your request I have therefore
enclosed a dozen Beans of each *pure* kind—(the exact variety used by me—)

4— The Beans forwarded in my last—Were *all* from the Negro Plant—Sown in
June *last*, and gathered *this* Autumn.

I have only to add that I shall feel it a pleasure to answer any further
questions—(should there be any) you may wish to ask—

With the greatest respect, | I am, Sir, | Your most obed.ᵗ serv.ᵗ | Henry Coe.— |
(Gardener)

C. Darwin. Esq.ᵉ | &c —

DAR 161: 193

CD ANNOTATIONS
Top of first page: 'done' *pencil, circled pencil*

[1] See letter from Henry Coe, 4 November 1857.

To J. D. Hooker 14 [November 1857][1]

Down Bromley Kent
14.ᵗʰ

My dear Hooker

On Tuesday I will send off from London, whither I go on that day,[2] Ledebour 3
remaining vols. Grisebach & Cybele, ie all that I have, & most truly am I obliged

to you for them.[3] I find the rule as yet of the species varying most in the large genera *universal* except in Miquels *very brief* & therefore imperfect *list* of Holland Flora,[4] which makes me very anxious to tabulate a fuller Flora of Holland.—

I shall remain in London till Friday *morning*, & if *quite* convenient to send me two vols of D. C. Prodromus, I c^d take them home & tabulate them:[5] I sh^d think a vol. with large best-known natural Family & vol. with several small broken Families w^d be best, *always supposing that the varieties are conspicuously marked in both.*— Have you the vol. published by Lowe on Madeira;[6] if so & if *any* varieties are marked I sh^d much like to see it, to see if I can make out anything about Habitats of vars. in so small an area,—a point on which I have become very curious. I fear there is no chance of your possessing Forbes & Hancock British shells, a grand work. which I much wish to tabulate.—[7]

very many thanks for seed of Adlumia cirrhosa, which I will carefully observe.[8] My notice in G. Ch. on kidney Beans has brought me a curious letter from intelligent Gardener,[9] with a most remarkable lot of Beans, crossed in marvellous manner **in 1st generation** like the Peas sent to you by Berkeley[10] & like those experimentised on by Gærtner & by Wiegmann.[11] It is a very odd case: I shall sow these seeds & see what comes up. How very odd that *pollen* of one form sh^d affect the outer coats & size of the Bean produced by pure species!

Many thanks for your letter about medals, which I burnt; I am glad about Lindley & very sorry about Lyell.[12] I hope he will never hear of the attempt: it is an old story that the very highest merit is undervalued.—

My week at Moor Park has done me wonderful good;[13] & has almost quite driven away the wretched feelings in my head: I only wanted rest, & that I got there in perfection; & took quite long walks & enjoyed the scenery like a gentleman at large.

Ever my dear Hooker | Yours most truly | C. Darwin

DAR 114.4: 215

[1] Dated by CD's intention to go to London and his stay at Moor Park (see nn. 2 and 13, below).
[2] According to Emma Darwin's diary, CD went to London on Tuesday, 17 November 1857 and returned to Down House on Friday, 20 November.
[3] CD had borrowed Ledebour 1842–53, Grisebach 1843–4, and Watson 1847–59 from Hooker (see letters to J. D. Hooker, 30 September [1857], 20 October [1857], and [23 October 1857]).
[4] See letter to J. D. Hooker, [23 October 1857].
[5] Candolle and Candolle 1824–73.
[6] Richard Thomas Lowe had recently published the first part of Lowe 1857[–72].
[7] E. Forbes and Hanley [1848–]1853.
[8] See letter to J. D. Hooker, 20 October [1857].
[9] Letter from Henry Coe, 4 November 1857. See also preceding letter.
[10] See letter to M. J. Berkeley, 29 February [1856], and letter from M. J. Berkeley, 7 March 1856.
[11] Gärtner 1849 and Wiegmann 1828.
[12] See letter to J. D. Hooker, 2 June [1857]. John Lindley, whom Hooker had proposed for a Royal Medal of the Royal Society, was to receive the award at the anniversary meeting of the society on

30 November 1857. Charles Lyell, who had been nominated for the Copley Medal, did not receive it in 1857 but was awarded the medal in 1858.

[13] CD spent the week of 5–12 November 1857 at Edward Wickstead Lane's hydropathic establishment ('Journal'; Appendix II).

From Williams & Norgate 17 November 1857

London, W.C. | *14, Henrietta-Street,* | *Covent-Garden.*
17/11 *1857.*

Messrs. Williams & Norgate[1] *present their compliments to* Mr. C. Darwin *and beg to inform* him, that „Heer geograph. Verbreitung der Käfer &c" has **not** been printed separately. The parts of the Mittheilungen aus dem Gebiete der theoret. Erdkunde, which contain this treatise—publ. in 1834. Vol I pt 1–4 cost about 10/–11/—[2]

DAR 91: 81

[1] Booksellers in London specialising in foreign scientific literature.
[2] Froebel and Heer 1834–6.

To J. D. Hooker 21 November [1857][1]

Down Bromley Kent
Nov. 21st

My dear Hooker

I was most sincerely sorry to receive your little note, saying how dangerously ill Mrs Henslow was.[2] I am sure Mrs Hooker & all at Hitcham have my most sincere sympathy. If the worst does happen & you make any stay at Hitcham do just say to my dear & most kind friend, Henslow, how truly I feel for his sorrow. I will not trouble him with writing to him.— To the last day of my life I shall feel under what deep obligations I lie to Henslow & Mrs Henslow for their extraordinary kindness to me at Cambridge. How many pleasant little dinners I have had with them; & how invariably kind poor Mrs Henslow was to me.[3] I do hope she has not suffered much. They were like the nearest & most affectionate relations to me.—

My dear Hooker | Yours affectionately | C. Darwin

I shd like much to have ever so short a scrap to tell me how Henslow is.—

DAR 114.4: 213

[1] Dated by the reference to the illness of Harriet Henslow (see n. 2, below).
[2] Harriet Henslow, wife of John Stevens Henslow and Hooker's mother-in-law, died on 20 November 1857 (*Gentleman's Magazine* n.s. 4 (1858) 1: 113).
[3] During his undergraduate days at Cambridge, CD was frequently invited to the Henslows' home. See *Correspondence* vol. 1.

To W. B. Tegetmeier 21 November [1857][1]

Down Bromley Kent
Nov. 21st

My dear Sir

Thank you for your note. I have been looking over my Catalogue of specimens,[2] & find to my surprise that I have no young chicken of Malay & without that for comparison, the skeleton of an old Bird w^{d} be comparatively useless to me; & therefore I think, if you could help me to purchase a few Malay eggs in the Spring, it would be *much* better for me than now getting a Cockrel; but I am sorry you sh^{d} have had the trouble of writing for nothing.

Next summer I shall probably give away *all* my Pigeons & you may rely on it, I will inform you.—[3]

I killed some young Jacobins the other day, which from being Bull-eyed I did not think worth sending you.

As soon as ever I can find time to go over my little disquisition on Pigeons, & compare the skeletons &c. I shall give up Pigeons.—[4]

I shall certainly attend, (health permitting) the Annual Show of Philoperisteron.[5] I am glad to hear you liked my little notice on Kidney Beans.—[6]

My dear Sir | Yours very sincerely | C. Darwin

Will you keep in mind Malay Eggs.—

New York Botanical Garden Library (Charles Finney Cox collection)

[1] Dated by the reference to CD's letter to the *Gardener's Chronicle* about kidney beans (see n. 6, below).
[2] CD's catalogue of poultry and pigeon specimens has not been located in the Darwin archive.
[3] See *Correspondence* vol. 7, letters to W. B. Tegetmeier, [21 April 1858] and 8 September [1858].
[4] CD had included a discussion of pigeons in chapter 1 of his species book (*Natural selection*, p. 25; see also Appendix III). He did not have time to revise or expand the manuscript until June 1858 (see *Correspondence* vol. 7, letter to W. B. Tegetmeier, 22 June [1858]).
[5] The annual show of the Philoperisteron Society, of which CD was a member, took place on 19 January 1858. The report of the event in the *Cottage Gardener*, 19 (1857–8): 256, noted: 'The visitors were numerous, fashionable, and scientific. It relieves Pigeon fancying from all charge of triviality when savans of such reputation as Messrs. Darwin and Waterhouse show, by their attendance and interest, that the changes capable of being produced in any species by domestication, are worthy of the deep attention of scientific inquirers; and in no species are these changes greater, or more varied, than in the Pigeon.' The notice was probably written by Tegetmeier, a regular contributor to the *Cottage Gardener*.
[6] Letter to *Gardeners' Chronicle*, 18 October [1857].

To John Lubbock [22 November 1857][1]

Down
Sunday Morning

Dear Lubbock

At the Philosoph. Club last Thursday I overheard Dr Sharpey speaking to Huxley in such high & warm praise of your paper & Huxley answering in same tone that it did me good to hear it.[2] And I thought I would tell you, for if you still wish to join Royal Soc.ᵞ, I sh.ᵈ think (Sharpey being influential in Council & Secretary) there c.ᵈ be no doubt of your admission. Even if you were not admitted the first year it cannot be thought the least disgraceful. I am not aware but perhaps you have been already proposed.—[3]

Will you be so kind as to give my cordial congratulations & most sincere good wishes to Miss Lubbock.—[4] I am very much obliged for my half of the invitation to the Breakfast: it will be a really beautiful sight, but I fear it would be too fatiguing for me so will not venture to accept it.

Believe me dear Lubbock | Yours sincerely | C. Darwin

Down House

[1] The date is established by CD's attendance at a meeting of the Royal Society Philosophical Club minutes (see n. 2, below) and by the reference to Miss Lubbock's engagement breakfast (see n. 4, below).

[2] CD attended a meeting of the Philosophical Club of the Royal Society on 19 November 1857 (Royal Society Philosophical Club minutes). William Sharpey, one of the secretaries of the society and a member of the Philosophical Club, was a noted physiologist. The paper under discussion was Lubbock's study of reproduction in *Daphnia* (Lubbock 1857).

[3] Lubbock was elected a fellow of the Royal Society in 1858.

[4] Mary Harriet Lubbock, John Lubbock's older sister, was married on 8 December 1857 to Robert Birkbeck of Keswick, Norfolk (*Gentleman's Magazine* n.s. 4 (1858): 98).

To Hugh Falconer 23 November 1857

Down
Nov. 23rd

My dear Falconer

You told me that you thought you could obtain information for me about the colours and markings of some Indian horses, from the very best authority, namely Col. E. Dickie.[1] If you could interest him to oblige me with answers to the following questions, it would be conferring a very great kindness on me. I have put the questions in full, though I do not suppose he could answer all; but if he would take the trouble to answer some of them, I should feel extremely much obliged to him, and to you for aiding me in getting information on a point on which I am very curious

My dear Falconer | Yours very sincerely | Ch. Darwin

Would you be so kind as to send off my queries pretty soon, as time is precious.[2]

[Enclosure]

1. Major Gwatkin in a letter published by Col. Hamilton Smith, speaking of the Kutch or Kahleawar breed of horses, says "Kutch is the country where the mares are bred; the sire is an Arab. They are generally greys or light duns, and almost invariably have the zebra-marks on the arms and thighs with list down the back."[3] I am very anxious to know what are the colours of the pure Kutch mares; for it

seems very surprising that the offspring, begotten by Arab sires, which I believe are never dun, should be *generally* dun or grey and have a dorsal stripe and transverse zebra-like marks on the legs. These latter characteristics, I hear, are generally lost in the Scandinavian duns, when crossed with any other colour.

2. In the Kutch or Kahleawar duns (whether or not a crossed breed), are the zebra-like leg-markings, and dorsal stripe very generally present? Does a transverse shoulder stripe, like that on the ass, ever occur? Is such shoulder stripe ever double, as sometimes, though very rarely, is the case with the ass.

3. Am I to understand from Major Gwatkin that the **Grey** Kutch horses have a dorsal stripe and zebra-like marks on the legs: I have never heard of this before.

4. Are any Kutch horses cream-coloured; and have such the dorsal, or shoulder, or leg stripes?

5. Chesnut horses often have a dorsal stripe: in India have they ever a shoulder or leg stripes?

6. Is the dun or other colour when accompanied by the above stripes, strongly inherited, when such coloured horses are crossed with others.

7. Are all the horses, whatever their colour may be, *when striped*, small and built like cobs.

8. What is the colour of the Dun or Cream or other colour, when accompanied by stripes *in the foal*, before first hair is shed? Are the stripes at this early age more or less conspicuous?[4]

C. Darwin

Down, | Nov. 22nd 1857.

Copy
DAR 144

[1] Edward John Dickey was the superintendent of the stud department of the Bengal Army from 1853.

[2] The copyist recorded that Falconer had written on the original: 'Wrote to Dr. Bruist Bombay 13 Jan^y, enclosing copy of the queries, as the best chance of having them well answered.' George Buist was the editor of the *Bombay times* and an authority on India (*DNB*).

[3] The letter from '*Major Gwatkin, Superintendent of the Hon. East India Company's Stud in Northern India*', received too late to be included in C. H. Smith 1843, was extracted in the 'Advertisement from the publisher' placed at the beginning of Charles Hamilton Smith's volume. The passage quoted comes from p. xi, and is marked in CD's copy of the work in the Darwin Library–CUL.

[4] In the manuscript of his species book, CD relied upon Smith's work for information on Indian horses (*Natural selection*, pp. 329–30). In *Variation* 1: 58–9, CD discussed the striped horses of India but did not mention either Dickey or Buist (see nn. 1 and 2, above).

To Asa Gray 29 November [1857][1]

Down Bromley Kent
Nov. 29th

My dear Gray

This shall be such an extraordinary note as you have never received from me, for it shall not contain *one* single question or request. I thank you for your impression on my views.[2] Every criticism from a good man is of value to me. What you hint at generally is very very true, that my work will be grievously hypothetical & large parts by no means worthy of being called inductive; my commonest error being probably induction from too few facts.— I had not thought of your objection of my using the term "natural Selection" as an agent; I use it much as a geologist does the word Denudation, for an agent, expressing the result of several combined actions. I will take care to explain, not merely by inference, what I mean by the term; for I must use it, otherwise I sh^d incessantly have to expand it into *some such* (here miserably expressed) formula as the following, "the tendency to the preservation (owing to the severe struggle for life to which all organic beings at some time or generation are exposed) of any the slightest variation in any part, which is of the slightest use or favourable to the life of the individual which has thus varied; together with the tendency to its inheritance".[3] Any variation, which was of no use whatever to the individual, would not be preserved by this process of "natural selection". But I will not weary you by going on; as I do not suppose I c^d make my meaning clearer without large expansion.— I will only add one other sentence: several varieties of Sheep have been turned out together on the Cumberland Mountains, & one particular breed is found to succeed so much better than all the others, that it fairly starves the others to death: I sh^d here say that natural selection picks out this breed, & would *tend* to improve it or aboriginally to have formed it.—[4]

Many thanks for seed & specimens of Adlumia:[5] I must confess from what I have seen in Bees whilst sucking Fumaria, I see no difficulty whatever in Bees crossing the individuals: & I would venture to predict that it has a nectary on *both* sides instead of on only one side, for the sort of cap of joined petals can be pushed with equal easiness both ways, but where there is only one nectary it can be pushed (as far as I have seen) only one way: Lecoq, I observe, brings forward Fumaria as a genus which c^d never be crossed by natural means,[6] whereas I *suspect* its structure is formed in direct relation to favour crossing!![7]

I sent you Gardeners Chronicle with little notice on Kidney Beans:[8] since writing it, I have received a most curious lot of Beans *naturally* crossed, & the seed-coats affected by the act of fertilisation like Gærtners Pea case.—[9] By the way I must tell you what I heard yesterday, though not in your line, but on subject of the crossing of individuals. Barnacles (Balanus) are hermaphrodite & with their well shut up shell offer as great a difficulty to crossing *as can well be conceived*: I found an individual with monstrous & *imperforate* penis, but yet with fertilised ova; but I did not know whether it might not be case of parthogenesis or a strange accident of some floating spermatozoa;[10] well yesterday I had an account by a man who watching some shells, saw one protrude its long prosciformed penis, & insert it in the shell of an adjoining individual![11] So here is a load off my mind.—

You speak of species not having any material base to rest on; but is this any greater hardship than deciding what deserves to be called a variety & be designated by a greek letter. When I was at systematic work, I know I longed to have no other difficulty (great enough) than deciding whether the form was distinct enough to deserve a name; & not to be haunted with undefined & unanswerable question whether it was a true species.[12] What a jump it is from a well marked variety, produced by natural cause, to a species produced by the separate act of the Hand of God. But I am running on foolishly.— By the way I met the other day Phillips, the Palæontologist,[13] & he asked me "how do you define a species?"— I answered "I cannot" Whereupon he said "at last I have found out the only true definition,—'any form which has ever had a specific name"!

I am infinitely obliged to you for your offer (if you can ever find time, & how much overworked you seem to be) of considering again a list of close species, such as Hooker would perhaps lump together: you could not do me a more essential service.[14]

If you do it, will you please take, if in your power, *large* & *small* orders as they come, for possibly there may be some difference in the rule in large natural & small broken families. I intend to go into this with Ledebour, as far as mere varieties are concerned.[15] In all Ledebour & *many* other Floras, I find the rule universal of the large genera presenting most varieties.[16] In the British Flora, by Mr Watsons aid,[17] I have struck out the most trifling varieties & I find the rule holds good, as it also does with the forms which most British Botanists rank as species, but which some one Botanist has considered a variety. This rule, as I must consider it of the large genera varying most, I look at as most important for my work & I believe it to be the foundation of the manner in which all beings are grouped in classes &c, together with what I rather vaguely call my principle of divergence ie the tendency to the preservation from extinction of the most different members of each group.—[18] But I am amusing myself by scribbling away all my notions without any mercy.

Forgive me, & believe | My dear Gray | Yours heartily obliged | C. Darwin

How I wish I knew what large, (for large it must be) Moth or Humble Bee visits & fertilises Lobelia fulgens in its native home: do you know any southern young Botanist who w^d look to this? I would cover a plant with a very coarse gauze cap, & then not a pod would set I believe. But by Jove I have broken my vow by a sort of question or request!

Gray Herbarium of Harvard University

[1] Dated by the relationship to the letter to Asa Gray, 5 September [1857].

[2] Gray's letter has not been found. It evidently responded to CD's explanation of his theory of natural selection, as put forward in letter to Asa Gray, 5 September [1857].

[3] CD included a definition of natural selection along these lines in *Natural selection*, p. 175, and *Origin*, p. 61.

[4] This case is also cited in *Natural selection*, p. 200.

[5] See letter from Asa Gray, [August 1857].

[6] Lecoq 1845, p. 61.

[7] See *Natural selection*, pp. 53–4, in which CD discussed the visiting of *Fumaria* by bees. See also letter to Asa Gray, 20 July [1857], in which CD gives other examples to illustrate his view.

[8] See letter to *Gardeners' Chronicle*, 18 October [1857].

[9] See letters from Henry Coe, 4 November 1857 and 14 November 1857. CD refers to the cases of seed-coats being affected by pollen from a different species reported in Gärtner 1849 (see letter to M. J. Berkeley, 29 February [1856]).

[10] CD described this case in *Living Cirripedia* (1854): 102 and referred to it in both *Natural selection*, p. 45, and *Origin*, p. 101. See also *Correspondence* vol. 4, letter to Louis Agassiz, 22 October 1848.

[11] See letter from Richard Bishop to Charles Spence Bate, 3 December 1857.

[12] See *Correspondence* vol. 4, letters to H. E. Strickland, 29 January [1849], [4 February 1849], and 10 February [1849]. See also *Correspondence* vol. 5, letter to J. D. Hooker, 25 September [1853].

[13] CD may have spoken to John Phillips at the meeting of the Philosophical Club of the Royal Society on 19 November 1857 at which both were present (Royal Society Philosophical Club minutes).

[14] Gray had already provided CD with a list of 'close species' (see *Correspondence* vol. 5, letter to Asa Gray, 8 June [1855], and letter from Asa Gray, 30 June 1855). The manuscript list is in DAR 165: 92/3.

[15] Ledebour 1842–53.

[16] See *Natural selection*, pp. 148–54.

[17] See letters from H. C. Watson, 14 December [1857] and 20 December [1857].

[18] See *Natural selection*, pp. 227–51, and *Origin*, pp. 111–26.

To George Bentham 1 December [1857][1]

Down Bromley Kent
Dec. 1st

My dear Sir

I thank you for so kindly taking the trouble of writing to me, on naturalised plants.[2] I did not know of or had forgotten the clover case. How I wish I knew what plants the Clover took the place of; but that would require more accurate knowledge of any one piece of ground than I suppose any one has. In the case of trees, being so long-lived, I sh^d think it would be extremely difficult to distinguish between true & new spreading of a species, & a rotation of crop.

With respect to your idea of plants travelling west, I was much struck by a remark of yours in the penultimate Linnean Journal on the spreading of plants from America near Behring St^s [3] Do you not consider so many more seeds & plants being taken from Europe to America, than in a reverse direction, would go some way to account for comparative fewness of naturalised American plants here. Though I think one might wildly speculate on European weeds having become well fitted for *cultivated* land, during thousands of years of culture, whereas cultivated land would be a new home for native American weeds, & they would not consequently be able to beat their European rivals when put in contest with them on cultivated land. Here is a bit of wild theory!

But I did not sit down intending to scribble thus; but to beg a favour of you: I gave Hooker a list of species of Silene, on which Gærtner has experimentised in crossing:[4] now I want **extremely** to be permitted to say that such & such are believed by Mr Bentham to be true species, & such & such to be only varieties.[5]

Unfortunately & stupidly Gærtner does not append authors' names to the species.—

Thank you heartily for what you say about my Book; but you will be greatly disappointed; it will be grievously too hypothetical. It will very likely be of no other service than collocating some facts; though I myself think I see my way approximately on the origin of species. But, alas, how frequent, how almost universal it is in an author to persuade himself of the truth of his own dogmas. My only hope is that I certainly see very many difficulties of gigantic stature.

Believe me | My dear Sir | Yours very sincerely | C. Darwin

If you can remember any cases of one introduced species beating out or prevailing over another, I sh^d be most thankful to hear it.—[6]

I believe the common Corn Poppy has been seen indigenous in Sicily; I sh^d like to know whether you suppose that seedlings of this wild plant, would stand a contest with our own Poppy: I sh^d almost expect that our Poppies were in some degree acclimatised & accustomed to our corn-fields. If this could be shown to be so in this & other cases I think we could understand why many not-trained American plants would not succeed in our agragrian habitats.

Royal Botanic Gardens, Kew (Bentham Letters: 682–3)

[1] The year is given by the reference to Bentham 1857 (see n. 3, below).
[2] The letter has not been located.
[3] Bentham 1857.
[4] See letter to George Bentham, 15 December [1857], and letter from George Bentham, [16 or 17 December 1857]. The list is in DAR 160.1: 151/3. The results of Karl Friedrich von Gärtner's experiments on *Silene* were described in Gärtner 1849, p. 722.
[5] See letter to George Bentham, 15 December [1857].
[6] Bentham apparently supplied CD with information about foreign seed having been sown near Montpellier. CD concluded that 'we must suppose the native plants in the long run beat the foreigners in the spots where both could grow.' (*Natural selection*, p. 193).

From J. D. Hooker [2 December 1857][1]

Kew
Wednesday

Dear Darwin

I arrived last Monday, having left my wife at Hitcham—[2] all were well & cheerful there—when I left— Old Henslow was very much cut up, though being wondrous stolid he did not show it much & he, in common with all the family, occupied themselves throughout industriously. Poor Mrs. H went off as peacefully as possible,—without a struggle, having had no pain for the last 12 hours.— She was a very gentle lady, who I had become very deeply attached to of late years. She inherited the calm self-possession of the Jenyns to the full;[3] had long known that any severe attack would certainly carry her off, & had led that kind of life which regarded death as going out of one room into another—happy soul!

Henslow was most sincerely pleased with the letters of his friends on the occasion, & I need not say with your's as much as any— all spoke lovingly of her, as you did. too, & the old gentleman was always repeating *that*.

I took down the most difficult genus of Indian plants I could think of to work at.—viz *Impatiens* of which there are just 100 Indian species! I have made the first rough draft of a monograph of them.[4] & was interested at the result as bearing on one of your problems, The genus is an extremely local one, there being both very few extra-Indian species, & the Indian species being extremely local! & you will be glad to know that they do not run into marked varieties at all, nor the species badly into one another. though somewhat.

I have still some of your questions to send you answers to & others I cannot answer I am sorry to find.[5]

Have you read Livingstone? he says that the proximity of wild bitter cucumbers to cultivated ditto, makes the fruit of the latter bitter too![6]

With best regards | Ever yours | J D Hooker

Lindley was received at R.S. with great applause[7]

DAR 104: 178-9

CD ANNOTATIONS
0.1 Kew . . . *that*. 2.3] *crossed pencil*
4.1 I have . . . applause 7.1] *crossed pencil*

[1] The Wednesday before letter to J. D. Hooker, 4 December [1857].
[2] Harriet Henslow had died in her home in Hitcham, Suffolk, on 20 November 1857 (see letter to J. D. Hooker, 21 November [1857]).
[3] Harriet Henslow was the daughter of George Leonard Jenyns of Bottisham Hall, Cambridgeshire, and the sister of Leonard Jenyns.
[4] In J. D. Hooker and Thomson 1859, p. 106, Hooker wrote of *Impatiens* that 'it would be difficult to indicate another genus in the vegetable kingdom, presenting amongst its species so many and such different modifications of structure, and of which the species are so universally and so excessively prone to vary.'
[5] See letter to J. D. Hooker, 20 October [1857].
[6] The crossing of sweet and bitter cucumbers and melons was reported in Livingstone 1857, pp. 48–9. Hooker identified the botanical specimens described in the work.
[7] John Lindley was awarded one of the Royal Society's Royal Medals at the society's anniversary meeting on 30 November 1857 (*Proceedings of the Royal Society of London* 9 (1857–9): 39–40).

From Richard Bishop to Charles Spence Bate[1] 3 December 1857

Plymõ
Decr 3. 57.

My dear friend,

I have much pleasure in answering, to the best of my power, the inquiries made by Mr Darwin respecting the impregnation of the Balanus;—had I known that the observation was new, or could possibly possess so much interest as your esteemed correpondent attaches to it, I would certainly have watched the process more *critically*.[2] I can only hope on some future occasion to be enabled to supply more precise information, if some other observer should not forestall me.

To take M[r] Darwins queries seriatim—

1st The *Species*—Balanus communis?

2nd Was the penis inserted into more than one individual? not *positive*, but I believe it was. It was extended to a length equal to about 3 times that of any single branch of the cirrus, & waved in every direction till it came in contact with the cirrus of a neighbouring balanus, when it was inserted, I believe into more than one. It should be observed that my group is not large, consisting of only 5 full sized animals & a few small ones.

3rd The insertion did not occupy more than 2 or 3 seconds.

4th Not inserted deeply as far as I could judge—at *which end* of valves was not noticed.

5 & 6. My *own* impression is *decidedly*, that the recipient individual was during the time exerting its cirri, with more than usual energy, & gave evidence of the intruder being a welcome guest, but on these points my memory is unfortunately not supported by that of a friend who joined me in noticing the act.

7. The specimens were under water at the time, I have not under other circumstances seen any movement of the cirri.

The particular instance referred to occurred on a fine bright day, about 3 months ago, when the sun was shining full on my glass, and the specimen had been recently taken & was in full vigour. The group still appears healthy, but I have not since observed a repetition of the indulgence of its amatory propensities. Is it a question of *season*?

I remain | yours sincerely | R[d] Bishop

Spence Bate.

DAR 160: 189

CD ANNOTATIONS
Top of first page: 'Ch 3'[3] *brown crayon*

[1] CD had corresponded with Charles Spence Bate when CD was preparing his monograph on the Cirripedia (see *Correspondence* vol. 5). Bate had apparently written to CD about Bishop's observations (see letter to Asa Gray, 29 November [1857]) and subsequently transmitted Bishop's response to CD's queries.

[2] CD had previously found only indirect evidence to suggest occasional cross-fertilisation between individuals of the hermaphrodite cirripede *Balanus balanoides* (*Living Cirripedia* (1854), p. 102).

[3] The annotation refers to CD's chapter 'On the possibility of all organic beings occasionally crossing' (*Natural selection*, pp. 35–91), in which CD discussed the possible pairing of hermaphrodite cirripedes like *Balanus balanoides*. The case given by Bishop was not, however, cited in CD's discussion here (*Natural selection*, p. 45) nor subsequently in *Origin*.

To J. D. Hooker 4 December [1857][1]

Down Bromley Kent
Dec. 4th

My dear Hooker

I send off this day D. C.[2] & the 2 pamphlets, for which very many thanks. I was particularly glad to have seen Thuret.—[3] I enclose the results of D. C. which, of course, I do not wish returned; but I sh[d] be very glad of your opinion on two points therein specified, whenever you have a little leisure, which probably will not be soon.—[4] I have to beg one other favour; viz name of any intelligent Curator (& permission to use your or Sir William's name as introduction) of any Botanic Garden in hot or hottish & *dryish* country, as Sydney or Cape of G. Hope, that I may enquire about temperate plants withstanding dryish *heat*, for my Glacial Chapter.[5] I know you once gave me reference to some remark of your own on tropical or temperate plants ascending or descending more on dry or damp mountains; but thinking it would never concern me, I did not keep it.

I sh[d] be very glad to hear sometime how Henslow & all the Family are after their dreadful loss.[6] Pray give my very kind remembrances & sympathy to M[rs] Hooker;[7] but perhaps she is yet at Hitcham, & indeed I do not know whether you are at Kew.—

My dear Hooker | Yours most truly | C. Darwin

The Parcel shall be paid to London.—

I hope you expressed my most sincere sympathy to poor dear Henslow.

Thank you very much for note just received telling me all I wanted to know about the Henslows—[8] Thanks.—

Endorsement: '1857'
DAR 114.4: 216

[1] The year is given by the reference to Harriet Henslow, who died on 20 November 1857.
[2] Candolle and Candolle 1824–73.
[3] Thuret 1854–5. CD cited this work in his chapter on hybridism, where he wrote: 'Mr. Thuret has shown that Fucus serratus could quite easily be fertilised by F. vesiculosus, whereas he never once could effect, after repeated trials, the reciprocal cross.' (*Natural selection*, p. 413).
[4] See following letter. The final results of CD's calculations derived from volumes of Candolle and Candolle 1824–73 are given in Table III of *Natural selection*, pp. 153–4.
[5] See following letter and letter to F. J. H. von Mueller, 8 December [1857].
[6] See letter from J. D. Hooker, [2 December 1857].
[7] Frances Harriet Hooker was the daughter of John Stevens and Harriet Henslow.
[8] Letter from J. D. Hooker, [2 December 1857].

From J. D. Hooker [6 December 1857][1]

Kew
Sunday

Dear Darwin

Your DC. results are very curious & suggestive.[2] I see no objection to your lumping all the orders (Polygon. Labiatæ, Scrophul. Borrag. together, but not

Proteaceæ, or if you do include *Proteaceæ* it should be also alluded to seperately in a foot note as well, because its distribution is so very local.[3]

Now with regard to Labiatæ there is a great deal to be said, though perhaps not much that will be satisfactory to you. In the first place it is not only as Natural an Order as any in the Vegetable Kingdom, but one of the largest & best limited & the most equably diffused. & is the best elucidated of any of its size both generically & specifically of any:—about this there are not two opinions amongst Botanists. It therefore forms a formidable obstacle to you & must be studied a little carefully.[4]

I would strongly wish to see you take more vols of DC., especially Compositæ, Ericeæ & others, for I cannot think that your case will be established upon any but such evidence as is afforded by plants spread over the whole globe— That in small areas, the species of large genera are more variable than of small, may not argue that the same holds good for same genera in large areas[5]

Such facts as *Rubus* being variable in Europe & not so in Himalaya prove this— It is also to be remembered that a form marked enough to be ranked as a variety in a local Flora, may not have that value—in a general Flora.

Benthams late researches into the British Flora, have so greatly modified his views of the limits of species, that in my eyes they invalidate the results of local Floras very materially. He has now completed the MSS of his British Flora,[6] having studied every species from all parts of the world, & most of them alive in Britain, France & other parts of Europe— Well—he has turned out as great a lumper as I am! **& worse**.[7] Then did you see a paper of Decaisne's on Pyrus, translated in Gard. Chron about 3 weeks ago[8]—in which he adopts Thomsons & my views of species & says that if he had to monograph. Plantago again he would reduce whole *sections* to one species:—& of course as many species (i.e. marked forms would then rank as varieties.— Now it was Decaisne (a most admirable Botanist) who on receiving the Fl. Indica wrote me most kindly & earnestly begging me to reconsider my mode of viewing species, & hinting that I was going to the devil.— All this does not directly affect your results, but it shows that you should draw them from materials of all kinds—local & general, & from systematists of all shades of opinion. A comparison of Babingtons & Benthams Floras[9] will be *invaluable*: the latter will be out at Easter

I find that Linum Catharcticum has occasionally the upper leaves alternate—:[10] normally all are opposite except the upper floriferous, which latter are normally always alternate— I find that in one or two cases an alternate non-floriferous upper pair, which may be theoretically accounted for by the *[illeg]* of the flower-buds.[11]

I send the Cucubalus &c list,[12] & will forward the remaining answers soon

Write to D.r Ferdinand Mueller Govt. Botanist Melbourne Victoria Australia & use your own name—& ours too if you like— he will be tremendously proud to hear from you—is a good active Botanist & will do your behests—[13] He is coming home in 1859 so put him up to all you want done *toute suite*— Also try Chas: Moore Esq Govt. Botanic Garden Sydney—a second-rate man, but quite capable of giving you good information—[14] I know of nobody at Cape of Good Hope.

I think the fact of moisture favouring extension of species is a very important one for you

AL incomplete
DAR 104: 195–6, DAR 47: 192

CD ANNOTATIONS
1.3 include . . . seperately] *scored brown crayon*
5.3 He has . . . Easter 5.16] *crossed pencil*
6.1 I find . . . flower-buds. 6.5] 'Ch. 7.'[15] *added pencil*
7.1 I send . . . for you 9.2] *crossed pencil*

[1] The date is the Sunday following the letter to J. D. Hooker, 4 December [1857].
[2] See preceding letter.
[3] In CD's final tabulation, the orders mentioned are treated separately (see *Natural selection*, pp. 153–4).
[4] See *Natural selection*, pp. 155 and 156 n. 1, where CD discussed the case of Labiatae, in which the smaller genera have more varieties than the larger genera, thus constituting one of the orders in which CD's 'rule' did not hold.
[5] See letter to J. D. Hooker, 9 December [1857], for CD's response to this criticism.
[6] George Bentham's *Handbook of the British flora* (Bentham 1858) was published in 1858.
[7] Hooker was known for his tendency to 'lump' together plants that other botanists might consider separate species.
[8] The *Gardeners' Chronicle and Agricultural Gazette*, 14 November 1857, p. 773, carried a translation of a paper by Joseph Decaisne originally published in the *Bulletin de la Société Botanique de France* (Decaisne 1857). Decaisne, botanist at the Jardin des plantes, wrote that 'it would be a great acquisition if describers of plants would condense their species by reducing them to really stable and natural types instead of dividing and multiplying them *ad infinitum*, as has been the custom for the last 30 years. This opinion is not exclusively my own; it is also that of my excellent friend Dr. J. D. Hooker (Flora Indica, Introductory Essay, &c.)' (Decaisne 1857, p. 773). The references are to J. D. Hooker and Thomson 1855 and to J. D. Hooker 1853–5.
[9] Babington 1851 and Bentham 1858.
[10] In October 1857, CD had asked Hooker to investigate this point (see CD note attached to letter to J. D. Hooker, 20 October [1857]). For Hooker's further comments, see n. 11, below.
[11] On CD's note questioning Hooker about the leaf arrangement in *Linum* (see letter to J. D. Hooker, 20 October [1857], CD note), Hooker answered: 'Upper leaves often so when inflorescence begins. Sometimes without inflorescence ie below inflorescence'. Hooker also drew a diagram to illustrate this point.
[12] This list is probably the one referred to in the letter to George Bentham, 1 December [1857]. It was a list of species that Karl Friedrich von Gärtner considered to be species of the genus *Silene* that were capable of crossing and generating hybrids (Gärtner 1849, p. 722). CD subsequently extracted a further list of species that Gärtner had hybridised and which he considered were true species of *Cucubalus* (see letter to George Bentham, 15 December [1857] and letter from George Bentham, [16 or 17 December 1857]). The limits of *Silene* and *Cucubalus* were in doubt and CD wanted Bentham's opinion on the validity of Gärtner's species.
[13] See following letter.

14 For CD's correspondence with Charles Moore, see *Correspondence* vol. 7, letter from Charles Moore, 11 August 1858. William Jackson Hooker, director of the Royal Botanic Gardens at Kew, had opposed the appointment in 1848 of Moore as director of the moribund botanic gardens in Sydney (Gilbert 1986, p. 72). In 1855, when the Legislative Council of New South Wales was investigating charges of wrong-doing against Moore, William Hooker was alleged to have expressed the view that: 'Mr Moore is not, in my opinion, a scientific botanist. I have known him for some time; and he is an excellent practical head-gardener, but not a *botanist*, in the sense in which I understand the term.' (Gilbert 1986, p. 92).

15 This refers to chapter 7 of CD's species book (*Natural selection*, pp. 279–338). The chapter was entitled 'Laws of variation: varieties and species compared'.

To Ferdinand Jakob Heinrich von Mueller 8 December [1857][1]

Down Bromley Kent
December 8th

Dear Sir

Sir William & Dr Hooker have told me that I might use their names as an introduction to you.[2] I do not know whether my name is known to you; but I have so often heard of interesting Botanical facts of your discovery, from Dr Hooker, that I feel as if I had been introduced to you. I am very anxious for a piece of information which possibly you may be able to give me. It is whether very many British or north-European, *perennial* plants can withstand, living & *seeding*, under the climate of S. Australia.—[3] I presuppose that the plants are kept in a well-weeded garden, free from the intrusion of the native vegetation. They might be supposed to be watered in very dry weather; my point being whether they could withstand the heat, & perfect their seeds. I presume all S. European plants could succeed well; but I want to know whether the greater number of British or still more north-ranging plants could survive & seed.— If you could answer me this question even approximately, & would take the trouble to write to me, I sh^d be *very much obliged.*— I believe there a good many British naturalised plants in S. Australia; (Have you ever enumerated them?) & this of course answers my question *most fully* as far as they are concerned, even under conditions not the most favourable, as they have to struggle with the native vegetation.—

With apologies for troubling you, & with much respect, I beg leave to remain, Dear Sir | Yours very faithfully | C. Darwin

DAR 92: 31–2

1 Dated by the relationship to the preceding letter.

2 Mueller was the director of the botanic gardens in Melbourne, Australia.

3 CD evidently received a reply from Mueller, for he cited him on a different point in *Natural selection*, p. 553. CD later asked Charles Moore the same question about the ability of European plants to set seed in Australia (*Correspondence* vol. 7, letter from Charles Moore, 11 August 1858). See also following letter.

To J. D. Hooker 9 December [1857][1]

Down Bromley Kent
Dec. 9th

My dear Hooker

I am infinitely obliged for your opinion on the results from D. C. *If you can spare* 2 or 3 more vols. of D. C. & can send them by enclosed address on 16th, I will *most* gladly have them worked out: please choose whatever vols. you think best to try the result.[2]

But I cannot agree with you *for my object*, that general monographs are best: (1st) I presume the varieties w[d] be best known in small country like ours; 2[d] a very large genus might have very few species in many separate countries & then according to my doctrine, *on average* it w[d] not be a numerically increasing or varying genus. Again a genus, though small for its order in a monograph, might be large in any one country, & then it ought to be there on average an increasing or varying genus. For such & other reasons, I rely more on local floras, but I am very anxious to see how rule goes in whole orders. Generally perhaps universally I sh[d] have expected, owing to great diffusion of plants, that same rule would hold in all cases, viz in local & mundane flora.—

I will see how Labiatæ are in Ledebour, Asa Gray & Koch;[3] & I will divide the species in D. C. into two more equal bodies.—

I quite see that this case is a great blow to me; but please, observe, I now rest on pretty large induction. Britain by 3 separate & very different men (I long to try Bentham),[4] France Germany. N. Italy, Roumelia, Ledebour (tried by separate volumes, & as whole) United States Canary Is[d], India,, & N. Zealand.— All tell one story.—

Please thank M[r] Bentham about Silene,[5] & thanks for all other points in your letter.—

I have written to Müller & Moore.—[6]

You speak of my having "so few aids"; why you yourself for years & years have aided me in innumerable ways, lending me books, giving me endless facts, giving me your invaluable opinion & advise on all sorts of subjects, & more than all, your kindest sympathy.

My dear Hooker | Yours affectionately | C. Darwin

Endorsement: '1857'
DAR 114.4: 217

[1] Dated by the relationship to the letter to J. D. Hooker, 4 December [1857], and the letter from J. D. Hooker, [6 December 1857].

[2] CD eventually used six volumes of Candolle and Candolle 1824–73 for his calculations. They covered the orders Leguminosae, Rosaceae, Borragineae, Scrophulariaceae, Acanthaceae, Verbenaceae, Labiatae, Solanaceae, Proteaceae, and Polygonaceae (*Natural selection*, pp. 153–4).

[3] Ledebour 1842–53, A. Gray 1856a, and Koch 1843–4.

[4] CD was not able to incorporate data from Bentham 1858 into his table on the number of species and varieties in large and small genera (see *Natural selection*, pp. 149, 152). By the time Bentham 1858 was published, CD was busy writing *Origin*.

⁵ See letter to George Bentham, 1 December [1857], and letter from J. D. Hooker, [6 December 1857].

⁶ See preceding letter. CD's letter to Charles Moore, director of the botanic gardens at Sydney, Australia, has not been located.

From H. C. Watson 14 December [1857]¹

Thames Ditton | S.W.
14 Decʳ

My dear Sir

I will return your list of varieties in British Plts with some notes in course of few days.² But it is to be feared that I can very imperfectly meet your object.

1. As to the *actual commingling*:—it is seldom I can say aught of this. Many of the varˢ or subspecies I have seen only in gardens & herbaria. And of those which I have myself picked wild,—in divers instances I cannot now recollect whether mingled with the type or not.

2— As to *difference of situation*:— This is pretty often the case. For instance, there are maritime & alpine forms of many inland & lowland plants. In their appropriate places (shores & hills) they remain so different, as usually to be retained in lists for distinct species. And where the differences are less obvious, or transition forms frequent, they are set down as species & vary.— Corresponding instances occur with plants of relatively dry & wet places. But as the cause of the variety is here usually obvious, it is less frequently deemed a species, & often passed by.

Thus, Polygonum amphibium

α natans

β terrestre

are two states of a species, which no botanist would have united into one species, if they had arrived in England from 2 different countries. They are not in your list, just because they are known to be varieties from situation;—the very fact you want to reach thus causing their exclusion.

3. *Area.* Within our limited Country the area or geographic space of varieties is usually an included & considerably smaller portion of the area of the type species. (On a wider view, there are varieties outside the area of the type species. For instance, there are boreal or arctic forms of temperate species.) Alpine forms with us are usually not included in the area of the type species;—but they overlap somewhat.—

I wish your list had been taken from the 1857 edition of the London Catalogue.³

In answer to your question about the Cybele Britannica,—I quite hope to commence printing Vol. 4 by or before Spring. But the result of so much labour is so often compressed into two or three figures (noˢ) that progress is slow.—⁴

Yours very truly | Hewett C. Watson

To | C. Darwin | Esq

DAR 98: 11–12

CD ANNOTATIONS

0.1 Thames . . . not. 2.4] *crossed pencil*

Top of first page: 'Remarks on Range [*over* '&'] & Stations of varieties.—' *ink*; 'All used' *pencil, circled pencil*

[1] Dated by the relationship to the letter from H. C. Watson, 20 December [1857].

[2] CD's list of varieties extracted from the fourth edition of Hewett Cottrell Watson and John Thomas Irvine Boswell-Syme's *London catalogue of British plants* (Watson and Syme eds. 1853) is in DAR 45: 9–15. His calculations relating to the list are in DAR 16.1: 128–33, 173–88. See also letter from H. C. Watson, 20 December [1857].

[3] CD used Watson and Syme eds. 1853 for his Table I (*Natural selection*, p. 149), but for the ranges of species he used the subsequent edition (Watson and Syme eds. 1857) (*Natural selection*, p. 168).

[4] Although the printing of the fourth volume of Waton's *Cybele Britannica* commenced in 1858, it was not published until 1859 (Watson 1847–59, 4: 3).

To George Bentham 15 December [1857][1]

> Down Bromley Kent
> Dec. 15th—

My dear Sir

I am in truth ashamed to trouble you so soon again; but since I sent you the list of Silene, I find that I very stupidly overlooked a list of Cucubalus, which I had forgotten that Gærtner ranked as quite distinct genus.—[2]

I have now copied a list in pairs of the crosses which he made in Cucubalus, & it would be a very great kindness, if you would look it over & mark whether *in each pair*, you consider the male & female as only varieties of the same species or as distinct species, & allow me to state that such is your opinion.—[3]

My object is to show that where the fertility of a cross makes a very close approach to perfect fertility there is often difference of opinion whether the forms are varieties or species. If any of the pairs are undoubtedly good & distinct species, the case does not concern me. If this would not cause you much trouble, it w^d be a great kindness, for I have already got a curious parallel list of graduated evidence from fertility & ordinary evidence on what to call species and what varieties.—

My dear Sir | Yours very sincerely | C. Darwin

Royal Botanic Gardens, Kew (Bentham Letters: 681)

[1] Dated by the relationship to the letter to George Bentham, 1 December [1857].

[2] See letter to George Bentham, 1 December [1857], and letter from J. D. Hooker, [6 December 1857]. The list is in DAR 160.1: 151/1. Both lists were returned to CD enclosed in the letter from George Bentham, [16 or 17 December 1857].

[3] In a discussion of the difficulty of using fertility as a test for distinguishing species and varieties, CD cited Bentham's view that the *Cucubalus* plants on which Karl Friedrich von Gärtner experimented were 'only varieties of Silene inflata.' (*Natural selection*, p. 404).

From George Bentham [16 or 17 December 1857][1]

My dear Sir

I return the lists.[2] But in questions where Gærtner may have obtained unexpected results in hybridising one would require to know whether he has got the right species under h⟨is⟩ names for the confusion in ⟨ ⟩ genera as Silene is very great in Continental botanic gardens One often sees Silene nutans italica inflata and a few others doing duty for dozens of tender ones which have died. Silene vespertina is one which we used to grow in our flower gardens ⟨as⟩ S. bipartita. It is a very common Mediterranean seacoast plant. S. noctiflora is a cornfield annual which we have sometimes in England S. viscosa (Cucubalus viscosus Linn) is in the way of Italica but I believe ⟨ ⟩ species from the ⟨ ⟩ central Asia

Yours very sincerely | George Bentham

DAR 160: 151

[1] The date is based on the assumption that the letter was written after the letter to George Bentham, 15 December [1857], and before the letter to George Bentham, 18 December [1857].
[2] CD had sent a list of *Silene* species enclosed in the letter to George Bentham, 1 December [1857]. Another list of *Cucubalus* species taken from Gärtner 1849 was sent in the letter to George Bentham, 15 December [1857]. Bentham added comments to both lists, which are preserved together with this letter.

To T. H. Huxley 16 December [1857][1]

Down Bromley Kent
Dec. 16th

My dear Huxley

In my opinion your Catalogue is simply the very best Resume by far, on the whole Science of Natural History, which I have ever seen.[2]

I really have no criticisms; I agree with every word. Your metaphors & explanations strikee me as *admirable*. In many parts it is curious how what you have written agrees with what I have been writing, only with the melancholy difference for me that you put everything in twice as striking a manner, as I do.

I append more for the sake of showing that I have attended to the whole, than for any other object, a few **most** trivial criticisms.

I was amused to meet with some of the arguments, which you advanced in talk with me, on classification; & it pleases me, as my long proses were so far not thrown away, as they led you to bring out here some good sentences.[3]

But on classification I am not quite sure that I yet wholly go with you, though I agree with every word you have here said.— The whole, I repeat, in my opinion is admirable & excellent—

Ever yours | C Darwin

I return by the same post your pamphlet.—

[Enclosure][4]

p. 2. *all* crust of earth is not formed of *strata*, as in injected Trap and ⁵ masses of the wider extent.

p. 6. *indefinite* series of alterations: according to our present knowledge is not thus too strong, though I believe in it.

15. Silex *chief* ingredient of flint; may it not be called sole ingredient.

18. Sexes never combined in same individual. Quatrefages seems to believe in Serranus being hermaphrodite;[6] I forget author's name.

20. Is it wise to give, *without any allusion to common view*, your view of Star-fish &c. being Annulosa; the rest of world looking at them as type of one of the grandest Divisions of animal kingdom.[7]

27. I think little expansion is wanted in middle paragraph to show how a fish can be morphologically more complex and physiologically less so than the highest molluscs. I do not quite understand.

Higher in page. Sentence beginning "the other sense" was not *at first* clear, as I did not see with what contrasted.

39. I do not understand how you can say that if only fossils existed there would be no difficulty in practically species—as there is variation amongst fossils, as with recent, there seems to be same difficulty in grouping.

40. Top sentence strikes me as too long.

44. Mud being deposited in all parts of sea, sufficient to imbed remains and to be preserved to future generations, I believe this to be the *gravest of errors*: but I cannot enter into argument much too long.

57. Is it not bold to say one foot never deposited in one year: think of floods after earthquake, at least add never a foot over wide area.

Imperial College of Science, Technology and Medicine Archives (Huxley 5: 151); DAR 145

[1] The year is given by the reference to Huxley's *Catalogue* (see n. 2, below).

[2] CD was at the time reading proof-sheets of Huxley's 'explanatory introduction' to *A catalogue of the collection of fossils in the Museum of Practical Geology* (see letter to J. D. Hooker, 25 December [1857]). The work, with Robert Etheridge as co-author, remained unpublished until 1865.

[3] CD had discussed classification in letters to T. H. Huxley, 15 September [1857], 26 September [1857], [before 3 October 1857], and 3 October [1857].

[4] It is likely that the following comments on Huxley's introduction, transcribed from a copy in DAR 145, were enclosed with this letter. The original manuscript of the enclosure is not with this letter in the Huxley papers at Imperial College.

[5] A space was left at this point by the copyist.

[6] Quatrefages de Bréau 1855–6, p. 80, referred to fishes of the genus *Serranus*. The case is cited in *Natural selection*, p. 44, where CD also mentioned that Jean Louis Armand de Quatrefage de Bréau's source was perhaps a memoir by Adolphe Dufossé (Dufossé 1856).

[7] Huxley, drawing upon Johannes Peter Müller's researches on the developmental history of echinoderms, had arrived at the conclusion that the structure of the larvae of echinoderms was founded on a plan similar to the idealised form of annelid larvae. He proposed the unorthodox view that echinoderms were most closely related to the annelids (see T. H. Huxley 1856–7, 13: 635–9, and Winsor 1976, pp. 113–21).

To W. D. Fox 17 December [1857]

Down Bromley Kent
Dec. 17th

My dear Fox

I am very much obliged for your Rabbit Letter: I have forgotten your having before mentioned to me this breed: I presume it is that called at Shows the "Himalaya breed".—[1] You call it black nosed & brown-tailed; I suppose the ears are black. If *not*, & you could get me an old accidentally dead specimen I sh^d be very glad, but I can hardly doubt it is the so-called Himalaya, one of which I have now skeletonising from Zoolog. Soc^y.— I am *particularly glad to hear about this variety being so true*: its history has been published, & I have it somewhere, but for the life of me I cannot tell where.[2] It is was produced, **I think**, from a cross of two *sub*-breeds of the French Silver Grey Rabbit.— I know it is *closely* connected with Silver Greys. If the ears are **not** black, will you tell me: all that I have seen had *red* eyes.—

There is one odd point in this breed, which will serve to identify it, the young are quite white & the black ears &c appear subsequently.—[3] I had forgotten that the **under** side of tail was brown.— By the way Silver-Grey Rabbits are quite black when born, a curious contrast with so called Himalaya.—

I have heard from Mr Woodd & will second him & if possible attend the voting.—[4]

In regard to Douche you will find the *same* water will last sweet, especially with charcoal floating on it, for a *long time*.[5]

Yours affectionately | C. Darwin

Postmark: 18 DEC 1857
Christ's College Library, Cambridge (Fox 105)

[1] CD mentioned the Himalayan breed of domestic rabbits in the letter to W. D. Fox, 14 June [1856].

[2] In *Variation* 1: 108 n. 15, CD cited an article in the *Cottage Gardener* 18 (1857): 141–2, on the origin of the Himalayan rabbits. The author believed the breed resulted from a cross between a tame silver gray doe and a silver gray buck that had 'one-eighth of wild blood in him.' (*Cottage Gardener* 18 (1857): 141).

[3] In *Variation* 1: 109, CD cited Fox as having found that on occasion in these rabbits 'the young are born of a very pale grey colour'.

[4] Charles Henry Lardner Woodd, Fox's brother-in-law and a geologist with whom CD had earlier corresponded (see *Correspondence* vol. 4, letter to C. H. L. Woodd, 4 March 1850), was seeking membership of the Athenæum Club (see *Correspondence* vol. 7, letters to George Bentham, 27 January [1858], and W. D. Fox, 31 January [1858]). Woodd is not listed in Francis Gledstanes Waugh, *Members of the Athenæum Club, 1824 to 1887* (London, [1888]).

[5] CD had erected a douche in the garden of Down House in 1849 to enable him to continue the cold-water treatment that he had begun at James Manby Gully's hydropathic establishment in Malvern (see *Correspondence* vol. 4, letters to Susan Darwin, [19 March 1849], and to W. D. Fox, 4 September [1850]). By 1853, CD no longer used the douche (see *Correspondence* vol. 5, letter to Edward Cresy, 29 April [1853]).

From J. D. Hooker [17–23 December 1857][1]

Dear Darwin

I am disgusted at remembering suddenly that I had forgotten to send the vols: of DC.—[2] I shall send on Thursday by carrier that with Leguminosæ (& others)—for I am really interested in that curious anomaly.[3]

Bentham I know thinks that he has created too many genera & others far more.—but whether enough to influence the whole Flora is more than I can tell—[4]

I have begun my Introd: Essay to Tasmania Flora.[5] I think I shall confine it to a clear exposition of all the main features of the Flora of Australia & leave all conclusion drawing to others. I find there are already upwards of 100 Europ. plants well naturalize⟨d⟩ in vicinity of Melbourne already.

I send by post 2 tracts from A Gray— I cannot follow Dana's metaphysical ideas on species.[6] What do you say to them

Ever yours | J D Hooker

DAR 104: 194

CD ANNOTATIONS
0.1 Dear . . . others. 3.3] *crossed pencil*
3.3 Europ.] *crossed pencil*
4.1 I send . . . Hooker 5.1] *crossed pencil*

[1] Dated by the relationship to the letter to J. D. Hooker, 25 December [1857], in which CD reported that the books had arrived at Down that morning. The weekly carrier service between London and Down operated on Thursdays. Hooker had been asked to send the books on 16 December (see letter to J. D. Hooker, 9 December [1857]). Hooker's mention of sending them 'on Thursday' makes Wednesday, 23 December the latest possible day on which this letter could have been written.
[2] Candolle and Candolle 1824–73. See letter to J. D. Hooker, 9 December [1857].
[3] Hooker refers to CD's calculations on the order Labiatae (see letter from J. D. Hooker, [6 December 1857]).
[4] Bentham 1858.
[5] J. D. Hooker 1860. The introductory essay of the *Flora Tasmaniæ* (J. D. Hooker 1855[–60]) was published separately.
[6] The 'two tracts' were probably Asa Gray's review of Arthur Henfrey's *Elementary course on botany* (London, 1857) and James Dwight Dana's 'Thoughts on species' (Dana 1857), both of which appeared in the November issue of the *American Journal of Science and Arts*. Copies of both papers are in the Darwin Pamphlet Collection–CUL. For CD's opinion of Dana 1857, see letter to J. D. Hooker, 25 December [1857].

To George Bentham 18 December [1857][1]

Down Bromley Kent
Dec. 18th

My dear Sir

I am very much obliged for answers which were just what I wanted & quite as explicit as I expected.—[2] I have had already occasion to put in a salvo about doubtful identification of plants experimentised on.—[3] Cucubalus viscosus & Italicus are extremely sterile together as might have been expected from what you say. All the other forms are extremely fertile, only one degree below normal fertility, & which one degree of lessened fertility may, I believe, be accounted for by the requisite manipulation &c.—

Anyhow Gærtner finds the same slight degree of lessened fertility (or rather a *greater* degree of infertility) between Cowslip & Primrose, & the Blue & Red common Anagallis—[4]

With very many thanks | Yours sincerely | C. Darwin

Royal Botanic Gardens, Kew (Bentham Letters: 700a)

[1] Dated by the relationship to the letters to George Bentham, 1 December [1857] and 15 December [1857], and the letter from George Bentham, [16 or 17 December 1857].
[2] See letter from George Bentham, [16 or 17 December 1857].
[3] CD perhaps refers to his remarks concerning Karl Friedrich von Gärtner's crosses of *Datura stramonium* and *D. tatula* that 'it may be questioned . . . whether these forms should be considered as anything but varieties.' (*Natural selection*, p. 404).
[4] See *Natural selection*, p. 405.

From Edward Hewitt 18 December 1857

Eden Cottage, Spark Brook, | nr. Birmm.
Decr. 18/57.

My dear Sir.

I am really sorry your very courteous letter has been so long without reply, but an *unusually severe* personal illness, from family bereavement, was the only cause.[1] Thank God, I am now better!

As to Hybrids, the product of intermixture between the different varieties of Pheasants and domestic fowls, for years, and years, I admit it was alike my "hobby" and "my folly". At one time I had 12 pairs of birds thus mated in my own possession, including attempts to cross the common English Pheasant, with the Chinese Silver Pencilled;—the English with the Chinese Golden, and also each variety of these Pheasants, with the Common Fowl. I obtained product from each "cross" in the Pheasants, inter se, but could never get even one solitary chick from either variety of *Chinese* Pheasants and a fowl, though frequently in the case of the English bird. This appeared to me *very remarkable* at the time, as copulation was *quite* as frequent in the one case as the other.[2]

In **all** these trials I admit the male bird was the Pheasant, were I to "try again", I should reverse the sexes *also*;—(why?) this year a friend of mine had a common English Pheasant, the survivor of a pair reared tamely, the cock having been lost by a mischance. The owner begged the loan of a cock Silver Sebright Bantam—a beautiful bird, but *a notoriously bad* **breeder** *with his* **own** *hens*—yet strange to say, they reared an Hybrid from him, and the hen Pheasant.

What is most curious, this offspring has both **wattles** *and* a **rosey**-comb!!—not the **slightest** vestige of which, did I ever see before in any *Hybrid*. The formation of the nostrils is (as *always* hitherto the case,) similar to the mother-pheasant's, but to prove how strangely Nature confutes us, if we force her from her accustomed habits, this bird is not even so large as the Sebright. The plumage will be eventually quite as beautiful as eccentric, closely approximating that of a Chinese Silver *cock* Pheasant, but the ground (particularly on the scapular feathers,) will show a clear (tho' slight) guilding.

The above is decidedly the **only** instance *I ever knew* of a male fowl, breeding with a hen Pheasant, *of any variety whatever*. I purpose (if all goes on smoothly,) to write full particulars of this "curiosity," for the general benefit of poultry friends, after its adult plumage is fully complete.

Hybrids are commonly very difficult "to tell" (as to the sex,) until *pretty well adult*, but commonly, the superiority of size, where several are in rearing from the same nest, will be conclusive; I myself do not know any other test, whilst yet immature.

In reply to your remaining query, I have opened hundreds of eggs, containing the partially formed embryo of a Hybrid.[3]

I once possessed a cock (english) pheasant, "netted" at "barley-time", that became quite tame, and almost every egg laid to him was fructified, yet only very few *hatched* comparatively. judging from appearances, they had progressed to the sixth or seventh day of incubation, but in only one instance, every egg (of eleven) hatched, although the foster-mother (a Game Hen) actually eat up in toto one of her chicks, (that showed a slight disposition to hemorrhage) even before it was well out of the shell; yet with that fatality *peculiar* (*I* **do** believe) to *all* "pets", not one of the ten remaining ones came to maturity;—rats, cats, and (at last) some timber "blowing down", all lending a helping hand.

In all instances within my own knowledge of Hybrids, for the male Ph! and common Hen fowl (of any variety) the produce has been immensely in advance as to size, of *either* parent; it is therefore worth reflection, that (as before narrated) the sexes opposed of the parents, a pigmy is the order of the day. There must be some occult cause for this, however indescribable, and such a result, I for one, *was not in anyway* prepared for. Again there appears no obvious reason, why a Chinese Ph!, should *not* breed with a fowl, when our common one *will*; they copulate, yet *not* **one** egg fructifies. It proves a limit is placed to *our* caprices, for they will very rarely breed a "cross" (if ever) at liberty.[4] however carefully devised, in the animal world, otherwise most probably it would be pre-occupied by monsters, to the ruin and exclusion, of all others.

I am indeed, on reperusal, afraid you will think my letter tedious and far too prolix, it is also anything but carefully indited, still it expresses *my* convictions, and with every good wish, it is freely rendered.

I am, My dear Sir, | Yours very faithfully, | Edward Hewitt.

DAR 166: 196

CD ANNOTATIONS

1.1 I am . . . better! 1.3] *crossed pencil*
2.9 as copulation . . . other. 2.10] *scored brown crayon*
3.4 The owner . . . Pheasant. 3.6] *scored brown crayon*
4.1 What . . . comb!!] *scored brown crayon*
4.2 The formation . . . mother-pheasant's, 4.3] *scored brown crayon*
6.1 Hybrids . . . conclusive; 6.3] *scored brown crayon*
7.1 In reply . . . eggs,] *double scored brown crayon*

[1] Hewitt had suffered the sudden death of his only sister in February 1857, leaving him with an invalid mother to care for (*Cottage Gardener* 17 (1856–7): 384–6). CD probably wrote to Hewitt soon after learning from William Bernhard Tegetmeier that Hewitt had made a number of crosses between pheasants and fowl (see letter to W. B. Tegetmeier, 6 February [1857]). Tegetmeier drew upon Hewitt's experiences in the relevant section of his *Poultry book* (Tegetmeier ed. 1856–7, pp. 165–7).
[2] CD reported this information in *Natural selection*, p. 435 n. 8.
[3] CD cited Hewitt on this point in *Natural selection*, p. 422.
[4] The passage 'for they . . . liberty.' was added at the bottom of the letter with a cross to indicate its position in the main text.

From H. C. Watson 20 December [1857][1]

Thames Ditton
20 Dec[r]

My dear Sir

I have taken up your list of British plts with subspecies or varieties several times during the week, & still with the same idea arising;—namely, that I am not prepared to go seriatim through it in the manner, & with the brevity, you desired.—[2] I do not think any botanist can. And thus, I fear, the only result of my beginning to write notes on your list, is, that the list is defaced & spoilt for any other use.

In writing the final volume of my Cybele Britannica, I find myself unable to carry out the ideas or inquiries originally intended.[3] And why?— Mainly, because the limits of species are so uncertain in nature,—so dissimilar in books. Add to this, the extreme inequality in the Structural value of groups designated by the same abstract term (ex. gr. Orders). The result is, I can only group & compare details,—not condense them into real generalisations.

This leads me to devote many pages to my own notions about species & classifications,—rather irrelevant in a book on local botany;—& perhaps somewhat limping over that same ground which will be better trod by yourself.[4] I cannot find the proof of species being definite & immutable, whatever they may seem to be at any one time & spot.—

Sincerely | Hewett C.[l] Watson

Charles Darwin | Esq

DAR 98: 13–14

CD ANNOTATIONS
0.3 My dear . . . use. 1.6] *crossed pencil*

[1] Dated by the relationship to the letter from H. C. Watson, 14 December [1857].
[2] See letter from H. C. Watson, 14 December [1857]. The list, with Watson's markings, is in DAR 45: 9–15.
[3] Watson was preparing the fourth volume of *Cybele Britannica* (Watson 1847–59).
[4] Watson paid considerable attention to the problem of defining species (Watson 1847–59, 4: 27–44).

From Edward Hewitt   22 December 1857

Eden Cottage, Spark Brook, | n.ʳ Birm.ᵐ
Dec.ʳ | 22/57.

My dear Sir

I duly rec.ᵈ your letter and at once to the best of my knowledge, reply to your queries.[1] In retrospect, I certainly do think my time might have been far more profitably employed than in breeding Hybrids; although at the time I was making these efforts, my attention (so far as poultry-matters went,) rested on this almost exclusively. To the subject—

In the year I tried so many, and so various "crosses"; certainly more than 300 Eggs were placed for incubation under foster-mothers, of which as certainly not a dozen hatched, the eggs that proved faulty were opened only when there was absolute certainty they would not produce chickens, and all "hope" was gone of their so doing, simply to see if they were fecundated, and *guess*, if possible, "why" they did not progress.[2] The result is in my last letter. This I invariably noticed, that eggs laid in April and May, (by fowls to Pheasants,) were the *only* ones with me productive, later in the season failure was *universal*.[3] I bred several at different times from a Golden Cock Pheasant and Common Hen *Pheasant*, but not from Golden or Silver cock Pheasants, and a fowl. It was only the *English* cock Ph.ˢ with the **fowl** that produced.

I raised three (all hens,) and singularly enough, very plain coloured birds, from a Golden cock Ph.ˢ and Silver Hen Ph.ᵗ, but never could produce a single bird if I reversed the sexes, of the parents.[4] Indeed with **best** luck, the product was most uncertain. As unfortunately at that time I did not keep a diary, I have no present direct reference, but I well remember from 55 eggs, *under very careful hens*, only 3 youngsters saw day-light, yet the parents copulated openly, without fear of bystanders, several times daily, many eggs being germinated.[5]

I never knew, but of the single instance mentioned in my former note, of a *cock* **fowl** breeding with a hen *pheasant*, in all other cases, the hen pheasant quite declined the amours of her appointed companion, and indeed simple aversion at the onset, always ended in perfect hate. In the pair now spoken of—cock silverlaced Sebright with hen common Ph.ˢ—copulation was frequent and daily, yet only the one solitary egg fertilized out of many, the remainder being what are commonly called "clear-eggs". The really small size of this product is easily supposed by its weight not 24 ounces, and looking even *considerably* less to the eye, but it "handles" well. Another remarkable caprice of Dame Nature, (for thus putting her out of her way,) is, in the legs of this bird, f.m hatching time the colour of the legs and feet until lately was deep-horn inclining to a shade of blue, but now they are become of a kind of sullied creamy white, together with the bill, the upper mandible however, as yet retains a narrow stripe of the original colour down the centre throughout, but it also appears to be fast fading away. The comb and wattles although perfect, are rudimental, and will never be developed as in the sire.

Again altho' I have seen Hybrids of 6 or seven years standing, I never yet myself saw one with well-formed spurs, available to do real duty in an affray,[6] but Hybrids do not want them particularly for this purpose, for with the most uncontrolable predeliction for "a row", or rather to oppress all weaker companions, even their own kind, they oft-times become perfect cannibals, and will tear and eat their victim (if not removed) peacemeal even while living. The last I ever kept destroyed and consumed nearly the whole of the flesh of its only companion, all along the back before I was aware of the fact, tho' the poor creature still lived, and in perfect horror, I eschewed Hybrids from that very day. They are recklessly cruel and capricious with all about them, tyrannous, and unusually wild, even where their parents were perfect "pets."[7]

The Sebright Hybrid at present seems more docile, but youth is still on its side, and time will most likely tell tales. In conclusion, tho' carefully & most *narrowly* watched, I never saw any salacious efforts on the part of the males towards their similarly bred female companions, or in favour of other poultry, nor yet the females manifest the slightest natural desires, my experience tells me, Hybrids like to be solitary, and the aversion on the part of other poultry is reciprocal.

These few hurried notes may possibly amuse you, and with my best wishes | I am, My D.r Sir, | Yours very faithfully, | Edward Hewitt.

DAR 166: 197

CD ANNOTATIONS
0.3 My dear . . . subject— 1.5] *crossed pencil*
3.7 many . . . germinated.] *underl pencil*

[1] CD's letter, written in response to the letter from Edward Hewitt, 18 December 1857, has not been located.
[2] Information from this letter was reported in CD's discussion of the sterility of hybrids (see *Natural selection*, pp. 422, 429). CD also cited Hewitt's view that the early death of embryos caused the sterility of such crosses in *Origin*, p. 264.
[3] Cited in *Natural selection*, p. 431.
[4] Cited in *Natural selection*, p. 435 n. 9.
[5] Cited in *Natural selection*, p. 422.
[6] Cited in *Natural selection*, p. 452.
[7] CD used this information in his discussion of the instincts of animals (*Natural selection*, p. 486).

To A. R. Wallace   22 December 1857

Down Bromley Kent.
Dec. 22/57

My dear Sir

I thank you for your letter of Sept. 27th.—[1] I am extremely glad to hear that you are attending to distribution in accordance with theoretical ideas. I am a firm believer, that without speculation there is no good & original observation. Few travellers have ⟨at⟩tended to such points as you are now at work on; & indeed the whole subject of distribution of animals is dreadfully behind that of Plants.— You say that you have been somewhat surprised at no notice having been taken of your paper in the Annals:[2] I cannot say that I am; for so very few naturalists care for anything beyond the mere description of species. But you must not suppose that your paper has not been attended to: two very good men, Sir C. Lyell & M[r] E. Blyth at Calcutta specially called my attention to it.[3] Though agreeing with you on your conclusion⟨s⟩ in that paper, I believe I go much further than you; but it is too long a subject to enter on my speculative notions.—

I have not yet seen your paper on distribution of animals in the Arru Isl[ds]:—[4] I shall read it with the **utmost** interest; for I think that the most interesting quarter of the whole globe in respect to distribution; & I have long been very imperfectly trying to collect data for the Malay archipelago.—

I shall be quite prepared to subscribe to your doctrine of subsidence: indeed from the quite independent evidence of the Coral Reefs I coloured my original map in my Coral volume of the Arru Isl[d] as one of subsidence, but got frightened & left it uncoloured.—[5] But I can see that you are inclined to go **much** further than I am in regard to the former connections of oceanic islands with continent:[6] Ever since poor E. Forbes propounded this doctrine, it has been eagerly followed; & Hooker elaborately discusses the former connections of all the Antarctic isl[ds] & New Zealand & S. America.— About a year ago I discussed this subject much with Lyell & Hooker (for I shall have to treat of it) & wrote out my arguments in opposition; but you will be glad to hear that neither Lyell or Hook⟨er⟩ thought much of my arguments: nevertheless for once in my life I dare withstand the almost preternatural sagacity of Lyell.—[7]

You ask about Land-shells on islands far distant from continents: Madeira has a few identical with those of Europe, & here the evidence is really good as some of them are sub-fossil.[8] In the Pacific isl[ds] there are cases, of identity, which I cannot at present persuade myself to account for by introduction through man's agency; although D[r] Aug. Gould has conclusively shown that many land-shells have there been distributed over the Pacific by man's agency.[9] These cases of introduction are most plaguing. Have you not found it so, in the Malay archipelago? it has seemed to me in the lists of mammals of Timor & other islands, that *several* in all probability have been naturalised.

Since writing before, I have experimentised a little on some land-mollusca & have found sea-water not quite so deadly as I anticipated.[10] You ask whether I shall discuss "man";—I think I shall avoid whole subject, as so surrounded with prejudices, though I fully admit that it is the highest & most interesting problem for the naturalist.— My work, on which I have now been at work more or less for 20 years, will *not* fix or settle anything; but I hope it will aid by giving a large collection of facts with one definite end: I get on very slowly, partly from ill-health, partly from being a very slow worker.— I have got about half written; but I do not suppose I shall publish under a couple of years. I have now been three whole months on one chapter on Hybridism![11]

I am astonished to see that you expect to remain out 3 or 4 years more:[12] what a wonderful deal you will have seen; & what interesting areas,—the grand Malay Archipelago & the richest parts of S. America!— I infinitely admire & honour your zeal & courage in the good cause of Natural Science; & you have my very sincere & cordial good wishes for success of all kinds; & may all your theories succeed, except that on oceanic islands, on which subject I will do battle to the death

Pray believe me. | My dear Sir | Yours very sincerely | C. Darwin

British Library (Add. 46434)

[1] Letter from A. R. Wallace, [27 September 1857].
[2] Wallace 1855.
[3] See *Correspondence* vol. 5, letter from Edward Blyth, 8 December 1855. Charles Lyell had read Wallace's paper in November 1855 and was much struck by it, for he began his scientific journals with notes on Wallace 1855 (Wilson ed. 1970, p. xli). CD had interleaved notes on Wallace 1855 in his copy of the journal in which it appeared. These are transcribed in *Correspondence* vol. 5, letter from Edward Blyth, 8 December 1855, n. 1.
[4] Wallace 1857.
[5] CD refers to a map showing the distribution of coral reefs throughout the world in *Coral reefs*, plate 3. The coloured areas indicated great zones of the sea-floor that were undergoing either elevation or subsidence. Wallace suggested in his paper that the Aru Islands once formed part of New Guinea, the intervening land having subsided (Wallace 1857, p. 479).
[6] To account for the similarity between many Australian genera and species and those found on Aru, Wallace suggested that these regions might have been connected at one time through New Guinea. He argued that other islands with rich floras similar to continental floras could be shown by geological evidence to have been recently united with continents, giving Britain and Sicily as examples. (Wallace 1857, pp. 478–9).
[7] See letters to Charles Lyell, 5 July [1856], and to J. D. Hooker, 5 July [1856].
[8] See *Correspondence* vol. 5, letter from R. T. Lowe, 19 September 1854.
[9] A. A. Gould 1852–6.
[10] See letter from T. V. Wollaston, [11 or 18 December 1856].
[11] According to CD's 'Journal' (Appendix II), he began the chapter on hybridism on 30 September and completed it on 29 December 1857.
[12] Wallace returned to England from Singapore in 1862, having spent eight years in the Malay Archipelago (*DNB*).

To J. D. Hooker 25 December [1857][1]

Down Bromley Kent
Dec. 25th

My dear Hooker.

The Books arrived quite safely this morning.[2]

Emma desires her thanks to Mrs Hooker for her note, but she is sorry she thought of troubling herself about the socks.

I will do D. Candolle first, as I guess that most likely to be wanted: the job will be a long one.—[3]

I am very glad to have R. Brown's Prodromus.—[4]

Whenever I try the Labiatæ in different ways, which will not be soon, you shall hear the result, as I shall wish much to tell it.

Thanks for Dana's pamphlet:[5] I am much disappointed in it. I have high respect for Dana, but in this brochure "I do not think he is much of a Baconian philosopher," as our friend J. E. Gray says.— I hate such far-fetched analogies, as comparing an atom of oxygen with a living being.— I believe, poor fellow, he believes in 1st Ch[r] of Genesis, so great allowances must be made for him.[6]

I have been reading in proof Sheets Huxley's "Explanatory Catalogue" & it seems to me quite an admirable, but very brief, resume on the Natural Sciences[7]

I am very sorry to hear that you do not intend to give generalisations in your Tasmanian Introduction, but I do not believe you will be able to resist: what is in the spirit must come out.[8]

I hope that poor dear Henslow is going on well.—

Ever yours | C. Darwin

I have just finished a tremendous job, my chapter on Hybridism: it has taken me 3 months to write, after all facts collected together![9]

Endorsement: '/57'
DAR 114.4: 218

[1] The year is established by CD's reference to having finished his chapter on hybridism, which was completed on 29 December 1857 ('Journal'; Appendix II).
[2] See letter from J. D. Hooker, [17–23 December 1857].
[3] CD did not conclude his statistical analysis of large and small genera until the spring of 1858 and rewrote his discussion of this work between April and June 1858 (J. Browne 1980, p. 87).
[4] Brown 1810.
[5] Dana 1857 (see letter from J. D. Hooker, [17–23 December 1857]).
[6] In 'Thoughts on species' (Dana 1857), James Dwight Dana tried to establish the immutability of biological species by analogy with chemical species or elements.
[7] T. H. Huxley and Etheridge 1865. See also letter to T. H. Huxley, 16 December [1857].
[8] J. D. Hooker 1860. See letter from J. D. Hooker, [17–23 December 1857] and n. 5.
[9] See preceding letter, n. 11.

From James Hunt [before 29 December 1857][1]

The Com[n] Song Thrush, & American Robin or thrush, have bred together, but the young did not live

The Black, or Globose Currassow, & Red Curassow have bred here, several years ago,[2] but are they not at present consided the same Species.

The different kind of Guinea Fowl, have never been cross^d here or the Peafowl.

I am very Sorry that I have not been able at present to find the Notes on the Crossbred Pintail Ducks. You wished in your Letter, and as I cannot speak possitively from memory, I have not stated any thing about them, but I will continue serching for them, and should they be found iwill send them to you instantly.[3]

And I shall be most happy to give you any information, on this or any other subject that lays in my Power.

I am Sir | Your most | Obedent Servant | James Hunt

Chas Darwin Esq.

DAR 166: 281

CD ANNOTATIONS
1.0 The . . . Esq. 7.1] *crossed pencil*
Verso of letter: 'Are not all rules false for Cultivated Plants' *pencil*

[1] CD used the information in this letter in his chapter on hybridism, which he recorded having completed on 29 December 1857 ('Journal'; Appendix II).

[2] Hunt was the head keeper at the zoological gardens of the Zoological Society of London until 1859 (Scherren 1905, p. 104). CD had consulted him previously (see Notebook M, p. 138 (*Notebooks*)).

[3] Hunt apparently did not find the notes. In the manuscript of his species book, CD stated that Hunt showed him 'several years ago a lot of young birds, which he knew to be the offspring of a pair of these hybrids inter se' (*Natural selection*, p. 439; see also p. 433 n. 1).

APPENDIX I
Translations of letters

From Victor de Robillard[1] 20 September 1856

Port Louis, Ile Maurice,
20 September 1856.

To Mr Charles Darwin.

Sir,

Mr. Ch. T. Beke[2] has communicated to the Natural Histroy Society here, of which I am a member, a letter received from you on various points which are the object of your research. As I have been engaged for many years in forming a collection of shells, which has given me the opportunity to go frequently to the sea shore, I will send you my remarks, although the subject that interests you did not especially attract my attention.

1. I have never seen trees thrown up on the shore by the sea; after hurricanes the flooding of rivers sometimes transports some to the estuaries, and then the sea brings them back to the shore, but then they are recognisable as trees of the island itself— After tidal waves I have often seen many seeds on the shore, but they came from marine plants— However, I recall having seen fairly large sized seeds, but I did not examine them enough to know whether they belonged to marine plants or to trees; if they were the seeds of trees, it could be that they reached the sea with river flood-waters and they could have come from the interior of the island, from those parts where the rivers cut through woods and forests— I am therefore unable to say anything precise on this point.

2. I am not aware of there having been here flocks of birds that have migrated and reached the islands; all those that live here have been introduced from other countries— Neither have I ever known of new species of birds having been found that have arrived on the island after losing their way while migrating from one land to another.

3. As far as domesticated animals and others are concerned, many species that have become acclimatised have been introduced from different countries— our horses have come from the Cape, Australia, Timor Islands, Pegu,[3] England and France, Buenos Aires—mules from Buenos Aires, France, the Red Sea, the Cape, the Persian Gulf. Cattle from the Cape, Australia, Madagascar, India—sheep and goats from India, Arabia, Australia, the Cape, Abyssinia; poultry from India, Madagascar, some species from Europe and the East Coast of Africa. Pigeons

from France and India; dogs from France, England, the Cape—pigs from France, England, India, the Malaysian Islands, China, Siam—guinea-fowl from Madagascar: Partridges from India—ducks and geese from the Cape, India, Madagascar—

Thus, Sir, before doing research

AL incomplete
DAR 205.3 (Letters)

[1] For the transcription of this letter in its original French and CD's annotations, see pp. 225-6. Robillard, a resident of Mauritius, was an active member of the natural history society of the island. He published several papers on Mauritian shells in the *Transactions of the Royal Society of Arts and Sciences of Mauritius*.

[2] Charles Tilstone Beke, an explorer of Abyssinia, had, in 1853, become a partner in a Mauritius mercantile house with the intention of opening up trade routes between Britain and Abyssinia. His second wife, Emily Alston, was from Mauritius. CD evidently contacted him in London, although Beke's name is not on CD's list of individuals who could provide him with overseas contacts (see *Correspondence* vol. 5, CD memorandum, [December 1855]).

[3] A district of Burma.

From Louis Sulpice Bouton to Charles Tilstone Beke[1] 24 September [1856][2]

My dear M[r] Beke

I shall do some research at leisure in the Museum and in the Transactions on the subject referred to in order to be able to send you later more precise details.[3]

Ever yours | L. Bouton

24 Sep

Extract from a letter of M Autard de Bragard— Savanne—[4] 17 Sept— 1856

In 1835—36—or 37, I sent to the Natural History Society a crested thrush from Bourbon[5] "Turdus cafer" taken at Morne Brabant a short time after a hurricane— Also another bird killed in the Savanne known to belong to a Madagascan species—

You will find these specimens in the Museum and details concerning them in the transactions of the Society—

DAR 205.3 (Letters)

[1] For the transcription of this letter in its original French and for CD's annotations, see p. 232.

[2] The year is given by the extract included in the letter.

[3] See letter from Victor de Robillard, 20 September 1856, in which it is mentioned that Charles Tilstone Beke had communicated a request from CD to the Société d'Histoire Naturelle of Mauritius. Bouton was secretary of the society until his death in 1878.

[4] Savanne is a district of south-west Mauritius.

⁵ Bourbon was the former name of Réunion Island, 130 miles south-west of Mauritius. The bird was possibly the now extinct Réunion huppe (see letter from Victor de Robillard, 26 February 1857).

From Victor de Robillard 26 February 1857[1]

Port Louis, | Mauritius
26 Feb /57.

My dear Sir,

I have received your letter of 7 December and I thank you for the approaches you have made to Mr Cuming for the shells;[2] I see from the written reply that he has made to you that he wants to sell rather than exchange; thus I have written to him and am sending him a list of the shells that I want in order to find out the prices.

I shall now give you the information that you asked me for, according to my view of things:

You say that in some cases, the same molluscs are found inhabiting different islands, although separated by the sea. In the islands that surround us, it is very rare to see the same land shells on different islands— Mauritius has many more land shells than Bourbon,[3] and they are not the same species, with the exception of an achatina in common and a small cyclostome of the woods— none of the land shells of the Seychelles exist on Mauritius nor do those of Madagascar, which would confirm the opinion of those who believe that the islands have risen in isolation from the depths of the ocean and have not been separated by volcanic upheaval and submarine eruptions, as some maintain—

I do not think that eggs could be transported by the sea from one island to another; they would encounter too many enemies to be able to reach another shore, where the shocks in any case could destroy the vital principle—

Several years ago a beautiful red-mouthed achatina of Madagascar was brought here, which has spread much and has quickly become acclimatised.

As for the freshwater molluscs, there are two or three species of Mauritius that are found on Bourbon; I think they are indigenous to the two islands without having been transported from one to the other. Their eggs must travel down to the sea through the overflowing of rivers but the passage from one island to another should not happen, they should be destroyed in the same manner as those of land molluscs—

I believe that now it is very easy to transport species from one island to another, either the molluscs themselves or their eggs, it only takes the will of man.

As for the most recent discoveries of birds on Mauritius or Bourbon, nothing new has been encountered since the Dodo and the Solitaire.[4]

On Bourbon there is a bird which lives secluded in the mountain woodlands, which is becoming very rare there, it is known by the name of Huppe;[5] it has never been met with on Mauritius—

Among the insects, I have noticed for a few years here, a blue butterfly, which is said to exist on Madagascar and a white one, which may be from the African coast— How have they arrived here? Presumably by means of plants underneath which the eggs happened to have been attached when the insects laid.

You ask me whether during hurricanes light objects could be transported from Bourbon to Mauritius? Following the accepted theory of the formation of this great disturbance of nature and from the manner in which they are created and move, it is a certain fact established by a number of observations, that is

AL incomplete
DAR 205.3 (Letters)

[1] For a transcription of this letter in its original French and CD's annotations, see pp. 347–9.
[2] Hugh Cuming had assembled one of the largest shell collections in Britain.
[3] Bourbon, now called Réunion, was a French colony in the Indian Ocean, 130 miles south-west of Mauritius.
[4] The dodo of Mauritius and solitaire of Bourbon were extinct (Strickland and Melville 1848, p. 27, 60).
[5] The Réunion huppe is now extinct (Staub 1976).

APPENDIX II
Chronology 1856–7

This appendix contains a transcription of Darwin's 'Journal' for the period 1856–7. Darwin commenced his 'journal' in August 1838 and continued to maintain it until December 1881. In this small notebook, measuring 3 inches by 4½ inches, Darwin recorded the periods he was away from home, the progress and publication of his work, and important events in his family life.

The version published by Sir Gavin de Beer as 'Darwin's Journal' (1959a) was edited before the original 'Journal' had been found and relied upon a transcription made by an unknown copyist. The original, now in the Darwin Archive, Cambridge University Library (DAR 158), reveals that the copyist did not clearly distinguish between the various types of entries it contains.

From 1845 onward, Darwin recorded all that pertained to his work (including his illnesses, since these accounted for time lost from work) on the left-hand pages of the 'Journal', while the periods he was away from home and family events were noted on the right-hand pages. In order to show clearly Darwin's deliberate separation of the types of entries he made in his 'Journal', the transcription has the left- and right-hand pages labelled, with the left-hand page of an opening preceding the right-hand page.

All alterations, interlineations, additions, and the use of a different ink or pencil have been noted. In addition, the editors have inserted relevant information throughout this transcription of the 'Journal' for 1856–7. These interpolations are enclosed in square brackets to distinguish them from Darwin's own entries.

———

[Left]
1856
[6 May. Read 'The action of sea-water on the germination of seeds' to the Linnean Society of London.][1]
May 14th Began by Lyells advice *writing* species sketch.—
Oct 13th Finished 2d Chapt. (& before part of[2] Geograph. Distr.)
Dec 16th '' 3d Chapt.[3]

[Right]
[8 January. Attended meeting of Philoperisteron Society.][4]
[31 January. Attended council meeting of the Royal Society.][5]

[10–14 March. Stayed in London. Attended council meeting of the Royal Society on 13 March.][6]
[13–16 April. Mary Elizabeth and Charles Lyell visited.][7]
[17 April. Attended council meeting of the Royal Society.]
[22–28 April. Frances Harriet and Joseph Dalton Hooker visited.][8]
[25–28 April. Thomas Vernon Wollaston visited.][9]
[26–28 April. Henrietta Anne and Thomas Henry Huxley visited.][10]
[26 April. John Lubbock dined.][11]
[5–8 May. Stayed in London. Attended council meeting of the Royal Society on 8 May.][11]
[29 May. Went to London with Henrietta Emma and George Howard Darwin. Attended council meeting of the Royal Society on 30 May.][13]
[12 June. Attended council meeting of the Royal Society.]
[18–21 June. Stayed in London and attended council meeting of the Royal Society on 19 June.][14]
[16 July. Dined at home of Philip Henry Stanhope.][15]
[29 July. Attended Anerley poultry show.][16]
[14 August. Visited London.][17]
[25 August. Took Henrietta to London to see Benjamin Collins Brodie.][18]
[8 September. Dined at Holwood.][19]
Sept 13th Leith Hill, returned 19th.—[20]
[15–18 October. Stayed in London and attended meeting of the Philosophical Club on 16 October.][21]
November[22] At. Sarah died[23]
[13 November. Attended meeting of the Philosophical Club.]
Dec 6th Charles Waring Darwin Born[24]

[Left]
1857
Jan 26th Finished Ch. 4. Var. Natural
March 3d.[25] Find Ch. 5 Struggle for Existence
March 31st finished Ch. 6. Nat. Selection
Sept. 29th finished Ch. 7 & 8; but one month lost at Moor Park.
Sept. 30th–to December 29th on Hybridism.—

[Right]
1857
[13–16 January. Went to London with William Erasmus Darwin. Attended meeting of the Philosophical Club and lecture at the Royal society on 15 January.][26]
[4–7 March. Stayed in London and attended meeting of the Philosophical Club on 5 March. Brought George home from school.][27]

April 22ᵈ Moor Park. Returned May 6ᵗʰ Did me astonishing good
[11 June. Attended meeting of the Philosophical Club.]
June 16ᵗʰ to Moor Park: returned on 30ᵗʰ— Etty there[28] On 27ᵗʰ went to Selbourne[29]
[10 August. Attended Crystal Palace poultry show.][30]
[13–14 August. John Stevens Henslow visited.][31]
In latter part of Sept. for about week, Lenny had very intermittent Pulse, but now Oct 6ᵗʰ seems quite well; latter part of Oct. occasionally poorly. Nov. 13ᵗʰ seems quite well[32]
[31 October. Joseph Dalton Hooker visited.][33]
(Nov. 5ᵗʰ–12ᵗʰ Moor Park)

[1] *Collected papers* 1: 264–73.
[2] 'before part of' *added in pencil above* 'Geograph. Distr.'
[3] 'Dec . . . Chapt.' *written in pencil;* '16' *over* '14'.
[4] *Cottage Gardener* 15 (1855–6): 301.
[5] Information on CD's attendance at council meetings of the Royal Society is taken from the Royal Society Council minutes.
[6] Emma Darwin's diary.
[7] Emma Darwin's diary.
[8] Emma Darwin's diary.
[9] Emma Darwin's diary.
[10] Emma Darwin's diary.
[11] Emma Darwin's diary.
[12] Emma Darwin's diary.
[13] Emma Darwin's diary.
[14] Emma Darwin's diary.
[15] Letter to J. D. Hooker, 13 July [1856].
[16] Letters to W. B. Tegetmeier, 14 August [1856], and to John Lubbock, [29 July 1856]. The Anerley show was held in Anerley Gardens, near Sydenham, Kent.
[17] Emma Darwin's diary.
[18] Emma Darwin's diary. Henrietta Emma Darwin's health began to fail in 1856 and worsened in the early months of 1857. On 9 April 1857, Emma Darwin took her to the seaside at Hastings, Sussex, where Henrietta remained until 12 May (Emma Darwin's diary).
[19] Emma Darwin's diary. Holwood House, 1½ miles from Down near Keston, Kent, was the seat of Robert Monsey Rolfe, Baron Cranworth (1790–1868), lord chancellor of England, 1852–8 (*DNB*).
[20] Leith Hill Place, Surrey, was the home of Emma Darwin's brother Josiah Wedgwood III and his wife Caroline, CD's sister.
[21] Letter to J. D. Hooker, 9 October [1856]. Information on CD's attendance at Philosophical Club meetings is taken from the Royal Society Philosophical Club minutes.
[22] 'November' *after del* 'Oct'.
[23] Sarah Elizabeth (Sarah) Wedgwood, the last surviving child of Josiah Wedgwood of Etruria and Emma and CD's aunt, died at Petleys, Down, on 6 November 1856 at the age of 80 (*Emma Darwin* 2: 161).
[24] 'Dec . . . Born' *written in pencil*.
[25] 'March 3ᵈ' *above del* 'Feb 27ᵗʰ'.
[26] Emma Darwin's diary. Thomas Henry Huxley's and John Tyndall's paper 'On the structure and motion of glaciers' was read to the Royal Society on 15 January (see letter to T. H. Huxley, 17 January [1857]).

[27] Emma Darwin's diary. George Howard Darwin, then aged 11, had entered Clapham Grammar School, in south London, in the autumn of 1856.

[28] 'Etty there' *added*. Emma Darwin had taken Henrietta Darwin to Edward Wickstead Lane's hydropathic establishment at Moor Park, near Farnham, Surrey, on 29 May, where she remained until 7 August. Henrietta returned to Moor Park for another extended stay on 22 August (Emma Darwin's diary). Recalling this period of ill-health, Henrietta Litchfield wrote: 'Both parents were unwearied in their efforts to soothe and amuse whichever of us was ill; my father played backgammon with me regularly every day, and my mother would read aloud to me.' (*Emma Darwin* 2: 163).

[29] According to Francis Darwin (*LL* 2: 67), during his stay at Moor Park CD 'made a pilgrimage to the shrine of Gilbert White at Selbourne [Hampshire].' For CD's reading of editions of White's *The natural history and antiqities of Selborne*, see *Correspondence*, vol. 4, appendix IV.

[30] Letter to John Lubbock, 11 August [1857]. The Crystal Palace poultry show, held twice yearly from 1857, was the largest in the London area.

[31] Letter to J. S. Henslow, 10 August [1857].

[32] For CD's fears about Leonard Darwin's failing health, see letter to J. D. Hooker, 30 September [1857].

[33] Letter to J. D. Hooker, [23 October 1857].

APPENDIX III

Dates of composition of Darwin's manuscript on species

Many of the dates of letters in this volume of correspondence have been based on or confirmed by reference to Darwin's manuscript on species (DAR 8–15.1, inclusive; transcribed and published as *Natural selection*). This manuscript, begun in May 1856, was nearly completed by June 1858. At that point Darwin was 'interrupted', as he put it, by a letter from Alfred Russel Wallace which provided a summary of Wallace's theory of transmutation (*Correspondence* vol. 7, letter to Charles Lyell, 18 [June 1858]). Darwin recorded in his 'Journal' the dates and headings of chapters as they were completed (see Appendix II). The following table gives the date of completion of each chapter as recorded in the 'Journal' and the chapter headings as supplied by Darwin, followed by the reference of the manuscript original in the Darwin Archive, Cambridge University Library. The page numbers of the published versions of the chapters (*Natural selection*) are also given. Chapter 1 is not extant nor was it recorded in Darwin's 'Journal'. Chapter 2 is not extant but was recorded in Darwin's 'Journal'.

[Notes to the table on the facing page:]

1 The title of this chapter has been taken from a table of contents to which CD added the names of chapters as he completed them (*Natural selection*, pp. 21–4). The manuscript of the table of contents is in DAR 8: 1–8. One folio from the chapter is extant (DAR 47: 95).

2 This title has also been taken from CD's table of contents (see n. 1, above).

3 In *Natural selection*, p. 95, Robert Stauffer suggests that the section intended to be added to chapter 4 was completed on 14 April 1858. Stauffer considers the alterations to chapter 6 were completed by 12 June. However, it seems more likely that CD worked on the revisions for both chapters during the period 14 April to 12 June.

Chapter number	Date completed	Chapter title and references
1	[Not known]	[Variation under domestication][1] (not extant)
2	13 October 1856	[Variation under domestication][2] (not extant)
11	13 October 1856	Geographical distribution (DAR 14; *Natural selection*, pp. 534–66)
3	16 December 1856	On the possibility of all organic beings occasionally crossing, & on the remarkable susceptibility of the reproductive system to external agencies (DAR 8; *Natural selection*, pp. 35–91)
4	26 January 1857	Variation under nature (DAR 9; *Natural selection*, pp. 95–171)
5	3 March 1857	The struggle for existence as bearing on natural selection (DAR 10.1; *Natural selection*, pp. 173–212)
6	31 March 1857	On natural selection (DAR 10.2; *Natural selection*, pp. 214–74)
7	29 September 1857	Laws of variation: varieties & species compared (DAR 11.1; *Natural selection*, pp. 279–338)
8	29 September 1857	Difficulties on the theory of natural selection in relation to passages from form to form (DAR 11.2; *Natural selection*, pp. 339–86)
9	29 December 1857	Hybridism (DAR 12; *Natural selection*, pp. 388–462)
10	9 March 1858	Mental powers and the instincts of animals (DAR 13; *Natural selection*, pp. 466–527)
[4]	12 June 1858[3]	[Discussion on large genera and divergence] (DAR 15.1; *Natural selection*, pp. 134–64)
[6]	12 June 1858	[Correcting chapter 6] (DAR 10.2: 26a–nn; *Natural selection*, pp. 226–50)

MANUSCRIPT ALTERATIONS AND COMMENTS

The alteration notes and comments are keyed to the letter texts by paragraph and line numbers. The precise section of the letter text to which the note applies precedes the square bracket. The changes recorded are those made to the manuscript by CD or his amanuensis; readers should consult the Note on editorial policy in the front matter for details of editorial practice and intent. The following terms are used in the notes as here defined:

del	deleted
illeg	illegible
interl	interlined, i.e., inserted between existing text lines
omitted	omitted in transcription
over	written over, i.e., superimposing

To John Maurice Herbert 2 January [1856]
1.4 regretted] 'ed' *over illeg*

To William Darwin Fox 3 January [1856]
2.1 Call Drake] *added*
2.2 elsewhere;] *semicolon altered from colon*
2.5 have as] *after del* 'had th'
2.6 dead] *interl*
3.1 dead] *interl*
3.1 Ducks] *after del* 'I'
3.1 Young] *added*
5.2 me)] *parenthesis over comma*
7.2 for] 'f' *over illeg*

To John Stevens Henslow 3 January [1856]
2.2 !] *over full stop*
3.2 Canada] *interl*

To John Edward Gray 14 January [1856]
1.1 helped] *over* '[me]'
3.2 *domestic*] *interl*
3.2 or Ducks] *interl*

To W. B. Tegetmeier 14 January [1856]
0.2 Jan] *over* 'Dec'
1.5 old] *over illeg*
1.5 Pulleine] 'P' *over* 'R'
3.4 to . . . collections.—] *added*; 'to' *over full stop*

3.5 the] *added*
5.1 &] *above del* 'but'
5.1 it] *over illeg*

To John Phillips 18 January [1856]
1.5 p. 162] *interl*
1.5 give] *altered from* 'gives'
2.1 & here] *after del illeg*
2.1 it] *interl*
2.2 apparent] *interl*
2.4 becoming] *altered from* 'become'
2.9 sometimes] *interl*
2.9 crystallized] *interl*
3.2 270,] *after del* '267,'
3.6 vertical] *after del* 'at'
7.2 in various parts of world] *interl*
7.5 together,] *interl*
7.6 these] *over* 'them'
7.6 subsequently] *interl*

To Walter Elliot 23 January 1856
3.3 *domestic*] *interl*
3.8 very] *interl*
3.8 old] *before omitted caret*
4.5 the bones] 'the' *above del* 'all'

To Katharine Murray Lyell 26 January [1856]
1.3 & ardent] '&' *added*

2.2 on] *over* 'con'
2.9 himself,] *comma over full stop*

To John Phillips 28 January [1856]
1.2 a] *added*

To W. B. Tegetmeier [1 February 1856]
1.2 tomorrow morning] *interl*

To William Yarrell 6 [February 1856]
0.2 6] *over* '5'
1.1 spirits.—] *full stop over comma*

To William Erasmus Darwin
    [26 February 1856]
1.7 chaps;] *semicolon over full stop*
1.7 but it] 'it' *altered from* 'is'
2.1 are] *interl*
3.5 of bother] 'of' *interl*

To Miles Joseph Berkeley
    29 February [1856]
2.1 experiments] *interl*
2.5 nearly] *interl*
2.5 & this] '&' *over illeg*
2.6 & Corn Salad] *interl*
2.7 was yours the tall or dwarf?] *added*
5.5 few] *interl*
5.8 sometimes] *interl*
5.9 change] *after del* 'vari'
6.3 close together 6.4] *interl*
6.4 Sweet] 'eet' *added*
6.4 Peas they] 't' *over* 'T'

To George Henry Kendrick Thwaites
    8 March 1856
1.3 then] *over illeg*
2.4 these] *altered from* 'this'
2.6 *perfectly*] *after del* 'certainly'
3.7 domestic] *interl*
3.7 or Rabbits] *interl*

To W. D. Fox 15 March [1856]
1.1 on] *above del* 'with'
1.2 however] *interl*
1.3 (can you find out for me)] *interl; square brackets in MS*
1.3 the real . . . Hounds] *interl after del* 'they'
3.2 in any way,] *interl*
3.5 I shall] *after del* 'we'
5.4 certain] *after del* 'cet'

To W. B. Tegetmeier 15 March [1856]
1.3 to purchase] 't' *over illeg*
1.5 already] *interl*
4.1 black] *interl*

To W. B. Tegetmeier 20 March [1856]
1.3 I sh^d . . . crosses. 1.4] *added*
5.1 quite] *interl*
5.2 (I think vol. 2.)] *interl*
5.3 on] *over* 'in'
6.3 makes] 's' *added*
6.4 but] 'b' *over* '&'
6.5 very] *interl*

To Thomas Henry Huxley 2 April [1856]
1.3 a] *over illeg*

To Walter Baldock Durrant Mantell
    3 April [1856]
2.1 , Latitude, 2.2] *interl*

To Joseph Dalton Hooker 8 April [1856]
1.4 not] *interl*
1.6 (I had . . . am),] *parentheses added*
2.5 fancy)] *parenthesis over comma*
2.5 on] *over* 'in'
2.6 Lindley] 'L' *over illeg*
2.12 & I fancy . . . Botanists. 2.13] *interl*
4.2 If . . . know.—] *added*
5.1 I have . . . wife.—] *added*
6.3 could] *after del* 'got'
6.4 9½ or] *interl*

To T. H. Huxley 9 April [1856]
3.3 miles,—] *comma over full stop*
3.4 best] *altered from* 'begin'

To W. B. D. Mantell 10 April [1856]
3.4 in] *over comma*
5.3 ours;] *semicolon over colon*
5.3 viz] *interl*
5.3 would] 'w' *over illeg*

To Henry Tibbats Stainton 13 April [1856]
2.1 (when I was not on council) 2.2] *interl*

To C. J. F. Bunbury 21 April [1856]
1.4 with] *above del* 'for'
4.1 persevere] *above del* 'pesevere'
4.7 & I think . . . plants.] *interl*
5.6 (existence . . . species)] *added*

6.2 exterminated.] *full stop altered from comma;*
before del 'because there were no other species *fitted*
to a colder [*above del* 'warmer'] *climate, &*
therefore able to seize their place.'
6.3 warmer] *after del illeg*
6.3 would] *after del* 'of the'
7.3 up] *before del* 'the'
7.5 (& European . . . Good Hope) 7.6] *added*
8.1 many] *after del* 'more'
8.3 pure] *above del* 'simple'
8.8 Abyssinia] 'Ab' *over illeg*
8.8 as the] *after del* 'of the'
8.9 as I] 'as' *added*
8.9 from Richard] *interl*
8.9 northern] *interl*
9.1 that] *altered from* 'what'
9.2 a] *over illeg*
9.6 identical] *interl*
13.1 diffusion] *after del illeg*
13.11 greater] *interl*

To Charles Lyell 21 April [1856]
1.5 a] *interl*
1.7 (I sh^d think)] *interl*
1.7 lava] *after del* 'flowing'
1.9 first] *interl*
2.2 very] *interl*
2.3 vertical] *interl*
2.3 in closely approximate planes] *interl*
3.1 **very** narrow,] **'very'** *interl*
3.1 (which] *after del* 'with'
4.1 coming] *after del* 'cm'
7.1 I supposed] *after del* 'I have seen horizontally
laminated basaltic lava passing by the finest
gradations into columns, but this, on reflection does not bear on your case,'
8.3 planes] *above del* 'lines'
9.2 The longitudinal . . . columns.— 9.4] *after
deleted diagram depicting longitudinal planes of
division in a lava-stream*

To Edward Sabine 23 April [1856]
1.2 who] *over* '&'
1.4 & of] 'of' *interl*
1.5 Several] *after del* 'But the word late.'
3.3 eminently] *first* 'e' *over* 'v'

To W. B. Tegetmeier 25 April [1856]
1.1 possibly] *interl*
1.5 (or any Cocks.)] *added*
2.1 all] *after del* 'but'

2.2 &] *over* 'but'
3.1 Poultry] *interl*
3.2 very well] 'very' *over* 'vy'
5.2 addressed] *after del* 'to'

To Asa Gray 2 May [1856]
1.12 list,] *before del* 'however short,'
1.12 (if any)] *interl*
1.17 (if ever so few)] *interl*
1.21 being] *after del* 'having speci'
1.22 (excluding] *after del* '(ie not necess'
1.26 the story] 'th' *over* 'a'
1.26 of relationship] *interl*
2.7 fraction] *before del* 'showing'
3.1 503 to 514] *above del* '465'
6.2 Hooker)] ')' *over comma*
6.4 now] *interl*
6.4 N. America] *above del* 'your country'
7.1 in De candolle] *after del* 'strikes'
7.5 they do] *after del* 'do'
7.5 either . . . publickly] *interl*
7.6 social,] *comma over full stop*
7.7 on the . . . range.] *added*
8.4 or . . . of U.S.] *interl*
8.7 have] *after del illeg*
8.8 other] *above del* 'one'
8.10 on] *over illeg*
9.1 simple] *interl*
9.3 me] *over illeg*
9.3 in introduction to 9.4] *interl before del* 'in'
12.3 Essay] *before del illeg*
12.3 desired,] *comma over full stop*
14.1 (Excuse . . . short)] *square brackets in MS*

To Laurence Edmondston 3 May [1856]
2.2 adult] *interl*
3.3 or Habits] *interl*

To Charles Lyell 3 May [1856]
1.1 very interesting] 'very' *interl*
2.3 possible] *interl*
2.10 opinions] *after del* 'not'
3.2 prejudices.] *above del illeg*
3.9 sympathy.] 't' *over illeg*
3.10 Thursday] *altered from* 'Tuesday'
5.3 pond] *above del* 'mud'

To C. J. F. Bunbury 9 May [1856]
2.8 many] *interl*
2.9 But] *interl*
2.12 on last] 'on' *over* 'at'
2.13 & on Lyell] *interl*

To J. D. Hooker 9 May [1856]
2.5 from] *over* '&'
3.4 positively] *interl*
4.2 dreadfully] 'd' *over* 'f'
4.3 of an] 'of' *over illeg*
4.4 at the] 'the' *altered from* 'this'

To Henry Ambrose Oldfield 10 May [1856]
1.5 Nepaul] *above del* 'Thibet'
2.2 Mastiffs,] *comma over semicolon*
2.8 or Rabbits] *interl*
2.8 Pigeons] *above del* 'latter'

To J. D. Hooker 11 May [1856]
1.2 several] *interl*
1.4 , ie] *added*
1.6 (By the way . . . been proposed.) 1.8] *square brackets in MS*
2.1 (Many thanks . . . in London.)] *square brackets in MS*
5.2 a separate] 'a' *interl*
5.2 i.e.] *interl*
5.9 preliminary] 'p' *over illeg*
5.10 details] *after del illeg*
6.3 error,] *after del* 'err'
6.3 should] *after del* 'w'
6.7 & drawing up a sketch] *interl*
7.1 an] *altered from* 'a'

To W. B. Tegetmeier 11 May [1856]
1.1 authentic] 'au' *over illeg*
6.2 Zoolog.] 'oolog.' *added over full stop*

To J. D. Hooker 21 [May 1856]
1.1 The Lectures . . . clever. 1.2] *added pencil*
1.7 (N.B I found . . . told Huxley so.) 1.8] *square brackets in MS*
3.1 there,] *comma added pencil*
3.1 Leptospermum &] *interl*; '&' *in pencil*
3.2 are confined] 'are' *in pencil over* 'is'
3.2 Australia,] *comma added pencil*
3.2 are they] *in pencil over* 'is it'
5.1 fossil] *interl*

To J. W. Lubbock 27 May [1856]
1.1 no] *over illeg*

To W. B. Tegetmeier 31 May [1856]
1.1 manifold] 'm' *over* 'a'
2.4 Could] *after del* 'But,'

To J. D. Hooker 1 June [1856]
1.3 Gardens!!] *first exclamation mark over comma*

To W. D. Fox 4 June [1856]
2.1 a grand] 'a' *over illeg*

To S. P. Woodward [after 4 June 1856]
10.1 (p. 299)] *after del illeg*
11.1 (in new map.)] *interl*
12.2 my] *over* 'the'
13.1 Is] *altered from* 'In'
13.1 page wrong] *added pencil after del* 'not f^{d}'
13.2 Modiolarca] *altered from* 'Modilioarca'
14.2 ranges.?] '?' *added pencil*
14.3 to rather distant isl^{ds} 14.4] *interl*
17.1 & p. 419—(bottom)] *added*
17.2 vertebrata] *over* 'animals'
18.1 names in] *interl*
19.2 in] *added*
20.1 (4th Par)] *interl*
20.1 Fulgur:] *before del interl* '(p. 421 4th Par)'
20.2 Bay?] '?' *over full stop; before del illeg*

To W. B. D. Mantell 5 June [1856–9]
0.2 5] *over* '4'

To T. V. Wollaston 6 June [1856]
1.5 all which] *interl*

To W. D. Fox 8 [June 1856]
2.5 & Fowls] *interl*
2.5 E. Africa.—] 'E' *over* 'W'
3.4 or Duck] *interl*
4.8 must I] *interl*
5.2 6.?] '?' *over* '—'

To Edgar Leopold Layard 8 June [1856]
1.2 have] *interl*
2.4 domestic] 'd' *over illeg*
3.6 domestic] *interl*
4.2 Ascension] *after del* 'the'
5.4 &] *over illeg*

To John Lubbock [8 June 1856]
1.1 are] *over* 'do'

To H. C. Watson [after 10 June 1856]
1.5 North] *after del illeg*

1.5 **N.**] *added*
1.6 from] *after del* 'ought'
1.13 old] *after del* 'Europe'
1.15 (at least of shells)] *interl*

To John Lubbock 12 [June 1856]
1.1 get] 'g' *over illeg*
1.4 having] 'h' *over illeg*
2.1 well] *after del* 'all'

To J. S. Henslow 16 June [1856]
2.2 nice] *interl*
3.3 publish,] *comma over colon*
3.5 have] *interl*

To Charles Lyell 16 [June 1856]
2.1 same)] *parenthesis over comma*
2.3 & P. Santo] *interl*
2.5 over] 'o' *over illeg*
3.2 a lower] 'a' *over illeg*
6.2 chart;] *semicolon over colon*

To J. D. Hooker 17–18 [June 1856]
1.1 subjects, to] 'to' *above del* 'on'
1.5 the knowledge] *after del* 'living be'
1.8 so called] *interl*
2.8 ultra!-honesty] '!' *added*
7.2 quite] *over* '.—'
8.1 Club,] *before del illeg*
8.1 *N.W.*] *before del interl* 'near Behring Str' *pencil*
8.1 portion] *added pencil above del* 'point'
8.2 of Asia;] 'of' *interl*
8.3 middle] *after del* 'other'

To J. D. Hooker 22 June [1856]
2.3 **about**] *after del illeg*
3.3 (as far as I can see)] *interl*

To W. B. Tegetmeier 24 June [1856]
3.1 they] *over* 'it'
3.5 yet] *interl*

To Charles Lyell 25 June [1856]
0.2 25] *over* '24'
2.3 Essay] 'E' *over* 'e'
2.5 my wrath 2.6] *inserted in gap*
2.8 European] *inserted in gap*
2.11 Raoũl Is^d] *inserted in gap*
2.11 America] *before del comma*
2.12 Acunha] *above del* 'Acunda'
2.12 making,] *comma added*

2.14 circumpolar] *inserted in gap above del* 'contin'
2.15 and if we . . . America; 2.16] *added*
2.19 perhaps] *inserted in gap*
2.23 organisms] 's' *added*
2.25 and central] *interl*
2.26 miocene] *inserted in gap*
2.26 eocene] *inserted in gap*
2.27 that] *altered from* 'the'
2.27 *fundamentally*] ²'n' *inserted*
2.27 America] *inserted in gap*
2.27 miocene] *inserted in gap*
2.28 shells] ²'s' *added*
2.30 in uninterrupted seas] *interl*
2.30 since] *before del* 'since'
2.31 the isthmus] 'the' *inserted in gap*
2.31 the Aralo-] *inserted in gap*
2.32 basin] *inserted in gap*
2.32 deposits,] ²'s' *added*
2.33 Fauna] *inserted in gap*
2.33 Marsupials] *inserted in gap*
2.36 level] *inserted in gap*
2.39 organisms] ²'s' *added*
2.39 our oceans] *altered from* 'one ocean'
2.39 , moreover] *inserted in gap*
2.40 vaster] *above del* 'faster'
3.1 as] *above del* 'and'
3.2 Consider that the] *interl*
3.3 Nevada,] *inserted in gap*
3.3 are] *interl*
3.4 Asia,] *comma added*
3.5 alps of Abyssinia] *inserted in gap*
3.5 of the 3.6] *interl*
3.8 alone] *interl*
3.9 Silla of Caraccas] *inserted in gap*
3.9 Itacolumi] *inserted in gap*
3.9 S. Ventana] *inserted in gap*
3.10 in] *interl*
3.10 these] *altered from* 'them'
3.10 peaks] *interl*
3.12 rock] *inserted in gap*
3.13 bank;] *semicolon altered from full stop*
3.13 New Caledonia, . . . exception. 3.14] *interl*
3.17 as continents,] 's,' *added*
3.18 oceans,] 's' *added*
3.19 continents,] *comma over colon*
3.19 forming continents;] *semicolon over comma*
3.20 volcanos] 's' *added*
3.21 cores] *inserted in gap*
3.21 syenite diorite] *inserted in gap*
3.21 porphyry] *inserted in gap*
3.25 my Coral volume] *inserted in gap*

3.25 volcanic] *after del* 'the'
3.25 outbursts] 's' *added*
3.27 -poured] *inserted in gap*
4.4 atolls] *inserted in gap*
4.5 coral] *inserted in gap*
4.6 Low Arch] *inserted in gap*
4.6 Arch.] *inserted in gap*
4.6 Laccadive] *inserted in gap*
4.7 heights] 's' *added*
4.11 the] *interl*
4.11 Silurian] *inserted in gap*
4.12 carboniferous formations,] *inserted in gap*
4.13 these formations] *altered from* 'their for-mation'
4.15 Hooker's] *apostrophe inserted*
4.17 with,] *comma added*
4.17 not] *interl*
4.18 nearly every] *above del* 'necessary'
4.19 island] *altered from* 'islands'
5.3 Batrachians] *inserted in gap*
5.4 Acaciæ] *inserted in gap*
5.4 Banksias &c] *inserted in gap*
5.4 paucity] *inserted in gap*
5.7 former] *above del* 'some'
6.3 Atlantis] 's' *over* 'c'
6.3 as checking] 'as' *altered from* 'a'
6.8 proved.] *above del* 'found.'

To W. B. Tegetmeier [July 1856]
1.1 one] *interl*
3.1 as one] 'as' *over* 'one'
3.3 not] *interl*
3.3 though] 't' *over* 'I'

To J. E. Gray 1 July [1856]
1.2 to give] *after del* 'if'
1.3 on your] 'on' *over* 'yo'
2.1 genera] *after del* 'ran'

To T. H. Huxley 1 July [1856]
1.2 are;] *semicolon over colon*
1.3 answered;] *semicolon over colon*
1.7 Ascidian] *interl*
1.7 & southern] *interl*
1.8 such] *after del* 'not'
3.1 Thanks] *over illeg*
6.2 almost] *interl*

To J. D. Hooker 5 [July 1856]
2.7 (& partially Pelargonium) 2.8] *interl*

To Charles Lyell 5 July [1856]
1.2 volcanic *Cosmogony*!] *inserted in gap*
1.5 every] *inserted in gap*
1.6 island] *altered from* 'islands'
2.1 looks] 's' *added*
2.4 &c] *above del* 'and'
2.4 strata;] 'strata' *above del* 'streams'
2.4 impresses the] 'the' *interl*
2.6 though the] *inserted in gap*
2.6 thicknesses] *altered from* 'thickness'
2.7 after writing coral-book] *interl*
2.8 notes] 's' *added*
2.8 the upper] 'the' *interl*
2.8 the Cordillera,] 'the' *interl*
2.10 the maritime] 'the' *interl*
2.11 (consequently] '(' *added*
2.11 of land) and] *above del* 'a law'
2.12 (as if . . . other)] *parentheses over commas*
2.13 the] *interl*
2.13 coral-map,—] ',—' *added*
2.14 conversations] *inserted in gap*
2.14 the] *added*
2.17 cracks,] *comma added*
2.17 the] *interl*
2.17 tilts,] *inserted in gap*
2.19 the cracks] *inserted in gap*
2.20 doubt.] *before omitted square brackets designating new paragraph*
3.2 local] *above del* 'coral'
3.2 elevations] *altered from* 'elevation'
3.2 (but of course . . . explained) 3.3] *above del* 'harshly'
3.5 piles] *inserted in gap*
3.6 shrinking] *inserted in gap*
3.7 lakes or] *interl*
3.7 Fresh] *inserted in gap*
3.9 is] *interl*
3.11 insisted] *inserted in gap*
3.12 Paper,] *inserted in gap*
3.14 large.] *before omitted square brackets designating new paragraph*
4.1 from there] 'there' *over* 'these'
5.1 (with absolute truth)] *parentheses inserted*
5.4 conceited.] *inserted in gap*

To J. D. Hooker 5 July [1856]
1.4 in my eyes 1.5] *interl*
3.1 to you] *after del interl* 'it'
3.1 for] *over* 'as'
4.2 translated] *after del comma*
5.3 I suspect] *after del* 'in'

6.2 very] *interl*
6.3 of] *over illeg*
7.1 sufficiently] *after del* 'po'
7.4 fertile] *interl*
9.1 two rivers of 9.2] *interl*
13.3 very] *interl*
13.6 nearly] *interl*

To J. D. Hooker 8 [July 1856]
1.4 looking into] *interl*
2.5 confined] 'c' *over illeg*
2.6 these] *altered from* 'this'
2.9 (like the Chonos plants)] *interl*
2.12 Do] *above del* 'Are'
2.13 Are . . . rangers?] *interl*

To T. H. Huxley 8 July [1856]
1.6 But I suspect . . . receptacle. 1.7] *added*
2.3 also] *interl*
2.4 unisexual] *above del* '*bisexual*'
2.5 Does the position . . . crossing? 2.6] *added*
2.5 unisexual] *interl*

To Charles Lyell 8 July [1856]
1.1 also . . . to lend me. 1.2] *added*
2.7 account] *altered from* 'accounted' *after del interl* 'have'
2.8 on the Azores] *interl*
3.2 about] *interl*
3.2 sudden] *interl*
5.2 week] *altered from* 'weeks'
7.3 there] *after del illeg*

To J. D. Hooker 13 July [1856]
1.7 me] 'm' *over* 'a'
1.14 of close] 'of' *interl*
2.1 your] *interl*
4.1 about] *interl*
4.7 Book?] '?' *above del colon*
4.7 a correct . . . oneself.— 4.8] *added*
4.8 ever] *interl*
4.8 without] *after del* 'ever'
6.1 on distribution,] *interl*
6.5 at second hand] *interl*
7.4 & faunas arctic] '& faunas' *interl*
7.4 & faunas & the] '& faunas' *interl*

To James Dwight Dana 14 July [1856]
1.1 you have] *interl*
2.2 any of] *interl*
2.4 (Has not Agassiz noticed the Fish?)] *added*

2.5 on the] 'on' *over* 'in'
3.3 temperate & arctic] *interl*
3.5 from my ignorance 3.6] *interl*
3.9 Decandolle] *after del* 'de'
3.11 same] *interl*
5.7 (our)] *interl*
6.6 if] *over* 'c^d'

To Asa Gray 14 July [1856]
1.5 the plants] 'the' *interl*

To J. D. Hooker 19 July [1856]
2.8 experimentise by] *above del* 'exp'
3.2 in terrestrial 3.3] *interl*
3.3 the semen is dry] *interl*
4.1 If you were . . . most difficult cases. 4.2] *added*
5.1 To] *before del* 'w'
5.4 mono- & diœcious 5.5] *above del illeg*
5.7 same stock,] 'same' *interl*
5.9 occasional] *interl*
5.10 have thought] 'have' *interl*
5.14 to prevent too much intermarriage] *interl*
5.14 I do not . . . of sexes.— 5.16] *added*
6.3 the French] *interl*
7.3 facts] *after del* 'evidence re'
8.1 extremely,] *comma over full stop*

To J. D. Hooker 26 [July 1856]
3.3 as much] 'as' *interl*
3.4 in] *after del* '&'
4.1 alive] *interl*
4.1 Kew] *after del caret*
4.3 I would gladly try.— 4.4] *added*
4.5 on] *after del* 'our'
4.6 Himalaya] *above del* 'Indian'
4.9 same genera] *above del* 'genus'
4.9 &] *before del* 'of'
5.1 in 17 genera] *interl*
5.2 Exactly . . . in length! 5.3] *added*

To John Lubbock [29 July 1856]
2.2 same] *over illeg*
6.1 Martin] *interl*

To J. D. Hooker 30 July [1856]
1.5 predicament] 'pre' *over* 'per'
1.5 mine] *above del* 'mine'
1.10 subsidence] *after del* 'changes'
1.12 both *identical* & allied] *interl*
1.14 most] *after del* 'my'
1.15 geographical] *interl*

1.15 believe] *above del* 'belief'
2.1 (not geographical)] *interl*
3.1 of Lyell's] *interl*
3.8 all] *after del* 'are'
3.8 species are] *interl*
3.9 by laws of change] *interl*
3.13 of Vestiges,] 'of' *added*

To J. D. Hooker 5 August [1856]
1.4 earnestly] *interl*
1.12 also] *interl*
1.17 distribution;] *semicolon over colon*
1.21 investigation] *after del illeg*
2.1 (I think) (I have got lists) 2.2] *interl*
2.4 weakens your] 'y' *over illeg*
2.4 not] *before del* 'quite'
2.6 to] *over semicolon*
2.7 taken] *above del* 'had'
2.7 trouble] 'bl' *crossed*
2.9 very] *above del* 'vey'
4.2 your] *over* 'th'
4.3 low] *after del illeg*
4.4 of origin of species] *interl*
5.2 finally] *interl*
5.4 in planes . . . surfaces] *interl*
6.3 obsidian] *interl*

To J. S. Henslow 6 August [1856]
2.1 sown in open ground] *above del* 'slightly forced
    planted out'

To J. D. Hooker 7 August [1856]
2.3 even] *interl*
2.4 in my note-book] *interl*
2.6 have] *above del* 'hope'

To W. B. Tegetmeier [15–22 August 1856]
1.3 Smyrna] 'm' *over* 'y'
1.4 object.—] *full stop over comma*

To W. B. Tegetmeier 23 August [1856]
1.1 & measurement. 1.2] *interl*
2.5 , & w^d advise you of their going.—] *added*;
    ', & w^d' *over* '.—'
3.1 About] 'A' *over illeg*

To Asa Gray 24 August [1856]
1.5 it] *interl*
1.6 (without] *parenthesis over comma*
1.9 an] *after del illeg*

1.11 (whether or no naturalised)] *parentheses over
    commas*
3.4 that] *interl*
3.5 only] *interl after del* 'I found'
3.5 did] *after del* 'which'
3.7 in] *added*
4.1 your] *over* 'the'

To T. C. Eyton 27 [August 1856]
1.4 is whether] 'is' *interl*
1.5 generally] *interl*

To W. B. Tegetmeier 30 August [1856]
1.1 am] *over* 'w'
2.1 dead] *interl*
2.2 (alive)] *interl*

To T. C. Eyton 31 August [1856]
2.4 (1849 p. 392)] *square brackets in MS*
2.9 commencement] *after del* 'origin'
4.1 What] 'W' *over illeg*
4.2 geese] *interl*
5.4 living] *interl*
6.1 (we have no ponds hereabouts) 6.2] *added*
8.4 of water plant] *interl*
10.2 request] *after del illeg*
11.1 have] *over illeg*

To John Lubbock 5 September [1856]
1.4 climat] *altered from* 'climate'
3.1 of] *over* '.—'

To J. D. Hooker 8 September [1856]
1.9 must] *interl*
1.9 come] *altered from* 'comes'
1.10 there be] *interl*
2.2 (floating . . . different rates) 2.3] *square
    brackets in MS*
2.2 in salt-water] *interl*
3.7 case] *interl*
3.8 Sceptics] ¹'c' *over* 'k'

To George Varenne Reed 8 September [1856]
1.3 nearly] *after del* 'so'

To Joseph Augustin Hubert de Bosquet
    9 September [1856]
2.1 strongly] *interl*
2.3 well] *interl*
4.3 on] *over illeg*
4.3 scales] ²'s' *added*

To Laurence Edmondston 11 September [1856]
3.2 & wild] *interl*

To W. B. Tegetmeier 21 September [1856]
2.3 I want much the Great Hare Rabbit.] *added*
4.1 with 5 toes 4.2] *interl*

To Philip Henry Gosse 22 September [1856]
1.3 leads] *altered from* 'lead'; *after del* 'has'
1.4 do *not*] *interl*
1.4 willingly] *interl*
2.6 wild] *interl*
3.3 or Fancy 3.4] *interl*
3.5 interest,] *comma over semicolon*

To John Lubbock 23 September [1856]
1.2 Quatrefages] 'r' *over* 'er'
1.3 &] *over* '—'
6.3 happiness.] *above del* 'marriage.'
6.3 them] *over illeg*

To J. D. Hooker 28 September [1856]
1.3 with *some* stalk & leaves] *interl*
3.1 any day,] *after del* 'early on a Wednesday, addressed'
4.1 In] *after del illeg*
4.1 on one . . . Distribution 4.2] *interl*
4.3 in regard to] *interl*
5.1 your] *over* 'vy'

To W. B. Tegetmeier 28 [September 1856]
1.7 red] *above del* 'read'
2.1 old] *interl*
2.2 or] *added*
4.1 what] *above del* 'that'
4.2 distinct] *interl*

To J. D. Dana 29 September [1856]
1.4 from your work] *interl*
1.6 for] *over illeg*
5.1 creation, in] 'in' *above del* 'to'
5.5 of any organisms] *interl*
6.5 (I think) 6.6] *interl*
8.2 &c,] *over comma*

To W. D. Fox 3 October [1856]
2.4 to Malvern] *interl*
3.3 School;] *semicolon altered from colon*
6.2 on] *over* 'in'
6.2 variation] *after del* 'the'

6.6 Rabbits] *after del* 'Pigeons'
6.8 it much] *after del* 'them'
7.2 how] *after del* 'w'

To J. D. Hooker 5 October [1856]
2.4 Such] 'S' *over* 'Th'

To J. D. Hooker 9 October [1856]
0.2 9] *over* '8'
1.7 the vitality] *interl*
4.2 Geography;] *semicolon altered from colon*

To Asa Gray 12 October [1856]
1.6 non-Botanist] *after del* 'man'
1.7 appreciate] 're' *over illeg*
1.9 much.] *full stop over comma; before del* '& va'
2.2 (ie ¼)] *interl*
3.5 & orders] *interl*
4.4 Is E. Asia . . . comparison? 4.6] *added*
5.1 in one point] *interl*
5.4 results] *altered from* 'results'
5.6 (nearly)] *interl*
5.6 world] 'w' *over illeg*
5.8 number of] *interl*
8.1 (or p. 229)] *interl*

To W. B. Tegetmeier 15 October [1856]
2.7 the Cock is] *interl*
2.7 having] 'h' *over* 'k'
2.7 it is] *interl*

To J. D. Hooker [16 October 1856]
1.1 Please read this, first,] *added pencil*
1.4 to save . . . writing.] *added*
2.1 my] *interl in pencil*
2.1 will be] *added pencil above del* 'is'
2.2 it is] *interl*
6.1 very] *interl pencil*

To J. D. Hooker [19 October 1856]
2.7 own] *added*

To W. B. Tegetmeier 19 October [1856]
1.2 Thursday] *above del blotted* 'Thursday'
1.4 the kindness of] *interl*
1.7 are very] 'are' *above del illeg*
1.7 of] *above del illeg*
1.13 have been] *after del* 'a'
1.14 eat] *above del illeg*

To W. D. Fox 20 October [1856]
1.2 Malvern,] *comma altered from full stop*

4.1 he] *over illeg*
4.5 I forget which] *interl*
5.4 much] *interl*
5.4 often happens] *above del* 'usual'
5.5 eaten] 'en' *added*
6.3 (or 10 days)] *interl*
6.6 after] *over* 'in'

To John Lubbock 27 October [1856]
1.10 (V. Back of Page)] *interl*
1.12 The] *altered from* 'These'
1.13 sorts of] *interl*
2.2 a priori] *interl*
2.3 have] *interl*
2.3 been] *altered from* 'be'
2.3 the] *added*
2.4 that] *over* '.—'
4.2 "darkened"] 'ed' *over* 'end'
5.1 your own] *interl*
5.2 times] *above del* 'generations'
8.2 worth] *after del* 'not'
9.1 fully] *after del* 'think'
11.1 not?—] '?' *over full stop*
12.1 successive . . . individual] *interl after del* 'generations'
12.2 but not . . . generations] *above del* ', & there is no reason to believ'
12.2 I suppose . . . improbable.— 12.3] *added*
13.2 necessarily] *interl*
13.3 ie] *interl*
13.4 seen] *interl*
13.5 view:] *colon over comma*
14.1 W^d it . . . evidence] *added pencil*

To John Lubbock [1 November 1856]
1.4 & largely] '&' *above del* 'I think'
1.5 But I cannot . . . using it daily. 1.6] *added over* '—'

To J. A. H. de Bosquet
  [before 3 November 1856]
1.1 &] *over full stop*
1.2 these] *altered from* 'them'
2.1 your] *over* 'the'
2.1 (old cock bird)] *interl*
3.1 authentically] *interl*
3.1 flown] *after del* 'ever'
3.2 or 400] *interl*

To W. B. Tegetmeier 3 November [1856]
1.2 cocks, have] 'have' *over illeg*

5.1 &] *over illeg*
5.3 to Carstangs . . . carriage free.— 5.4] *added*

To Charles Lyell 10 November [1856]
4.1 than] *altered from* 'that'
6.5 during the] 'the' *interl*

To J. D. Hooker 11–12 November [1856]
1.4 my M.S] *above del* 'it'
1.7 general] *interl*
2.4 resist,] *comma altered from full stop*
4.2 some kinds] *above del* 'many'
4.3 up!] '!' *over colon*
4.3 not] *interl*
4.6 damp with gastric juice] *interl*
5.1 during few walks] *interl*
5.4 possible] *interl*
5.5 very tenacious] *interl*
11.2 anything] *above del* 'much'
11.4 19'] *interl*
11.6 my Book's] *above del* 'work'

To George Howard Darwin and W. E. Darwin
  13 [November 1856]
1.10 Martha] *altered from* 'Marthat'

To J. D. Hooker 15 November [1856]
1.9 Some of the] 'Some of' *interl*; 'the' *altered from* 'The'
1.10 long] *after del illeg*
1.11 so] *below del* 'that'
1.12 they] *altered from* 'these'
1.14 warm temperate] *interl*
1.14 &] *after del* 'than'
1.15 comparing] *after del* 'the'
1.18 Mountains] *after del illeg*
1.19 on the] 'the' *over* 'an'
1.19 islands] *altered from* 'island'

To John Murray 17 November [1856–7]
1.8 of] *over* 'D'
2.3 altogether] 'al' *added*

To J. D. Hooker 18 November [1856]
2.4 when] *before del* 'they some'
2.4 the few] 'the' *altered from* 'those'
2.4 got] *after del* 'that'
2.5 tropical species] *above del* 'equatorials'
2.15 Arctic] *altered from* 'Arctics'
2.15 species] *added*
3.1 21] '1' *in pencil over* '3'
3.3 an owl] *above del* 'it'

To John Higgins 19 November [1856]
2.2 payment] *below del illeg*

To W. B. Tegetmeier 19 November [1856]
2.3 shall] *altered from* 'sh^d'
2.4 it,] *interl*
3.2 thanks] *before del* 'also'
4.1 large] *interl*
6.1 seen] *after del* 'not'
6.1 only] *after del* 'the late nos'

To J. D. Hooker 23 November [1856]
2.1 M.S.] *after del* 'sket'
2.1 strongly] *after del* 'that the'
2.8 my conclusion] *interl*
2.11 (No doubt . . . trace).— 2.12] *square brackets in MS*
2.13 the] *interl*
2.14 variations] *after del* 'variability'
2.17 (even . . . animals) 2.18] *parentheses over commas*
2.20 Time +] '+' *over comma*

To Asa Gray 24 November [1856]
4.4 but are] *after del* 'except'
5.1 the only] 'the' *interl*

To W. E. Darwin 25 [November 1856]
2.3 to] *above del* 'home'

To George Bentham 26 November [1856]
1.5 mostly] *above del* 'almost all'
1.7 timber] *above del* 'large'
1.7 a keel] *above del* 'flowers'
1.8 & pistil,] *before del* 'within the Keel'
1.9 the trees] *interl*
1.12 shut up] *interl*
1.14 (but this . . . question)] *interl*
1.15 as] *over illeg*
2.1 myself] *interl*
2.1 (I . . . summer.) 2.2] *square brackets in MS*
2.1 to see if the keel opens] *interl*

To John Thompson? 26 November [1856]
2.1 to you] *interl*
2.5 old] *interl*
2.9 &] *over illeg*
3.1 forgive me] 'me' *over* 'the'

To W. B. Tegetmeier 29 November [1856]
1.1 (handsome)] *interl*
3.3 it)] ')' *over comma*

To George Bentham 30 November [1856]
2.2 "were] *after del* 'had their'
2.6 than the ordinary flowers. 2.7] *interl*
2.11 lead] *before del* 'to'
2.15 an] *interl*
2.15 as they do] *over* '[as to in]'
2.16 apetalous] *after del* 'the'

To J. D. Hooker [early December 1856]
1.1 return] *after del* 'alter this &'
1.1 this] *over* 'it'
1.1 with any remarks:—] *added*
2.2 rapid] *interl*
3.1 But] *after del* 'All the members of'
3.4 the pollen] 'the' *interl*
3.4 under water 3.5] *interl*

To J. D. Hooker 1 December [1856]
2.2 sexes,] *before del* '—'
3.2 be worth your] 'be' *interl*

To George Bentham 3 December [1856]
1.5 a] *interl*

To T. H. Huxley 9 December [1856]
1.6 of Bees;] *interl*
2.9 on the sides low down] *interl*

To J. D. Hooker 10 December [1856]
1.1 much] *interl*
1.2 (I find . . . structure) 1.4] *square brackets in MS*
1.3 in exactly] 'in' *interl*
1.3 the trees] *interl*
1.4 systematic] *interl*
1.13 condense,] *comma over colon*
2.1 Give . . . kind note.— 2.2] *added*
4.7 (as yet)] *interl*
6.5 (Leguminosæ, . . . sea-water)] *interl*
6.6 have] *altered from* 'has'
6.8 dead] *interl*
8.1 a] *over illeg*

To T. H. Huxley 13 [December 1856]
1.2 I fear . . . good species. 1.3] *added*
2.5 the unbranched part] *interl*
8.1 connection,] *after del* 'cases of'
8.3 nearly] *interl*

To Thomas William St Clair Davidson
  23 December [1856]
1.3 for] *interl*
1.3 give] *above del* 'answer'

1.6 or is not] *interl*
1.9 equally] *interl*
2.3 in order] *interl*
2.5 or caused by] *interl*
2.6 many] *after del illeg*

To Asa Gray 1 January [1857]
1.6 species of] *interl*
4.2 one of] *interl*
4.3 distribution] *after del* 'statistics'
7.6 &c] *interl*
7.13 geological] *interl*

To J. D. Hooker 17 January [1857]
1.4 &c.] *over full stop*

To T. H. Huxley 17 January [1857]
1.8 a talus] *after del* 'the'
1.11 push out] *interl*
1.14 a pile] 'a' *above del illeg*
1.14 pile of] *interl*
1.16 freezing] *above del* 'cohesion'
1.16 brittle] *interl*
2.3 temperature] *above del* 'point'

To J. D. Hooker 20 January [1857]
5.1 greedily] *interl*
5.1 millet] *above del* 'some'
5.2 will] *over* 'with'

To J. D. Hooker [after 20 January 1857]
1.2 assistance from] *interl*
3.6 time] 't' *over illeg*
4.3 very] *interl*
4.5 & Australia] *interl*
4.8 a] *after del* 'the'
4.9 now] *interl*
4.12 might] 'm' *over illeg*
11.2 is] *after del* 'are'
11.2 has] *altered from* 'have'
11.3 strongly] *after del caret*
11.3 very] *added*
11.5 affinities] *after del* '&'
11.6 latter] *interl*
11.7 how] *after del illeg*
11.9 the best] 'the' *added*

To William Sharpey, Secretary, Royal Society
  24 January [1857]
1.2 special] *altered from* 'especial'
1.5 of the] 'the' *over* 'c'

To Henry Doubleday [before 5 February 1857]
1.4 have] *interl*
1.4 been] *over* 'be'

To W. B. Tegetmeier 6 February [1857]
2.2 (ie before)] *interl*
2.4 you of] 'you' *interl*
3.3 noticed] *after del* 'f'
7.1 By] *after del* 'When you next see Carstang, I
  wish you would do a little commission for me,
  ask him the next time he sees a common *Dutch*
  or German Pouter in market, short-legged &
  altogether inferior to our Pouters, to buy one
  for me if he can get it, as I have been told, for
  a few shillings & send it by enclosed
  address.—'

To W. D. Fox 8 February [1857]
5.2 &c;] *semicolon over colon*
9.1 at Oulton Hothouses] *interl*
9.2 eject] *after del* 'when touched'
9.4 always] *interl*

To Charles Lyell 11 February [1857]
1.4 power] *before del* 'left'
1.4 this w^d be very desirable] *interl*
2.1 especially] *interl*
3.1 Cocos] *over illeg*
3.5 (& Volcanic islands)] *interl*

To W. B. Tegetmeier 11 February [1857]
4.1 Scanderoons;] *semicolon over comma*

To John Innes [after 16 February 1857]
1.3 in this Book] *interl*

To W. E. Darwin [17 February 1857]
1.3 me] 'm' *over illeg*
1.6 any] *interl*
1.8 greatest] *altered from* 'great'
4.4 aloud to Meeting] *interl*
4.6 I felt] 'I' *interl*

To W. D. Fox 22 February [1857]
1.1 (1st)] *interl*
2.2 she] 's' *over* 't'
4.3 side] *altered from* 'sides'
6.1 heartily] *above* 'heatily'

To J. D. Hooker 15 March [1857]
1.1 *which please return.*] *interl*

2.1 in U. States] *interl*
3.8 more or less] *interl*
3.8 must] *interl before del interl* 'can'
3.9 were] *interl*
3.9 separation on mainlands] *interl*
3.10 island] *altered from* 'islands'
4.1 remark] *after del illeg*
4.1 on] *over* 'of'
4.3 on the mainland] *interl*
4.6 & feebly . . . d'Acunha. 4.7] *interl*
5.3 & island] *interl*
5.3 main-land] 'main-' *added*
6.1 this part of] *interl*

To Asa Gray [after 15 March 1857]
0.2 (I have . . . Postage)] *added*
1.1 has been] 'has' *altered from* 'have'
1.9 to your] 'to' *interl*
1.14 careful] *after del* 'the'
1.16 overvalued] 'va' *interl*
2.1 on] *over illeg*
2.1 observations] *after del illeg*
3.5 always] *interl*
3.8 the] *over* 'its'
3.10 of] *over* 'on'
3.16 rule] *after del illeg*
4.2 has led] *after del* 'which'
4.3 doctrine] *after del* 'case'
5.2 in them 5.3] *interl*
5.6 (I have . . . N. York) 5.7] *interl; square brackets in MS*
6.1 with distinct individual 6.2] *interl*
6.3 obstacle] *interl*
6.5 Keel-petals] *after del* 'flowers'
6.5 when ready for fertilisation; 6.6] *interl*
6.6 that] *interl*
6.9 look] *interl*
7.2 the greatest] 'the' *over* 'my'
7.4 it is apt to be] *interl*
7.4 (I am] '(' *over full stop*
7.7 with list] *after del* 'on them'

To Edward Sabine 16 March [1857]
1.2 to attend] *interl*
1.3 Committee.] *full stop over comma*
2.4 state] *after del* 'effect'
2.6 also?] *question mark over semicolon*
2.7 are stones . . . ice?] *interl*
2.8 able to] *interl*
2.8 between] *interl*
3.3 the shingle] 'the' *interl*

To J. D. Hooker [21 March 1857]
1.2 or list] *interl*
1.5 range] *after del* 'on'
1.5 the species in] *interl*
1.7 these] *over* 'they'
1.7 way.] *full stop over semicolon*
1.8 as I asked him] *interl*

To J. D. Dana 5 April [1857]
1.5 cooler] *interl after del interl* 'col'
2.5 Crust.] *interl*
3.2 some work] 'work' *interl*
4.3 have] *altered from* 'has'

To J. D. Hooker 8 April [1857]
0.2 8th] '8' *over* '7'
2.1 point; &] ';' &' *added*
2.1 to do this 2.2] *interl*
3.3 in England & India.] *interl*
6.1 British plants in] *interl;* 'plants' *after del* 'pant'
6.1 N. Zealand] *before del* 'plants'
6.1 or] *before del* 'any'
7.1 One] *altered from* 'one'
7.3 is] *after del* 'are'

To J. D. Hooker 12 April [1857]
3.3 the fur] 'the' *interl*
3.4 range of] 'of' *over illeg*
3.7 hairy] *after del* 'wi'
3.16 *varieties*] *after del* 'plants'
3.17 proportionally] *interl*
4.5 almost exclusively] *interl*
4.5 destruction] *after del* 'almost exclusive'
6.2 plants] *after del* '**dryed**'
6.2 of a country] *interl*
6.3 when dryed] *interl*
6.5 the Azores] 'the' *interl*
7.1 endemic to] *above del* 'from'
7.3 is] *interl*
7.4 manner] *above del* 'a way'
7.5 the suckers of] *interl*
10.3 seeds] *after del* 'pl'
10.3 after] *added*

To Laurence Edmondston 19 April [1857]
2.1 of] *over* '.—'
2.7 near Hythe has] 'near Hythe' *interl*

To P. H. Gosse 27 April [1857]
1.3 of same species 1.4] *interl*
1.7 a] *interl*

1.8 *very*] *interl*
1.8 *small*] *interl*
1.11 that] *altered from* 'than'
2.3 to Europe] *interl*
2.4 of] *after del illeg*
2.7 to a birds foot 2.8] *interl*

To J. D. Hooker [29 April 1857]
2.5 plants] *above del* 'paper'
2.6 in Alps.] *interl*
2.6 writing] *after del illeg*
2.8 saying] *interl*
2.8 cold] *after del caret*
2.9 Are arctic . . . apetalous?] *added*
3.2 direct] *interl*
3.2 of climate] *interl*
3.3 for me,] *interl*
3.4 *varieties*] *after del* 'any'
3.4 in] *above del* 'under'
4.5 he did] *after del* '&'
5.2 (quite] *after del* 'if'
10.1 a] *interl*

To W. D. Fox [30 April 1857]
2.4 you] *interl*
2.6 after] *after del* 'in'

To Alfred Russel Wallace 1 May 1857
1.1 received a few days ago] *interl*
1.6 every] *altered from* 'evey'
1.8 different] *after del* 'own'
3.6 backed] 'b' *over illeg*
3.7 truth] *after del* 'very'
3.11 & Gærtner,] '&' *interl*
5.2 domestic] *interl*
5.4 with the] 'the' *interl*
5.5 of his] *interl*
5.6 & **Cats** skins] *interl*
6.1 are believed] *interl*
6.2 to] *interl*
6.5 had] *altered from* 'have'
7.1 on which] 'on' *interl*

To J. D. Hooker [2 May 1857]
2.5 it] *after del* 'sometimes'
2.5 & again.] *interl*
2.10 apelatous] *above del* 'no'
2.10 flowers"!] *before del* 'at all'
3.1 but false] *interl*
3.2 & Dreges] *interl*
3.7 Does] *over* '—'

To J. D. Hooker [3 May 1857]
2.2 (now] *parenthesis over comma*
3.3 & Adamson's] *interl*
4.2 &] *over* '.—'

To Asa Gray 9 May [1857]
1.6 species of] *interl*
1.7 feel] *over* 'be'
1.8 whole] *interl*
1.9 is] *over* 'are'
1.9 disjoined] *altered from* 'disjointed'
1.12 species] *interl*
1.12 on average] *above del* '*generally*'
1.13 again] *interl*
1.13 from want of knowledge;] *interl*
1.14 out] *interl*
3.6 looking . . . of some] *above del* 'notice, for'
4.2 (as I thought)] *interl*
6.2 all was] 'was' *interl*
6.2 or errors.] *interl*

To W. B. Tegetmeier 12 [May 1857]
2.4 very likely] *interl*
4.1 thrown] *over illeg*
5.1 your note] *above del* 'it'
5.1 it)] *above del* 'your note'
5.2 tell] *before misplaced caret*
5.2 briefly] *interl*
6.5 Silver] 'S' *over* 's'
7.3 ever] *after del* 'any'
8.2 to breed from;] *interl*
8.2 which is] *interl*
10.1 for I think I know which is which.—] *interl*

To W. E. Darwin 13 May [1857]
1.2 am very sorry to say that] *interl*
5.1 Aunt] *interl*
5.2 tomorrow] 't' *over illeg*
6.1 Hand] 'H' *over* 'h'
6.1 in] *interl*

To J. D. Hooker 16 [May 1857]
2.2 (ie not monstrosity)] *interl*
2.3 or organ] *interl*
2.5 is very] 'is' *interl*
2.7 is *very*] 'is' *interl*
2.7 in length of legs] *interl*
2.8 in all classes;—] *interl*
2.10 or two] *interl*
2.10 group] *above del* 'genus'
2.11 have expected] 'have' *interl*

2.11 to have] 'have' *interl*
2.11 been] *altered from* 'be'
2.11 very] *interl*
2.11 but I do . . . point, for] *added after del* 'for'
2.14 part or organ] *above del* 'respect'
2.14 the] *interl*
2.14 to it] *interl*
2.15 such part or organ] 'or organ' *interl*
2.16 your] *interl*
4.4 part] *altered from* 'parts'
4.5 condense the] 'the' *altered from* 'them'
4.5 hairs] *interl*
5.1 positively] *interl*

To W. B. Tegetmeier 18 May [1857]
1.1 interesting] *before omitted comma*
2.4 (sent by the Rajah Sir J. Brooke)] *interl*
3.1 the] *over* 'they'
4.3 & long-tailed] *interl*
4.5 it: I] ': I' *over* '.—'
5.5 Is your Edition completed yet?] *added*
6.2 the Meave . . . Breed.—] *added*
7.1 me to see the] *interl*
7.3 I work] 'I' *added*
8.1 the] *added*
8.1 Polands] *added*

To J. D. Dana 25 May [1857]
2.2 had heard] 'had' *altered from* 'have'

To J. D. Hooker [June 1857]
3.1 differ in] *before del* 'one or'

To J. D. Hooker 2 June [1857]
5.3 (No No I see it is local *work*!!!)] *added; square brackets in MS*
5.7 in *Zoology*] *after del* 'which'
5.9 Hancock] *above del* 'him'
5.11 *Private*] *interl*
5.11 Huxley] *after del* 'Pretwi'
5.12 a] *interl*
5.16 distinct] *after del* 'rather'
5.18 mentioned] *after del comma*
5.22 to take different views] *interl*
5.22 a near] 'a' *interl*
8.3 Moor] 'M' *over* 'm'
12.1 and] *after del* 'as I thought'

To William Sharpey 2 June [1857]
1.4 expressing . . . Botanist,] *interl*
1.6 to me] *interl*

1.10 one of] *interl*
1.16 seem] *after del* 'always'
1.17 without] *after del* 'my'
2.5 to a Body] 'to' *interl*

To J. D. Hooker 3 June [1857]
1.7 thinking of this I] *above del* 'I then I'
1.13 in female] *interl*
1.14 of head] *before del interl* 'of fem'
1.14 *astonishingly*] 'h' *over illeg*
1.17 hermaphrodite] *interl*
1.19 my] *interl*
2.2 see] *interl*
2.4 more than] *interl*
2.5 with plant not seedlings 2.6] *interl*
2.18 3] *over illeg*
6.1 I believe . . . Chile!!!] *added pencil*
6.2 Chile] *after del* 'Ja'

To J. D. Hooker 5 June [1857]
1.2 (but a much interested one)] *interl*
2.2 a] *interl*
2.9 but] *interl*
2.10 Crown] 'C' *over* 'c'
2.14 to] *interl*
2.16 frightfully] *interl*
2.16 work,] *before omitted comma*
2.18 the] *interl*
2.19 under all circumstances] *interl*
2.22 He] 'H' *over* 'h'
2.23 it] *over illeg*
2.24 that] *interl*
2.25 Bunsen] *before omitted point*
3.1 by] *over* 'on'
3.2 (or rather Waterhouse's)] *interl*
3.4 in plants] *interl*
3.10 instead] 'i' *over* '&'
3.11 being] *interl*
4.3 most] 'm' *over* 's'
8.1 I am . . . said.] *pencil*

To T. C. Eyton 9 June [1857]
2.2 regret] *over illeg*
5.2 colts] *over* 'cots'
5.2 me,] *comma over full stop*
6.1 same] *underl pencil*

To *Gardeners' Chronicle* [before 13 June 1857]
1.6 Is] *over illeg*

To Asa Gray 18 June [1857]
1.4 note] *interl*

1.5 he says] 'he' *interl*
2.4 only] *interl*
2.7 (*conversely . . . wide ranges.*)] *square brackets in MS*
2.8 & Families] *interl*
2.8 (ie from Extinction)] *interl*
2.12 & as nature does not lie] *interl*
3.1 naturalised] *interl*
5.2 led] *before del interl* 'to them'
5.3 animals'] 's' *added*
5.5 often] *interl*
5.8 favourable] *after del* 'eminently'
5.9 & C. C. Sprengels'] *interl*
5.10 that the] 'the' *altered from* 'these'
5.10 Compositæ] *interl*
5.10 (I am . . . pollen.)] *square brackets in MS*
5.14 in various genera] *interl*
5.17 in] *above del* 'of'
5.17 early] *after del* 'developm'
5.19 regular & normal] *interl*
5.20 under] *after del caret*
5.20 (& Leguminosæ)] *interl*
5.21 me, as] ', as' *over* '.—'
6.1 **very**] *interl*
7.4 most!] '!' *over* '—'

To W. B. Tegetmeier 23 June [1857]
1.1 Of] *over* 'On'
1.5 about] *interl*
1.7 about] *after del* 'bef'

To J. D. Hooker 25 June [1857]
1.3 feebly] *interl*
1.5 you] *interl*

To W. B. Tegetmeier 25 [June 1857]
1.1 to Down] *interl*
1.2 chips the] 'the' *over* 'its'
4.1 Carcase] *after interl del illeg*

To T. C. Eyton 26 [June 1857]
1.4 before] *before omitted point*
2.2 in] 'i' *over* 'o'
3.2 Ireland] *above del* 'Scotland'
3.3 the ass.] 'the' *interl*

To J. D. Hooker 1 July [1857]
1.1 the] *over* 'are'
1.2 summer,] *comma over full stop*
1.3 make] *after del* 'pu'
1.6 affair] *interl*

2.2 that] *interl*
2.11 in] *after del illeg*
3.1 relation] *after del* 'more distinct'
3.1 distinct] *after del* 'than have true leaves'
3.2 of the mature] *after del* 'have'; 'of' *over* 'to'
3.2 have to the conditions,] *interl*
3.3 the] *altered from* 'these'
3.3 of the cotyledons] *interl*

To J. D. Hooker 5 July [1857]
2.1 I] *over illeg*
6.1 Fresh] *interl*
6.5 & then] *above del* 'it was'

To T. H. Huxley 5 July [1857]
1.3 & in which way] *interl*
1.8 then] *interl*
1.13 consists of greatly] *above del* 'is only'
1.13 early] *before del* 'or middle'
2.5 an Elephant or] 'an' *altered from* 'a'; 'Elephant or' *interl*

To T. H. Huxley 9 July [1857]
1.4 (I state . . . myself)] *square brackets in MS*
1.4 that] *added*
2.3 man] *interl*
3.1 cannot] *interl*
3.3 have] *altered from* 'has'
3.4 begin] *altered from* 'begins'
3.10 distinct] *before del* 'embryos'
4.1 we] *after del* 'y'
6.3 usually] *above del* 'almost'
7.2 to] *before del* 'me'

To J. D. Hooker 14 July [1857]
0.3 Please Read last part. soon.] *added*
1.4 a)] *after del illeg captions*
2.1 As] *after del illeg line of text*
4.2 (N.B I once . . . peloric)] *square brackets in MS*
7.1 talking] 'l' *over* 'k'
7.4 what] *after del* 'whether'
7.6 England] *after del* 'in'

To John Lubbock 14 [July 1857]
1.2 muzzy] *altered from* 'muzzey'
2.4 (as 339:323 :: 51.:48.5)] *square brackets in MS*
2.5 strong] *after del* 'nearly'
2.5 general] *after del* 'very'
7.2 to you.—] *added*; 'to' *over* '.—'

To W. B. Tegetmeier [19 July 1857]
5.3 forget.—] *before del added* 'I wish I could get chick'

To Asa Gray 20 July [1857]
1.1 disjunction] *after del* 'extincti'
1.10 his] *after del* 'the'
1.11 best worth keeping] *interl*
1.14 find] *interl*
1.14 made] *over* 'make'
1.23 will] *after del illeg*
1.29 (but . . . grounds)] *interl*
2.3 of the works of] *interl*
3.1 *possibility* of] *interl*
3.2 fertilisation in] 'in' *over* 'of'
4.2 varieties] *above del* 'they'
4.3 & Diclytra] *interl*
4.6 scarlet] *interl*
4.7 honey] *after del* 'nectary'
4.8 *one*] *added*
4.8 of the flower] *interl*
5.4 *occasionally*] *interl*

To W. E. Darwin 21 [July 1857]
2.2 in deciphering] *interl*
4.9 write to] *before del* 'him.—'
4.9 the Tutor . . . vacations?] *added*; 'the' *over* '—'

To W. B. Tegetmeier 27 July [1857]
5.2 by stamps] *interl*
5.3 your] *altered from* 'you'

To J. D. Hooker 1 August [1857]
2.2 11ᵗʰ] '11' *over* '12'
2.4 to search for it.] *interl*
3.1 mere] *above del* 'mere'
3.2 varieties] *after del* 'marked'
3.2 (as far as I have yet gone] *interl*
3.6 well-worked] *interl*
3.16 ——Fragment of Indian Flora] *added*
4.3 known] *after del* 'marked'

To Laurence Edmondston 2 August [1857]
2.3 offer] *after del* 'so'
2.4 it is] 'it' *interl*
3.1 (Eel-backs)] *interl*
3.4 dark] *interl*
3.7 geographically] *interl*
4.1 stripe] *after del illeg*
4.1 with the duns] *interl*
4.3 when they occur] *interl*

4.5 and] *above del* 'but'
4.6 has been] *interl*
5.2 it,] *interl*
5.3 one] *after del* 'are'
5.3 When duns . . . results?] *added*
5.6 Something] 'S' *over illeg*
5.7 colours] *interl*

To J. S. Henslow 10 August [1857]
2.1 (N. Kent Division)] *square brackets in MS*

To John Lubbock 11 August [1857]
1.1 not] *interl*
1.4 a miserable] 'a' *over* 'as'

To John Lubbock 12 [August 1857]
2.3 &] *over* 'I'

To J. D. Hooker 22 August [1857]
1.3 that] *interl*
2.2 Schlagenheit] 'Sch' *before del* 'a'
2.2 right?)] '?' *interl*
3.5 be] *interl*
3.5 some time] *interl*
3.6 do not] *above del* 'hardly'
3.28 Centre] 'C' *over* 'F'
4.1 plants] *interl*

To Asa Gray 5 September [1857]
2.6 not strongly] 'not' *interl*
2.13 in] *above del illeg*
3.7 with its pollen . . . by Birds] *interl*
3.19 Vestiges] 'V' *over* 'v'
3.21 perhaps] *interl*
4.3 forming] *after del* 'in'
4.4 naturalised & of] *interl*
4.6 the large] *after del* 'in'
4.6 $\frac{88}{1000}$] '88' *above del* '61'
4.7 $\frac{50}{1000}$] '50' *before del* '37'
4.8 your] *above del* 'the'
4.9 $\frac{48}{1000}$] '48' *above del* '35'
4.9 $\frac{46}{1000}$] '46' *above del* '33'
5.1 have] *interl*
5.2 the larger] 'the' *interl*
6.7 to the large] 'to the' *interl*
6.8 Are the close . . . difference?] *added*
7.2 Bees] *after del* 'vis'
7.3 the wing] 'the' *interl*
7.4 the same] 'the' *interl*
9.2 thanks &] '&' *over full stop*
9.2 & your Lesson . . . reading.] *added*

9.4 (which is not likely)] *parentheses over commas*
11.2 desired] 's' *added for clarity*
11.5 half] *interl*
11.10 "rogueing"] *inserted in gap*
11.16 is not] 'not' *inserted in gap*
11.17 he] *interl*
11.18 &c.—] *added over full stop*
12.5 a] *interl*
12.5 change] *altered from* 'changed'
12.6 in the conditions] 'in the' *interl*
13.1 unerring] *inserted in gap*
13.1 or] *over* 'on'
13.6 years, or] 'or' *above del* 'and'
13.7 not] *interl*
14.11 again] *interl*
15.2 non] *over illeg*
16.12 true] *above del* 'twin'
17.2 many] *above del* 'my'
18.2 of] *altered from* 'on'

To J. D. Hooker 6 September [1857]
6.1 wild] *interl*

To W. E. Darwin [before 11 September 1857]
1.6 school] *interl*
1.10 wood] *interl*

To J. D. Hooker 11 September [1857]
5.1 the] *above del* 'my'
5.1 Books] *after del* 'work'
10.2 tank] *after del* 'hot-water'
13.2 the shoulder] 'the' *interl*
15.3 when] *after del* 'aft'
16.1 on] *after del* 'of'

To T. H. Huxley 15 September [1857]
1.7 much] *interl*

To T. H. Huxley 26 September [1857]
1.6 or] *over* '&'
1.7 & judging] *interl*
2.1 have written] *after del* '& Brulle'
2.2 anatomy of] *interl*
2.2 14] *over* '13'
3.1 1836,] *before del colon*
3.1 work] *altered from* 'worth'
8.1 anatomical] *after del* 'strictly'

To W. B. Tegetmeier 29 September [1857]
1.2 Head] *after del* 'Body'
1.4 have done] 'v' *over illeg*

1.7 a] *interl*
2.1 young] *above del* 'old'

To J. D. Hooker 30 September [1857]
1.3 like to] 'to' *over* '&'
2.10 France &] *interl*
3.1 2] *over* 'or'

To T. H. Huxley 3 October [1857]
0.2 Oct.] 'O' *over* 'Sa'
2.3 element] *interl*
3.2 to look] *after del* 'not'
4.3 do] *after del* 'wo'
4.5 stood] *above del* 'been'
4.5 if allocated] 'if' *over illeg*
4.6 we may] *after del* 'of cour'

To James Buckman 4 October [1857]
5.4 Or . . . free.—] *added*
5.6 Birds] *interl*

To J. S. Henslow 14 October [1857]
1.1 you] *interl*
1.2 sport] *after del* 'suff'
1.3 separate] *after del* 'spe'

To J. S. Henslow 18 October [1857]
2.1 *apparently*] *before del* 'on'
2.1 only certain] *interl*
2.3 10 sepals] *after del* '8'

To J. D. Hooker 20 October [1857]
1.3 the carriage] 'the' *interl*
1.3 will return] 'will' *above del* 'can'
1.6 soon as] *interl*
2.3 of the above Books] *interl*
2.4 Ledebour will] 'Ledebour' *interl*
3.2 entirely] *interl*
4.1 with varieties marked.] *interl*
11.1 is mentioned] *interl*
11.1 not] *after del illeg*

To J. D. Hooker [23 October 1857]
1.3 the] *interl*
2.2 per Parcel Delivery] *interl*
5.4 ie,] 'i' *over* 'e'

To W. D. Fox 30 October [1857]
1.6 &] *over* 'but'
1.7 & I] 'I' *added*
2.5 our] *interl*
6.8 visitors,] 'v' *over* 'w'

To W. E. Darwin [November 1857]
2.7 small] *interl*

To T. C. Eyton 2 November [1857]
1.2 common] *above del illeg*

To T. H. Huxley [before 12 November 1857]
1.4 (whatever he may be)] *interl*
1.8 new] *interl*

To J. D. Hooker 14 [November 1857]
1.3 as yet] *interl*
2.1 quite] *interl*
2.9 which I much] 'which' *over* '—'
3.4 **in 1ˢᵗ generation**] *interl*
3.5 by Wiegmann] 'by' *interl*

To J. D. Hooker 21 November [1857]
1.7 many] *interl*

To W. B. Tegetmeier 21 November [1857]
1.2 to my surprise] *interl*
1.2 young] *interl*
1.4 purchase] *interl*
5.1 (health] *parenthesis over comma*

To John Lubbock [22 November 1857]
1.1 I] *after del illeg*
1.4 (Sharpey] *parenthesis over comma*
2.3 Breakfast] 'k' *over* 'f'

To Asa Gray 29 November [1857]
1.3 on] *over* 'of'
1.8 expressing] 're' *over illeg*
1.10 for] 'f' *over illeg*
1.11 (here miserably expressed)] *interl; square brackets in MS*
1.12 preservation] ¹'r' *over illeg; before del caret*
1.19 several] *after del* 'man can if'
1.19 of Sheep] *interl*
1.19 have] 'h' *over illeg*
1.23 or] *after del* '('
2.2 seen in] 'in' *above del* 'of'
2.2 whilst sucking] *interl after del* 'on'
3.2 the seed-coats] 'the' *over illeg*
3.3 act] *after del* 'first'
3.5 (Balanus)] *interl*
3.6 up] *added*
3.6 offer] *altered from* 'offers'
4.4 (great] *after del* 'tha'
4.10 I have] *after del double quotation mark*

4.11 definition] *above del* 'one'
5.2 again] *interl*
6.4 all] *interl*
6.9 for my work] *interl*
6.11 in classes &c,] *interl*
6.12 ie] *above del* 'or'
6.12 tendency to the] *interl*
6.12 from extinction] *interl*
6.14 mercy] *altered from* 'mery'
8.1 Moth or Humble Bee] *above del* 'insects'
8.1 visits] *altered from* 'visit'
8.2 fertilises] *altered from* 'fertilise'
8.2 young] *added*

To George Bentham 1 December [1857]
1.2 know of] 'of' *interl*
1.3 Clover] *interl*
1.4 has.] *above del* 'knows.'
2.6 wildly] *interl*
2.7 well] *interl*
2.7 land,] *after del* 'pl'
2.9 consequently] *interl*
2.9 European] *after del* 'old'
2.9 rivals] *interl*
2.10 land.] *after del* 'plants'
3.5 not] *before del* 'say'
3.5 names] *altered from* 'name'
4.5 in] *interl*
6.1 introduced] *interl*
7.5 in this & other cases] *interl*
7.5 many] *above del* 'many'
7.5 trained] *above del* 'accustomed'

To J. D. Hooker 4 December [1857]
1.1 day] *interl*
1.7 Garden] *interl*
1.7 or hottish] *interl*

To Ferdinand Jakob Heinrich von Mueller
     8 December [1857]
1.5 very] *interl*
1.12 north-ranging] *above del* 'northern'
1.15 (Have you ever enumerated them?)] *interl*

To J. D. Hooker 9 December [1857]
2.4 a numerically] *interl*
2.4 genus] *above del* 'species'
2.6 on average] *interl*
2.8 perhaps universally] *interl*
2.9 same] *over illeg*
2.9 hold] *after del illeg*

4.1 observe] *after del* 'I'
4.4 United States] *interl*

To George Bentham 15 December [1857]
1.3 ranked] *after del* 'made'
2.3 of the same species] *interl*
3.2 often] *interl*
3.5 parallel] *interl*

To T. H. Huxley 16 December [1857]
1.1 on] *over illeg*
2.2 explanations] *after del illeg*
2.3 been writing] *above del* 'written'
2.4 you put] 'you' *interl*
4.1 meet with] *above del* 'see'

To W. D. Fox 17 December [1857]
1.1 have] *over* 'had'
1.3 brown-tailed;] *semicolon over colon*
1.9 I know . . . Silver Greys.] *added*

2.3 that the] 'the' *interl*

To George Bentham 18 December [1857]
2.1 (or . . . infertility) 2.2] *parentheses over commas*

To A. R. Wallace 22 December 1857
1.10 specially] *interl*
3.3 in my Coral volume] *interl*
3.6 poor E.] *interl*
3.10 be] *after del illeg*
4.2 some of them] *above del* 'they'
4.4 man's] *after del* 'savage'
4.7 ?] *over colon*
5.4 the highest] 'the' *interl*
5.6 by] *over* 'in'
6.9 have] *over* 'now'

To J. D. Hooker 25 December [1857]
3.1 that] *interl*
7.2 the] *after del illeg*

BIBLIOGRAPHY

The following bibliography contains all the books and papers referred to in this volume by author–date reference or by short title. Short titles are used for some standard reference works (e.g., *DNB*, *OED*), for CD's books and papers, and for editions of his letters and manuscripts (e.g., *Descent*, *LL*, *Notebooks*). Other references are given in author–date form.

Adams, Henry and Adams, Arthur. [1853–]1858. *The genera of recent mollusca; arranged according to their organisation.* 3 vols. London.

ADB: *Allgemeine deutsche Biographie.* Under the auspices of the Historical Commission of the Royal Academy of Sciences. 56 vols. Leipzig: Duncker and Humblot. 1875–1912.

Agassiz, Louis. 1847. *Nouvelles études et expériences sur les glaciers actuels, leur structure, leur progression et leur action physique sur le sol.* Paris.

——. 1849. On the differences between progressive, embryonic, and prophetic types in the succession of organized beings through the whole range of geological times. *Proceedings of the American Association for the Advancement of Science. First meeting, held at Philadelphia, September, 1848,* pp. 432–8.

——. 1850. De la classification des animaux dans ses rapports avec leurs développement embryonaire et avec leur histoire paléontologique. *Archives des Sciences Physiques et Naturelles. Supplément à la Bibliothèque Universelle de Genève* 15: 190–204.

——. 1851. Observations on the blind fish of the Mammoth cave. *American Journal of Science and Arts* 2d ser. 11: 127–8.

——. 1857. On some young gar pikes from Lake Ontario. *American Journal of Science and Arts* 2d ser. 23: 284.

Ajasson de Grandsagne, Jean Baptiste François Étienne. 1829–33. *Histoire naturelle de Pline.* 20 vols. Paris.

Albin, Eleazar. 1731–8. *A natural history of birds.* 3 vols. London.

Alder, Joshua and Hancock, Albany. 1844. Report on the British Nudibranchiate Mollusca. *Report of the 14th meeting of the British Association for the Advancement of Science held at York,* pp. 24–9.

——. 1845–55. *A monograph of the British Nudibranchiæ Mollusca: with figures of all the species.* 7 pts. in 1 vol. London.

Aldrovandi, Ulysse. 1599–1603. *Ornithologiae. De avibus historiae libri . . . cum indice septendecim linguarum copiosissimo.* 3 vols. Bologna.

Allen, William and Thomson, Thomas Richard Heywood. 1848. *A narrative of the expedition sent by Her Majesty's Government to the River Niger, in 1841. Under the command of Captain H. D. Trotter, R.N.* 2 vols. London.

Allman, George James. 1850. On the present state of our knowledge of the freshwater Polyzoa. *Report of the 20th meeting of the British Association for the Advancement of Science held at Edinburgh,* pp. 305–37.

Alum. Cantab.: *Alumni Cantabrigienses. A biographical list of all known students, graduates and holders of office at the University of Cambridge, from the earliest times to 1900.* Compiled by J. A. Venn. Part II. From 1752 to 1900. 6 vols. Cambridge: Cambridge University Press. 1940–54.

Alum. Oxon. : *Alumni Oxonienses: the members of the University of Oxford, 1715–1886.* By Joseph Foster. 4 vols. Oxford. 1888.

Andral, Gabriel et al. 1829–35. *Dictionnaire de médicine et de chirurgie pratiques.* 15 vols. Paris.

[Asso y del Rio, Ignacio Jordán de]. 1784. *Introductio in oryctographiam, et zoologiam Aragoniæ accedit enumeratio stirpium in eadem regione noviter detectarum.* [Madrid].

Astley, Thomas. [Publisher]. 1745–7. *A new general collection of voyages and travels . . . in Europe, Asia, Africa, and America.* 4 vols. London.

Atkins, Hedley. 1974. *Down: the home of the Darwins; the story of a house and the people who lived there.* London: Royal College of Surgeons.

Audouin, Jean Victor and Brullé, Gaspard Auguste. 1834–7. *Histoire naturelle des insectes, traitant de leur organisation et de leurs mœurs en général, par M. V. Audouin . . . et comprenant leur classification et la description des espèces, par M. A. Brullé.* Vols. 4, 5 and 6: Coléoptères. Vol. 9: Orthoptères, Hémiptères. By G. A. Brullé. (No more published.) Paris.

Audouin, Jean Victor. 1824. Recherches anatomiques sur la femelle du *Drile jaunatre,* et sur le mâle de cette espèce. *Annales des Sciences Naturelles* 2: 443–62.

Aust. dict. biog.: *Australian dictionary of biography: 1788–1850; 1851–1890.* Edited by Douglas Pike and Bede Nairn. 6 vols. Melbourne: Melbourne University Press. 1966–76.

Autobiography: *The autobiography of Charles Darwin 1809–1882. With the original omissions restored.* Edited and with appendix and notes by Nora Barlow. London: Collins. 1958.

Azara, Félix d'. 1801. *Essais sur l'histoire naturelle des quadrupèdes de la province du Paraguay.* 2 vols. Paris.

Babington, Charles Cardale. 1851. *Manual of British botany, containing the flowering plants and ferns arranged according to the natural orders.* 3d ed. London.

Backhouse, James. 1856. *A monograph of the British Hieracia.* York.

Baird, William. 1850. *The natural history of the British Entomostraca.* London.

Barbot, John. 1732. *A description of the coasts of North and South-Guinea; and of Ethiopia Inferior, vulgarly Angola: being a new and accurate account of the western maritime countries of Africa.* Vol. 5 of Churchill, Awnsham, *A collection of voyages and travels.* London.

Barnéoud, François Marius. 1846. Mémoire sur le développement de l'ovule, de l'embryon et des corolles anomales. *Annales des Sciences Naturelles. Botanique* 3d ser. 6: 268–96.

Barry, Martin. 1838. Researches in embryology. First series. *Philosophical Transactions of the Royal Society of London*, pp. 301–41.

Barth, Heinrich. 1857–8. *Travels and discoveries in North and Central Africa: being a journal of an expedition undertaken under the auspices of H.B.M.'s government, in the years 1849–1855*. 5 vols. London.

Barton, Benjamin Smith. 1809. *Specimen of a geographical view of the trees and shrubs, and many of the herbaceous plants of North-America*. Philadelphia.

Bartram, William. 1791. *Travels through North and South Carolina, Georgia, East and West Florida, the Cherokee country . . . containing an account of the soil and natural productions of those regions*. Philadelphia.

Baudement, Émile. 1847. Observations sur les analogies et les différences des arcs viscéraux de l'embryon dans les deux sous-embranchements des vertébrés. *Annales des Sciences Naturelle. Zoologie* 3d ser. 7: 73–86.

Bechstein, Johann Matthäus. [1789–95]. *Gemeinnützige Naturgeschichte Deutschlands nach allen drey Reichen*. 4 vols. Leipzig.

Bell, Thomas. 1841. Some account of the Crustacea of the coasts of South America, with descriptions of new genera and species; founded principally on the collections obtained by Mr. Cuming and Mr. Miller. *Transactions of the Zoological Society of London* 2: 39–66.

Bénézit, Emmanuel. 1976. *Dictionnaire critique et documentaire des peintres, sculpteurs, dessinateurs et graveurs de tous les temps et de tous les pays, par un groupe d'écrivains spécialistes français et étrangers*. Revised edition. 10 vols. Paris: Librairie Gründ.

Bentham, George. 1855. Historical notes on the introduction of various plants into the agriculture and horticulture of Tuscany: a summary of a work entitled *Cenni storici sulla introduzione di varie piante nell'agricoltura ed orticultura Toscana*. By Dr. Antonio Targioni Tozzetti. Florence 1850. *Journal of the Horticultural Society of London* 9: 133–81.

——. 1857. Synopsis of the genus *Clitoria*. [Read 3 March 1857.] *Journal of the Proceedings of the Linnean Society. Botany* 2 (1858): 33–44.

——. 1858. *Handbook of the British flora; a description of the flowering plants and ferns indigenous to, or naturalized in, the British Isles. For the use of beginners and amateurs*. London.

Berghaus, Heinrich Karl Wilhelm. 1845–8. *Physikalischer Atlas*. 2 vols. Gotha.

Berkeley, Miles Joseph. 1857. *Introduction to cryptogamic botany*. London.

BHGW: *Biographisch-literarisches Handwörterbuch zur Geschichte der exacten Wissenschaften*. By Johann Christian Poggendorff. Vols. 1–4, Leipzig: Johann Ambrosius Barth. 1863–1904. Vol. 5, Leipzig and Berlin: Verlag Chemie. 1926.

BLKO: *Biographisches Lexikon Des Kaiserthums Oesterreich, enhaltend die Lebensskizzen der denkwürdigen Personen, welche seit 1750 in den österreichischen Krönlandern geboren wurden oder darin gelebt und gewirkt haben*. By Constant von Wurzbach. 60 vols. Vienna. 1856–1890.

Blyakher, L. Y. 1982. *History of embryology in Russia from the middle of the eighteenth century to the middle of the nineteenth century.* Translated from the Russian by H. I. Youssef and B. A. Malek, with an introduction by Jane Maienschein. Washington, D.C.: Smithsonian Institution and the National Science Foundation.

Blyth, Edward. 1837. On the counterfeiting of death, as a means to escape from danger, in the fox and other animals. *Magazine of Natural History* n.s. 1: 566–74.

——. 1855a. Report on a zoological collection from the Somáli country. *Journal of the Asiatic Society of Bengal* n.s. 24 (1856): 291–306.

——. 1855b. Further remarks on the different species of orang-utan. *Journal of the Asiatic Society of Bengal* n.s. 24 (1856): 518–28.

——. 1859. On the different animals known as wild asses. *Journal of the Asiatic Society of Bengal* n.s. 28: 229–53.

Bochart, Samuel. 1675. *Opera omnia.* 2 vols. Leiden.

Bonaparte, Charles Lucien. 1850–7. *Conspectus generum avium.* 2 vols. Leiden.

——. 1855a. Coup d'œil sur les pigeons (quatrième partie). *Comptes Rendus Hebdomadaires des Séances de l'Académie des Sciences* 40: 15–24.

——. 1855b. Remarques de S. A. Monseigneur le Prince Bonaparte au sujet de cette communication ['Sur deux chevaux sauvages, d'une espèce nouvelle (*Equus hemippus*) . . . par M. Is. Geoffroy-Saint-Hilaire']. *Comptes Rendus Hebdomadaires des Séances de l'Académie des Sciences* 41: 1219–20.

Bonney, Thomas George. 1919. *Annals of the Philosophical Club of the Royal Society, written from its minute books.* London: Macmillan and Co.

Boreau, Alexandre. 1840. *Flore du centre de la France; ou, description des plantes qui croissent spontanément dans le région centrale de la France.* 2 vols. Paris.

Bosquet, Joseph Augustin Hubert de. 1854. *Monographie des Crustacés fossiles du terrain Crétacé du Duché de Limbourg.* Haarlem.

——. 1857. Notice sur quelques cirripèdes récémment découverts dans le terrain crétacé du Duché de Limbourg. *Natuurkundige Verhandelingen van de Hollandsche Maatschappij der Wetenschappen té Haarlem* 2d ser. 13, pt 3: 1–36.

Brehm, Christian Ludwig. 1831. *Handbuch der Naturgeschichte aller Vögel Deutschlands.* Ilmenau.

British Museum (Natural History). 1904–6. *The history of the collections contained in the natural history departments of the British Museum.* 2 vols. London.

Brockway, Lucile H. 1979. *Science and colonial expansion. The role of the British Royal Botanic Gardens.* New York and London: Academic Press.

Brown, Robert. 1810. *Prodromus floræ Novæ Hollandiæ et insulæ Van-Diemen, exhibens characteres plantarum quas annis 1802–1805.* Vol. 1. London.

——. 1814. General remarks, geographical and systematical, on the botany of Terra Australis. Appendix 3, pp. 533–613 of vol. 2 of Flinders, Matthew, *A voyage to Terra Australis.* 2 vols. and atlas. London.

Browne, Janet. 1978. The Charles Darwin–Joseph Hooker correspondence: an analysis of manuscript resources and their use in biography. *Journal of the Society for the Bibliography of Natural History* 8: 351–66.

Browne, Janet. 1980. Darwin's botanical arithmetic and the "principle of divergence", 1854–1858. *Journal of the History of Biology* 13: 53–89.

——. 1983. *The secular ark: studies in the history of biogeography*. New Haven and London: Yale University Press.

——. 1990. Spas and sensibilities: Darwin at Malvern. In Porter, Roy and Bynum, W. F., eds., *The medical history of waters and spas. Medical History*, supp. 10. London: Wellcome Institute for the History of Medicine.

Browne, William George. 1799. *Travels in Africa, Egypt, and Syria, from the year 1792 to 1798*. London.

Brullé, Gaspard Auguste. 1844. Recherches sur les transformations des appendices dans les Articulés. *Annales des Sciences Naturelles. Zoologie* 3d ser. 2: 271–373.

Buckman, James. 1856. Notes on experiments in the botanical garden of the Royal Agricultural College. *Report of the 26th meeting of the British Association for the Advancement of Science held at Cheltenham*, Transactions of the sections, p. 83.

——. 1857. Report on the experimental plots in the botanical garden of the Royal Agricultural College at Cirencester. *Report of the 27th meeting of the British Association for the Advancement of Science held at Dublin*, pp. 200–15.

Buffon, Georges Louis Leclerc, Comte de. 1793. *The natural history of birds. From the French of the Count de Buffon. Illustrated with engravings; and a preface, notes, and additions, by the translator*. 9 vols. London.

Bunbury, Charles James Fox. 1848. *Journal of a residence at the Cape of Good Hope; with excursions into the interior, and notes on the natural history, and the native tribes*. London.

Bunbury, Frances Joanna, ed. 1891–3. *Memorials of Sir C. J. F. Bunbury, Bart. Early life* 1 vol.; *Middle life* vols. 1–3; *Later life* vols. 1–5. Mildenhall.

Burckhardt, John Lewis. 1819. *Travels in Nubia*. London.

——. 1822. *Travels in Nubia*. 2d ed. London.

——. 1829. *Travels in Arabia*. 2 vols. London.

——. 1830. *Notes on the Bedouins and Wahábys, collected during his travels in the east*. 2 vols. London.

Burke's landed gentry: Burke's genealogical and heraldic history of the landed gentry. By John Burke. 1–18 editions. London: Burke's Peerage Ltd. 1833–1972.

Burke's peerage: Burke's peerage and baronetage. 1–105 editions. London: Burke's Peerage (Genealogical Books). 1826–1980.

Burnett, Waldo I. 1854. Researches on the development of viviparous aphids. *American Journal of Science and Arts* 2d ser. 17: 62–78.

Butter, Donald. 1839. *Outlines of the topography and statistics of the southern districts of Oud'h, and of the cantonment of Sultanpur-Oud'h*. Calcutta.

Candolle, Alphonse de. 1855. *Géographie botanique raisonnée; ou exposition des faits principaux et des lois concernant la distribution géographique des plantes de l'époque actuelle*. 2 vols. Paris and Geneva.

Candolle, Augustin Pyramus de and Candolle, Alphonse de. 1824–73. *Prodromus systematis naturalis regni vegetabilis, sive enumeratio contracta ordinum generum specierumque plantarum huc usque cognitarum, juxta methodi naturalis normas digesta.* 17 vols. Paris.

Candolle, Augustin Pyramus de. 1820. Géographie botanique. In vol. 18 of Cuvier, Frédéric, ed. *Dictionnaire des sciences naturelles.* 61 vols. and 12 vols. of plates. Strasbourg and Paris. 1816–45.

Carpenter, William Benjamin. 1856. Researches on the Foraminifera. Pt 1. Containing general introduction, and monograph of the genus Orbitolites; Pt 2. On the genera Orbiculina, Alveolina, Cycloclypeus and Heterostegina. *Philosophical Transactions of the Royal Society of London* 146: 181–236, 547–69.

Carroll, P. Thomas. 1976. *An annotated calendar of the letters of Charles Darwin in the library of the American Philosophical Society.* Wilmington, Delaware: Scholarly Resources.

Cassini, Henri. 1813. Observations sur le style et le stigmate des synanthérées. *Journal de Physique, de Chemie, d'Histoire Naturelle et des Arts* 76: 97–128, 181–201, 249–75.

Catalogue de la Bibliothèque nationale: Catalogue général des livres imprimés de la Bibliothèque nationale. 231 vols. Paris: Bibliothèque nationale. 1897–1981.

Cautley, Proby Thomas. 1840. On the fossil remains of Camelidæ of the Sewaliks. *Journal of the Asiatic Society of Bengal* n.s. 9, pt 1: 620–4.

CDEL: A critical dictionary of English literature and British and American authors, living and deceased. By S. Austin Allibone. 3 vols. Philadelphia and London. 1859–71. Supplement by John Foster Kirk. 2 vols. Philadelphia and London. 1891.

Chalmers-Hunt, J. M. 1976. *Natural history auctions 1700–1972. A register of sales in the British Isles.* London and New York: Sotheby Parke Bernet Publications.

[Chambers, Robert]. 1844. *Vestiges of the natural history of creation.* London.

Chamisso, Adelbert von. 1831. De Plantis in Expeditione Romanzoffiana. Observatis disserere pergitur. *Linnaea* 6: 545–98.

Chesney, Francis Rawdon. 1850. *The expedition for the survey of the rivers Euphrates and Tigris.* 2 vols. London.

Clergy list: The clergy list . . . containing an alphabetical list of the clergy. London. 1841–.

Collected papers: The collected papers of Charles Darwin. Edited by Paul H. Barrett. 2 vols. Chicago and London: University of Chicago Press. 1977.

Colp, Ralph, Jr. 1977. *To be an invalid: the illness of Charles Darwin.* Chicago and London: University of Chicago Press.

Complete peerage: The complete peerage of England Scotland Ireland Great Britain and the United Kingdom extant extinct or dormant. By G. E. Cokayne. New edition, revised and much enlarged. Edited by the Hon. Vicary Gibbs and others. 12 vols. London: The St Catherine Press. 1910–59.

Conolly, Arthur. 1834. *Journey to the north of India, overland from England, through Russia, Persia, and Affghaunistaun.* 2 vols. London.

Coral reefs: The structure and distribution of coral reefs. Being the first part of the geology of the voyage of the Beagle, under the command of Capt. FitzRoy, R.N. during the years 1832 to 1836. By Charles Darwin. London. 1842.

Coral reefs 2d ed.: The structure and distribution of coral reefs. Second edition revised. London. 1874.

Correspondence: The correspondence of Charles Darwin. Edited by Frederick Burkhardt and Sydney Smith. Cambridge: Cambridge University Press. 1985–.

Corse, John. 1799a. An account of the method of catching wild elephants at Tipura. *Asiatic Researches* 3: 229–48.

——. 1799b. Observations on the different species of Asiatic elephants, and their mode of dentition. *Philosophical Transactions of the Royal Society of London*, pp. 205–36.

[Cowell, F. R.] 1975. *The Athenaeum: club and social life in London, 1824–1974*. London: Heinemann.

Crawfurd, John. 1828. *Journal of an embassy from the governor-general of India to the courts of Siam and Cochin China; exhibiting a view of the actual state of those kingdoms*. London. Reprint ed. with an introduction by D. K. Wyatt. Kuala Lumpur, London and New York: Oxford University Press. 1967.

Cuvier, Georges. 1834–6. *Recherches sur les ossemens fossiles, où l'on rétablit les caractères de plusieurs animaux dont les révolutions du globe ont détruit les espèces*. 4th ed. 10 vols. and 2 atlases. Paris.

DAB: Dictionary of American biography. Under the auspices of the American Council of Learned Societies. 20 vols., index, and 7 supplements. New York: Charles Scribner's Sons; London: Oxford University Press. 1928–81.

Dana, James Dwight. 1852[–53]. *Crustacea*. Vol. 13 of *United States Exploring Expedition during the years 1838–1842. Under the command of Charles Wilkes, U.S.N.* 2 pts. Philadelphia. With atlas. London and Philadelphia. [1861].

——. 1853. *On the classification and geographical distribution of Crustacea: from the report on Crustacea of the United States Exploring Expedition, under Captain Charles Wilkes, U.S.N., during the years 1838–1842*. Philadelphia.

——. 1855. Anniversary address of the president. *Proceedings of the American Association for the Advancement of Science. Ninth meeting, held at Providence, Rhode Island*, pp. 1–36.

——. 1857. Thoughts on species. *American Journal of Science and Arts* 2d ser. 24: 303–16.

Daniell, William Freeman. 1849. *Sketches of the medical topography and native diseases of the gulf of Guinea western Africa*. London.

Darwin pedigree: Pedigree of the family of Darwin. Compiled by H. Farnham Burke. Privately printed. 1888. Reprinted in Freeman, Richard Broke, *Darwin pedigrees*. London: printed for the author. 1984.

Darwin, Charles and Wallace, Alfred Russel. 1858. On the tendency of species to form varieties; and on the perpetuation of varieties and species by natural means of selection. [Read 1 July 1858.] *Journal of the Proceedings of the Linnean Society* 3 (1859): 45–62.

Davidson, Thomas William St Clair. 1851–86. *British fossil Brachiopoda*. Vol. 1: *Tertiary, cretaceous, oolitic, and liasic species. With a general introduction.* Vol. 2: *Permian and carboniferous species.* 6 vols. and 2 vols. plates. London.

DBF: Dictionnaire de biographie française. Under the direction of J. Balteau, M. Barroux, M. Prevost, and others. 17 vols. (A–Humann). Paris: Librairie Letouzey et Ané. 1933–89.

DBI: Dizionaria biografica degli Italiani. Edited by Alberto M. Ghisalberti. 33 vols. (A–De Foresta). Rome: Istituto della Enciclopedia Italiana. 1960–87.

DBL: Dansk biografisk leksikon. Founded by Carl Frederick Bricka, edited by Povl Engelstoft and Svend Dahl. 26 vols. and supplement. Copenhagen: J. H. Schultz. 1933–44.

de Beer, Gavin, ed. 1959a. Darwin's Journal. *Bulletin of the British Museum (Natural History) Historical series 2: 3–21.*

——. 1959b. Some unpublished letters of Charles Darwin. *Notes and Records of the Royal Society of London* 14: 12–66.

Decaisne, Joseph. 1857. On the development of the floral organs in the pear. *Gardeners' Chronicle and Agricultural Gazette*, 14 November 1857, p. 773.

Demas, Don Sinbaldo. 1840. On the Egyptian system of artificial hatching. *Journal of the Asiatic Society of Bengal* n.s. 8: 38–48.

Descent: The descent of man, and selection in relation to sex. By Charles Darwin. 2 vols. London, 1870, 1871.

Deshayes, Gérard Paul. 1824–37. *Description des coquilles fossiles des environs de Paris.* 2 vols. and atlas. Paris.

——. 1853–4. *Catalogus Concharum Bivalvium, quæ in Musæo Britannico asservantur.* Pt 1: Veneridæ; pt 2: Petricoladæ (concluded); Corbiculadæ. [Edited by J. E. Gray.] London.

Desmond, Adrian. 1982. *Archetypes and ancestors: palaeontology in Victorian London, 1850–1875.* London: Blond & Briggs.

Desmond, Ray. 1977. *Dictionary of British and Irish botanists and horticulturists, including plant collectors and botanical artists.* 3d ed. London: Taylor and Francis.

Dictionary of Mauritian biography: Dictionary of Mauritian biography. Société de l'Histoire de l'Ile Maurice. Port Louis, Mauritius. [1941–].

di Gregorio, Mario Aurelio Umberto. 1984. *T. H. Huxley's place in natural science.* New Haven and London: Yale University Press.

Dixon, Edmund Saul. 1848. *Ornamental and domestic poultry: their history and management.* London.

——. 1851. *The dovecote and the aviary: being sketches of the natural history of pigeons and other domestic birds in a captive state, with hints for their management.* London.

DNB: Dictionary of national biography. Edited by Leslie Stephen and Sidney Lee. 63 vols. and 2 supplements (6 vols.). London: Smith, Elder and Co. 1885–1912. *Dictionary of national biography 1912–80.* Edited by H. W. C. Davis,

J. R. H. Weaver, and others. 7 vols. London: Oxford University Press. 1927–86.

DNZB: *A dictionary of New Zealand biography.* Edited by G. H. Scholefield. 2 vols. Wellington: Department of Internal Affairs New Zealand. 1940.

Don, David. 1841. An account of the Indian species of *Juncus* and *Luzula*. *Transactions of the Linnean Society of London* 18: 317–26.

Drège, Jean François. 1837–39. *Catalogus plantarum exsiccatarum Africæ australioris quas emturis offert J. F. D.* 3 pts. Königsberg.

——. 1840. *Catalog südafrikanischer getrockneter Pflanzen welche unter folgenden Bedingungen zu haben sind.* Hamburg.

——. 1843. Zwei pflanzengeographische Dokumente. With an introduction by Ernst Friedrich Heinrich Meyer. *Flora, oder allgemeine botanische Zeitung.* Suppl. to n.s. 1: 1–200.

——. [1847]. *Herbarien der südafrikanischen aussertropischen Flora.* Hamburg.

DSAB: *Dictionary of South African biography.* Edited by W. J. de Kock, D. W. Krüger, and C. J. Beyers. Vol. 1, Cape Town: Nasionale Boekhandel Beperk; vols. 2–3, Cape Town: Tafelberg Publishers; vol. 4, Durban and Pretoria: Butterworth & Co. 1968–1981.

DSB: *Dictionary of scientific biography.* Edited by Charles Coulston Gillispie. 14 vols., supplement, and index. New York: Charles Scribner's Sons. 1970–80.

Dufossé, Adolphe. 1856. De l'hermaphodisme chez certains vertébrés. *Annales des Sciences Naturelles. Zoologie* 4th ser. 5: 295–332.

Dupree, Anderson Hunter. 1959. *Asa Gray, 1810–1888.* Cambridge, Mass.: Belknap Press of Harvard University Press.

Dureau de la Malle, Adolphe Jules César Auguste. 1855. Des transformations opérées lors du retour des diverses variétés de nos animaux et de nos oiseaux domestiques à l'état sauvage, et du passage de la servitude à l'indépendance et à la liberté. *Comptes Rendus Hebdomadaires des Séances de l'Académie des Sciences* 41: 688–92.

Duval, (jardinier). 1852. *Histoire du pommier et sa culture.* Paris.

East-India register and army list: *The East-India register and army list. Compiled from the official returns, by permission of the Honourable the East-India Company.* 16 vols. London. 1845–60.

East-India register and directory: *The East-India register and directory. Compiled, by permission of the Honourable the East-India Company, from the official returns received at the East-India House.* 42 vols. London. 1803–44.

Eaton, John Matthews. 1852. *A treatise on the art of breeding and managing tame, domesticated, and fancy pigeons.* London.

EB: *Encyclopaedia Britannica.* 11th ed. 29 vols. Cambridge: Cambridge University Press. 1910–11.

Ecklon, Christian Friedrich and Zeyher, Carl. 1835–7. *Enumeratio plantarum Africæ Australis extratropicæ.* 3 pts. Hamburg.

Edwards, George. 1758–64. *Gleanings of natural history, exhibiting figures of quad-rupeds, birds, insects, plants, &c. Most of which have not, till now, been either figured or described.* 3 pts. London.

EI: *Enciclopedia Italiana di scienze, lettere ed arti.* 35 vols. Rome: Istituto della Enciclopedia Italiana. 1929–39.

Eights, James. 1852. Description of a new animal belonging to the Crustacea, discovered in the Antarctic Seas, by the author. *Transactions of the Albany Institute* 2 (1833–52): 331–4.

——. 1856. Description of an Isopod crustacean from the Antarctic seas, with observations on the New South Shetlands. *American Journal of Science and Arts* 2d ser. 22: 391–7.

Emma Darwin (1904): *Emma Darwin, wife of Charles Darwin: a century of family letters.* Edited by Henrietta Litchfield. 2 vols. Privately printed. Cambridge: Cambridge University Press. 1904.

Emma Darwin: *Emma Darwin: a century of family letters, 1792–1896.* Edited by Henrietta Litchfield. Revised edition. 2 vols. London: John Murray. 1915.

Encyclopaedia of New Zealand: *An Encyclopaedia of New Zealand.* Edited by A. H. McLintock. 3 vols. Wellington, New Zealand: R. E. Owen, Government Printer. 1966.

Eve, A. S. and Creasey, C. H. 1945. *Life and work of John Tyndall.* London: Macmillan and Co.

Everest, Robert. 1834. On the climate of the fossil elephant. *Journal of the Asiatic Society of Bengal* n.s. 3: 18–24.

Eyton, Thomas Campbell. 1837a. Some osteological peculiarities in different skeletons of the genus *Sus*. *Proceedings of the Zoological Society of London* pt 5: 23.

——. 1837b. Some remarks upon the theory of hybridity. *Magazine of Natural History* n.s. 1: 357–9.

——. 1840. Remarks on the skeletons of the common tame goose, the Chinese goose, and the hybrid between the two. *Magazine of Natural History* n.s. 4: 90–2.

——. 1846[–53]. *The herd book of Hereford cattle.* 2 vols. London.

——. 1856. *A catalogue of the species of birds in his possession.* Wellington, Salop.

——. 1867. *Osteologia avium; or, a sketch of the osteology of birds.* 2 vols. Wellington, Salop.

Fabre, Esprit. 1854. On the species of Ægilops of the south of France, and their transformation into cultivated wheat. Translated from the French. *Journal of the Royal Agricultural Society of England* 15: 167–80.

Falconer, Hugh. 1856. On Prof. Huxley's attempted refutation of Cuvier's laws of correlation, in the reconstruction of extinct vertebrate forms. *Annals and Magazine of Natural History* n.s. 17: 476–93.

——. 1857. Description of two species of the fossil mammalian genus Plagiaulax from Purbeck. *Quarterly Journal of the Geological Society of London* 13: 261–82.

——. 1863. On the American fossil elephant of the regions bordering the Gulf of Mexico, (*E. Columbi*, Falc.); with general observations on the living and extinct species. *Natural History Review* n.s. 3: 43–114.

558   Bibliography

Farley, John. 1982. *Gametes & spores: ideas about sexual reproduction, 1750–1914.*
    Baltimore and London: Johns Hopkins University Press.
Férussac, Jean Baptiste Louis d'Audebard de and Deshayes, Gérard Paul.
    1820–51. *Histoire naturelle générale et particulière des mollusques terrestres et fluviatiles.*
    4 vols. Paris.
Forbes, Edward. 1845. On the distribution of endemic plants, more especially
    those of the British Islands, considered with regard to geological changes.
    *Report of the 15th meeting of the British Association for the Advancement of Science held at
    Cambridge,* Transactions of the sections, pp. 67–8.
——. 1846. On the connexion between the distribution of the existing fauna and
    flora of the British Isles, and the geological changes which have affected their
    area, especially during the epoch of the northern drift. *Memoirs of the Geological
    Survey of Great Britain, and of the Museum of Economic Geology in London* 1: 336–432.
——. 1854. On the discovery by Dr. Overweg of Devonian rocks in North Africa.
    *Report of the 21st meeting of the British Association for the Advancement of Science held at
    Ipswich,* Transactions of the sections, p. 58.
——. 1854. Zoology and Botany. Pp. 80–92 of Forbes, Edward and Latham,
    R. G., *The natural history department of the Crystal Palace described.* London.
——. 1856. Map of the distribution of marine life, illustrated chiefly by fishes,
    molluscs and radiata; showing also the extent & limits of the homoiozoic belts.
    Pp. 99–102 of vol. 4 of Johnston, Alexander Keith, ed., *The physical atlas of
    natural phenomena.* 2d ed. 4 vols. Edinburgh and London.
Forbes, Edward and Godwin-Austen, Robert Albert Cloyne. 1859. *The natural
    history of the European seas. By the late Prof. Edw. Forbes . . . edited and continued by
    Robert Godwin-Austen.* London.
Forbes, Edward and Hanley, Sylvanus. [1848–]53. *A history of British Mollusca and
    their shells.* 4 vols. London.
Forbes, James David. 1845. *Travels through the Alps of Savoy and other parts of the
    Pennine chain* 2d ed. Edinburgh.
Fossil Cirripedia (1851): *A monograph of the fossil Lepadidæ, or pedunculated cirripedes of
    Great Britain.* By Charles Darwin. London. 1851.
Fossil Cirripedia (1854): *A monograph of the fossil Balanidæ and Verrucidæ of Great
    Britain.* By Charles Darwin. London. 1854.
Foster, Michael and Lankester, E. Ray, eds. 1898–1903. *The scientific memoirs of
    Thomas Henry Huxley.* 4 vols. and supplement. London: Macmillan and Co.
*Foundations: The foundations of the Origin of Species. Two essays written in 1842 and 1844
    by Charles Darwin.* Edited by Francis Darwin. Cambridge: Cambridge Univer-
    sity Press. 1909. Reprint. New York: Kraus Reprint Co. 1969.
Freeman, Richard Broke. 1978. *Charles Darwin: a companion.* Folkestone:
    W. Dawson & Sons; Hamden, Conn.: Archon Books, Shoe String Press.
Fries, Elias Magnus. 1846. *Summa vegetabilium Scandinaviæ.* Stockholm and Leipzig.
Froebel, Julius and Heer, Oswald. 1834–6. *Mittheilungen aus dem Gebiete der
    theoretischen Erdkunde.* Vol. 1 (no more published). Zurich.

Fryer, John. 1698. *A new account of East-India and Persia, in eight letters. Being nine years travels, begun 1672. And finished 1681.* London.

Fürnrohr, August Emanuel. 1839. *Flora Ratisbonensis, oder Uebersicht der um Regensburg wildwachsenden Gewächse.* Vol. 2 of Fürnrohr, August Emanuel, ed. *Naturhistorische Topographie von Regensburg.* 3 vols. in 2. Regensburg. 1838–40.

Gärtner, Karl Friedrich von. 1844. *Versuche und Beobachtungen über die Befruchtungsorgane der vollkommeneren Gewächse und über die natürliche und künstliche Befruchtung durch den eigenen Pollen.* Pt 1 of *Beiträge zur Kenntniss der Befruchtung der vollkommeneren Gewächse.* Stuttgart.

——. 1849. *Versuche und Beobachtungen über die Bastarderzeugung im Pflanzenreich. Mit Hinweisung auf die ähnlichen Erscheinungen im Thierreiche.* Stuttgart.

Gage, Andrew Thomas. 1938. *A history of the Linnean Society of London.* London: Linnean Society.

Gage, Andrew Thomas and Stearn, William Thomas. 1988. *A bicentenary history of the Linnean Society of London.* London and San Diego: Academic Press for the Linnean Society of London.

Gascoigne, Robert Mortimer. 1984. *A historical catalogue of scientists and scientific books: from the earliest times to the close of the nineteenth century.* New York and London: Garland Publishing.

Gaskell, Elizabeth Cleghorn 1857. *The life of Charlotte Brontë.* 2 vols. London.

Gay, Claude. 1845[–53]. *Historia física y política de Chile: Botanica.* 8 vols. in 4. Paris and Santiago.

Geoffroy Saint-Hilaire, Isidore. 1855a. Sur deux Chevaux sauvages, d'une espèce nouvelle (*Equus hemippus*), donnés par S. M. L'Impératrice à la Ménagerie du Muséum d'Histoire naturelle. *Comptes Rendus Hebdomadaires des Séances de l'Academie des Sciences* 41: 1214–19.

——. 1855b. Sur le genre Cheval, et en particulier sur l'Hémione et l'Onagre. *Comptes Rendus Hebdomadaires des Séances de l'Academie des Sciences* 41: 1220–4.

Geology of 'Beagle': Geological observations on coral reefs, volcanic islands, and on South America. By Charles Darwin. London. 1851.

Gérard, Frédéric. 1845. Géographie zoologique. Vol. 6, pp. 112–92, of Orbigny, Alcide Charles Victor Dessalines d', ed., *Dictionnaire universel d'histoire naturelle.* 13 vols. and 3 vols. atlas. Paris. 1841–9.

Gerber, Friedrich. 1842. *Elements of the general and minute anatomy of man and the Mammalia, chiefly after original researches.* Notes and an appendix by George Gulliver. London.

Gilbert, Lionel. 1986. *The Royal Botanic Gardens, Sydney. A history, 1816–1985.* Melbourne: Oxford University Press.

Gilbert, Pamela. 1977. *A compendium of the biographical literature on deceased entomologists.* London: British Museum (Natural History).

Gladwin, Francis, trans. 1783–6. *Ayeen Akbery; or, the institutes of the Emperor Akber.* Translated from the original Persian. 3 vols. Calcutta.

Godwin-Austen, Robert Alfred Cloyne. 1856. On the possible extension of the coal-measures beneath the south-eastern part of England. *Quarterly Journal of the Geological Society of London* 12: 38–73.

Goodsir, John. 1857. On the morphological constitution of the skeleton of the vertebrate head. *Edinburgh New Philosophical Journal* n.s. 5: 123–78.

Gosse, Edmund. 1890. *The life of Philip Henry Gosse F.R.S.* London.

Gosse, Philip Henry. 1847. *The birds of Jamaica*. Assisted by Richard Hill. London.

——. 1851. *A naturalist's sojourn in Jamaica*. Assisted by Richard Hill. London.

——. 1856. *Tenby: a sea-side holiday*. London.

Gould, Augustus Addison. 1852–6. *Mollusca and shells*. Vol. 12 and atlas of *United States Exploring Expedition during the years 1838–42. Under the command of Charles Wilkes, U.S.N.* Philadelphia.

Gould, John. 1848. *The birds of Australia*. 7 vols. London.

——. 1850–83. *The birds of Asia*. 7 vols. London.

——. 1855. On a new species of the genus *Prion. Proceedings of the Zoological Society of London* 23: 87–8.

Graba, Karl Julian. 1830. *Tagebuch, geführt auf einer Reise nach Färö im Jahre 1828*. Hamburg.

Gray, Asa. 1842. Notes of a botanical excursion to the mountains of North Carolina, &c.; with some remarks on the botany of the higher Alleghany mountains. In a letter to Sir W. J. Hooker. *London Journal of Botany* 1 (1842): 1–14, 217–37; 2 (1843): 113–25; 3 (1844): 230–42.

——. 1848. *A manual of the botany of the northern United States*. Cambridge, Mass.

——. 1856–7. Statistics of the flora of the northern United States. *American Journal of Science and Arts* 2d ser. 22: 204–32; 23: 62–84, 369–403.

——. 1856a. *Manual of the botany of the northern United States: second edition; including Virginia, Kentucky, and all east of the Mississippi: arranged according to the natural system*. New York.

——. 1856b. Review of Hooker, Joseph Dalton and Thomson, Thomas, *Flora Indica. American Journal of Science and Arts* 2d ser. 21: 134–137.

——. 1857a. *First lessons in botany and vegetable physiology*. New York.

——. 1857b. Account of the botanical specimens. List of dried plants collected in Japan, by S. Wells Williams, Esq., and Dr. James Morrow. In vol. 2, pp. 305–32, of *Narrative of the expedition of an American squadron to the China Seas and Japan, performed in the years 1852, 1853, and 1854, under the command of the Commodore M. C. Perry, United States Navy*. 3 vols. New York, Washington, D.C., and London. 1856–7.

——. 1859. Diagnostic characters of new species of phænogamous plants, collected in Japan by Charles Wright, botanist of the U.S. North Pacific Exploring Expedition. (Published by request of Captain John Rodgers, commander of the expedition.) With observations upon the relations of the Japanese flora to that of North America, and of other parts of the northern temperate zone. With an appendix by N. J. Andersson. *Memoirs of the American Academy of Arts and Sciences* n.s. 6: 377–452.

Gray, George Robert. 1849. Notice of two examples of the genus Gallus. *Proceedings of the Zoological Society of London,* pp. 62–3.

Gray, John Edward. [1830–5]. *Illustrations of Indian zoology; chiefly selected from the collection of Major-general Hardwicke.* 2 vols. London.

——. 1855. *Catalogue of Pulmonata or air-breathing Mollusca in the collection of the British Museum.* Pt 1. London.

Griffith, Edward, *et al.* 1827–35. *The animal kingdom arranged in conformity with its organization, by the Baron Cuvier, . . . with additional descriptions of all the species hitherto named, and of many not before noticed.* 16 vols. London.

Griffith, Richard. 1857. On the remains of fossil plants discovered in the Yellow Sandstone strata, situate at the base of the Carboniferous limestone series of Ireland, in connexion with a communication on that subject from M. Adolphe Brongniart. *Journal of the Royal Dublin Society* 1 (1856–7): 313–25.

Grisebach, August Heinrich Rudolph. 1843–4. *Spicilegium florae Rumelicae et Bithynicae exhibens synopsin plantarum quas aest. 1839 legit.* 2 vols. Brunswick.

——. 1849. Report on the progress of geographical botany, during the year 1844. Pp. 317–413 of *Reports and papers on botany.* Edited by Arthur Henfrey. London.

Grote, Arthur. 1875. Introduction, being a memoir of the late Mr. Ed. Blyth, C.M.Z.S., and Hon. Member Asiatic Soc. of Bengal. *Journal of the Asiatic Society of Bengal* n.s. 43 (pt 2) (extra number, August 1875): iii–xxiv.

Gulliver, George, ed. 1846. *The works of William Hewson, F.R.S.* London.

Gunther, Albert E. 1975. *A century of zoology at the British Museum through the lives of two keepers, 1815–1914.* Folkestone: William Dawsons and Sons.

Gyllenhaal, Leonardo. 1808–27. *Insecta Suecica.* 1 vol. in 4. Skara and Leipzig.

Hall, Marie Boas. 1984. *All scientists now: the Royal Society in the nineteenth century.* Cambridge: Cambridge University Press.

Hamilton, William John. 1856. Anniversary address of the president. *Quarterly Journal of the Geological Society of London* 12: xxvi–cxix.

Hancock, Albany. 1848. On the boring of the Mollusca into rocks, &c.; and on the removal of portions of their shells. *Annals and Magazine of Natural History* 2d ser. 2: 225–48.

——. 1849. On the excavating powers of certain sponges belonging to the genus *Cliona*; with descriptions of several new species, and an allied generic form. *Annals and Magazine of Natural History* 2d ser. 3: 321–48.

——. 1850. On the anatomy of the freshwater Bryozoa, with descriptions of three new species. *Annals and Magazine of Natural History* 2d ser. 5: 173–204.

——. 1858. On the organization of the *Brachiopoda*. [Read 14 May 1857.] *Philosophical Transactions of the Royal Society of London* 148: 791–869.

Hancock, John. 1886. Letters from C. Darwin, Esq., to A. Hancock, Esq., communicated by John Hancock, Esq. *Natural History Transactions of Northumberland, Durham, and Newcastle-upon-Tyne* 8: 250–78.

Harcourt, Edward William Vernon. 1851. *A sketch of Madeira; containing information for the traveller or invalid visitor.* London.

——. 1855. Notes on the ornithology of Madeira. *Annals and Magazine of Natural History* 2d ser. 15: 430–8.

Harte, Negley. 1986. *The University of London, 1836–1986: an illustrated history.* London and Atlantic Highlands, New Jersey: The Athlone Press.

Hartung, Georg. 1864. *Geologische Beschreibung der Inseln Madeira und Porto Santo.* Leipzig.

Harvey, William Henry. 1849. *The sea-side book.* London.

——. 1854. *The sea-side book; being an introduction to the natural history of the British coasts.* 3d ed. London.

Haughton, Samuel. 1853–4. On the depth of the sea deducible from tidal observations. *Proceedings of the Irish Academy* 6: 354–5.

——. 1856. Discussion of tidal observations made by direction of the Royal Irish Academy in 1850–51. [Read 24 April 1854.] *Transactions of the Royal Irish Academy* 23: 35–140.

——. 1859. On *Cyclostigma*, a new genus of fossil plants from the Old Red Sandstone of Kiltorcan, Co. Kilkenny; and on the general law of phyllotaxis in the natural orders. *Journal of the Royal Dublin Society* 2 (1858–9): 407–20.

Haworth, Adrian Hardy. 1803–28. *Lepidoptera Britannica: sistens digestionem novam insectorum Lepidopterorum.* London.

Heer, Oswald. 1855. Ueber die fossilen Pflanzen von St. Jorge in Madeira. *Neue Denkschriften der allgemeinen schweizerischen Gesellschaft für die gesammten Naturwissenschaften* 15 (1857): 1–40.

Henslow, John Stevens. 1835. *A catalogue of British plants, arranged according to the natural system, with the synonyms of De Candolle, Smith, Lindley, and Hooker.* 2d ed. Cambridge.

——. 1856. On the triticoidal forms of Ægilops and on the specific identity of Centaurea nigra and C. nigrescens. *Report of the 26th meeting of the British Association for the Advancement of Science held at Cheltenham*, Transactions of the sections, pp. 87–8.

Herbert, William. 1846. Local habitation and wants of plants. *Journal of the Horticultural Society of London* 1: 44–9.

——. 1847. On hybridization among vegetables. *Journal of the Horticultural Society of London* 2: 1–28, 81–107.

Hérétieu, ——. 1841. Note sur une variété assez rare du *Lepus timidus*. *Revue Zoologique, par la Société Cuvierienne* 4: 33–5.

Herschel, John Frederick William. 1831. *A preliminary discourse on the study of natural philosophy.* London.

——, ed. 1849. *A manual of scientific enquiry; prepared for the use of Her Majesty's Navy: and adapted for travellers in general.* London.

Hincks, Thomas. 1852. On a peculiar organ which occurs on some of the marine Bryozoa and which appears to indicate a difference of sex. *Report of the 22nd meeting of the British Association for the Advancement of Science held at Belfast*, Transactions of the sections, pp. 75–6.

Hodgson, Brian Houghton. 1833. Description of the wild dog of the Himalaya. *Asiatic Researches* 18, pt 2: 221–37.

——. 1856. On a new perdicine bird from Tibet. *Journal of the Asiatic Society of Bengal* 25: 165–6.

Hooker, Joseph Dalton. 1844–7. *Flora Antarctica.* Pt 1 of *The botany of the Antarctic voyage of H.M. Discovery Ships* Erebus *and* Terror *in the years 1839–1843, under the command of Captain Sir James Clark Ross.* 2 vols. in 1. London.

——. 1851. On the vegetation of the Galapagos Archipelago, as compared with that of some other tropical islands and of the continent of America. *Transactions of the Linnean Society of London* 20: 235–62.

——. 1852. On the climate and vegetation of the temperate and cold regions of East Nepal and the Sikkim Himalaya Mountains. *Journal of the Horticultural Society of London* 7: 69–131.

——. 1853–5. *Flora Novæ-Zelandiæ.* Pt 2 of *The botany of the Antarctic voyage of H.M. Discovery Ships* Erebus *and* Terror, *in the years 1839–1843, under the command of Captain Sir James Clark Ross.* 2 vols. London.

——. 1854. *Himalayan journals; or, notes of a naturalist in Bengal, the Sikkim and Nepal Himalayas, the Khasia mountains, &c.* 2 vols. London.

——. 1855[-60]. *Flora Tasmaniæ.* Pt 3 of *The botany of the Antarctic voyage of H.M. Discovery Ships* Erebus *and* Terror, *in the years 1839–1843, under the command of Captain Sir James Clark Ross.* 2 vols. London.

[——]. 1856. Review of Alphonse de Candolle, *Géographie botanique raisonnée. Hooker's Journal of Botany and Kew Garden Miscellany* 8: 54–64, 82–8, 112–21, 151–7, 181–91, 214–19, 248–56.

——. 1857. On the botany of Raoul Island, one of the Kermadec group in the South Pacific Ocean. *Journal of the Linnean Society. Botany* 1: 125–9.

——. 1862. Outlines of the distribution of Arctic plants. [Read 21 June 1860.] *Transactions of the Linnean Society of London* 23: 251–348.

Hooker, Joseph Dalton and Thomson, Thomas. 1855. *Flora Indica: being a systematic account of the plants of British India, together with observations on the structure and affinities of their natural orders and genera.* London.

——. 1858. *Præcursores ad Floram Indicam*: being sketches of the natural families of Indian plants, with remarks on their distribution, structure, and affinities. *Journal of the Proceedings of the Linnean Society of London. Botany* 2: 1–29, 54–103, 163–80.

——. 1859. Præcursores ad Floram Indicam.— Balsamineæ. [Read 16 June 1859]. *Journal of the Proceedings of the Linnean Society of London. Botany* 4 (1860): 106–57.

[Hooker, William Jackson]. 1830a. On the species of the genus Colletia, of the natural order Rhamneæ, discovered by Dr. Gillies in South America. *Botanical Miscellany* 1: 150–9.

——. 1830b. *The British flora.* London.

——. 1840. *Flora Boreali-Americana; or, the botany of the northern parts of British America.* 2 vols. London.

Hooker, William Jackson and Arnott, George Arnott Walker. 1855. *The British flora: comprising the phænogamous or flowering plants, and the ferns.* 7th ed. London.

Hooker, William Jackson and Hooker, Joseph Dalton. 1847. Botany of the Niger expedition: notes on Madeira plants. *London Journal of Botany* 6: 125–39.

Hooper, Robert. 1831. *Lexicon medicum; or medical dictionary.* 6th ed. London.

Hopkins, Annette Brown. 1952. *Elizabeth Gaskell: her life and work.* London: John Lehmann.

Hopkins, William. 1835. Researches in physical geology. [Read 4 May 1835.] *Transactions of the Cambridge Philosophical Society* 6 (1838): 1–84.

——. 1836. An abstract of a memoir on physical geology; with a further exposition of certain points connected with the subject. *London and Edinburgh Philosophical Magazine and Journal of Science* 8: 227–36, 272–81, 357–66.

——. 1847. Report on the geological theories of elevation and earthquakes. *Report of the 17th meeting of the British Association for the Advancement of Science held at Oxford,* pp. 33–92.

Hornschuch, Christian Friedrich. 1848. Ueber Ausartung der Pflanzen. *Flora, oder allgemeine botanische Zeitung* n.s. 6: 17–28, 33–44, 50–64, 66–86. Horsfield, Thomas. 1824. *Zoological researches in Java, and the neighbouring islands.* London.

Hudson, William. 1762. *Flora Anglica: exhibens plantas per regnum Angliæ sponte crescentes, distributas secundum systema sexuale.* London.

Humboldt, Alexander von. 1814–29. *Personal narrative of travels to the equinoctial regions of the New Continent, during the years 1799–1804, by Alexander de Humboldt and Aimé Bonpland . . . translated into English by Helen Maria Williams.* 7 vols. London.

——. 1846–58. *Cosmos: sketch of a physical description of the universe.* Translated [by Elizabeth Juliana Sabine] under the superintendence of Edward Sabine. 4 vols. London.

Hummel, Arvid-David. 1821–29. *Essais entomologiques.* 7 pts. St Petersburg.

Hutchinson, Horace Gordon. 1914. *Life of Sir John Lubbock, Lord Avebury.* 2 vols. London: Macmillan and Co.

Hutton, Thomas. 1846. Rough notes on the zoology of Candahar and the neighbouring districts . . . with notes by Ed. Blyth, curator of the Asiatic Society's museum. *Journal of the Asiatic Society of Bengal* n.s. 15: 135–70.

——. 1848. Notes on the nidification of Indian birds. (Communicated by E. Blyth, Esq.) *Journal of the Asiatic Society of Bengal* n.s. 17, pt 2: 3–13, 681–96.

——. 1850. *The chronology of creation; or, geology and scripture reconciled.* Calcutta.

Huxley, Leonard, ed. 1900. *Life and letters of Thomas Henry Huxley.* 2 vols. London: Macmillan and Co.

——, ed. 1918. *Life and letters of Sir Joseph Dalton Hooker . . . based on materials collected and arranged by Lady Hooker.* 2 vols. London: John Murray.

Huxley, Thomas Henry. 1849. On the anatomy and the affinities of the family of the *Medusæ. Philosophical Transactions of the Royal Society of London,* pt 2: 413–34. Reprinted in Foster and Lankester, eds. 1898–1903, 1: 9–32.

——. 1851. Zoological notes and observations made on board H.M.S. Rattle-snake during the years 1846–50. *Annals and Magazine of Natural History* 2d ser. 7: 304–6, 370–4; 8: 433–42. Reprinted in Foster and Lankester, eds. 1898–1903, 1: 80–95.

——. 1852. Researches into the structure of the Ascidians. *Report of the 22d meeting of the British Association for the Advancement of Science held at Belfast*, Transactions of the sections, pp. 76–7. Reprinted in Foster and Lankester, eds. 1898–1903, 1: 194–6.

——. 1855. On certain zoological arguments commonly adduced in favour of the hypothesis of the progressive development of animal life in time. [Read 20 April 1855.] *Notices of the Proceedings at the meetings of the members of the Royal Institution of Great Britain* 2 (1854–8): 82–5. Reprinted in Foster and Lankester, eds. 1898–1903, 1: 300–4.

——. 1856a. On natural history, as knowledge, discipline, and power. *Notices of the Proceedings at the meetings of the members of the Royal Institution of Great Britain* 2 (1854–58): 187–95. Reprinted in Foster and Lankester, eds. 1898–1903, 1: 305–14.

——. 1856b. On the method of palaeontology. *Annals and Magazine of Natural History* 2d ser. 18: 43–54. Reprinted in Foster and Lankester, eds. 1898–1903, 1: 432–44.

——. 1856–7. Lectures on general natural history. *Medical Times & Gazette* n.s. 12: 429–32, 481–4, 507–11, 563–7, 618–23; 13: 27–30, 131–4, 157–60, 278–81, 383–6, 462–3, 537–8, 586–8, 635–9; 14: 133–5, 181–3, 255–7, 353–5, 505–8, 638–40; 15: 159–62, 186–9, 238–41, 467–71.

——. 1857a. On the present state of knowledge as to the structure and functions of nerve. *Notices of the Proceedings at the meetings of the members of the Royal Institution of Great Britain* 2 (1854–8): 432–7. Reprinted in Foster and Lankester, eds. 1898–1903, 1: 315–20.

——. 1857b. Observations on the structure of glacier ice. *Philosophical Magazine* 4th ser. 14: 241–60. Reprinted in Foster and Lankester, eds. 1898–1903, 1: 482–501.

——. 1857–8. On the agamic reproduction and morphology of *Aphis*. *Transactions of the Linnean Society of London* 22 (1859): 193–220, 221–36. Reprinted in Foster and Lankester, eds. 1898–1903, 2: 26–80.

——. 1858. On the theory of the vertebrate skull. [Read 17 June 1858.] *Proceedings of the Royal Society of London* 9 (1857–59): 381–457. Reprinted in Foster and Lankester, eds. 1898–1903, 1: 538–606.

——. 1859. *The oceanic Hydrozoa; a description of the Calycophoidæ and Physophoridæ observed during the voyage of H.M.S. 'Rattlesnake', in the years 1846–1850*. London.

Huxley, Thomas Henry and Etheridge, Robert. 1865. *A catalogue of the collection of fossils in the Museum of Practical Geology, with an explanatory introduction*. London.

IBN: *Index bio-bibligraphicus notorum hominum*. Part C: *Corpus alphabeticum sectio generalis*. Edited by Jean-Pierre Lobies. 45 vols. (A–Czyżyk). Osnabrück: Biblio Verlag. 1974–89.

Jackson, Benjamin Daydon. 1913. *Catalogue of the Linnean specimens of Amphibia, Insecta, and Testacea, noted by Carl von Linné.* London: printed for the Linnean Society.

Janson, Edward Westley. 1848. Notice of the occurrence of rare Coleopterous insects, with observations on their habits; to which is appended the description of a species hitherto unrecorded as British. *Zoologist* 6: 2108–10.

Jenyns, Leonard. 1856. On the variation of species. *Report of the 26th meeting of the British Association for the Advancement of Science held at Cheltenham,* Transactions of the sections, pp. 101–5.

——. 1862. *Memoir of the Rev. John Stevens Henslow.* London.

Jerdon, Thomas Claverhill. 1867. *The mammals of India; a natural history of all the animals known to inhabit continental India.* Roorkee, India.

Jones, William, trans. 1796. *Institutes of Hindu law: or, the ordinances of Menu, according to the gloss of Cullúca. Comprising the Indian system of duties, religious and civil.* Calcutta and London.

Jordan, Alexis. 1846[–9]. *Observations sur plusieurs plantes nouvelles rares ou critiques de la France.* 6 pts. Paris and Leipzig.

Journal of researches: Journal of researches into the geology and natural history of the various countries visited by H.M.S. Beagle. By Charles Darwin. London. 1839.

Kinahan, John Robert. 1856. Remarks on the habits and distribution of marine Crustacea on the eastern shores of Port Philip, Victoria, Australia; with descriptions of undescribed species and genera. *Journal of the Royal Dublin Society* 1 (1856–7): 111–34.

Koch, Wilhelm Daniel Joseph. 1843–4. *Synopsis florae Germanicae et Helveticae, exhibens stirpes phanerogamas rite cognitas.* 2d ed. 2 vols. Frankfurt and Leipzig.

Kölreuter, Joseph Gottlieb. 1761–6. *Vorläufige Nachricht von einigen das Geschlecht der Pflanzen betreffenden Versuchen und Beobachtungen.* 4 pts. Leipzig.

Krauss, Ferdinand. 1844–6. Pflanzen des Cap- und Natallandes. *Flora, oder allgemeine botanische Zeitung* n.s. 2: 261–74, 277–307, 346–59, 423–32, 551–6, 819–35; n.s. 3: 81–93, 305–14, 337–44, 753–64; n.s. 4: 113–21, 129–38, 209–219.

——. 1846. *Beiträge zur Flora des Cap- und Natallandes.* Regensburg.

Krohn, August David. 1859. Beobachtungen über den Cementapparat und die weiblichen Zeugungsorgane einiger Cirripedien. *Archiv für Naturgeschichte* 25: 355–64.

Lacaze-Duthiers, Félix Joseph Henri. 1852a. Recherches sur l'armure génitale femelle des insectes Orthoptères. *Annales des Sciences Naturelles. Zoologie* 3d ser. 17: 207–51.

——. 1852b. Recherches sur l'armure génitale femelle des insectes Hémiptères. *Annales des Sciences Naturelles. Zoologie* 3d ser. 18: 337–90.

Lacordaire, Jean Théodore. 1854–75. *Histoire naturelle des insectes. Genera des Coléoptères, etc.* 11 vols. in 12. Vols. 10 and 11 by F. Chapuis. Paris.

Lane, Edward Wickstead. 1857. *Hydropathy; or, the natural system of medical treatment. An explanatory essay.* London.

Lange, Johan Martin Christian. 1857. Oversigt over Grønlands planter. Vol. 2, Appendix 6, pp. 106–35, of Rink, Hinrich Johannes, *Grønland, geographisk og statistik beskrevet*. 2 vols. Copenhagen.

Larousse XX: *Larousse du XXe siècle*. Under the direction of Paul Augé. 6 vols. Paris: Librairie Larousse. 1928–33.

Latham, John. 1821–8. *A general history of birds*. 11 vols. Winchester.

Lecoq, Henri. 1845. *De la fécondation naturelle et artificielle des végétaux et de l'hybridation, considéré dans ses rapports avec l'horticulture, l'agriculture et la sylviculture*. Paris.

Ledebour, Karl Friedrich von. 1842–53. *Flora Rossica sive enumeratio plantarum in totius imperii Rossici provinciis Europaeis, Asiaticis et Americanis hucusque observatarum*. 4 vols. Stuttgart.

Lichtenstein, Martin Heinrich Karl. 1812–15. *Travels in southern Africa, in the years 1803–6*. Translated from the original German, by Anne Plumtre. 2 vols. London.

Lightfoot, John. 1777. *Flora Scotica; or, a systematic arrangement in the Linnæan method, of the native plants of Scotland and the Hebrides*. 2 vols. London.

Lindley, John. 1830. *An introduction to the natural system of botany; or, a systematic view of the organisation, natural affinities, and geographical distribution, of the whole vegetable kingdom*. London.

——. 1846. *The vegetable kingdom; or, the structure, classification, and uses of plants, illustrated upon the natural system*. London.

——. 1850. Memorandum concerning a remarkable case of vegetable transformation. *Journal of the Horticultural Society of London* 5: 29–32.

——. 1854. *School botany, and vegetable physiology; or, the rudiments of botanical science. A new edition*. London.

Linnaeus (Carl von Linné). 1761. *Fauna Suecica sistens animalia Sueciæ regni: mammalia, aves, amphibia, pisces, insecta, vermes*. 2d ed. Stockholm.

——. 1783. *Philosophia botanica, in qua explicantur fundamenta botanica cum definitionibus partium, exemplis terminorum, observationibus rariorum*. 2d ed. Vienna.

Little, J. 1840. On the different breeds of cattle on the western side of the peninsula of India, and what has been done there for the improvement of sheep and wool:—with observations on the American plough. *Transactions of the Agricultural and Horticultural Society of India* 7: 111–29.

Littleton, Adam. 1678. *Linguæ latinæ liber dictionarius quadripartitus. A Latine dictionary, in four parts*. London.

Living Cirripedia (1851): *A monograph of the sub-class Cirripedia, with figures of all the species. The Lepadidæ; or, pedunculated cirripedes*. By Charles Darwin. London. 1851.

Living Cirripedia (1854): *A monograph of the sub-class Cirripedia, with figures of all the species. The Balanidæ (or sessile cirripedes); the Verrucidæ, etc.* By Charles Darwin. London. 1854.

Livingstone, David. 1857. *Missionary travels and researches in South Africa; including a sketch of sixteen years' residence in the interior of Africa, and a journey from the Cape of Good Hope to Loanda on the west coast; thence across the Continent, down the river Zambesi, to the Eastern Ocean.* London.

Loiseleur Deslongchamps, Jean Louis Auguste. 1842–3. *Considérations sur les céréales et principalement sur les froments. Partie historique* (1842); *Partie pratique et expérimentale* (1843). 2 pts. Paris.

London and provincial medical directory: The London and provincial medical directory. London. 1851–.

London encyclopædia: The London Encyclopædia or universal dictionary of science, art, literature and practical mechanics. 22 vols. London. 1829.

Lopes de Lima, José Joaquim. 1844–62. *Ensaios sobre a statistica das possessões portuguezas na africa occidental e oriental; na Asia occidental; na China, e na Oceania.* 5 vols. Lisbon.

Loudon, John Claudius. 1842. *An encyclopaedia of trees and shrubs; being the arboretum et fruticetum Britannicum abridged.* London.

Lowe, Richard Thomas. 1856. Species *Plantarum Maderensium* quædam novæ, vel hactenus ineditæ, breviter descriptæ. *Hooker's Journal of Botany and Kew Garden Miscellany* 8: 289–302.

———. 1857[–72]. *A manual flora of Madeira and the adjacent islands of Porto Santo and the Dezertas.* 2 vols. London.

Lubbock, John. 1855. On the freshwater *Entomostraca* of South America. [Read 7 May 1855.] *Transactions of the Entomological Society of London* n.s. 3: 232–40.

———. 1857. An account of the two methods of reproduction in *Daphnia*, and of the structure of the ephippium. [Read 29 January 1857.] *Philosophical Transactions of the Royal Society of London* 147: 79–100.

———. 1859. On the arrangement of the cutaneous muscles of the larva of *Pygæra bucephala*. *Transactions of the Linnean Society of London* 22: 173–91.

Lucas, Prosper. 1847–50. *Traité philosophique et physiologique de l'hérédité naturelle.* 2 vols. Paris.

Lurie, Edward. 1954. Louis Agassiz and the races of man. *Isis* 45: 227–42.

Lyell, Charles. 1830–3. *Principles of geology, being an attempt to explain the former changes of the earth's surface, by reference to causes now in operation.* 3 vols. London.

———. 1850. On craters of denudation, with observations on the structure and growth of volcanic cones. *Quarterly Journal of the Geological Society of London* 6: 207–34.

———. 1853. *Principles of geology; or, the modern changes of the earth and its inhabitants considered as illustrative of geology.* 9th ed. London.

———. 1855. *A manual of elementary geology; or, the ancient changes of the earth and its inhabitants as illustrated by geological monuments.* 5th ed. London.

———. 1857a. *Supplement to the fifth edition of 'A manual of elementary geology'.* London.

———. 1857b. *Supplement to the fifth edition of 'A manual of elementary geology'.* 2d ed. London.

——. 1858. On the structure of lavas which have consolidated on steep slopes; with remarks on the mode of origin of Mount Etna, and on the theory of 'craters of elevation'. *Philosophical Transactions of the Royal Society of London* 148: 703–86.

Lyell, Katharine Murray, ed. 1881. *Life, letters and journals of Sir Charles Lyell, Bart.* 2 vols. London.

McAndrew, Robert. 1854. On the geographical distribution of Testaceous Mollusca in the north-east Atlantic and neighbouring seas. *Proceedings of the Literary and Philosophical Society of Liverpool* 8: 8–56.

Macgillivray, William. 1837–52. *A history of British birds, indigenous and migratory.* 5 vols. London.

Malherbe, Alfred. 1846. *Catalogue raisonné d'oiseaux de l'Algerie, comprenant la description de plusieurs espèces nouvelles.* Metz.

Mammalia: Pt 2 of *The zoology of the voyage of H.M.S. Beagle.* By George Robert Waterhouse. Edited and superintended by Charles Darwin. London. 1838–9.

Martins, Charles. 1849. On the vegetable colonisation of *the British Islands, Shetland, Feroe, and Iceland. Edinburgh New Philosophical Journal* 46: 40–52.

Maury, Matthew Fontaine. 1855a. *The physical geography of the sea.* London.

——. 1855b. *Explanations and sailing directions to accompany the wind and current charts, approved by Commodore Charles Morris.* 7th ed. Philadelphia.

Meyen, Franz Julius Ferdinand. 1834[–5]. *Reise um die Erde ausgeführt auf dem Königlich Preussischen Seehandlungs-Schiffe Prinzess Louise, commandirt von Capitain W. Wendt, in den Jahren 1830, 1831 und 1832.* 3 vols. Berlin.

Milne-Edwards, Henri. 1834–40. *Histoire naturelle des Crustacés, comprenant l'anatomie, la physiologie et la classification de ces animaux.* 3 vols. Paris.

——. 1844. Considérations sur quelques principes relatifs a la classification naturelle des animaux, et plus particulièrement sur la distribution méthodique des mammifères. *Annales des Sciences Naturelles. Zoologie* 3d ser. 1: 65–99.

——. 1845. Observations sur le développement des Annélides. *Annales des Sciences Naturelles. Zoologie* 3d ser. 3: 145–82.

Miquel, Frederich Anton Wilhelm. 1837. *Disquisitio geographico-botanica de plantarum Regni Batavi distributione.* Leiden.

Mitra, Rajendralala. 1885. History of the Society. Pt 1 of *Centenary review of the Asiatic Society of Bengal. From 1784 to 1883.* Calcutta.

Modern English biography: *Modern English biography: containing many thousand concise memoirs of persons who have died since the year 1850.* By Frederic Boase. 3 vols. and supplement (3 vols.). Truro: printed for the author. 1892–1921.

Montagu, George. 1803–8. *Testacea Britannica; or, natural history of British shells.* 2 pts. and supplement. London.

Moore, James R. 1977. On the education of Darwin's sons: the correspondence between Charles Darwin and the Reverend G. V. Reed, 1857–64. *Notes and Records of the Royal Society of London* 32: 51–70.

Moore, James R. 1985. Darwin of Down: the evolutionist as squarson-naturalist. In Kohn, David, ed., *The Darwinian heritage*. Princeton: Princeton University Press in association with Nova Pacifica.

Moore, John. 1735. *Columbarium; or, the pigeon-house. Being an introduction to a natural history of tame pigeons*. London.

[Moore, John]. 1765. *A treatise on domestic pigeons; comprehending all the different species known in England . . . Carefully compiled from the best of authors. To which is added, a most ample description of that celebrated and beatiful pigeon called the almond tumbler*. London.

Moore, Thomas. 1855. *The ferns of Great Britain and Ireland*. Edited by John Lindley. London.

Moquin-Tandon, Horace Bénédict Alfred. 1841. *Éléments de tératologie végétale, ou histoire abrégée des anomalies de l'organisation dans les végétaux*. Paris.

Murchison, Roderick Impey. 1847. Additional remarks on the deposit of Œningen in Switzerland. *Quarterly Journal of the Geological Society of London* 3: 54.

——. 1849. On the geological structure of the Alps, Apennines, and Carpathians, more especially to prove a transition from secondary to tertiary rocks, and the development of Eocene deposits in southern Europe. *Quarterly Journal of the Geological Society of London* 5: 157–312.

Natural selection: *Charles Darwin's Natural Selection; being the second part of his big species book written from 1856 to 1858*. Edited by R. C. Stauffer. Cambridge: Cambridge University Press. 1975.

Navy list: Admiralty. *By Authority. The Navy list*. London. 1814–.

NBU: *Nouvelle biographie universelle*. Edited by Jean Chrétien Ferdinand Hoefer. 46 vols. Paris. 1852–66.

NDB: *Neue deutsche Biographie*. Under the auspices of the Historical Commission of the Bavarian Academy of Sciences. 15 vols. (A–Maltzan). Berlin: Duncker and Humblot. 1953–86.

Neumeister, Gottlob. 1837. *Das Ganze der Taubenzucht*. Weimar.

Newport, George. 1851. On the impregnation of the ovum in the Amphibia. (First series.) *Philosophical Transactions of the Royal Society of London* pt 1, pp. 169–242.

Nicholson, B. A. R. 1851. Note on a new species of Francolin. *Proceedings of the Zoological Society of London* pt 19: 128.

Niebuhr, Carsten. 1779. *Description de l'Arabie, d'après les observations et recherches fait dans le pays même*. [Translated from the German by F. L. Mourier. New edition revised by Joseph de Guignes.] Paris.

NNBW: *Nieuw nederlandsch biografisch woordenboek*. Edited by P. C. Molhuysen, P. J. Blok, and K. H. Kossmann. 10 vols. Leiden: A. W. Sijthoff. 1911–37.

Nordmann, Alexander von. 1839. Polype nouveau de la Mer-Noire (extrait abrégé). *L'Institut* 7: 95.

Notebooks: *Charles Darwin's notebooks*. Edited by Paul H. Barrett, Sandra Herbert, David Kohn, Sydney Smith, and Peter Gautrey. London: British Museum (Natural History); Ithaca: Cornell University Press. 1987.

NUC: *The national union catalog*. Pre-1956 imprints. 685 vols. and supplement (vols. 686–754). London and Chicago: Mansell. 1968–81.

O'Byrne, William R. 1849. *A naval biographical dictionary: comprising the life and services of every living officer in Her Majesty's Navy, from the rank of admiral of the fleet to that of lieutenant, inclusive*. London.

OED: *The Oxford English dictionary. Being a corrected re-issue with an introduction, supplement, and bibliography of A new English dictionary*. Edited by James A. H. Murray, Henry Bradley, W. A. Craigie, and C. T. Onions. 12 vols. and supplement. Oxford: Clarendon Press. 1970. Supplement. Edited by R. W. Burchfield. 4 vols. Oxford: Clarendon Press. 1972–86.

Ogilby, William. 1835. Observations on several rare and undescribed species of Mammalia and birds, brought from the Gambia. *Proceedings of the Zoological Society of London* pt 3: 97–105.

Orbigny, Alcide Charles Victor Dessalines d'. 1835–47. *Voyage dans l'Amérique Méridionale (le Brésil, la République orientale de l'Uruguay, la République Argentine, la Patagonie, la République du Chili, la République de Bolivia, la République du Pérou), exécuté pendant les années 1826 . . . 1833*. 6 vols. in 7 and 4 atlases. Paris and Strasbourg.

Origin: *On the origin of species by means of natural selection, or the preservation of favoured races in the struggle for life*. By Charles Darwin. London. 1859.

Orton, Reginald. 1855. *On the physiology of breeding. Two lectures, delivered to the Newcastle Farmers' Club*. 2d ed. Sunderland.

Osbeck, Per. 1771. *A voyage to China and the East Indies . . . Translated from the German, by John Reinhold Forster*. 2 vols. London.

Ospovat, Dov. 1981. *The development of Darwin's theory: natural history, natural theology, and natural selection, 1838–59*. Cambridge and New York: Cambridge University Press.

Owen, Richard. 1846. Report on the archetype and homologies of the vertebrate skeleton. *Report of the 16th meeting of the British Association for the Advancement of Science held at Southampton*, pp. 169–340.

——. 1849. *On parthenogenesis, or the successive production of procreating individuals from a single ovum*. London.

——. 1855a. On the anthropoid apes, and their relations to man. [Read 9 February 1855.] *Notices of the Proceedings at the meetings of the members of the Royal Institution of Great Britain* 2 (1854–8): 26–41.

——. 1855b. *Lectures on the comparative anatomy and physiology of the invertebrate animals, delivered at the Royal College of Surgeons*. 2d ed. London.

——. 1856. Description of a fossil cranium of the musk-buffalo . . . from the 'lower-level drift' at Maidenhead, Berkshire. [Read 19 December 1855.] *Quarterly Journal of the Geological Society of London* 12: 124–31.

——. 1857a. On the affinities of the *Stereognathus ooliticus* (Charlesworth), a mammal from the Oolitic Slate of Stonesfield. [Read 5 November 1856.] *Quarterly Journal of the Geological Society of London* 13: 1–11.

Owen, Richard. 1857b. On the characters, principles of division, and primary groups of the class Mammalia. *Journal of the Proceedings of the Linnean Society of London. Zoology* 2 (1858): 1–38.

Pallas, Pyotr Simon. 1767–80. *Spicilegia zoologica quibus novae imprimis et obscurae animalium species iconibus, descriptionibus atque commentariis illustrantur. 14 pts in 2 vols.* Berlin.

——. 1774. Equus Hemionus, Mongolis Dshikketaei Dictus. *Novi Commentarii Academiae Scientiarum Imperialis Petropolitanae* 19: 394–417.

——. 1798. Description of the *Equus Hemionus* or *Dshiggetai* of the eastern deserts of Middle Asia. [Translation]. *Philosophical Magazine* 2: 113–21, 234–40.

Pan, Jixing. 1984. Charles Darwin's Chinese sources. *Isis* 75: 530–4.

Perry, Matthew Calbraith. 1856–7. *Narrative of the expedition of an American squadron to the China Seas and Japan, performed in the years 1852, 1853, and 1854, under the command of Commodore M. C. Perry, United States Navy. Compiled from the original notes and journals of Commodore Perry by Francis L. Hawks.* 3 vols. New York, Washington, D.C., and London.

Persoon, Christiaan Henrik, ed. 1797. *Systema vegetabilium.* By C. Linnaeus. 15th ed. Gottingen.

——. 1805–7. *Synopsis Plantarum, seu enchiridium botanicum, complectens enumerationem systematicam specierum hucusque cognitarum.* 2 pts. Paris and Tubingen.

Pfeiffer, Ludwig George Karl. 1848. *Monographia heliceorum viventium. Sistens descriptiones systematicas et criticas omnium huiuas familiae generum et specierum hodie cognitarum.* 2 vols. Supplement, 4 vols. in 6. 1853–77. Leipzig.

[Philippart, John]. 1823–26. *The East India military calendar; containing the services of general and field officers of the Indian army.* 3 vols. London.

Phillips, John. 1855. *The rivers, mountains and sea-coast of Yorkshire. With essays on the climate, scenery, and ancient inhabitants of the country.* 2d ed. London.

——. 1856. Report on cleavage and foliation in rocks, and on the theoretical explanations of these phænomena.— Part I. *Report of the 26th meeting of the British Association for the Advancement of Science held at Cheltenham*, pp. 369–96.

Physicians: The roll of the Royal College of Physicians of London. Vol. 3: (1801–25). By William Munk. 2d ed., revised and enlarged. London. 1878. Vol. 4: *Munk's roll, vol. 4: Lives of the fellows of the Royal College of Physicians* (1826–1925). Compiled by G.H. Brown. London: Royal College of Physicians. 1955.

Pickering, Charles. [1848]. *The races of man: and their geographical distribution.* Vol. 9 of the United States Exploring Expedition during the years 1838–42 under the command of Charles Wilkes, U.S.N. London.

——. 1850. *The races of man; and their geographical distribution.* New edition, to which is prefixed, an analytical synopsis of the natural history of man, by John Charles Hall. (Bohn's Illustrated Library.) London.

Pingree, Jeanne. 1968. *Thomas Henry Huxley: A list of his scientific notebooks, drawings and other papers, preserved in the college archives.* London: Imperial College of Science and Technology.

Porter, Robert Ker. 1821–2. *Travels in Georgia, Persia, Armenia, ancient Babylonia, &c. &c. during the years 1817, 1818, 1819 and 1820.* 2 vols. London.

Post Office directory of Durham, Northumberland, Cumberland and Westmorland: The Post Office directory of the county of Durham, and the principal towns and adjacent places in Northumberland, Cumberland & Westmorland. London. 1873–.

Post Office directory of Lancashire: Post Office directory of Lancashire. London. 1858–.

Post Office directory of Lincolnshire: Post Office directory of Lincolnshire. London. 1849–.

Post Office directory of the six home counties: Post Office directory of the six home counties, viz., Essex, Herts, Kent, Middlesex, Surrey and Sussex. London. 1845–.

Post Office London directory: Post Office London directory. London. 1802–.

Praeger, William E. 1935. Six unpublished letters of Charles Darwin. *Papers of the Michigan Academy of Science, Arts, and Letters* 20: 711–15.

Prestwich, Joseph. 1840. On the geology of Coalbrook Dale. [Read 1834.] *Transactions of the Geological Society of London* 2d ser. 5, pt 3: 413–95.

———. 1854. On the distinctive physical and palæontological features of the London Clay and the Bracklesham Sands; and on the independence of these two groups of strata. *Quarterly Journal of the Geological Society of London* 10: 435–54.

Prichard, James Cowles. 1843. *The natural history of man; comprising inquiries into the modifying influence of physical and moral agencies on the different tribes of the human family.* London.

Provincial medical directory: The provincial medical directory. London. 1847.

Quatrefages de Bréau, Jean Louis Armand de. 1855–6. Physiologie comparée. Les métamorphoses. *Revue des deux Mondes* 10 (1855): 90–116, 275–314; 3 (1856): 496–519, 859–83; 4 (1856): 55–82.

Quoy, Jean René Constant and Gaimard, Joseph Paul. 1830[–4]. *Zoologie.* 4 vols. In Dumont d'Urville, Jules Sébastien César, *Voyage de découvertes de l'Astrolabe. Exécuté par ordre du Roi, pendant les années 1826-1827-1828-1829, sous le commandement de J. Dumont d'Urville.* 15 vols. Paris. 1830[–5].

Rathke, Martin Heinrich. 1840. Zur Entwickelungsgeschichte der Dekapoden. *Archiv für Naturgeschichte* 6: 241–9.

Rehbock, Philip F. 1979. The early dredgers: 'naturalizing' in British seas, 1830–1850. *Journal of the History of Biology* 12: 293–368.

Reid, Mayne. 1857. *The war-trail; or, the hunt of the wild horse.* London.

Richard, Achille. [1847]. *Tentamen Floræ Abyssinicæ seu enumeratio plantarum.* Pt 3, vols. 4 and 5, of Lefebvre, Charlemagne Théophile, *Voyage en Abyssinie exécuté pendant les années 1839, 1840, 1841, 1842, 1843 par une commission scientifique.* 4 pts in 6 vols. Paris. [1845–54].

Richardson, Edward William. 1916. *A veteran naturalist; being the life and work of W. B. Tegetmeier.* London: Witherby & Co.

Richardson, John. 1829–37. *Fauna Boreali-Americana; or, the zoology of the northern parts of British America.* Assisted by William Swainson and William Kirby. 4 vols. London and Norwich.

Richardson, John. 1836. Report on North American zoology. *Report of the 6th meeting of the British Association for the Advancement of Science held at Bristol*, pp. 121–224.

——. 1845. Report on the ichthyology of the seas of China and Japan. *Report of the 15th meeting of the British Association for the Advancement of Science held at Cambridge*, pp. 187–320.

——. 1852. *Vertebrals, including fossil mammals*. In Forbes, Edward, ed., *The zoology of the voyage of H.M.S. Herald, under the command of Captain Henry Kellett, R.N., C.B., during the years 1845–51*. London. 1852–4.

Ross, James Clark. 1826. Zoology. Appendix, pp. 91–120, of Parry, William Edward, *Journal of a third voyage for the discovery of a north-west passage from the Atlantic to the Pacific; performed in the years 1824–25, in His Majesty's Ships Hecla and Fury*. London. 1826.

Rowlinson, J. S. 1971. The theory of glaciers. *Notes and records of the Royal Society of London* 26: 189–204.

Royal Gardens, Kew. 1899. *Catalogue of the library of the Royal Botanic Gardens, Kew*. (Royal Gardens, Kew, Bulletin of Miscellaneous Information, Additional Series 3). London.

Royal Society of London. 1839. *Catalogue of the scientific books in the library of the Royal Society*. London.

Royal Society of London. 1940. *The record of the Royal Society of London for the promotion of natural knowledge*. 4th ed. London: Royal Society.

Rudwick, Martin J. S. 1974. Darwin and Glen Roy: a 'great failure' in scientific method? *Studies in History and Philosophy of Science* 5: 97–185.

Rugby School register: *Rugby School register. Volume 1, from April 1675 to October 1857*. Revised and annotated by Godfrey A. Solly. Rugby: George Over (Rugby). 1933.

Rüppel, Wilhelm Peter Eduard Simon. 1845. *Systematische Uebersicht der Vögel Nord-Ost-Afrikas . . . Fortsetzung der neuen Wirbelthiere, zu der Fauna von Abyssinien gehoerig*. Frankfurt am Main.

Russell-Gebbett, Jean. 1977. *Henslow of Hitcham: botanist, educationalist and clergyman*. Lavenham, Suffolk: Terence Dalton.

Saint-Hilaire, Auguste de. 1847. *Leçons de botanique: comprenant principalement la morphologie végétale*. Paris.

Salter, James. 1857. On the vitality of seeds after prolonged immersion in the sea. [Read 6 May 1856.] *Journal of the Proceedings of the Linnean Society. Botany* 1: 140–2.

Salter, Thomas Bell. 1852. On the fertility of certain hybrids. *Phytologist* 4, pt 2: 737–42.

Sarjeant, William A. S. 1980. *Geologists and the history of geology: an international bibliography from the origins to 1978*. 5 vols. London: Macmillan.

SBL: *Svenskt biografiskt lexikon*. Edited by Bertil Boëthius, Erik Grill, and Birgitta Lager-Kromnow. 26 vols. (A–Nilsson). Stockholm: Albert Bonnier and P. A. Norstedt. 1918–89.

Scherren, Henry. 1905. *The Zoological Society of London. A sketch of its foundation and development and the story of its farm, museum, gardens, menagerie and library.* London, Paris, New York, and Melbourne: Cassell & Co.

Scherzer, Karl von. 1861–3. *Narrative of the circumnavigation of the globe by the Austrian frigate Novara . . . in the years 1857, 1858, & 1859.* 3 vols. London.

Schiödte, Jörgen C. [1849]. Bidrag til den underjordiske Fauna. *Det Kongelige Danske Videnskabernes Selskabs Skrifter. Naturvidenskabelig og Mathematisk* n.s. 2 (1851): 1–39.

Schlagintweit, Hermann Rudolph Alfred and Schlagintweit, Robert. 1857. Notes on some of the animals of Tibet and India. *Report of the 27th meeting of the British Association for the Advancement of Science held at Dublin*, Transactions of the sections, pp. 106–8.

Schlagintweit, Hermann Rudolph Alfred, Schlagintweit, Adolphe, and Schlagintweit, Robert. 1861–6. *Results of a scientific mission to India and high Asia, undertaken between the years 1854 and 1858, by order of the court of directors of the Honourable East India Company.* 4 vols. Leipzig and London.

Schreber, Johann Christian Daniel von. 1778–1846. *Die Säugethiere in Abbildungen nach der Natur mit Beschreibungen.* Continued by Johann Andreas Wagner. 7 vols. Erlangen.

Secord, James A. 1981. Nature's fancy: Charles Darwin and the breeding of pigeons. *Isis* 72: 163–86.

Senebier, Jean. [1800]. *Physiologie végétale, contenant une description des organes des plantes, et une exposition des phénomènes produits par leur organisation.* 5 vols. Geneva.

Seringe, Nicolas Charles. 1845–9. *Flore des jardins et des grandes cultures.* 3 vols. Paris and Lyon.

Serres, Pierre Toussaint Marcel de. 1809a. Mémoire sur les yeux composés et les yeux lisses des Orthopteres, et sur la manière dont ces deux espèces d'yeux concourent a la vision. *Journal de Physique, de Chimie, et de l'Histoire Naturelle* 68: 277–98.

——. 1809b. Comparaison des organes de la mastication des Orthoptères avec ceux des autres animaux. *Annales du Muséum d'Histoire Naturelle* 14: 56–73.

——. 1811. De l'odorat, et des organes qui paroissent en être le siège, chez les Orthoptères. *Annales du Muséum d'Histoire Naturelle* 17: 426–41.

Sharpe, Daniel. 1855. On the structure of Mont Blanc and its environs. [Read 15 November 1854.] *Quarterly Journal of the Geological Society of London* 11: 11–26.

——. 1856. On the last elevation of the Alps; with notices of the heights at which the sea has left traces of its action on their sides. *Quarterly Journal of the Geological Society of London* 12: 102–23.

Sheridan, Richard Brinsley Butler. 1781. *The school for scandal, a comedy; as it is performed at the Theatre-Royal in Drury-Lane.* Dublin.

Siebold, Karl Theodor Ernst von. 1856. *Wahre Parthenogenesis bei Schmetterlingen und Bienen. Ein Beitrag zur Fortpflanzungsgeschichte der Thiere.* Leipzig.

——. 1857. *On a true parthenogenesis in moths and bees; a contribution to the history of reproduction in animals.* Translated by William S. Dallas. London.

Silliman, Benjamin, Jr. 1851. On the Mammoth cave of Kentucky. *American Journal of Science and Arts* 2d ser. 11: 332–9.

Sloane, Hans. 1707–25. *A voyage to the islands Madera, Barbados, Nieves, S. Christophers and Jamaica, with the natural history . . . of the last of those islands.* 2 vols. London.

Smith, Charles Hamilton. 1843. *The natural history of horses. The Equidæ or genus equus of authors.* Vol. 12 of Jardine, William, ed., *The naturalist's library.* 40 vols. Edinburgh. 1833–43.

Smith, Crosbie. 1989. William Hopkins and the shaping of dynamical geology: 1830–60. *British Journal for the History of Science* 22: 27–52.

Smith, Frederick. 1853. Monograph of the genus *Cryptocerus*, belonging to the group *Cryptoceridae*—Family *Myrmicidae*—Division *Hymenoptera Heterogyna*. *Transactions of the Entomological Society of London* n.s. 2 (1852–3): 213–28.

Smith, James Edward. 1824–36. *The English flora.* 5 vols. in 6. Vol. 5, pts 1 and 2, by William Jackson Hooker. London.

Smith, John. 1841. Notice of a plant which produces perfect seeds without any apparent action of pollen. *Transactions of the Linnean Society of London* 18: 509–12.

Smith, Thomas. 1822. On certain species of Carduus and Cnicus which appear to be dioecious. *Transactions of the Linnean Society of London* 13: 592–603.

Sorby, Henry Clifton. 1856. On the microscopical structure of mica-schist. *Report of the 26th meeting of the British Association for the Advancement of Science held at Cheltenham,* Transactions of the sections, p. 78.

South America: Geological observations on South America. Being the third part of the zoology of the Beagle, under the command of Capt. Fitzroy, R.N. during the years 1832 to 1836. By Charles Darwin. London. 1846.

Spallanzani, Lazzaro. 1769. *An essay on animal reproductions.* Translated from the Italian [by M. Maty]. London.

Spencer, Herbert. 1855. *The principles of psychology.* London.

Sprengel, Christian Konrad. 1793. *Das entdeckte Geheimniss der Natur im Bau und in der Befruchtung der Blumen.* Berlin.

[Stainton, Henry Tibbats]. 1856. Why did Mr. Westwood get the Royal Medal? *Entomologist's Weekly Intelligencer,* 12 April 1856, pp. 9–10.

Stenton, Michael. 1976. *Who's who of British members of Parliament, 1832–1885.* Brighton: Harvester Press.

Stephens, James Francis. 1828–46. *Illustrations of British entomology; or, a synopsis of indigenous insects: containing their generic and specific distinctions.* 11 vols. and supplement. London.

——. 1829. *A systematic catalogue of British insects: being an attempt to arrange all the hitherto discovered indigenous insects in accordance with their natural affinities.* London.

Steudel, Ernst Gottlieb. 1840–1. *Nomenclator botanicus seu: synonymia plantarum universalis, enumerans ordine alphabetico nomina atque synonyma tum generica tum specifica.* 2d ed. 2 pts. Stuttgart and Tübingen.

Straus-Durckheim, Hercule Eugène. 1819. Mémoire sur les Daphnia, de la classe des Crustacés. *Mémoires du Muséum d'Histoire Naturelle* 5: 380–425.

Strickland, Hugh Edwin and Melville, A. G. 1848. *The dodo and its kindred; or the history, affinities, and osteology of the dodo, solitaire, and other extinct birds of the islands Mauritius, Rodriguez, and Bourbon.* London.

Sulivan, Henry Norton, ed. 1896. *Life and letters of the late Admiral Sir Bartholomew James Sulivan, K.C.B. 1810–1890.* London.

Swayne, George. 1824. On fertilizing the blossoms of pears. *Transactions of the Horticultural Society of London* 5: 208–13.

Tabler, Edward C. 1977. *Pioneers of Natal and southeastern Africa 1552–1878.* Cape Town and Rotterdam: A. A. Balkema.

Targioni Tozzetti, Antonio. 1850. *Cenni storici sulla introduzione di varie piante nell' agricoltura ed orticoltura toscana.* Florence.

Tegetmeier, William Bernhard, ed. 1856–7. *The poultry book: including pigeons and rabbits; comprising the characteristics, management, breeding and medical treatment of poultry, being the results of personal observation and practice of the Rev. W. Wingfield, G. W. Johnson, Esq.* Re-arranged and edited by W. B. Tegetmeier. London.

——. 1856. On the very remarkable peculiarities existing in the skulls of the feather-crested variety of the domestic fowl, now known as the Polish. *Proceedings of the Zoological Society of London* pt 24: 366–8.

——. 1858. On the formation of the cells of bees. *Report of the 28th meeting of the British Association for the Advancement of Science held at Leeds*, Transactions of the sections, pp. 132–33.

——. 1867. *The poultry book: comprising the breeding and management of profitable and ornamental poultry, their qualities and characteristics.* London.

——, ed. 1879. *Columbarium; or, the pigeon-house. Being an introduction to a natural history of tame pigeons. By John Moore. Reprinted verbatim et literatim from the original edition of 1735, with a brief notice of the author by W. B. Tegetmeier.* London.

Tegetmeier, William Bernhard and Weir, Harrison. 1868. *Pigeons: their structure, varieties, habits and management.* London.

Tellkampf, Theodor G. 1844a. Ueber den blinden Fisch der Mammuthhöhle in Kentucky. *Archiv für Anatomie, Physiologie, und wissenschaftliche Medicin* 9: 381–95.

——. 1844b. Beschreibung einiger neuer in der Mammuth-Höhle in Kentucky aufgefundener Gattungen von Gliederthieren. *Archiv für Naturgeschichte* 10: 318–22.

Temminck, Coenraad Jacob and Knip, Pauline de Courcelles. 1811. *Les pigeons.* 2 vols. (Vol. 1 by C. J. Temminck; vol. 2 by Florent Prévost.) Paris.

Thompson, William. 1849–56. *The natural history of Ireland.* 4 vols. (Vol. 4 edited by Robert Patterson.) London.

Thomson, Keith Stewart. 1975. H.M.S. *Beagle*, 1820–1870. *American Scientist* 63: 664–72.

Thunberg, Carl Peter. 1784. *Flora Japonica.* Leipzig.

——. 1794. Botanical observations on the Flora Japonica. *Transactions of the Linnean Society* 2: 326–342.

Thuret, Gustave Adolphe. 1854–5. Recherches sur la fécondation des Fucacées, suivies d'observations sur les anthéridies des Algues. *Annales des Sciences Naturelles. Botanique.* 4th ser. 2: 197–214; 3: 5–28.

Thwaites, George Henry Kendrick. 1847. On conjugation in the Diatomaceæ. *Report of the 17th meeting of the British Association held at Oxford,* Transactions of the sections, p. 87.

Torrey, John and Gray, Asa. 1838[–43]. *A Flora of North America: containing abridged descriptions of all the known indigenous and naturalized plants growing north of Mexico: arranged according to the natural system.* 2 vols. New York and London.

Tuckerman, Edward. 1849. Observations on American species of the genus Potamogeton, L. *American Journal of Science and Arts* 7: 347–60.

Tulasne, Louis René. 1852. *Monographia Podostemacearum.* Paris.

Tyndall, John. 1856. Comparative view of the cleavage of crystals and slate rocks. *Notices of the Proceedings at the meetings of the members of the Royal Institution of Great Britain* 2 (1854–8): 295–308.

——. 1860. *The glaciers of the Alps. Being a narrative of excursions and ascents, an account of the origin and phenomena of glaciers, and an expositon of the physical principles to which they are related.* London.

Tyndall, John and Huxley, Thomas Henry. 1857. On the structure and motion of glaciers. *Philosophical Transactions of the Royal Society of London* 147: 327–46. Reprinted in Foster and Lankester, eds. 1898–1903, 2: 1–25.

Tytler, Robert C. 1854. Miscellaneous notes on the fauna of Dacca, including remarks made on the line of march from Barrackpore to that station. *Annals and Magazine of Natural History* 2d ser. 14: 168–77.

Variation: The variation of animals and plants under domestication. By Charles Darwin. 2 vols. London. 1868.

Vaux, William Sandys Wright. 1851. *Handbook to the antiquities in the British Museum: being a description of the remains of Greek, Assyrian, Egyptian, and Etruscan art preserved there.* London.

Vicq d'Azyr, Félix. 1787–1830. *Médicine.* 13 vols. In *Encyclopédie methodique.* 190 vols. Paris.

Visiani, Roberto de. 1842–52. *Flora Dalmatica, sive enumeratio stirpium vascularium quas hactenus in Dalmatia lectas et sibi observatas descripsit digessit rariorumque iconibus illustravit.* 3 vols. Leipzig.

Volcanic islands: Geological observations on the volcanic islands visited during the voyage of H.M.S. Beagle, together with some brief notices of the geology of Australia and the Cape of Good Hope. Being the second part of the geology of the voyage of the Beagle, under the command of Capt. FitzRoy, R.N. during the years 1832 to 1836. By Charles Darwin. London. 1844.

Wahlenberg, Göran. 1824–6. *Flora Suecica.* 2 pts. Upsala.

Wallace, Alfred Russel. 1853. *A narrative of travels on the Amazon and Rio Negro.* London.

——. 1855. On the law which has regulated the introduction of new species. *Annals and Magazine of Natural History* 2d ser. 16: 184–96.

——. 1857. On the natural history of the Aru Islands. *Annals and Magazine of Natural History* 2d ser. 20, suppl.: 473–85.

——. 1869. *The Malay Archipelago: the land of the orang-utan, and the bird of paradise. A narrative of travel, with studies of man and nature.* 2 vols. London.

——. 1905. *My life: a record of events and opinions.* 2 vols. London: Chapman & Hall.

Wallich, Nathaniel. 1851. Specimen faunæ subterraneæ, being a contribution towards the subterranean fauna, by J. C. Schiödte. Translated from the Danish. *Transactions of the Entomological Society of London* n.s. 1 (1850–1): 134–57.

Waterhouse, George Robert. 1846–8. *A natural history of the Mammalia.* 2 vols. London.

——. 1858. *Catalogue of British Coleoptera.* London.

Watson, Hewett Cottrell. 1832. *Outlines of the geographical distribution of British plants.* Edinburgh.

——. 1835. *Remarks on the geographical distribution of British plants; chiefly in connection with latitude, elevation, and climate.* London.

——. 1843. *The geographical distribution of British plants.* 3d ed. Pt 1 (no more published). London.

——. 1847–59. *Cybele Britannica; or British plants, and their geographical relations.* 4 vols. London.

Watson, Hewett Cottrell and Syme, John Thomas Irvine Boswell, eds. 1853. *The London catalogue of British plants.* 4th ed. London.

——, eds. 1857. *The London catalogue of British plants: published under the direction of the Botanical society of London adapted for marking desiderata in exchanges of specimens, &c.* 5th ed. London.

Waugh, Francis Gledstanes. [1888]. *Members of the Athenaeum Club, 1824 to 1887.* London.

Webb, Philip Barker and Berthelot, Sabin. 1836–50. *Histoire naturelle des Iles Canaries.* 3 vols. in 7 and atlas. Paris.

Wedgwood, Barbara and Wedgwood, Hensleigh. 1980. *The Wedgwood circle, 1730–1897.* London: Studio Vista.

Wedgwood, Hensleigh. 1859–65. *A dictionary of English etymology.* 3 vols. in 4. London.

Wellesley Index: *The Wellesley index to Victorian periodicals 1824–1900.* Edited by Walter E. Houghton and others. 5 vols. Toronto: University of Toronto Press. London: Routledge and Kegan Paul. 1966–89.

Wellsted, James Raymond. 1840. *Travels to the city of the Caliphs, along the shores of the Persian Gulf and the Mediterranean. Including a voyage to the coast of Arabia, and a tour on the island of Socotra.* 2 vols. London.

Westwood, John Obadiah. 1839–40. *An introduction to the modern classification of insects; founded on the natural habits and corresponding organisation of the different families.* 2 vols. London.

[Widdrington, Samuel Edward] (Cook, Captain S. E., *pseud.*). 1834. *Sketches in Spain during the years 1829, 30, 31, & 32; containing notices of some districts very little known; of the manners of the people, government, recent changes, commerce, fine arts, and natural history.* 2 vols. London.

——. 1844. *Spain and the Spaniards, in 1843.* 2 vols. London.

Wiegmann, Arend Friedrich August. 1828. *Ueber die Bastarderzeugung im Pflanzenreiche.* Brunswick.

Wight, Robert. 1841. Remarks on the fruit of the natural order Cucurbitaceæ. *Journal of Botany* 3: 387–92.

Williams, Herbert William. 1971. *A dictionary of the Maori language.* 7th ed. Wellington, New Zealand: A.R. Shearer.

Wilson, Leonard Gilchrist, ed. 1970. *Sir Charles Lyell's scientific journals on the species question.* New Haven and London: Yale University Press.

Wiltshear, F. G. 1913. The botany of the Antarctic voyage. *Journal of Botany. British and Foreign* 51: 355–8.

Wimmer, Christian Friedrich Heinrich. 1840. *Flora von Schlesien, preussischen und österreichischen Antheils, oder vom oberen Oder- und Weichsel-Quellen-Gebiet. Nach natürlichen Familien, mit Hinweisung auf das Linnéische System.* Breslau, Ratibor, and Pless.

Winsor, Mary Pickard. 1976. *Starfish, jellyfish, and the order of life: issues in nineteenth-century science.* New Haven and London: Yale University Press.

Wollaston, Thomas Vernon. 1854. *Insecta Maderensia; being an account of the insects of the islands of the Madeiran group.* London.

——. 1856. *On the variation of species, with special reference to the Insecta; followed by an inquiry into the nature of genera.* London.

——. 1857. *Catalogue of the coleopterous insects of Madeira in the collection of the British Museum.* London.

——. 1860. Review of *On the origin of species by means of natural selection*, by Charles Darwin. *Annals and Magazine of Natural History* 3d ser. 5: 132–43.

——. 1861. On the Atlantic Cossonides. *Transactions of the Entomological Society of London* n.s. 5 (1858–61): 362–407.

Woodward, Samuel Pickworth, 1851–6. *A manual of the Mollusca; or, a rudimentary treatise of recent and fossil shells.* 3 pts. London.

——. 1856a. On an *Orthoceras* from China. *Quarterly Journal of the Geological Society of London* 12: 378–81.

——. 1856b. On the land and freshwater shells of Kashmir and Tibet, collected by Dr. T. Thomson. *Proceedings of the Zoological Society of London* pt 24: 185–7.

WWW: *Who was who*. A companion to *Who's who*. 7 vols. and cumulated index, 1897–1980. London: Adam & Charles Black. 1920–81.

Youatt, William. 1845. *The dog.* London.

BIOGRAPHICAL REGISTER
AND INDEX TO CORRESPONDENTS

This list includes all persons mentioned in the letters and notes that the editors have been able to identify and all correspondents. Dates of letters to and from correspondents are given in chronological order. Letters to the correspondents are listed in roman type; letters from correspondents in italic type. Following the register, a list of all biographical sources referred to in the entries is given. These works are also listed in the main bibliography.

Academia Caesarea Leopoldino-Carolina Naturae Curiosorum
8 September [1857]

Acton, Samuel Poole. Wine and spirit merchant and postmaster in Bromley, Kent. (*Post Office directory of the six home counties* 1851.)

Adams, Arthur (1820–78). Naval surgeon, conchologist, and author. (*Modern English biography, NUC.*)

Adams, Henry (1813–77). Conchologist and author. (*NUC.*)

Adanson, Michel (1727–1806). French botanist who studied the flora of Senegal. (*DBF.*)

Agassiz, Jean Louis Rodolphe (Louis) (1807–73). Swiss geologist and zoologist. Professor of natural history, Neuchâtel, 1832–46. Emigrated to the United States in 1846. Professor of natural history, Harvard University, 1847–73. Established the Museum of Comparative Zoology, Harvard, 1859. Foreign member, Royal Society, 1838. (*DAB, DSB.*)

Airy, George Biddell (1801–92). Plumian professor of astronomy and director of the Cambridge Observatory, 1828–35. Astronomer Royal, 1835–81. FRS 1836. (*DNB, DSB.*)

Ajasson de Grandsagne, Jean Baptiste François Étienne (1802–45). French naturalist and writer. Studied and translated works of ancient science. Devoted himself to publishing a popular encyclopaedia. This enterprise exhausted his fortune and he died ruined in Lyon. (*DBF.*)

Albert Francis Charles Augustus Emmanuel, Prince Consort of England (1819–61). Married Queen Victoria in 1840. A patron of the arts and sciences. Established a model farm at Windsor in 1840. Chancellor of Cambridge University, 1847. Put forward the idea of the Great Exhibition and was actively involved in its organisation, 1849–51. Supported many philanthropic causes. (*DNB.*)

Albin, Eleazar (*fl.* 1713–59). Naturalist and water-colour painter. (*DNB.*)

Alder, Joshua (1792–1867). Zoologist who specialised in the Mollusca. A member of the Newcastle Literary and Philosophical Society, 1815. Collaborated with Albany Hancock on the *Monograph of British nudibranchiate Mollusca* (1845–55). (*DNB.*)

Aldrovandi, Ulisse (1522–1605). Italian naturalist. Professor of natural history at Bologna University, 1561. Established the botanic garden and public museum of Bologna. (*DBI, DSB.*)

Alexander III (356–323 B.C.). King of Macedon, known as 'The Great', who extended his realm into Asia Minor and northern India. (*EB.*)

Allen, William (1793–1864). Naval officer. Took part in the Niger expeditions of 1832 and 1841–2. Rear-admiral, 1862. Published books on travel. FRS 1841. (*DNB.*)

Allman, George James (1812–98). Botanist and zoologist. Professor of botany, Dublin University, 1844; professor of natural history, Edinburgh University, 1855–70. FRS 1854. (R. Desmond 1977, *DNB.*)

Andersson, Charles John (Carl Johann) (1827–67). Swedish-born ornithologist, hunter, and explorer. Accompanied Francis Galton to South Africa in 1849. Organised a successful expedition to Lake Ngami and published an account of the expedition in 1856. Continued to explore the South African interior and to carry out ornithological studies. (*DSAB, Modern English biography, SBL.*)

Andersson, Nils Johan (1821–80). Swedish botanist who explored the Americas on the *Eugenie* expedition, 1851–3. (*SBL.*)
   [6 April 1856]

Andral, Gabriel (1797–1876). French physician. Author of many medical works. (*DBF.*)

Aristophanes (*c.* 448–385 B.C.). Comic dramatist and poet of Athens. (*EB.*)

Aristotle (384–322 B.C.). Greek philosopher. Author of many works, including *Historia animalium.* (*EB.*)

Arnott, George Arnott Walker (1799–1868). Botanist. Worked with William Jackson Hooker in Glasgow from 1830 to 1840. Described, with Hooker, the plants collected on F. W. Beechey's voyage of 1825–8. Regius professor of botany, Glasgow, 1845–68. (R. Desmond 1977, *DNB.*)

Arnott, Neil (1788–1874). Physician and meteorologist. Surgeon in the East India Company's service in China, 1807–9 and 1810–11. Practised as a physician in London, 1811–55. Gave lectures at the Philomathic Institution, which he published as *Elements of physics* in 1827. FRS 1838. (*DNB.*)

Asso y del Rio, Ignacio Jordán de (1742–1814). Spanish orientalist, jurist, historian, and bibliographer. Consul general in Holland and in Guyana. (*IBN.*)

Atkinson, William Henry (1821–76). Director of public instructions for Bengal. Secretary of the Asiatic Society of Bengal, 1855–63; vice-president, 1865. (*Modern English biography.*)

Audouin, Jean Victor (1797–1841). French zoologist. Began publication of the *Annales des Sciences Naturelles*, 1824. Studied the marine invertebrates of the coasts of Brittany and Normandy with Henri Milne-Edwards. Professor of zoology, Muséum d'histoire naturelle, 1833. (*DBF, DSB.*)

Autard de Bragard, Gustave Adolphe (d. 1876). Magistrate of Mauritius. (*DBF.*)

Azara, Felix d' (1746–1811). Spanish explorer and army officer. Sent to South America to survey Spanish and Portuguese territories. Published works on the fauna of Paraguay. (*NBU.*)

Babington, Charles Cardale (1808–95). Botanist and archaeologist. BA, St John's College, Cambridge, 1830. Professor of botany, Cambridge University, 1861–95. An expert on plant taxonomy. FRS 1851. (*DNB, DSB.*)
   22 November 1856

Backhouse, James (1825–90). Botanist who collected plants in Norway, 1851; Ireland, 1854; and Scotland, 1859. Also collected the flora around Teesdale, England. (R. Desmond 1977.)

Baer, Karl Ernst von (1792–1876). Estonian naturalist. Professor of anatomy at Königsberg University, 1817; of zoology, 1826–34. An active member of the Academy of Sciences in St Petersburg, 1834–67. Carried out embryological researches and discovered the mammalian egg, 1826. Propounded the influential view that development proceeds from the general to the specific. (*DSB, NDB.*)

Baily, John. Poulterer and dealer in live birds at 113 Mount Street, Berkeley Square, London. Author of works on the management of domestic and game fowl. (Freeman 1978, *Post Office London directory* 1853.)

Baily, John, Sr. Father of John Baily. Poultry judge.

Baird, William (1803–72). Scottish physician who practised in London, 1833–41. Assistant in the zoological department of the British Museum, 1841–72. FRS 1867. (*DNB, Modern English biography.*)

Baker, Samuel C. and Charles N. Dealers in ornamental poultry and live wild fowls at 3 Half Moon Passage, Chelsea. (*Post Office London directory* 1851.)

Bakewell, Robert (1725–95). Farmed at Dishley, Leicestershire. Notable stock-breeder. (*DNB.*)

Barbot, Jacques (James). Brother of John Barbot. Emigrated from France to Britain with his brother and became a British subject in 1688. Participated in voyages to the African coast, 1698 and 1699. (*DBF.*)

Barbot, Jean (John) (d. 1720). French agent in the ivory and slave trade off the coast of Guinea. Became a British subject in 1688. Travelled in Sierra Leone and western Africa. (*DBF.*)

Barnéoud, François Marius (b. 1821). French botanist and geologist. (*NUC.*)

Barnes, James (1806–77). Gardener to Lady Louisa Barbara Rolle at Bicton House, Devon, 1839–69. (R. Desmond 1977.)

Barrande, Joachim (1799–1883). French-born palaeontologist and stratigrapher. Employed as a tutor by the French royal family until their exile in 1830. Railway engineer in Prague where he became interested in fossils. Collected and described the fossils of the Bohemian basin, 1840–83. The results of his studies were published in 30 volumes entitled *Système Silurien du centre de la Bohême* (1852–1902). (*DSB.*)

Barry, Martin (1802–55). Studied medicine in London, Paris, Erlangen, and Berlin. MD, Edinburgh, 1833; did not practise as a physician. Carried out researches in embryology and histology in Britain and Germany; discovered the presence of spermatozoa in ova, 1843. FRS 1840. (*DNB, DSB.*)

Barth, Heinrich (1821–65). German geographer and traveller. Member of a British expedition to establish commercial relations with the states of central and western Africa, 1850–5. Published an account of his travels in five volumes (1857–8). (*EB, NDB.*)

Bartlett, Abraham Dee (1812–97). Taxidermist in London, 1834–52. Superindendent of the natural history department of the Crystal Palace, 1852–9; of the Zoological Society's gardens, Regent's Park, 1859–97. (*Modern English biography.*)

Barton, Benjamin Smith (1766–1815). American physician and naturalist. Professor of natural history and botany at the College of Pennsylvania, 1790. (*DAB.*)

Bartram, William (1739–1823). American traveller and naturalist. Explored the south-eastern part of the United States, 1773–8, and published an account of his travels (1791). (*DAB.*)

Bashford, F. Of Surdah Factory, Lower Bengal. Silkworm breeder. Friend of Edward Blyth.

Bate, Charles Spence (1819–89). Dentist and scientific writer. Practised dentistry in Swansea, 1841–51, then Plymouth. Studied the Crustacea. FRS 1861. (*DNB.*)
    3 December 1857

Bates, Henry Walter (1825–92). Naturalist. Undertook a joint expedition to the Amazon with Alfred Russel Wallace, 1848–50; continued to explore the area alone until 1859. Developed the concept of mimicry. Published *The naturalist on the River Amazons* in 1863. Assistant secretary, Royal Geographical Society, 1864–92. FRS 1881. (*DNB, DSB.*)

Baudement, Émile (1810–64). French embryologist and agriculturist. (*NUC.*)

Beaufort, Francis (1774–1857). Naval officer; retired as rear-admiral in 1846. Hydrographer to the Admiralty, 1832–55. One of the founders of the Royal Astronomical Society and of the Royal Geographical Society. FRS 1814. (*DNB.*)

Bechstein, Johann Matthäus (1757–1822). German ornithologist. (*NDB.*)

Beckles, Samuel Husband (1814–90). Barrister. Discovered in the Purbeck beds the oldest known Mammalian fossils. FRS 1859. (*Modern English biography*, Sarjeant 1980.)

Beke, Charles Tilstone (1800–74). Explorer and antiquary. Travelled in Abyssinia, 1840–3; Syria and Palestine, 1861–2. A business associate of Blyth Brothers, Mauritius, 1853–9. (*DNB*.)
24 September [1856]

Beke, Emily Alston. Resident of Mauritius. Second wife of Charles Tilstone Beke. (*CDEL*.)

Bell, Thomas (1792–1880). Dental surgeon at Guy's Hospital, London, 1817–61. Professor of zoology, King's College, London, 1836. President of the Ray Society, 1843–59. One of the secretaries of the Royal Society, 1848–53. President of the Linnean Society, 1853–61. Described the reptiles from the *Beagle* voyage. FRS 1828. (R. Desmond 1977, *DNB*.)

Bennett, John Joseph (1801–76). Botanist. Secretary of the Linnean Society, 1840–60. Assistant keeper, department of botany, British Museum, 1827; keeper, 1859–70. FRS 1841. (R. Desmond 1977, *DNB*.)

Benson, William Henry (1803–70). Entomologist. (*IBN*.)

Bentham, George (1800–84). Botanist. Honorary secretary of the Horticultural Society, 1829–40. President of the Linnean Society, 1861–74. Published *Genera plantarum* (1862–83) with J. D. Hooker. FRS 1862. (*DNB, DSB*.)
26 November [1856], 30 November [1856], 3 December [1856], 1 December [1857], 15 December [1857], [16 or 17 December 1857], 18 December [1857]

Berghaus, Heinrich Karl Wilhelm (1797–1884). German geographer and cartographer. Lecturer at the Academy of Architecture in Berlin. (*IBN*.)

Berkeley, Miles Joseph (1803–89). Clergyman and botanist. Perpetual curate of Apethorpe and Wood Newton, Northamptonshire, 1833–68. An expert on British fungi. FRS 1879. (*DNB, DSB*.)
29 February [1856], 7 March 1856

Berthelot, Sabin (1794–1880). French naturalist and traveller who studied in the Canary Islands, 1820–30, and directed the botanic garden at La Orotava, Tenerife. Agent in Santa Cruz, Tenerife, 1847; consul, 1867. (*DBF*.)

Birch, Samuel (1813–85). Egyptologist. Assistant keeper of the department of oriental antiquities in the British Museum, 1844–61; keeper, 1861–85. Translator of old Chinese classics; archaeologist. (*DNB*.)
6 February [1856], [12 March 1856]

Birkbeck, Robert. Married Mary Harriet Lubbock in 1857. (*Burke's landed gentry* 1898.)

Bishop, Richard. Of Plymouth. Provided CD with observations on *Balanus*.
3 December 1857

Blyth, Edward (1810–73). Unsuccessful druggist in Tooting. Wrote zoological articles and edited zoological works before being appointed curator of the museum of the Asiatic Society of Bengal in 1841. He retired from this post in 1862 and returned to England where he continued to write on zoology and on the question of the origin of species. (*DNB, DSB*.)

Blyth, Edward, cont.
  [before 8 January 1856], 8 January [1856], 23 January 1856,
  23 February 1856, 26 February 1856, [c. 22 March 1856], [3 April 1856]
Blyth, Mrs Edward (d. 1857). Married Edward Blyth in 1854. (Grote 1875.)
Bochart, Samuel (1599–1667). French biblical scholar. Pastor of a Protestant
  church in Caen. Published *Geographia Sacra* (1646 and 1651) and *Hierozoicon*
  (1663), which treats of the animals of Scripture. His works were published
  together as Bochart's *Opera omnia* in 1675. (*DBF, EB.*)
Bomare, Jacques Christophe Valmont de. *See* Valmont de Bomare, Jacques
  Christophe.
Bonaparte, Charles Lucien, Prince of Canino, (1803–57). French zoologist.
  Lived in America, 1822–8. Returned to Italy in 1828 where he was politically
  active. Attempted to enter France in 1848 but was expelled. Lived in England
  before being allowed to settle in Paris, 1850. Devoted the rest of his life to work
  on taxonomy. (*DBF, DSB.*)
Boott, Francis (1792–1863). Physician and botanist. Had a successful medical
  practice in London, 1825–32. Lecturer on botany at the Webb Street School of
  Medicine, London, 1825. Secretary of the Linnean Society, 1832–9; treasurer,
  1856–61. (R. Desmond 1977, *DNB.*)
Boreau, Alexandre (1803–75). French botanist and pharmacist. Director of the
  botanic garden and of the natural history museum, Angers, 1838. (*DBF.*)
Bosquet, Joseph Augustin Hubert de (1814–80). Pharmacist in Maestricht.
  Stratigrapher and palaeontologist who studied the Cretaceous and Tertiary
  strata of Holland and Belgium. Member of the Royal Academy of Sciences of
  Amsterdam. (*NNBW*, Sarjeant 1980.)
  9 September [1856], [before 3 November 1856]
Boswell-Syme, John Thomas Irvine (1822–88). Scottish botanist. Curator of
  the Botanical Society, Edinburgh, 1850. (R. Desmond 1977, *Modern English
  biography.*)
Bouchard-Chantereaux, Nicolas Robert (1802–64). French naturalist who
  studied the Crustacea and Mollusca. (*DBF.*)
Bouton, Louis Sulpice (1799–1878). Botanist of Mauritius. Author of works on
  the agricultural and medicinal plants of the island. Secretary of the Natural
  History Society of Mauritius, 1839–78. (*NUC.*)
  24 September [1856]
Brehm, Christian Ludwig (1787–1864). German clergyman and ornitholo-
  gist. His bird collection formed the basis of the Rothschild museum in Tring.
  (*NDB.*)
Brent, Bernard Philip (d. 1867). Pigeon fancier and author. Studied pigeon
  breeding in France and Germany. (*CDEL.*)
  23 October 1857
Bright, John (1811–89). Manufacturer and statesman. Leader, with Richard
  Cobden, of the Anti-Corn Law League. MP for Durham, 1843; Manchester

1847; Birmingham, 1858. Radical reformer on the issues of taxation, education, free trade, and non-intervention. (*DNB*.)

Bristowe, Henry Fox (1824–93). Barrister. Called to bar, 1847; incorporated at Lincoln's Inn, 1854. Nephew of William Darwin Fox. Married Selina Bridgeman, 1850. (*Alum. Cantab.*)

Brodie, Benjamin Collins (1783–1862). Surgeon. Professor of comparative anatomy and physiology, Royal College of Surgeons, 1816. Sergeant-surgeon to Queen Victoria. President of the Royal Society, 1858–61. Created baronet, 1834. FRS 1810. (*DNB, DSB*.)

Brongniart, Adolphe Théodore (1801–76). French palaeobotanist and taxonomist. A founder of the *Annales des Sciences Naturelles*, 1824. Professor of botany, Muséum d'histoire naturelle, 1833. Foreign member, Royal Society, 1852. (*DBF, DSB*.)

Brooke, James (1803–68). Raja of Saráwak, Borneo, 1841–63. British commissioner and consul-general, 1847. (*DNB*.)

Brown, Robert (1773–1858). Botanist. Naturalist on a voyage to Australia, 1801–5; published descriptions of the plants he collected there. Librarian to Joseph Banks, 1810–20. Keeper of the botanical collections, British Museum, 1827–58. FRS 1811. (*DNB, DSB*.)

Browne, William George (1768–1813). Traveller in Africa and the Middle East. (*DNB*.)

Brullé, Gaspard Auguste (1809–73). French entomologist. Professor of entomology and comparative anatomy, Dijon University, 1839–73. (*DBF*, P. Gilbert 1977.)

Buch, Christian Leopold von (1774–1853). A widely-travelled German geologist and geographer. Foreign member, Royal Society, 1828. (*DSB, NDB*.)

Buchanan, James (1791–1868). Minister to Great Britain, 1853–6. Fifteenth president of the United States, 1856–60. (*DAB*.)

Buckman, James (1814–84). Geologist and agriculturist. Curator of the Birmingham Philosophical Institute, 1842–7. Professor of geology and botany at the Royal Agricultural College, Cirencester, 1848–63. Farmed according to scientific principles in Dorset, 1863–84. (*DNB*, Sarjeant 1980.)
  4 October [1857]

Buffon, Georges Louis Leclerc, Comte de (1707–88). French naturalist and scientific administrator. (*DBF, DSB*.)

Buist, George (1805–60). Anglo-Indian journalist and scientific observer. Editor of the *Bombay Times*, 1839–59. Inspector of the astronomical, magnetic, and meteorological observatories of Bombay, 1842. Author of numerous papers on the geology, geography, and agriculture of India. (*DNB*.)

Bult, Benjamin Edmund. Secretary to the Office of the Charitable & Provident Society for the Aged & Infirm Deaf & Dumb. Bookseller and newsagent at 25 New Quebec Street, London. Pigeon fancier and member of the Philoperisteron Society. (*Post Office London directory* 1858.)

Bunbury, Charles James Fox, 8th Baronet (1809–86). Botanist and palaeobotanist. Collected plants in South America, 1833–4; South Africa, 1838–9. Accompanied Charles Lyell to Madeira in 1853. Married Frances Joanna Horner, Mary Lyell's sister, in 1844. FRS 1851. (R. Desmond 1977, Sarjeant 1980.)

> *7 February 1856*, *16 April 1856*, 21 April [1856], [before 9 May 1856], 9 May [1856]

Bunbury, Frances Joanna (1814–94). Daughter of Leonard Horner. Married Charles James Fox Bunbury in 1844. (*Burke's peerage* 1980, Freeman 1978.)

Bunsen, Robert Wilhelm Eberhard (1811–99). German chemist. Professor of chemistry, University of Marburg, 1842; University of Heidelberg, 1852–89. Carried out important work in spectroscopy in the 1860s. Foreign member, Royal Society, 1858. (*DSB*, *NDB*.)

Burckhardt, John Lewis (Johann Ludwig) (1784–1817). Swiss traveller in the Middle East and Africa. His accounts of his travels were published posthumously. (*Alum. Cantab.*, *DNB*.)

Burnett, Waldo Irvin (1828–54). American entomologist. (P. Gilbert 1977.)

Butler, Louisa Jane. *See* Galton, Louisa Jane.

Butler, Thomas (1806–86). BA, St. John's College, Cambridge, 1829. Rector of Langar with Barnston, Nottinghamshire, 1834–76. (*Alum. Cantab.*)

Butter, Donald (1799–1877). Surgeon in the Bengal army from 1833. Superintending surgeon in Benares from 1854 to 1859. (*Modern English biography.*)

Buttler, John Olive. Lieutenant in the 49th regiment native infantry of Madras. (*East-India register and army list, for 1848.*)

Buxton, Charles (1823–71). Partner in the brewery of Truman, Hanbury, Buxton, & Co. MP for Newport, 1857; Maidstone, 1859; East Surrey, 1865–71. (*DNB.*)

Caesar, Gaius Julius (102–44 B.C.). Roman soldier and statesman. (*EB.*)

Camden, Marquis of. *See* Pratt, George Charles.

Campbell, Andrew. Superintendent of the Darjeeling station, 1840, with responsibility for political relations between the British and the Sikkim government. Wrote many papers on Himalayan geography. Travelled with Joseph Dalton Hooker in Sikkim and was imprisoned with Hooker by the Sikkim rajah. (J. D. Hooker 1854, L. Huxley ed. 1918.)

Candolle, Alphonse de (1806–93). Swiss botanist whose home was a centre of botanical activity. Professor of botany and director of the botanic gardens, Geneva, 1835–50. Son of Augustin Pyramus de Candolle. Foreign member, Royal Society, 1869. (*DSB.*)

Candolle, Augustin Pyramus de (1778–1841). Swiss botanist. Professor of natural history, Academy of Geneva, 1816–35. Foreign member, Royal Society, 1822. (*DSB.*)

Canino, Prince of. *See* Bonaparte, Charles Lucien.

Canning, Charles John, Earl (1812–62). Conservative MP for Warwick, 1836–7. Under secretary of state for foreign affairs, 1841–6. Postmaster general, 1853–5. Governor-general of India, 1855–62. In 1858, following the Indian mutiny, he became the first viceroy when the government of India was transferred from the East India Company to the crown. Created earl in 1859. (*Complete peerage, DNB.*)

Canning, Charlotte (1817–61). Married Charles John Canning in 1835. A lady of the bedchamber, 1842–55. Died of jungle fever in Calcutta. (*Complete peerage.*)

Carpenter, Philip Pearsall (1819–77). Conchologist and Presbyterian minister. In 1855 he bought a collection of shells from Mazatlan, Mexico, that weighed 14 tons, and devoted the rest of his life to classifying and describing these shells. (*DNB.*)

Carpenter, William Benjamin (1813–85). Physician and naturalist. Fullerian professor of physiology at the Royal Institution, 1844–56; professor of forensic medicine at University College London and lecturer on physiology at the London Hospital, 1845–56. Registrar of the University of London, 1856–79. FRS 1844. (*DNB, DSB.*)

Cassini, Alexandre Henri Gabriel (Henri), Vicomte de (1781–1832). French botanist. (*DBF.*)

Cautley, Proby Thomas (1802–71). Served in the Bengal artillery. Constructed the Ganges canal works, 1843–54. Member of the council of India, 1858–68. Made an extensive collection of fossil Mammalia from the Sewalik Hills in the north-west provinces of India, which he gave to the British Museum. FRS 1846. (*DNB.*)

Chambers, Robert (1802–71). Publisher, writer, and geologist. Partner with his brother, William Chambers, of Chambers's publishing company in Edinburgh. Joint editor of *Chambers's Journal* from 1832. Anonymous author of *Vestiges of the natural history of creation* (1844). (*DNB, DSB.*)

Chamisso, Adelbert von (1781–1838). French-born poet and botanist whose family fled to Germany during the French Revolution. Member of the expedition to the Pacific under Otto von Kotzebue, 1815–18. Adjunct curator of the botanical gardens, Berlin, 1819; curator, 1833. (*DSB, NDB.*)

Chesney, Francis Rawdon (1789–1872). Officer in the Royal Artillery. Visited Turkey, 1829; surveyed the isthmus of Suez to assess the practicability of a canal, 1830; explored the valley of the Euphrates, 1831; and navigated the lower Euphrates and explored the Tigris, 1835–7. Stationed at Hong Kong, 1843–7. Major-general, 1855; general, 1868. Published accounts of his surveys. (*DNB.*)

Chevreul, Michel Eugène (1786–1889). French chemist. Professor of chemistry, Muséum d'histoire naturelle, 1830; director, 1864–79. Director of dyeing

Chevreul, M. E., cont.
  at the Manufactures royales des Gobelins, 1824–83. Foreign member, Royal Society, 1826. (*DBF, DSB.*)
Chitty, Edward (1804–63). Legal reporter. Called to the bar at Lincoln's Inn, 1829. Went to Jamaica in 1840. Author, in addition to several works on equity law, of *The fly-fisher's text book* (1841), written under the pseudonym Theophilus South. (*DNB.*)
Chitty, Thomas (1802–78). Older brother of Edward Chitty. Special pleader and legal writer at 1 King's Bench Walk, London. (*DNB.*)
Christy, Alfred. Resided in Surrey Square, Old Kent Road, London. Supplied CD with information on the flight of carrier pigeons. (*Post Office London directory* 1857.)
  11 February 1857
Churchill, Awnsham (d. 1728). Bookseller and publisher. (*DNB.*)
Clive, Marianne. Daughter of George Tollet. Married William Clive in 1829. (*Burke's landed gentry* 1846.)
Clive, William (1795–1883). Vicar of Welshpool, Montgomeryshire, 1819–65. Married Marianne Tollet, daughter of George Tollet, in 1829. (*Alum. Cantab., Modern English biography.*)
Cobden, Richard (1804–65). Manufacturer and statesman. Leading member of the Anti-Corn Law League, 1838–46. MP for Stockport, 1841–7; for West Riding, Yorkshire, 1847–57; for Rochdale, 1859–65. Vocal advocate for free trade and for non-intervention. (*DNB.*)
Coe, Henry. Gardener to the Lunatic Asylum at Fareham, Hampshire.
  4 November 1857, 14 November 1857
Colling, Charles (1751–1836). Stock-breeder. One of the earliest and most successful improvers of the breed of shorthorn cattle. Had a farm near Darlington, Durham. (*DNB.*)
Colling, Robert (1749–1820). Stock-breeder. Sold to his brother, Charles Colling, the bull from which the improved stock of shorthorns was bred. Farmed at Barmpton, Durham. (*DNB.*)
Columella, Lucius Junius Moderatus (*fl.* 100 A.D.). Agricultural writer of Gades. Author of *De re rustica* and *De arboribus*. (*EB.*)
Conolly, Arthur (1807–42?). Captain in the East India Company's service. Travelled in central Asia, 1829–31, and published an account of his journey. Joined the British envoy's staff in Afghanistan in 1840. Imprisoned in Bokhara and eventually killed while on a diplomatic mission. (*DNB.*)
Cook, James (1728–79). Commander of several voyages of discovery. Circumnavigated the world, 1768–71 and 1772–5. FRS 1776. (*DNB, DSB.*)
Corker, Edward Langley. Tracing-paper manufacturer, stationer, account book manufacturer, engraver, and printer at 11 Queen Street, Cheapside, London. Pigeon fancier. President of the Southwark Columbarian Society. (*Post Office London directory* 1858.)

Covington, Syms (1816?–61). Became CD's servant in the *Beagle* in 1833 and remained with him as assistant, secretary, and servant until 1839. Emigrated to Australia, 1839. (Freeman 1978.)

9 March 1856, 22 February 1857

Crawfurd, John (1783–1868). Orientalist and linguist. Held several civil and political posts in Java, India, Siam, and Cochin China. Returned to England in 1827 and promoted the study of Indo-China. FRS 1818. (*DNB.*)

Cresy, Edward (1823–70). Principal assistant clerk at the Metropolitan Board of Works and architect to the fire brigade. (*DNB.*)

Crisp, Edwards (1806–82). Physician. Fellow of the Zoological Society of London. (*Modern English biography.*)

4 April 1857

Crowe, John Rice (1792–1877). British consul-general in Norway, 1843–75. Knighted, 1875. (*Modern English biography.*)

Crump, Charles Wade. Lieutenant in the Madras artillery. (*East-India register and army list, for 1856.*)

[before 8 January 1856]

Cuming, Hugh (1791–1865). Naturalist and traveller. Collected shells and living orchids in the Pacific, on the coast of Chile, and in the Philippine Islands. Returned to England in 1839. (R. Desmond 1977, *DNB.*)

Curtis, John (1791–1862). Entomologist and artist. Executed engravings for many eminent naturalists. President of the Entomological Society, 1855. (*DNB.*)

Cuvier, Georges (1769–1832). French systematist, comparative anatomist, palaeontologist, and administrator. Professor of natural history, Collège de France, 1800–32; professor of comparative anatomy, Muséum d'histoire naturelle, 1802–32. Permanent secretary of the Académie des sciences from 1803. Foreign member, Royal Society, 1806. (*DBF, DSB.*)

Dallas, William Sweetland (1824–1900). Clerk in the City of London. Prepared lists of insects for the British Museum, 1847–58. Curator of the Yorkshire Philosophical Society's museum, 1858–68. Assistant secretary to the Geological Society of London, 1868–1900. Translated German works for CD and prepared the index for *Variation* and the glossary for the 6th edition of *Origin*. (Freeman 1978, *Modern English biography*, Sarjeant 1980.)

Dana, James Dwight (1813–95). American geologist and zoologist. Naturalist to the United States Exploring Expedition to the Pacific, 1838–42; described the geology and the zoophytes and Crustacea from the voyage. An editor of the *American Journal of Science and Arts* from 1846. Appointed professor of geology, Yale University, 1849; assumed duties, 1856–90. Foreign member, Royal Society, 1884. (*DAB, DSB.*)

14 July [1856], *8 September 1856*, 29 September [1856], *8 December 1856*, 5 April [1857], *27 April 1857*, 25 May [1857]

Daniell, William Freeman (1818–65). Surgeon and botanist. Assistant surgeon to the British army in West Africa, 1841–53. Sent plants to the Royal Botanic Gardens, Kew. (R. Desmond 1977, *DNB.*)

8 October – 7 November 1856, 14 November 1856

Darwin, Anne Elizabeth (Annie) (1841–51). CD's oldest daughter. (*Darwin pedigree.*)

Darwin, Catherine. *See* Darwin, Emily Catherine.

Darwin, Charles Waring (1856–8). Youngest child of CD. Died of scarlet fever. (*Darwin pedigree.*)

Darwin, Elizabeth (1847–1926). CD's daughter. (*Darwin pedigree.*)

Darwin, Emily Catherine (1810–66). CD's sister. Resided at The Mount, Shrewsbury, until she became Charles Langton's second wife in 1863. (*Darwin pedigree.*)

Darwin, Emma (1808–96). Youngest daughter of Josiah Wedgwood II. Married CD, her cousin, in 1839. (*Emma Darwin.*)

Darwin, Erasmus (1731–1802). Physician, botanist, and poet. Advanced an evolutionary theory similar to that subsequently expounded by Lamarck. CD's grandfather. FRS 1761. (*DNB, DSB.*)

Darwin, Erasmus Alvey (1804–81). CD's brother. Attended Shrewsbury School, 1815–22. Matriculated Christ's College, Cambridge, 1822; MB, 1828. At Edinburgh University, 1825–6. Qualified but never practised as a physician. Lived in London from 1829 to his death. (*Alum. Cantab.*)

Darwin, Francis (Frank) (1848–1925). CD's son. BA, Trinity College, Cambridge, 1870. Collaborated with CD on several botanical projects, 1875–82. Lecturer in botany, Cambridge University, 1884; reader, 1888–1904. Edited CD's letters. FRS 1882. (*DNB, DSB.*)

Darwin, George Howard (1845–1912). CD's son. BA, Trinity College, Cambridge, 1868; fellow, 1868–78. Called to the bar in 1872 but did not practise law. Plumian professor of astronomy and experimental philosophy, Cambridge University, 1883–1912. FRS 1879. (*DNB, DSB.*)

13 [November 1856]

Darwin, Henrietta Emma (1843–1927). CD's daughter. Married Richard Buckley Litchfield (*Alum. Cantab.*) in 1871. (*Burke's landed gentry* 1952.)

[2 August 1857]

Darwin, Horace (1851–1928). CD's son. BA, Trinity College, Cambridge, 1874. Apprenticed to an engineering firm in Kent before returning to Cambridge in 1875 to design and make scientific instruments. Formed the Cambridge Scientific Instrument Company in 1885 and served as chairman. Mayor of Cambridge, 1896–7. FRS 1903. (*Alum. Cantab., DNB.*)

Darwin, Leonard (1850–1943). CD's son. Attended the Royal Military Academy, Woolwich; entered the Royal Engineers, 1871; major, 1889. Served on several scientific expeditions, including those for the observation of the transit of Venus, 1874 and 1882. Instructor in chemistry, School of Military Engineering, Chatham, 1877–82. Staff, Intelligence Department, War Office, 1885–90. Liberal Unionist MP, Lichfield division of Staffordshire, 1892–5. President, Royal Geographical Society, 1908–11; Eugenics Education Society, 1911–28. Chairman, Bedford College, London University, 1913–20. (Sarjeant 1980, *WWW*.)

Darwin, Marianne. *See* Parker, Marianne.

Darwin, Robert Waring (1766–1848). Physician. MD, Leiden, 1785. Had a large practice in Shrewsbury and resided at The Mount, which he built *c.* 1796–8. Third son of Erasmus Darwin by his first wife, Mary Howard. Married Susannah, daughter of Josiah Wedgwood I, in 1796. CD's father. FRS 1788. (Freeman 1978.)

Darwin, Susan Elizabeth (1803–66). CD's sister. Lived at The Mount, Shrewsbury, until her death. (*Darwin pedigree.*)

Darwin, Susannah (1765–1817). Daughter of Josiah Wedgwood I. Married Robert Waring Darwin in 1796. CD's mother. (*Darwin pedigree.*)

Darwin, William Erasmus (1839–1914). CD's oldest child. Attended Rugby School. BA, Christ's College, Cambridge, 1862. Banker in Southampton. (*Alum. Cantab.*)

[26 February 1856], 13 [November 1856], 25 [November 1856],
10 [December 1856], [17 February 1857], 13 May [1857],
21 [July 1857], [before 11 September 1857], [before 29 October 1857],
29 [October 1857], [November 1857]

Davidson, Thomas William St Clair (1817–85). Artist and palaeontologist. Expert on fossil brachiopods. FRS 1857. (*DNB*, Sarjeant 1980.)

23 December [1856], *29 December 1856*

Davy, Humphry (1778–1829). Professor of chemistry at the Royal Institution, 1802–13. President of the Royal Society, 1820–7. FRS 1803. (*DNB*, *DSB*.)

Davy, John (1790–1868). Physiologist and anatomist. Brother of Humphry Davy. FRS 1834. (*DNB*.)

10 January 1856

Dawson, John Frederick (1803–70). Rector of St Peter's and vicar of All Saints', Toynton, Lincolnshire, 1827–70. Resided at the Woodlands, near Bedford. (*Alum. Cantab.*)

Decaisne, Joseph (1807–82). French botanist at the Jardin des plantes, Paris, 1824. Professor of statistical agriculture, Collège de France, 1845. Foreign member, Royal Society, 1877. (*DBF.*)

Demas, Don Sinbaldo. *See* Mas y Sans, Sinbaldo de.

Derby, Earl of. *See* Stanley, Edward Smith.

Deshayes, Gérard Paul (1797–1875). French palaeontologist and conchologist. Collected and studied living and fossil molluscs. Prepared a catalogue (1853–4) of the bivalve molluscs in the British Museum. (*DBF, DSB.*)

Desjardins, Julien François (1799–1840). Born in Mauritius, he studied in Paris at the Muséum d'histoire naturelle. He returned to Mauritius and established a colonial museum in 1826 and a natural history society in 1829, to which he served as secretary. (*DBF.*)

Dickey, Edward John (1804–79). Army officer in India. Superintendent of the stud department of the Bengal Army from 1853. (*Modern English biography.*)

Dickie, George (1812–82). Scottish botanist. Professor of natural history, Belfast, 1849–60. Professor of botany, Aberdeen, 1860–77. Specialised in the study of marine algae and described many of the collections sent to Kew Gardens. FRS 1881. (R. Desmond 1977, *DNB.*)

1 December 1856

Dieffenbach, Ernst (1811–55). German physician, naturalist, and geologist. Surgeon and naturalist to the New Zealand Company, 1839–41. Supernumerary professor of geology at Giessen, 1850–5. Translated CD's *Journal of researches* into German (1844). (*DNZB*, Sarjeant 1980.)

Dixon, Edmund Saul (1809–93). Rector of Intwood with Keswick, Norfolk, 1842–93. Author of books on the history and management of poultry. Published under the pseudonym Eugene Sebastian Delamer from 1854. (*Alum. Cantab.*, *Modern English biography.*)

Don, David (1800–41). Professor of botany, King's College, London, 1836–41. Linnean Society librarian, 1822–41. (R. Desmond 1977, *DNB.*)

Doubleday, Henry (1808–75). Quaker tradesman in Epping. Naturalist and entomologist noted for his work on the systematics of Lepidoptera. (R. Desmond 1977, *DNB.*)

26 January 1857, [before 5 February 1857], *5 February 1857*

Douglas, George, 16th Earl of Morton, (1761–1827). Reported to the Royal Society in 1821 an apparent case of telegony that became the most famous example of the phenomenon. FRS 1785. (*Burke's peerage* 1980.)

Drège, Johann Franz (Jean François) (1794–1881). German botanical collector, traveller, and horticulturist. Collected extensively in South Africa, 1826–33. (*DSAB.*)

Drysdale, Lady (d. *c.* 1882). Mother-in-law of Edward Wickstead Lane. According to Henrietta Litchfield, she was 'a great reader, a great whist-player, and the active capable housekeeper of the great [hydropathic] establishment.' (*Emma Darwin* (1904), 2: 184.)

Dufossé, Adolphe (1805–77). French doctor who served in a ship sailing between Marseilles and the Levant until 1851. Professor of chemistry at the medical school in Marseilles, 1854. Studied marine organisms. (*DBF.*)

Dureau de la Malle, Adolphe Jules César Auguste (1777–1857). French geographer, naturalist, and historian. (*DBF.*)

Duval. French gardener known as 'jardinier Duval'. Published works on gardening.

Eaton, John Matthews. Author of works on pigeon breeding. (*NUC.*)

Ecklon, Christian Friedrich (1795–1868). Danish botanical collector and apothecary. Collected in South Africa, 1823–8 and 1829–32, with Karl Ludwig Philipp Zeyher, with whom he published a catalogue of South African plants (1835–7). (*DSAB.*)

Edmondston, Laurence (1795–1879). Physician and naturalist. Had a medical practice in Unst, the most northerly of the Shetland Islands. Made numerous additions to the list of British birds. Father of Thomas Edmondston. (*DNB.*)
> *[before 3 May 1856]*, 3 May [1856], 11 September [1856], 19 April [1857], 2 August [1857]

Edmondston, Thomas (1825–46). Botanist. Elected professor of botany and natural history, Andersonian Institution, Glasgow, 1845. Naturalist aboard HMS *Herald*, 1845–6. Accidentally shot in Peru, 1846. (R. Desmond 1977, *DNB.*)

Edwards, George (1694–1773). Naturalist and artist. Librarian of the Royal College of Physicians, 1733. FRS 1757. (*DNB.*)

Egerton, Philip de Malpas Grey- (1806–81). Of Oulton Park, Cheshire. Tory MP for South Cheshire, 1835–68. Palaeontologist who specialised in fossil fish. FRS 1831. (*DNB*, Sarjeant 1980.)

Ehrenberg, Christian Gottfried (1795–1876). German naturalist, microscopist, and traveller. Studied the development of coral reefs and worked extensively on the Infusoria. Foreign member, Royal Society, 1837. (*DSB*, *NDB.*)

Eights, James. American naturalist and geologist. Collected specimens in the South Shetland Islands near Antarctica. (*NUC.*)

Élie de Beaumont, Jean Baptiste Armand Louis Léonce (1798–1874). French mining engineer and geologist. Propounded a catastrophist theory of mountain elevation. Foreign member, Royal Society, 1835. (*DBF, DSB.*)

Elliot, Walter (1803–87). Indian civil servant and archaeologist. Commissioner for the administration of the Northern Circars, 1845–54. Member of the council of the governor of Madras, 1854–60. Wrote articles on Indian natural history and culture. FRS 1878. (*DNB.*)
> 23 January 1856

Engelmann, George (1809–84). German botanist, physician, and meteorologist. Emigrated to America in 1832. Settled in Illinois where he prospected and botanised. Established a successful medical practice in 1835 and continued his botanical researches. (*DAB.*)

Etheridge, Robert (1819–1903). Palaeontologist. Curator of the museum of the British Philosophical Institution, 1850–7. Palaeontologist to the Geological Survey, 1863. Assistant keeper in geology at the British Museum (Natural History), 1881–91. FRS 1871. (*DNB.*)

Everest, Robert (b. 1799). Clergyman and geologist in India. (*Alum. Oxon., Clergy list.*)
> 18 June [1856]

Eyton, Elizabeth Frances (d. 1870). Wife of Thomas Campbell Eyton.

Eyton, Thomas Campbell (1809–80). Shropshire naturalist. Friend and Cambridge contemporary of CD. Author of several works on natural history. On coming into possession of the family estate at Eyton, Shropshire, he built a museum in which he formed one of the finest collections of skins and skeletons of the birds of Europe. (*DNB*.)

21 August [1856], 27 [August 1856], 31 August [1856], 5 October [1856], 9 June [1857], 26 [June 1857], 22 [July 1857], 2 November [1857]

Fabre, Esprit. French botanist and agricultural writer. (*Catalogue de la Bibliothèque nationale.*)

Fairbeard, William and Julius. Nurserymen at Teynham, Sittingbourne, Kent. (*Post Office directory of the six home counties* 1855.)

Falconer, Hugh (1808–65). Palaeontologist and botanist. Superintendent of the botanic garden, Saharanpur, India, 1832–42. Superintended the arrangement of Indian fossils for the British Museum, 1844. Superintendent of the botanic garden, Calcutta, and professor of botany, Calcutta Medical College, 1848–55. FRS 1845. (*DNB, DSB*.)

8 March [1857], 23 November 1857

Faraday, Michael (1791–1867). Apprentice to a bookbinder, 1804. Attended scientific lectures, including some given by Humphry Davy. Appealed to Davy for patronage and became his assistant at the Royal Institution in 1812. Began his own researches in 1816 and contributed extensively to the fields of electrochemistry, magnetism, and electricity. FRS 1824. (*DNB, DSB*.)

Fenwick, Ralph (d. 1873). Firework maker in Lambeth from 1839. Supplied the fireworks on the occasion of Queen Victoria's accession, 1837, and at the end of the Crimean War, 1856. Killed, with seven others, by an explosion at his premises. (*Modern English biography*.)

Férussac, Jean Baptiste Louis d'Audebard, Baron de (1745–1815). French geologist and palaeontologist. Carried out work on molluscs and crustaceans which was published by his son André. (*DBF*, Sarjeant 1980.)

Fisher, Francis (1810–63). Fellow of Jesus College, Cambridge. (*Alum. Cantab.*)

Fitton, William Henry (1780–1861). Physician and geologist. President of the Geological Society, 1827–9; vice-president, 1831–46. FRS 1815. (*DNB, DSB*.)

FitzRoy, Maria. Second wife of Robert FitzRoy.

FitzRoy, Mary Henrietta (d. 1852). Married Robert FitzRoy in 1836. (*Burke's peerage* 1980.)

FitzRoy, Robert (1805–65). Naval officer, hydrographer, and meteorologist. Commander of the *Beagle*, 1828–36. Tory MP for Durham, 1841–3. Governor of New Zealand, 1843–5. Superintendent of the dockyard at Woolwich, 1848–50. Chief of the meteorological department of the Board of Trade, 1854. Vice-admiral, 1863. FRS 1851. (*DNB, DSB*.)

Flinders, Matthew (1774–1814). Naval officer, hydrographer, and explorer. Surveyed a large part of the Australian coast. (*Aust. dict. biog., DNB*.)

Forbes, David (1828–76). Geologist and philologist. Travelled in South America, 1857–60, in search of ores of nickel and cobalt for the firm of Evans & Askins, nickel-smelters of Birmingham. FRS 1856. (*DNB.*)

Forbes, Edward (1815–54). Zoologist, botanist, and palaeontologist. Naturalist on board HMS *Beacon*, 1841–2. Professor of botany, King's College, London, 1842. Palaeontologist with the Geological Survey, 1844–54. Professor of natural history, Edinburgh University, 1854. FRS 1845. (*DNB, DSB.*)

Forbes, James David (1809–68). Professor of natural philosophy, Edinburgh University, 1833–60. Secretary of the Royal Society of Edinburgh, 1840–51. Principal, University of St Andrews, 1860–8. FRS 1832. (*DNB, DSB.*)

Forster, Johann Reinhold (1729–98). German traveller, naturalist, and geographer. Accompanied James Cook on his circumnavigation of the globe, 1772–5. Foreign member, Royal Society, 1772. (*DSB, NDB.*)

Fox, Edith Darwin (b. 1857). Daughter of Ellen Sophia and William Darwin Fox. (*Darwin pedigree.*)

Fox, Ellen Sophia (1820–87). Second wife of William Darwin Fox, 1846. (*Darwin pedigree.*)

Fox, Frederick William (b. 1855). Son of Ellen Sophia and William Darwin Fox. (*Darwin pedigree.*)

Fox, Henry Stephen (1791–1846). Diplomat. Minister plenipotentiary to Buenos Aires, 1831–2; Rio de Janeiro, 1833–6; Washington, D.C., 1836–44. Uncle of Charles James Fox Bunbury. Formed a herbarium at Rio de Janeiro. (R. Desmond 1977, *DNB.*)

Fox, William Darwin (1805–80). Clergyman. BA, Christ's College, Cambridge, 1829. CD's second cousin. A close friend at Cambridge who shared CD's enthusiasm for entomology. Maintained an active interest in natural history throughout his life and provided CD with much information. Rector of Delamere, Cheshire, 1838–73. Spent the last years of his life at Sandown, Isle of Wight. (*Alum. Cantab.*)

   3 January [1856], *8 March [1856]*, 15 March [1856], 4 June [1856], 8 [June 1856], 14 June [1856], 3 October [1856], 20 October [1856], 8 February [1857], 22 February [1857], [30 April 1857], 30 October [1857], 17 December [1857]

Frankland, Edward (1825–99). Professor of chemistry at Owens College, Manchester, 1851–7. Lecturer in chemistry, St Bartholomew's Hospital, London, 1857–64. FRS 1853. (*DNB, DSB.*)

Franklin, John (1786–1847). Naval officer and Arctic explorer. Lieutenant-governor of Van Diemen's Land (Tasmania), 1837–43. Leader of the 1845 expedition in search of a north-west passage during which all hands perished. FRS 1823. (*DNB.*)

Frémont, John Charles (1813–90). American explorer, politician, and soldier. Served in the United States Topographical Corps and participated in many exploring expeditions. Republican candidate in the 1856 presidential elections in which he was defeated by James Buchanan. (*DAB.*)

Fries, Elias Magnus (1794–1878). Swedish botanist. Professor of botany at Uppsala University, 1835. Foreign member, Royal Society, 1875. (*DSB, SBL.*)

Fröbel, Carl Ferdinand Julius (Julius) (1805–93). German author and politician. Lecturer at the technical school, later university, in Zurich, 1833. (*NDB.*)

Fryer, John (d. 1733). MB, Pembroke College, Cambridge, 1671. Traveller in India and Persia in the interests of the East India Company, 1672–82. FRS 1697. (*DNB.*)

Fürnrohr, August Emanuel (1804–61). German botanist. (*NUC.*)

Gabriel, Edmund (*fl.* 1850s). English commissioner for the suppression of the slave trade and consul for Angola. (Livingstone 1857, pp. 389, 398.)

Gaimard, Joseph Paul (1796–1858). French naval surgeon and naturalist. Participated in many voyages of scientific exploration. A colleague of Jean Quoy. (*DBF, DSB.*)

Gainsborough, Thomas (1727–88). Painter of landscapes and portraits. One of the original members of the Royal Academy of Art, 1768. (*DNB.*)

Galton, Francis (1822–1911). Traveller and scientific writer. Explored unknown regions of south-western Africa, 1850–2. Carried out various statistical inquiries. Founder of the eugenics movement. CD's cousin. FRS 1860. (*DNB, DSB.*)

7 July [1857]

Galton, Louisa Jane (d. 1897). Married Francis Galton in 1853. (*Darwin pedigree.*)

Gardeners' Chronicle.

[before 6 December 1856], [after 28 February 1857],
[before 13 June 1857], [before 25 July 1857], 18 October [1857],
[before 12 November 1857]

Garnett, Thomas (1799–1878). Manufacturer and naturalist. Several times mayor of Clitheroe, Yorkshire. (*DNB.*)

Garrett, James R. (1820–55). Irish solicitor and botanist in County Down. Fern fancier. (R. Desmond 1977.)

Gärtner, Karl Friedrich von (1772–1850). German physician and botanist. From 1824, he devoted his time solely to the study of plant hybridisation. (*DSB, NDB.*)

Gassiot, John Peter (1797–1877). Wine merchant and scientific writer. Chairman of Kew Observatory. Carried out electrical experiments. FRS 1840. (*DNB, DSB.*)

Gay, Claude (1800–73). French naturalist and traveller who surveyed the flora and fauna of Chile. Professor of physics and chemistry at Santiago College, 1828–42. (*DBF.*)

Geoffroy Saint-Hilaire, Isidore (1805–61). French zoologist. Professor of zoology, Muséum d'histoire naturelle, 1841. (*DBF, DSB.*)

Gérard, Frédéric. French writer on botanical and horticultural topics, a follower of Étienne Geoffroy Saint-Hilaire.

Gerber, Friedrich (1797–1872). German anatomist. (*NUC.*)

Gilbert, Henry. Dentist at 3 Suffolk Place, Pall Mall East, London. Pigeon fancier. (*Post Office London directory* 1857.)

Gillies, John (1792–1834). Naval surgeon. Collected plants in South America. (R. Desmond 1977.)

Gisborne, Thomas (1794–1852). Whig politician and essayist. (*DNB.*)

Gladwin, Francis (d. 1813?). Orientalist. Served in the Bengal army. Translated the *Ayeen Akbery; or, the institutes of the Emperor Akber* (1783–6). (*DNB.*)

Glover, Thomas. Of Smedley Hill, Cheetham, near Manchester. Breeder of cactuses. (*Post Office directory of Lancashire* 1858.)
  26 October 1857

Gmelin, Johann Georg (1709–55). Naturalist and explorer. Professor of chemistry and natural history at the Academy of Sciences, St Petersburg, 1731–47. Professor of medicine, botany, and chemistry at Tübingen University, 1749. (*DSB, NDB.*)

Godron, Dominique Alexandre (1807–80). French botanist, zoologist, and ethnologist. Head of the faculty of sciences, Nancy, 1854, where he established a natural history museum and a botanic garden. (*DBF.*)

Godwin-Austen, Robert Alfred Cloyne (1808–84). Geologist. Frequently a member of the council of the Geological Society, 1841–76; secretary, 1843–4 and 1853–4. FRS 1849. (*DNB, DSB.*)

Goodsir, John (1814–67). Curator of the Edinburgh University museum and demonstrator in anatomy, 1843–6. Professor of anatomy at Edinburgh, 1846–67. FRS 1846. (*DNB, DSB.*)

Gosse, Edmund William (1849–1928). Poet and writer. Clark lecturer in English literature, Trinity College, Cambridge, 1885–90. Librarian to the House of Lords, 1904–14. Son of Philip Henry Gosse. (*DNB.*)

Gosse, Philip Henry (1810–88). Naturalist, traveller, and writer. Sent to Jamaica by the British Museum to collect birds and insects, 1844–6. Author of numerous works on natural history and marine biology. FRS 1856. (*DNB.*)
  22 September [1856], 28 September 1856, 27 April [1857]

Gould, Augustus Addison (1805–66). American physician and conchologist. Practised medicine in Massachusetts. Co-author, with Louis Agassiz, of the *Principles of zoology* (1848). In 1846, began describing the shells collected during the United States Exploring Expedition, 1838–42. An active member of the Boston Society of Natural History. (*DAB, DSB.*)

Gould, John (1804–81). Self-taught ornithologist and artist. Taxidermist to the Zoological Society of London, 1826–81. Travelled in Australia, 1838–40. Described the birds collected on the *Beagle* and *Sulphur* expeditions. FRS 1843. (*DNB, DSB.*)

Graba, Karl Julian. German traveller and ornithologist. (*NUC.*)

Graham, Robert (1786–1845). Physician and botanist. Regius professor of botany, Edinburgh University, 1820–45. Physician to the Edinburgh Infirmary. (*DNB.*)

Gray, Asa (1810–88). American botanist. Fischer professor of natural history, Harvard University, 1842–73. Devoted much time to the Harvard botanic garden and herbarium. Foreign member, Royal Society, 1873. (*DAB, DSB.*)

2 May [1856], 14 July [1856], *[early August 1856]*, 24 August [1856], *23 September 1856*, 12 October [1856], *4 November 1856*, 24 November [1856], 1 January [1857], *16 February 1857*, *13 March 1857*, [after 15 March 1857], 9 May [1857], *[c. 24 May 1857]*, *1 June 1857*, 18 June [1857], *7 July 1857*, 20 July [1857], *[August 1857]*, 5 September [1857], 29 November [1857]

Gray, George Robert (1808–72). Zoologist; an expert on insects and birds. Assistant in the zoological department of the British Museum, 1831–72. Brother of John Edward Gray. FRS 1865. (*DNB*, P. Gilbert 1977.)

Gray, John Edward (1800–75). Naturalist. Assistant zoological keeper at the British Museum, 1824; keeper, 1840–74. FRS 1832. (R. Desmond 1977, *DNB.*)

14 January [1856], 19 January [1856], 1 July [1856]

Greg, William Rathbone (1809–81). Essayist who contributed to many of the leading quarterlies. Comptroller of the stationery office, 1864–77. (*DNB.*)

Griffith, Edward (1790–1858). Naturalist and solicitor. Translated Georges Cuvier's *Règne animal* making considerable additions (1827–35). (*DNB.*)

Griffith, Richard (1784–1878). Irish geologist and civil engineer. (*DNB.*)

Griffith, William (1810–45). Botanist. Appointed assistant surgeon to the East India Company, 1832. Travelled extensively in India making natural history collections with the aim of compiling a flora of India. Superintendent of the Calcutta botanic garden and professor of botany at the Calcutta Medical College, 1842–4. Died of hepatitis in Malacca. (*DNB, DSB.*)

Grisebach, August Heinrich Rudolf (1814–79). German botanist. Travelled through the Balkan peninsula and north-western Asia Minor, 1839–40, studying the flora of these regions. Professor of botany, Göttingen University, 1847. (*DSB, NDB.*)

Grove, William Robert (1811–96). Judge and physical scientist. Professor of experimental philosophy, London Institution, 1841–6. An active member of the Royal Society; treasurer and chairman of the executive committee of the Philosophical Club, 1847. FRS 1840. (*DNB, DSB.*)

Gulliver, George (1804–82). Anatomist and physiologist. Surgeon to the Royal Horse Guards. Member of the council of the Royal College of Surgeons, 1852. Hunterian professor of comparative anatomy and physiology, 1861. FRS 1839. (*DNB.*)

20 *January [1856]*

Gully, James Manby (1808–83). Physician. Studied medicine in Paris and Edinburgh. Practised in London, 1830–42. In 1842 he set up a hydropathic establishment in Malvern and became a successful practitioner of the water treatment until his retirement in 1872. His reputation suffered when, in 1876, his affair with Mrs Bravo, who was accused of poisoning her husband, was revealed. (*DNB, Modern English biography*.)

Gurney, John Henry (1819–90). Norfolk financier and ornithologist. MP, King's Lynn, Norfolk, 1854–65. (*Modern English biography*, Stenton 1976.)
2 July 1856

Guyot, Arnold Henri (1807–84). Swiss geographer and glacial geologist who emigrated to the United States in 1848. Studied the distribution of erratic boulders in Switzerland, 1840–7. Professor of physical geography and geology at Princeton, 1854–84, where he founded a museum. (*DAB, DSB*.)

Gwatkin, Edward. Superintendent of the East India Company's stud at Haupper in northern India. Entered the Bengal Army in 1804. Colonel in the 31st Bengal Native Infantry, 1853–5. (*Modern English biography, East-India register and directory 1828*.)

Gyllenhaal, Leonhard (1752–1840). Swedish entomologist; an expert on Coleoptera. (*DSB*.)

Hamilton, William John (1805–67). Geologist. MP for Newport, Isle of Wight, 1841–7. Secretary of the Geological Society of London, 1832–54; President 1854 and 1865. (*DNB*.)

Hancock, Albany (1806–73). Zoologist and palaeontologist in Newcastle upon Tyne. Collaborated with Joshua Alder on the *Monograph of British nudibranchiate Mollusca* (1845–55). Contributed several papers on the boring apparatus of sponges, molluscs, and cirripedes to the Tyneside Naturalists' Field Club, which he had helped to found in 1846. (*DNB*, Sarjeant 1980.)
25 May [1856]

Hanley, Sylvanus Charles Thorp (1819–1900). Conchologist. Author with Edward Forbes of *A history of British Mollusca, and their shells* (1848–53). (*Modern English biography*.)

Harcourt, Edward William Vernon (1825–91). Author of *A sketch of Madeira* (1851). An active member of artillery volunteers from 1862. Sheriff of Oxfordshire, 1875. MP for Oxfordshire, 1878–85; for South Oxfordshire, 1885–6. (*Modern English biography*.)
31 May 1856, 19 August [1856], 23 August [1856]

Hardinge, Henry, Viscount Hardinge of Lahore, (1785–1856). Army officer and Tory politician. Governor-general of India, 1844–7. Field-marshal, 1855. (*DNB*.)

Hardwicke, Thomas (d. 1835). Joined the Bengal Military Establishment in 1778. Rose to the rank of major-general in the Indian Army. Commanded the regiment of Bengal artillery from 1816. (*East-India register* 1835, [Philippart] 1823–6.)

Hardy, Francis. Farmer in Beesby, Lincolnshire. CD's tenant. (*Post Office directory of Lincolnshire* 1849.)

Harris, John Williams (1808–72). Established the first shore whaling station on the east coast of New Zealand. Storekeeper and sheep farmer. Discovered bones of *Dinornis* in 1837 which were described by Richard Owen. (*DNZB*.)

Hartung, Georg (1822?–91). German geologist who in 1852–3 made investigations on Madeiran geology with Charles Lyell. Author of numerous travel books. (Freeman 1978, *NUC*.)

Harvey, William Henry (1811–66). Irish botanist. Colonial treasurer in Cape Town, 1836–42. Keeper of the herbarium, Trinity College, Dublin, 1844. Professor of botany, Trinity College, Dublin, 1856–66. Described the algae specimens collected by CD on the *Beagle* voyage. (*DNB, DSB*.)
   3 January 1857

Haughton, Samuel (1821–97). Clergyman and palaeobotanist. Professor of geology at Dublin University, 1851–81. MD, Dublin, 1862; registrar of the medical school. FRS 1858. (R. Desmond 1977, *DNB*, Sarjeant 1980.)

Haworth, Adrian Hardy (1767–1833). Entomologist, botanist, and author. Formed a collection of 40,000 insects and a large herbarium. (R. Desmond 1977, *DNB*.)

Hayne, Watson Ward. Solicitor at 25 Red Lion Square, Holborn, London. Pigeon fancier and member of the Philoperisteron Society. (*Post Office London directory* 1858.)

Heer, Oswald (1809–83). Swiss botanist, palaeontologist, and entomologist. An expert on Tertiary flora. Professor of botany and entomology and director of the botanical garden in Zurich, 1834. Went to Madeira for his health in 1850. (*DSB, NDB*.)

Heineken, C. (d. 1829). German physician and naturalist who resided in Madeira, 1826–9. (*Zoological Journal* 5 (1830): 191–6.)

Hemmings, Henry. A servant of Sarah Elizabeth (Sarah) Wedgwood. (Freeman 1978.)

Henfrey, Arthur (1819–59). Botanist. An advocate of German physiological botany in Britain. Lecturer on botany, St George's Hospital, London, 1847. Professor of botany, King's College, London, 1854–9. FRS 1852. (*DNB, DSB*.)

Henry, Prince (1394–1460). Third son of John I of Portugal. Navigator who explored the coast of West Africa. (*EB*.)

Henslow, Ann (*c.* 1767–1857). Aunt of John Stevens Henslow. (*Gentleman's Magazine* n.s. 3 (1857) 2: 574.)

Henslow, Frances Stevens (1775–1856). Mother of John Stevens Henslow. (Jenyns 1862.)

Henslow, George (1835–1925). Clergyman and teacher. BA, Christ's College, Cambridge, 1858. Headmaster, Hampton Lucy Grammar School, Warwick, 1861–5; Grammar School, Store Street, London, 1865–72. Lecturer in botany at St Bartholomew's Medical School, 1886–90. Younger son of John Stevens Henslow. (*Alum. Cantab.*, R. Desmond 1977.)

Henslow, Harriet (1797–1857). Daughter of George Leonard Jenyns and sister of Leonard Jenyns. Married John Stevens Henslow in 1823. (*Burke's landed gentry* 1879.)

Henslow, John Stevens (1796–1861). Clergyman, botanist, and mineralogist. Professor of mineralogy, Cambridge University, 1822–7; professor of botany, 1825–61. Extended and remodelled the Cambridge botanic garden. Curate of Little St Mary's Church, Cambridge, 1824–32; vicar of Cholsey-cum-Moulsford, Berkshire, 1832–7; rector of Hitcham, Suffolk, 1837–61. CD's teacher and friend. (*DNB, DSB.*)

> 3 January [1856], 22 January [1856], 16 June [1856], *2 August 1856*, 6 August [1856], [after 6 December 1856], 10 August [1857], 25 September [1857], 14 October [1857], 18 October [1857]

Herbert, John Maurice (1808–82). BA, St John's College, Cambridge, 1830; fellow, 1832–40. Barrister, 1835. County court judge, South Wales, 1847–82. (*Alum. Cantab., Modern English biography.*)

> 2 January [1856]

Herbert, Mary Anne. Poetess. Wife of John Maurice Herbert.

Herbert, William (1778–1847). Naturalist, classical scholar, linguist, politician, and clergyman. Noted for his work on plant hybridisation. Rector of Spofforth, Yorkshire, 1814–40. Dean of Manchester, 1840–7. (*DNB, DSB.*)

Herring, William. Bird and animal dealer at 34 FitzRoy Terrace, New Road, London. (*Post Office London directory* 1855.)

Herschel, Alexander Stewart (1836–1907). Son of John Frederick William Herschel. Attended Clapham Grammar School, London, 1851–4. BA, Trinity College, Cambridge, 1859. Professor of physics at Glasgow University, 1868–71; University of Durham College of Science, Newcastle upon Tyne, 1871–86. (*Alum. Cantab., DNB.*)

Herschel, John Frederick William (1792–1871). Astronomer, mathematician, chemist, and philosopher. Member of many learned societies. President of the Royal Society, 1827–9, 1839–41, 1847–9. Carried out astronomical observations at the Cape of Good Hope, 1834–8. Master of the mint, 1850–5. Created baronet, 1838. FRS 1813. (*DNB, DSB.*)

Hewitt, Edward. Judge of poultry exhibitions. Resided in Sparkbrook, Birmingham. (Tegetmeier ed. 1856–7, p. 89.)

> *18 December 1857, 22 December 1857*

Hewson, William (1739–74). Surgeon and physiologist. FRS 1770. (*DNB, DSB.*)

Higgins, John (1796–1872). Land agent. Born in Shrewsbury but settled in Alford, Lincolnshire in 1819. Agent for Robert Waring Darwin's and Susan Darwin's estates in that county and for CD's farm at Beesby.

> 19 November [1856]

Hill, Richard (1795–1872). Jamaican-born magistrate and planter who was educated in England. Active in the anti-slavery movement. Naturalist who specialised in the ornithology of Jamaica. Assisted Philip Henry Gosse in the preparation of works on the natural history of the island.

> *10 January 1857, 12 March 1857*

Hill, Rowland, 2d Viscount (1800–75). MP for Shropshire, 1821–32; North Shropshire, 1832–42. Lord lieutenant of Shropshire, 1845–75. (*Alum. Oxon., Complete peerage.*)

Hincks, Thomas (1818–99). Unitarian minister in Leeds, 1855–69. Zoologist who studied Polyzoa. FRS 1872. (*DNB.*)

Hodgson, Brian Houghton (1800–94). Orientalist and ethnologist. In the service of the East India Company from 1818; assistant resident in Nepal, 1820; resident, 1833–43. Acquired an important collection of Sanskrit and Tibetan manuscripts. Wrote extensively on the geography, ethnography, and natural history of India and the Himalayas. Lived in Darjeeling, 1845–58, thereafter settled in England. FRS 1877. (*DNB.*)

Holland, Edward (1806–75). CD's second cousin. BA, Trinity College, Cambridge, 1829. Liberal MP for East Worcestershire, 1835–7; Evesham, 1855–68. President of the Royal Agricultural Society. (*Alum. Cantab.*, Stenton 1976.)

Holland, Henry (1788–1873). Physician. Distant cousin of the Darwins and Wedgwoods. Travelled extensively. Physician in ordinary to Queen Victoria, 1852. President of the Royal Institution. Created baronet, 1853. FRS 1815. (*DNB, Physicians.*)

Holland, Miss.
  [May 1856]

Homer. Epic poet of Greece. (*EB.*)

Hooker, Frances Harriet (1825–74). Daughter of John Stevens Henslow. Married Joseph Dalton Hooker in 1851. (R. Desmond 1977.)

Hooker, Joseph Dalton (1817–1911). Botanist. Accompanied James Clark Ross on the Antarctic expedition, 1838–43, and published the botanical results of the voyage. Botanist to the Geological Survey, 1845. Travelled in the Himalayas, 1848–50. Assistant director, Royal Botanic Gardens, Kew, 1855–65; director, 1865–85. Worked chiefly on taxonomy and plant geography. Son of William Jackson Hooker. Friend and confidant of CD. FRS 1847. (*DNB, DSB.*)
  8 April [1856], *7 May 1856*, 9 May [1856], 11 May [1856],
  21 [May 1856], 1 June [1856], 17–18 [June 1856], 22 June [1856],
  [26 June or 3 July 1856], 5 [July 1856], 5 July [1856], 8 [July 1856],
  10 July 1856, 13 July [1856], 19 July [1856], 26 [July 1856],
  30 July [1856], *4 August 1856*, 5 August [1856], 7 August [1856],
  8 September [1856], 28 September [1856], 5 October [1856],
  9 October [1856], [16 October 1856], [19 October 1856],
  9 November 1856, 11–12 November [1856], 15 November [1856],
  [16 November 1856], 18 November [1856], *22 November 1856*,
  23 November [1856], [early December 1856], *[early December 1856]*,
  1 December [1856], *7 December 1856*, 10 December [1856],
  24 December [1856], 17 January [1857], 20 January [1857],

[after 20 January 1857], 15 March [1857], [21 March 1857], 8 April [1857], *[11 April 1857]*, 12 April [1857], [29 April 1857], [2 May 1857], [3 May 1857], 16 [May 1857], [June 1857], 2 June [1857], 3 June [1857], 5 June [1857], 25 June [1857], *[27] June 1857*, 1 July [1857], 5 July [1857], 14 July [1857], 1 August [1857], 22 August [1857], 6 September [1857], 11 September [1857], 30 September [1857], 20 October [1857], [23 October 1857], 14 [November 1857], 21 November [1857], *[2 December 1857]*, 4 December [1857], *[6 December 1857]*, 9 December [1857], *[17–23 December 1857]*, 25 December [1857]

Hooker, Marie Elizabeth (b. 1857). Daughter of Frances Harriet and Joseph Dalton Hooker.

Hooker, William Jackson (1785–1865). Botanist. Regius professor of botany, Glasgow University, 1820. Established the Royal Botanic Gardens at Kew, 1841, and served as its first director. Father of Joseph Dalton Hooker. FRS 1812. (*DNB, DSB.*)

Hooper, Robert (1773–1835). Medical writer. Practised medicine in Savile Row, London, specialising in pathology. (*DNB.*)

Hopkins, William (1793–1866). Mathematician and geologist. A highly successful mathematics tutor at Cambridge. President of the Geological Society, 1851–3. Specialised in quantitative studies of geological and geophysical questions. FRS 1837. (*DNB, DSB.*)

Horace [Quintus Horatius Flaccus] (65–8 B.C.). Roman poet. (*EB.*)

Hordern, Ellen Frances (d. 1879). Married John Lubbock in 1856. (Freeman 1978.)

Hordern, Peter (1797–1836). Perpetual curate of Chorlton-cum-Hardy, Staffordshire, 1816. (*Alum. Oxon.*)

Hornby, William Windham (b. 1812). Captain in the Royal Navy. Noted breeder of Dorking fowl. (O'Byrne 1849.)

Horner, Anne Susan (1789–1862). Married Leonard Horner in 1806. (Freeman 1978.)

Horner, Leonard (1785–1864). Geologist and educationist. Warden of University College London, 1827–31. Inspector of factories, 1833–56, and a promoter of a science-based education at all social levels. President of the Geological Society, 1845–7 and 1860–2. Father-in-law of Charles Lyell and Charles James Fox Bunbury. FRS 1813. (*DNB, DSB.*)

Hornschuch, Christian Friedrich (1793–1850). German naturalist. Extraordinary, later ordinary, professor of natural history and botany and director of the botanic garden, University of Greifswald, 1820. (*ADB.*)

Horsfield, Thomas (1773–1859). American naturalist. Served the Dutch government in Java and Sumatra, 1799–1811, after which he transferred his services to the British and was sent to Bangka to study the natural history of the island. Keeper of the museum of the East India Company, Leadenhall Street, London, 1820–59. (R. Desmond 1977, *DNB.*)

Hotham, Beaumont, 3d Baron (1794–1870). Army officer. Tory MP for Leominster, 1820–41; East Riding, Yorkshire, 1841–68. (*DNB*.)

Hudson, William (1733–93). Botanist. Sub-librarian at the British Museum, 1757–8. Author of *Flora Anglica* (1762). Director and botanical demonstrator at the Apothecaries Garden, Chelsea, 1765–71. FRS 1761. (*DNB, DSB*.)

Humboldt, Friedrich Wilhelm Heinrich Alexander (Alexander) von (1769–1859). Eminent Prussian naturalist and traveller. Official in the Prussian mining service, 1792–6. Explored equatorial South America, 1799–1804. Travelled in Siberia, 1829. Foreign member, Royal Society, 1815. (*DSB, NDB*.)

Hummel, Arvid David (d. 1836). Russian entomologist. (P. Gilbert 1977.)

Hunt, James. Head keeper at the Zoological Society's gardens.
  [before 29 December 1857]

Hunt, Thomas Carew (d. 1886). Consul at Archangel, 1832; Azores, 1839–48; Bordeaux, 1866. Collected plants for the Botanical Society of London. (R. Desmond 1977.)

Hutton, Thomas. Army officer and naturalist in India. Author of works on natural history and scriptural geology.
  8 March 1856

Huxley, Henrietta Anne (Nettie) (1825–1915). Lived in Australia, where she met Thomas Henry Huxley in 1847. Came to England and married Huxley in 1855. (Freeman 1978.)

Huxley, Leonard (1860–1933). Son of Thomas Henry Huxley. Biographer. Assistant editor of the *Cornhill Magazine*, 1901; editor, 1916–33. (*DNB*.)

Huxley, Noel (1856–60). First son of Henrietta and Thomas Henry Huxley. Died of scarlet fever. (L. Huxley ed. 1900.)

Huxley, Thomas Henry (1825–95). MB, London, 1845. Assistant surgeon in HMS *Rattlesnake*, 1846–50, when he investigated hydrozoa; wrote up results of voyage, 1850–4. Lecturer on natural history, Royal School of Mines, 1854; naturalist to the Geological Survey, 1855. Served on several Royal commissions. Hunterian professor of comparative anatomy, Royal College of Surgeons, 1863–9; Fullerian professor of physiology, Royal Institution, 1863–7. President of the Royal Society, 1883–5. FRS 1851. (*DNB, DSB*.)
  2 April [1856], 9 April [1856], 4 May [1856], 27 May [1856],
  1 July [1856], 8 July [1856], 9 December [1856], 13 [December 1856],
  4 January [1857], 17 January [1857], 3 February [1857],
  5 July [1857], *7 July 1857*, 9 July [1857], 15 September [1857],
  26 September [1857], *[before 3 October 1857]*, 3 October [1857],
  [before 12 November 1857], 16 December [1857]

Innes, John (1817–94). BA, Trinity College, Oxford, 1839; MA, 1842. Curate of Farnborough, Kent, 1842. Perpetual curate of Down, later vicar, 1846–69. Left Down in 1862 after inheriting property in Scotland; changed his name to Brodie Innes *c*. 1860. (J. R. Moore 1985.)
  [after 16 February 1857]

Janson, Edward Westley (1822–91). Businessman and entomologist. Authority on British Coleoptera. Curator of the Entomological Society, 1850–63; librarian, 1863–74. Natural history agent, publisher, and bookseller, 1867. (*Entomologist* 24 (1891): 252, P. Gilbert 1977.)

Jardine, William, 7th Baronet (1800–74). Naturalist. A founder of the *Annals and Magazine of Natural History*, 1841. Commissioner on salmon fisheries of England and Wales, 1860. FRS 1860. (*DNB*.)

Jenyns, George Leonard (1763–1848). Vicar of Swaffham Prior, Cambridgeshire, 1787–1848; prebendary of Ely, 1802–48. Inherited Bottisham Hall, Cambridgeshire, from his second cousin in 1787. (*Alum. Cantab.*)

Jenyns, Leonard (1800–93). Naturalist and clergyman. Brother-in-law of John Stevens Henslow. Vicar of Swaffham Bulbeck, Cambridgeshire, 1828–49. Moved to South Stoke, near Bath, 1850; Swainswick, near Bath, 1852; settled in Bath, 1860. Founded the Bath Natural History and Antiquarian Field Club, 1855. Member of many scientific societies. Described the *Beagle* fish specimens. Adopted the name Blomefield in 1871. (R. Desmond 1977, *DNB*.)

Jerdon, Thomas Claverhill (1811–72). Zoologist. Assistant surgeon in the Madras service, 1835–64. Author of works on the birds and mammals of India. (*DNB*.)

John I (1357–1433). King of Portugal. (*EB*.)

Johnson, George William (1802–86). Writer on gardening. Professor of moral and political economy, Calcutta, 1839–42. Founded the *Cottage Gardener* in 1848. (R. Desmond 1977, *DNB*.)

Johnston, Alexander Keith (1804–71). Geographer. Published numerous atlases and maps. Published the first English atlas of physical geography, 1848. Constructed the first globe of physical geography for the Great Exhibition, 1851. (*DNB*.)

Jones, William (1746–94). Orientalist and jurist. Judge of the high court in Calcutta, 1783–94. Founded the Asiatic Society of Bengal, 1784. FRS 1772. (*DNB*.)

Jordan, (Claude Thomas) Alexis (1814–97). French botanist. (*DSB*.)

Kamandáki (Kamandaki Pandita). Indian author of Sanskrit treatise on political economy. (*NUC*.)

Kelaart, Edward Frederick (*c.* 1818–60). Physician and botanist. Born in Ceylon and returned there in 1841 and 1849. Served in the Ceylon Medical Service. In Gibraltar, 1843–5. An active member of the Ceylon branch of the Royal Asiatic Society. (R. Desmond 1977.)

Kennedy, Lord David. Lieutenant in the 1st regiment light infantry of Madras. (*East-India register and army list, for 1848.*)

Kinahan, John Robert (1828–63). Irish physician. Professor of natural history in the department of science and art, Museum of Industry, Dublin. (*Modern English biography.*)

Kippist, Richard (1812–82). Botanist. Librarian of the Linnean Society, 1842–81. (R. Desmond 1977, *DNB*.)
  23 February [1857]
Knip, Pauline (de Courcelles) (1781–1851). French artist. (*NUC*.)
Koch, Wilhelm Daniel Joseph (1771–1849). German botanist. (*ADB*.)
Kölreuter, Joseph Gottlieb (1733–1806). German botanist. Professor of natural history and director of the botanic gardens in Karlsruhe. Carried out hybridisation experiments on plants. (*DSB, NDB*.)
Krauss, Christian Ferdinand Friedrich von (1812–90). German naturalist and collector who travelled in South Africa, 1838–40. Joined the staff of the natural history museum in Stuttgart, 1840; director, 1856. (Sarjeant 1980, Tabler 1977.)
Krohn, August David (1803–91). Russian invertebrate anatomist and embryologist who travelled extensively. (Blyakher 1982.)
Kynnersley, Mary (d. 1886). Née Mary Sneyd. Married her cousin, Clements John Sneyd-Kynnersley of Walton House, in 1830. (*Burke's landed gentry* 1952.)
Lacaze-Duthiers, Félix Joseph Henri de (1821–1901). French invertebrate zoologist. Assistant to Henri Milne-Edwards. Professor of zoology in Lille, 1854–64; at the Muséum d'histoire naturelle, Paris, 1856–9; at the faculty of sciences, Paris, from 1869. (*DSB*.)
Lacordaire, Jean Théodore (1801–70). French traveller and naturalist. Professor of zoology, University of Liège, Belgium, 1835. (P. Gilbert 1977, *NBU*.)
Læstadius, Lars Levi (1800–61). Swedish botanist. (*SBL*.)
Lamarck, Jean Baptiste Pierre Antoine de Monet de (1744–1829). French naturalist. Held various botanical posts at the Jardin du roi, 1788–93. Professor of zoology, Muséum d'histoire naturelle, 1793. Believed in spontaneous generation and the progressive development of animal types; propounded a theory of transmutation. (*DSB*.)
Lane, Edward Wickstead (1823–89). Proprietor of a hydropathic establishment at Moor Park, near Farnham, Surrey. (Freeman 1978.)
Lange, Johan Martin Christian (1818–98). Danish botanist. (*DBL*.)
Langton, Charles (1801–86). Rector of Onibury, Shropshire, 1832–40. Left the Church of England in 1840. Resided at Maer, Staffordshire, 1840–6; Hartfield Grove, Sussex, 1847–62. Married Charlotte Wedgwood in 1832. After her death, he married Emily Catherine Darwin in 1863. (*Alum. Oxon.*, *Emma Darwin*.)
Langton, Charlotte (1797–1862). Emma Darwin's sister. Married Charles Langton in 1832. Resided at Maer, Staffordshire, 1840–6; Hartfield Grove, Sussex, 1847–61. (*Emma Darwin*.)
Langton, Edmund (1841–75). Son of Charles and Charlotte Langton. BA, Trinity College, Cambridge, 1864. Admitted at Lincoln's Inn, 1864. (*Alum. Cantab.*, *Emma Darwin*.)

La Saussaye, Jean François de Paule Louis Petit de (1801–78). French antiquary who devoted himself to the study of numismatology and archaeology. (*NBU.*)

Laslett, Isaac. Bricklayer in Down, Kent. (*Post Office directory of the six home counties* 1845.)

Latham, John (1740–1837). Physician and ornithologist. A founder of the Linnean Society, 1788. FRS 1775. (*DNB.*)

Layard, Edgar Leopold (1824–1900). Served in the Ceylon civil service, 1846–54; in the Cape of Good Hope civil service, 1855–70. Studied the birds and shells of Ceylon. Founded the South African museum. Published on the birds of South Africa (1867). (*Modern English biography, DSAB.*)
    8 June [1856], *[September–October 1856]*

Leadbeater, John. Bird dealer and taxidermist at 19 Brewer Street, Golden Square West, London. Ornithologist to Queen Victoria. (*Post Office London directory* 1858.)

LeConte, John Lawrence (1825–83). American entomologist and physician. Contributed to the subject of geographical distribution. (*DAB.*)

Lecoq, Henri (1802–71). French naturalist. Professor of natural history and director of the botanic garden and mineralogical cabinet at Clermont-Ferand. Published many geological and botanical works. (*NBU*, Sarjeant 1980.)

Ledebour, Karl Friedrich von (1786–1851). German botanist. Assistant professor of natural history and director of the botanic gardens, Dorpat, 1812–36. Undertook a botanical excursion in Siberia, 1826–7. (*NDB.*)

Lemann, Charles Morgan (1806–52). Botanist and physician. Collected plants on Madeira, 1837–8, and Gibraltar, 1840–1. Retired from his medical practice in 1845. (*Alum. Cantab.*, R. Desmond 1977.)

Lewes, George Henry (1817–78). Writer. Studied physiology from 1855 and published several works on the subject. Common law husband of Mary Ann Evans (George Eliot). (*DNB.*)

Lewis, John. Carpenter in Down, Kent. (*Post Office directory of the six home counties* 1845.)

Lichtenstein, Martin Heinrich Karl (1780–1857). German naturalist and traveller. Visited the Cape Colony, 1804–6, and published an account of his travels (1810–11). Professor of zoology in Berlin, 1811. Director of the university zoological museum, 1813. (*NDB.*)

Liénard, Elizée (1808–69?). Notary and naturalist of Mauritius. Studied the fish, annelids, and mollusca of the island. Member of the Natural History Society of Mauritius. (*Dictionary of Mauritian biography.*)

Lightfoot, John (1735–88). Botanist and clergyman. FRS 1781. (R. Desmond 1977, *DNB.*)

Lindley, John (1799–1865). Botanist and horticulturist. Assistant secretary to the Horticultural Society, 1822–41; vice-secretary, 1841–58. Professor of botany at London University (later University College London), 1828–60. Editor of the *Gardeners' Chronicle* from 1841. FRS 1828. (*DNB, DSB.*)

Linley, Elizabeth Ann (1754–92). Singer. Married Richard Brinsley Sheridan in 1772. Sister of Maria Linley. (*DNB*.)

Linley, Maria (d. 1784). Singer in Bath concerts and oratorios. (*DNB*.)

Linnaeus (Carl von Linné) (1707–78). Swedish botanist and zoologist. Proposed a system for the classification of the natural world and reformed scientific nomenclature. (*DSB*.)

Litchfield, Henrietta Emma. *See* Darwin, Henrietta Emma.

Littleton, Adam (1627–94). Lexicographer. Chaplain to Charles II, 1670. Headmaster of Westminster School. Author of a Latin dictionary. (*DNB*.)

Livingstone, David (1813–73). Explorer and missionary in Africa, 1841–56. Published an account of his travels in 1857. Consul on the east coast of Africa and commander of an expedition to explore east and central Africa, 1858–64. Returned to explore the Nile basin, 1866–73. FRS 1858. (*DNB*.)

Lobb, William (1809–64). Plant collector for the Veitch Nursery in Brazil, Chile, Patagonia, Peru, Ecuador, and Colombia, 1840–8, and in Calfornia and Oregon, 1849–57. (R. Desmond 1977.)

Logan, Robert Francis (1827–87). Scottish entomologist, specialising in Lepidoptera. (P. Gilbert 1977.)

Loiseleur Deslongchamps, Jean Louis Auguste (1775–1849). French botanist and physician. Studied the medicinal qualities of plants. (*NBU*.)

Loudon, John Claudius (1783–1843). Landscape gardener and horticultural writer. Travelled in northern Europe, 1813–15; in France and Italy, 1819–20. Founded and edited the *Gardener's Magazine*, 1826–43, and *Magazine of Natural History*, 1828–36. (R. Desmond 1977, *DNB*.)

Lowe, Richard Thomas (1802–74). Clergyman and naturalist. Chaplain on Madeira, 1832–54. Rector of Lea, Lincolnshire, 1854–74. (R. Desmond 1977, *DNB*.)

12 April 1856

Loyd, Samuel Jones, 1st Baron Overstone, (1796–1883). Banker and preeminent authority on banking and finance; partner in the London and Westminster Bank. Whig MP for Hythe, 1819–26. A leading promoter of the Great Exhibition of 1851. (*DNB*.)

Lubbock, Diana Hotham (1836–1917). John Lubbock's sister. Married William Powell Rodney in 1856. (*Burke's peerage* 1980.)

Lubbock, Ellen Frances. *See* Hordern, Ellen Frances.

Lubbock, Harriet (1810–73). Married John William Lubbock in 1833. Mother of John Lubbock. (Freeman 1978.)

[8 December 1856]

Lubbock, John (1834–1913). Banker, politician, and naturalist. Son of John William Lubbock and a neighbour of CD in Down. Studied entomology and anthropology. An active supporter of CD's theory of natural selection. Created Baron Avebury in 1900. FRS 1858. (*DNB, DSB*.)

[14 January 1856], 24 April [1856], [8 June 1856], 12 [June 1856],

[29 July 1856], 5 September [1856], 23 September [1856],
27 October [1856], [1 November 1856], [6 March 1857],
14 [July 1857], 11 August [1857], 12 [August 1857],
[22 November 1857]

Lubbock, John William, 3d Baronet (1803–65). Astronomer, mathematician, and banker. Treasurer and vice-president of the Royal Society, 1830–5 and 1838–47. First vice-chancellor of London University, 1837–42. CD's neighbour in Down. FRS 1829. (*DNB, DSB.*)
27 May [1856], 18 November [1856]

Lubbock, Mary Harriet (d. 1910). John Lubbock's older sister. Married Robert Birkbeck in 1857. (*Burke's landed gentry* 1898.)

Lucas, Prosper (1805–85). French physician and medical writer interested in heredity. (*NUC.*)

Lyell, Charles (1797–1875). Uniformitarian geologist whose *Principles of geology* and *Elements of geology* appeared in many editions. Professor of geology, King's College, London, 1831–3. President of the Geological Society, 1834–6 and 1849–50. Travelled widely and published accounts of his trips to the United States. Knighted, 1848; created baronet, 1864. Scientific mentor and friend of CD. FRS 1826. (*DNB, DSB.*)
21 April [1856], *1–2 May 1856*, 3 May [1856], 16 [June 1856],
17 June 1856, 25 June [1856], *[1 July 1856]*, 5 July [1856],
8 July [1856], 10 November [1856], *[16 January 1857]*,
11 February [1857], 13 April [1857]

Lyell, Henry (1804–75). Army officer in India. Married Katharine Murray Horner in 1848. Brother of Charles Lyell. (*Burke's peerage* 1980.)

Lyell, Katharine Murray (1817–1915). Daughter of Leonard Horner. Married Henry Lyell, brother of Charles Lyell, in 1848. Collected plants in India. Edited *Life, letters and journals of Sir Charles Lyell* (1881) and memoirs of Charles James Fox Bunbury and Leonard Horner. (R. Desmond 1977.)
26 January [1856]

Lyell, Leonard, 1st Baron (1850–1926). MP for Orkney and Shetland, 1885–1900. Created Baron Lyell of Kinnordy in 1914. Oldest son of Henry and Katharine Murray Lyell. (*Burke's peerage* 1980, *WWW.*)

Lyell, Mary Elizabeth (1808–73). Oldest child of Leonard Horner. Married Charles Lyell in 1832. An accomplished conchologist. (Freeman 1978.)

Macarthur, John (1767–1834). Lieutenant in the New South Wales corps, 1789; arrived in Sydney in 1790. Captain, 1795–1804. Established a farm at Parramatta, where he introduced merino sheep. Cultivated vines and olives at Camden, New South Wales. A member of the first legislative council of New South Wales, 1825–31. (*Aust. dict. biog., DNB.*)

Macarthur, William (1800–82). Australian horticulturist, viniculturist, and amateur botanist. Son of John Macarthur. Member of the legislative council of New South Wales, 1849–55 and 1864–82. Commissioner for New South Wales

Macarthur, William, cont.
  to the Paris Exhibition, 1855. Knighted in 1856 and made an officer in the Legion of Honour. (*Aust. dict. biog.*, *DNB.*)
Macaulay, Thomas Babington, 1st Baron (1800–59). Historian and politician. Created Baron Macaulay of Rothley, 1857. FRS 1849. (*DNB.*)
Macgillivray, William (1796–1852). Conservator of the museum of the Royal College of Surgeons, Edinburgh, 1831–41. Professor of natural history, Aberdeen University, 1841. (*DNB.*)
Maclaren, Charles (1782–1866). Established the *Scotsman* in 1817; editor from 1820 to *c*. 1848. Author of geological and topographical works. (*DNB*, Sarjeant 1980.)
Malherbe, Alfred (d. 1866). French naturalist and ornithologist. (*NUC.*)
Mantell, Walter Baldock Durrant (1820–95). Geologist and naturalist. Emigrated to New Zealand, 1840. Made important studies of the *Dinornis* beds. (*DNZB*, Sarjeant 1980.)
  3 April [1856], 10 April [1856], 5 June [1856–9]
Martha. A servant of Sarah Elizabeth (Sarah) Wedgwood. (Freeman 1978.)
Martins, Charles Frédéric (1806–89). French meteorologist and botanist. Professor of botany, Montpellier, 1847. (*NBU.*)
Mason, Philip Brookes (1842–1903). Entomologist and botanist. (R. Desmond 1977, P. Gilbert 1977.)
Massereene, Viscount. *See* Skeffington, John.
Mas y Sans, Sinbaldo de (1809–68). Author of works on China, the Philippine Islands, and Spain. (*NUC.*)
Maury, Matthew Fontaine (1806–73). American naval officer, hydrographer, and meteorologist. Head of the National Observatory, 1844–61. Published a series of wind and current charts with sailing instructions for all the oceans. (*DAB*, *DSB.*)
Maximilian Joseph of Este (1782–1863). Archduke of Austria. (*ADB.*)
Mayor, Robert Bickersteth (1820–98). Mathematical master at Rugby School, 1845–63. Fellow of St John's College, Cambridge, 1845–64. Ordained deacon in 1845; priest, 1850. (*Alum. Cantab.*)
McAndrew, Robert (1802–73). Liverpool merchant, yachtsman, and naturalist. Member of the dredging committee of the zoological section of the British Association for the Advancement of Science, 1844. FRS 1853. (Rehbock 1979.)
McGrigor, Charles Roderic, 2d Baronet (1811–90). Author and army agent with his brother, Walter McGrigor, at 17 Charles Street, St James's Square, London, 1844–69; at 25 Charles Street, 1869–90. Succeeded to the baronetcy in 1858. (*Modern English biography.*)
Melville, Alexander Gordon. Co-author of *The dodo and its kindred* (1848) with Hugh Edwin Strickland.

Meyen, Franz Julius Ferdinand (1804–40). Prussian botanist and physician. Naturalist on board the *Prinzess Louise*, 1830–2. Published the results of the voyage, 1834–43. Carried out microscopic investigations of plants and animals. (*ADB, DSB.*)

Meyer, Ernst Heinrich Friedrich (1791–1858). German botanist. Director of the botanical garden and associate professor of botany in Königsberg, 1826; professor, 1829. (*ADB.*)

Miers, John (1789–1879). Botanist and engineer. Travelled and worked in South America, 1819–38. Author of many papers describing South American plants. FRS 1843. (R. Desmond 1977, *DNB.*)

Miller, William Hallowes (1801–80). Mineralogist and crystallographer. Professor of mineralogy, Cambridge University, 1832–80. FRS 1838. (*DNB, DSB.*)

Milne-Edwards, Henri (1800–85). French zoologist. Professor of hygiene and natural history, École centrale des arts et manufactures, 1832. Professor of entomology, Muséum d'histoire naturelle, 1841, with responsibility for the collections of crustaceans, myriapods, and arachnids as well as insects; professor of mammalogy, 1861. Foreign member, Royal Society, 1848. (*DSB.*)

Miquel, Friedrich Anton Wilhelm (1811–71). Dutch botanist. Described the flora of the Dutch East Indies. Director, Rotterdam botanic garden, 1835–46. Professor of botany, Amsterdam, 1846–59; Utrecht, 1859–71. (*DSB.*)

Mitchell, David William (1813–59). Secretary of the Zoological Society of London, 1847–59; director of the Jardin d'acclimatation, Paris, 1859. (*Modern English biography*, Scherren 1905.)

Montagu, George (1751–1815). Writer on natural history. His collections of birds and other animals were purchased by the British Museum after his death. (*DNB.*)

Moore, Charles (1820–1905). Botanist. Director, botanic garden, Sydney, 1848–96. (*Aust. dict. biog.*, R. Desmond 1977.)

Moore, Frederic. Assistant in the East India Company's museum, Leadenhall Street, London.

Moore, John (d. 1737). Apothecary in London noted for his worm powders. Author of a treatise on domesticated pigeons. (Tegetmeier ed. 1879.)

Moquin-Tandon, Horace Bénédict Alfred (1804–63). French naturalist. Professor of zoology, Marseilles, 1829; professor of botany, Toulouse, 1833; professor of natural history, faculty of medicine, Paris, 1853. (*Larousse XX, NBU.*)

Morrey, Mrs. Cook to Sarah Wedgwood at Petleys, Down, until 1856. (Freeman 1978.)

Morris, John (1810–86). Geologist. Originally a pharmaceutical chemist in Kensington. Professor of geology, University College London, 1854–77. (*DNB*, Sarjeant 1980.)

1 March 1856

Morrow, James (1820–65). Agriculturist to the United States expedition to China and Japan, 1852–4. Collected botanical specimens. (*NUC.*)

Morton, Lord. *See* Douglas, George.

Mueller, Ferdinand Jakob Heinrich von (1825–96). German-born explorer and botanist who emigrated to Australia in 1847. Government botanist in Victoria, 1853. Botanist to the North-West Australia Expedition, 1855–7. Director of the botanic garden, Melbourne, 1857–73. FRS 1861. (*Aust. dict. biog.*, R. Desmond 1977.)

    8 December [1857]

Müller, Johannes Peter (1801–58). German comparative anatomist, physiologist, and zoologist. Professor of anatomy and physiology, Berlin University, 1833. (*DSB.*)

Murchison, Roderick Impey (1792–1871). Geologist noted for his work on the Silurian system. A leading figure in the Geological Society, British Association for the Advancement of Science, and Royal Geographical Society. Director of the Geological Survey of Great Britain, 1855. Knighted, 1846; created baronet, 1866. FRS 1826. (*DNB, DSB.*)

Murray, Charles Augustus (1806–95). Diplomatist and author. Travelled in North America, 1834–5, and wrote an account (1839) that went through three editions. Consul-general in Egypt, 1846–53. Envoy and minister plenipotentiary to the Court of Persia, 1854–9. (*DNB.*)

Murray, John (1808–92). Publisher and author of guide-books. CD's publisher from 1845. (*DNB.*)

    17 November [1856–7], 20 November [1856–7]

Neumeister, Gottlob. Pigeon fancier and author. (*NUC.*)

Newport, George (1803–54). Naturalist and surgeon. Practised medicine in London, 1837–47. President of the Entomological Society, 1844–5. Received a pension in 1847, after which he devoted himself to natural history. FRS 1846. (*DNB, DSB.*)

Niebuhr, Carsten (1733–1815). German cartographer to a Danish expedition to Arabia, 1761–7. Travelled through Persia, Palestine, and Constantinople on his return, making geographical observations and exact copies of cuneiform transcriptions at Persepolis. (*DSB.*)

Nordmann, Alexander von (1803–66). Finnish zoologist and palaeontologist. Professor of natural history at the Lyceum Richelieu, Odessa, Russia, 1830–45; professor of zoology in Helsinki, 1848. (Sarjeant 1980.)

Norman, Ebenezer. Schoolmaster in Down whom CD employed as a copyist. (Freeman 1978.)

Ogilby, William (1808–73). Irish barrister and naturalist. Practised as a barrister in London, 1832–46, before settling in Ireland. Honorary secretary of the Zoological Society, 1839–46. (*Alum Cantab.*, *Quarterly Journal of the Geological Society* 45 (1888–9): *Proceedings*, p. 46.)

Oken, Lorenz (1779–1851). German naturalist and leading exponent of Naturphilosophie. Professor of natural history, Zurich University, 1832–51. (*ADB, DSB, EB.*)

Oldfield, Henry Ambrose. Surgeon and author. Assistant surgeon in the Bengal Army. Surgeon to the British Residency in Nepal, 1850–63. (*NUC.*) 10 May [1856]

Orbigny, Alcide Charles Victor Dessalines d' (1802–57). French palaeontologist who travelled widely in South America, 1826–34. Professor of palaeontology, Muséum d'histoire naturelle, 1853. (*DSB.*)

Orr, William Somerville (d. 1873). Publisher at 2 Amen Corner, St Paul's Churchyard, London, 1837–59. (*Modern English biography.*)

Orton, Reginald (1810–62). Surgeon in Sunderland. (*DNB.*)

Osbeck, Per (1723–1805). Swedish clergyman and traveller. (*NUC.*)

Overstone, Lord. *See* Loyd, Samuel Jones.

Owen, Richard (1804–92). Comparative anatomist. Assistant-conservator of the Hunterian Museum, Royal College of Surgeons, 1827; Hunterian professor, 1836–56. Superintendent of the natural history departments, British Museum, 1856–84. Described the *Beagle* fossil mammal specimens. FRS 1834. (*DNB, DSB.*)

Paget, James (1814–99). Assistant surgeon to St Bartholomew's Hospital, London, 1847; surgeon, 1861–71. Surgeon-extraordinary to Queen Victoria, 1858. FRS 1851. (*DNB.*)

Pallas, Pyotr Simon (1741–1811). German naturalist and geographer who travelled extensively in the Russian empire. Foreign member, Royal Society, 1763. (*ADB, DSB.*)

Palliser, John (1807–87). Geographer and explorer. Undertook a hunting expedition in the western and north-western districts of North America, 1847, and published in 1853 a popular account of his experiences. Commanded the British North American Exploring Expedition, 1857–60, charged with surveying the boundary territories of British North America between Lake Superior and the Pacific. (*DNB.*)

Palmerston, Lord. *See* Temple, Henry John.

Parker, Henry (1788–1856). MD, Edinburgh, 1814. Physician to the Shropshire Infirmary. Married Marianne Darwin in 1824. (*Darwin pedigree, Provincial medical directory.*)

Parker, Marianne (1798–1858). CD's oldest sister. Married Henry Parker in 1824. (*Darwin pedigree.*)

Parry, William Edward (1790–1855). Rear-admiral and explorer. Led three voyages between 1819 and 1824 in search of a north-west passage. FRS 1821. (*DNB.*)

Parslow, Joseph (1809/10–98). CD's manservant at 12 Upper Gower Street *c.* 1840 and butler at Down House until 1875. (Freeman 1978.)

Patterson, Robert (1802–72). Naturalist and merchant in Belfast. Author of zoological works. A founder of the Natural History Society of Belfast, 1821, and served as its president for many years. An active member of the British Association for the Advancement of Science. FRS 1859. (*DNB*.)
    10 March [1857], 12 November [1857]

Penfold, James. BA, Christ's College, Cambridge, 1830. Rector of Thorley, Isle of Wight, 1841–50. (*Alum. Cantab.*)

Pennant, Thomas (1726–98). Traveller and naturalist. FRS 1767. (*DNB, DSB*.)

Perry, Matthew Calbraith (1794–1858). American naval officer who extended the commercial and naval interests of America in the Pacific. (*DAB*.)

Persoon, Christiaan Henrik (1761–1836). Dutch botanist born in South Africa and educated in Germany. Primarily known as a mycologist, he was also the author of *Synopsis plantarum* (1805–7), which attempted to describe all the phanerogams then known. (*DSB*.)

Petit de la Saussaye, Jean François de Paule Louis de. *See* La Saussaye, Jean François de Paule Louis Petit de.

Pfeiffer, Ludwig Georg Karl (1805–77). German physician and naturalist. Author of a number of botanical and conchological works. (*ADB*.)

Phillips, John (1800–74). Geologist. Keeper of the museum of the Yorkshire Philosophical Society, 1825–40. Assistant secretary of the British Association for the Advancement of Science, 1832–62. Deputy reader in geology, Oxford University, 1853; professor, 1860–74. FRS 1834. (*DNB, DSB*.)
    18 January [1856], 28 January [1856]

Pickering, Charles (1805–78). American physician and naturalist. Curator of the Academy of Natural Sciences, Philadelphia, 1833–7. Chief zoologist to the United States Exploring Expedition, 1838–42. Published on the geographical distribution of plants and animals. (*DAB*.)

Plieninger, Wilhelm Heinrich Theodor von (1795–1879). German theologian, schoolteacher, and palaeontologist. (Sarjeant 1980.)

Pliny, the Elder [Gaius Plinius Secundus] (*c.* 23–79). Roman advocate and soldier. Author of the *Naturalis historia*. (*EB*.)

Porter, Robert Ker (1777–1842). Painter and traveller. Published and illustrated accounts of his journeys. British consul in Venezuela, 1826–41. (Bénézit 1976, *DNB*.)

Portlock, Joseph Ellison (1794–1864). Major-general in the Royal Engineers. Engaged in the geological and productive economical sections of the Irish survey, 1824–43. FRS 1837. (*DNB*.)

Pratt, George Charles, 2d Marquis of Camden, (1799–1866). MP for Ludgershall, Wiltshire, 1821–6; Bath, 1826–30; Dunwich, 1831–2. A lord of the Admiralty, 1828–9. (*Alum. Cantab., Modern English biography*.)

Prestwich, Joseph (1812–96). Geologist. Wine merchant in London. An expert on the Tertiary geology of Europe. Discovered evidence for the antiquity of man. Professor of geology at Oxford, 1874–88. FRS 1853. (*DNB, DSB*.)

Prichard, James Cowles (1786–1848). Physician and ethnologist who maintained that the races of man are varieties of one species. A commisioner of lunacy in London, 1845. FRS 1827. (*DNB, DSB.*)

Pritchard, Charles (1808–93). Clergyman and astronomer. Headmaster of Clapham Grammar School, 1834–62, where he erected an observatory. Hulsean lecturer at Cambridge, 1867. Savilian professor of astronomy at Oxford, 1870. FRS 1840. (*DNB, DSB.*)

Pugh, Miss. Governess of the Darwin children, 1856–7. Later became insane and was confined to an asylum. (Freeman 1978.)

Pulleine, Robert (1806–68). BA, Emmanuel College, Cambridge, 1829. Rector of Kirkby-Wiske, 1845–68. (*Alum. Cantab.*)

Quatrefages de Bréau, Jean Louis Armand de (1810–92). French zoologist and anthropologist. Undertook a study of invertebrates and, in 1844, accompanied Henri Milne-Edwards on a zoological excursion off the coast of Sicily. Awarded the chair of natural history at the Lycée Henri IV, Paris, 1850; professor of the natural history of man, Muséum d'histoire naturelle, 1855. Foreign member, Royal Society, 1879. (*DSB.*)

Quoy, Jean René Constant (1790–1869). French naval surgeon and naturalist. Participated in many voyages of scientific exploration. (*DSB.*)

Radlkofer, Ludwig (1829–1927). German botanist. (*NUC.*)

Ramsay, Andrew Crombie (1814–91). Geologist with the Geological Survey, 1841; senior director of the Geological Survey for England and Wales, 1862; director-general, 1871–81. Professor of geology, University College London, 1847–52; lecturer at the Government School of Mines, 1852–71. Knighted, 1881. FRS 1849. (*DNB, DSB.*)

Ransome, James Allen (1806–75). Partner, then head, of the firm of agricultural machinery makers, Ransome and Sons, in Ipswich, 1829–75. Established the allotment system for labourers. Founder of the Working Men's College in Ipswich. (*Modern English biography.*)

Ransome, Robert (1795–1864). Iron founder. Senior partner in the firm Ransome and Sons, Ipswich. (*Modern English biography.*)

Rathke, Martin Heinrich (1793–1860). German embryologist and anatomist. Professor of zoology and anatomy in Königsberg, 1835–60. Studied marine organisms and the embryonic development of sex organs. (*ADB, DSB.*)

Rawlinson, Henry Creswicke (1810–95). Assyriologist. Army officer in the East India Company's service, 1827–44. Consul at Baghdad, 1844; consul-general, 1851–5. Deciphered a Persian cuneiform. (*DNB.*)

Rawson, Arthur (1818–91). Perpetual curate of Trinity Church, Bromley, Kent, 1843–82. Married Charlotte Elizabeth Clay, 1848. (*Alum. Cantab.*)

Réaumur, René Antoine Ferchault de (1683–1757). French mathematician, physicist, and natural historian. Invented the thermometer scale that bears his name. Foreign member, Royal Society, 1738. (*DSB.*)

Reed, George Varenne (1816–86). BA, Jesus College, Cambridge, 1837. Curate of Hayes, 1837–9; of Tingewick, Buckinghamshire, 1839–54. Rector of Hayes, 1854–86. Tutor to George Howard Darwin. (*Alum. Cantab.*)
    8 September [1856]

Reissek, Siegfried (1819–71). German botanist. (*NUC.*)

Richard, Achille (1794–1852). French physician and botanist. (*NUC.*)

Richardson, John (1787–1865). Arctic explorer, physician, and naturalist. Physician to the Royal Hospital at Haslar, 1838. Inspector of hospitals, 1840. Commanded an expedition in search of John Franklin, 1847–9. FRS 1825. (*DNB*, Sarjeant 1980.)
    17 July 1856

Rink, Hinrich Johannes (1819–93). Danish naturalist who explored Greenland, 1848–51; member of a Danish commission to investigate trade with Greenland, 1851. Resided in Greenland from 1853. (*DBL.*)

Robillard, Victor de. Resident of Mauritius. Active member of the Natural History Society of Mauritius.
    20 September 1856, 26 February 1857

Rodney, William Powell (1829–68). Married Diana Hotham Lubbock in 1856. (*Burke's peerage* 1980.)

Rolle, Louisa Barbara (1796–1885). Widow of Baron Rolle of Stevenstone. Resided at Bicton House, Devon, until her death. (*Complete peerage.*)

Ross, James Clark (1800–62). Naval officer and polar explorer. Discovered the north magnetic pole in 1831. Commander of an expedition to the Antarctic, 1839–43. In command of a search expedition for John Franklin, 1848–9. Rear-admiral, 1856. FRS 1828. (*DNB, DSB.*)

Royal Society
    18 July 1856, 24 January [1857]

Royle, John Forbes (1799–1858). Surgeon and naturalist in the service of the East India Company. Superintendent of the botanic garden in Saharanpur, 1823–31. Professor of materia medica, King's College, London, 1837. In charge of the East India House correspondence relating to vegetable productions, 1838. A founder of the Philosophical Club of the Royal Society, 1847. FRS 1837. (R. Desmond 1977, *DNB.*)

Rüppell, Wilhelm Peter Eduard Simon (1794–1884). German naturalist, traveller, and invertebrate palaeontologist. (*ADB*, Sarjeant 1980.)

Sabine, Edward (1788–1883). Geophysicist and army officer. General secretary of the British Association for the Advancement of Science, 1838–59. Foreign secretary of the Royal Society, 1845–50; treasurer, 1850–61; president, 1861–71. FRS 1818. (*DNB, DSB.*)
    23 April [1856], 16 March [1857]

Saint-Hilaire, Augustin François César Prouvençal (Auguste de) (1779–1853). French naturalist. Surveyed the flora and fauna of Brazil, 1816–22. (*DSB.*)

Salter, Thomas Bell (1814–58). Physician and botanist. Practised medicine in Ryde, Isle of Wight. Published papers on botany. (R. Desmond 1977.)

Santarem, Juan de (*fl.* 1450). Portuguese navigator. (*EB.*)

Say, Thomas (1787–1834). American entomologist and conchologist. A founder of the Academy of Natural Sciences of Philadelphia, 1812. Curator of the American Philosophical Society, 1821–7; professor of natural history, University of Pennsylvania, 1822–8. Became a member of the utopian community at New Harmony, Indiana, where he carried out his scientific work, 1825–34. (*DAB, DSB.*)

Scherzer, Karl von (1821–1903). Viennese scientific traveller and diplomat. Principal scientist of the *Novara* expedition. Austrian consul in London, 1875–8. (*BLKO.*)

Schiödte, Jörgen Matthias Christian (1815–84). Danish zoologist. (Gascoigne 1984.)

Schlagintweit, Adolphe (1829–57). German geologist and explorer. Lecturer in geology, University of Münich, 1853. Accompanied his brothers on a scientific expedition to India and Tibet, during which he was put to death by the amir of Kashgar. (*ADB, EB.*)

Schlagintweit, Hermann Rudolph Alfred (1826–82). German geographer and explorer. Lecturer in physical geography at the University of Berlin, 1851–4. With his brothers Adolph and Robert, undertook a scientific expedition, sponsored by the East India Company and supported by the Prussian government, to India and Tibet, 1854–7. (*ADB, BHGW.*)

Schlagintweit, Robert (1833–85). German traveller and explorer. Undertook with his brothers a scientific expedition to India and Tibet, 1854–7. (*ADB, BHGW.*)

25 September 1857

Schlegel, Hermann (1804–84). German naturalist. Director of the National Museum of the Netherlands in Leiden. (*ADB*, Sarjeant 1980.)

Schmidt, Ferdinand Joseph (1791–1875). Austrian businessman and entomologist. Published a list in the *Laibacher Zeitung* (no. 146, 1852) of all the species he had collected since 1832. (*BLKO.*)

Schouw, Joachim Frederik (1789–1852). Danish botanist. Extraordinary professor of botany, Copenhagen University, 1820; professor, 1845–53. Curator of the Copenhagen botanic gardens, 1841. (*DBL, DSB.*)

Schreber, Johann Christian Daniel von (1739–1810). German naturalist. Foreign member, Royal Society, 1795. (*ADB.*)

Scott, David. Assistant surgeon in the Bengal Medical Establishment. (*East-India register and army list, for 1856.*)

Sebright, John Saunders, 7th Baronet (1767–1846). Whig politician and agriculturist. Published works on animal breeding. (*DNB.*)

Sedgwick, Adam (1785–1873). Geologist and clergyman. Woodwardian professor of geology, Cambridge University, 1818–73. Canon of Norwich, 1834–73. FRS 1821. (*DNB, DSB.*)

Seemann, Berthold Carl (1825–71). German-born traveller and botanist. Went as gardener to the Royal Botanic Gardens, Kew, 1844. Naturalist to HMS *Herald*, 1847–51; published the *Botany* of the voyage, assisted by Joseph Dalton Hooker (1852–7). (R. Desmond 1977, *DNB*.)

Seemann, Wilhelm E. G. (d. 1868). Co-editor, with his brother Berthold Carl Seeman, of *Bonplandia*, the botanical journal of the Academia Caesarea Leopoldino-Carolina Naturae Curiosorum, 1853–62.

Senebier, Jean (1742–1809). Swiss plant physiologist. Was the first to understand correctly the nature of gas exchange in photosynthesis. Librarian, Republic of Geneva, 1773. (*DSB*.)

Seringe, Nicolas Charles (1776–1858). French botanist. (*NUC*.)

Serres, Marcel Pierre Toussaint de (1780–1862). French palaeontologist and zoologist. Professor of mineralogy and geology, University of Montpellier, 1811. Magistrate in Montpellier, 1814–52. (*DSB*, Sarjeant 1980.)

Sharpe, Daniel (1806–56). Geologist. Merchant in Portugal, 1835–8. Studied slaty cleavage and wrote several papers on the cause of cleavage and foliation. President of the Geological Society, 1856. FRS 1850. (*DNB*, Sarjeant 1980.)

Sharpey, William (1802–80). Physiologist. Professor of anatomy and physiology at University College London, 1836–74. Secretary of the Royal Society of London, 1853–72. FRS 1839. (*DNB*, *DSB*.)
2 June [1857], 24 January [1857]

Sheridan, Richard Brinsley (1751–1816). Dramatist and parliamentary orator. (*DNB*.)

Siebold, Karl Theodor Ernst von (1804–85). German invertebrate zoologist. Professor of zoology and comparative anatomy, University of Munich, 1853–83. Co-editor with Rudolf Albert von Koelliker of the *Zeitschrift für wissenschaftliche Zoologie*. (*DSB*, *NDB*.)

Silliman, Benjamin (1779–1864). American chemist, geologist, and mineralogist. Professor of chemistry and natural history, Yale University, 1802–53. Founder and first editor of the *American Journal of Science and Arts*, 1818. (*DAB*, *DSB*.)

Silliman, Benjamin, Jr (1816–85). American chemist. Teaching assistant to his father Benjamin Silliman, 1837. An editor of the *American Journal of Science and Arts*, 1838–85. Professor of practical chemistry, Yale, 1846. Succeeded his father as professor of chemistry and natural history, 1853. Carried out researches on petroleum. (*DAB*, *DSB*.)

Skeffington, John, Viscount Massereene, (1812–63). Of Antrim Castle, Northern Ireland. (*Complete peerage*.)

Sloane, Hans (1660–1753). Physician and naturalist. Visited Jamaica, 1687–9, as physician to the Duke of Albemarle; published an account of his travels in two volumes (1707 and 1725). FRS 1685. Succeeded Isaac Newton as president of the Royal Society, 1727, a position he held until 1741. (*DNB*.)

Smethurst, Thomas (b. *c.* 1803). Practised as a surgeon. Author of a work on the water cure, 1843. Proprietor of a hydropathic establishment at Moor Park, 1850–5, which he sold to Edward Wickstead Lane. Accused and acquitted of murdering a woman with whom he had entered into a bigamous relationship, 1859. Imprisoned for bigamy, 1859–60. (*Modern English biography.*)

Smith, Charles Hamilton (1776–1859). Army officer and writer on natural history. FRS 1824. (*DNB.*)

Smith, Frederick (1805–79). Entomologist. (Gilbert 1977.)
    10 November 1857

Smith, George (1815–71). Missionary and explorer in China. Bishop of Victoria, Hong Kong, 1849–65. (*DNB.*)

Smith, James Edward (1759–1828). Botanist. Founded the Linnean Society in 1788 and held the office of president until his death. FRS 1785. (*DNB, DSB.*)

Smith, John (1798–1888). Curator at the Royal Botanic Gardens, Kew, 1842–64. (R. Desmond 1977.)

Smith, Thomas (d. *c.* 1825). Microscopist. Friend of Robert Brown. FRS 1816. (R. Desmond 1977.)

Smith and Beck. Partnership of James Smith and Richard Beck, manufacturers of optical instruments.

Smith, Elder, and Company. Partnership of George Smith (1789–1846) and Alexander Elder (1790–1876). Publishers and East India agents. (*DNB, Modern English biography.*)

Smyth, Charles Piazzi (1819–1900). Assistant astronomer at the Royal Observatory, Cape of Good Hope, 1835–45. Astronomer Royal of Scotland and professor at Edinburgh University, 1846–88. FRS 1857. (*DNB, DSB.*)

Snow, George. Carrier between Down and London. (*Post Office directory of the six home counties* 1858.)

Soane, John (1753–1837). Architect. Founder of Soane Museum. Erected the picture gallery at Dulwich College, 1812. (*DNB.*)

Sorby, Henry Clifton (1826–1908). Geologist who pioneered microscopic petrology. President of the Geological Society, 1878–80. Established the chair of geology at Sheffield University. FRS 1857. (*DNB, DSB.*)

Sowerby, George Brettingham (1788–1854). Conchologist and artist. Described the *Beagle* fossil shells. (*DNB*, Sarjeant 1980.)

Sowerby, George Brettingham, Jr (1812–84). Conchologist and scientific illustrator. (*DNB*, Sarjeant 1980.)

Spach, Édouard (1801–79). French botanist. (*NUC.*)

Spallanzani, Lazzaro (1729–99). Italian priest and natural scientist, best known for his disproof of spontaneous generation and his elucidation of the process of digestion. Professor in Modena, 1763–9; Pavia, 1769–99. (*DSB.*)

Spencer, Herbert (1820–1903). Civil engineer on the railways, 1837–41 and 1844–6. Supported the complete suffrage movement and in 1844 became sub-editor of the *Pilot*, the newspaper devoted to the movement. Sub-editor of the

Spencer, Herbert, cont.
> Economist, 1848–53. From 1852, author of papers on evolution and works on philosophy and social science. (*DNB, DSB.*)
> 11 March [1856]

Sprengel, Christian Konrad (1750–1816). German botanist who studied insect-aided fertilisation of flowers. (*ADB, DSB.*)

Stainton, Henry Tibbats (1822–92). Entomologist. Founder and editor of the *Entomologists' Annual* (1855–74) and of the *Entomologist's Weekly Intelligencer*, 1856–61. FRS 1867. (*DNB.*)
> 13 April [1856]

Stanhope, Philip Henry, 5th Earl (1805–75). Historian. Tory MP for Wootton Bassett, 1830–2; Hertford, 1832–3 and 1835–52. President of the Society of Arts, 1846–75. Succeeded to the earldom in 1855. FRS 1827. (*DNB.*)

Stanley, Edward Smith, 13th Earl of Derby, (1775–1851). Whig MP for Preston, 1796–1812; Lancashire, 1812–32. President of the Linnean Society, 1828–33. President of the Zoological Society, 1831–51. Formed a private menagerie at Knowsley. (*DNB.*)

Stephens, Henry (1795–1874). Agricultural writer. Studied agricultural practice in Berwickshire, and then made an agricultural tour of the Continent, 1818–19. Farmed in Forfarshire, 1820–30. (*DNB.*)

Stephens, James Francis (1792–1852). Entomologist and zoologist. Employed in the Admiralty office, Somerset House, 1807–45. Assisted in arranging the insect collection of the British Museum. (*DNB*, P. Gilbert 1977.)

Steudel, Ernst Gottlieb (1783–1856). German botanist. (*ADB.*)

Stevens, John Crace. Auctioneer and natural history dealer at 38 King Street, Covent Garden, London. (*Post Office London directory* 1855.)

Stevens, Samuel. Natural history agent and collector of British Coleoptera and Lepidoptera. Brother of John Crace Stevens. (Wallace 1905, 1: 266.)

Stimpson, William (1832–72). American naturalist and conchologist. Naturalist to the North Pacific Exploring Expedition, 1853–6. Worked in the invertebrate department of the Smithsonian Institution, 1856–65. Director of the Chicago Academy of Sciences, 1865–72. (*DAB, DSB.*)

Stokes, John Lort (1812–85). Naval officer. Served in HMS *Beagle* as midshipman, 1826–31; mate and assistant surveyor, 1831–7; lieutenant, 1837–41; commander, 1841–3. Admiral, 1877. (R. Desmond 1977, *DNB.*)

Storey, John (*c.* 1801–59). Botanist specialising in the flora of Northumberland and Durham. Secretary of the Tyneside Naturalists' Field Club, 1849–57. (R. Desmond 1977.)

Straus-Durckheim, Hercule Eugène Grégoire (1790–1865). French comparative anatomist. (*NUC.*)

Strickland, Hugh Edwin (1811–53). Geologist and zoologist. An advocate of reform in zoological nomenclature. FRS 1852. (*DNB*, Sarjeant 1980.)

Strzelecki, Paul Edmund de (1797–1873). Polish-born explorer and geologist. Explored the Australian interior, 1839–40. Became a British subject in 1845. FRS 1853. (*Aust. dict. biog.*, *DNB*.)

Sulivan, Bartholomew James (1810–90). Naval officer and hydrographer. Lieutenant in HMS *Beagle*, 1831–6. Surveyed the Falkland Islands in HMS *Arrow*, 1838–9. Commander of HMS *Philomel*, 1842–6. Resided in the Falkland Islands, 1848–51. Commanded HMS *Lightning* in the Baltic, 1854–5. Naval officer of the marine department of the Board of Trade, 1856–65. Admiral, 1877. (*DNB*.)

Swainson, William (1789–1855). Naturalist and illustrator who collected in Sicily and Brazil. Adopted the quinary system of classification. Emigrated to New Zealand in 1837. FRS 1820. (*DNB*, *DSB*.)

Swayne, George (*c.* 1746–1827). Botanist and clergyman. (R. Desmond 1977.)

Sykes, William Henry (1790–1872). Naturalist and military officer in the East India Company. Statistical reporter to the Bombay government, 1824–31. Elected to the board of directors of the East India Company, 1840; deputy chairman 1855; chairman, 1856. FRS 1834. (R. Desmond 1977, *DNB*.)

Syme, John Thomas Irvine Boswell. *See* Boswell-Syme, John Thomas Irvine.

Targioni Tozzetti, Antonio (1785–1856). Italian naturalist. (*EI*.)

Tegetmeier, William Bernhard (1816–1912). Editor, journalist, lecturer, and naturalist. Pigeon fancier and an expert on fowls. (E. W. Richardson 1916.)
    1 January [1856], 14 January [1856], [1 February 1856],
    15 March [1856], 20 March [1856], 25 April [1856], 11 May [1856],
    31 May [1856], 4 June [1856], 24 June [1856], [July 1856],
    14 August [1856], [15–22 August 1856], 23 August [1856],
    30 August [1856], [18 September 1856], 21 September [1856],
    28 [September 1856], 15 October [1856], 19 October [1856],
    3 November [1856], 19 November [1856], 29 November [1856],
    4 December [1856], 6 February [1857], *11 February 1857*,
    11 February [1857], 18 February [1857], 12 [May 1857],
    18 May [1857], [18 June 1857], 23 June [1857], 25 [June 1857],
    [19 July 1857], 27 July [1857], 29 September [1857],
    21 November [1857]

Tellkampf, Theodor G. German entomologist. Brother of the economist Johann Louis Tellkampf (*ADB*). Studied cave fauna in the United States. (*NUC.*)

Temminck, Coenraad Jacob (1778–1858). Dutch ornithologist. (*NNBW*.)

Temple, Frederick (1821–1902). Fellow and tutor of Balliol College, Oxford, 1842–8. Headmaster of Rugby School, 1857–69. Bishop of Exeter, 1869–85; of London, 1885–96; Archbishop of Canterbury, 1896–1902. (*DNB*.)

Temple, Henry John, 3d Viscount Palmerston, (1784–1865). Statesman. MP for Cambridge University, 1811–31. Foreign secretary, 1830–41 and 1846–51. Home secretary, 1852–5. Prime minister, 1855–8 and 1859–65. (*DNB*.)

Tenant, James. Keeper of the aquarium at the Zoological Society's gardens. (Woodward 1851–6, 3: 331 n.)
31 March 1857, 23 April 1857

Thompson, John (1810/11–59). Manager of the Earl of Derby's menagerie at Knowsley until 185?. Superintendent of the gardens of the Zoological Society of London, 1852–9. (*Modern English biography.*)
26 November [1856]

Thompson, William (1805–52). Irish naturalist. President of the Belfast Natural History Society, 1843–52. (R. Desmond 1977, *DNB.*)

Thomson, Thomas (1817–78). East India Company surgeon and botanist in India. Collaborated with Joseph Dalton Hooker on various publications and joined him on his Himalayan expedition in 1849. Curator of the Asiatic Society's museum, Calcutta, 1840–1. Superintendent of the Calcutta botanic garden and professor of botany at the Calcutta Medical College, 1854–61. FRS 1855. (R. Desmond 1977, *DNB.*)

Thomson, Thomas Richard Heywood. Ethnologist. (*NUC.*)

Thorley, Miss. Governess to the Darwin children, 1850–6. (Freeman 1978.)

Thunberg, Carl Peter (1743–1828). Swedish botanist. A protégé of Linnaeus, he accompanied a Dutch merchant vessel to Japan, 1772–9, spending three years in the Cape Colony to learn Dutch and later visiting India and Ceylon en route to Japan. Botanical demonstrator, Upsala University, 1779; professor of botany, 1784. (*DSB, SBL.*)

Thuret, Gustave Adolphe (1817–75). French lawyer and attaché in Constantinople, 1840, who gave up his diplomatic career to devote himself to botanic researches. Made the seaweed *Fucus* the special object of his study, on which he published a number of papers. Resided in Cherbourg, 1852–6, and later at Cap d'Antibes on the Mediterranean in order to carry out researches on living algae. (*DSB.*)

Thwaites, George Henry Kendrick (1811–82). Botanist and entomologist. Superintendent of the Peradeniya botanic garden, Ceylon, 1849; director, 1857–80. FRS 1865. (R. Desmond 1977, *DNB.*)
8 March 1856

Tippoo Sahib (1753–99). Sultan of Mysore. (*EB.*)

Tollet, George (1767–1855). Of Betley Hall, Staffordshire. Justice of the peace and deputy lieutenant of Staffordshire. Agricultural reformer. Close friend of Josiah Wedgwood II. (*Burke's landed gentry* 1879.)

Tollet, Marianne. *See* Clive, Marianne.

Torrey, John (1796–1873). American botanist and chemist. Collected and described plants from various government explorations. Professor of chemistry at the College of Physicians and Surgeons, New York City, 1827–55. (*DAB, DSB.*)

Tuckerman, Edward (1817–86). American botanist. Lecturer in history at Amherst College, Massachusetts, 1854; professor of botany, 1858–86. An expert on lichens. (*DAB.*)

Tulasne, Louis René (1815–85). French mycologist and philanthropist. (*DSB.*)

Tyndall, John (1820–93). Irish physicist, lecturer, and populariser of science. Professor of natural philosophy at the Royal Institution, London, 1853; superintendent of the Royal Institution, 1867–87. FRS 1852. (*DNB, DSB.*) 4 February [1857]

Tytler, Robert C. Of the 38th Regiment of Bengal Native Light Infantry.

Valmont de Bomare, Jacques Christophe (1731–1807). French mineralogist and naturalist. Travelled extensively as naturalist for the government before starting annual courses of public lectures at the Jardin des plantes. Author of the *Dictionnaire raisonné universal d'histoire naturelle* (1764). (*DSB, NBU.*)

Van Voorst, John (1804–98). Employed by Longmans, 1826–33. Publisher in Paternoster Row, 1833–86. Fellow of the Microscopical Society of London. (*Modern English biography.*)

Vaux, William Sandys Wright (1818–85). Antiquary. Employed in the department of antiquities of the British Museum, 1841. Keeper of the department of coins and medals, 1861–70. (*DNB.*)

Vicq d'Azyr, Félix (1748–94). French anatomist. Elected to the Académie des sciences in 1774. Physician to the queen, 1788. Author of several works and a pioneer of veterinary studies in France. (*DSB.*)

Victoria, Queen of Great Britain and Empress of India (1819–1901). Succeeded to the throne on the death of her uncle, William IV, 20 June 1837. (*DNB.*)

Vigors, Nicholas Aylward (1785–1840). Irish zoologist and politician. FRS 1826. (*DNB.*)

Vilmorin, Pierre Philippe André Levêque de (1816–60). French botanist. Director of the family seed firm, Vilmorin-Andrieux et Cie of Paris, 1843–60. Editor of *Bon jardinier*, 1844–60. Carried out extensive breeding experiments and set up an experimental farm at Verrières-le-Buisson. (*DSB.*)

Vine, William (1814–92). Officer in the Madras cavalry. General, 1888. (*Modern English biography.*)

Visiani, Roberto de (1800–78). Italian botanist. (*NUC.*)

Vivian, Edward (1808–93). Banker in Torquay. Author of several scientific papers. (*Modern English biography.*)

Wahlenberg, Göran (Georg) (1780–1851). Swedish botanist and homoeopathist. Travelled extensively in Europe, 1799–1814, studying the geographical distribution of plants. Professor of medicine and botany at the University of Upsala, 1828. (*DSB, SBL.*)

Wallace, Alfred Russel (1823–1913). Collector in the Amazon, 1848–52; in the Malay Archipelago, 1854–62. Independently discovered natural selection in 1858. Lecturer and author of works on protective coloration, mimicry, and zoogeography. Became a Spiritualist in 1865 and was considered controversial because of his views on Spiritualism, vaccination, land nationalisation, women's rights, and socialism. FRS 1893. (*DNB, DSB.*) 1 May 1857, [*27 September 1857*], 22 December 1857

Wallace, Peter. Farm superintendent on Ascension Island.
    10 September 1856
Wallich, Nathaniel (1786–1854). Danish-born botanist. Superintendent of the Calcutta botanic garden, 1815–41. FRS 1829. (R. Desmond 1977, *DNB*.)
Waterhouse, George Robert (1810–88). Naturalist. A founder of the Entomological Society, 1833. Curator, Zoological Society, 1836–43. On the staff of the mineralogical and geological branch of the natural history department of the British Museum, 1843–80. Described CD's Mammalia and entomological specimens from the *Beagle* voyage. (*DNB*, Gilbert 1977.)
    14 April 1857
Watson, Hewett Cottrell (1804–81). Botanist, phytogeographer, and phrenologist. Published various guides on the distribution of British plants. Edited the *Phrenological Journal*, 1837–40. Collected plants in the Azores, 1842. An early supporter of the idea of the progressive development of plant species. (*DNB*, *DSB*.)
    5 June 1856, 10 June 1856, [after 10 June 1856], *20 June 1856,*
    10 November 1856, 19 November 1856, 26 November 1856,
    [28 December 1856], 10 March 1857, 13 March 1857, 14 June [1857],
    14 December [1857], 20 December [1857]
Webb, Philip Barker (1793–1854). Botanist and traveller. Visited Madeira and the Canary Islands, 1828–30, and, with Sabin Berthelot, was the author of *Histoire naturelle des Iles Canaries* (1836–50). (R. Desmond 1977, *DNB*.)
Wedgwood, Allen. *See* Wedgwood, John Allen.
Wedgwood, Caroline (Carry) (b. 1836). Daughter of Jessie and Harry Wedgwood. (*Emma Darwin*.)
Wedgwood, Caroline Sarah (1800–88). CD's sister. Married Josiah Wedgwood III, her cousin, in 1837. (*Darwin pedigree*.)
Wedgwood, Cecily Mary (1837–1917). Daughter of Francis and Frances Mosley Wedgwood. (*Darwin pedigree*.)
Wedgwood, Elizabeth. *See* Wedgwood, Sarah Elizabeth (Elizabeth).
Wedgwood, Ernest Hensleigh (1838–98). BA, Trinity College, Cambridge, 1860. Clerk in the Colonial Office. Son of Hensleigh and Frances Mackintosh Wedgwood. (*Alum. Cantab.*)
Wedgwood, Frances Julia (Snow) (1833–1913). Writer. Daughter of Hensleigh and Frances Mackintosh Wedgwood. (*Burke's peerage* 1980.)
Wedgwood, Frances Mackintosh (Fanny Hensleigh) (1800–89). Married Hensleigh Wedgwood in 1832. (*Burke's peerage* 1980, Freeman 1978.)
    18 [August 1856 – January 1858]
Wedgwood, Frances Mosley (Fanny Frank) (1808–74). Married Francis Wedgwood in 1832. (*Burke's peerage* 1980.)
Wedgwood, Francis (Frank) (1800–88). Master-potter and partner in the works at Etruria until 1876. Emma Darwin's brother. Married Frances Mosley in 1832. (*Alum. Cantab.*)

Wedgwood, Godfrey (1833–1905). Son of Francis and Frances Mosley Wedgwood. (Freeman 1978.)
    21 April [1856]

Wedgwood, Harry. *See* Wedgwood, Henry Allen.

Wedgwood, Henry Allen (Harry) (1799–1885). BA, Jesus College, Cambridge, 1821. Barrister. Emma Darwin's brother. Married Jessie Wedgwood in 1830. (*Alum. Cantab.*)

Wedgwood, Hensleigh (1803–91). Philologist and barrister. Fellow, Christ's College, Cambridge, 1829–30. Metropolitan police magistrate at Lambeth, 1832–7. Registrar of metropolitan carriages, 1838–49. An original member of the Philological Society, 1842. Published his *Dictionary of English etymology* in 1857. Emma Darwin's brother. Married Frances Mackintosh in 1832. (*DNB*.)
    [before 29 September 1857]

Wedgwood, Jessie (1804–72). CD and Emma Darwin's cousin. Married Henry Allen Wedgwood in 1830. (*Emma Darwin.*)

Wedgwood, John Allen (Allen) (1796–1882). CD and Emma Darwin's cousin. Vicar of Maer, Staffordshire, 1825–63. (*Alum. Cantab.*)

Wedgwood, Josiah, I (1730–95). Master-potter. Founded the Wedgwood pottery works at Etruria, Staffordshire. CD's grandfather. FRS 1783. (*DNB, DSB.*)

Wedgwood, Josiah, II (1769–1843). Of Maer Hall, Staffordshire. Master-potter of Etruria. Whig MP for Stoke-on-Trent, 1832–4. Emma Darwin's father. (*Burke's peerage* 1980, *Emma Darwin.*)

Wedgwood, Josiah, III (Jos) (1795–1880). Partner in the Wedgwood pottery in Staffordshire until 1841, when he moved to Leith Hill Place, Surrey. Emma Darwin's brother. Married Caroline Sarah Darwin, his cousin, in 1837. (*Burke's peerage* 1980.)

Wedgwood, Katherine Elizabeth Sophy (Sophy) (1842–1911). Daughter of Caroline and Josiah Wedgwood III. CD's niece. (*Emma Darwin.*)

Wedgwood, Louisa Frances (1834–1903). Daughter of Jessie and Harry Wedgwood. (*Emma Darwin.*)

Wedgwood, Lucy Caroline (1846–1919). Daughter of Caroline and Josiah Wedgwood III. (*Darwin pedigree.*)

Wedgwood, Margaret Susan (1843–1937). Daughter of Caroline and Josiah Wedgwood III. Married Arthur Charles Vaughan Williams in 1869. Mother of Ralph Vaughan Williams (*DNB*). (*Emma Darwin.*)

Wedgwood, Sarah Elizabeth (Eliza) (1795–1857). Cousin of CD and Emma Darwin. Sister of Jessie Wedgwood. (*Emma Darwin.*)

Wedgwood, Sarah Elizabeth (Elizabeth) (1793–1880). Emma Darwin's sister. Resided at Maer Hall, Staffordshire, until 1847, then at The Ridge, Hartfield, Sussex, 1847–62. She moved to London before settling in Down in 1868. (*Burke's peerage* 1980, *Emma Darwin.*)

Wedgwood, Sarah Elizabeth (Sarah) (1778–1856). CD and Emma Darwin's aunt. Resided at Camp Hill, Maer Heath, Staffordshire, 1827–47, then moved to Petleys, Down. (*Emma Darwin.*)

Wedgwood, Sophy. *See* Wedgwood, Katherine Elizabeth Sophy.

Weir, Harrison William (1824–1906). Animal painter and author. Illustrator for the *Illustrated London News*, 1842–1906, as well as other illustrated papers. An experienced breeder and judge of poultry and pigeons. (*DNB.*)

Weld, Charles Richard. Librarian of the Royal Society of London.

Wellsted, James Raymond (1805–42). Officer in the East India Company's surveying ship in the Red Sea, 1830–3. Examined Socotra Island, 1834. Travelled in Oman, 1835 and 1837. Retired from service, 1839. FRS 1837. (R. Desmond 1977, *DNB.*)

Westwood, John Obadiah (1805–93). Entomologist and palaeographer. An active member of the Entomological Society. Hope professor of zoology, Oxford University, 1861–93. Entomological referee for the *Gardeners' Chronicle*. (*DNB*, P. Gilbert 1977.)
   23 November 1856

Whewell, William (1794–1866). Mathematician and historian and philosopher of science. Master of Trinity College, Cambridge, 1841–66. Professor of moral philosophy, Cambridge University, 1838–55. FRS 1820. (*DNB, DSB.*)

White, Walter (1811–93). Cabinet-maker in Reading until 1834; in America, 1834–9. Employed as an attendant in the library of the Royal Society, 1844; assistant secretary and librarian, 1861–84. Author of many travel books. (*DNB, Modern English biography.*)
   26 October 1857

Wicking, Matthew. Brewer of Jenner, Wicking and Jenner, 153 Southwark Bridge Road, London. Breeder of fancy pigeons. Member of the Philoperisteron Society. (*Post Office London directory* 1856.)

Widdrington, Samuel Edward (d. 1856). Naval officer. Author of book on Spain. FRS 1842. (*DNB.*)

Wiegmann, Arend Friedrich August (1802–41). German naturalist. Extraordinary professor, University of Berlin. Founder and editor of the *Archiv für Naturgeschichte*, known as 'Wiegmann's Archiv', 1835–41. (*NUC.*)

Wight, Robert (1796–1872). Botanist. Assistant surgeon, then naturalist, to the East India Company, 1829–31 and 1834–53. Published numerous works on the botany of India. FRS 1855. (R. Desmond 1977, *DNB.*)

Willdenow, Karl Ludwig (1765–1812). German botanist and apothecary in Berlin. Gave up his apothecary shop when he became professor of natural history at the Berlin Medical-Surgical College in 1798. Curator of the Berlin botanic garden, 1801. (*ADB, DSB.*)

Williams & Norgate. London booksellers and publishers who specialised in foreign scientific literature.
   17 November 1857

Wilson, Horace Hayman (1786–1860). Orientalist. Secretary of the Asiatic Society of Bengal, 1811. Professor of Sanskrit, Oxford University, 1832. Director of the Royal Asiatic Society of London, 1837–60. Published a Sanskrit-English dictionary. FRS 1834. (*DNB.*)

Wilson, William Greive (1819–96). BA, St. John's College, Cambridge, 1842; fellow, 1844–9. Rector of Forncett, Norfolk, 1847–96. Tutor to William Erasmus Darwin, 1858. (*Alum. Cantab.*)

Wimmer, Christian Friedrich Heinrich (1803–68). German botanist and educationist. (*ADB.*)

Wingfield, William Wriothesley (b. 1815). Vicar of Gulval, Cornwall. Co-author with George William Johnson of a work on poultry breeding and management. (*Alum. Oxon., NUC.*)

Wolff, Caspar Friedrich (1734–94). German embryologist. Field doctor in the Prussian army, 1761, and lecturer in anatomy at the Breslau Military Hospital. Became academician in anatomy at the St Petersburg Academy of Sciences, 1766. Successfully refuted the theory of preformation. (*ADB, DSB.*)

Wollaston, Thomas Vernon (1822–78). Entomologist and conchologist. Collected insects and shells on Madeira. (*DNB*, P. Gilbert 1977.)
    [February 1856], 6 June [1856], *[c. 27 June 1856]*, *[27 June 1856]*, *[early November 1856]*, *[11 or 18 December 1856]*, *[12 April 1857]*, *[November–December 1857]*

Wood, Searles Valentine (1798–1880). Geologist. Studied the fossils of the the East Anglian Crag, specialising in Eocene fossil Mollusca. (*DNB*, Sarjeant 1980.)

Woodd, Charles Henry Lardner (1821–93). Of Oughtershaw Hall, Yorkshire. Justice of the peace. Elected fellow of the Geological Society in 1846. Brother of Ellen Sophia Fox. (*Burke's landed gentry* 1952.)

Woodward, Samuel Pickworth (1821–65). Naturalist. Sub-curator, Geological Society, 1839–45. Professor of geology and natural history at the Royal Agricultural College, Cirencester, 1845. Assistant in the department of geology and mineralogy, British Museum, 1848–65. (*DNB*, Sarjeant 1980.)
    2 May 1856, 15 May [1856], 27 May [1856], 3 June [1856], *4 June 1856*, [after 4 June 1856], *[after 4 June 1856]*, *15 July 1856*, *[15 July 1856]*, 18 July 1856

Wooler, William A. Landowner in Dinsdale, near Darlington, Durham. (*Post Office directory of Durham, Northumberland, Cumberland and Westmorland* 1873.)

Wright, Charles (1811–85). American botanical collector. Botanist to the United States North Pacific Expedition, 1853–6. Investigated the botany of Cuba, 1856–67. (*DAB.*)

Yarrell, William (1784–1856). Zoologist. Newspaper agent and bookseller in London. Author of standard books on British birds and fishes. (*DNB.*)
    6 [February 1856]

Youatt, William (1776–1847). Veterinary surgeon in London. Gave private lectures on veterinary science; from 1830 he delivered these lectures at London University. Founded the *Veterinarian* in 1828. Author of a series of handbooks on the breeds, management, and diseases of farm animals. (*DNB.*)

Zeyher, Karl Ludwig Philipp (1799–1858). German botanist and plant collector. Collected extensively in South Africa. Author with Christian Friedrich Ecklon of a descriptive catalogue of South African plants (1835–7). (*DSAB.*)

Unidentified.

9 October [1856]

Bibliography of biographical sources

ADB: *Allgemeine deutsche Biographie*. Under the auspices of the Historical Commission of the Royal Academy of Sciences. 56 vols. Leipzig: Duncker and Humblot. 1875–1912.

Alum. Cantab.: *Alumni Cantabrigienses. A biographical list of all known students, graduates and holders of office at the University of Cambridge, from the earliest times to 1900.* Compiled by J. A. Venn. Part II. From 1752 to 1900. 6 vols. Cambridge: Cambridge University Press. 1940–54.

Alum. Oxon. : *Alumni Oxonienses: the members of the University of Oxford, 1715–1886.* By Joseph Foster. 4 vols. Oxford. 1888.

Aust. dict. biog.: *Australian dictionary of biography: 1788–1850; 1851–1890.* Edited by Douglas Pike and Bede Nairn. 6 vols. Melbourne: Melbourne University Press. 1966–76.

Bénézit, Emmanuel. 1976. *Dictionnaire critique et documentaire des peintres, sculpteurs, dessinateurs et graveurs de tous les temps et de tous les pays, par un groupe d'écrivains spécialistes français et étrangers.* Revised edition. 10 vols. Paris: Librairie Gründ.

BHGW: *Biographisch-literarisches Handwörterbuch zur Geschichte der exacten Wissenschaften.* By Johann Christian Poggendorff. Vols. 1–4, Leipzig: Johann Ambrosius Barth. 1863–1904. Vol. 5, Leipzig and Berlin: Verlag Chemie. 1926.

BLKO: *Biographisches Lexikon Des Kaiserthums Oesterreich, enthaltend die Lebensskizzen der denkwürdigen Personen, welche seit 1750 in den österreichischen Krönlandern geboren wurden oder darin gelebt und gewirkt haben.* By Constant von Wurzbach. 60 vols. Vienna. 1856–1890.

Blyakher, L. Y. 1982. *History of embryology in Russia from the middle of the eighteenth century to the middle of the nineteenth century.* Translated from the Russian by H. I. Youssef and B. A. Malek, with an introduction by Jane Maienschein. Washington, D.C.: Smithsonian Institution and the National Science Foundation.

Burke's landed gentry: *Burke's genealogical and heraldic history of the landed gentry.* By John Burke. 1–18 editions. London: Burke's Peerage Ltd. 1833–1972.

Burke's peerage: *Burke's peerage and baronetage.* 1–105 editions. London: Burke's Peerage (Genealogical Books). 1826–1980.

Catalogue de la Bibliothèque nationale: *Catalogue général des livres imprimés de la Bibliothèque nationale*. 231 vols. Paris: Bibliothèque nationale. 1897–1981.

CDEL: *A critical dictionary of English literature and British and American authors, living and deceased*. By S. Austin Allibone. 3 vols. Philadelphia and London. 1859--71. Supplement by John Foster Kirk. 2 vols. Philadelphia and London. 1891.

Clergy list: *The clergy list . . . containing an alphabetical list of the clergy*. London. 1841–.

Complete peerage: *The complete peerage of England Scotland Ireland Great Britain and the United Kingdom extant extinct or dormant*. By G. E. Cokayne. New edition, revised and much enlarged. Edited by the Hon. Vicary Gibbs and others. 12 vols. London: The St Catherine Press. 1910–59.

DAB: *Dictionary of American biography*. Under the auspices of the American Council of Learned Societies. 20 vols., index, and 7 supplements. New York: Charles Scribner's Sons; London: Oxford University Press. 1928–81.

Darwin pedigree: *Pedigree of the family of Darwin*. Compiled by H. Farnham Burke. Privately printed. 1888. Reprinted in Freeman, Richard Broke, *Darwin pedigrees*. London: printed for the author. 1984.

DBF: *Dictionnaire de biographie française*. Under the direction of J. Balteau, M. Barroux, M. Prevost, and others. 17 vols. (A Humann). Paris: Librairie Letouzey et Ané. 1933–89.

DBI: *Dizionaria biografica degli Italiani*. Edited by Alberto M. Ghisalberti. 33 vols. (A–De Foresta). Rome: Istituto della Enciclopedia Italiana. 1960–87.

DBL: *Dansk biografisk leksikon*. Founded by Carl Frederick Bricka, edited by Povl Engelstoft and Svend Dahl. 26 vols. and supplement. Copenhagen: J. H. Schultz. 1933–44.

Desmond, Ray. 1977. *Dictionary of British and Irish botanists and horticulturists, including plant collectors and botanical artists*. 3d ed. London: Taylor and Francis.

Dictionary of Mauritian biography: *Dictionary of Mauritian biography*. Société de l'Histoire de l'Ile Maurice. Port Louis, Mauritius. [1941–].

DNB: *Dictionary of national biography*. Edited by Leslie Stephen and Sidney Lee. 63 vols. and 2 supplements (6 vols.). London: Smith, Elder and Co. 1885–1912. *Dictionary of national biography 1912–80*. Edited by H. W. C. Davis, J. R. H. Weaver, and others. 7 vols. London: Oxford University Press. 1927–86.

DNZB: *A dictionary of New Zealand biography*. Edited by G. H. Scholefield. 2 vols. Wellington: Department of Internal Affairs New Zealand. 1940.

DSAB: *Dictionary of South African biography*. Edited by W. J. de Kock, D. W. Krüger, and C. J. Beyers. Vol. 1, Cape Town: Nasionale Boekhandel Beperk; vols. 2–3, Cape Town: Tafelberg Publishers; vol. 4, Durban and Pretoria: Butterworth & Co. 1968–1981.

DSB: *Dictionary of scientific biography*. Edited by Charles Coulston Gillispie. 14 vols., supplement, and index. New York: Charles Scribner's Sons. 1970–80.

East-India register and army list: *The East-India register and army list*. Compiled from the official returns, by permission of the Honourable the East-India Company. 16 vols. London. 1845–60.

East-India register and directory: The East-India register and directory. Compiled, by permission of the Honourable the East-India Company, from the official returns received at the East-India House. 42 vols. 1803–44.

EB: *Encyclopaedia Britannica*. 11th ed. 29 vols. Cambridge: Cambridge University Press. 1910–11.

EI: *Enciclopedia Italiana di scienze, lettere ed arti*. 35 vols. Rome: Istituto della Enciclopedia Italiana. 1929–39.

Emma Darwin: *Emma Darwin: a century of family letters, 1792–1896*. Edited by Henrietta Litchfield. Revised edition. 2 vols. London: John Murray. 1915.

Emma Darwin (1904): *Emma Darwin, wife of Charles Darwin: a century of family letters*. Edited by Henrietta Litchfield. 2 vols. Privately printed. Cambridge: Cambridge University Press. 1904.

Freeman, Richard Broke. 1978. *Charles Darwin: a companion*. Folkestone: W. Dawson & Sons; Hamden, Conn.: Archon Books, Shoe String Press.

Gascoigne, Robert Mortimer. 1984. *A historical catalogue of scientists and scientific books: from the earliest times to the close of the nineteenth century*. New York and London: Garland Publishing.

Gilbert, Pamela. 1977. *A compendium of the biographical literature on deceased entomologists*. London: British Museum (Natural History).

Grote, Arthur. 1875. Introduction, being a memoir of the late Mr. Ed. Blyth, C.M.Z.S., and Hon. Member Asiatic Soc. of Bengal. *Journal of the Asiatic Society of Bengal* n.s. 43 (pt 2) (extra number, August 1875): iii–xxiv.

Hooker, Joseph Dalton. 1854. *Himalayan journals; or, notes of a naturalist in Bengal, the Sikkim and Nepal Himalayas, the Khasia mountains, &c.* 2 vols. London.

Huxley, Leonard, ed. 1900. *Life and letters of Thomas Henry Huxley*. 2 vols. London: Macmillan and Co.

Huxley, Leonard, ed. 1918. *Life and letters of Sir Joseph Dalton Hooker . . . based on materials collected and arranged by Lady Hooker*. 2 vols. London: John Murray.

IBN: *Index bio-bibligraphicus notorum hominum*. Part C: *Corpus alphabeticum sectio generalis*. Edited by Jean-Pierre Lobies. 45 vols. (A–Czyżyk). Osnabrück: Biblio Verlag. 1974–89.

Jenyns, Leonard. 1862. *Memoir of the Rev. John Stevens Henslow*. London.

Larousse XX: *Larousse du XXᵉ siècle*. Under the direction of Paul Augé. 6 vols. Paris: Librairie Larousse. 1928–33.

Livingstone, David. 1857. *Missionary travels and researches in South Africa*. London.

Modern English biography: *Modern English biography: containing many thousand concise memoirs of persons who have died since the year 1850*. By Frederic Boase. 3 vols. and supplement (3 vols.). Truro: printed for the author. 1892–1921.

Moore, James R. 1985. Darwin of Down: the evolutionist as squarson-naturalist. In Kohn, David, ed., *The Darwinian heritage*. Princeton: Princeton University Press in association with Nova Pacifica.

NBU: *Nouvelle biographie universelle*. Edited by Jean Chrétien Ferdinand Hoefer. 46 vols. Paris. 1852–66.

NDB: *Neue deutsche Biographie*. Under the auspices of the Historical Commission of the Bavarian Academy of Sciences. 15 vols. (A–Maltzan). Berlin: Duncker and Humblot. 1953–86.

NNBW: *Nieuw nederlandsch biografisch woordenboek*. Edited by P. C. Molhuysen, P. J. Blok, and K. H. Kossmann. 10 vols. Leiden: A. W. Sijthoff. 1911–37.

NUC: *The national union catalog*. Pre-1956 imprints. 685 vols. and supplement (vols. 686–754). London and Chicago: Mansell. 1968–81.

O'Byrne, William R. 1849. *A naval biographical dictionary: comprising the life and services of every living officer in Her Majesty's Navy, from the rank of admiral of the fleet to that of lieutenant, inclusive*. London.

[Philippart, John]. 1823–26. *The East India military calendar; containing the services of general and field officers of the Indian army*. 3 vols. London.

Physicians: *The roll of the Royal College of Physicians of London*. Vol. 3: (1801–25). By William Munk. 2d ed., revised and enlarged. London. 1878. Vol. 4: *Munk's roll, vol. 4: Lives of the fellows of the Royal College of Physicians* (1826–1925). Compiled by G.H. Brown. London: Royal College of Physicians. 1955.

Post Office directory of Durham, Northumberland, Cumberland and Westmorland: *The Post Office directory of the county of Durham, and the principal towns and adjacent places in Northumberland, Cumberland & Westmorland*. London. 1873–.

Post Office directory of Lancashire: *Post Office directory of Lancashire*. London. 1858–.

Post Office directory of Lincolnshire: *Post Office directory of Lincolnshire*. London. 1849–.

Post Office directory of the six home counties: *Post Office directory of the six home counties, viz., Essex, Herts, Kent, Middlesex, Surrey and Sussex*. London. 1845–.

Post Office London directory: *Post Office London directory*. London. 1802–.

Provincial medical directory: *The provincial medical directory*. London. 1847.

Rehbock, Philip F. 1979. The early dredgers: 'naturalizing' in British seas, 1830–1850. *Journal of the History of Biology* 12: 293–368.

Richardson, Edward William. 1916. *A veteran naturalist; being the life and work of W. B. Tegetmeier*. London: Witherby & Co.

Sarjeant, William A. S. 1980. *Geologists and the history of geology: an international bibliography from the origins to 1978*. 5 vols. London: Macmillan.

SBL: *Svenskt biografiskt lexikon*. Edited by Bertil Boëthius, Erik Grill, and Birgitta Lager-Kromnow. 26 vols. (A–Nilsson). Stockholm: Albert Bonnier and P. A. Norstedt. 1918–89.

Scherren, Henry. 1905. *The Zoological Society of London. A sketch of its foundation and development and the story of its farm, museum, gardens, menagerie and library*. London, Paris, New York, and Melbourne: Cassell & Co.

Stenton, Michael. 1976. *Who's who of British members of Parliament, 1832–1885*. Brighton: Harvester Press.

Tabler, Edward C. 1977. *Pioneers of Natal and southeastern Africa 1552–1878*. Cape Town and Rotterdam: A. A. Balkema.

Tegetmeier, William Bernhard, ed. 1856–7. *The poultry book: including pigeons and rabbits; comprising the characteristics, management, breeding and medical treatment of poultry, being the results of personal observation and practice of the Rev. W. Wingfield, G. W. Johnson, Esq.* Re-arranged and edited by W. B. Tegetmeier. London.

Tegetmeier, William Bernhard, ed. 1879. *Columbarium; or, the pigeon-house. Being an introduction to a natural history of tame pigeons. By John Moore. Reprinted verbatim et literatim from the original edition of 1735, with a brief notice of the author by W. B. Tegetmeier.* London.

Wallace, Alfred Russel. 1905. *My life: a record of events and opinions.* 2 vols. London: Chapman & Hall.

Woodward, Samuel Pickworth, 1851–6. *A manual of the Mollusca; or, a rudimentary treatise of recent and fossil shells.* 3 pts. London.

WWW: Who was who. A companion to *Who's who.* 7 vols. and cumulated index, 1897–1980. London: Adam & Charles Black. 1920–81.

INDEX

The dates of letters to and from Darwin's correspondents are listed in the Biographical register and index to correspondents and are not repeated here. Darwin's works are indexed under the short titles used throughout this volume and listed in the bibliography.

Aberdeen, Scotland, 293, 294
abortive organs: CD's portfolio of notes on, 253 & 254 n.7
Abyssinia: animals of, 9; botanical relations of, 226 n.2, 260, 262; flora of, 78 & 80 nn.4 & 5, 81 & 82 n.8
Acacia, 198, 353, 372 & 373 n.7, 417
Academia Caesarea Leopoldino-Carolina Naturae Curiosorum, 451 & nn.1 & 2, 452 & 453 nn.1 & 5
Acanthaceae, 502 n.2
Acronycta, 328
actinians, 162 n.6
Acton, Samuel Poole, 233, 288
Adams, Arthur, 128 n.12, 182 & 185 n.6
Adams, Henry, 128 n.12, 182 & 185 n.6
Adanson, Michel, 390
Adelops, 283
Adkins, Mr, 211 & n.5
Adlumia, 437, 492 & 493 n.5; *cirrhosa*, 447 & 450 n.16, 469 & n.12, 487 & n.8, 492 & 493 n.5
Aegilops, 176, 190 & 191 n.8, 197 & nn.3 & 6, 199, 202 & 203 n.3
Africa: birds of, 7, 116 & 120 n.6; botanical relations of, 78–9, 80–2 & nn.8 & 9, 167 n.6; cats of, 241, 270 & 271 n.5; diseases among white men in, 241–2 & 243 n.4, 322 & n.6; dogs of, 137, 241, 243, 410 & 411 n.5, 435 & nn.1 & 2; fish of, 188; fowls of, 9–10, 26, 28 & 30 nn.23 & 24, 69 & 70 n.7, 135 & 136 n.7, 243, 246–7 & n.2, 248–9, 290 & n.3; geology of, 80–2, 153, 154, 164 & 165 n.3; mammals of islands of, 242 & 243 n.5, 269–70 & nn.5 & 7, 272 & n.3; pigs of, 206 & n.3, 207 n.5, 211 & 212 n.2, 240 & n.6, 410 & 411 n.3; zoological relations of, 78–9, 80

Agassiz, Louis, 106 & 107 n.5, 228; and blind cave fauna, 180 & 181 n.5, 215–16 & nn.2 & 5, 299 & 300 n.1; CD's opinion of, 456; Asa Gray's opinion of, 340 & 342 n.8; on fish of North America, 187, 216 & 217 n.7; on law of parallelism, 216, 235 & 237 n.6, 299–300; on races of man, 184 & 185 n.13; and scientific methodology, 315, 360 & 362 n.6, 363 & 364 n.4; sends collector to Pacific, 381 & 382 n.9, 400 n.3; and use of negative evidence in geology, 237; view of species, 90, 236
Agrilus darwinii, 374 & 376 n.8
Agrostis, 50 & n.8, 342
Aira antarctica, 261 & 263 n.11
Airy, George Biddell, 238 & 239 n.6
Ajasson de Grandsagne, Jean Baptiste François Étienne, 29 n.13
Akis, 374
Albert, Prince, 434 n.2
Albin, Eleazar, 26 & 29 n.8
Alder, Joshua, 405 n.10
Aldrovanda, 157, 171 & n.13
Aldrovandi, Ulysse, 14 & 18 n.57, 59 n.5
Alexander the Great, 10
algae, 105, 112 & n.6, 266, 305 & n.9, 316–17 & nn.2 & 3. *See also Fucus*
Allen, William, 30 n.23, 61
Allman, George James, 428 & n.9
Alopecurus, 423
Alps: geology of, 48 & 49 n.5, 144 & 146 n.2
Alsine media, 370
Amadina, 13
American Association for the Advancement of Science, 195 & 197 n.4, 215 & 216 n.1
American Journal of Science and Arts, 181 n.3, 196 & 198 n.8, 246, 258, 362 n.3, 403, 412 & 414 n.3, 422 & 423 n.2, 437 & n.3, 508 n.6

Amherstia, 38
Amsterdam Island, 337 & 338 n.7
Anacharis, 401
Anagallis arvensis, 369
Anas: boschas, 352; *maxima*, 352 & 355 n.2, 353; *moschata*, 9. *See also* ducks
Ancylus fluviatilis, 89 & 91 n.6
Andersson, Charles John, 71 & 72 nn.1 & 2
Andersson, Nils Johan, 444 & 445 n.16, 452 & 453 n.8
Andral, Gabriel, 474 & n.6
Anerley, Kent: CD attends poultry show at, 160 & n.6, 192 & 193 n.2, 204 & n.5, 205 & n.3; poultry show at, 110 & 111 nn.3–5, 234
Angerona, 332
animals: blind cave fauna, 180 & 181 nn.4 & 5, 216 & nn.2–5, 235, 282–4 & n.1, 299 & 300 n.1; blood differences of, 24; cross-fertilisation among, 161 & 162 n.2, 173, 178 & 179 n.3, 190 & 191 nn.3 & 5, 302 & 303 n.14, 307 & 308 n.8, 428 & n.9, 433, 492 & 494 nn.10 & 11, 496–7 & nn.1–3; illustrations of on ancient monuments, 13, 107–8 & n.3, 137 & 138 n.8; instincts of, 378 n.1, 513 & n.7; markings of, 8; naturalised species, 220–2 & 203 n.3, 225–6, 241 & 243 n.3, 243, 514 & 515 n.9; prolificness of hybrids, 40–1 & 42 n.20, 387 & 388 n.6; variability of highly developed organs, 395–6 & nn.1, 2 & 3, 407 & 408 nn.3 & 4, 409 & 410 n.9
Annals and Magazine of Natural History, 387 & 388 n.3, 514 & 515 n.3
Annals des Sciences Naturelles, 317 n.2
annelids, 506 & n.7
Annobon Island, 61 & 65 n.3, 242 & 243 n.5, 243
Anophthalmus, 283
Anser mareca, 119
Antarctic, 261; flora of, 166–7 & n.3, 170 & 171 n.2, 172 & n.2, 174, 177, 179 & 180 n.11, 191–2 & n.4, 203 & 204 nn.1, 2 & 5; zoology of, 262, 380
Antennaria, 341
Anthemis, 231
Anthyllis, 388 & 390 n.2
Antilles, 9, 10
Antiquarian Society of London, 110 n.4
Aphis, 457 n.10, 484–5 & nn.3 & 4
Aplecta, 328
Appalachian Mountains, 231 n.7
Aquilegia, 309 & 310 n.6
Arabia: animals of, 14, 27, 40 & 42 n.15, 64; fowls of, 10 & 17 n.39, 61; pigeons of, 88 n.2

Arbutus, 294
Arctic, 261, 263 n.12; flora of, 140–1, 151 & 152 n.7, 166–7 & n.3, 179 & 180 n.11, 384 & 385 n.7, 390 & n.2; fossil musk-ox of, 61 & 66 n.13; geology of, 263 n.12; molluscs of, 151 & 152 n.7, 186; zoology of, 187–8 & n.1, 262
Arctic and alpine species, 81–2, 113, 182 & n.3, 186, 198–9 & 200 nn.3 & 4, 257, 258, 262, 390; apetalous flowers in, 384 & 385 n.7, 390 & n.3; dwarfing in, 113, 371, 396; migration of, 81–2, 186, 262 & 264 n.14, 271, 272–3, 274, 340; woolliness of, 384 & 385 nn.3, 7 & 8, 388–9 & n.2, 390 & nn.6 & 7
Arenaria, 132, 311, 312
Aristida, 102
Aristolochia, 156, 170, 176 & 177 n.15
Aristophanes, 10, 13, 14
Aristotle, 9
Arnott, George Arnott Walker, 175 & 177 n.4
Arnott, Neil, 73 n.7
Arrhenatherum, 423
Artemisia, 356–7
Arthrochortus, 75 & n.6
Articulata, 421 n.3
Aru Islands, 514 & 515 nn.4–6
Ascension Island, 126 n.4; fauna of, 220–2 & 223 n.3; pigeons of, 137 & 138 n.7, 214 & 215 n.5; plants of, 102 & 103 n.2
ascidians, 161 & 162 nn.2–4, 173, 178
Asclepias, 341, 445 & 449 n.5
Asiatic Society of Bengal, 15 n.2, 18 n.66, 38 & 41 nn.2 & 7, 43 & 44 n.2
Asida, 374
asparagus, 371
asses, 8; coloration of, 409 & 410 n.10, 411, 417; of Arabia, 40 & 42 n.15; wild forms of, 6 & 16 n.6, 14, 27, 64 & 66 nn.24, 30 & 31
Asso y del Rio, Ignacio Jordán de, 116 & 120 n.7
Assyria, 62, 107–8 & nn.2 & 3
Astacus, 216, 302 n.7, 425 & 427 n.3
Astley, Thomas, 242 & 243 n.5
Astragalus hypoglottis, 132
Astronomical Society of London, 110 n.4
Athenæum, 3 & n.7, 7, 455 & n.2, 464 & n.5
Athenæum Club, 1 & 2 n.3, 467, 507 & n.4; W. E. Darwin proposed for membership of, 45 & 46 n.9, 349 & 350 n.2; T. H. Huxley's membership of, 103 & 104 n.13, 106 & 107 n.3, 109 & 110 nn.7 & 8, 112 & n.3; John Lubbock proposed for membership of, 46 n.9, 349–50 & n.3

Atkinson, William Henry, 38 & 42 n.5, 43 & 45 n.3

Atlantis, 375

aubergines, 46, 49

Audouin, Jean Victor, 456 & 457 nn.7 & 9

Aukland Island, 153

Australia, 317 n.3; birds of, 14; botanical relations of, 81, 153, 171 n.2, 176 & 177 n.11, 198–9, 260, 262, 271, 325; fish of, 187–8; fowl of, 56; geology of, 153, 154; plants of, 105, 112 & n.7, 178, 501 & n.3, 508 & n.5; wines of, 345 & 346 n.5; zoological relations of, 125, 154, 271, 515 n.6

Austria: scientific expedition of, 337–8 & nn.1 & 7

Autard de Bragard, Gustave Adolphe, 232

Ava, Burma, 260

Ayeen Akberi, 31 & 32 n.4

Azara, Félix d', 7 & 16 n.12, 354 & 355 n.4

Azores Islands, 123 n.2, 153, 154, 156, 174; flora of, 80, 372

Babington, Charles Cardale: and aquatic plants, 277 & 280 n.1; *Manual of British botany*, 175 & 177 n.3, 197, 438 & 439 n.7, 443 & 444 n.12, 499 & 500 n.9

Backhouse, James, 355 & 358 n.4

Baer, Karl Ernst von, 461 & 462 n.4

Baikal, Russia, 156, 157

Baily, John, Jr, 2 & 3 n.4, 21 & n.2, 152 & 153 n.5, 234 & 235 n.2

Baily, John, Sr, 2 & 3 n.5, 235 n.2

Baird, William, 251 & 252 n.8

Baker, Charles N., 2 & 3 n.3, 160, 234

Baker, Samuel C., 2 & 3 n.3, 160, 234

Bakewell, Robert, 201 & 202 n.5

Balanus, 492 & 494 nn.10 & 11, 496 & 497 nn.2 & 3; *balanoides*, 318 & n.3, 497 nn.2 & 3; *communis*, 497

Bali, 290 n.4

Bangkok, Thailand, 9

Banksia, 112 & n.7

barbel, 364, 365

barberry, 228–9, 372, 417

Barbot, James, 243 n.5

Barbot, John, 28 & 30 n.24, 242 & 243 n.5

Barclay, James, 98 & 99 n.4

Barlaston, Staffordshire, 269 & n.10, 304 n.3, 419 n.2

Barmouth, Wales, 1 & 2 n.6

Barnéoud, François Marius, 420 & 421 n.3, 427 & 428 nn.4 & 6

Barnes, James, 37 nn.2 & 7, 80 & 82 n.3

Barrande, Joseph, 366 & 367 n.11

Barry, Martin, 193 & n.4

Barth, Heinrich, 421 & 422 n.3

Bartlett, Abraham Dee, 39 & 42 n.13, 288 n.3

Barton, Benjamin Smith, 93 & 95 n.14

Bartram, William, 182 & n.5

Bashford, F., 7 & 16 n.8

Bate, Charles Spence, 494 n.11, 496–7 & nn.1 & 2

Bateleur, 248 n.2

Bates, Henry Walter, 481 & 482 n.3

Batrachium, 131, 351, 356

bats, 75 n.4, 421

Baudement, Émile, 462 & n.7

Beagle, HMS, 55 & 56 n.4, 98, 223 n.3, 305 n.9

bears, 147 & 148 n.3

Beaufort, Francis, 404 & 405 n.15

Bechstein, Johann Matthäus, 211 & 213 n.3, 240 & n.6

Beckenham, Kent: railway extension to, 73 & n.17, 286 & n.8, 335 & n.5, 471

Beckles, Samuel Husband, 321 n.2

bees, 343 & n.4, 452 & 453 n.9; boring of flowers, 466–7 & n.5; cells of, 377–8 & nn.1–3; and fertilisation of plants, 291 & n.4, 295 & n.3, 335, 346 & 347 n.5, 361, 433 & nn.5 & 6, 437 & n.5, 446–7 & 449 n.12, 465–7 & nn.4 & 6, 487 & n.9, 489 & n.6, 492 & 494 n.7, 493; hybrids among, 301 & 302 n.6, 307 & 308 n.3; parthenogenesis in, 156 & 158 n.8, 171 & n.12, 301 & 302 nn.2 & 3, 343 & nn.1 & 2

Beesby, Lincolnshire, 275 n.2

beetles, 32 & 33 n.1, 90, 252–3 & nn.2–4

Beh, Mr, 439

Beke, Charles Tilstone, 225 & 226 n.2, 232 & n.3

Beke, Emily Alston, 226 n.2

Belcher, Edward, 188

Bell, Thomas, 72 & 73 nn.3 & 9, 110 n.4, 381 & 382 n.8

Bennett, John Joseph, 321 & 322 n.4

Benson, William Henry, 184 & 185 n.10

Bentham, George, 102 & 103 n.4, 324, 464, 467 n.3; and apetalous flowers, 291, 295 & n.2; and cross-fertilisation in Leguminosae, 286–7 & n.2, 291, 361 & 362 n.7; and *Cucubalus*, 504, 505 & n.2, 508–9 & n.2; and Cucurbitaceae, 326; and CD's species theory, 495; and laburnum, 217; and naturalised plants, 494 & 495 n.6; on flora of Britain, 499 & 500 nn.6 & 9, 502 & n.4, 508 & n.4; on primroses and cowslips, 103; on vitality of ancient seeds,

Bentham, George, cont.
3 & n.8; and plants of north-west North America, 494 & 495 n.3; reputation as species lumper, 37; reviews botanical works, 103 & 104 n.10, 299 n.6; and Royal Medal, 85 & n.5; and *Silene*, 494 & 495 n.5, 500 n.12, 502 & 503 n.5, 504 & nn.2 & 3, 505 & n.2

Berghaus, Heinrich Karl Wilhelm, 97 n.8

Berkeley, Miles Joseph, 266 n.1; *Introduction to cryptogamic botany*, 49 & 50 n.5; and seed-coats of hybrid peas, 47 & n.7, 49, 487 & n.10, 494 n.9; seed-salting experiments of, 46–7 & n.3, 49 & 50 n.1

Berthelot, Sabin, 116 & 120 n.5, 117, 119, 144 & n.6, 443 & 444 n.7, 450 & n.2, 460 & 461 n.6

Bessel's Green, Riverhead, Kent, 4 n.11, 470 & 471 n.2

Bible: fowls mentioned in, 13 & 18 n.55

Birch, Samuel, 19 & 20 n.3, 35, 57–8 & n.3; CD arranges to visit, 57 & nn.1 & 2

birds: blood corpuscles of, 24; blown by hurricanes, 232, 336; and dispersal of molluscs, 385 & n.9, 453 & n.10, 455 & n.4; and dispersal of seeds, 100 & 101 n.10, 212 & 213 nn.12 & 13, 233 & n.4, 238, 239–40 & n.5, 248 & n.2, 250 & n.6, 266, 267 & 268 n.5, 274 & n.5, 294 & 295 n.7, 305 & n.8, 309 & 310 n.7, 354, 365 n.2, 382–3 & nn.3 & 5, 389 & 390 n.6; distribution of, 206 & n.7; extinct species of, 232 n.5, 336 & 338 n.4, 348; of Africa, 7, 116 & 120 n.6; of Canary Islands, 116–19; of Germany, 206 & nn.2, 3 & 5; of Jamaica, 233 n.3, 319, 352 & 355 n.2; of Madeira, 27, 115–20 & nn.1, 3, 4 & 12, 206 & n.7, 320 n.2; of Mauritius, 225, 226; of Réunion, 232 n.5; of Scandinavia, 8; of Spain, 116 & 120 nn.7 & 8; osteology of, 417 & n.3; variability in, 407, 411

Birkbeck, Robert, 490 n.4

Bishop, Richard, 494 n.11, 496–7 & nn.1 & 2

Black Mountains, North Carolina, 230 & 231 n.7

blackbirds, 206

Blatta, 456 & 457 nn.5–10 & 14

Blyth, Edward, 32 n.4, 388 n.3; asks CD's assistance with petition for increase in salary, 38 & 41 n.2, 43 & 44 n.2; asks CD to send him animals for sale, 39–40; on a zoological collection from Somáli, 61 & 65 n.5; on Asiatic lions, 6 & 16 n.6; on cabbage family, 9 & 17 n.28; on camels, 7–8 & 16 nn.18–20; on canaries, 7; on cultivated varieties of fish, 26 & 29 n.4; on deer, 28 & 30 n.29, 40; on

fertility of hybrids, 40–1 & 42 nn.20 & 21; on geese, 53 n.1, 213 n.11; on guinea-fowls, 9–11, 65 n.5; on hatching eggs artificially, 43; on hybrids of birds, 7, 8 & 16 n.23; on instinct, 7 & 16 n.9; on markings in animals, 8; on natural history of Seychelles, 338 n.5; on orang-utans, 6 & 16 n.5, 25 & 29 n.1; on origin of domestic fowls, 13–14, 26–7, 28, 69 & 70 n.7; on pheasants, 11–12, 26; on pigeons, 28, 39, 62–4 & 66 n.21, 67–9; on rabbits, 11; on turkeys, 12–13, 27, 28; on wild asses, 6 & 16 n.6, 8, 27 & 30 n.19, 40 & 42 n.16, 64 & 66 nn.23 & 28; on wild cattle, 4 & 5 n.1, 6 & 15 n.3, 12 & 18 n.47, 28; on wild dogs, 40; sends CD skins of jungle fowl, 26; and A. R. Wallace on species, 514 & 515 n.3

Blyth, Mrs Edward, 38 & 42 n.9

boas, 319

Bochart, Samuel, 14 & 18 n.58, 61 & 65 n.2

Bodleian Library, Oxford, 23 & n.5

Bogota, Columbia, 340

Boltenia, 162 n.4

Bomare, Jacques Christophe Valmont de. *See* Valmont de Bomare, Jacques Christophe

Bombay, India, 38, 61

Bombyx mori, 302 n.2

Bonaparte, Charles Lucien, 7 & 16 n.15, 8 & 17 n.24, 64 & 66 n.25

Bonin Islands, 400 & n.3

Bonplandia, 453 n.5

Boott, Francis, 195 & 197 n.5, 398 & 399 n.4

Boreau, Alexandre, 429 & n.6, 438 & 439 n.5, 444 & 445 n.14, 469 & n.14

Borneo: animals of, 199, 201, 388 & n.9; botanical relations of, 81; cattle of, 12; fowls of, 136 & n.5, 296 & n.4, 344–5 & n.2, 414 & 415 n.4

Borragineae, 498 & 500 n.3, 502 n.2

Bos, 12. *See also* buffalos, cattle

Bosquet, Joseph Augustin Hubert de: and CD's study of pigeons, 255–6 & n.1, 265 & n.6; and fossil cirripedes, 219 & nn.2 & 3, 237 & n.12, 257 & n.5

Boswell-Syme, John Thomas Irvine, 149–50 & 151 n.3, 209, 277 & 279 nn.3 & 4, 438 & 439 n.8, 443, 504 nn.2 & 3

Bouchard-Chantereaux, Nicolas Robert, 314 & n.5

Bougainvillea, 38

Bourbon Island. *See* Réunion Island

Bouton, Louis Sulpice, 232 & n.3

brachiopods, 308 & nn.2 & 3, 312–14 & nn.1 & 2, 404 & 405 n.11, 454 n.3

bramble, 221
Brazil, 154, 318 & 320 n.2
Brehm, Christian Ludwig, 206 & nn.2, 3, & 5, 207 & n.2
Brent, Bernard Philip, 3 & 4 n.11, 58 n.5, 60 & n.6, visits Down House, 247; and hybrid canaries, 470 & 471 n.1
Bridge, Mr, 414 & 415 n.5, 416
Bright, John, 375 & 376 n.14
Bristowe, Henry Fox, 58 & 59 n.10
Bristowe, Selina, 58 & 59 n.10
British Association for the Advancement of Science, 55 n.2, 202; James Buckman's papers for, 464 & nn.5 & 7; Cheltenham meeting of, 197 & nn.1 & 3; dredging committee of, 91 n.11; Dublin meeting of, 317 & n.4, 455 & n.1; Glasgow meeting of, 30 & 31 n.1; Oxford meeting of, 485 n.6
British Museum, 161 & 162 n.3, 320 n.3, 322 n.4, 373 & 376 n.4, 377, 410, 482 & n.1; CD plans to donate collections to, 21 & n.4, 485; CD visits, 57 & 58 n.3
Brock, Dr, 473
Brodie, Benjamin Collins, 73 n.7, 205 n.4, 369 n.5
Brongniart, Adolphe Théodore, 123 n.3
Brooke, James, 12 & 18 n.51, 135 & 136 nn.5 & 12, 296 & n.4, 344 & 345 n.2, 388 & n.9, 393 & 394 n.7, 397
Brown, Robert, 178, 322, 325 & 326 n.5, 400, 516 & n.4
Browne, William George, 69 & 70 n.6
Brullé, Gaspard Auguste, 420 & 421 nn.3 & 5, 424–6 & 427 nn.4 & 9, 427–8 & nn.3 & 5, 456 & 457 n.7
Brussels, Belgium, 336
bryony, 281 n.6
Bryozoa, 173 & nn.4 & 5, 404 & 405 n.11
Buccinum antarcticum, 127
Buch, Christian Leopold von, 84 n.2
Buchanan, James, 237 n.11
Buckman, James, 96 & 97 n.9, 463, 464 & nn.5, 7 & 8, 484 n.6
buffalos, 14, 61–2
Buffon, Georges Louis Leclerc, Comte de, 10 & 17 n.37, 14
Buist, George, 491 & nn.2 & 4
Bult, Benjamin Edmund, 21 & nn.5 & 6, 34 & 35 n.2
Bult, Samuel, 257 & n.6
Bunbury, Charles James Fox, 84 n.4, 321 n.2; and CD's species theory, 36 & 37 n.8, 78 & 80 n.3, 80–1, 91 n.7; and flora of Cape Colony,

36–7 & n.10, 78–9; and migration of plants, 80–2, 105 & 106 n.3; on new variety of *Colletia*, 36–7, 78 & 80 n.2; on plants common to Cape Colony and Europe, 104 & nn.1 & 3, 105 & 106 nn.2 & 3
Bunbury, Frances Joanna, 84 n.4
Bunsen, Robert Wilhelm Eberhard, 409 & 410 n.7
Bunyan, John, 376 n.11
Buprestidae, 374
Burckhardt, John Lewis, 8 & 16 n.20, 40 & 42 n.15
Burlington House, Piccadilly, 107 & 110 n.4, 347 & n.1, 429 n.8
Burma, 225 & 226 n.3, 459
Burnett, Waldo Irvin, 484 & 385 n.5
bustards, 10
Butea, 287
Butler, Thomas, 2 n.6
Butter, Donald, 12 & 18 n.49
butterflies and moths, 88, 493; classification of, 329 n.3, 332; copper, 454 & 455 n.3, 471 & n.3; double broods of, 332, 333; geographical distribution of, 215 & nn.2 & 4; interest of Darwin boys in collecting, 32 & 33 n.1, 454 & 455 n.2; of Mauritius, 348; parthenogenesis in, 156 & 158 n.8, 171 & n.12, 302 n.2; variation in, 327–8, 331 & n.2, 331–3 & n.3
Butterworth, Mr, 439
Buttler, John Olive, 5 & 6 n.5
Buxton, Charles, 166 & n.1
Buxus sempervirens, 294

cabbage family, 9 & 17 n.28, 178 & 179 n.9, 201, 291, 433
Caccabis, 11
cactuses, 460 & 461 n.2, 472–3 & nn.1 & 2
Caesalpineae, 286, 287
Caesar, Gaius Julius, 14
Cairina moschata, 9, 352
Calcutta Sporting Review, 6 & 16 n.6, 26
Calcutta, India, 38, 43 & 44 n.2, 157 & 158 nn.11, 12, & 13, 171 n.13; botanic gardens, 263, 280 & 281 n.2
Callithamnia, 317
Callitriche verna, 192, 202 & 203 n.5, 233, 244
Camarões, Rio dos. *See* Cameroons, estuary of
Cambarus, 216
Cambridge University, 250 & n.5, 335 & n.6, 475 & 476 n.2
Camden, Marquis of. *See* Pratt, George Charles
camels, 7–8 & 16 nn.18–20, 14

Cameroons, 322 n.2; estuary of, 270 & 271 n.6
Campanulaceae, 190, 309 & n.3
Campbell, Andrew, 469 & n.13
Campeche, Bay of (Mexico), 129 n.27
Canada, 132, 327 & n.2, 362–3 & n.1
canaries, 7, 27, 470 & 471 n.1; hybrids of, 318; of Jamaica, 228 & n.6, 318 & 320 n.2; of Madeira, 228, 320 n.2
Canary Islands: birds of, 13, 27, 116 & 120 n.5; flora of, 80, 372, 443 & 444 n.7, 502; geology of, 84 n.4, 163–4, 168; and land-bridge theory of distribution, 164–5 & nn.3 & 8, 174; molluscs of, 90 & 91 n.11, 146; T. V. Wollaston visits, 479 n.1
Candolle, Alphonse de, 82 & n.10, 94 & 95 n.18, 170 & n.4, 172, 281, 282 & n.3; and alpine species, 396; and botanical statistics, 93 & 95 nn.9–11, 245 & 246 n.9, 316 n.3, 363; and disjoined species, 339–40 & 342 nn.5 & 7, 392 n.3; *Géographie botanique raisonnée*, 25, 75 n.3, 107 n.8, 200 & 202 nn.2–4, 203 & 204 n.6, 244, 245 & 246 n.7, 246 & n.11, 298 & 299 n.6, 316 n.3, 325 & 326 n.4, 358 & 359 n.7, 363 & 364 n.3; on crossing, 190 & 191 n.4; on means of dispersal, 105, 201 & 202 n.3; on naturalised plants, 92–3 & 94 nn.5 & 13, 103 & 104 n.11, 201 & 202 n.4, 246 & n.7, 325; on representative species, 180 & 181 n.7; on social plants, 93 & 95 n.16; *Podromus*, 452 & 453 n.7, 460 & 461 n.10, 468 & 469 n.6, 487 & n.5, 498 & nn.2 & 4, 499, 502 & n.2, 508 & n.2, 516
Candolle, Augustin Pyramus de, 171 & n.13, 448; *Podromus*, 452 & 453 n.7, 460 & 461 n.10, 468 & 469 n.6, 487 & n.5, 498 & nn.2 & 4, 499, 502 & n.2, 508 & n.2, 516
Canino, Prince of. See Bonaparte, Charles Lucien
Canis, 27
Canning, Charles John, 38 & 42 n.8
Canning, Charlotte, 38 & 42 n.8
Cantal, France, 215 & n.3
Cape Colony, South Africa, 271, 498, 500; animals of, 9, 137 & 138 nn.2, 4 & 7, 296; crustaceans of, 380; fish of, 81, 188; flora of, 36–7 & n.10, 78–9 & 80 nn.4–6, 80–2 & n.9, 95 n.10, 104 & nn.1–3 & 7, 105, 167 & n.6, 363 & 364 nn.1 & 2, 389 & 390 n.4; geology of, 80–2 & n.9; molluscs of, 185–6; pigeons of, 137
Cape Town, South Africa, 138 n.1
Carabus, 374
Caracas, Venezuela, 353, 354

Carcinus maenas, 381
Cardamine hirsuta, 369, 370
Carex, 351, 357, 391 & 392 n.3, 398 & 399 n.4, 401, 446
carnations. See *Dianthus*
carp family, 26 & 29 n.4, 212, 216, 364 & 365 n.2, 379; dace, 212; tench, 379
Carpenter, Philip Pearsall, 182 & 184 n.6
Carpenter, William Benjamin, 73 nn.7 & 14, 114 & n.2, 189 & n.1, 260
Carpophaga oenea, 8
carrots, 450
Carstang, Mr, 246 & 247 n.3, 248 & 249 n.2, 256, 334 & n.3, 338, 339, 344
Cassia obovata, 353
Cassini, Henri, 413 & 414 n.12
Cassiope hypnoides, 230
Casuarina, 112 & n.7
caterpillars, 8
cats, 8; civet, 270 & 271 n.5; correlation of colour and deafness in, 136 & n.11, 141 & 142 n.1, 307; hybrids of, 137 & 138 n.4; of Africa, 241, 270 & 271 n.5; of Ascension Island, 221–2; of Borneo, 388 & n.9; of Singapore, 388 & n.9; of West African islands, 270 & 271 n.5; osteological comparison of, 212
cattle, 270, 422 & n.4; blood corpuscles of, 24 & n.4; and dispersal of plants, 353, 354; Ghor-Khur, 64 & 66 n.29; Hereford, 210 & n.2, 211 & 213 nn.5 & 7; hybrid crosses in, 455 & n.2; musk-ox, 61 & 66 n.13, 62; Niata, 7 & 16 n.11; of Arabia, 14; of India, 4–5 & 6 n.7, 12 & 18 n.50, 65; short-horn, 201 & 202 n.5
Cautley, Proby Thomas, 12 & 18 n.48
Cayman Islands, 352
Celebes, 290 n.4, 387
Centaurea, 197, 202 & 203 n.2
Cereus, 460 & 461 n.2, 472–3 & n.2
Cervus: elephus, 14 & 18 n.62; *virginianus*, 6. See also deer
Cetonidae, 374
Ceylon, 317 n.3; flora of, 105, 260, 262 & 264 n.18, 293; poultry and pigeons of, 54 & 55 nn.4 & 6; vegetation of, 54 & 55 n.3
Chacura, 11 & 18 n.43
chaffinches, 27
Chambers, Robert, 178 & 179 n.9, 194, 200 n.8, 236 & 237 n.9, 449 n.6
Chambers's Journal, 40
chameleons, 27
Chamisso, Adelbert von, 132 & 133 n.9
chamois, 262

Charadrius hiaticula, 118

charlock, 3 n.8, 483

charr, 19

Chasmatopterous nigrocinctus, 374

Chatham Island, 145 & 146 n.10

Cheltenham, Gloucestershire, 197 & nn.1 & 3, 202

Chesney, Francis Rawdon, 8 & 16 n.18, 14 & 18 nn.56 & 59, 40 & 42 n.17, 43 & 45 n.5, 61 & 65 n.6, 64 & 66 n.26

Chevreul, Michel Eugène, 410 n.7

Chevreulia, 167

Chile, 22, 125, 172 & n.3, 176 & 177 n.11, 178, 385 n.4

Chiloe, Chile, 125

chimpanzees, 419

China: ancient accounts of pigeons and fowls of, 57 n.2, 57–8 & n.3; animals of, 7, 11, 19–20 & n.4, 26, 57 n.2, 62 & 66 n.20, 345, 386, 442, 510; artificial hatching of ducks' eggs in, 43; flora of, 19–20 & n.4, 92, 156 & 157 n.3

Chiococca racemosa, 372 & 373 n.8

Chitty, Edward, 318 & 320 n.3

Chitty, Thomas, 318

chloroform, 334 & 335 n.2, 438 & 439 n.3

Chonos Archipelago, Chile, 22

chooboos, 455 & n.2

Chordarieae, 317

Christy, Alfred, 336 & 337 nn.2 & 3

Chrosomitris, 7 & 16 n.15

Chrysanthemum, 75 & n.7

Chrysosplenium, 132

Chthamalus, 219 & n.3, 265 & n.6, 300 n.3

Churchill, Awnsham, 242 & 243 n.5

Cicindelidae, 374

Circaea, 311

cirripedes, 130, 399; antennae of, 161–2 & nn.8 & 10; cement glands and ovaria of, 301–2 & nn.7–10, 307 & 308 n.5; classification of, 194 & 195 n.7, 454 & n.3; cross-fertilisation in, 492 & 494 nn.10 & 11, 496–7 & nn.1–3; CD's study of, 55 & 56 nn.7–9; T. H. Huxley's study of, 454 & nn.2 & 3; Tertiary fossils, 219 & nn.1–3 & 5–7, 237 & n.12, 256 n.1, 265 & n.6, 300 & n.3; variability in highly developed organs, 407 & 408 n.4

Cirsium, 357

Cladrastis, 423

Clapham Grammar School, 219 nn.2 & 3, 238 & 239 n.6, 250 n.4, 269 n.1, 346 & 347 nn.6 & 7, 476 & 477 n.9

classification: affinities of isolated genera and families, 326; CD's portfolio of notes on, 462 & n.5; genealogical basis of, 309, 456 & 457 nn.12 & 13; T. H. Huxley's view of, 461–2, 505; lumping or splitting of species, 37, 182–4, 206 & n.5, 277–8, 398, 438 & 439 n.6, 499 & 500 n.7; of Lepidoptera, 329 n.3, 332; of man, 367 & n.12, 419 & 420 n.5; parallels with etymology, 458 & 459 n.2; and principle of divergence, 443; species definitions, 96, 194 & nn.6 & 7, 195–6, 309, 492–3 & 494 n.12; use of embryonic characters in, 417–18, 420 & 421 nn.3–5, 424–6 & 427 nn.5–8, 427–8 & nn.3, 5 & 7, 463 & n.3

Clausilia, 96, 184

cleavage and foliation: CD's view of, 22–3 & n.2, 200 n.9, 201–2 & n.7, 217 & 218 n.6, 331 n.7; John Tyndall on, 200 & n.9, 201, 217 & 218 n.4, 330 & 331 n.7

Clematis, 298

Clive, Marianne, 25 n.3

Clive, William, 25 & n.3

clover, 494

Cnestis, 102, 400

coal deposits, 198 & 200 n.2, 363

Cobden, Richard, 375 & 376 n.14

Coco Island (Pacific Ocean), 337

coconuts: coco-de-mer, 27, 337 & 338 n.5

Coe, Henry, 480–1 & nn.2 & 3, 486 & n.1

Coleoptera, 134 n.3, 283, 306; geographical distribution of, 215 n.4, 488 & n.2; loss of tarsi, 375 & 376 n.12; of Madeira, 158 & 159 n.3, 159, 373 & 376 n.4, 374–5 & 376 nn.8–10 & 13, 478–9 & 480 nn.1, 2, 6 & 7; variation in, 158 & 159 n.3, 252–3 & nn.2–4, 407 & 408 n.3

Colletia, 36, 36 & nn.1 & 2, 37 n.3, 78 & 80 n.2, 80

Colling, Charles, 202 n.5

Colling, Robert, 202 n.5

Columba: intermedia, 53 & 54 n.4, 62 & 66 n.14; *leucocephala*, 227 & 228 n.2; *livia*, 8, 62 & 66 n.14, 204 n.3, 236, 394 n.8; *oenas*, 117. *See also* pigeons

Columella, Lucius Junius Moderatus, 9–10 & 17 n.33

Colymbetes, 90, 99 & 100 n.5

Combes, Mr and Mrs, 440

Compositae, 391, 413 & 414 n.12, 424 & n.9

Conchoderma, 302 & 303 n.9, 307

Coniferae, 258, 281, 452

Connarus, 400

Conolly, Arthur, 12 & 18 n.46

Conus, 95

Convallaria majalis, 230, 257

Cook, James, 14, 451 n.2

Cook-Flower, Mr, 470

Coracia, 7, 8, 41

Coral reefs, 109 & 110 n.10, 146 n.4, 162 n.6, 163 & 165 n.2, 168 & 169 nn.3 & 4, 514 & 515 n.5

coral reefs and islands: CD's view of, 154–5 & 156 n.7, 167–8 & 169 nn.3 & 4, 514 & 515 n.5; Charles Lyell on, 145 & 146 n.4, 163 & 165 n.2, 165 & n.7

Cordilleras, 105, 168, 384 & 385 n.4

Corker, Edward Langley, 226 & 227 n.4

corn-salad, 46, 49

Corse, John, 9 & 17 n.27

Corydalis, 423

Cottage Gardener, 58 & n.7, 226 & 227 n.6, 257, 489 n.5, 507 n.2

Cotton, Mrs, 35 & n.2

Covington, Syms, 55 & 56 n.7, 345

cowslips. *See Primula*

Crangon, 380

Crawford, Mr, 439

Crawfurd, John, 9 & 17 n.29, 30 n.26, 106 & 107 n.4, 109

Cresy, Edward, 507 n.5

Crimean War, 55 & 56 n.5, 122 & 123 n.5

Crisp, Edwards, 365

Crithagra braziliensis, 318 & 320 n.2

crossbills, 396 & n.2

cross-fertilisation: CD's portfolio of notes on, 190 & 191 n.6, 424 & n.11; in animals, 161 & 162 n.2, 173 & nn.3–7, 178 & 179 n.3, 190 & 191 nn.3 & 5, 302 & 303 n.14, 307 & 308 n.8, 428 & n.9, 433, 492 & 494 nn.10 & 11, 496–7 & nn.1–3; in aquatic plants, 280 n.1, 289, 291 & n.3, 291–2 & nn.2 & 5, 292–3 & n.4, 305, 412 & n.4, 413; in Compositae, 413 & 414 nn.13 & 14, 424 & n.9; in *Fucus*, 317 & n.2, 469 & n.10, 498 & n.3; in Fumariaceae, 423 & 424 n.6, 433 & n.5, 437, 492 & 493 nn.5–7; in grasses, 178 & 179 n.5, 413 & 414 n.11, 423 & 424 n.8, 432–3 & n.7, 437 & n.4; in invertebrates, 173 & nn.3, 5–7, 178 & 179 n.3, 302 & 303 n.14; in Leguminosae, 286–7 & n.2, 295, 296 & n.2, 309, 346 & 347 n.5, 361, 413, 423, 433; in plant species, 170–1 & n.12, 176–7, 190 & 191 nn.5 & 7, 437, 446–7 & 449 n.12, 465–6 & 467 nn.4 & 6; in trees, 190 & 191 nn.5 & 7, 286–7 & n.2, 291, 294 & 295 nn.2, 5 & 6, 297–8, 304 & 305 nn.4 & 5, 315 & 316 nn.8 & 9,

340 & 342 n.10, 358 & 359 n.4, 360–1, 402–3, 413; in *Viola*, 295, 298; role of insects in, 170–1, 176, 287 n.2, 291 & n.4, 295 & n.3, 346 & 347 n.5, 361, 414 & n.16, 433 & nn.5 & 8, 446 & 449 nn.12 & 13, 493

Crowe, John Rice, 106 & 107 n.2, 107, 373 n.7, 441 n.5

Crucianella, 407

Cruciferae, 291, 433

Crump, Charles Wade, 4–5

Crustacea, 180 & 181 nn.3 & 6, 381 & 382 n.8; blind forms from caves, 216 & n.5; classification of, 399 & 400 n.4, 454 & n.3; dispersal of, 319, 352, 354; embryonic development of, 421 n.5, 425–6 & 427 nn.2, 3 & 9; fighting of males for females, 383 & n.8; geographical distribution of, 180 & 181 n.6, 235 & 237 nn.3 & 5, 262 & 264 n.17, 366 & 367 nn.2 & 5, 380–1 & 382 nn.4–6, 399 & 400 n.2; T. H. Huxley's study of, 301–2 & nn.7, 8 & 10, 307 & 308 n.4, 399 & 400 n.4, 420 & 421 n.9, 425 & 427 n.2; John Lubbock's study of, 181 & n.10, 193 n.4, 231 & 232 n.7; of Britain, 6; of US North Pacific Expedition, 381 & 382 n.10, 399 & 400 n.3; parthenogenesis in, 252 n.2

Cryptocerus, 481 & 482 nn.2 & 4

Crystal Palace, 218 & 219 n.2, 390 n.6, 442 & n.3

Cuba, 353

Cucubalus, 499 & 500 n.12, 504 & nn.2 & 3, 509

cucumbers. *See* Cucurbitacea

Cucurbitacea, 326 & n.12, 496 & n.6

Cumberland Mountains, Cumbria, 492

Cuming, Hugh, 96 & 97 n.5, 126 & 128 n.5, 129 n.28, 182 & 184 n.5, 347 & 348 n.2

curassows, 517

Curculionidae, 375

Curtis, John, 90

Cutleria multifida, 317

Cuvier, Georges, 27 & 29 n.13, 43 & 45 n.6, 106 & 107 n.5, 112 n.4, 147 & 148 nn.4, 5 & 10, 461, 463

Cyclograpsus, 381

Cyclostigma, 123 n.3

Cynthia cardui, 41

Cyperus, 102

Cyprinodonts, 216

Cyprinus carpio, 26

Cyprus, 13

Cyrena, 183, 189

Cystoseirae, 317

Cytisus, 152 nn.3 & 4, 240 & 241 n.3

Dacca, 28

dace. *See* carp family

Dactylis, 423

Dahuria, Manchuria, 156 & 157 n.3

Dalbergia, 287

Dallas, William Sweetland, 303 n.13, 343 n.1

Dana, James Dwight, 299; and blind cave animals, 180 & 181 n.3, 215–16 & n.2, 235, 299 & 300 n.1; and coral reefs, 169 n.4; and CD's species theory, 235–6; and geographical distribution of Crustacea, 180 & 181 n.6, 235 & 237 nn.3 & 5, 366 & 367 nn.2, 5 & 6, 380–1 & 382 nn.3 & 6, 399 & 400 n.2; and John Lubbock's study of crustaceans, 231 & 232 n.7; on geology, 145 & 146 n.7, 235 & 237 n.5, 381; on parallelism between fossil record and embryogenesis, 299–300 & n.3; on species, 236 & 237 n.7, 508 & n.6, 516 & nn.5 & 6

dandelions, 102, 341

Daniell, William Freeman, 239 & 240 n.2, 247 n.2, 248 & 249 n.3, 257, 435 & n.2; on animals of Africa, 241–2 & nn.2–4, 269–70; and botanical collecting, 321–2 & nn.2, 3 & 7, 324 & nn.2 & 4

Daphnia, 193 n.4, 231 n.2, 251–2 & nn.2–5 & 7–10, 254–5 & nn.2–4, 302 n.4, 489–90 & n.3

Darwin, Anne Elizabeth, 238 & 239 n.2, 347 n.8

Darwin, Charles Robert: declines reviewing books, 205 & n.2; photographs of, 195 & 197 n.6;

—CHARACTER AND OPINIONS: of Louis Agassiz, 340 & 342 n.8, 456; of J. D. Dana, 516 & nn.5 & 6; of W. F. Daniell, 321, 324 & n.4; of Edward Forbes, 178; of W. D. Fox, 335 & n.6; of Albany Hancock, 403–4 & 405 nn.4 & 9–12; of T. H. Huxley, 404; of John Lindley, 406; of John Lubbock, 250–2 & n.2, 254–5; of Joseph Prestwich, 404, 409; of publishing species theory, 100 & n.7, 106–7 & n.8, 109, 133 n.1, 135–6 & n.10, 142, 169 & nn.7 & 8, 201, 236, 346; of A. R. Wallace, 387 & 388 n.3, 515; of T. V. Wollaston, 134, 147 & 148 n.7

—CHILDREN AND FAMILY: advises W. E. Darwin, 343, 344; birth of C. W. Darwin, 297 & n.5, 301, 334 & 335 n.2; death of mother, 25 & n.6; education of sons, 32, 219 n.3, 238 & 239 n.6, 334, 346 & 347 n.6, 476 & 477 n.9; fears hereditary ill-health of children, 460 & 461 n.13; illnesses of, 205 & n.4, 368 & 369 n.5,

460 & 461 n.13; opinion of Clapham Grammar School, 346 & 347 n.6, 476 & 477 n.9; opinion of Rugby School, 476; opinion of value of collecting for children, 32; professions for sons, 334, 343, 345

—FINANCES: rent from farm, 275 & n.2; shares in railways, 215 n.6, 335 n.5; subscriptions to societies, 101 & 102 n.3, 171 & 172 n.14, 347 & n.1

—HEALTH, 2, 3; abandons water cure, 335; delays work on species book, 267, 366, 417; hydropathic treatment, 238 & 239 n.4, 377 & n.5, 383 & n.6, 384, 385 & 386 nn.2–4, 6 & 9, 394–5 & n.8, 476, 487 & 488 n.13, 507 n.5; improvement in, 32; opinion of medical treatment, 238; poor state of, 20 & n.1, 21, 87, 285, 373, 377, 393, 404, 417, 444, 460, 475, 476; suffers from overwork, 346 & 347 n.8, 397, 476; tries mineral acids, 335 & 336 n.7

—PUBLICATIONS: nomination of John Richardson for Royal Medal, 85 n.4; 'On certain areas of elevation and subsidence in the Pacific and Indian Oceans, as deduced from the study of coral formations', 109 & 110 n.10; 'On the action of sea-water on the germination of seeds', 46 & 47 n.3, 49 & 50 n.2, 203 n.6, 277 & 278 n.2; 'On the connexion of certain volcanic phenomena', 168 & 169 n.6; 'On the distribution of the erratic boulders . . . of South America', 174 & 175 n.7; 'On the tendency of species to form varieties; and on the perpetuation of varieties and species by natural means of selection', 450 n.17

—READING: Asa Gray, *First lessons in botany*, 452 & 453 n.6; E. W. V. Harcourt on Madeiran ornithology, 120 nn.3 & 4, 206 n.7; J. D. Hooker's review of A. de Candolle *Géographie botanique raisonnée*, 203, 244; Reginald Orton, *Physiology of breeding*, 211 n.3; K. T. E. von Siebold, *True parthenogenesis*, 303 n.13. *See also Correspondence* vol. 4, Appendix IV

—SCIENTIFIC VIEWS: of Alphonse de Candolle's *Géographie botanique raisonnée*, 325 & 326 n.4; of cleavage and foliation, 22–3 & n.2, 200 n.9, 201–2 & n.7, 217 & 218 n.6, 331 n.7; of connection between cement glands and ovaria of cirripedes, 301–2 & nn.7–10, 307 & 308 n.5; of controversy between T. H. Huxley and Hugh Falconer, 147 & 148 nn.2 & 10; of coral reefs and islands, 154–5 & 156 n.7, 167–8 & 169 nn.3 & 4, 514 & 515 n.5; of elevation and subsidence, 115 &

n.6, 153–5, 165 n.2, 168–9; of glacial phenomena, 322–3, 324, 325 & 326 n.10, 329 & n.2; of Asa Gray's statistics of US flora, 244–6 & nn.5 & 6; of J. D. Hooker's *Flora Novæ-Zelandiæ*, 178–9; of imperfection of nature, 178, 452 & 453 n.9; of land-bridge theory of distribution, 115, 143–4, 147, 153–5, 170 & 171 n.2, 174, 191, 193 & 194 nn.2 & 3, 199, 200–1, 203 & 204 n.5, 236, 267, 325 & 326 n.8, 358 & 359 nn.5 & 6, 376, 514–15 & nn.5–7; of medals of Royal Society, 403–4 & 405 nn.3, 4 & 19, 405–6 & n.2, 408–9, 487 & n.12; of Richard Owen's classification of man, 367 & n.12, 419 & 420 n.5; of parthenogenesis, 484; of phrenology, 32; of scientific methodology, 105, 109, 315, 325, 360 & 362 n.6, 363 & 364 n.4, 404; of significance of transmutation laws for natural history, 194; of telegony, 214 n.4; of use of negative evidence in geology, 219 & 220 n.4, 236–7 & n.12, 265 & n.6; of volcanos and volcanic phenomena, 83–4 & n.2, 154–5, 167–8 & 169 nn.3, 4 & 6

—SPECIES BOOK: George Bentham's opinion of, 495; C. J. F. Bunbury's opinion of, 36 & 37 n.8, 78, 80–1 & 82 n.9; Hugh Falconer's opinion of, 445; Asa Gray's opinion of, 437, 445, 445–6, 492; J. D. Hooker reads, 179 & 180 n.11, 190 & 191 n.2, 194, 202, 234 & n.3, 247 & nn.1 & 2, 259–62 & nn.1–3 & 9–20, 266–7, 271, 432 & 433 nn.3 & 4; J. D. Hooker's opinion of, 89 & 91 n.7, 99–100 & n.6, 199–200, 201, 259–60, 280–1; T. H. Huxley's opinion of, 89 & 91 n.7, 99–100 & n.6; Charles Lyell's opinion of, 89–90 & 91 nn.7 & 10, 100 & n.7, 134 & n.4, 179 & 180 n.12, 194 & nn.5, 6, 8 & 9, 199 & 200 n.8, 236 & 237 n.10; plans to give facts for and against theory, 54, 80–1, 178; progress on manuscript, 133 n.4, 179, 207 & n.3, 234, 254 & 255 n.5, 304 & 305 n.5, 308 n.3, 335, 368, 377, 387, 417, 461 & n.14, 476 & 477 n.8, 515 & n.11, 516; publication of, 89 & 91 n.10, 100 & n.7, 106–7 & n.8, 130–1 & 133 n.1, 135–6 & n.10, 142, 169 & nn.7 & 8, 235–6, 238, 265 & n.5, 346; H. C. Watson's opinion of, 130–1; T. V. Wollaston's opinion of, 89 & 91 n.7, 99–100 & n.6, 134 & nn.8 & 10

—SPECIES WORK: abortive organs, 159 & n.3; adaptation, 445; bees' cells, 377 & 378 n.1; botanical statistics, 92–3, 94 & 95 n.18, 245

& 246 n.9, 316 n.3, 325 & 326 n.6, 363, 429 & n.9, 430 & n.3, 446; classification, 449, 456 & 457 n.13, 462 & n.6, 462–3 & n.3, 467, 492–3 & 494 n.12, 505 & 506 n.3; correlation between organs or parts, 136 & n.11, 141 & 142 n.1, 307 & 308 n.9; effect of external conditions on variation, 149, 281–2, 384, 387 & 388 n.7, 389 & 390 n.4, 445; embryology, 58 & n.8, 416, 418–19, 420 & 421 nn.2–5, 424–6 & 427 nn.5–8, 427–8 & nn.3, 5 & 7; experiments on breaking the constitution of plants, 25 n.7, 143 & n.5, 198 n.6, 202 & 203 nn.3 & 4, 465 & n.2, 467 & n.3; experiments on cross-breeding pigeons, 54, 60 & n.3, 66 & n.21, 142, 210, 431; experiments on cross-fertilising plants, 190, 217–18, 287, 291, 295, 413, 447 & 450 n.16, 469 & n.12, 487 & n.8, 492 & 493 n.5; experiments on *Dianthus*, 349 & n.3; experiments on dispersal of aquatic plants, 81 & 82 n.5, 192 & n.5, 202 & 203 nn.5 & 6, 233 & 234 n.2, 244 & n.2; experiments on dispersal of molluscs, 239, 243 & 244 nn.2 & 3, 253 & n.5, 304–5 & n.7, 305–6 & n.2, 335 & 336 n.8, 338 & n.8, 346, 382–3 & n.3, 385 & n.9, 515 & n.10; experiments on dispersal of seeds by birds, 81 & 82 n.5, 100, 212 & 213 n.14, 248 & n.2, 250 & n.6, 267 & 268 nn.4 & 5, 274 & n.5, 281 & nn.3 & 6, 282 & n.3, 294 & 295 n.7, 309 & 310 n.7, 389 & 390 n.6, 453 & n.11, 455 & n.4; experiments on dispersal of seeds by fish, 212, 324 & n.5, 364 & 365 n.2, 365, 379 & 380 n.3; experiments on dispersal of seeds by trees, 99, 100 & n.9, 105 & 106 n.4; experiments on floating plants in salt-water, 217, 371–2 & 373 nn.5–7, 408; experiments on germinating seeds from pond mud, 81 & 82 n.5, 100 & 101 n.10, 212 & 213 n.14, 389 & 390 n.6, 453 & n.11; experiments on immersion of seeds in salt-water, 46–7 & nn.3 & 4, 92 n.15, 103 n.3, 105 & 106 n.4, 192 & n.5, 212, 233, 265–6 & n.1, 277 n.2, 371–2, 419 & 420 n.4, 429 & n.5; experiments on number of species in small area, 192 & n.6; experiments on role of bees in fertilisation, 414 & n.16, 446 & 449 n.12, 465–6 & 467 n.4; experiments on seedling survival, 363 & 364 n.5, 371 & 372 n.4, 407 & 408 nn.6 & 7, 418; extinction, 407 & 408 n.8, 413, 422 & 423 nn.4, 5 & 10, 431–2 & 433 n.2, 448; hybrids, 178, 210 & 211 n.3, 334, 409 & 410 n.10, 509–10 & 511 nn.1–3, 512–13 & nn.2–7, 515 & n.11, 516–17 & n.3;

instinct, 378 n.1, 513 & n.7; natural selection, 272 & 273 n.3, 282, 370, 407 & 408 nn.6–8, 447–8, 449, 464 n.8, 492 & 493 nn.3 & 4, 494; neuter insects, 301 & 302 n.6, 302, 481–2 & nn.1, 4 & 5; pangenesis theory of inheritance, 484 & 485 n.7; parthenogenesis, 301 & 302 nn.2 & 3; portfolio of notes, 50 & n.7; portfolio of notes on abortive organs, 253 & 254 n.7; portfolio of notes on classification, 462 & n.5; portfolios of notes on crossing, 190 & 191 n.7, 424 & n.11; portfolio of notes on dichogamy, 190 & 191 n.6; portfolio of notes on embryology, 418 & n.4; portfolio of notes on geographical distribution of animals, 34 & n.5, 126 & n.9, 184 & 185 n.14, 186 & 187 n.4, 188 & n.5, 216 & 217 n.7, 226 & n.4, 270 & 271 n.7, 284 & n.17, 300 & n.4, 306 & 307 n.7, 348 & 349 n.6; portfolio of notes on geographical distribution of plants, 37 & n.13, 103 & 104 n.15, 157 & 158 n.15, 196 & 197 n.9; portfolios of notes on hybridism, 51 & n.9, 53 & 54 n.5, 166 & n.5, 456 & n.4; portfolio of notes on means of dispersal, 49 & n.6, 80 & n.9, 90 & 92 n.17, 98 & n.5, 223 & n.6, 243 & n.7, 266 & n.2, 319 & 320 n.7, 336 & 337 n.4, 355 & n.5, 365 & n.4, 365 & 366 n.2, 379 & 380 n.2; portfolios of notes on palaeontology and extinction, 130 & n.3, 300 & n.4, 321 & n.5, 424 & n.10; portfolios of notes on variation and varieties, 31 n.2, 53 & 54 n.5, 97 & n.10; principle of divergence, 420 & 421 nn.3 & 5, 427–8 & n.5, 443, 448–9, 493 & 494 n.18; rudimentary organs, 333 & 334 n.2, 339, 415 n.3; selective pressure, 272 & 273 n.4; sexual selection, 72 n.2, 75 & n.5; study of large and small genera, 93 & 95 nn.10 & 11, 314 & 316 nn.3 & 4, 325 & 326 n.6, 363, 391 & 392 n.4, 429 & n.9, 430 & nn.2 & 3, 438 & 439 nn.7–13, 443–4 & nn.12–14, 446 & 449 nn.8–10, 452 & 453 nn.3 & 4, 460, 471 & 472 n.7, 478–9 & 480 nn.2, 3, 6 & 7, 486 7 & nn.3–7, 493 & 494 n.16, 498 & nn.2–4, 498–9 & 500 nn.2–5, 502 & nn.2–4, 516 & nn.2 & 3; struggle for existence, 113, 230 & 231 n.8, 483, 494; time and species' change, 145, 240, 271, 272, 280; transitions between organs, 307 & 308 nn.4 & 9; vitality of ancient seeds, 45 & 46 n.6. *See also* cross-fertilisation, dispersal, geographical distribution, species, variation and varieties
—TRIPS AND VISITS: to B. E. Bult, 34 & 35 n.2; declines attending Miss Lubbock's engagement breakfast, 490; declines visit to J. D. Hooker, 191; to Hugh Falconer, 350, 366, 384 & 385 n.3; to J. D. Hooker, 244 & n.3, 246 & n.11, 247 n.2, 248 & n.1, 260 & 263 nn.5 & 7, 267, 269 n.9; invited to visit United States, 195, 209, 230; to Charles Lyell, 100 & n.8, 106 & 107 n.7, 179 & 180 n.13; plans to visit London, 267–8 & n.3, 484 & 385 n.8; plans to visit Malvern, 385 & n.5; plans to visit Moor Park, 404, 475 & 476 n.6; plans to visit Tenby, 110 n.9, 110 & 111 n.4, 135 & 136 n.3, 346; to British Museum, 57 & 58 n.3; to Dulwich College, 434 & 435 n.4; to Leith Hill Place, 224 & 225 n.1, 227 & n.1; to London, 34 & n.1, 45, 57 & n.1, 100 & n.8, 112 n.5, 122 & n.3, 123 & n.4, 127 n.1, 141 & n.3, 148 & n.12, 234 & n.5, 244 & n.3, 247 n.2, 248 & n.1, 260 & 263 nn.5 & 7, 264–5, 267 & 268 n.3, 268 & 269 n.9, 322 n.2, 322 & 323 n.2, 349, 369 n.5; to Moor Park, 238 & 239 n.4, 347 n.8, 377 & n.5, 383 & n.6, 384, 385 & 386 n.5, 388, 390, 393 n.1, 394 & 395 n.5, 407 & 408 n.8, 413 & 414 n.10, 414 & 415 n.1, 476 & 477 n.2, 483 & n.3, 484 & 485 n.2; to Philosophical Club meetings, 112 & n.5, 148 & n.12, 181 & n.9, 234 & n.5, 263 n.5, 267–8 & n.3, 269 n.9, 358 & 359 n.5, 485 n.8, 489 & 490 n.2, 494 & n.13; to Philoperisteron Society, 1 & n.2, 21 n.5, 51 n.7, 489 & n.5; to poultry shows, 160 & n.6, 192 & 193 n.2, 204 & n.5, 205 & n.3, 442 & n.3; to Royal Society council meetings, 34 n.1, 57 n.1, 109 & 110 nn.4 & 8, 112, 123 n.5; to P. H. Stanhope, 179 & 180 n.13; to Matthew Wicking, 45 & 46 n.7, 142 & n.5

Darwin, Charles Waring, 87 n.4, 218 n.13, 250 n.3, 297 & nn.1 & 5, 302 & n.1, 303, 306 & 307 n.3, 334 & 335 n.2

Darwin, Emily Catherine, 51 & n.8, 395 & n.9, 477 & n.10, 478 & n.5

Darwin, Emma, 66 & 67 n.2, 88 & 89 n.5, 107, 440 & n.2, 477, 516; birth of C. W. Darwin, 301, 302 & n.1, 303 & 304 n.5, 304, 335 n.2; health of, 3, 87 & n.4, 111 n.4, 138, 298; lip surgery of, 343 & 344 n.5; nurses Eliza Wedgwood, 442 & 443 n.3, 451 & 452 n.4; pregnancy of, 135 & 136 n.3, 151 & 152 n.8, 191, 218 & n.13, 238, 249 & 250 n.3, 264, 265, 267, 275 & 276 n.1, 285; use of chloroform in childbirth, 438 & 439 n.3; visits Barlaston, 418 & 419 n.2; visits Hartfield, 45 & 46 n.3; visits Hastings, 368 & 369 n.5, 369,

Darwin, Emma, cont.
372 & 373 n.9, 386 n.5, 394 & 395 n.4; visits Moor Park, 386 n.5, 394 & 395 n.5, 404 & 405 n.17; visits the Tollets, 475 & 476 n.4

Darwin, Erasmus, 134 & 135 n.9

Darwin, Erasmus Alvey, 34 n.1, 57 n.1, 58 & 59 n.10, 122 n.3, 244 & n.4, 335, 350, 393, 471 n.2, 477

Darwin, Francis, 45 & 46 n.3, 344 & n.7; assists CD's experiments, 305 & n.8, 365 n.2; education of, 219 n.3; interest in Lepidoptera, 32 & 33 n.1

Darwin, George Howard, 46 n.3; education of, 218 & 219 nn.2 & 3, 238 & 239 n.6, 249 & 250 n.4, 346 & 347 n.6, 476 & 477 n.9; interest in entomology, 32 & 33 n.1, 138 & 139 n.2, 215 & n.5, 238 & 239 n.5; interest in heraldry, 238 & 239 n.5; school holidays of, 218 & 219 n.1, 285–6 & n.4, 303 & 304 nn.2 & 4; and Sarah Wedgwood's funeral, 267–9 & n.2

Darwin, Henrietta Emma, 269 n.8; attends Ellen Frances Lubbock's wedding, 231 & 232 n.6; health of, 205 n.4, 285, 301, 368 & 369 n.5, 369 & 370 n.2, 385 & 386 n.5, 389 & 390 n.5, 394 & 395 nn.4, 5 & 7, 416 & n.2, 440, 468 & 469 n.7, 476; returns home from Moor Park, 471 & n.4, 475 & 476 n.5; visits Hastings, 368 & 369 n.5, 385–6 & n.5, 389 & 390 n.5, 390; visits Moor Park, 386 n.5, 394 & 395 nn.5 & 7, 404 & 405 n.17, 416 & n.2, 440 & nn.1 & 2, 469 n.7, 476 & 477 n.6

Darwin, Horace, 45 & 46 n.3; education of, 219 n.3; interest in Lepidoptera, 32 & 33 n.1

Darwin, Leonard, 344 & n.7, 451 & 452 n.6; childhood sayings of, 45 & 46 n.4; education of, 219 n.3; health of, 460 & 461 n.13, 468 & 469 n.8, 476 & 477 n.7; interest in Lepidoptera, 32 & 33 n.1

Darwin, Susan Elizabeth, 51 & n.8, 386 n.9, 395 & n.9, 507 n.5

Darwin, Susannah, 25 n.6

Darwin, William Erasmus: allowance of, 394 & 395 n.3; clothing for, 285; education of, 45, 343 & 344 n.2, 345, 434 & 435 n.8, 475 & 476 nn.2 & 3, 477–8 & n.2; interest in Lepidoptera, 32 & 33 n.1, 454 & 455 n.2; interest in painting, 478 & n.6; interest in photography, 451 & 452 n.3; membership in debating society, 343; profession for, 343, 345; proposed for membership of Athenæum Club, 45

& 46 n.9, 349 & 350 n.2; school holidays of, 303 & 304 n.4, 394 & 395 n.6; visits Manchester, 434 & n.1; visits Tenby, 136 n.3; and Sarah Wedgwood's funeral, 267–9 & n.2

Datura, 509 n.3

Davidson, Thomas William St Clair, 308 & nn.2 & 3, 312–14 & nn.1 & 2

Davy, John, 19

Dawson, John Frederick, 479 & 480 n.8

Decaisne, Joseph, 499 & 500 n.8

deer, 335; elks, 28; fallow, 6, 14, 28 & 30 n.28; Irish, 40; red, 6, 40; reindeer, 28 & 30 n.29, 61, 62

Demas, Don Sinbaldo. *See* Mas y Sans, Sinbaldo de

Dendrocygna arborea, 354, 355

Denmark, 132, 133

Derby, Earl of. *See* Stanley, Edward Smith

Deshayes, Gérard Paul, 48 & 49 n.1, 96 & 97 nn.3–5, 182

Desjardins, Julien François, 27 & 30 n.17

Dezerta Grande, Madeira, 75, 158, 377 n.2

Dianthus, 4, 25, 36–7 & n.11, 82 n.9, 349 & n.3, 372

Diatomaceae, 55 n.2, 484 & 485 n.6

dichogamy: CD's portfolio of notes on, 190 & 191 n.6

Dickie, Edward John, 490 & 491 n.1

Dickie, George, 289 & 290 n.4, 293–4 & n.2

Diclytra, 433

Dictyoteae, 317

Dieffenbach, Ernst, 128 n.15

Dinornis, 75 n.2

Diptera, 399, 419

Dipterix, 287

dispersal: by hurricanes, 89, 232, 274 & n.5, 319, 337, 348, 352–3, 354; by icebergs, 70–1 & n.2, 174 & 175 nn.4 & 6, 203 & 204 n.4, 260, 261 & 263 nn.10 & 11, 263; by insects, 89, 90 & 92 n.15, 99 & 100 nn.4 & 5; CD's portfolios of notes on means of, 49 & n.6, 80 & n.9, 90 & 92 n.17, 98 & n.5, 223 & n.6, 243 & n.7, 266 & n.2, 319 & 320 n.7, 336 & 337 n.4, 355 & n.5, 365 & n.4, 365 & 366 n.2, 379 & 380 n.2; land-bridge theory of, 89 & 91 n.8, 115, 130 n.2, 139 & 140 n.1, 140 & 141 n.3, 143–4 & nn.2–5, 147 & 148 n.9, 150, 153–5 & 156 n.3, 159 & n.3, 167–8, 170 & 171 nn.2 & 7, 174, 191, 193 & 194 nn.2 & 3, 198–200 & nn.3 & 4, 200–1, 203 & 204 n.5, 236, 261, 325 & 326 n.8, 358 & 359 n.5, 376 & 377 nn.2 & 3, 383 & n.4, 514 & 515 nn.5–7; of aquatic plants, 81, 82, 192, 202; of lizards, 250, 270 & 271 nn.5 & 7, 346 & 347 n.3; of molluscs, 225, 233 & n.4, 243 & 244 nn.2 & 3, 253 & n.5, 304–5 & n.8, 305–6 & 307 n.2, 335 & 336 n.8, 338 & n.8, 346, 382–3 & n.3, 385 & n.9, 388, 515 & n.10; of plants, 198 & 200 n.4, 201 & 202 n.3, 325, 358 & 359 nn.5 & 6, 379 & 380 n.3; of plants by birds, 100 & 101 n.10, 238–9, 250 & n.6, 274, 281 & nn.3 & 6, 282 & n.3, 309 & 310 n.7, 336–7, 389 & 390 n.6; of plants by sea-currents, 122 & 123 n.2, 156, 162, 170 & 171 n.11, 192 & n.5, 200 n.4, 202, 233, 244, 260, 266, 276, 348, 372, 419 & 420 n.4, 429 & n.5; of seeds by algae, 266; of seeds by birds, 212 & 213 nn.12–14, 248 & n.2, 267 & 268 n.5, 389 & 390 n.6, 453; of seeds by fish, 212, 324 & nn.5 & 6, 364 & 365 n.2, 365; of seeds by trees, 98, 99, 100 & n.9, 105 & 106 n.4, 162, 225, 233, 276 & 278 n.1, 319; specific centres theory of, 78, 89, 194 & 195 n.9, 199 & 200 n.8, 201

divergence, 420 & 421 nn.3 & 5, 427–8 & n.5, 443, 448–9, 493 & 494 n.18

Dixon, Edmund Saul, 13 & 18 n.55, 26 & 29 n.9, 62 & 66 n.20, 69 & 70 nn.3–5, 393 & 394 n.12

dodos, 337 & 338 n.4, 348 & 349 n.4

dogs: blood corpuscles of, 24 & n.5; bulldogs, 149 & n.5; correlation between hair and teeth in, 307; degeneracy of European breeds in foreign lands, 52, 149 & nn.1, 4–6, 241; dingos, 24; T. C. Eyton's study of, 410 & 411 n.4; greyhounds, 214; hybrids of, 40 & 42 n.20, 51, 52, 137; illustrations of on ancient monuments, 107–8 & n.3, 137 & 138 n.8; mongrels, 50–1, 57, 213; of Africa, 137, 239, 241, 243, 410 & 411 n.5, 435 & nn.1 & 2; of Tibet, 107 & n.3, 350 & n.2; Scottish deerhounds, 50, 57 & 58 nn.1 & 2; setters, 149 n.5

Don, David, 368 & n.3, 369

Doubleday, Henry, 327–8, 331 & n.2, 331–3 & n.1

Douglas, George, 214 n.4, 301 & 302 n.5

doves. *See* pigeons

Down House: beech tree of, 45 & 46 n.1; B. P. Brent visits, 247; Emily Catherine Darwin visits, 395 & n.9; Susan Elizabeth Darwin visits, 395 & n.9; douche of, 507 & n.5; extension to, 451 & 452 n.5, 460, 475 & n.2, 475 & 476 n.8, 477; Robert FitzRoy visits, 345; grey mare of, 286 & n.6, 303, 478 & n.4; J. S. Henslow visits, 438 & 439 n.4, 441 & n.1, 444 & 445 n.15; J. M. Herbert invited to, 2; J. D. Hooker visits, 66 & 67 n.2, 72 n.16, 85 & n.1, 87 & n.3, 89 & 91 n.7, 100 & n.6, 103 n.1, 471, 475 & 476 n.7; T. H. Huxley visits, 66 & 67 n.2, 73, 74 & n.2, 87 & n.3, 89 & 91 n.7, 100 & n.6, 101; kitchen garden of, 143 n.2; John Lubbock dines at, 87 & n.3; Charles Lyell visits, 34 n.4, 75 n.1, 84 n.2, 89 & 90 n.1 & 10; pigeon house of, 45 & 46 n.8; postal address of, 189 n.1; H. C. Watson invited to, 66 & 67 n.2; Caroline and Josiah Wedgwood visit, 344 & n.10; Frances Julia (Snow) Wedgwood visits, 344 & n.9; Frances Mosley Wedgwood visits, 395 & n.9; Harry Wedgwood visits, 343 & 344 n.3; Harry and Jessie Wedgwood visit, 395 & n.9; well of, 435–6 & n.2; T. V. Wollaston visits, 66 & 67 n.2, 87 & n.3, 88 & 89 n.2, 89 & 91 n.7, 100 & n.6

Down, Kent, 141; Friendly Club, 4 n.2; national school, 444 n.5

Drège, Jean François, 102 & 104 n.7, 104 & n.2, 107 & 109 n.3, 363 & 364 nn.1 & 2, 368 & n.2, 389 & 390 n.4

Drosera, 157, 469

Drysdale, Lady, 416

Dublin, Ireland, 455 & n.1; Royal Society, 353

ducks, 9, 100 & 101 n.10; as agents of dispersal, 100 & 101 n.10, 319, 352–3, 354, 382–3 & n.3, 385 & n.9; black, 334, 338; Buenos Ayres, 135, 226; call, 2 & 3 n.1, 135, 142, 238, 257; CD's acquisition of, 2, 54, 56, 135 & 136 n.5, 137 & 138 n.9, 224, 226, 238, 288, 334 & n.4, 338; hook-billed, 135, 257; hybrids of, 53, 142, 166 & n.4, 214, 257 & n.7, 480 & n.3; mallards, 166 n.4; musk-, 3, 9, 15, 214, 480 & n.3; of China, 19–20 & n.4; of Jamaica, 352–3, 354 & 355 n.2; penguin, 257, 431; pintails, 166 n.4, 517 & n.3; Rouen, 226; tufted, 135; variation in, 334 & n.4

Dufossé, Adolphe, 506 n.6

Dulwich College, 434 & 435 n.4

Dureau de la Malle, Adolphe Jules César Auguste, 61 & 65 n.3

Durvilloea, 317

Duval, (jardinier), 37 & n.12

Dyaks: of Borneo, 393

Dytiscus marginalis, 90

eagles, 250 & n.6, 267

East India Company, 18 n.65, 38 & 41 nn.2 & 3, 191 & n.9; museum, 6 & 15 n.4, 158 n.13

Eaton, John Matthews, 256 & 257 n.2, 394 nn.5 & 9
echinoderms, 160–1 & n.2, 506 & n.7
Ecklon, Christian Friedrich, 105 n.3
Edinburgh Review, 298 & 299 n.6
Edinburgh, Scotland, 412 & n.3
Edmondston, Laurence, 97, 98, 223–4 & n.5, 378–9 & n.2, 440–1 & n.4
Edmondston, Thomas, 224 n.7
Edwards, George, 9 & 17 n.32, 13 & 18 n.52
Edwardsia, 172, 200 n.4, 408
eels, 75
Egerton, Philip de Malpas Grey-, 72 & 73 n.7, 109 & 110 n.8, 336 n.9
Egypt, 43
Ehrenberg, Christian Gottfried, 106 & 107 n.5
Eights, James, 128 n.11, 180 & 181 n.8, 380 & 382 n.4
Elateridae, 375
Elatineacea, 294
elephants, 9 & 17 n.27, 186 & 187 n.3, 448
Elephas primigenius, 264 n.19
elevation and subsidence, 115 & n.6, 125, 140 & 141 n.3, 174, 199, 514 & 515 n.5; CD's view of, 153–5, 167–8 & 169 nn.3, 4 & 6, 174, 193, 514 & 515 n.5; Edward Forbes's view of, 140 & 141 n.3; Charles Lyell's view of, 83–4 & n.2, 99 & 100 n.2, 144–6, 163–5 & nn.2 & 8; effect of on climate, 260 & 263 n.8
Élie de Beaumont, Jean Baptiste Armand Louis Léonce, 84 n.2
Elliot, Walter, 30–31 & nn.2, 3, 5 & 6, 256, 257, 443 & 444 n.4, 460 n.5
embryology: Brullé's law of, 420 & 421 nn.2, 3 & 5, 424–6 & 427 nn.4 & 9, 427–8 & nn.3 & 5; CD's portfolio of notes on, 418 & n.4; facts of support CD's species theory, 445; and fossil record, 216, 235 & 237 n.6, 299–300; of Crustacea, 421 n.5, 425–6 & 427 nn.2, 3 & 9; of fish, 425, 427; of fowls, 58 & n.8, 489; of pigeons, 58 & n.8, 91 n.10, 152; of plants, 416, 417–18, 418–19, 428–9; use of in classification, 420 & 421 nn.3–5, 424–6 & 427 nn.5–7, 427–8 & nn.3, 5 & 7, 428–9 & n.2, 461, 506 n.7
Engelmann, George, 228 & 230 n.1
England: brachiopods of, 313; butterflies of, 328; Coleoptera of, 134 n.3; flora of, 113 & n.3, 131–3 & nn.3 & 4, 139, 149–50 & 151 n.1, 360, 369, 438 & 439 nn.7–9, 443 & nn.10 & 12, 460, 493 & 494 n.17, 499 & 500 n.6, 502 & n.4, 503 & 504 nn.2 & 3, 511 & 512 n.2;

fossil mammals of, 61 & 66 n.13; fowls of, 10, 11–12, 13; molluscs of, 48, 89 & 90 n.1, 145–6, 185–6; naturalised plants of, 201 & 202 n.4, 208 & 209 n.4, 391–2 & nn.8 & 9, 413 & 414 n.9
Entada, 156, 170 & 171 n.11, 175 & 177 n.5
Entomological Society of London, 375 & 376 n.11, 378 & n.2, 451
Entomologists' Annual, 77 n.2
Entomologist's Weekly Intelligencer, 75 & 77 n.2, 85 n.9
Entomostraca, 181 & n.10, 231 & 232 n.7
Epacrideae, 198, 201
Epilobium, 175 & 177 nn.1 & 2, 176
Erhardt, S., 136 n.7
Erica, 311, 312
Ericaceae, 150, 389
Erigeron, 341
Eriocaulon septangulare, 78
Eriogaster lanestris, 88 & 89 n.3
Erodium cicutarium, 229
erratic boulders, 174 & 175 n.7, 327 & n.5, 362 & 363 nn.4 & 5; of New Zealand, 70 & 71 nn.1 & 2, 74, 130, 337
Erythrina, 287, 361
Escobar, Pedro, 242, 270
Etheridge, Robert, 506 n.2, 516 n.7
Ethiopia. *See* Abyssinia
Etna, Mount, 165
Etruria, Staffordshire, 269 n.10
Eucalyptus, 198, 201, 417
Eupatorium, 398
Euphorbia, 102, 316
Eurypodius, 380
Everest, Robert, 148–9 & nn.1 & 6, 223 n.1
extinction, 407, 448; CD's portfolio of notes on, 130 & n.3, 300 & n.4, 321 & n.5, 424 & n.10; effect on plant distribution, 413, 422–3, 431–2 & 433 n.2
Eyton, Elizabeth Frances, 207
Eyton, Thomas Campbell: and coloration of horses, 411 & n.6, 417 & n.4; and CD's study of means of dispersal, 212; and Hereford cattle, 210 & n.2, 211 & 213 nn.5 & 8; and hybrid geese, 480 & n.2; and hybrid pigs, 206–7 & n.5, 211 & 212 n.2, 410 & 411 n.3, 417 & n.2; study of birds, 212 & 213 n.9, 410 & 411 n.2, 417 & n.3; study of dogs, 211, 239, 410 & 411 n.4, 435 & n.1; study of fowls, 210–11, 240 & n.7; study of pigs, 206 & 207 nn.2 & 3, 211 & 212 n.2

Fabre, Esprit, 177 n.6

Faeroe Islands, 132, 260, 262, 368, 379 & n.4

Fairbeard, Julius, 48 n.9

Fairbeard, William, 48 n.9

Falconer, Hugh, 72 & 73 n.15, 157, 263; and coloration of horses, 490–1 & n.2; controversy with T. H. Huxley, 112 & n.4, 147 & 148 n.2, 175 & 177 n.8, 176; CD visits, 350, 366, 384 & 385 n.3; and dogs of India, 148 & n.3, 149 & n.5, 350 & nn.1 & 2; and fossil mammals, 321 n.2, 366 & 367 nn.8 & 9; and woolly alpine plants and animals, 384 & 385 n.3, 389

Falkland Islands, 22, 56 n.3, 125, 154, 188

Faraday, Michael, 112

Fareham, Hampshire, 480 & 481 n.1

Farnborough, Kent, 189 n.1, 234 & 235 n.5

Fennell, James H., 166 n.4

Fenwick, Ralph, 123 n.5

Fernando Po Island, 242, 243, 270

ferns, 195

Férussac, Jean Baptiste Louis d'Audebard de, 48 & 49 n.3

Festuca, 131, 401, 423

Fiennes, Twiselton, 166 & n.4

Fiji Islands, 317 n.3

finches, 7, 13, 27, 206, 396, 470

fish, 216 & 217 n.7, 235, 364, 365 379; blind forms from caves, 180 & 181 n.5, 216 & n.2; cultivated varieties of, 26; and dispersal of seeds, 212, 319, 324 & nn.5 & 6, 355, 364 & 365 n.2, 365, 379 & 380 n.3; electrical organs of, 307; embryology of, 425, 427; geographical distribution of, 187–8 & n.4; of Africa, 81, 188; of China, 26; of Jamaica, 320 n.5, 353; of Madeira, 264 n.16; of northern and southern hemispheres, 187–8 & n.4, 366 & 367 n.3; vitality of ova, 19. *See also* carp family

Fisher, Francis, 297 & n.2

Fitton, William Henry, 409 & 410 n.6

FitzRoy, Maria, 345 n.4

FitzRoy, Mary Henrietta, 345 n.4

FitzRoy, Robert, 345 & n.4

flamingos, 14

Flinders Island, 112 & n.7

Flinders, Matthew, 325 & 326 n.5

Flora, oder allgemeine botanische Zeitung, 151 & 152 n.2, 157 & 158 n.9, 217 & 218 n.11, 363 & 364 n.2

Florideae, 317

Flustra, 173 & nn.4 & 6, 302 & 303 n.14

Foraminifera, 189 & n.1

Forbes, David, 338 & n.9

Forbes, Edward, 115, 128 n.10, 224 n.7, 264 n.15; *A history of British Mollusca*, 89 & 90 n.2, 487 & n.7; and geographical distribution, 78 & 80 n.4, 125 & 126 n.3, 261; land-bridge theory of distribution, 139 & 140 n.1, 143 & 144 n.2, 147 & 148 n.9, 154, 159 n.3, 164, 376, 383 & n.4, 514 & 515 n.7; map of distribution of marine life, 103 & 104 n.8, 107 & 110 n.6, 127 & 128 n.14; on natural history of European seas, 175 & 177 n.7, 178, 191 & 192 nn.1 & 3; and plant migration during worldwide cold period, 81 & 82 n.7, 140 & 141 n.3, 262; poor opinions of, 139 n.4, 178; and variability in brachiopods, 308 & 309 n.5

Forbes, James David, 84 & 85 n.5, 202 n.7; glacial theory of, 218 n.3, 323 n.2, 324, 330 & nn.3 & 6

Forncett, Norfolk, 478 n.2

Forster, Johann Reinhold, 451 n.2

Fossil Cirripedia (1851), 219 & 220 n.5, 237 & n.12

fossils: of birds, 130 n.3, 337 & 338 n.4, 348 & n.4; of brachiopods, 308 & n.2, 312–14 & nn.1 & 2; of cirripedes, 219 & nn.1–3, 5–7, 237 & n.12, 256 n.1, 265 & n.6, 300 & n.3; of lizards, 366; of mammals, 82, 126, 130 n.3, 154, 186 & 187 n.3, 320–1 & nn.2–4, 338, 366 & 367 nn.8 & 9; of plants, 89 & 91 n.8, 112 & n.7, 122 & 123 n.3, 200 n.2, 321 & n.4, 327 & n.4, 338, 363; of shells, 48, 95–6, 126–7, 129, 189, 315, 431–2; of trees, 180 & 181 n.8; variation in, 308 & n.2, 312–14 & nn.1 & 2, 506

fowls: ancient descriptions and illustrations of, 13; bantams, 20 & n.3, 28 & 30 n.26, 50 & 51 n.3, 58, 231 & 232 n.9; black-skinned, 7 & 16 n.12; Cochin China, 2 & 3 n.1, 3, 50, 58, 60; coloration of, 393 & 394 n.13, 397; crève-coeur, 288 & n.6; CD's acquisition of, 2–3 & nn.2 & 4, 21 & n.1, 31 & 32 n.5, 56, 107, 121, 135, 137 & 136 n.9, 269, 288, 296 & n.4, 388 & n.9, 443 & 444 n.4; Dorking, 2 & 3 n.1, 50, 124, 210, 227; double spurs in, 28 & 30 n.25; embryonic characters of, 58 & n.8, 489; frizzled, 345; game, 510; Ghoondook, 436 & n.3; guinea-, 9–11 & 17 nn.31 & 32, 61 & 65 nn.4, 7 & 8, 69 & 70 n.6, 220–1, 320 n.6, 517; Hamburgh, 210, 334 n.6, 393, 397; hybrids of, 39 & 42 n.11, 41 & 42 n.21, 286 n.6, 334 & n.6, 509–10 & 511 nn.1–3, 512–13 & nn.2–7; jungle, 1 n.3, 14, 61 & 65 n.6, 256, 416 & n.3, 459; Malay, 2 & 3 n.1, 121, 247, *fowls, cont.*

256, 459, 489; of Africa, 9–10, 26, 28 & 30 nn.23 & 24, 69 & 70 n.7, 135 & 136 n.7, 243, 246–7 & n.2, 248–7, 269 & 270 nn.3 & 4, 290 & n.3, 296; of Annobon Island, 61 & 65 n.3; of Ascension Island, 220–1; of Borneo, 296 & n.4, 344–5 & n.2, 388 & n.9, 393 & 394 n.7, 397, 414 & 415 n.4; of Burma, 459; of Ceylon, 54 & 55 n.6; of China, 7, 19–20 & n.4, 57 n.2, 345, 442, 510; of the Gambia, 26, 238; of India, 62 & 66 nn.16–18, 256, 443 & 444 n.4; of Jamaica, 10; of Japan, 57 n.2; of Persia, 256, 290 & n.2, 339 & nn.2 & 3; of Sierra Leone, 241 & 243 n.2, 246 & 247 n.2, 248 & 249 n.3, 256; of Singapore, 345, 388 & n.9; origin of domesticated breeds of, 9–11, 13–14, 26–7, 135, 387; osteological variation in, 210–11 & n.4, 234 & 235 n.4, 247, 256, 436 & n.3; Polish, 13, 20 n.3, 27, 60 & 61 n.8, 87 & 88 n.5, 110, 121 & 122 n.2, 152 & 153 n.1, 160 n.7, 210 & 211 n.4, 235 & n.6, 276 n.2, 392 & 393 nn.2 & 3, 397, 436 n.3; rudimentary characters of, 333 & 334 n.2, 339; rumpless Polish, 333 & 334 n.2, 415 & n.3, 416, 431 & n.4, 459; Sebright bantams, 509, 512–13; silky, 7, 110, 339, 344–5; Spanish, 21; variation in, 210–11 & n.4, 288

Fox, Edith Darwin, 335 n.4, 346 & n.3

Fox, Ellen Sophia, 136 nn.1 & 2, 250, 335 n.4, 346 & n.3

Fox, Frederick William, 4 n.12

Fox, Henry Stephen, 36 & 37 n.6

Fox, William Darwin: account of long-buried seeds, 3 & 4 n.9; children of, 3 & 4 n.12, 334 & 335 nn.3 & 4, 476 & 477 n.9; and correlation of colour and deafness in cats, 136 & n.11, 141 & 142 n.1; and CD's species theory, 135, 238, 335; and CD's study of dispersal, 238–9, 324 n.6; and CD's study of domestic animals, 2–3 & n.2, 3, 50, 124, 135, 238; entomological excursions with CD, 335 & n.6; and fruit-trees, 477 & n.11; health of, 135 & 136 nn.1 & 2, 237–8, 249 & 250 n.2, 346 & n.2, 386 & n.7; and hybrid crosses between pigs, 386 & n.8; and lizards' eggs, 250 & n.7, 346 & 347 n.4; and origin of Scottish deerhound, 50–1, 57 & 58 nn.1 & 2; and rabbits, 288 nn.2 & 4, 507 & nn.1 & 3

France: brachiopods of, 313; butterflies of, 215 & n.3; flora of, 438 & 439 n.5, 444 & 445 n.14, 460, 499 & 500 n.6, 502

Franciscea, 473

Francolinus, 61

Frankenia laevis, 36 & 37 n.11

Frankland, Edward, 410 n.8

Franklin, John, 405 n.14

Freemason's Tavern, London, 1 n.2

Frémont, John Charles, 229 & 231 n.4, 237 n.11

Friendly Islands, 317 n.3

Fries, Elias Magnus, 132 & 133 n.12

Fringillidae. *See* finches

Fröbel, Julius, 488 n.2

frogs, 33 & 34 n.4, 130 & n.4, 270, 319

Fry, Mr, 137 & 138 nn.6 & 7, 213, 214, 223 n.2

Fryer, John, 13 & 18 n.53

Fucoideae, 316

Fucus, 112 & n.6, 317 & n.2, 368 & n.3, 369 & 370 nn.3 & 6, 469 & n.10, 498 n.3

Fumaria, 433 & n.5, 492 & 494 n.7

Fumariaceae, 423 & 424 n.6, 437

Fürnrohr, August Emanuel, 429 & n.7, 438 & 439 n.5

furze. *See* gorse

Gabriel, Edmund, 269 & 271 n.3, 290 & n.3

Gaimard, Joseph Paul, 128 n.4

Gainsborough, Thomas, 434 & nn.3 & 4

Galápagos Islands, 92 n.16, 96, 153, 337, 381

Galium, 131, 356

galls, 450

Gallus: bankiva, 29 n.6, 416 & n.3; *oeneus*, 39 & 42 n.11, 61; *sonneratii*, 41; *temminckii*, 61 & 65 nn.10 & 11; *varius*, 38–9 & 42 n.11. *See also* fowls

Galton, Francis, 421–2 & n.2

Galton, Louisa Jane, 422 & n.5

Gambia, 26, 223, 238

ganoids, 300 & n.2

Gardeners' Chronicle, 3 n.8, 297 n.3, 499 & 500 n.8; CD writes concerning bees and fertilisation of kidney beans, 465–7 & n.4, 468 & 469 n.9, 480–2 & n.2, 487 & n.9, 489 & n.6, 492 & 494 n.8; CD writes concerning colouring of horses, 411 & n.2; CD writes concerning cross-fertilisation in Leguminosae, 296 & n.2; CD writes concerning deep wells, 435–6 & n.2; CD writes concerning hybrids of *Dianthus*, 349; CD writes concerning prolificness of foreign seed, 482–3 & nn.2, 5 & 6

garfish, 216 & 217 n.7, 235

Garnett, Thomas, 480 n.3

gar-pike. *See* garfish

Garrett, James R., 351 n.3

Gärtner, Karl Friedrich von: *Beiträge zur Kenntnis der Befruchtung*, 254 & 255 n.4; and cross-fertilisation of plants, 296 n.2; on *Cucubalis*, 494 & 495 n.4, 495, 500 n.12, 504 & n.3, 505 & n.2, 509 & n.3; on *Dianthus*, 387 & 388 n.6; on hybrid sterility, 349 & n.4; on laburnum, 151 & 152 n.3; on parthenogenesis in plants, 254 & 255 n.4; on wheat, 413 & 414 n.15; and seed-coats of hybrid peas, 47 & 48 n.8, 49 & 50 n.4, 487 & n.11, 492 & 494 n.9

Gaskell, Elizabeth Cleghorn, 458 & 459 n.6

Gaskell, William, 459 n.6

Gassiot, John Peter, 238 & 239 n.6

Gay, Claude, 384 & 385 n.4

geese: hybrids of, 53, 212 & 213 nn.10 & 11, 240 & n.7, 480 & n.2; of China, 480 & n.2; of India, 53 & n.3; seed in craw of, 4 & n.3

generation: in algae, 112 & n.6, 316–17 & n.2; in animals, 190, 433; in bees, 453 & n.9; in cirripedes, 492 & 494 nn.10 & 11; in Compositae, 424 & n.9; in *Daphnia*, 193 n.4, 251–2 & nn.2–5 & 9; in grasses, 423 & 424 n.8, 432–3; in hermaphrodites, 178, 190, 433; in molluscs, 173 & n.7; in parthenogenetic species, 231 & nn.2 & 3; in plants, 433 & n.8; in trees, 294 & 295 n.6, 304; separation of sexes, 190, 294, 304 & 305 n.3, 402 & 403 n.2; and speciation, 262; and telegony, 213 & 214 n.4; theories of, 210 & 211 n.3, 484 & 485 n.7

gentians, 262 & 264 n.14

Geoffroy Saint-Hilaire, Étienne, 461 & 462 n.3

Geoffroy Saint-Hilaire, Isidore, 64 & 66 nn.24 & 29

geographical distribution, 75 n.1; CD's manuscript on, 179 & 180 n.11, 190 & 191 n.2, 193 & 194 n.3, 247 & nn.1 & 2, 432 & 433 nn.3 & 4; CD's portfolio of notes on animals, 34 & n.5, 126 & n.9, 184 & 185 n.14, 186 & 187 n.4, 188 & n.5, 216 & 217 n.7, 226 & n.4, 270 & 271 n.7, 284 & n.17, 300 & n.4, 306 & 307 n.7, 348 & 349 n.6; CD's portfolio of notes on plants, 37 & n.13, 103 & 104 n.15, 157 & 158 n.15, 196 & 197 n.9; effect of climate on, 245, 258, 262 & 264 n.20, 498 & n.5, 500, 501 & n.3; effect of extinction on, 407 & 408 nn.6 & 7, 413, 422–3 nn.4, 5 & 10, 431–2 & 433 n.2, 448; lines of migration, 125, 260, 261, 262, 263, 264 n.20, 271; mapping of, 125; migration of animals during worldwide cold period, 125, 151, 161 & 162 n.4, 262 & 264 nn.16, 17 & 20; migration of aquatic plants, 78, 82, 192; migration of birds, 116, 119 & 120 n.12; migration of insects, 215 n.4, 376 &

377 n.2; migration of molluscs, 89 & 91 n.5, 186; migration of plants during worldwide cold period, 81–2 & n.9, 104 n.3, 105, 131, 139 & 140 n.3, 140 & 141 n.4, 150 & 151 n.2, 151 & 152 n.6, 156 & 157 nn.2 & 4, 162, 166–7 & n.3, 170, 174 & 175 n.4, 179 & 180 n.11, 180 & 181 n.7, 182, 198–9 & 200 n.4, 201, 203, 259–60, 262 & 264 n.14, 271–2, 272–3, 273–4, 280, 314–15 & 316 nn.6 & 7, 325 & 326 n.8, 340; migration of sub-Arctic species, 81, 260, 271–2, 272–3, 274; of ascidians, 160–1; of butterflies, 215 & nn.2 & 4; of Coleoptera, 215, 374 & 376 nn.10 & 13, 488 & n.2; of Crustacea, 180 & 181 n.6, 235 & 237 nn.3 & 5, 366 & 367 nn.2 & 5, 380–1 & 382 nn.3 & 6, 399 & 400 n.2; of disjoined species, 229, 339–40 & 342 nn.4 & 5, 391 & 392 nn.3 & 4, 401 & 403 n.1, 413 & 414 n.5, 422–3, 431–2 & 433 n.2; of echinoderms, 160–1; of fish, 187–8 & n.4, 366 & 367 n.3; of insects, 325 & 326 n.9, 374–5 & 376 n.13, 376; of insular flora and fauna, 125, 153–5, 166–7, 270, 272 & n.3, 325 & 326 nn.7–9, 348, 358–9, 388; of lizards, 346 & 347 n.4; of mammals, 14, 78–9, 80, 125, 193, 199, 242 & 243 n.5, 269–70 & nn.5 & 7, 272 & n.3; of molluscs, 89, 90 & 92 n.15, 111 n.2, 123 & nn.2 & 3, 125 & 126 n.3, 126–7, 129 & 130 n.2, 145–6 & n.11, 151 & 152 n.7, 152–3, 183–4 & n.1, 185–6 & 187 nn.1–3, 189 & n.2, 243 & 244 n.2, 304–5 & n.7, 305–6 & n.2, 348, 388; of representative species of animals, 8–9, 11, 160–1 & n.2, 161 & 162 n.4, 180 & 181 n.7, 262 & 264 nn.16 & 19, 306, 366 & 367 nn.3 & 4, 380–1 & 382 n.3, 478–9 & n.2; of representative species of plants, 36–7 & n.11, 81, 92 & 95 n.7, 105, 130, 180, 192, 262 & 264 nn.15–17, 271, 306, 316 n.7, 339, 340, 366 & 377 n.3, 380–1 & 382 n.3, 452, 478–9 & n.2, 480 n.7; plants common to Asia and United States, 148, 150, 151 & 152 nn.5 & 6, 156 & 157 nn.2 & 4, 245 & 246 n.8; plants common to England, Ireland, and Europe, 151; plants common to Europe and Asia, 339; plants common to Europe and Australia, 176 & 177 n.12; plants common to Europe and Cape Colony, 78–9, 81–2 & nn.8 & 9, 104 & nn.1–3, 105, 151; plants common to Europe and United States, 131–3 & nn.3 & 4, 139 & 140 n.3, 140–1 & n.4, 148, 149–50 & 151 n.2, 182 & nn.3, 5 & 6, 209 & nn.5–8, 228–9, 245 & 246 n.5, 258 & 259 n.1, 340 & 342 n.9; plants common to

South America and New Zealand, 172 & n.3, 176 & 177 n.11; provinces of, 96 & 97 n.8; A. R. Wallace's study of, 514 & 515 nn.5 & 6. *See also* dispersal

Geological Society of London, 110 n.4, 145 & 146 n.7, 156 n.8, 199 & 200 n.2, 260 & 263 n.6, 344 & n.8, 451

geology, 506; antiquity of continents, 115, 125, 143–4, 147, 153–5, 163–5 & n.2, 167–8, 170 & 171 nn.5 & 8, 199; denudation, 492; and geographical distribution, 198, 260, 261; methodology similar to that of etymology, 458 & 459 n.3; of Cape Colony, 80; of coal deposits, 198 & 200 n.2; of India, 153; of Madeira, 83–4 & nn.2 & 4, 99 & 100 n.2, 164, 168, 265 & n.4, 376 & 377 n.4; of mountain systems, 154; of Pacific islands, 337 & 338 n.7; of South America, 22–3; use of negative evidence in, 219 & 220 n.4, 236–7 & n.12, 265 & n.6, 366. *See also* cleavage and foliation, elevation and subsidence, glaciers and glacial phenomena, ice and icebergs, volcanos and volcanic phenomena

Geology of the 'Beagle', 23 & n.6, 33

Geoneria, 473

George Inn, Borough, 234 & 235 n.5

Gérard, Frédéric, 215 & nn.2 & 4

Gerber, Friedrich, 24 & n.2

Germany, 132; flora of, 229, 289 & 290 n.3, 438 & 439 n.14, 502; fossils of, 321 & n.3; natural history of, 206 & nn.2, 3 & 5, 213 n.3; scientific methodology of naturalists of, 461 & 462 n.4

Geum, 176, 177 n.1, 310

Gilbert, Henry, 431 & n.2

Gillies, John, 36 & 37 n.5

Gisborne, Thomas, 211 & 213 n.7.

glaciers and glacial phenomena, 83 & 85 n.5, 174 & 175 n.4, 262, 325 & 326 n.10, 327 & n.5, 362–3 & n.5; and geographical distribution, 70 & 71 n.2, 81–2 & n.6, 150; J. D. Hooker's study of, 217 & 218 n.3; T. H. Huxley's and John Tyndall's study of, 322–3 & nn.2–6, 329 & nn.3 & 4, 330 & nn.5–7, 456 & 457 n.3; in New Zealand, 70 & 71 nn.1 & 2, 337; Charles Lyell's views on, 144–6; theories of, 218 n.3, 323 n.2, 324, 330 & nn.5–7. *See also* ice and icebergs, erratic boulders

Gladiolus, 82 n.9

Gladwin, Francis, 32 n.4

Glasgow, Scotland, 30 & 31 n.1

Glover, Thomas, 460 & 461 nn.2 & 3, 472–3 & n.1

Gloxinia, 473

Glyceria fluitans, 102 & 103 n.5

Glyptonotus antarcticus, 380 & 382 n.4

Gmelin, Johann Georg, 64 & 66 n.30

gneiss, 217

goats, 8, 222, 270; blood corpuscles of, 24 & n.4; hybrid crosses in, 51 & 53 n.2; of Tibet, 350 n.2

Godron, Dominique Alexandre, 444 & 445 n.14

Godwin-Austen, Robert Alfred Cloyne, 177 n.7, 178, 192 n.3, 198 & 200 n.2, 201

Golden Cross, Charing Cross, London, 59 & n.4

goldfinches, 7, 470

goldfish, 364, 379

Goodsir, John, 420 & 421 nn.7 & 8

Goodward, Surrey, 336 & 337 n.3

gorillas, 184

gorse, 416, 417 & 418 n.2

Gosse, Edmund William, 383 n.5

Gosse, Philip Henry, 10 & 17 n.36, 11 & 17 n.40, 109 & 110 n.9, 320 nn.1 & 2; and birds as means of dispersal, 233; *Birds of Jamaica*, 9 & 17 nn.30 & 31, 352 & 355 n.2; and CD's experiments on dispersal of molluscs, 382–3 & nn.3 & 5; and pigeons and rabbits of Jamaica, 227–8 & nn.2, 3 & 5, 232–3 & n.1

Gould, Augustus Addison, 127 & 128 nn.11, 20 & 21, 129, 514 & 515 n.9

Gould, John, 12 & 18 n.45, 119 & 120 nn.3 & 11; *The birds of Asia*, 14 & 18 n.64; *The birds of Australia*, 14 & 18 n.63, 41 & 42 n.22

Graba, Karl Julian, 379 & n.4

Graham, Mrs, 434 & n.3

Graham, Robert, 412 & n.3

Gramineae. *See* grasses

grasses, 78, 82, 102 & 103 n.5, 178, 263 n.11, 390 n.6; cross-fertilisation in, 413, 423 & 424 n.8, 432–3

grasshoppers, 283

Gray, Asa, 132 & 133 n.10, 248, 280, 281, 299, 367; and A. de Candolle *Géographie botanique raisonnée*, 258 & 259 n.5, 325 & 326 n.3; and cross-fertilisation of plants, 423–4 & nn.5 & 11, 432–3 & nn. 6 & 7, 437 & nn.4 & 5, 492 & 493 nn.5–7; and CD's species theory, 432, 437 & n.2, 445–6 & 449 n.4, 447–9 & 450 n.17, 492–3 & n.2; and CD's study of close species, 443 & 444 n.11, 446 & 449 nn.9 & 10, 450 & n.5, 493 & 494 n.14; and CD's study of large and small genera, 314 & 316 nn.3–6, 363, 391, 446 & 449 nn.7–10; describes flora of Japan, 339 & 342 nn.2–4, 340 & 342 n.7; and disjoined species, 391 & 392 n.3, 401 &

403 n.1, 413 & 414 n.5, 422–3, 431–2 & 433 n.2; *First lessons in botany*, 339 & 342 n.1, 422 & 423 n.3, 447 & 449 n.14, 452 & 453 n.6; and intermediates between varieties, 182 & n.7, 195–6, 208 & 209 n.2; invites CD to visit United States, 195, 230; *A manual of the botany of the northern United States*, 150 & 151 n.5, 195 & 196 nn.2 & 3, 285 & n.5, 391 & 392 nn.7 & 8, 438 &439 n.11, 443 & 444 n.10, 446 & 449 nn.7 & 8, 502 & n.3; and naturalised plants, 196, 208, 228, 229, 258 & 259 nn.3 & 4, 285, 315 & 316 n.10, 341–2, 391 & 392 n.6, 402 & 403 n.3, 413, 447 & 450 n.15, 483 & n.3; on alpine flora, 182 & n.3, 257, 258, 262, 314 & 316 n.6, 390; opinion of Louis Agassiz, 340 & 342 n.8; and plants common to Asia and United States, 156 & 157 n.4, 258; and plants common to Europe and United States, 92–3, 151 n.2, 182 & n.4, 230 & 231 n.6, 257–8 & 259 n.1, 274 & n.1, 284–5 & nn.2 & 3, 314–15 & 316 nn.6 & 7, 340 & 342 n.9, 359–60; and protean genera, 315, 341 & 342 n.11, 352 & n.1, 355–7 & 358 nn.1, 5, 8 & 9, 361 & 362 nn.9–11, 398 & 399 n.2; and representative species, 315 & 316 n.7; reviews botanical works, 508 & n.6; and separation of sexes in trees, 315 & 316 n.8, 340 & 342 n.10, 358 & 359 n.4, 360–1 & 362 n.5, 391 & 392 n.5, 402 & 403 n.2, 413, 419 & n.4; and social plants, 228–9 & 230 n.2; statistics of United States flora, 92–4 & nn.3–8, 11 & 12, 196 & 198 n.8, 209 & n.8, 229 & 231 n.5, 244–6 & nn.3 & 6, 258 & 259 n.2, 285 & n.6, 314 & 316 n.2, 325 & 326 n.6, 360 & 362 n.3, 391 & 392 n.2, 399 & n.6, 412 & 414 n.3, 422 & 423 n.2, 447 & 450 n.15; and variation in highly developed organs, 396 n.1

Gray, George Robert, 61 & 65 n.10, 352

Gray, Jane Loring, 195

Gray, John Edward, 19 & 20 nn.2 & 3, 23, 32, 128 n.15, 373 & 376 n.5, 516; and lumping or splitting of species, 182; poor reputation of, 90 & 91 n.13, 96 & 97 n.6; and ranges of echinoderms, 160–1 & n.2

Grays, Essex, 48 & 49 n.4

grebes, 41

Greece, 13–14

Greenland, 140, 258, 260, 264 n.13, 314 & 316 n.6, 340, 400 & 401 n.4

Greg, William Rathbone, 205 & n.2

Griffith, Edward, 7 & 16 n.12

Griffith, Richard, 123 n.3

Griffith, William, 157 & 158 n.12

Grisebach, August Heinrich Rudolf, 80 n.7; *Spicilegium florae Rumelicae*, 443 & 444 n.9, 460 & 461 n.8, 468 & 469 n.4, 471, 486 & 487 n.3

Grove, William Robert, 238 & 239 n.6

guava, 354, 355

Guernsey, 473

Guiana, 422

Guinea, 17 n.34, 26

Gulliver, George, 24 & n.4

Gulliver, Mr, 152, 153, 211 & n.6, 226 & 227 n.5, 334, 397 & 398 n.3, 415, 431, 436

gulls, 41 & 42 n.22

Gully, James Manby, 238 & 239 n.2, 346 & n.2, 376 n.3, 385 & 386 nn.5, 7 & 9, 477 n.3, 507 n.5

Gurney, John Henry, 166 & n.1

Guyot, Arnold Henri, 230 & 231 n.7

Gwatkin, Edward, 490 & 491 n.3, 491

Gyllenhaal, Leonhard, 478 & 480 nn.5 & 6

Hadena adusta, 328

Hadrus, 374

Halicarcinus, 380

Hamilton, William John, 146 n.7

Hampshire County Lunatic Asylum, 480 & 481 n.1

Hancock, Albany, 72 & 73 n.12, 85 & n.7, 113 & n.3, 403–4 & 405 nn.4 & 9–12, 405 & 406 nn.2 & 4, 487

Hanley, Sylvanus Charles Thorp, 89 & 90 n.2, 487 & n.7

Harcourt, Edward William Vernon: lends book to CD, 206 & n.2, 207 & n.2; on Madeiran birds, 115–20 & nn.1, 3 & 4, 206 & nn.6 & 7

Hardinge, Henry, 38 & 42 n.3

Hardwicke, Thomas, 11

Hardy, Francis, 275 & n.3

Harpalus vividus, 253 & n.4

Harpalyce, 328, 332

Harris, John Williams, 74 & 75 n.2

Hartfield, Sussex, 458 & 459 n.4

Hartung, Georg, 84 nn.2 & 4, 265 n.4

Harvey, William Henry, 305 & n.9, 309 & n.5, 368 & n.3, 369; *Seaside book*, 184 & 185 n.11, 189; and sexes of algae, 316–17 & nn.2 & 3

Hastings, Sussex, 368 & 369 n.5, 386 & n.5, 389 & 390 n.5, 394 & 395 n.4

Haughton, Samuel, 123 n.3, 145 & 146 n.8

hawks, 212, 238, 239, 248 & n.2, 266

Haworth, Adrian Hardy, 328 & 329 n.4, 332 & 333 n.1

Hayne, Watson Ward, 227 n.4, 334

heaths. *See* Ericaceae

Hedyotis, 102

Heer, Oswald, 89 & 91 n.8, 144 & n.6, 488 & n.2; on fossil plants of Madeira, 99 & 100 n.3, 105 & 106 n.1

Hegeter, 374

Heinecken, C., 374 & 376 n.9

Heiracium, 355

Helianthemum, 357

Helix, 89–90 & 91 n.13, 96 & 97 n.5, 253 & n.5, 305; antiquity of species of, 186 & 187 n.3, 189; *aspersa*, 336 n.8; *monodon*, 27; *pomatia*, 243 & 244 n.3, 335 & 336 n.8, 338 & n.8, 346; *pulvinata*, 306 n.2. *See also* molluscs

Helops, 374

Helosciadium, 371

Hemionus, 409 & 410 n.10

Hemmings, Henry, 267 & 269 n.7, 269, 286, 303 & 304 n.3

Henfrey, Arthur, 508 n.6

Henry (Prince of Portugal), 17 n.34

Henslow, Ann, 467 & 468 n.5

Henslow, Frances Stevens, 25 & n.5

Henslow, George, 297 & n.4, 418 & 419 n.3

Henslow, Harriet, 488 & n.2, 495 & 496 nn.2 & 3

Henslow, John, 468 n.5

Henslow, John Stevens, 260, 267 & 268 n.3, 418 n.3, 453, 516; *A catalogue of British plants*, 438 & 439 n.8, 443 & 444 n.10; and CD's attempts to break the constitution of plants, 465 & n.2, 467 & n.3; and CD's children, 297 & n.5; death of mother, 25 & n.6; death of wife, 488 & nn.2 & 3, 495 & 496 n.2, 498 & nn.1, 7 & 8; gives butterfly specimen to W. E. Darwin, 454 & 455 n.3, 471 & n.3; gives CD his photograph, 467; and Hitcham Horticultural Society, 467 & 468 n.4; interest in botanical education, 142–3 & n.3, 297 & n.3; provides CD information on insurance clubs, 4 & nn.1 & 2; and seeds for CD's experiments, 25 & n.7, 142 & 143 n.2, 197 & 198 n.6; study of *Aegilops*, 175 & 177 n.6, 179, 197 & n.3, 202; visits Down, 438 & 439 n.4, 441 & n.1, 442 & n.2, 442 & 443 n.2, 444 & 445 n.15, 477 & n.12

Herbert, John Maurice, 1 & 2 n.4

Herbert, Mary Anne, 1 & 2 n.4

Herbert, William, 152 n.3, 368 & n.3, 387 & 388 n.6, 448

heredity, 192 n.2, 370 & n.6, 484 & 485 n.7. *See also* hybrids and hybridism

Hérétieu, M., 29 n.3

hermaphrodites, 161 & 162 n.2, 170–1 & n.12, 173 & nn.3, 5 & 7, 178 & 179 n.2, 190, 292 & n.5, 428; in animals, 307 & 308 n.9, 433; in fishes, 506 & n.6; in plants, 190, 280 n.1, 292 & n.5, 295 n.6, 298, 433 & n.8

herons, 212, 383

Herring, William, 40 & 42 n.14

Herschel, Alexander Stewart, 346 & 347 n.7

Herschel, John Frederick William, 112, 238 & 239 n.6, 327 n.3, 338 n.2, 347 n.7, 360 & 362 n.4

Hewitt, Edward, 334 n.6, 509–10 & 511 nn.1–3, 512–13 & nn.2–7

Hewson, William, 24 & n.3

Hibiscus populneus, 319

Hieracium, 315, 341, 351

Higgins, John, 275 & n.2

High Elms (home of the Lubbocks), 87 & n.5

Hill, Richard, 17 n.31, 233 & n.3, 318–19 & 320 nn.2–5 & 7, 352–4 & 385 n.2, 383 & n.7

Hill, Rowland, 206 & 207 n.4, 207, 240 & n.6, 410 & 411 n.3

Himalayas, 54, 262 & 264 n.18; flora of, 292, 293, 315, 340, 368 & n.4, 392 n.3, 499; thistle of, 363–4, 408, 469

Hincks, Thomas, 173 & nn.4 & 5, 302 & 303 n.14, 307 & 308 n.6

Hippolyte, 380

Hitcham, Suffolk, 25 & n.7, 143 n.4, 297 n.3, 488, 495 & 496 n.2, 498; Horticultural Society, 467 & 468 n.4

Hodgson, Brian Houghton, 11 & 18 n.42, 27 & 29 n.14, 40

Holland: flora of, 438 & 439 n.10, 450 & n.4, 460 & 461 n.11, 468, 469, 471 & 472 n.7, 487 & n.4

Holland, Edward, 463 & 464 n.4

Holland, Henry, 166 n.1

Holland, Miss, 88 & n.1

holly, 414 & n.16, 424

hollyhocks, 190

Homer, 13

Homoptera, 253

honeysuckle, 371

Hooker, Frances Harriet, 66 & 67 n.2, 72 & 73 n.16, 87 & n.3, 107, 109, 122, 151, 202, 217, 234, 248, 298, 304, 369, 468, 471 & n.5, 488, 498 & n.7, 516; birth of children, 372 & 373 n.9, 438 & 439 n.3, 443 & 444 n.2; visits Hastings, 390; visits Yarmouth, 408

Hooker, Joseph Dalton, 31 n.1, 244, 254, 273, 327 n.2, 412, 473, 501; birth of children, 438 & 439 n.3, 443 & 444 n.2; and controversy between Hugh Falconer and T. H. Huxley, 147 & 148 n.10, 175–6 & 177 n.8, 178; and cross-fertilisation in plants, 178 & 179 n.2, 190 & 191 n.4, 194 & 195 n.10, 291–2 & nn.2 & 5, 292–3 & n.4, 309 & n.2, 315 & 316 n.9; and cross-fertilisation in trees, 294 & 295 nn.3 & 4, 297–8 & 299 nn.1 & 3, 304 & 305 n.1; and Cucurbitaceae, 96 & 97 n.7; and CD's attempts to break the constitution of plants, 467 n.3; CD believes too critical of poor observers, 384, 389 & 390 n.3; CD borrows books from, 443 & 444 nn.6–8, 450 & n.2, 452 & 453 n.2, 468 & 469 nn.2–6, 471 & n.2, 486–7 & nn.3, 5 & 6, 498 & nn.2 & 3, 502 & n.2, 508 & nn.1 & 2, 516 & nn.2 & 4; and CD's experiments on dispersal of molluscs, 324 & n.6; and CD's experiments on plant dispersal, 46 & 47 n.2, 49, 105, 122 & 123 n.2, 192 & n.5, 203 n.5, 233 & 234 n.2, 240 & 241 n.2, 244 & n.2, 248, 274, 294 & 295 n.7, 305 & n.8, 309 & 310 n.7, 324 & n.6; and CD's species theory, 89 & 91 n.7, 99–100 & n.6, 106–7 & n.8, 109 & 110 n.11, 201, 259 n.6, 281–2; and CD's tabulation of large and small genera, 363, 429, 438, 443–4 & nn.12–14, 450 & nn.4 & 6, 452 & 453 nn.2–4, 460 & 461 nn.4–12, 468 & 469 n.2, 471 & nn.2 & 7, 486–7 & nn.3–7, 498 & nn.2–4, 498–9 & 500 nn.2–5, 502 & nn.2–4, 508 & n.1, 516 & nn.2 & 3; and death of Harriet Henslow, 488, 495 & 496 n.2, 498 & nn.1, 7 & 8; definition of species, 194 & n.5; and disjoined species, 340; and embryonic characters of plants, 416, 417 & 418 n.2, 418–19, 428–9 & n.2; examiner in botany for East India Company, 191 & n.9; and fish of India, 187; *Flora Indica*, 298 & 299 n.6, 309 & 310 n.6, 419 & 420 n.2, 423 & 424 n.7, 438 & 439 n.12, 452 & 453 n.4, 499 & 500 n.8; *Flora Novæ-Zelandiæ*, 147 & 148 n.8, 172 & n.3, 179, 294, 297, 337 & 338 n.3, 444 & n.13; *Flora Tasmaniæ*, 508 & n.5, 516 & n.8; and glaciers, 217 & 218 n.3, 323 n.3, 324, 329 n.3; and Asa Gray's letters, 273 & 274 n.2, 280 & 281 n.1, 325 & 326 n.3; *Himalayan journals*, 413 & 414 n.6; and T. H. Huxley's membership of the Athenæum Club, 103 & 104 n.13, 106 & 107 n.3, 109 & 110 nn.7 & 8, 111–12 & n.3; and land-bridge theory of distribution, 143–4 &

n.3, 147 & n.9, 153, 154 & 156 n.4, 159 n.3, 170 & 171 nn.2 & 8, 191, 193 & 194 n.2, 198–200 & n.4, 200–1, 261, 325 & 326 n.8, 358 & 359 n.5, 514 & 515 n.7; and medals of Royal Society, 72 & 73 nn.2 & 11, 77, 85, 403–4 & 405 nn.2 & 3, 405, 408–9 & 410 n.2, 487 & n.12, 496 & n.7; and migration during worldwide cold period, 151 & 152 n.7, 156 & 157 n.2, 170 & 172 n.2, 174, 259–62 & 263–4 nn.1–3 & 9–20, 272, 280; and North American exploring expedition, 362 & 363 n.3; and numbers of species in small area, 192 & n.6; on flora of Antarctic islands, 96 & 97 n.7, 166–7 & nn.3 & 6, 172 & nn.2 & 3, 174, 176, 191–2 & n.4, 203; on flora of Arctic, 390 & n.2; on flora of Ascension Island and St Helena, 102 & 103 n.2; on flora of Galápagos, 172 & n.4; on flora of Himalayas, 54 & 55 n.3, 81–2, 293, 315, 499; on flora of India, 363 & 364 n.6, 450 & n.6, 496 & n.4; on flora of New Zealand, 92 & 95 n.15, 102 & 103 n.1, 178, 190, 245 & 246 n.5, 297–8 & 299 nn.2, 3 & 6, 430 & n.2, 438 & 439 n.12, 444 & n.13, 483 & n.4, 500 n.8; on flora of Raoul Island, 145 & 146 n.10, 267 & 268 n.6, 400 n.3; on flora of Tasmania, 199 & 200 n.5, 298 & 299 n.5, 304; on flora of Tristan d'Acunha, 203 & 204 nn.1 & 2; on geological evidence in explaining geographical distribution, 198, 261; on highness of conifers, 258 & 259 n.5, 281; on laburnum, 151 & 152 n.4, 217 & 218 nn.9 & 12, 240 & 241 n.3; on new variety of *Colletia*, 78; on primroses and cowslips, 103, 107; on scientific methodology, 109; on vitality of ancient seeds, 3 & n.8; on *Wahlenbergia*, 419 & 420 n.3; opinion of CD's species theory, 89 & 91 n.7, 99–100 & n.6, 199–200, 201, 259–60, 280–1; opposes W. F. Daniell as botanical collector, 321–2 & nn.2, 3 & 7, 324 & n.2; plans to visit Down, 453, 454, 468, 471, 475 & 476 n.7; and plants common to Asia and United States, 151, 156 & 157 nn.2 & 4; and polymorphic species, 368 & n.4, 369–70 & n.3, 371 & 372 n.2; and protean plant genera, 315, 358 & 359 nn.3 & 4; reads CD's species manuscript, 179 & 180 n.11, 190 & 191 n.2, 194, 202, 234 & n.3, 247 & nn.1 & 2, 259–62 & 263–4 nn.1, 2, 9, 10, 12–20, 266–7, 271, 280–1, 432 & 433 nn.3 & 4; reputation as species lumper, 182 & 184 n.8, 195, 277, 398, 493, 499 & 500 n.7; review of A. de Candolle *Géographie botanique raisonnée*, 122 & 123 n.6,

Hooker, J. D., cont.
200 & 202 n.2, 203 & 204 n.6, 244 & n.6, 246 & n.11, 258 & 259 n.5, 325 & 326 n.4; and separation of sexes in trees, 190, 294 & 295 nn.3 & 4, 297–8 & 299 nn.1 & 3, 305 n.1, 419 & n.4; and *Silene*, 494; and variability of highly developed organs, 395–6 & nn.1, 4 & 5, 400 & nn.1 & 6, 407 & 408 nn.2 & 5, 409 & 410 n.9; and variation due to external causes, 280, 369–70 & n.6, 371, 384, 388–9 & n.1, 390 & nn.1 & 4, 396; visits Down, 66 & 67 n.2, 72 & 73 n.16, 85 & n.1, 87 & n.3, 89 & 91 n.7, 103 n.1; visits Hastings, 369, 372 & n.1; visits Manchester, 460; visits Switzerland, 194 & 195 n.11, 196, 202, 217, 246

Hooker, Marie Elizabeth, 373 n.9, 440 n.2, 471 & n.5

Hooker, William Jackson, 36 & 37 n.4, 132 & 133 n.8, 175 & 177 n.4, 176, 289 & n.2, 324, 326 n.3, 327 n.2, 338 n.7, 498, 501 n.14

Hooper, Robert, 474 & n.4

Hopkins, William, 83 & 84 n.3, 84 & 85 n.5, 168 & 169 nn.5 & 6, 330 n.6, 409

Horace, 383 n.8

Hordern, Peter, 87 n.2

Hormoseira, 317

Hornby, William Windham, 50 & 51 n.2

Horner, Anne Susan, 264 & 265 n.2

Horner, Leonard, 367 n.11, 457 n.4

Hornschuch, Christian Friedrich, 151 & 152 n.2, 157 & 158 n.9, 170 & 171 n.10, 218 n.11, 240 & 241 n.3

horses, 40, 270; Arabian, 214; chestnut, 440–1, 491; colouring of, 409 & 410 n.10, 411 & nn.2 & 6, 417 & n.4, 490–1 & nn.3 & 4; dun-coloured, 440–1 & nn.4 & 5, 444 & 445 n.17, 491; of India, 490–1 & nn.3 & 4; wild, 14, 64 & 66 n.26

Horsfield, Thomas, 6 & 15 n.4, 9 & 17 n.26, 41

Hotham, Beaumont, 350 n.3

Hudson, William, 355 & 358 n.3

Humboldt, Alexander von, 38 & 42 n.10, 214 & 215 n.7

Hummel, Arvid David, 456 & 457 n.14

Hunt, James, 516–17 & n.3

Hunt, Thomas Carew, 122 & 123 n.2

hurricanes: as means of dispersal, 89, 232, 319, 337, 348, 352–3, 354

Hutton, Thomas, 8 & 16 nn.17 & 19, 51–3 & nn.1–4, 213 n.11

Huxley, Henrietta Anne, 66 & 67 n.2, 73, 74, 87 & n.3, 101 & 102 n.5, 176, 307 & 308 n.7, 318, 454

Huxley, Noel, 318 & n.2, 454 & n.6

Huxley, Thomas Henry: appointed examiner to London University, 113 & 114 n.2, 114 & n.3, 161 & 162 n.7; and Athenæum Club membership, 103 & 104 n.13, 106 & 107 n.3, 109 & 110 nn.7 & 8; birth of children, 318 & n.2; and *Blatta*, 456 & 457 nn.5–10 & 14; and Brullé's law of embryonic development, 420 & 421 nn.2, 3 & 5, 424–6 & 427 nn.4 & 9, 427–8 & nn.3 & 5; and candidates for Royal Medal, 85 & n.8; *Catalogue* for the Museum of Practical Geology, 505–6 & n.2, 516 & n.7; and classification, 175 & 177 n.9, 454 & n.3, 456 & 457 nn.12 & 13, 461–2 & n.2, 462–3, 505 & 506 nn.3 & 7; controversy with Hugh Falconer, 147 & 148 n.10, 175 & 177 n.8, 178; controversy with Richard Owen, 106 & 107 n.4, 260 & 263 n.6, 329 & 330 n.5, 330 & n.4, 484 & 485 nn.3 & 4; criticises doctrines of distinguished naturalists, 106 & 107 n.5, 111–12 & nn.2 & 4, 147 & 148 nn.3–5, 330 n.4; and cross-fertilisation in invertebrates, 161 & 162 nn.2–4, 173 & nn.3, 5 & 7, 178 & 179 n.2, 302 & 303 n.14, 307 & 308 n.8, 428 & n.9; and CD's species theory, 89 & 91 n.7, 99–100 & n.6; lectures at Royal Institution, 66 & 67 n.3, 74 & n.3; lectures at School of Mines, 106 & 107 n.5, 111–12 & nn.2 & 4, 113 & 114 n.3, 161 & 162 nn.5 & 6, 231 & n.4, 302 & nn.7 & 12, 330 n.4, 399 & 400 n.4, 420 & 421 n.9, 454 & nn.2 & 4; and John Lubbock's study of *Daphnia*, 251, 489–90 & n.2; on nerves, 454 & n.4; on scientific methodology, 461 & 462 n.4; opinion of Albany Hancock, 403–4; and parthenogenesis, 301 & 302 n.2, 484–5 & nn.3 & 4; praises CD's cirripede monograph, 454 & n.2; publications of, 101 & 102 n.4, 106 & 107 n.6; receives Royal Medal, 106 & 107 n.6, 114; study of aphids, 484–5 & nn.3 & 4; study of Crustacea, 301–2 & nn.7, 8 & 10, 307 & 308 n.4, 318 & n.3, 399 & 400 n.4, 454 & nn.2 & 3; study of glaciers, 218 n.3, 322–3 & n.2, 324 & n.3, 329 & nn.2 & 3, 454 & n.5, 456 & 457 nn.3 & 11; visits Down, 66 & 67 n.2, 73, 74 & n.2, 87 & n.3, 89 & 91 n.7, 101; visits Switzerland, 218 n.3, 323 n.2, 454 n.5, 457 n.11; visits Tenby, 162 & n.9

hyaenas, 40

hybrids and hybridism, 176, 178; Edward Blyth on, 6–14, 26–8; CD's experiments on, 47 & n.6, 60 & n.3, 64 & 66 n.22; CD's manuscript on, 516 & nn.1 & 9; CD's portfolios of notes on, 51 & n.9, 54 n.5, 166 & n.5, 456 & n.4; degeneracy of, 52, 149 & nn.1, 4, & 6; fertility of, 40–1, 206 & 207 n.5, 211 & 212 n.2, 212 & 213 n.11, 387 & 388 n.6, 512 & 513 n.2; of algae, 498 & n.3; of beans, 481 & n.3, 486 & n.1; of bees, 301 & 302 n.6, 307 & 308 n.3; of birds, 516–17 & n.3; of cactuses, 460 & 461 n.3, 472–3 & n.2; of canaries, 318, 470 & 471 n.1; of cats, 137; of cattle, 4–5 & 6 n.7, 6 & 15 n.3; of *Dianthus*, 349; of dogs, 40, 50–1, 51, 57 & 58 nn.1 & 2, 137, 213; of ducks, 142, 166 & n.4, 214, 257, 352, 480 & n.3, 517 & n.3; of *Epilobium*, 175 & 177 n.1; of fowls, 53, 509–10 & 511 nn.1–3, 512–13 & nn.2–7; of geese, 212 & 213 n.11, 240 & n.7, 480 & n.2; of *Gloxinia*, 473; of goats, 51 & 53 n.2; of horses, 213–14, 409 & 410 n.10; of laburnum, 217 & 218 n.11; of peacocks, 474 n.7; of peas, 47 & nn.7, 8 & 10, 49, 492 & 494 n.9; of pheasants, 166 & n.3, 334 & n.6, 509–10 & 511 nn.1–3, 512–13 & nn.2–7; of pigeons, 98 & 99 nn.2 & 4, 142, 210; of pigs, 206–7 & n.5, 211 & 212 n.2, 240 & n.6, 386 & n.8; of yaks and cows, 455 & n.2, 469 & n.13; and origin of species, 191 n.4; theories of, 210 & 211 n.3. *See also* cross-fertilisation, Karl Friedrich von Gärtner, William Herbert, and Joseph Gottlieb Kölreuter

Hydrobius piceus, 89
Hymenicus, 380
Hymenoptera, 326, 456 & 457 nn.8 & 9
Hypsiprymnus, 366 & 367 n.9
Hythe, Kent, 379 & n.5

ice and icebergs, 263, 327 & n.5, 362–3 & n.5; effects of pressure and freezing on, 322–3 & nn.3–7, 329 & n.4, 330 & nn.5 & 7; and plant dispersal, 174 & 175 nn.4 & 6, 200 n.3, 203 & 204 n.4, 260, 261 & 263 n.10. *See also* glaciers and glacial phenomena
Iceland, 139, 260, 262 & 264 n.13
ichthyology, 188 n.4
Illustrated London News, 7 & 16 n.10, 13 & 18 n.54
Impatiens, 496 & n.4
India: botanical relations of, 260, 262; camels of, 7–8 & 16 nn.16–20; cattle of, 4–5 & 6 n.7, 12 & 18 nn.47–50, 28, 65; chooboos of, 455 & n.2; dogs of, 149 & nn.4 & 6; flora of, 182, 292 & n.5, 363 & 364 n.6, 369, 438 & 439 n.12, 496 & n.4, 499 & 500 n.8, 502; fowls of, 1 n.3, 7, 10–11, 12–14 & 18 nn.52–9, 256, 339 n.2; geese of, 53 & n.3, 212 & 213 n.11; geology of,

153; pigeons of, 8, 28, 31 & 32 nn.5 & 6, 53 & 54 n.4, 67–9, 87, 223; rabbits of, 28 & 30 n.27, 88 n.3 *See also* Himalayas, Khasia Mountains, Nilgiri Hills, Siwalik Hills
India House, 31 & 32 n.4
Indian Sporting Review, 40 & 42 n.16, 67
Innes, John, 240 n.5, 268 & 269 n.6, 343
insects: alpine, 262; as agents of dispersal, 89, 90 & 92 n.15, 99 & 100 nn.4 & 5, 376; blind cave forms, 180 & 181 n.3, 216, 282–4 & n.1; CD's study of large and small genera of, 478 & 480 n.5, 488; fighting of males for females, 375; instincts of, 7; intermediates between varieties of, 171 & 172 n.15, 177 n.16, 208 & 209 n.3; of Madeira, 158 & 159 n.2, 159, 325 & 326 n.9, 374–5 & 376 nn.10 & 13, 376 & 377 n.2; of Mauritius, 348; of Sweden, 478 & 480 n.5; parthenogenesis in, 252 n.2; role in fertilisation of plants, 170, 176, 287 n.2, 291 & n.4, 295 & n.3, 335, 346 & 347 n.5, 361, 414 & n.16, 433 & nn.5 & 6, 446 & 449 nn.12 & 13, 469 n.12, 493; social species, 301 & 302 n.6, 481–2 & nn.1 & 4; variability in, 134, 158, 283, 419
Ipswich, Suffolk, 453 n.10, 455 & n.4
Ireland: plants of, 139, 150; rabbits of, 11, 351 & n.2, 485 & nn.2 & 3
Ischia, Italy, 165
islands: absence of frogs on, 34 n.4, 130 n.4, 154, 271 n.7; antiquity of, 115 & n.6, 125 & 126 n.4, 144, 147, 154–5, 163–5 & nn.2 & 8, 167–8, 193, 199 & 200 n.7; colonisation of, 270 & 271 n.8; distribution of mammals on, 193 & 194 nn.3 & 4, 199, 201, 242 & 243 n.5, 269–70 & nn.5 & 7, 270 & 271 n.7, 514; insular flora and fauna, 75–6, 80, 90 & 92 n.16, 96 & 97 n.7, 158, 166–7 & nn.3–6, 260 & 263 n.4, 337, 374–5, 400 & n.3, 423, 448. *See also* dispersal, geographical distribution
Isle of Wight, 3 & 4 n.9
Isnardia palustris, 132
Italy, 14, 502, 515 n.6

jaguars, 388 & n.10, 457 & 458 n.3
Jamaica: animals of, 227–8 & nn.2, 3 & 5, 233 & nn.2 & 3; birds blown to, 336; ducks of, 354; fish of, 320 n.5, 353; fowls of, 10, 320 n.6; natural history of, 9 & 17 nn.30 & 31, 318–19 & 320 nn.2, 4 & 5, 352–4 & 385 n.2; rabbits of, 485
Janson, Edward Westley, 134 & n.3

Japan: fish of, 188, 264 n.16; flora of, 92, 339 & 342 nn.2–4 & 7, 392 n.3; fowls of, 57 n.2

Jardine, William, 119

Java: animals of, 9, 12 & 17 n.26, 199, 201; fowls of, 39; plants of, 105

jellyfish, 173 & nn.3, 6 & 7, 178 & 179 n.3

Jenyns, George Leonard, 496 n.3

Jenyns, Leonard, 496 n.3; on variation of species, 197 & n.2, 202

Jerdon, Thomas Claverhill, 5 & 6 n.7

John I (King of Portugal), 10 & 17 n.34

Johnson, George William, 398 n.5

Johnston, Alexander Keith, 104 n.8

Jones, William, 62 & 66 n.17

Jordan, Alexis, 132 & 133 n.11

Journal de la Conchyliologie, 129 n.28

Journal de Medicin, 474

Journal of researches, 72 n.3, 126 n.4, 130 n.3, 273 & n.2, 277 & n.1, 337 & 338 n.6, 363 n.3

Juan Fernandez Island, 153, 337

Juncus, 312, 357, 368 & n.3, 369 & 370 n.6

juniper, 294, 372

Juno, 14

Kaffraria, South Africa, 214 & 215 n.6

Kamandáki (Kamandaki Pandita), 62 & 66 n.16

Kashmir, 182 & 184 n.2, 186

Keeling Islands, 192

Kelaart, Edward Frederick, 54 & 55 n.6

Kennedy, Lord David, 5 & 6 n.6

Kerguelen Land, 144, 153, 293, 337, 359 n.6; botanical relations of, 104 n.9, 192, 203 & 204 n.5; cabbage of, 178 & 179 n.9, 191

Khasia Mountains, India, 262, 292 n.5, 293

kidney-beans, 46, 49, 291 n.4, 433 & n.6, 437 & n.5, 446, 465–7 & n.4, 480–1 & nn.2 & 3, 489 & n.6, 492 & 494 n.9; effect of hybrid crosses on seed-coats of, 480–1 & nn.2 & 3, 486 & n.1, 487 & n.9, 492 & 494 n.9

Kinahan, John Robert, 380 & 382 n.5, 381

Kinder, Mr, 463

King, Colonel, 379 & n.5

Kippist, Richard, 347 & n.1, 429 & n.8

Kirkby-Wiske, Yorkshire, 3 n.5

Knip, Pauline (de Courcelles), 228 n.2

Knole House, Kent, 434 & 435 n.5

Knowsley Hall, Lancashire, 166 & n.2

Koch, Wilhelm Daniel Joseph, 438 & 439 n.14, 443 & 444 n.6, 450 & n.2, 460 & 461 n.5, 502 & n.3; on *Subularia*, 277 & 280 n.2, 289 & n.3, 293

kohlrabi, 17 n.28

Kölreuter, Joseph Gottlieb, 387 & 388 n.6, 413 & 414 n.13

Krauss, Christian Ferdinand Friedrich von, 79 & 80 n.7

Krohn, August David, 303 n.10

Kynnersley, Mary, 394 & 395 n.2

La Plata, Argentina, 271

La Saussaye, Jean François de Paule Louis Petit de, 129 n.28

Labiatae, 391, 498 & 500 n.3, 499 & 500 n.4, 502 n.2, 508 n.3, 516

Labrador, 258, 259, 285, 316 n.6

laburnum, 151 & 152 nn.2–4, 157 & 158 n.9, 170 & 171 n.10, 217 & 218 nn.9 & 12, 240 & 241 n.3, 429 & n.4

Lacaze-Duthiers, Félix Joseph Henri de, 456 & 457 n.8

Laccadive Islands (Indian Ocean), 155

Lacordaire, Jean Théodore, 283 & 284 n.2

Læstadius, Lars Levi, 357

Lagopus, 8

Laibacher Zeitung, 284 & n.4

Lamarck, Jean Baptiste Pierre Antoine de Monet de, 90, 445

Lane, Edward Wickstead, 239 n.4, 377 n.5, 414 n.10, 440 n.1, 471 n.4, 476 n.5, 488 n.13; CD's opinion of, 385 & 386 nn.2, 4 & 9, 416 & n.2, 476 & 477 nn.2, 4 & 5

Lange, Johan Martin Christian, 400 & n.4

Langton, Charlotte, 46 n.3, 335, 459 n.4

Langton, Edmund, 365 n.2

Laparocerus, 375

Lapland, 260

Laslett, Isaac, 452 n.5

Latham, John, 10 & 17 n.35

Lathyrus, 347 n.5, 356, 428, 433, 466

laurel, 281 n.6, 353

Layard, Edgar Leopold, 136 & n.12, 137 & 138 nn.1, 2, 4, 6 & 7, 213 & 215 n.6, 223 nn.1 & 2

Leadbeater, John, 27 & 29 n.12

LeConte, John Lawrence, 216 & 217 n.3

Lecoq, Henri, 492 & 493 n.6

Ledebour, Karl Friedrich von, 133 & n.13, 438 & 439 n.15, 443 & 444 n.8, 450 & n.2, 460 & 461 n.9, 468 & 469 n.3, 471, 486 & 487 n.3, 493 & 494 n.15, 502 & n.3

Leguminosae, 198, 305, 391, 401, 417, 433, 502 n.2, 508; apetalous, 291 & n.3, 295 & n.2; cross-fertilisation in, 286–7 & n.2, 291, 296 & n.2, 309 & n.2, 346 & 347 n.5, 361, 413. *See also* clover, gorse, kidney-beans, laburnum, peas

Leith Hill Place, Surrey, 45 & 46 n.4, 227 & n.1, 344 n.10

Lemann, Charles Morgan, 178 & 179 n.7

leopards, 43–4, 388 & n.10

Lepidoptera. *See* butterflies and moths

Leptospermum, 112

Lepus, 8, 11, 29 n.3, 351 & n.2. *See also* rabbits

Lewes, George Henry, 192 n.2

Lewis, John, 46 n.8, 267 & 269 n.5, 475

Leyden, Mr, 318

Liatris, 398

Lichtenstein, Martin Heinrich Karl, 9, 137 & 128 n.5

Liénard, Elizée, 27 & 30 n.17

Lightfoot, John, 289 & n.2

Limas, Juan de. *See* Lopes de Lima, José Joaquim

Limax, 306

Limnaea auricularis, 128

Limosella, 192, 202 & 203 n.5, 266, 280, 289, 293, 294, 412 & n.4

Linaria, 311

Lindley, John, 146 n.9, 222, 260, 267 & 268 n.3; on new variety of *Colletia*, 36 & 37 nn.1–3, 78 & 80 n.2, 80; and Royal Medal, 72 & 73 nn.3, 11 & 14, 403 & 405 n.3, 404, 405–6 & n.4, 409 & 410 nn.5 & 8, 487 & n.12, 496 & n.7; *School botany*, 143 & n.4; *The vegetable kingdom*, 46, 49 & 50 n.2, 72 & 73 n.5, 292 & n.4, 406 & n.3

Lingula, 313

Linley, Elizabeth Ann, 435 n.4

Linley, Maria, 435 n.4

Linnaeus (Carl von Linné), 191 n.6, 195, 277, 356, 357; and apetalous flowers, 384 & 385 n.7; cabinet of, 327 & 329 n.2; *Fauna Suecica*, 327 & 329 n.2; on hairiness of alpine plants, 390 & n.3

Linnean Society of London, 103, 171 & 172 n.14, 266 n.1, 322 n.4, 329 n.2, 364 n.6, 451, 467, 485 n.3; CD's papers read to, 105 & 106 n.4, 450 n.17; journal of, 203 & n.6, 281 & n.3, 282, 419 & 420 nn.2 & 5, 494 & 495 n.3; library of, 429 & n.8; moves to Burlington House, 107 & 110 n.4, 347 & n.1, 429 n.8; transactions of, 254 & n.2

linnets, 206, 470

Linum catharticum, 469 & n.14, 499 & 500 nn.10 & 11

lions, 4 & 6 n.2, 6 & 16 n.6, 8, 71

Lipozygis, 291

Lister, E., 50 & 51 n.4

Lithodes, 380

Lithotryà, 219 & 220 n.7

Little, J., 18 n.47

Littleton, Adam, 18 n.52

Living Cirripedia (1851) and (1854), 55 & 56 nn.7–9, 101 & n.3, 158 n.10, 171, 219 & 220 n.6, 302 n.8, 314 & n.4, 318 n.3, 494 n.10

Livingstone, David, 496 & n.6

lizards, 250 & n.7, 270 & 271 nn.5 & 7, 346 & 347 n.4, 366

Lobb, William, 146 n.9

Lobelia, 420 n.2, 446, 466, 493

Logan, Robert Francis, 332

Loiseleur Deslongchamps, Jean Louis Auguste, 178 & 179 n.5, 413 & 414 n.11, 424 n.8

Loiseleuria procumbens, 230

Lombok, Indonesia, 290 n.4

London, Brighton, and South Coast Railway, 286 n.8

The London encyclopaedia, 17 n.38

London Journal of Botany, 182 & n.3

London University, 114 nn.1–3, 161 & 162 n.7

Lopes de Lima, José Joaquim, 270 & 271 n.8

Loudon, John Claudius, 286 & 287 n.3, 287, 291 & n.2, 294 & 295 n.5, 315 & 316 n.8

Lowe, Richard Thomas: and birds of Madeira, 116, 117, 118, 119 & 120 n.2; and fish of Madeira, 264 n.16; and land shells of Madeira, 90, 306 & 307 n.5, 376 n.9, 514 & 515 n.8; on flora of Madeira, 75–6 & nn.3 & 5–7, 80 & 82 n.4, 487 & n.6; sends CD specimens from Madeira, 33 & 34 n.3, 224 n.6

Loyd, Samuel Jones, 45 & 46 n.10, 103 & 104 n.12, 109, 113 & 114 n.2

Lubbock, Diana Hotham, 231 & 232 n.6

Lubbock, Ellen Frances, 87 & n.2, 231 & 232 n.5, 442

Lubbock, Harriet, 301

Lubbock, John, 454; birth of children, 232 n.5; business interests hinder scientific pursuits, 400 & n.6; CD asks for Sebright bantam from, 20, 51 n.3; CD asks to borrow fly-pincers from, 138, 141; and CD's botanical statistics, 429 & n.9, 430 & n.3; CD proposes for Athenæum Club, 349–50 & n.3; discovers fossil musk-ox, 66 n.12; interest in insects, 215; interest in invertebrates, 114 & n.4; invited to dine at Down, 74 n.2, 87 & n.3, 442 & 443 n.2; marriage of, 87 & n.2, 181 & n.10, 231; and membership in Royal Society, 490 & n.3; on generation in *Daphnia*, 250–2 & nn.2–5 & 7–10, 254–5 & nn.2–4, 489–90 &

Lubbock, John, cont.
　n.2; and parthenogenesis, 192–3 & nn.3 & 4,
　231 & n.2, 254 & 255 n.2, 301 & 302 n.4;
　study of crustaceans, 181 & n.10, 193 n.4, 231
　& 232 n.7; study of larval anatomy, 399 &
　400 n.5
Lubbock, John William, 113 & 114 n.2, 114, 215
　n.6, 274–5, 335 n.5
Lubbock, Mary Harriet, 490 & n.4
Lucas, Prosper, 474 n.7
Lukis, Mr, 473
Luzula, 311, 312
Lyallia, 103 & 104 n.9, 178
Lychnis, 143 & n.5, 371
Lyell, Charles, 32, 191, 212, 264, 281, 298; and
　antiquity of continents, 125 & 126 n.4, 170 &
　171 nn.5 & 8, 174, 176 & 177 n.14; and
　Austrian scientific expedition, 337–8 & nn.7
　& 10; and Joseph Barrande's 'colonies', 366
　& 367 n.11; and cleavage and foliation, 218
　n.6; and Copley Medal, 403 & 405 n.8, 409 &
　410 n.8, 487 & n.12; and CD's species theory,
　89–90 & 91 nn.7 & 10, 100 & n.8, 106–7 &
　n.7, 109 & 110 n.11, 134 & n.4, 135–6 &
　n.10, 145, 169 & n.7, 179 & 180 n.12, 194 &
　nn.5, 6, 8 & 9, 199 & 200 n.8, 236 & 237 n.10,
　265 & n.5; CD visits, 100 & n.8, 106 & 107
　n.7, 179 & 180 n.13; and effect of elevation
　and subsidence on climate, 260 & 263 n.8;
　experiments on immersion of seeds in salt
　water, 92 n.15; and land-bridge theory of
　distribution, 89 & 91 n.8, 143–4, 144–6, 147
　& 148 nn.9 & 11, 153–5, 159 n.3, 164 & 165
　n.3, 167–8, 170 & 171 n.7, 172, 174, 193 &
　194 nn.2 & 5, 198 & 200 n.1, 200, 376 & 377
　nn.2 & 3, 514 & 515 n.7; and letter to CD
　from R. T. Lowe, 75 n.1; on fossil mammals,
　320–1 & nn.2–4, 366 & 367 nn.8 & 10; on
　geology of Madeira, 83–4 & nn.2 & 4, 99 &
　100 n.2, 265 & n.4, 376 & 377 n.4; on
　migration, 90 & 92 n.15, 99–100, 174 & 175
　n.4, 203 & 204 n.4; on molluscs, 89 & 90 n.1,
　99; opinion of Joseph Prestwich, 409; opinion
　of Louis Agassiz, 315; *Principles of geology*, 168,
　169 & 170 n.8, 174 & 175 n.6; and review of
　J. D. Hooker's work, 304 & 305 n.6; and
　species, 134 & n.7, 182 & 184 n.8, 194 & nn.5
　& 6, 321 & nn.3 & 4; and struggle for
　existence, 448; study of glaciers, 456 & 457
　n.4; supplement to *Elements of geology*, 366 &
　367 nn.9 & 10, 381 & 382 n.12, 400 & n.7;
　visits Continent, 89 & 91 n.9, 366 & 367 n.11;

　visits Down, 34 n.4, 75 n.1, 84 n.2, 89 & 90
　nn.1 & 10; and volcanoes and volcanic
　phenomena, 83–4 & n.2, 99 & 100 n.2,
　163–5 & nn.1, 2 & 8, 167–8; and
　A. R. Wallace on species, 514 & 515 n.3
Lyell, Henry, 32 & 33 n.5
Lyell, Katharine Murray, 32 & 33 n.2
Lyell, Leonard, 32 & 33 n.4
Lyell, Mary Elizabeth, 84 n.4, 90 nn.1 & 9, 136
　n.10, 264, 298, 457 n.4
Lythrum hyssopifolium, 133

McAndrew, Robert, 90 & 91 n.11, 146 & n.11
Macarthur, John, 346 n.5
Macarthur, William, 345 & 346 n.5, 361 & 362
　n.8
Macaulay, Thomas Babington, 251
Macgillivray, William, 98 & n.4, 98 & 99 n.3,
　128 n.9, 223 & 224 n.3, 396 n.2
McGrigor, Charles Roderic, 269 & 270 n.1
McGrigor, Walter, 269 & 270 n.1
Maclaren, Charles, 156 n.7
Madagascar, 10, 137 & 138 n.3, 153, 155, 232,
　260, 293, 348
Madeira, 270; absence of certain insects on, 154,
　374 & 376 n.10 & 13; birds blown by gales to,
　336; birds of, 27, 115–20 & nn.1, 3, 4 & 12,
　206 & n.7, 320 n.2; Coleoptera of, 374–5 &
　376 nn.6–10 & 13, 376 & 377 n.2, 407 & 408
　n.3, 478–9 & 480 nn.1, 2, 6 & 7; distinctness
　of fauna of, 92 n.16; fish of, 76, 264 n.16; flora
　of, 75–6 & nn.3 & 5–7, 80, 178 & 179 n.7,
　372, 487 & n.6; fossil plants of, 89 & 91 n.8;
　frogs of, 33; geology of, 83–4 & nn.2 & 4, 99
　& 100 n.2, 163–4 & 165 n.8, 167–8, 168, 265
　& n.4, 376 & 377 n.4; insects of, 158, 159, 253
　& n.4, 325 & 326 n.9, 374–5, 438 & 439 n.13;
　and land-bridge theory of distribution, 144,
　153, 164–5 & nn.3 & 8, 167–8, 174, 376 &
　377 nn.2 & 3, 383 & n.4; molluscs of, 75 n.1,
　89–90 & 91 nn.6 & 11, 96, 145–6 & n.11,
　185–6, 253 & n.4, 306, 514 & 515 n.8;
　naturalised species of, 478 & 479 n.2; pigeons
　of, 223; rabbits of, 485 & n.3; zoological
　relations of, 325 & 326 n.9. *See also* Dezerta
　Grande, Porto Santo
Madras, India, 31 & nn.1, 2 & 5, 38
Maestricht, Belgium, 256 n.1, 257 & n.5, 265 &
　n.6, 320
Malay Archipelago, 7, 8–9, 290 & n.4, 387 &
　388 n.5, 411, 514, 515 & n.12
Maldonado, Uruguay, 36

Malherbe, Alfred, 116 & 121 n.6

Malvern, Worcestershire, 238 & 239 nn.2 & 3, 249, 346 & nn.2 & 8, 373 & 376 n.3, 385 & 386 nn.3, 5, 7 & 9, 394, 476 & 477 n.3, 507 n.5

mammals: classification of, 367 & n.12, 419 & 420 n.5; embryology of, 425, 462 & n.7; fossil, 82, 126, 186 & 187 n.3, 320-1 & nn.2-4, 366 & 367 nn.7 & 8; geographical distribution of, 14, 79, 80, 193, 199, 201; naturalised, 514; of West African islands, 242 & 243 n.5, 269-70 & nn.5 & 7, 272 & n.3

Mammoth Cave, Kentucky, 216 & n.2, 299

man: classification of, 367 & n.12, 419 & 420 n.5, 463; interracial crosses in, 51-2; origins of, 125 & 126 n.8; races of, 184 & 185 nn.12 & 13, 463; relation of complexion to susceptibility to diseases, 241-2 & 243 n.4

Manchester, 460; Art-Treasures Exhibition, 434 & nn.2 & 3, 459 n.6

mangetout, 49 & 50 n.3

Mantell, Walter Baldock Durrant: and erratic boulders of New Zealand, 70 & 71 nn.1 & 2, 74, 130; and sexual selection in man, 72 n.2, 75 & n.5

Manual of scientific enquiry, 327 & n.3, 336 & 338 n.2

Marianne Islands, 400 & n.3

Marshall Island (Pacific Ocean), 155

marsupials, 125, 154, 367 n.9

Martha (servant of Sarah Wedgwood), 267 & 269 n.7

Martins, Charles Frédéric, 260 & 263 n.4, 264 n.13

Maruta cotula, 229 & 231 n.3

Mas y Sans, Sinboldo de, 45 n.8

Mason, Philip Brookes, 373 & 376 n.6

Massereene, Lord. *See* Skeffington, John

mastodons, 126, 186 & 187 n.3

Maull and Fox (photographers), 197 n.6

Mauritius, 138 n.3; birds blown to, 232, 336; flora and fauna of, 225-6, 347-8; Natural History Society of, 27 & 30 n.17, 225 & 226 n.1, 232 & n.3

Maury, Matthew Fontaine, 144 & n.6, 164 & 165 n.3, 174 & 175 nn.2 & 3

Maximilian, Archduke, 338 n.7

Mayor, John, 393 & 394 n.9

Mayor, Miss, 451

Mayor, Robert Bickersteth, 285 & 286 n.3, 434 & 435 n.6, 451 & 452 n.2, 475 & 476 n.2, 478 & n.3

medusae, 107 n.6

Melbourne, Australia, 499 & 500 n.13, 508

Meleagris, 9, 10, 13, 29 n.12. *See also* fowls, guinea-

Melville, Alexander Gordan, 338 n.4

Mentha, 131, 196, 315, 341, 356

Menyanthes, 294

Meyen, Franz Julius Ferdinand, 384 & 385 n.4

Meyer, Ernst Heinrich Friedrich, 104 & n.2, 389 & 390 n.4

Microlestes, 321 & n.3

microscopes, 193 & n.3

Mid-Kent Railway, 73 & n.17, 215 & n.6, 286 n.8, 335 n.5

Miers, John, 72 & 73 nn.3 & 7

Miller, William Hallowes, 181 & n.9

millet, 324 & n.5, 364

Milne-Edwards, Henri, 251 & 252 n.8, 399 & 400 n.4, 405 n.10; on classification, 420 & 421 n.4, 425-6 & 427 nn.5-7, 428 & n.8; receives Copley Medal, 73 n.14

Mimoseae, 286, 287

minnows, 364 & 365 n.2

Miquel, Frederick Anton Wilhelm, 438 & 439 n.9, 450 & n.4, 460 & 461 n.11, 471 & 472 n.7, 487 & n.4

mistletoe, 445, 448

Mitchell, David William, 69 & 70 n.7, 288 & n.4

Mitra, Rajendralala, 66 n.16

moas, 130 n.3

molluscs, 253 & n.5; circulatory system of, 404 & 405 n.10; common to Europe and America, 151 & 152 n.7; disjoined species of, 431-2; dispersal of, 89 & 91 nn.5 & 6, 233 & n.4, 236 & 237 n.8, 239, 253, 304-5 & n.7, 305-6 & n.2, 335 & 336 n.8, 338 & n.8, 346, 382-3 & n.3, 383 & n.4, 385 & n.9, 388, 453 & n.10, 455 & n.4, 514 & 515 nn.8-10; geographical distribution of, 96, 111 n.2, 123 & nn.2 & 3, 125 & 126 n.3, 126-7, 129 & 130 n.2, 140 & 141 n.4, 146 & n.11, 151 & 152 n.7, 183-4 & n.1, 185-6, 189 & n.2, 514 & 515 nn.8-10; geological history of, 48 & 49 n.5, 95-6 & 97 nn.1 & 2, 127 & 129 n.24; intermediates between varieties of, 171 & 172 n.15; naturalised species of, 514 & 515 n.9; of Africa, 78, 154; of America, 154; of Canary Islands, 90 & 91 n.11; of England, 48, 89 & 90 n.1, 146, 185-6; of Jamaica, 318 & 320 n.3; of Madeira, 75 n.1, 89 & 91 nn.6 & 11, 90 & 91 n.11, 96, 146 & n.11, 185-6, 253 & n.4, 306, 514 & 515 n.8; of Mauritius, 225 & 226 n.1, 347; of New Zealand, 185-6; protean genera of, 315; reproduction in, 173 & n.7; variation in, 96, 183-4

Mombasa, Kenya, 138 n.3

Monizia edulis, 75 & n.5

monkeys, 270 & 271 n.7, 375

Monotropa uniflora, 340

Montagu, George, 89 & 90 n.3

Montia fontana, 192, 202 & 203 n.5, 233, 244

Montpellier, France, 495 n.6

Moor Park, Surrey, 238 & 239 n.4, 347 n.8, 377 n.5, 383 & n.6, 384, 385, 388, 390, 394 & 395 n.5, 404, 407 & 408 n.8, 413 & 414 n.10, 414 & 415 n.1, 416 & n.1, 417, 440 n.1, 461 n.14, 469 n.7, 471 n.4, 475 & 476 n.5, 476 & 477 nn.2, 5 & 6, 483, 484 & 485 nn.1 & 2, 487 & n.13

Moore, Charles, 499–500 & 501 n.14, 501 n.3, 502 & 503 n.6

Moore, Frederic, 6 & 15 n.4, 15

Moore, John, 59 n.5, 394 n.9

Moore, Thomas, 299 n.6

Moquin-Tandon, Horace Bénédict Alfred, 121 n.5, 384 & 385 nn.6 & 8, 389 n.2, 396 & nn.6 & 7

Moray, Mrs. *See* Morrey, Mrs

Moreton Bay, 127

Mormodes, 335

Morrey, Mrs, 267 & 269 n.7

Morris, John, 48

Morrow, James, 342 n.2

Morton, Lord. *See* Douglas, George

moths. *See* butterflies and moths

Mueller, Ferdinand Jakob Heinrich von, 176 & 177 n.13, 178, 499 & 500 n.13, 501 & nn.2 & 3, 502 & 503 n.6

mules, 53, 214

Müller, Johannes Peter, 506 n.7

mullets, 353

Murchison, Roderick Impey, 48 & 49 n.5, 338 n.7, 362 & 363 n.3

Murray, Charles Augustus, 135 & 136 n.6, 290 & n.2, 339 & n.3

Murray, John, 273 & nn.1 & 2, 277 & nn.1 & 2

Musari, Sayzid Mohammed, 31 n.3

Museum of Practical Geology, 330 n.5

musk-ox. *See* cattle

Musophaga, 117

Mustela, 8

Mydaus, 9 & 17 n.26, 41

Myosotis, 142 & 143 n.2, 202 & 203 n.4, 315, 341, 465 & n.2, 467 & n.2

Myosurus, 172

Myrtaceae, 198

Myrtus, 354, 355

Mysis, 420 & 421 n.9, 425 & 427 n.2

Nag's Head, Borough, 88, 240, 273, 436 & n.4

Nanina, 90 & 91 n.13

National Review, 205 n.2

natural selection, 272 & 273 n.3, 282, 370, 407 & 408 nn.6–8, 447–8, 492 & 493 nn.4 & 5

Natural selection. See Darwin, Charles Robert: species book

Nebria, 374

Necrophorus, 374

Nepal, 107 & n.3

Nepenthes, 396 & n.4

Neumeister, Gottlob, 60 & n.5

Neuroptera, 419

New Guinea, 515 nn.5 & 6

New Holland. *See* Australia

New Zealand: birds of, 10, 14; botanical relations of, 92 & 94 n.7, 145, 153, 154, 155, 171 n.2, 172 & n.3, 176 & 177 n.11, 178, 198–9 & 200 nn.3 & 4, 267, 325, 358 & 359 n.6; crustaceans of, 366 & 367 n.5, 380; erratic boulders of, 70 & 71 n.2, 74, 130, 337; fish of, 188; flora of, 70–1 & n.3, 105, 112 n.7, 190, 294 & 295 n.3, 297–8 & 299 n.3, 304, 315 & 316 n.9, 358 & 359 n.6, 368 & n.4, 369, 413, 430 & n.2, 438 & 439 n.12, 444 & n.13, 483 & n.4, 502; geology of, 70 & 71 n.2, 145, 337; indigenous animals of, 34 n.4, 74 & 75 nn.2 & 4, 130 & nn.3 & 4; molluscs of, 185–6; trees of, 294 & 295 n.3, 297–8 & 299 nn.1–4, 304, 358, 360; zoological relations of, 188

Newcastle upon Tyne, 405 n.9

Newfoundland, 316 n.6

Newport, George, 77 & n.4

Nicholson, B. A. R., 61 & 65 n.8

Nicobar Islands, 8

Niebuhr, Carsten, 10 & 17 n.39, 14 & 18 n.61, 61 & 65 n.7

Niger Expedition, 28 & 30 n.23, 322 n.7

Nigeria, 241–2

Nilgiri Hills, India, 5, 8, 54, 81, 105, 262 & 264 n.18

Nordmann, Alexander von, 173 & n.6

Norman, Ebenezer, 443 & 444 n.5, 450 n.17, 452 & 453 n.3, 460, 480 n.7

North American Exploring Expedition, 327 & n.2, 362–3 & n.1

North Pacific Exploring Expedition, 400 n.3

Norway, 132; dun-coloured ponies of, 411, 440 & 441 n.5, 444 & 445 n.17; seeds washed up on coast of, 107 n.2

Norwich, Norfolk, 477 & 478 n.2
Numenius arquatus, 118
Numida, 26, 69
nuts, 372, 419 & 420 n.4, 429 & n.5
Nymphaea, 102 & 103 n.1

oats, 379
Oconee, Mount, South Carolina, 182 & n.5
Ocymum basilicon, 469
Oedemeridae, 375
Oenothera, 341
Ogilby, William, 26 & 29 n.7
Oken, Lorenz, 461 & 462 n.3
Oldfield, Henry Ambrose, 107 & 108 n.3
Ononis, 291
orang-utans, 6, 25 & 29 n.1, 125, 184 & 185 n.12
Orbigny, Alcide Charles Victor Dessalines d',
    128 nn.4, 8 & 13
orchids, 152 n.3, 190, 325, 335, 449 n.5
Oregonia, 380
Orenburg, [USSR], 27
Organ Mountains, Brazil, 81
Origin, 196 & 197 n.9
Orkney Islands, 132, 260, 379 n.5
Ornithorhynchus, 419
Orr, William Somerville and Co. (publishers),
    398 n.4
Orthoptera, 456 & 457 n.6
Orton, Reginald, 210 & 211 n.3
Osbeck, Per, 474 & n.8
Osmia atricapilla, 377
Otion, 301 & 303 n.9
Otiorhynchus, 374
Otis tarda, 10–11
Oulton Park, Cheshire, 335 & 336 n.9
Overstone, Baron. *See* Loyd, Samuel Jones
Owen, Richard, 38 & 42 n.3, 404; controversy
    with T. H. Huxley, 106 & 107 n.4, 109, 112
    n.2, 260 & 263 n.6, 329 & 330 n.5, 330 & n.4,
    484 & 485 nn.3 & 4; and fossil mammals, 321
    n.2; lectures of, 330 n.5; and methodology in
    anatomy, 147 & 148 n.5, 461 & 462 n.3; on
    Bryozoa, 173 n.6; on classification of man,
    367 & n.12, 419 & 420 n.5; on *Dinornis*, 75
    n.2; on musk-ox, 61 & 66 n.13; on orang-
    utans, 25 & 29 n.1; and parthenogenesis, 112
    n.2, 158 n.8, 231 & n.2, 484 & 485 nn.3 & 4;
    reputation of, 298; theory of vertebral origin
    of the skull, 330 & n.4
owls, 24, 250 & n.6, 274, 281, 282 & n.3
Oxalis, 200 n.4, 368, 370

Oxford: meeting of British Association in, 54 &
    55 n.2
oxlips. *See Primula*

Paget, James, 344 n.5
Palaeontographical Society, 126 n.2, 314 nn.4 &
    6
palaeontology: Joseph Barrande's theory of
    colonies in, 366 & 367 n.11; CD's portfolio of
    notes on, 130 & n.3, 300 & n.4, 321 & n.5,
    424 & n.10; methodology in, 147 & 148 nn.2,
    4, 5 & 10; parallel between record of and
    embryogenesis, 216, 235 & 237 n.6, 299–300.
    See also fossils
Pallas, Pyotr Simon, 60 & 61 n.8, 64 & 66 n.30,
    187
Palliser, John, 327 n.2
Palmerston, Lord. *See* Temple, Henry John
palms, 38, 79
Panicum clandestinum, 423
panthers, 43–4
Papilionaceae, 286 & n.2, 291, 423, 433. *See also*
    Leguminosae
Paraguay, 355 n.4
Parker, Henry, 58 & 59 n.11
parrots, 14, 354, 355, 388 & n.11
Parry, William Edward, 262 & 264 n.17
Parslow, Joseph, 141 & n.4, 192, 240 n.5, 303,
    343 & 344 n.6, 394 & 395 n.3, 478
parthenogenesis, 156 & 157 nn.7 & 8, 171 &
    n.12, 231 & nn.2 & 3, 301 & 302 nn.2–4;
    T. H. Huxley's study of, 301 & 302 n.2,
    484–5 & nn.3 & 4; in Articulata, 252 n.2; in
    bees, 156 & 158 n.8, 171 & n.12, 301 & 302
    nn.2 & 3, 343 & nn.1 & 2; in butterflies and
    moths, 156 & 158 n.2, 171 & n.12, 302 n.2; in
    Daphnia, 251–2 & nn.3–5 & 9; in plants, 157
    n.7, 254 & 255 nn.2 & 3; Richard Owen's
    study of, 112 n.2, 158 n.8, 231 n.2, 484 & 485
    nn.3 & 4
partridges, 10, 11, 27 & 29 n.15, 221–2, 239–40
    & n.5, 248 & n.3, 267; *chukors*, 11–12
Patterson, Robert, 351 & nn.2 & 3, 485 & n.2
peacocks, 474 n.7
pear-trees, 477 & n.11
peas: cross-fertilisation in, 47 & 48 nn.9, 10 & 12,
    346 & 347 n.5; effect of hybrid crosses on
    seed-coats of, 47 & nn.7 & 8, 487 & nn.10 &
    11, 492 & 494 n.9. *See also Lathyrus*, mange-
    tout, sweetpeas
Pelargonium, 167

Penfold, James, 250 & n.5

Pennant, Thomas, 10 & 17 n.38

Peradeniya, Ceylon, 55 n.1

Perdix, 11, 27 & 29 n.15

Peronea, 327, 328, 331, 332 & 333 n.1

Perry, Matthew Calbraith, 342 nn.2 & 3, 400 n.3

Persia: fowls of, 135, 256 & 257 n.19, 290 & n.2, 339 & nn.2 & 3

Persoon, Christiaan Henrik, 190 & 191 n.7, 304 & 305 n.2, 315 & 316 n.8

Petleys, Down (home of Sarah Wedgwood), 238 & 239 n.7, 267 & 269 n.3

Pezophaps solitarius, 338 n.4

Pfeiffer, Ludwig Georg Karl, 96 & 97 n.5, 127 & 129 n.21

Phalaris, 423

Phaseolus, 361

pheasants, 8 & 17 n.25, 9–10, 14, 26, 61 & 65 n.6, 166 n.3, 221–2; golden, 512 & 513 n.4; hybrids of, 166 & n.3, 334 & n.6, 509–10 & 511 nn.1–3, 512–13 & nn.2–7; of China, 509 & 511 n.2; silver, 512 & 513 n.4

Phillips, John, 22–23 & nn.1 & 2, 33 & n.2, 493 & 494 n.13

Philoperisteron Society, 59 n.2, 257 n.4, 275 & 276 n.2, 276; CD attends annual show of, 1 & n.2, 51 n.7, 489 & n.5; CD attends meeting of, 21 n.5

Philosophical Club of the Royal Society, 247 n.2, 260 & 263 nn.5 & 7, 263 n.5, 302 & n.2, 321 n.2, 338 n.7, 408; CD attends meetings of, 112 & n.5, 148 & n.12, 181 & n.9, 234 & n.5, 267–8 & n.3, 269 n.9, 302 & 303 n.11, 358 & 359 n.5, 404 & 405 n.16, 485 n.8, 489 & 490 n.2, 494 n.13

Phleum, 423

phoenix, 27 & 29 n.13, 29

phrenology, 32

Phryma, 340 & 342 nn.5 & 7, 401

Phylica, 167

Pickering, Charles, 125 & 126 n.8, 184 & 185 n.12

Pieris, 332

pigeons, 39, 62–4 & 66 n.21, 67–9, 305; ancient references to, 31 & nn.3 & 4, 57–8 & n.3, 236; bald-pates, 63–4, 67, 69 & 70 n.5; barbs, 69 & 70 n.3, 223, 393, 431, 459; black wing bars in, 204 n.3, 214, 223, 379; Brunswicker, 276; carriers, 31 & 32 n.6, 59, 60, 62, 63, 67, 69, 87 & 88 n.4, 226 & 227 n.4, 234, 235, 255, 334, 336 & 337 nn.1 & 7; coloration of, 137 & 138 n.7, 210, 276 & n.3, 379 & n.5, 393 & 394

n.10, 397, 409, 441 n.4; CD's acquisition of, 2–3 & n.4, 31 & 32 nn.5 & 6, 35 & n.1, 45 & 46 n.8, 54 & 55 n.4, 56, 58, 59, 60, 87 & 88 nn.2 & 4, 98–9 & n.5, 107, 124 & nn.3 & 4, 135 & 136 n.5, 223, 226, 228, 288, 335, 378–9 & n.2; CD's breeding of, 54, 60 & n.3, 66 n.21, 142, 210, 431; death of from overeating salt, 60 & 61 n.9; dragons, 60; embryonic characters of, 58 & n.8, 91 n.10, 152; fantails, 63, 64, 67, 68, 69, 226; finnikins, 226 & 227 n.6, 256 n.1, 257 & n.4; flight of, 336 & 337 n.7; helmets, 397, 414; hybrids of, 98 & 99 nn.2 & 4, 397; jacobins, 64, 69 & 70 n.3, 431, 436, 489; laughing, 87 & 88 n.2, 110 & 111 n.2, 122, 124 & nn.3 & 4, 152 & 153 nn.1 & 2, 160 & n.2, 210; leg feathers of, 63; *lotun* or *lotan* (ground-tumblers), 32 n.6, 63, 67; magpies, 397, 414; nuns, 45, 67, 69 & 70 nn.3 & 4, 431; of Ascension Island, 137 & 138 n.7, 214 & 215 n.5, 220, 222; of Borneo, 388 & n.9; of Cape Colony, 137; of China, 19–20 & n.4, 57 n.2; of the Gambia, 223, 238; of India, 8, 28 & 30 n.30, 31 & 32 nn.5 & 6, 53 & 54 n.4, 62–4 & 66 n.20, 67–9, 87 & 88 n.3, 223; of Jamaica, 227 & 228 nn.2 & 3, 233; of Madeira, 223; of Persia, 276 & n.4; of Shetland Islands, 223 & nn.2 & 3; of Sierra Leone, 241 & 243 n.2; of Singapore, 388 & n.9; of West Africa, 276 & n.5; owls, 436; passenger, 24; pouters, 45, 58 & n.6, 63, 67, 68, 69, 257 n.6, 334 & n.7, 339; reversion in, 204 n.3, 393 & 394 n.8, 397; rock, 378–9 & n.2, 440 & 441 n.3; runts, 63, 67, 68, 69, 152 & 153 nn.4 & 5, 204 & n.3, 204–5 & n.2, 208 & n.2, 210 & 211 n.2, 211, 226, 234 & 235 n.2, 246 & 247 n.1, 334, 397 & 398 n.3, 414–15 & n.6, 416, 431 & nn.1 & 3, 436; Scanderoons, 60, 152, 160 & n.4, 204 & n.4, 208 & n.3, 210, 224, 249 & n.4, 339, 392–3 & 394 n.4, 397, 459; smiters, 459 & nn.1 & 4, 463 & 464 n.3, 464; trumpeters, 45, 210, 431; tumblers, 31, 54 & 55 n.5, 67, 68, 276; turbits, 45, 64, 210, 393, 397, 431, 436; turners, 226 & 227 n.6, 257 n.4; variation in, 58 & n.6, 160 & n.3, 204, 207, 208 & n.2, 223, 236, 238, 387–8, 414–15; wild, 62, 97–8, 223

pigs: CD's study of jaws of, 211, 240 & n.6; hybrids of, 206–7 & nn.3 & 5, 211 & 212 n.2, 386 & n.8, 410 & 411 n.3; Neapolitan, 386; of Africa, 240, 410 & 411 n.3; of China, 386; of Jamaica, 354, 355; variation in, 206 & 207 nn.2 & 3, 211 & 212 nn.2 & 3

pike, 187–8
Pilumnus hirtellus, 381
Pimelia, 374
Piscidia, 287
Plagiaulax, 366 & 367 nn.8 & 9
Plantago, 499
plants: agrarian species, 483, 495; aquatic species, 78, 82, 157, 171 & n.13, 192, 233, 244, 279 & 280 n.1, 289, 291, 372, 412 & n.4, 413; embryology of, 416, 417–18, 418–19, 421 n.3, 427–8 & n.6, 428–9 & n.2; fossil forms of, 122 & 123 n.3, 200 n.2, 321 & n.4, 327 & n.4, 338, 363; hardiness of endemic species, 372; hermaphroditic species, 190, 280 n.1, 292 & n.5, 295 n.6, 298, 433 & n.7; monotypic genera, 401; naturalised, 103 & 104 nn.10 & 11, 196, 201 & 202 n.4, 208 & 209 n.4, 228, 229, 245 & 246 n.6, 273, 285, 315 & 316 n.10, 319, 325, 341–2, 353, 368 & n.4, 391–2 & nn.8 & 9, 402 & 403 n.3, 413 & 414 n.9, 447 & 450 n.15, 482–3 & nn.2, 5 & 6, 494 & 495 n.6; number of in small area, 192 & n.6; parthenogenesis in, 157 n.7, 254 & 255 nn.2 & 3; pelorism in, 429 & n.4; polymorphic species, 368 & nn.3 & 4, 369–70 & n.3, 371 & 372 n.3, 409; secondary sex characters in, 407 & 408 n.2; social species, 93 & 95 n.16, 228–9 & 230 n.2, 245; struggle for existence among, 113 & n.3, 407 & 408 nn.6 & 7, 418. *See also* Arctic and alpine species, cross-fertilisation, dispersal, geographical distribution
Pleurotoma, 96
Plieninger, Wilhelm Heinrich Theodor von, 321 n.3
Pliny, 29 n.13
plovers, 396 & n.3
plum-trees, 477 & n.11
Po River, 171 n.13
Poa, 131
Podostemaceae, 291–2 & n.5, 292–3 & nn.2 & 4, 413
Podostemon, 293 & n.4, 294
Poedisca sordidana, 328
Polygala, 368 & n.3, 369
Polygonaceae, 498 & 500 n.3, 502 n.2
Polygonum, 131 & 132 n.6, 356, 503
Polyzoa, 173 & n.5, 303 n.14, 428 & n.9
Pongamia, 287
Poole, Dorset, 267 n.7
poppies, 495
Populus, 398

Porphyrio alleni, 119
Port Philip, Australia, 381
Porter, Robert Ker, 64 & 66 n.31
Portlock, Joseph Ellison, 309 & n.4
Porto Santo, Madeira: Coleoptera of, 375; fish of, 76; flora of, 75–6, 80 & 82 n.4, 144, 165 n.8, 377 n.2; molluscs of, 146 & n.11, 253 nn.1 & 5, 305–6 & nn.2 & 4; rabbits of, 33 & 34 n.2, 135 & 136 n.8, 224 & n.6
Portunus, 381
Potamogeton, 131, 357, 380 n.3, 398 & 399 n.5
potatoes, 37 & n.12
Potentilla, 312, 401
Poultry Chronicle, 3 & 4 n.11, 288 & n.6
poultry. *See* fowls
Prague, Czechoslovakia, 367 n.11
Pratt, George Charles, 345 & n.3
Prentice, C., 89
Prestwich, Joseph, 85 & n.7, 366, 403, 404 & 405 nn.4 & 13, 405 & 406 nn.2 & 4, 409
Prichard, James Cowle, 26–7 & 29 n.10
primroses. *See* Primula
Primula, 103, 311, 401, 464 n.8, 509 & n.3
Principe Island, 242 & 243 n.5, 243, 269, 270
Pringlea, 178 & 179 n.9, 191
Prion, 119 & 120 nn.3 & 11
Prionus, 119
Pritchard, Charles, 238 & 239 n.6, 476 & 477 n.9
Prosopis juliflora, 353
Proteaceae, 499 & 500 n.3, 502 n.2
Pselaphidae, 374
Psychidae, 302 n.2
Pterocarpus, 287
Pteromys petaurista, 9
Ptini, 375
Puget Sound, Washington, 380
Pugh, Miss, 301 n.3
Pulleine, Robert, 2 & 3 n.5, 21 & n.2
pumas, 71 & 72 n.3
Punch, 375 & 376 n.14
Purbeck beds, Dorsetshire, 320, 321 n.2, 338 & n.10, 366 & 367 nn.7 9, 381 & 382 n.11, 400 n.7
Pyrenees, 384 & 385 n.8, 388 & 389 n.2
Pyrgaera bucephala, 400 n.5
Pyrus, 312, 499 & 500 n.8

quaggas, 214 n.4, 409 & 410 n.10
quails, 267, 354, 355
Quarterly Review, 211 & 213 n.7
Quatrefages de Bréau, Jean Louis Armand de, 231 & n.3, 405 n.10, 506 & n.6

Quercus, 357

Quoy, Jean René Constant, 128 n.4

rabbits, 8, 11; alpine, 11, 262; Angora, 121, 124
& n.2, 135, 226; CD's acquisition of, 2, 54,
121, 124 & n.2, 135, 142 & n.3, 223 & 224
n.5, 226, 227–8, 234, 238, 287 & 288 nn.1 &
3, 351, 485 & n.2; Himalayan, 28 & 30 n.27,
88 n.3, 142 & n.3, 288 nn.1, 3 & 4, 459 & n.3,
507 & n.2; of Ascension Island, 221, 223; of
Ireland, 11, 351 & n.2, 485 & nn.2 & 3; of
Jamaica, 227 & 228 n.5; of Madeira, 485 &
n.3; of Porto Santo, 33 & 34 n.2, 135, 224 &
n.6; of Sandon, 85 & n.2; of Shetland Islands,
223, 379 & n.2, 440 & 441 n.2, 485 & n.3;
silver grey, 288 nn.1 & 3, 507 & nn.2 & 3;
variation in, 11, 121, 135 & 136 n.8, 142 n.3,
217 & 218 nn.7 & 8, 223–4 & n.5, 236, 238

Radlkofer, Ludwig, 156 & 157 n.7

railways: CD's shares in, 215 n.6, 335 n.5; extension to Beckenham, 73 & n.17, 215 & n.6, 286
& n.8, 335 & n.5, 471

Ramsay, Andrew Crombie, 154 & 156 n.8

Ransome, James Allen, 453 n.10, 455 & n.4

Ransome, Robert, 453 n.10, 455 & n.4

Ranunculus, 131, 277–8, 351, 356, 372, 398, 401,
469 & n.15

Raoul Island, 145 & 146 n.10, 153, 267 & 268
n.6, 359 & n.8

raspberry, 281 & n.6, 282 & n.3

Rathke, Martin Heinrich, 421 n.5, 425 & 427
n.3, 461 & 462 n.4

rats, 130 & n.3, 235 & 237 n.4, 299 & 300 n.1

Rawlinson, Henry Creswicke, 62 & 66 n.19

Rawson, Arthur, 49 & 50 n.6

Rawson, Charlotte Elizabeth, 49 & 50 n.6

Ray Society, 101 & nn.2–4, 157 & 158 n.10, 171,
314 n.4

Réaumur, René Antoine Ferchault de, 353

Reed, George Varenne, 218 & 219 n.3

Reid, Mayne, 440 & 440 n.3

reindeer. *See under* deer

Reissek, Siegfried, 152 nn.2–4, 217 & 218 n.11,
240 & 241 n.3

reproduction. *See* cross-fertilisation, generation,
hermaphrodites, parthenogenesis

Réunion Island, 232 & n.5, 348 & 349 n.3

reversion: in horses, 409 & 410 n.10, 411 & nn.2
& 6, 417 & n.4, 440–1 & n.4, 490–1 & nn.3
& 4; in pigeons, 409, 441 n.4; in rabbits, 85
n.2

Revillagigedo Islands (Pacific Ocean), 338

rhinoceros, 186 & 187 n.3, 264 n.19

Rhododendrons, 54 & 55 n.3

Rhône River, 171 n.13

Ribes, 356

Richard, Achille, 81 & 82 n.8

Richardson, John, 362 & 363 n.4, 366 & 367 n.3,
404 & 405 n.19, 409 & 410 n.3; on geographical distribution of fish, 187–8 & n.4; receives
Royal Medal, 73 n.14, 85 & n.4, 107 & 110
n.5

roaches, 365

Robillard, Victor de, 225–6, 232 n.3, 347–8

Robinia, 287, 361

robins, 516

Rodney, William Powell, 232 n.6

Rodriguez Island, 337 & 338 n.7

Roe, Mr, 459, 463

Rolle, Louisa Barbara, 36 & 37 n.2

Rosa, 131, 203, 315, 341, 356, 444

Rosaceae, 502 n.2

Ross, James Clark, 262 & 264 n.17, 263 n.8, 404
& 405 n.14

Royal Agricultural College, Cirencester, 464
nn.2 & 5

Royal Botanic Gardens, Kew, 123 n.2, 180 n.7,
322 n.3, 364 n.1, 372, 447 & 450 n.16

Royal Dublin Society, 380 & 382 n.5

Royal Geographical Society of London, 422 n.2,
451

Royal Institution of Great Britain, 67 n.3, 107
n.5, 112 & n.4, 114, 200 n.9

Royal Society of London, 252 n.2, 451, 467;
committee on North American Exploring
Expedition, 327 & n.2, 362–3 & nn.1 & 2;
council of, 72 & 73 nn.2 & 7, 76–7 & n.5, 86;
CD attends council meetings of, 34 n.1, 57
n.1, 108 & 110 nn.4 & 8, 112 n.5, 123 n.5; CD
referees papers for, 189 & n.1; CD vice-
president of, 73 n.2, 110; funds for scientific
research, 101 & 102 n.4, 321 & 322 n.2, 324
& nn.2 & 4; lectures at, 322 & 323 n.2; library
of, 474 & n.1; medals of, 55 & 56 n.8, 72 & 73
nn.3, 4 & 11–14, 77 & nn.3–5, 85–6 &
nn.4–9, 106 & 107 n.6, 108 & 110 n.5, 403–4
& 405 nn.2–4 & 8, 408–9 & 410 n.2, 487 &
n.12; moves to Burlington House, 110 n.4. *See
also* Philosophical Club of the Royal Society

Royle, John Forbes, 263

Rubus, 131, 152 n.3, 171, 277, 315, 341, 351, 355,
357, 371, 444, 499

rudimentary organs, 253, 333 & 334 n.2, 339,
415 n.3

Rugby School, 46 n.2, 269 n.1, 286 n.3, 304 n.3, 344 n.2, 345, 434 & 435 nn.1 & 6–8, 452 n.2, 454 & 455 n.2, 476 n.2, 476 & 477 n.9, 478 n.3

Rumelia, 443 & 444 n.9, 502

Rüppell, Wilhelm Peter Eduard Simon, 7 & 16 n.14, 11 & 18 n.43, 61

Russia: fishes of, 187; flora of, 131, 132, 133, 156, 438 & 439 n.15; molluscs of, 185–6

Sabine, Edward, 72 & 73 nn.10 & 14, 85–6 & n.4, 106 & 107 n.2, 403 & 405 nn.4 & 6; and North American Exploring Expedition, 362–3 & nn.1 & 2

Saint Helena, 96, 102 & 103 n.2, 126 n.4, 154

Saint Paul Island, 336 & 338 n.7

Saint Thomas Island. *See* São Tomé Island

Saint Vincent Island (Lesser Antilles), 319

Saint-Hilaire, Auguste de, 469 & n.15

Salix, 131, 171, 277, 315, 341, 398, 444

salmon, 187–8

Salter, James, 267 & 268 n.7

Salter, Thomas Bell, 175 & 177 nn.1 & 2

Salvages Islands, Madeira, 90 & 91 nn.11 & 13, 376 n.7

Sandon Park, Staffordshire, 85 & n.2

Sandwich Islands, 381

Santa Domingo, 353

Santa Fé de Bogota, New Granada. *See* Bogota, Columbia

Santarem, Juan de, 242, 270

São Tomé Island, 242 & 243 n.5, 269–70 & 271 nn.5 & 7

Saráwak, Borneo, 296 n.4, 345 n.2

Sarcophycus, 317

Sargassa, 317

Sarraceniaceae, 422–3

Saxifraga, 150, 176, 311, 315, 341, 351

Say, Thomas, 48 & 49 n.1, 128 n.19

Scabiosa, 102

Scandinavia, 8, 131, 140, 258, 262, 285

Scherzer, Karl von, 338 nn.1 & 7

Schiödte, Jörgen Matthias Christian, 283 & 284 n.3

Schlagintweit, Adolphe, 456 n.3

Schlagintweit, Hermann Rudolph Alfred, 443 & 444 n.3, 455 nn.1–3

Schlagintweit, Robert, 444 n.3, 455 & nn.1–3

Schlegel, Hermann, 61 & 65 n.12

Schmidt, Ferdinand Joseph, 284 & n.4

Schouw, Joachim Frederik, 97 n.8

Schreber, Johann Christian Daniel von, 66 n.29

School of Mines, 231 n.4, 400 n.4

Schrophulariaceae, 502 n.2

Sciurus maximus, 9

Scotland, 11, 260; flora of, 289 & 290 n.4, 293–4

seaweeds. *See* algae, *Fucus*

Scott, David, 67–8, 69 & 70 n.2

Scourge, HMS, 269 & 270 n.2

Scrophulariaceae, 401, 498 & 500 n.3

Sebright, John Saunders, 20 n.3

sedges, 78; 82

Sedgwick, Adam, 145, 199 & 200 n.6, 409

seeds: albumen in, 102 & 103 n.3; cast up on Azores, 123 n.2; cotyledons, 418; CD's experiments on dispersal of, 100 & nn.9 & 10, 212 & 213 n.14, 248 & n.2, 250, 419 & 420 n.4; CD's experiments on immersion of in salt-water, 36 & 37 n.9, 46–7 & nn.3 & 4, 49 & 50 n.2, 92 n.15, 103 n.3, 192 & n.5, 202 & 203 nn.5 & 6, 244 & n.2, 248, 265–6 & n.1; dispersal of, 105 & 106 n.4, 122 & 123 n.2, 200 nn.3 & 4, 212 & 213 nn.12 & 13, 233, 267 & 268 n.5, 305 & n.8, 324 & nn.5 & 6; dispersal of by birds, 274 & n.5, 281 & nn.3 & 6, 282 & n.3, 294 & 295 n.7, 309 & 310 n.7, 365 n.2, 389 & 390 n.6, 445; dispersal of by fish, 319, 353, 355, 364 & 365 n.2, 379 & 380 n.3; dispersal of by icebergs, 263 n.10; dispersal of by seacurrents, 276, 319; dispersal of by trees, 100 & n.9, 276, 319, 348; effect of pollen on, 47 & nn.7 & 8, 49 & 50 n.4, 480–1 & nn.2 & 3, 486 & n.1, 487 & nn.9 & 11, 492 & 494 n.9; failure of to float, 46 & 47 n.4; prolificness of foreign, 482–3 & nn.2, 5 & 6, 495 & n.6; vitality of ancient, 3 & n.8, 45 & 46 n.6

Seeman, Berthold Carl, 452 & 453 n.5

Seeman, Wilhelm E. G., 453 n.5

Selenia, 332

Senebier, Jean, 396 & n.7

Senecio, 357

Sequoia, 145 & 146 n.9

Seringe, Nicolas Charles, 474 & n.3

Serinus melanocephalus, 7 & 16 n.14

Serolis, 380

Serranus, 506 & n.6

Serres, Pierre Toussaint Marcel de, 456 & 457 n.6

Sevastopol, Crimean peninsula, 56 n.5

sexual selection, 71 & 72 nn.1 & 2, 75 & n.5

Seychelles Islands, 27, 154, 337 & 338 n.5, 348

Shakspear, Mr, 318 & 320 n.2

Sharpe, Daniel, 122 & 123 n.4, 125 & 126 n.2, 144 & 146 nn.1 & 2, 218 n.6, 331 n.7

Sharpey, William, 72 & 73 n.6, 324 & n.4, 327, 489–90 & n.2; and medals of Royal Society, 403 & 405 nn.5 & 7, 404 & 405 n.18, 405–6 & n.2, 410

sheep, 3, 24 & n.4, 270, 492 &493 n.4

Sheridan, Robert Brinsley, 13 & 18 n.53

Shetland Islands, 98 & n.5, 132, 180 & 181 n.8, 260; pigeons of, 223, 379 & n.2, 440 & 441 n.3; rabbits of, 223–4 & n.5, 379 & n.2, 440 & 441 n.2, 485 & n.3

Shrewsbury, 419 n.2

Siberia, 258

Sicily, Italy, 495, 515 n.6

Siebold, Karl Theodor Ernst von: on parthenogenesis, 156 & 158 n.8, 301 & 302 n.2; *Wahre Parthenogenesis*, 302 & 303 n.13, 307 & 308 n.2, 343 nn.1 & 2

Sierra Leone, 322 n.4

Silene, 356, 401, 494 & 495 nn.4 & 5, 500 n.12, 504 n.3

silkworms, 16 n.8

Silliman, Benjamin, 181 & nn.3 & 11, 246

Silliman, Benjamin, Jr, 180 & 181 n.3

Silpha, 374

Sinai, Mount, 11

Sinapis nigra, 483 & n.5

Singapore, 339 n.2, 345, 388 & n.9, 515 n.12

siskins, 7

Siwalik Hills, India, 73 n.15

Skeffington, John, 485 & n.4

Sloane, Hans, 15 & 18 n.67

Smethurst, Thomas, 386 n.2

Smith, Mr (of Uyea Sound), 99 n.4

Smith, Charles Hamilton, 490 & 491 n.3

Smith, Frederick, 481–2 & nn.1 & 4

Smith, George, 20 & n.2

Smith, James Edward, 289 & n.2, 329 n.2

Smith, John, 255 n.2

Smith, Thomas, 414 n.12

Smith and Beck (scientific instrument makers), 193 n.3

Smith, Elder & Co., 23 & n.6

Smithsonian Institute, 381 & 382 n.10

Smyth, Charles Piazzi, 79 & 80 n.8

snakes, 307

Snow, George, 88, 273

Snowdon (mountain), Wales, 198

Soane, John, 435 n.4

Société de biologie de Paris, 405 n.10

Solanaceae, 502 n.2

Solanum, 37 & n.12, 469

Solidago, 398

solitaires, 337 & 338 n.4, 348 & 349 n.4

Solomon, King, 10, 14

Sonchus, 368 & n.4, 370, 483 & n.4

Sorby, Henry Clifton, 217 & 218 nn.5 & 6

South Africa. *See* Cape Colony

South America: botany of, 167 & n.6, 171 n.2, 172 & n.3, 176 & 177 nn.10 & 11, 204 n.5

South America, 22 & 23 n.2, 128 n.16

South Eastern Railway, 335 n.5

South Shetland Islands, 261 & 263 n.11, 366 & 367 n.6, 380 & 382 n.4

Sowerby, George Brettingham, 127 & 128 n.16

Sowerby, George Brettingham, Jr, 101 & n.3

Spach, Edouard, 357 & 358 n.7

Spain, 139, 150

Spallanzani, Lazzaro, 254 & 255 n.3

species: definitions of, 96, 194 & nn.6 & 7, 195–6, 309, 492–3 & 494 n.12, 499 & 500 n.8; duration of, 127 & 129 n.22; polymorphic, 357 & 358 n.6, 368 & n.4, 370 & n.6, 371 & 372 n.2; single or multiple creation of, 78, 90, 93, 122, 125, 153 & 156 n.3, 170, 172, 176, 178, 190, 194 n.3, 236 & 237 n.7, 340 & 342 nn.5 & 7; transmutation of, 36 & 37 n.8, 78, 81, 145, 194 & n.6, 197, 199–200, 201, 236, 271, 274, 280, 432, 446, 447–9. *See also* classification; Darwin, Charles Robert: species work; geographical distribution; varieties and variation

Spencer, Herbert, 56 & 57 nn.2 & 3

spiders, 7, 216, 283

Spiraea aruncus, 258

Spirifera rostrata, 313

Spitsbergen, 260, 262

Splachnidium, 317

Sprengel, Christian Konrad, 309 & n.3, 413 & 414 n.14

Stachys, 311, 312

Stainton, Henry Tibbats, 75–7 & nn.2 & 3, 85 n.9

Stanhope, Philip Henry, 179 & 180 n.13

Stanley, Edward Smith, 166 n.2

starlings, 222

Statice, 37 n.11

steinbocks, 262

Stephens, Henry, 211 & 213 n.7

Stephens, James Francis, 328 & 329 n.4, 332 & 333 n.1, 478 & 480 n.4

Stereognathus ooliticus, 263 n.6

Steudel, Ernst Gottlieb, 413 & 414 n.7

Stevens, John Crace, 21 & n.1, 59 & n.3, 60, 87, 111 & n.6, 224, 249 n.5, 257 & n.6

Stevens, Samuel, 290 n.4
Stimpson, William, 381 & 382 n.10, 399 & 400 n.3
Stockport, 336
Stokes, John Lort, 55 & 56 n.6
Stonesfield, Oxfordshire, 148 n.5, 320
Storey, John, 113 & nn.2 & 3
storks, 324 & n.5, 380
Straus-Durckheim, Hercule Eugène Grégoire, 251 & 252 n.8
Strickland, Agnes, 440 n.4
Strickland, Hugh Edwin, 13, 62 & 66 n.14, 338 n.4, 494 n.12
Strzelecki, Paul Edmund de, 103 & 104 n.14, 106 & 107 n.4
Stuttgart, Germany, 321
Stylidum, 112
Subularia, 277 & 280 n.1, 289, 293–4 & n.2, 412 & n.4, 413
Sudbrook Park, Surrey, 477 n.5
sugar peas. *See* mangetout
Sulivan, Bartholomew James, 22 & 23 n.4, 55 & 56 nn.3 & 5, 345 & n.2, 471 & 472 n.6
Surat, India, 13
Sveaborg, Gulf of Finland, 55 & 56 n.5
Swainson, William, 13
Swanage, Dorset, 400 n.7
Swayne, George, 465 & 467 n.2, 466
Sweden, 132, 185–6, 478 & 480 n.5
sweet briar, 372
sweetpeas, 47, 190, 225, 291 & n.4, 346 & 347 n.5, 356, 428, 433, 466
Switzerland: butterflies of, 215 & n.2; J. D. Hooker visits, 195 n.11, 196, 202, 217; T. H. Huxley visits, 454 & n.5, 456 & 457 n.11
Sydenham, Kent, 111 n.3
Sydney, Australia, 362 n.8, 498, 500 & 501 n.14, 503 n.6
Sykes, William Henry, 14 & 18 n.65, 38 & 42 n.3, 43
Sylochelidon, 41
Sylvicola petechia, 318
Syme, John Thomas Irvine Boswell. *See* Boswell-Syme, John Thomas Irvine
Syria, 10–11, 14

Taraxacum, 370
tare, 305
Targioni Tozzetti, Antonio, 103 & 104 nn.10 & 11
Tarphi, 375

Tasmania: animals of, 125; botanical relations of, 176 & 177 n.11, 198–9 & 200 nn.3 & 4; flora of, 413, 508 & n.5; trees of, 304, 315
Tegetmeier, William Bernhard: and Anerley poultry show, 110 & 111 nn.4 & 5, 160 & n.6, 204 & nn.2 & 5, 205, 234; and barbs, 393, 397; and bees' cells, 378 & n.3; and carriers, 226, 234, 336 & 337 n.1; claims to be working with CD, 51 & n.7, 58 & n.7; and correlation between colour and down in pigeons, 276 & n.3, 393 & 394 nn.10 & 13, 397, 414 & 415 n.3; and crève-coeur fowls, 288 n.6; and CD's acquisition of ducks, 334, 338; and CD's acquisition of fowls, 21 & n.1, 87–8 & n.5, 110, 121, 152, 247, 256; and CD's acquisition of pigeons, 1, 59, 60, 87–8 & nn.2 & 4, 110, 124 & nn.3 & 4, 152, 160 & n.2; and CD's acquisition of rabbits, 121, 124 & n.2, 226, 459 & n.3; CD sends pigeons to, 431, 436 & n.2; exhibits fowl at Zoological Society, 160 & n.7, 339 & n.2, 344 & 345 n.3; and fantails, 226; and fowls from Africa, 243 n.2, 246 & 247 n.2, 248, 256, 290; and fowls from Borneo, 344–5 & nn.2 & 3, 392 & 394 n.7, 414 & 415 n.4; and fowls from India, 295 & 296 n.2; and fowls from Persia, 290 & n.2; health of, 296; and hybrid crosses, 60 & n.3, 210, 334 & n.6, 397, 414 & 415 n.2; and jungle fowls, 1 & n.3, 416 & n.3, 459; and Malay fowls, 489; on Polish fowls, 60 & 61 n.8, 160 & n.7, 210 & 211 n.4, 275 & 276 n.2; and Philoperisteron Society, 1 & n.2, 51 n.7, 257 n.4, 275 & 276 n.2, 489 & n.5; *Poultry book*, 60 n.7, 87 & 88 n.6, 227 & n.8, 276 & n.6, 295 & 296 n.3, 393 & 394 n.11, 397 & 398 nn.4–6, 416 & n.2, 431 & n.6; and rumpless Polish fowls, 333, 392 & 393 nn.2 & 3, 397, 415 & n.3, 416, 431 & n.4; and runts, 204, 204–5, 208 & n.2, 210 & 211 n.2, 226, 234 & 235 n.2, 246 & 247 n.1, 431 & nn.1 & 3, 436; and Scanderoons, 152, 160, 204, 208 & n.3, 210, 224, 249 & n.4, 392–3 & 394 n.6, 397; visits Mr Bult with CD, 34
telegony, 213 & 214 n.4, 301 & 302 n.5
Telephoridae, 374
Tellkampf, Theodor G., 282 & 284 n.1
Temminck, Coenraad Jacob, 29 n.15, 39, 69 & 70 n.4, 227 & 228 n.2, 233 n.2
Temple, Frederick, 434 & 435 n.7
Temple, Henry John, 376 n.14
Tenant, James, 364 & 365 nn.1 & 2, 379 & 380 n.1

Tenby, Wales, 162 & n.9, 346; CD plans to visit, 110 n.9, 110 & 111 n.4, 135 & 136 n.3

tench. *See* carp family

Tentyria, 374

Terebratella, 313

Terebratula, 96 & 97 n.9, 115

Testacella, 306

Tetrao, 26

Thalassidroma, 41, 119

Thames, river, 200 n.2

thistle, 363–4, 408, 469

Thlaspi alpestre, 132

Thompson, John, 287 & 288 nn.1 & 3

Thompson, William, 351 & nn.2 & 3

Thomson, Thomas, 157 & 158 n.10, 185 n.8, 280 & 281 n.2; *Flora Indica*, 298 & 299 n.6, 310 n.6, 364 n.6, 423 & 424 n.7, 438 & 439 n.12, 453 n.4, 499 & 500 n.8; '*Præcursores ad Floram Indicam*', 420 n.2

Thomson, Thomas Richard Heywood, 30 n.23

Thorley, Miss, 301 & n.3, 394 & 395 n.4

thrushes, 516

Thunberg, Carl Peter, 339 & 342 n.4

Thuret, Gustave Adolphe, 112 n.6, 317 & n.2, 498 & n.3

Thwaites, George Henry Kendrick, 54 & 55 nn.1–3 & 5, 485 n.6

Tibet: chooboos of, 455 & n.2; dogs of, 107 & n.3, 350 & n.2; molluscs of, 182 & 184 n.2, 186; partridges of, 11

Tierra del Fuego, 125, 359 n.6; botanical relations of, 93 & 95 n.15, 105, 166–7 & nn.3, 5 & 6, 172 & n.3, 192, 204 n.5, 369; crustaceans of, 380

tigers, 6, 12, 31 & n.2

Timor Island, 514

Tippoo Sahib, 5 & 6 n.4, 12

tits, 8

toads, 270

Tollet, George, 25 & n.3

Tollet, Georgina, 475 & 476 n.4

Tollet, Marianne. *See* Clive, Marianne

Torrey, John, 132 & 133 n.10

tortoises, 270

Tortrius, 327

Tottanus glottoides, 41

Toulouse, France, 389 n.2

Toynton, Lincolnshire, 480 n.8

Travancore, India, 5 & 6 n.4

trees, 226; dispersal of, 99, 353, 354; and dispersal of plants, 100 & n.9, 105 & 106 n.4, 319, 348; fossil, 180 & 181 n.8; separation of sexes in,

190 & 191 n.7, 287 & n.2, 291, 294 & 295 nn.2, 5 & 6, 297–8 & 299 n.4, 304 & 305 n.3, 315 & 316 n.8, 340 & 342 n.10, 358 & 359 n.4, 360–1 & 362 n.5, 391 & 392 n.5, 402 & 403 n.2, 413, 419 & n.4

Treron, 7, 8

Tribulus, 353

Trifolium, 464 n.7

Tristan d'Acunha, 153, 170 & 171 n.2, 172 & n.2, 176 & 177 n.10, 336, 358 & 359 n.6; botanical relations of, 166–7 & nn.3, 5 & 6, 191–2 & n.4, 203 & 204 nn.1, 2 & 5

Triticum, 341, 423

Triura, 216

trout, 187–8

Tuckerman, Edward, 398 & 399 n.5

Tulasne, Louis René, 292 & 293 n.2

Turkey, 13

turkeys, 9 & 17 n.30, 10–11, 12–13, 14, 26, 27, 28 & 30 n.24

turnstones, 41

Turtur, 8

Twofold Bay, New South Wales, 56 n.2

Tyndall, John, 260, 267 n.3; on cleavage and foliation, 200 & n.9, 201–2 & n.7, 217 & 218 n.4, 330 & 331 n.7; study of glaciers, 218 n.3, 322–3 & nn.2–6, 329 & nn.2–4, 330 & nn.5–7, 454 n.5, 457 n.3

Tytler, Robert C., 28 & 30 n.22

Umbelliferae, 391, 401

Unitarianism, 134 & 135 n.9, 205 n.2

United States: agrarian plants of, 182 & n.8, 209, 229, 285, 495; alpine flora of, 182 & nn.3 & 5; blind cave animals of, 235, 282–4, 299; botanical relations of, 92–4, 209, 245 & 246 nn.5, 6 & 8, 314–15 & 316 nn.3, 6 & 7; flora of, 131, 132, 133, 148, 150, 151 & 152 nn.5 & 6, 156 & 157 n.2, 182 & n.3, 391 & 392 nn.2, 3 & 8, 422–3, 438 & 439 n.11, 446 & 449 nn.7 & 8, 460, 502; geology of, 145 & 146 n.7, 154, 235 & 237 n.5; naturalised plants of, 228–9, 245 & 246 n.6, 258 & 259 nn.3 & 4, 285, 341, 391 & 392 nn.6 & 8, 402 & 403 n.3, 494; plants common to Asia, 245 & 246 n.8, 258, 259, 339–40 & 342 nn.3, 4, 6 & 7, 391 & 392 n.3; plants common to Europe, 92–3, 131–3 & nn.3 & 4, 139 & 140 nn.2 & 3, 140–1, 148, 149–50 & 151 n.2, 182, 209 & nn.5–8, 245 & 246 n.5, 257–8 & 259 n.1, 261, 272, 273–4, 285, 340 & 342 n.9, 391 & 392 n.3, 401; presidential elections of, 231 n.4, 236 & 237 n.11; social plants of, 93, 228–9 & 230 n.2, 245; topographic map of, 231 n.7; trees of, 93 & 95 n.14, 182, 358 & 359 n.4, 402 & 403 n.2, 413, 419 & n.4, 423; zoology of, 180, 187, 216 & 217 n.7, 235, 299

United States Exploring Expedition, 129 n.20

United States North Pacific Expedition, 342 n.6, 381 & 382 n.10

Valmont de Bomare, Jacques Christophe, 474 & n.7

Valparaiso, Chile, 380, 381

Van Voorst, John, 175

variation and varieties: 197, 281–2, 447, 448, 460 & 461 n.12; as incipient species, 430 n.3, 432; CD's attempts to induce, 465 & n.2, 467 & n.3; CD's portfolios of notes on, 31 n.2, 37 & n.13, 54 n.5, 97 & n.10; effect of isolation on, 262; in agrarian plants, 131, 133, 149, 182 & n.8, 208–9, 285 & n.7, 341, 464 & nn.5, 7 & 8; in allied species and genera, 419 & 420 n.3; in bees, 343 n.4; in birds, 206, 411; in brachiopods, 308 & n.2, 312–14 & nn.1 & 2; in cattle, 211 & 212 n.3; in Coleoptera, 252–3 & nn.2–4, 407 & 408 n.3, 478–9 & 480 nn.2–5; in fossils, 308 & n.2, 312–14 & nn.1 & 2, 506; in fowls, 210–11 & n.4, 288; in *Fucus*, 368, 369 & 370 n.6; in highly developed organs, 395–6 & n.1, 400 & nn.1 & 6, 407 & 408 n.2, 409 & 410 n.9; in insects, 134, 158, 283, 419; in large and small genera, 93 & 95 nn.10 & 11, 314 & 316 nn.3 & 4, 325 & 326 n.6, 363, 391 & 392 n.4, 429 & n.9, 430 & nn.2 & 3, 438 & 439 nn.7–13, 443–4 & nn.12–14, 446 & 449 nn.8–10, 452 & 453 nn.3 & 4, 460, 471 & 472 n.7, 478–9 & 480 nn.2, 3, 6 & 7, 486 7 & nn.3–7, 493 & 494 n.16, 498 & nn.2–4, 498–9 & 500 nn.2–5, 502 & nn.2–4, 516 & nn.2 & 3; in Lepidoptera, 327–8, 331–3 & n.3; in molluscs, 96, 183–4; in naturalised plants, 54, 182, 196, 228–9, 284, 368 & n.4, 369; in pigeons, 58 & n.6, 160 & n.3, 204, 207, 208 & n.2, 223, 236, 238, 387–8, 414–15; in pigs, 206 & 207 n.3, 211 & 213 nn.3 & 4; in polymorphic species, 368 & n.4, 369–70, 370 & n.6, 371 & 372 n.2; in protean genera, 315, 341 & 342 n.11, 351–2 & n.1, 355–7, 358 & 359 n.3, 361 & 362 nn.9–11, 398 & 399 n.2; in rabbits, 11, 121, 135 & 136 n.8, 142 n.3, 217 & 218 nn.7 & 8, 223–4 & n.8, 226, 236, 238, 351 & n.2; intermediate forms between, 171 & 172 n.15, 176 & 177 n.16, 182 & n.7, 195–6, 208 & 209 & n.3,

310–12 & nn.2 & 3, 363–4; laws of, 281–2, 371, 387 & 388 n.7, 437, 449; limits of, 134, 147 & 148 n.6, 190 & 191 n.4; of British plants, 493 & 494 n.17, 503 & 504 n.2, 511 & 512 n.2; of *Colletia*, 36 & 37 nn.1–3, 78 & 80 n.2; of species through time or space, 94 & 95 n.17, 95–6, 115, 240, 308, 313–14

Vaux, William Sandys Wright, 107 & 108 n.2

Verbascum thapsus, 229

Verbenaceae, 502 n.2

Vestiges of the natural history of creation, 36 & 37 n.8, 178 & 179 n.9, 194, 200 n.8, 236 & 237 n.9, 446 & 449 n.6

Vibrio, 50 & n.8

Viburnum, 294, 339 & 342 n.4

Vicq d'Azyr, Félix, 474 & n.5

Victoria, Queen, 29 n.12, 434 n.2

Vigors, Nicholas Aylward, 41

Vilmorin, Pierre Philippe André Lévêque de, 47 & 48 n.11

Vilmorin-Andrieux et Cie (seedsmen), 244

Vine, William, 28 & 30 n.30

Viola, 311; cross-fertilisation in, 295 & n.3, 298; varieties of, 131, 312, 351

Virgil, 376 n.2

Viscaria oculata, 372

Viscum, 156 & 157 n.6, 170–1, 176

Visiani, Roberto de, 460 & 461 n.7

Vitrinae, 306

Viverra civetta, 270

Vivian, Edward, 288 & n.5

Volcanic islands, 337 & 338 n.6

volcanos and volcanic phenomena, 83–4 & nn.2 & 4, 154–5, 330 n.3, 337; CD's views on, 83–4 & n.2, 99 & 100 n.2, 167–8 & 169 nn.3, 4 & 6; Charles Lyell's views on, 163–5 & nn.1, 2 & 8

Wahlenberg, Göran, 94 & 95 n.17

Wahlenbergia saxicola, 419 & 420 n.2

Wallace, Alfred Russel, 450 n.17, 482 n.3; and CD's acquisition of ducks and fowl, 290 & n.4, 387 & 388 n.8; and CD's view of species, 387–8; and jaguars, 388 & n.10, 457 & 458 n.3; and land-bridge theory of distribution, 514 & 515 nn.5 & 6; *A narrative of travels on the Amazon*, 388 & n.11; on species, 457 & 458 n.2, 514 & 515 n.3

Wallace, Peter: on fauna of Ascension Island, 220–2 & 223 n.1

Wallich, Nathaniel, 157 & 158 n.13, 283 & 284 n.2

Warren, Captain, 50 & 51 n.5

Washingtonia, 145 & 146 n.9

Waterhouse, George Robert, 32 & 33 n.3, 489 n.5; and geographical distribution, 125 & 126 n.3; on bees' cells, 377–8 & n.1; on rabbits, 217 & 218 n.8; on rodents, 235 & 237 n.4; and variation in highly developed organs, 409 & 410 n.9

waterlilies, 37, 380 n.3

Watson, Hewett Cottrell, 176, 280 n.2; and aquatic plants, 289, 412 & n.2; *Cybele Britannica*, 391 & 392 n.9, 450 & n.3, 468 & 469 n.5, 471, 486 & 487 n.3, 503 & 504 n.4, 511 & 512 n.3; and CD's experiments on seed dispersal, 265–6 & n.1, 276 & 278 n.2; and CD's species theory, 130–1; and CD's study of close species, 443 & 444 n.11, 493 & 494 n.17; and intermediates between varieties, 208 & 209 n.3, 310–12 & nn.2 & 3; invited to Down, 66 & 67 n.2; *London catalogue of British plants*, 277 & 279 nn.3 & 4, 438 & 439 n.8, 443, 503 & 504 n.3; and naturalised plants, 208 & 209 n.4; on distribution of British plants, 131–3 & nn.2–4, 139, 149–50 & 151 nn.1 & 2, 259 n.1; on species, 277–8, 511 & 512 n.4; on variation and varieties, 131, 149, 396 n.1, 503 & 504 nn.2 & 3, 511 & 512 n.2; and plant migration during worldwide cold period, 140 & 141 n.4; and plants common to Europe and United States, 139 & 140 nn.2 & 3, 140, 149–50 & 151 n.2, 209 & nn.5 & 7; poor opinion of Edward Forbes, 139 & 140 n.4; and protean genera, 342 n.13, 351–2 & n.1, 355–7 & 358 n.1, 358 & 359 n.3, 361 & 362 nn.9 & 10, 398 & 399 n.2, 403 & n.5, 412 & 414 n.4, 422 & 423 n.1, 447 & 449 n.13; publications of, 131 & 133 n.2, 149 & 151 n.1

Webb, Philip Barker, 119 & 120 n.5, 144 & n.6, 443 & 444 n.7, 450 & n.2, 460 & 461 n.6

Wedgwood, Caroline (Carry), 395 n.9

Wedgwood, Caroline Sarah, 46 n.4, 51 & n.8, 225 n.1, 227 n.1, 344 n.10

Wedgwood, Cecily Mary, 395 n.9

Wedgwood, Ernest Hensleigh, 303 & 304 n.3

Wedgwood, Frances Julia (Snow), 344 & n.9

Wedgwood, Frances Mackintosh, 205 & n.2, 239 n.3, 303 & 304 nn.2–4, 458 & 459 n.5, 478 n.5

Wedgwood, Frances Mosley, 395 & n.9

Wedgwood, Francis (Frank), 267 & 269 n.4, 269 n.10, 419 n.2

Wedgwood, Godfrey, 85 & n.2

Wedgwood, Henry Allen (Harry), 268 & 269 n.4, 343 & 344 n.3, 395 & n.9

Wedgwood, Hensleigh, 205 n.4, 267 & 269 n.4, 304 nn.2 & 3, 458 & 459 nn.1–3, 478 n.5

Wedgwood, Jessie, 395 & n.9

Wedgwood, John Allen (Allen), 267 & 269 n.4

Wedgwood, Josiah, II, 25 n.3, 269 n.3

Wedgwood, Josiah, III (Jos), 46 n.3, 225 n.1, 267 & 269 n.4, 275 n.3, 344 n.10

Wedgwood, Katherine Elizabeth Sophy (Sophy), 344 & n.10

Wedgwood, Louisa Frances, 395 n.9

Wedgwood, Lucy Caroline, 344 & n.10

Wedgwood, Margaret Susan, 344 & n.10

Wedgwood, Sarah Elizabeth (Eliza), 443 n.3, 451 & 452 n.4

Wedgwood, Sarah Elizabeth (Elizabeth), 46 n.3, 267 & 269 n.4, 286 & n.7, 459 n.4

Wedgwood, Sarah Elizabeth (Sarah), 238 & 239 n.7, 275 nn.2 & 3, 286 & n.2, 303 & 304 nn.3 & 6; death of, 265 & n.3, 267–9 & n.2

Wedgwood, Sophy. *See* Wedgwood, Katherine Elizabeth Sophy (Sophy)

Weir, Harrison William, 59 & n.2, 226, 276 & n.3, 431 & n.5

Weld, Charles Richard, 474 & n.1

Wellingtonia, 146 n.9

Wellsted, James Raymond, 40, 64 & 66 n.27

Wenham Ice Company, 323 n.5

West Indies, 372 & 373 n.8

Westerham, Kent, 390 n.6

Westminster Review, 191 & 192 n.2

Westwood, John Obadiah, 73 n.3, 77 & nn.3 & 5, 85 & n.8, 456 & 457 nn.5 & 14; and bees' cells, 378 & n.2; on cave insects, 181 n.4, 282–4 & nn.5–16

wheat, 176, 178 & 179 n.5; fertilisation of, 178 & 179 n.5, 413, 424 n.8; J. S. Henslow's study of, 175 & 177 n.6, 179, 197 & n.3, 202; ingestion of by fish, 364, 365; origins of, 197 & n.3, 199

Whewell, William, 125 & 126 n.7

White Mountains, New Hampshire, 259, 261

White, Walter, 474 & n.1

Whittlesea, Cambridgeshire, 335 & n.6

Wicking, Matthew, 21 & n.5, 45 & 46 n.7, 142 & n.5, 226, 339 & n.4, 397

Widdrington, Samuel Edward, 27 & 29 n.15, 116 & 120 n.8

Wiegmann, Arend Friedrich August, 296 n.2, 487 & n.11

Wight, Robert, 326 & n.12

Willdenow, Karl Ludwig, 196

Williams & Norgate (booksellers), 488 & n.1

willows, 171

Wilson, Horace Hayman, 28 & 30 n.21, 62 & 66 n.19

Wilson, William Greive, 477 & 478 n.2

Wimmer, Christian Friedrich Heinrich, 384 & 385 n.5

Wingfield, William Wriothesley, 398 n.5

Wolff, Caspar Friedrich, 461 & 462 n.4

Wollaston, Thomas Vernon, 75, 91 n.6; and Coleoptera of Madeira, 373–4 & 376 n.4, 478–9 & nn.1 & 2; and CD's experiments on dispersal of land molluscs, 305–6 & n.2; and CD's species theory, 89 & 91 n.7, 99–100 & n.6, 134 & nn.8 & 10; and CD's study of aberrant genera, 480 n.3; and CD's study of large and small genera, 438 & 439 n.13, 478–9 & 480 nn.3, 6 & 7; forwards Charles Lyell's letter to CD, 134 & n.2; *Insecta Maderensia*, 134 & n.6, 253 & n.3, 438 & 439 n.13; and insects of Madeira, 158–9 & n.2, 325 & 326 n.9, 374–5 & 376 nn.10 & 13; and intermediates between varieties, 171 & 172 n.15, 177 n.16, 208 & 209 n.3; and land-bridge theory of distribution, 144 & n.4, 147, 159 & n.3, 253 & n.5, 376 & 377 nn.2 & 3; names beetle after CD, 374 & 376 n.8; *On the variation of species*, 134 & n.5, 138 & 139 n.3, 147 & 148 n.6, 182 n.7; on the wastefulness of nature, 253; opinion of entomologists, 375 & 376 n.11; sends CD specimens from Madeira, 33 & 34 n.2, 224 n.6; variation in Coleoptera, 252–3 & nn.2–4, 407 & 408 n.3, 478–9 & nn.1 & 2; view of species, 89 & 91 n.7, 99–100 & n.6, 135 & n.8, 306 & 307 n.6; visits Down, 66 & 67 n.2, 87 & n.3, 88 & 89 nn.2 & 3, 89 & 91 n.7

Wood, Searles Valentine, 48 & 49 n.1

Woodd, Charles Henry Lardner, 507 & n.4

woodpeckers, 445, 448

Woods, A., 275 & n.3

Woodward, Samuel Pickworth, 89 & 91 n.4; and antiquity of islands, 115 & n.6, 125 & 126 n.4, 144 & n.5, 147 & 148 n.9, 153 & 156 n.5; *A manual of the Mollusca*, 111 & n.2, 115 & n.5, 123 & 124 n.2; on *Orthocera*, 124 & 126 n.1; on ranges of land and freshwater shells, 126–7 & n.1, 129, 182 & 184 nn.2, 7, 8 & 10, 185–6 & 187 nn.1–3, 189 & n.2; on species, 182–4, 189 & n.2; on variability in brachiopods, 308 & 309 n.5; on variability of species through time, 95–6, 115 & n.2

Wooler, William A., 87 & 88 n.3

Wright, Charles, 340 & 342 n.6

Xanthium, 356

yaks, 62, 455 & n.2, 469 & n.13

Yarrell, William, 35 & n.1, 51 n.7, 119 & 120 n.11, 224 & 225 n.3, 249 & n.5

yews, 281 n.6

Youatt, William, 149 n.4

Yuccatan, Mexico, 352, 354

Zanzibar, Tanzania, 138 n.3

zebras, 14, 409 & 410 n.10, 417

Zeyher, Karl Ludwig Phillip, 105 n.3

Zizania aquatica, 102 & 103 n.6

Zoological Society of London, 61 n.8, 70 n.7, 166 n.4, 276 n.2, 451; gardens, 142 & n.3, 239 & 240 n.4, 248 n.2, 250 & n.6, 288 & nn.1–3, 364 & 365 n.1, 379 & 380 n.1, 507, 517 n.2; meetings of, 182 & 184 n.2; W. B. Tegetmeier exhibits fowls at, 160 & n.7, 339 & n.2, 344 & 345 n.3

Zostera, 294

Zua, 185, 186, 189

Zylophasia polyodon, 328

Robert Darwin = Elizabeth Hill
1682-1754   1702-97

William Alvey Darwin = Jane Brown
1726-83   1746-1835

Elizabeth Collier =² Erasmus Darwin ¹= Mary
Pole   1731-1802
1747-1832

Samuel Fox = Ann
1765-1851   1777-1859

Charles
1758-1778

Erasmus
1759-1799

Rob
War
1766-

Samuel Tertius = Frances Anne
Galton   Violetta
1783-1844   1783-1874

Edward
1782-1829

Emma
1784-1818

Francis = Jane Harrie
Sacheverel Ryle
1786-1859   1794-1866

John
1787-1818

Harriot = Thomas Jam
1790-1825   Maling
            1778-1849

Elizabeth Ann   Henry Parker = Mari
(Bessy) 1808-1906 1788-1856   1798-

Mary Ann = Samuel   Lucy Harriot
1800-29   Ellis Bristowe 1809-48
           1800-55

Eliza   Milicent Adèle
1801-86   1810-83

Su
Eliz
18

Emma   Emma Sophia
1803-85   1811-1904

Ellen Sophia =² William ¹= Harriet Fletcher   Darwin
Woodd   Darwin   1799-1842   1814-1903
1820-87   1805-80

Frances Jane = John Hughes   Erasmus
b. 1806   1794-1873   1815-1909

Erasmus
1804

Julia   Francis = Louisa Jane   Emily Cath
b. 1809   1822-1911 Butler d.1897   1810